Dust Devils

Participants in the Workshop on Dust Devils on Earth and Mars, held at the International Space Science Institute, Bern, Switzerland, February 16–20, 2015. Not in this picture: Jon Merrison and Angelo Rossi

1. Patrick Whelley
2. Jim Murphy
3. Ralph Lorenz
4. Giles Harrison
5. Erika Barth
6. Naomi Murdoch
7. Scot Rafkin
8. Martina Klose
9. Béatrice Marticorena
10. Frank Montmessin
11. Gregoire Déprez
12. Lynn Neakrase
13. Aymeric Spiga
14. Dennis Reiss
15. Matt Balme
16. Bradley Jemmet-Smith
17. Manish Patel
18. John Zarnecki
19. Mark Lemmon
20. Zhaolin Gu
21. Tetsuya Takemi
22. Michael Kurgansky
23. Stephen Lewis
24. Henrik Kahanpää
25. Junshi Ito
26. Francesca Esposito
27. Peter Knippertz
28. Dan Tyler
29. Michael Zimmerman
30. Lori Fenton
31. Kathryn Steakley
32. Thorben Kelling
33. Gerhard Wurm

Dennis Reiss • Ralph Lorenz • Matthew Balme •
Lynn Neakrase • Angelo Pio Rossi • Aymeric Spiga •
John Zarnecki
Editors

Dust Devils

Chapters 2–12 were previously published in *Space Science Reviews* Volume 203, Issues 1–4, 2016

Springer

Editors
Dennis Reiss
Münster, Germany

Ralph Lorenz
Space Department
JHU Applied Physics Lab
Laurel, MD, USA

Matthew Balme
Dept. of Physical Sciences
Open University
Milton Keynes, UK

Lynn Neakrase
Astronomy
New Mexico State University
Las Cruces, NM, USA

Angelo Pio Rossi
Jacobs University Bremen
Bremen, Germany

Aymeric Spiga
Laboratoire de Météorologie Dynamique
Université Pierre et Marie Curie
Paris, France

John Zarnecki
International Space Science Institute
Bern, Switzerland

ISSN 1385-7525 Space Sciences Series of ISSI
ISBN 978-94-024-1133-1

Library of Congress Control Number: 2017946702

Cover illustration: Orbital view of a large dust devil in Amazonis Planitia (Mars) acquired by the High Resolution Imaging Science Experiment (HiRISE) with a horizontal ground resolution of up to 60 cm/pixel. The vortex is about 140 m in diameter and reaches a maximum height of approximately 20 km (estimated from the shadow length). The dust column at height is leaned and twisted by high altitude winds.
Image credit: NASA/JPL/University of Arizona (http://hirise.lpl.arizona.edu/ESP_026394_2160)

Printed on acid-free paper

This Springer imprint is published by Springer Nature
The registered company is Springer Science+Business Media B.V.
The registered company address is: Van Godewijckstraat 30, 3311 GX Dordrecht, The Netherlands

Contents

Special Issue on Dust Devils

Dennis Reiss[1] · Ralph D. Lorenz[2] · Matthew Balme[3] · Lynn D. Neakrase[4] · Angelo P. Rossi[5] · Aymeric Spiga[6] · John Zarnecki[7]

Dust devils have captivated humankind since antiquity, and yet also assume importance in that most modern of activities, planetary exploration. The close investigation of the planet Mars by an armada of spacecraft in the last couple of decades has stimulated extensive

✉ D. Reiss
dennis.reiss@uni-muenster.de

R.D. Lorenz
ralph.lorenz@jhuapl.edu

M. Balme
matt.balme@open.ac.uk

L.D. Neakrase
lneakras@nmsu.edu

A.P. Rossi
an.rossi@jacobs-university.de

A. Spiga
aymeric.spiga@upmc.fr

J. Zarnecki
zarnecki@issibern.ch

1 Institut für Planetologie, Westfälische Wilhelms-Universität, Wilhelm-Klemm-Str. 10, 48149 Münster, Germany

2 Applied Physics Laboratory, Johns Hopkins University, 11100 Johns Hopkins Road, Laurel, MD 20723, USA

3 Dept. of Physical Sciences, Open University, Walton Hall, Milton Keynes MK7 6AA, UK

4 Department of Astronomy, New Mexico State University, Las Cruces, NM, USA

5 Department of Physics and Earth Sciences, Jacobs University Bremen, Campus Ring 1, 28759, Bremen, Germany

6 Laboratoire de Météorologie Dynamique, Université Pierre et Marie Curie, Paris, France

7 International Space Science Institute, 3012 Bern, Switzerland

research on these desert whirlwinds, supported by improvements in field instrumentation on Earth, laboratory experiments and powerful numerical simulations. This convergence of capability and interest stimulated the workshop 'Dust Devils on Mars and Earth', conceived by Dennis Reiss, Ralph Lorenz, Matt Balme, Lynn D. Neakrase, Angelo P. Rossi, Aymeric Spiga and John Zarnecki, held under the auspices of the International Space Science Institute in Bern, Switzerland, during the week of February 16–20, 2015. The workshop drew not only planetary scientists, for whom terrestrial dust devils are a convenient proxy for their Martian cousins, but also scientists studying the impact of dust on the terrestrial climate.

Chapter 2 by Lorenz et al. (2016) surveys the historical development of dust devil studies, noting a vivid description of one in Sophocles' writings 25 centuries ago, the beginnings of scientific investigation in the 19th century and the systematic improvements in field research until the early 1970s. Mars exploration seems then to have been an additional stimulus to dust devil studies, particularly following the Pathfinder and Mars Global Surveyor missions of 1997. This chapter also reviews the occasionally damaging nature of dust devils, responsible for a few deaths per decade, and the role of dust devils in culture.

Field observations, by researchers on Earth and robots on Mars, are reviewed in Chap. 3 (Murphy et al. 2016). Remarkably, it is only in recent years that high-time-resolution field meteorological data for dust devil sites on Earth have become widely available; generally field data from Mars have been better! Terrestrial field measurements have quantified an extensive range of vortex parameters such as rotational, vertical and translational speeds, dust loads, and pressure deficits. Dust devil parameters obtained on Earth are more temporally limited compared to the long-term operating robotic missions on Mars. The latter have enabled the quantification of daily and seasonal dust devil occurrence on Mars. Recent developments measuring dust devils with electric field, infrasound, acoustic and seismic sensors on Earth and its application in future missions on Mars are discussed.

Dust devils were first seen on Mars from above, by the Viking orbiter cameras (the Viking landers' slow-scan facsimile cameras were not suitable for observing moving objects) and these and other orbital observations since are reviewed in Chap. 4 (Fenton et al. 2016). Orbital imaging at Mars has progressed dramatically in coverage and resolution, and useful dust devil statistics have emerged, noting a number of regions of particularly enhanced activity. Recent work has shown that dust devil height and spacing can be related to large-scale ambient properties, such as the thickness of the boundary layer, and dust devil tracking provides much-needed insight into Mars' near-surface winds. Dust raising by other processes (notably dust storms) is important on both Mars and Earth, and orbital observations of dust raising generally on both worlds is reviewed in this chapter.

Some of the most numerous and prominent ephemeral surface features on Mars are dark or bright surficial tracks left by the passages of dust devils, commonly named 'dust devil tracks'. In the last decades dust devil tracks have also been identified on Earth in high-resolution satellite imagery and have been subsequently studied *in situ*. Their different morphological types, formation mechanisms, and spatial and temporal distribution on Mars and Earth are reviewed in Chap. 5 (Reiss et al. 2016). Dust devil tracks as proxies for dust devil activity, ambient wind conditions, dust devil lifetimes, and solar panel clearing events are discussed. Their impact on the climate on Mars due to large-scale surface albedo changes and the current state-of-the-art in automatic detection of dust devil tracks is more generally discussed.

The conditions under which dust devils form are reviewed in Chap. 6 (Rafkin et al. 2016). The association of dust devils with strong surface heating is well-established, but the anecdotal observation of an 'optimum' wind speed range for dust devil occurrence was not so obvious. Datasets and numerical simulations supporting this effect are reviewed, and the generation of vorticity more generally is discussed.

The nature of dust devils as discrete, self-sustaining convective vortices has long attracted the attention of workers in fundamental fluid mechanics. Various idealizations of a vortex equating the centripetal acceleration of air parcels to that due to a pressure gradient, with various analytic descriptions of wind and pressure fields since the original ‘Rankine vortex’ model of the 19th century are reviewed in Chap. 7 (Kurgansky et al. 2016). A thermodynamical theory that explains how the vortices are maintained and provides an order-of-magnitude estimate of their principal parameters is also overviewed, as well as discussed in this chapter are the effects of vortex translational motion, of surface roughness and of lofted dust on the steady-state structure of dust devils.

Among the most powerful tools in contemporary dust devil studies is the computer, through the framework of (most aptly-named in this case) Large-Eddy Simulation (LES) described in Chap. 8 (Spiga et al. 2016). These simulations describe the three-dimensional fields of pressure, velocity etc. in the planetary boundary layer, and explicitly resolve the vortex structures that emerge spontaneously therein. This approach allows the visualization of flow structure that is generally impossible to see in the field, and brings a perspective of dust devils not so much as discrete entities, but as part of a convecting flow system. Simulations, which have been performed for both Earth and Mars conditions, allow the tracking of the lifetime of individual structures, and their growth and sometimes merging. These numerical experiments, which nonetheless often involve many idealizations, allow systematic evaluation of the influence of environmental parameters such as ambient windspeed, on dust devil populations.

Chapter 9 (Lorenz and Jackson 2016) assesses the challenge of systematically evaluating dust devil populations, given their sporadic nature and extremely skewed size and intensity distributions: like earthquakes, there are many more small ones than large ones. Statistical techniques for comparing and portraying the level of activity, as determined by a variety of observational techniques, are reviewed. It is concluded that the widely-held assumption that ‘Martian dust devils are larger than those on Earth’ is not strongly supported by data, given the very different techniques used to survey their populations (different scales of ob servation sample different parts of the size distribution), but the best comparisons do indeed suggest a factor of about 3 in median size, and it is certainly true that the largest dust devils documented are on Mars. The chapter also notes the nonstationary statistics of dust devils, that they occur in ‘bursts’ when conditions are favorable, and may do so in a pseudoperiodic manner.

The electrical properties of dust devils have been studied since the 1860s, and charge generation in dust is reviewed in Chap. 10 (Harrison et al. 2016). Planetary atmospheric electricity research has been stimulated by instrumentation development around the launch of an electric field sensor to Mars on the Schiaparelli lander in 2016, and field experiments on dust charging associated with that investigation. In this chapter the causes of atmospheric dust electrification in general are discussed and future instrumentation for electric field measurements on Mars are reviewed.

The particle-lifting process is reviewed at a microscopic level in Chap. 11 (Neakrase et al. 2016): a fundamental challenge for both Earth and Mars, but especially the latter, is that modeled flow speeds to launch dust are improbably high due to the strong cohesion between very small dust grains. Various physical processes to circumvent this apparent ‘lifting threshold speed’ barrier are reviewed, illuminated especially by laboratory experiments. Among possible contributors are sand-blasting (using easier-moved large grains to dislodge smaller ones), the suction or ‘Delta-P’ effect, wherein the vortex pressure drop causes the surface to exhaust pore gas upwards, the break-up of large dust aggregates, the role of electrostatic force, and even exotic light-driven processes. It is concluded that all these effects may play a role, and the dominant ones may vary from location to location.

The problem of how to scale up dust lifting observed in individual devils, to the dust devil input to the overall aerosol content of the atmosphere, is reviewed in Chap. 12 (Klose et al. 2016). Field observation and numerical simulation results can inform the parameterizations used in global models to estimate dust fluxes. Since global models cannot hope to resolve individual devils, population statistics and functional relationships between dust fluxes and ambient parameters must be devised. On Mars the role of dust devils is considered to be important, contributing tens of per cent to the total, whereas on Earth it is small at the global scale.

This synthesis of the state-of-the-art in dust devil studies from two planets, Earth and Mars, serves as a basis for a better understanding of dust devil processes and quantifying their global dust entrainment contribution into the atmospheres of both planets. Unresolved questions and future directions in dust devil research are outlined in this book to stimulate further research and future robotic missions on Mars.

Acknowledgements The Convenors of the workshop "Dust Devils on Earth and Mars" would like to express their sincere appreciation to ISSI, in particularly to Maurizio Falanga, Jennifer Fankhauser-Zaugg, Saliba Saliba, Andrea Fischer, and Silvia Wenger for their very kind and excellent organization of this workshop as well as their technical and editorial help. The Editors would also like to thank all participants of the workshop who contributed to this volume and all reviewers who contributed their time and expertise in reviewing the manuscripts.

References

L. Fenton, D. Reiss, M. Lemmon, B. Marticorena, S. Lewis, B. Cantor, Orbital observations of dust lofted by daytime convective turbulence. Space Sci. Rev. (2016, this issue). doi:10.1007/s11214-016-0243-6

R.G. Harrison, E. Barth, F. Esposito, J. Merrison, F. Montmessin, K.L. Aplin, C. Borlina, J.J. Berthelier, G. Déprez, W.M. Farrell, I.M.P. Houghton, N.O. Renno, K.A. Nicoll, S.N. Tripathi, M. Zimmerman, Applications of electrified dust and dust devil electrodynamics to Martian atmospheric electricity. Space Sci. Rev. (2016, this issue). doi:10.1007/s11214-016-0241-8

M. Klose, B.C. Jemmett-Smith, H. Kahanpää, M. Kahre, P. Knippertz, M.T. Lemmon, S.R. Lewis, R.D. Lorenz, L.D.V. Neakrase, C. Newman, M.R. Patel, D. Reiss, A. Spiga, P.L. Whelley, Dust devil sediment transport: from lab to field to global impact. Space Sci. Rev. (2016, this issue). doi:10.1007/s11214-016-0261-4

M.V. Kurgansky, R.D. Lorenz, N.O. Renno, T. Takemi, Z. Gu, W. Wei, Dust devil steady-state structure from a fluid dynamics perspective. Space Sci. Rev. (2016, this issue). doi:10.1007/s11214-016-0281-0

R.D. Lorenz, M.R. Balme, G. Zhaolin, H. Kahanpää, M. Klose, M.V. Kurgansky, M.R. Patel, D. Reiss, A.P. Rossi, A. Spiga, T. Takemi, W. Wei, History and applications of dust devil studies. Space Sci. Rev. (2016, this issue). doi:10.1007/s11214-016-0239-2

R.D. Lorenz, B.K. Jackson, Dust devil populations and statistics. Space Sci. Rev. (2016, this issue). doi:10.1007/s11214-016-0277-9

J. Murphy, K. Steakley, M. Balme, G. Deprez, F. Esposito, H. Kahanpää, M. Lemmon, R. Lorenz, N. Murdoch, L. Neakrase, M. Patel, P. Whelley, Field measurements of terrestrial and martian dust devils. Space Sci. Rev. (2016, this issue). doi:10.1007/s11214-016-0283-y

L.D.V. Neakrase, M.R. Balme, F. Esposito, T. Kelling, M. Klose, J.F. Kok, B. Marticorena, J. Merrison, M. Patel, G. Wurm, Particle lifting processes in dust devils. Space Sci. Rev. (2016, this issue). doi:10.1007/s11214-016-0296-6

S. Rafkin, B. Jemmett-Smith, L. Fenton, R. Lorenz, T. Takemi, J. Ito, D. Tyler, Dust devil formation. Space Sci. Rev. (2016, this issue). doi:10.1007/s11214-016-0307-7

D. Reiss, L. Fenton, L. Neakrase, M. Zimmerman, T. Statella, P. Whelley, A.P. Rossi, M. Balme, Dust devil tracks. Space Sci. Rev. (2016, this issue). doi:10.1007/s11214-016-0308-6

A. Spiga, E. Barth, Z. Gu, F. Hoffmann, J. Ito, B. Jemmett-Smith, M. Klose, S. Nishizawa, S. Raasch, S. Rafkin, T. Takemi, D. Tyler, W. Wei, Large-eddy simulations of dust devils and convective vortices. Space Sci. Rev. (2016, this issue). doi:10.1007/s11214-016-0284-x

Space Sci Rev (2016) 203:5–37
DOI 10.1007/s11214-016-0239-2

History and Applications of Dust Devil Studies

Ralph D. Lorenz[1] · Matthew R. Balme[2] · Zhaolin Gu[3] · Henrik Kahanpää[4] · Martina Klose[5,6] · Michael V. Kurgansky[7] · Manish R. Patel[2,8] · Dennis Reiss[9] · Angelo Pio Rossi[10] · Aymeric Spiga[11] · Tetsuya Takemi[12] · Wei Wei[13]

Received: 12 September 2015 / Accepted: 2 February 2016 / Published online: 4 April 2016

Abstract Studies of dust devils, and their impact on society, are reviewed. Dust devils have been noted since antiquity, and have been documented in many countries, as well as on the planet Mars. As time-variable vortex entities, they have become a cultural motif. Three major stimuli of dust devil research are identified, nuclear testing, terrestrial climate studies, and perhaps most significantly, Mars research. Dust devils present an occasional safety hazard to light structures and have caused several deaths.

✉ R.D. Lorenz
ralph.lorenz@jhuapl.edu

1 Johns Hopkins University Applied Physics Laboratory, Laurel, MD 20723, USA

2 Open University, Walton Hall, Milton Keynes, MK7 6AA, UK

3 Xi'an Jiaotong University, Xi'an, Shaanxi 710049, China

4 Finnish Meteorological Institute, 00101 Helsinki, Finland

5 Institute for Geophysics and Meteorology, University of Cologne, Pohligstr. 3, 50969 Cologne, Germany

6 Present address: USDA-ARS Jornada Experimental Range, Las Cruces, NM 88003, USA

7 A.M. Obukhov Institute of Atmospheric Physics, Russian Academy of Sciences, Pyzhevsky 3, 119017 Moscow, Russia

8 Space Science and Technology Department, STFC Rutherford Appleton Laboratory, Didcot, Oxfordshire OX11 0QX, UK

9 Institut für Planetologie, Westfälische Wilhelms-Universität, Wilhelm-Klemm-Str. 10, 48149 Münster, Germany

10 Department of Physics and Earth Sciences, Jacobs University Bremen Campus Ring 1, 28759 Bremen, Germany

11 Laboratoire de Meteorologie Dynamique (LMD), 4 place Jussieu, 75252, Paris cedex 5, France

12 Disaster Prevention Research Institute, Kyoto University, Uji, Kyoto 611-0011, Japan

13 Wuhan University of Technology, Wuhan, Hubei 430063, China

1 Introduction

"*And so the hours dragged by until the sun stood dead above our heads, a huge white ball in the noon sky, beating, blazing down, and then it happened—suddenly, a whirlwind! Twisting a great duststorm up from the earth, a black plague of the heavens filling the plain, ripping the leaves off every tree in sight, choking the air and sky. We squinted hard and took our whipping from the gods.*"

Thus speaks a sentry in the play 'Antigone', written by Sophocles in Greece around 450 BCE (e.g. Bowker 2011), capturing a scene that many dust devil researchers two and a half millennia later will doubtless appreciate. Dust devils are therefore a phenomenon that has been documented for as long as there has been documentation of anything, and short of finding a prehistoric sketch of a dust devil on a cave wall, appears to be the first meteorological report in the West of this phenomenon.

In the present paper, we review the history of dust devil research in various countries, and its motivations, and then the history of research on dust devils on Mars. We then survey the applications of dust devil research, and finally note the role of dust devils as a motif in culture. First, however, we offer some definitions and distinctions against other vortex structures, and note the designation of dust devils in different languages.

Dust devils are dry convective vortices, rendered visible by lofted dust (Fig. 1). The definition offered by Oke et al. (2007a) is succinct: "*A vortex rotating in either direction faster than its motion downwind. It must have a coherent columnar or cone shape extending in a vertical direction above the surface. It must carry dust to a height of at least 1 m and last for at least 10 s. It is distinct from a dust blow in that it maintains a fast rotating coherent structure with vertical extension and distinct from a small dust whirl in its size and duration.*" Usually the instability is generated by strong solar heating, although cold air masses over warm ground may cause them. Vortices generated by purely mechanical forcing (that can for example also loft snow or leaves in conditions without strong heating) such as those caused by outflows from mesoscale convective systems (e.g. gustnados) are not usually considered to be dust devils, although this chapter will occasionally stray beyond the narrowest definition. Where such outflows are cold (as is often the case in desert storms due to evaporative cooling), however, the distinction is less clear cut. Related phenomena include fire whirls associated with forest fires and the like, where flames may be collimated in narrow rotating columns.

Fig. 1 Image of a dust devil in Arizona, observed in a NASA field campaign. The large dust devil has a small 'little brother' to the right. Credit: NASA/U. of Michigan

Table 1 Terms for dust devils. Note that the Navajo word should not be used in casual conversation as it often has evil connotations in that culture. Note also some remarks in the Culture section at the end of this paper

Location/language	Name	Etymology, Other remarks
Australia	Willy-willy	Aboriginal/onomatopoeic? sometimes 'Whirly-whirl'
Arabic	Djinn	Spirit, devil (qv the anglicized form 'genie')
Japan	Jin-sen-pu	Dust whirlwind. Also tsumuji-kaze 'whirlwind'
France	Tourbillon de poussière	Dust whirl
	Diable de poussière	Dust devil
	Trombe de sable	Sand trumpet (archaic)
Germany	Staubwirbel	Dust whirl
	Kleintrombe	Small trumpet
	Sandhose	Sand hose
	Wirbelwind	Whirlwind
	Staubteufel	Dust devil (most common)
Russia	pyl'niy (peschaniy) vikhr' пыльный вихрь	Dust (sand) whirl
China	尘卷风	Dust whirl
India	Suntargalis or devva	Devil wind
Navajo	Chiindii	Spirit
Spanish (inc. Americas)	diablo de polvo	Dust devil
	demonio de polvo	Dust demon
	remolino de polvo	Dust whirl
	burro	(Donkey)
Italian	diavolo di sabbia (diavolo di polvere)	Dust Devil (Sand Devil)

Tornados, while of interest because of some dynamical vortex similarities (see e.g. the chapter by Kurgansky et al. in this volume) are quite distinct from dust devils in that their driving energy is from latent heat in moisture rather than sensible heat. Further, tornados are invariably associated with clouds, from which the vortex descends, whereas dust devils most usually occur in clear sky conditions and the visible structure grows upwards from the ground. Tornados and dust devils may have similar aspect ratios (height/diameter) but tornados are almost always much larger than dust devils, and are usually much more intense (in terms of core pressure drop and wind speed). Waterspouts are essentially tornado-like, in being usually associated with cloud, although they are often similar to dust devils in size.

Hurricanes (and other cyclonic storms such as typhoons) have very much larger horizontal and vertical scales than dust devils, and rather smaller aspect ratios. The large horizontal scale leads to significant effects of planetary rotation (i.e. the Coriolis effect) which is negligible for dust devils. While associated with moist convection, hurricanes do have the feature that like dust devils, much of their energy is supplied at their base (and similar heat engine formalisms can be applied to the generation of mechanical energy in each), and there may be some analogies in the migration paths and vortex structure.

1.1 Dust Devil Terminology

Table 1 summarizes the terms used for dust devils in different languages. Clearly some terms are simply descriptive of the phenomenon ('dust whirl(wind)'), some are onomatopoeic, and

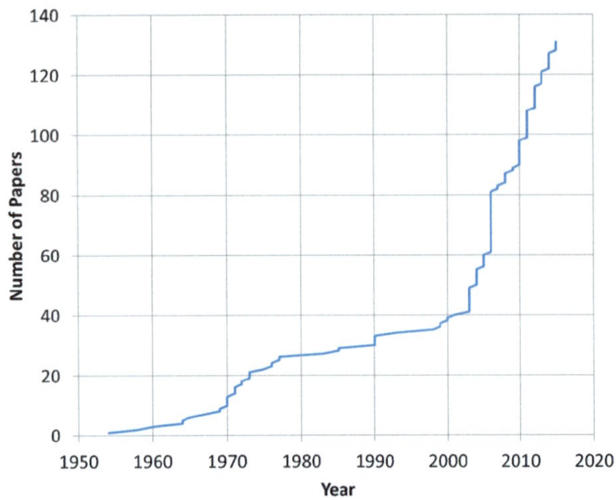

Fig. 2 The cumulative growth of the corpus of dust devil work in refereed journals in the ADS abstract database (http://adsabs.harvard.edu/). Note the slope change around 2003, and the jump circa 1970

some invoke the ephemeral/aethereal nature of them with reference to spirits or devils. It is of interest that the shape-inspired term 'trombe' appears in early French and German literature (sharing a root with the English 'trombone') but has largely fallen into disuse.

1.2 Scientific Significance of Dust Devil Studies

It is the general impression of the authors that the public finds dust devils 'interesting', a qualitative endorsement of the utility of their study. However, the scientific level of interest can be at least crudely quantified. An informal bibliometric study, searching for refereed journal papers with "Dust devil" in the title in the Smithsonian Physics/Astronomy database found some 131 papers 1930–2015. It is recognized that this database will not capture all the literature (particularly that in languages other than English), but it can be considered representative.

Figure 2 shows the cumulative number of papers versus time—it is seen that papers accumulate at the rate of about 1.5/yr until the early 21st century when there is an abrupt change in slope—essentially reflecting studies of dust devils on Mars as an ongoing enterprise—to ~4/yr. There is also a flurry of activity in the late 1960s/early 1970s reflecting work in the USA (see later). The particular jump in 2006 largely reflects a special issue of the journal Geophysical Research Letters that published papers discussed at a workshop devoted to dust devils hosted in Flagstaff, Arizona by Ronald Greeley in August 2005.

An inspection of the most-cited papers in this corpus (Table 2) attests to the importance of Mars discoveries, with half of the 'top 10' papers referring to Mars directly.

2 History of Dust Devil Studies—Earth

2.1 Overview

Dust devils are found (Fig. 3) in many parts of the world, although predominantly in desert regions. While dust devils have been reported from antiquity, their serious study began in the mid-19th century with the onset of global scientific exploration and growing commercial activity by the European powers: see the following sections on Britain, France and Germany.

Table 2 The most widely-cited dust devil papers (Harvard-SAO Database Title Search 'Dust Devil')

Year	Authors	Citations	Title
1985	Thomas & Gierasch	120	Dust Devils on Mars
1998	Renno et al.	93	A Simple Thermodynamical Theory for Dust Devils
1999	Metzger et al.	85	Dust Devil Vortices Seen by the Mars Pathfinder Camera
1983	Ryan & Lucich	79	Possible Dust Devils—Vortices on Mars
1960	Freier	71	The Electric Field of a Large Dust Devil
1970	Kaimal & Businger	71	Case Studies of a Convective Plume and a Dust Devil
1973	Sinclair	71	The Lower Structure of Dust Devils
2006	Balme & Greeley	67	Dust Devils on Earth and Mars
1969	Sinclair	61	General Characteristics of Dust Devils
2003	Greeley et al.	58	Martian Dust Devils: Laboratory Simulations

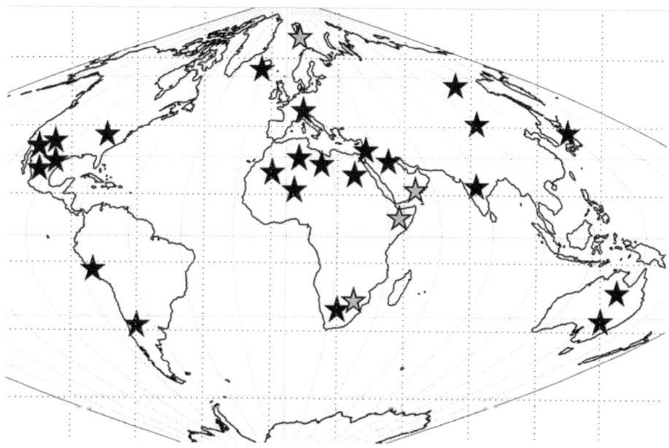

Fig. 3 Map of reported dust devil studies or sightings (*black stars*, not all historical reports included) in reports cited in this paper—note the predominance around 30° latitude, where the downwelling branch of the Hadley circulation leads to generally dryer conditions. *Grey stars* denote undocumented sightings by the present authors (UAE and Ethiopia) and their correspondents (Svalbard, Botswana)

In the 20th century, most scientific activity on dust devils was in the USA, a situation that broadly holds today. More recently, interest in dust devils has grown, in part due to environmental concerns, in Japan and China.

Dust devils would likely have remained a minor topic in meteorology, but for the Cold War. Specifically, nuclear weapons testing is typically conducted in desert areas for the obvious reason that these have low populations. Since dust devils might transport radioactive dust from these remote sites, some work on dust devils may have been sponsored by relevant agencies to assess this hazard (for example, Snow and McLelland's work sponsored by the US Defense Nuclear Agency) (Snow and McClelland 1990).

A second impetus to dust devil studies has been the exploration of Mars. While we review Mars dust devil work as an integrated whole in the next section, the impact of Mars exploration on dust devil research generally in different countries is noted here—work circa 1970 seems in part to have been associated with Mars and the US Viking program. The growing role of Europe in planetary exploration, with Mars Express, Beagle 2, and various proposed missions and involvements in NASA missions, has similarly stimulated dust devil research in other countries. In what follows we arrange the countries approximately by chronology of their first contributions to dust devil studies.

Note that it is not the purpose of this paper to comprehensively review all publications on dust devils. Some representative references from various countries and on various topics are identified only as a starting point. Other chapters in the present volume explore the literature more fully, and a useful overall review is Balme and Greeley (2006).

2.2 Italy and the Roman Empire

As our opening quotation suggests, the often dry and hot conditions around the Mediterranean Sea can yield dust. In military campaigns a cloud of dust kicked up by a marching or galloping army was often the earliest indication of its approach. Julius Caesar's various chronicles make only scant mention of dust raised by the wind, but Lucan's Pharsalia (an epic poetical account of the Roman Civil War, Lucanus, 65 C.E., 1926) has the following graphic account which offers a number of meteorological insights:

"*Libya has no mountains to oppose the rising wind and break its force, no cliffs to drive it back and scatter it, or to turn its hurricane to clear breezes; it does not fall upon forests and wear itself out by bending ancient oaks: all the land is level, and the wind travels freely and wreaks the fury of Aeolus all over the desert. There is no rain in the cloud of whirling dust which it drives furiously in circles; most of the land is lifted up by it and is suspended in the air, as the eddying motion is continuous. The needy Nasamonian sees his possessions flying in the wind and his dwelling blown to pieces; the Garamantian is laid bare, and his hut, beginning with the roof, is snatched away and flies aloft. Fire does not carry what it seizes to a greater height: as high as smoke may rise up and mar the face of day, so great is the dust that fills the air. And now, even fiercer than its wont, the wind attacked the Roman column; and the staggering men can find no footing on the sand, when even the spot they tread on is carried away. If Africa had a solid framework, so that the heavy weight of its cliffs might confine the South wind within hollow caverns, the wind would overset the whole world and wrench it from its foundations; but because the soil is easily driven about with its drifting sands, it remains stable by offering resistance at no point, and the lower stratum stands fast because the upper is dispersed. Driving furiously, the blast snatched up the men's helmets and shields and javelins, and rushed on, carrying them through the mighty void of heaven.*"

Clearly, this account may conflate dust/sand storms with dust devils, but it clearly notes the vortical motion characteristic of dust devils, and the fact that these strong motions are not associated with rainfall. Active saltation is described, and in addition to helmets and javelins being lofted, damage to structures is noted. Beyond these dramatic observations, some strikingly perceptive thoughts can be perceived—there is a hint that smoke plumes and dust reach the same height (both capped by the Planetary Boundary Layer, perhaps), and an argument that the low aerodynamic roughness of the African continent may enhance the ferocity of its winds.

After Roman times, the most notable studies are those by Leonardo da Vinci on vortices in general. He suggests colliding winds generate vorticity, and observed the following (Atalay and Wamsley 2009), which could perhaps be a tornado or waterspout rather than a dust devil:

"*It often happens that, when one wind meets another at an obtuse angle, these two winds circle around together and twine themselves into the shape of a huge column; and having thus condensed, the air acquires wright. I once saw such winds, raging around together, produce a hollow in the sand of the seashore as deep as the height of a man, removing from it stones of considerable size, and carrying sand and seaweed through the air for the space of a mile and dropping them in the water, whirling them around and transforming them into a dense column, which formed dark thick clouds at its upper extremity.*"

In connection with waterspouts in Italy, it may be noted that the young British scientist Michael Faraday wrote an interesting account of them while traveling with Humphrey Davy

Fig. 4 Dust devil depicted in Chinese ancient Poem thirteen centuries ago (drawn by Quanzong Zhang). Note that the dust column flares into a *plume at the top*, and occurs rather near sunset

in that country—his correspondence with Baddeley on dust devils is reported in that author's 1860 book.

As elsewhere, Mars exploration has stimulated terrestrial dust devil studies in Italy, and Rossi and Marinangeli (2004) reported the first observation of terrestrial counterparts to Martian dust devil tracks, in Niger.

2.3 China

In the year 737 C.E. (Kaiyuan 25th year, Tang Dynasty), the well-known ancient poet Wei Wang traveled to the frontier at the emperor's order. He wrote the famous poem *On Mission to the Frontier* after viewing the spectacular scenery on the way. In the poem, the sentence "*A plume of smoke rises up into the evening sky of the Great Desert, while the Yellow River dimmed as the setting sun goes down*" depicted the view of vast boundless desert and endless river (as shown in Fig. 4). In fact, the "smokes" mentioned in the sentence may be interpreted as the dust and sand lofted in the swirling winds of a dust devil.

China has extensive desert regions, especially in its north and west (see Fig. 5), and aeolian studies and efforts to mitigate the effects of sand dune migration on agricultural areas and infrastructure such as railway lines have become extensive, with research institutes devoted to aeolian studies. Sandstorms frequently occur in northern China (as many as 20 days a year in Mongolia), especially in Spring. In one recent example, in March in 2000, large sandstorms attacked Inner Mongolia and the Beijing district, with winds reaching force 8 or 9 (on the Beaufort scale, i.e. 17–24 m/s). Seven workers who worked in a two-floor building were blown down and two of them were killed.

As for dust devils, specifically, before the 21st century, there were only cursory studies. Le (1983) compiled the diurnal variation of 193 dust devils that occurred in Taklimakan Desert from 1964 to 1970. According to the data of Lei et al. (2005), during 1971–2000, 594 dust devils occurred in Shaanxi Province. Dust devils in these two places had some common characteristics. They began appearing after 9 a.m., became frequent in 12:00–15:00 (local time) and then declined. Before 8 a.m. And after 18:00, no dust devils occurred.

More recently, theoretical and simulation work has been pursued: e.g. Gu et al. (2006, 2010) simulated the formation process of dust devils with LES (Large Eddy Simulation) methods as a function of swirl ratio, and used an Euler-Lagrange approach to simulate the gas-solid two-phase flows of dust devils in the convective boundary layer. Gu et al. (2013)

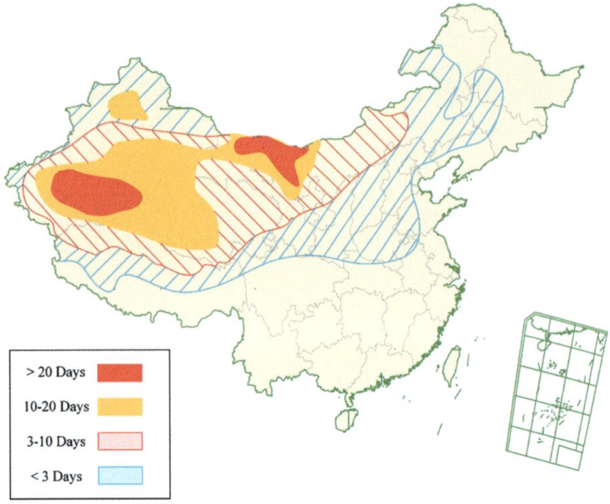

Fig. 5 Spatial distribution and the occurrence days per year of sandstorms in China

demonstrated a charging mechanism of sand particles with the adsorbed water on microporous surfaces of wind-blown sand, recognizing the mobility of water ions H^+/OH^- between the contacting sand grains with a temperature difference. Much of this work was aggregated in a book on these topics (Gu 2010). Additionally, the Turpan desert in Western China was the site of the first in-situ examination of dust devil tracks on Earth (Reiss et al. 2011a, 2011b).

2.4 Britain and the British Empire

While the cold and damp United Kingdom does not commonly have circumstances favorable to dust devil formation, many territories formerly under British administration have desert areas that do. Most notably these include North Africa, the Middle East, and Australia. The dispersal of soldiers, officials and other professionals across the globe, together with a strengthening tradition of reporting observations, created the opportunity for careful studies of dust devils beginning in the mid-1800s. It is of note that in the same period, the same motivations prompted the Royal Navy to commission "A Manual of Scientific Enquiry; prepared for the use of Officers in Her Majesty's Navy and Travellers in General", edited by the astronomer John Herschel (and including a chapter on meteorology by him, and one on geology by Charles Darwin). The Meteorology chapter (Herschel 1851) instructs readers on the care and use of meteorological instruments, and the careful recording of their measurements. It gives particular guidance on observations of whirlwinds:

'*All circumstances from the first trace or prognostic to their final dissipation should be minutely noticed... At what distance is the whirling motion of the air perceptible? What are the indications of the barometer during their approach and recess? Do any and what electrical phenomena accompany them.*'

A century and a half later, this remains good advice for dust devil study, on any planet.

Baddeley (1860), a surgeon in the Bengal Army, compiled an entire book, much of which is on the topic of dust devils—"Whirlwinds and Dust Storms of India: An Investigation into the Law of Wind and Revolving Storms at Sea". This appears to be one of the first detailed investigations published on dust devils.

The book struggles to develop a unified theory for vortical winds (such as cyclones as well as dust devils), and noting a number of electrical phenomena associated with dust devils makes the suggestion (erroneously) that electric forces are in fact the cause of the winds. This notion aside, the book usefully documents many visual observations of dust devils and storms in the vicinity of Lahore (now in modern Pakistan) including their ascent to more than 2000 ft, and the frequent presence of multiple cores. He notes the striking similarity of the multiple co-rotating cores to the motion of dancing Dervishes (an analogy repeated by others), and hints at a possible origin of this religious dance from the motion of dust devils:

"This remarkable sight gives the idea of a fairy dance round a ring; and the motions are from all accounts, exactly imitated by the dancing Dervishes of Turkey; one of their holy exercises being to whirl round and round like a top; singly, or in company with several others, performing at the same time a gyration round in a circle, as if their dance originated in the very phenomenon now described."

He describes following on foot one dust devil for a mile, then being able to penetrate the centre and walk along inside it—he notes "The centre was perfectly calm, while round about, the winds were blowing in every direction".

A casual sentence in the book (p. 81), noting electrical activity associated with dust devils observed with a gold-leaf electroscope and other means, makes the observation (13th October 1850) that a solution of starch and potassium iodide rapidly become discolored. Although the report does not offer any explanation or interpretation, it seems clear in retrospect that this is evidence of atmospheric oxidants, and specifically ozone, and anticipates by almost 150 years the consideration of dust devil oxidants in atmospheric chemistry on Mars. It may be noted that ozone was only discovered in 1840 (and so it may have been a 'hot topic' at the time) and that the starch-iodide reaction was developed soon thereafter as a specific test.

Various reports appear in the scientific and popular media of the period, documenting dust devils and other storms in various parts of the empire, notably India and South Africa. In the Straits Times, "Dust devils in India, Observations Made of Curious Phenomena", a report is made of observations of dust devils near Agra. Weather reports had predicted thunderstorms in the local area, which never manifested, and it was observed that "Numerous very large dust devils were forming during the middle of the day" instead. These dust devils were categorized to range in size from 20–100 m in diameter, in some cases exceeding 800 m in height. A curious observation was that of one dust devil that was noted to pause over a village for approximately 10 minutes, with no perceptible translational movement. Such stability and lack of movement is rare in dust devils and implies extremely quiescent ambient wind conditions, and though the reported static duration may potentially be somewhat exaggerated, it represents a rare observation of a stable, non-moving dust devil. The journalist also reported observing varying rates of rotation and dust vertical flux (assumed to be from visual observation), along with irregular masses of dust being raised in the vicinity of the base of the dust devil, resembling a "dust curtain" extending some 50 m from the dust devil itself, which was attributed to a descending current from just outside the limit of the upwelling dust column.

While not a meteorologist, it is worth noting William Rankine, a Scottish engineer who contributed significantly to the development of thermodynamics and its application to steam engine performance in the mid-late 1800s. Vortices figure prominently in his thinking, and his idealized model of a vortex (which like a temperature scale, and an engine cycle, bear his name) is often used in dust devil studies.

Little further British work on dust devils is of note until an important contribution by Flower (1936), a meteorologist working in Egypt for the Air Ministry (it is tempting to speculate whether Flower ever met or worked with Ralph Bagnold, in the same part of the empire at the same time; Bagnold was a desert soldier and pioneer in aeolian studies, his

1941 book 'Physics of Blown Sand and Desert Dunes' remains required reading today). Flower's tabulations of dust devil diameters, heights, longevity etc. remain one of the most comprehensive visual datasets, and were recently used to develop analytic correlations of diameter and longevity (Lorenz 2013).

As evidenced above, the UK has a long tradition in meteorological research, but while dust-raising in general and its impact on climate has been a current research topic, dust devils in particular have not received much 'mainstream' interest. Only a few reports of dust devil vortices appear in British journals, often from other parts of the (former) empire (e.g. Libya and Cyprus, McGinnigle 1966, 1970).

However, dust devil work in the 21st century at UK institutions was stimulated by the development of the Beagle 2 Mars lander, and the prospect of detecting dust devils in-situ on that mission. Some examples include a re-analysis of Viking data (Ringrose et al. 2003), consideration of the 'Delta-P effect' (Balme and Hagermann 2006), instrumentation trials and field measurements in Nevada (Mason et al. 2014), and analysis of Mars orbital imaging (Towner 2009).

2.5 Germany

The interdisciplinary research of Alexander von Humboldt opened new avenues in many branches of the natural sciences. From one of Humboldt's journeys to Southern America (Orinoco area), he offers what is likely one of the earliest 'scientific' reports on dust devils (von Humboldt 1849):

"*When under the vertical beam of the never-cloudy sun the charred grass layer has crumbled into dust, the hardened surface gapes open as if it would be shaken by strong quakes. When it is then touched by opposing air flows, whose quarrel balances through rotating movement, the plain allows a peculiar view. In the form of funnel shaped clouds with their tips sliding across the surface, sand rises vapor-like through the air-thin, electrically charged center of the whirl: similar to the rushing waterspouts that are feared by experienced skippers.*"

Other early reports about dust devils in German literature include observations in what is now Iraq by Schläfli (1870) who noted that small dust devils are frequent especially during summer. Early dust devil observations in Germany were also reported by Less (1891) and Budde (1883). The inventor of the photographic flash, Adolf Miethe (Miethe 1899) reported about ash devils (Aschehosen) in the Luneburg Heath in Lower Saxonia. These ash devils formed due to manmade peat fires; sometimes 20–50 per hour were observed with heights of 200–300 m. The author notes that the ash devils travel at speeds and in directions of the prevailing winds, and the formation ceased at wind speeds above $\sim$3.5 ms^{-1}.

While better-known for his theory of continental drift, the meteorologist, polar explorer and geoscientist Alfred Wegener made an interesting report (Wegener 1914) of dust devils on a lava plain, Odàđahraun, in Iceland, noting that he had previously only once seen a $\sim$20 m high dust devil on a hot summer day in Marburg, Germany. During their stay in Iceland in mid-June 1912, Wegener and his company saw 20–30 dust devils at a time on the wide plain with typical diameters of about 5–10 m and heights of more than 100 m. Wegener got the notion that dust devil height would be of the same order as the stratocumulus cloud base. In addition to dust devils, Wegener and his colleague frequently observed non-rotating convective plumes, which developed into dust devils and gave some insight into dust devil formation. Wegener notes that the dark volcanic sand would effectively absorb the solar heat, driving the formation of convective vortices. This report may be one of the highest-latitude field reports of dust devils and is quite widely cited thereafter, notably in Geiger's textbook

'The Climate Near the Ground' (1950). In this book, Geiger describes dust devils formed near Erlangen in southern Germany, where air flowing over a cliff edge induces vorticity causing devils on a heated chalky plain. Geiger notes an audible whistling of the devils, and goes on to summarize Flower's (1936) observations.

Letzmann (1939) aimed to facilitate a comprehensive investigation of dust devils by suggesting guidelines for the observations of atmospheric vortices, including dust devils. Among the focuses he suggests are shape, duration, track, atmospheric conditions, pressure changes and surface conditions, many of which are still used to characterize dust devils today.

Lammert (1947) summarizes an intensive period of dust devil occurrence in advance of a thunderstorm on a summer day in northern central Germany (Salzderhelden, Lower Saxony). Few minutes before the thunderstorm began, numerous (50–60 in total, 4–7 at a time) dust devils with 3–5 m diameter and heights of hundreds of meters occurred and lifted "whistling and howling corkscrew-like" considerable amounts of sediment. According to eye-witness reports, the dust devils were superseded by a dust storm generated by the convective downdrafts associated with the thunderstorm. The event came to an end with the starting rain.

Apart from natural dust devil occurrences in Germany (e.g. Koschmieder 1951), Maurer (1934) observed an interesting man-made dust devil. In May 1934, a group of about 30 young girls met on a sandy sports field in eastern Germany (Schwarza, Thuringia). As exercise, they were instructed to run in circles of about 15 m diameter. After four rounds, a dust devil began to form at the inside of the circle, grew to about 5 m height, left the field, and ceased quickly. Maurer emphasized the strong heating over sun exposed surfaces such as the sports field.

Recent papers on dust devils by German scientists have included some work on Large Eddy Simulation (e.g. Raasch and Franke 2011), and field measurements of dust-lifting in the Morroccan desert (Ansmann et al. 2009). The development of the High-Resolution Stereo Camera (HRSC), and its arrival at the Red Planet in 2003 on Mars Express stimulated significant work on remote observation of Martian dust devils (e.g. Stanzel et al. 2008; Reiss et al. 2011b) as well as terrestrial dust devil tracks (e.g. Reiss et al. 2010, 2011b, 2013).

2.6 France

The situation of historical dust devil reports in France is rather similar to Germany and the United Kingdom. Most of France itself is not prone to the formation of dust devils; convective vortices might occur on very hot summer days, especially in Southern France, but dusty material to be carried within the vortices are rarely available to form proper dust devils. As with the United Kingdom, opportunities to report dust devils arose when France administered colonies in North Africa and Central Asia, and soldiers, merchants, and travellers used to report observations of desert environments they were exploring in the middle of the nineteenth century. Other reports were based on observations in arid regions in America (e.g. Mexico, reports by M. Virlet d'Aoust) not under French administration.

Dust devils were usually reported as either anecdotical features within a broader exploration report, or as an additional dusty example in the taxonomy of natural vortices (as in the book by Zurcher and Margoller (1876) 'Trombes et Cyclones', which cites Baddaley's work). However, following the development of the French meteorological network in the late 1850s and 1860s, the interest for meteorological phenomena became more and more prominent amongst French scientists. Reports about dust devils started to be a matter of scientific interest, with questions arising about their properties and formation.

Fig. 6 A French woodcut from the 19th century 'Trombes de Sable dans la Steppe', showing the threatening aspect of large dust devils, perhaps like those reported by Prejevalski

For instance, in a June 1883 report to the Comptes Rendus de l'Académie des Science, Colonel Prejevalski[1] writes: "The various morphologies of dust devils are reminiscent of marine vortices: dust is transported upward, often in a spiraling motion". This description of a dust devil within an exploration report written by a military traveler (likely in Central Asia—see Fig. 6) is typical of what was proposed at the time. However, it is notable that this report is not simply describing the occurrence of dust devils and their morphology: it proposes an analogy with another natural whirling phenomena, and describes the dust transport within the dust devil. This report led M. Faye, who authored several other discussions about dust devils in Comptes Rendus de l'Académie des Sciences and Annuaire du Bureau des Longitudes since 1877, to publish a report in the same journal to severely criticize Colonel Prejevalski's observations.

According to him, those "simply cannot be regarded as a scientific document worth being published in the Comptes Rendus"! M. Faye then explains that dust devils, similarly to tornados, find their swirling motion originating from regional-scale vorticity (thus comes from above the boundary layer rather than from the surface), and this motion when it finally gets down to the surface would cause dust lifting. He concludes that what Colonel Prejevalski reports about the apparent motion of dust within a dust devil is an illusion, one of the same kind that led sea men to conclude that marine vortices would pump water from the surface to the higher atmosphere. M. Faye might be right for marine tornados, but we now know he had a wrong interpretation about dust devils, while Colonel Prejevalski is completely in line with modern observations of those phenomena. This discussion in the French literature parallels analogous debates about cyclones and other weather systems in (at least) Britain and America in the 19th century.

Recent dust devil research in France, like that in Germany, the UK and Italy, has been stimulated by climate studies and interest in dust-lifting in general, and by studies of the meteorology of Mars. The emphasis there has been on numerical simulation, e.g. Spiga and Forget (2009), although the French development of a seismometer instrument for the InSight Mars lander to be launched in 2016 may stimulate new observational studies.

[1] Although mentioned here since the discussion is in the French literature, in fact Nikolai Przhevalsky (transliterations of the Cyrillic vary) was a Russian geographer who explored Central and East Asia, which is presumably where his familiarity with dust devils arose. A wild horse species in that region is named after him.

2.7 United States of America

Although widespread familiarity with dust devils awaited the modern settlement of the desert Southwest, dust devils were recognized among the pantheon of rotating winds in William Ferrel's[2] discussion of cyclones and tornados (Ferrel 1881), where he describes 'sand spouts'. That work is notable in recognizing that the centripetal acceleration of the sand and dust must be generated somehow, and he ascribes it to the radial drag of the in-rushing winds. Further, he suggests that since the drag on each particle varies as its diameter squared, while the required force varies as diameter cubed, 'particles of sand of different sizes arrange themselves at different distances from the center', an important insight.

Although the Eastern part of the USA does not often encounter dust devils, having a more damp and temperate climate than the west, one of the earliest reports of dust devils was made there by Benjamin Franklin in Maryland (Franklin 1755). The report is remarkable not only in its observational detail, but also for Franklin's experiment in modifying the devil.

'*It appeared in the form of a sugar-loaf, spinning on its point, moving up the hill towards us… its smaller part near the ground appeared no bigger than a common barrel, but widening upwards, it seemed at 40 or 50 feet high, to be 20 or 30 feet in diameter.… I followed it, riding close by its side, and observed its licking up, in its progress, all the dust that was under its smaller part. As it is a common opinion that a shot, fired through a water-spout, will break it, I tried to break this little whirlwind, by striking my whip frequently through it, but without any effect.… Soon after, it quitted the road and took into the woods, growing every moment larger and stronger, raising, instead of dust, the old dry leaves with which the ground was thick covered, and making a great noise with them and the branches of the trees, bending some tall trees round in a circle swiftly and very surprisingly, though the progressive motion of the whirl was not so swift but that a man on foot might have kept pace with it; but the circular motion was amazingly rapid. By the leaves it was now filled with, I could plainly perceive that the current of air they were driven by, moved upwards in a spiral line; and when I saw the trunks and bodies of large trees invelop'd in the passing whirl, which continued intire after it had left them I no longer wondered that my whip had no effect on it in its smaller state. I accompanied it about three quarters of a mile, till some limbs of dead trees, broken off by the whirl, flying about and falling near me, made me more apprehensive of danger; and then I stopped, looking at the top of it as it went on, which was visible, by means of the leaves contained in it, for a very great height above the trees. Many of the leaves, as they got loose from the upper and widest part, were scattered in the wind; but so great was their height in the air, that they appeared no bigger than flies.*

My son, who was by this time come up with me, followed The whirlwind till it left the woods, and crossed an old tobacco-field, where, finding neither dust nor leaves to take up, it gradually became invisible below as it went away over that field. The course of the general wind then blowing was along with us as we travelled, and the progressive motion of the whirlwind was in a direction nearly opposite, though it did not keep a strait line, nor was its progressive motion uniform, it making little sallies on either hand as it went, proceeding sometimes faster and sometimes slower, and seeming sometimes for a few seconds almost stationary, then starting forward pretty fast again.'

In addition, the following observation by Thomas Jefferson, a farmer and careful recorder of meteorological phenomena as well as principal author of the Declaration of Independence, is of interest as being a likely description of dustless vortices (Jefferson 1787):

[2]The meridional circulation 'Ferrel cell' is named after this meteorologist.

Going out into the open air, in the temperate, and in the warm months of the year, we often meet with bodies of warm air, which, passing by us in two or three seconds, do not afford time to the most sensible thermometer to seize their temperature. Judging from my feelings only, I think they approach the ordinary heat of the human body. Some of them perhaps go a little beyond it. They are of about 20 or 30 feet diameter horizontally. Of their height we have no experience; but probably they are globular volumes wafted or rolled along with the wind. But whence taken, where found, or how generated? They are not to be ascribed to Volcanos, because we have none. They do not happen in the winter when the farmers kindle large fires in clearing up their grounds. They are not confined to the spring season, when we have fires which traverse whole counties, consuming the leaves which have fallen from the trees. And they are too frequent and general to be ascribed to accidental fires. I am persuaded their cause must be sought for in the atmosphere itself, to aid us in which I know but of these constant circumstances; a dry air; a temperature as warm at least as that of the spring or autumn; and a moderate current of wind. They are most frequent about sun-set; rare in the middle parts of the day; and I do not recollect having ever met with them in the morning.

Dust devils occur in dry areas of the USA outside the Southwest, as in this observation of multiple vortices rotating around a common column in South Carolina (Anonymous 1902):

"*It consisted of four separate whirlwinds which followed each other to the left around the center of a circle 10 or 15 feet in diameter, like horses going around a horse power thrashing machine. The whole circle also seemed to be moving to the left and around the center of an enlarging coil. The motion was made apparent by dust taken up from the soil, and it could not well be seen above 10 or 15 feet from the ground. Sometimes one or more of the small whirls would rise so as not to be visible, but presently it would touch the soil again in its regular place in the procession. This beautiful and curious motion continued for five minutes or more over a spot only about 100 feet in diameter. It then advanced northward the four whirls enlarging their circle to about 75 yards and then vanishing.*"

Ives (1947) wrote a review of dust devils, recording his own observation of dust devils in the Bonneville salt flats in Utah, one lasting some seven hours and traveling 40 miles. He also notes that dust devils can be started sometimes by driving a jeep across, or firing a shell into, the 'lake' in a mirage. Mirages require strongly superheated air, which is unstable to convection, so these anecdotes make sense. He speculates that activity of small animals might therefore trigger many dust devils—although this was obviously before the discovery of dust devils on Mars.

He also summarizes some in-situ measurements of meteorological quantities, and noted reports that rabbits and hares can be lofted by dust devils, implying updrafts of 35 mph or more. He notes that such encounters were usually fatal to the animals involved, although he observed himself kangaroo rats falling from devils (and surviving). He then deduced (by dropping such rats from a tower and measuring their terminal velocity) the corresponding updraft velocities as 25 mph, recording memorably that the animal was "apparently unhurt, although usually very angry".

Wyett (1954) reported a dust devil encounter at the airport in St. George, Utah, wherein the pressure drop was recorded by a barograph (see Fig. 7). Several papers emerge in the 1950s and 1960s on electric fields in dust devils, as well as on analytic models of dust devil vortices and considerations of their energetics. There is not space for a full review here—the reader is referred to other papers in this volume, and to an earlier review by Balme and Greeley (2006).

Many recognize the pivotal role of Peter C. Sinclair in dust devil research. Sinclair was a research student in the Atmospheric Sciences department at the University of Arizona in

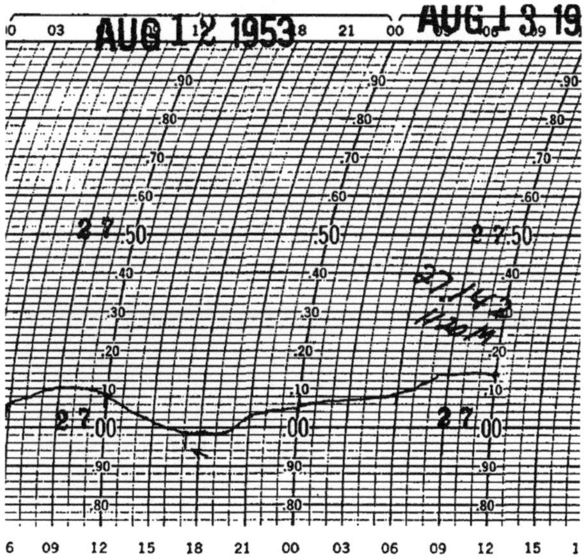

Fig. 7 Perhaps one of the first instrumental records of a dust devil encounter (arrowed), after Wyett (1954). The barograph scale is in inches of mercury—the 0.04 inch drop corresponds to ~1.4 hPa

Tucson, and generated some of the first usefully-robust statistics on dust devil occurrence and properties, as well as pioneering the active "chase and sample" methodology. A preliminary report (Sinclair 1964) captures the highlights of his work, initially using a hand-carried mobile observing station (Fig. 8).

Around the time of Sinclair's early work, an important survey was conducted by Lambeth (1966) of the Army Electronics Command at the White Sands Missile Range in New Mexico. This survey lasted May-September, using a linear array of six recording stations. When a human operator saw a dust devil, chart recorders were activated. Only 21 events were recorded in the five months of operation, a rather modest haul. Lambeth concludes with the suggestion that "dust devils can be measured in adequate numbers only by using mobile sensors".

Sinclair had by this time started to make many encounters using an impressive jeep-borne meteorology tower (Fig. 9), as well as measurements in a sailplane of the dust plumes atop dust devil. Sinclair's papers (1965, 1969, 1973) and thesis are well worth reading even today, noting such challenges as the high-speed data acquisition required on a hot, dusty, bouncing jeep in the desert (a cine camera recorded film of instrument readings!). Sinclair also performed a couple of visual surveys from fixed stations, data still referred to today.

It is of note that Sinclair's work was funded by the Office of Naval Research (ONR), although it is not obvious whether this had any connection with nuclear testing, or was merely one of many areas of fundamental research supported by the US Defense Department.

At this point space exploration and dust devil meteorology begin to intersect. J.A. Ryan, at the Missile and Space Systems Division of the Douglas Aircraft Corporation, speculated (Ryan 1964) that transient winds associated with 'a cyclonic system (somewhat analogous to a terrestrial dust devil)' might be responsible for the yellow clouds, presumed to be of dust, that had been seen telescopically on Mars. Later, Ryan and J. Carroll (then of the 'Space Sciences Division' of McDonnell-Douglas: the space program now allowed the progressively-consolidating aerospace industry to employ their own scientists, a situation that unfortunately is not widespread today) with NASA funding published several papers on field observations of dust devils in the Mojave desert in California, with a fixed meteorological station (e.g. Ryan and Carroll 1970). They appear to have been more lucky than

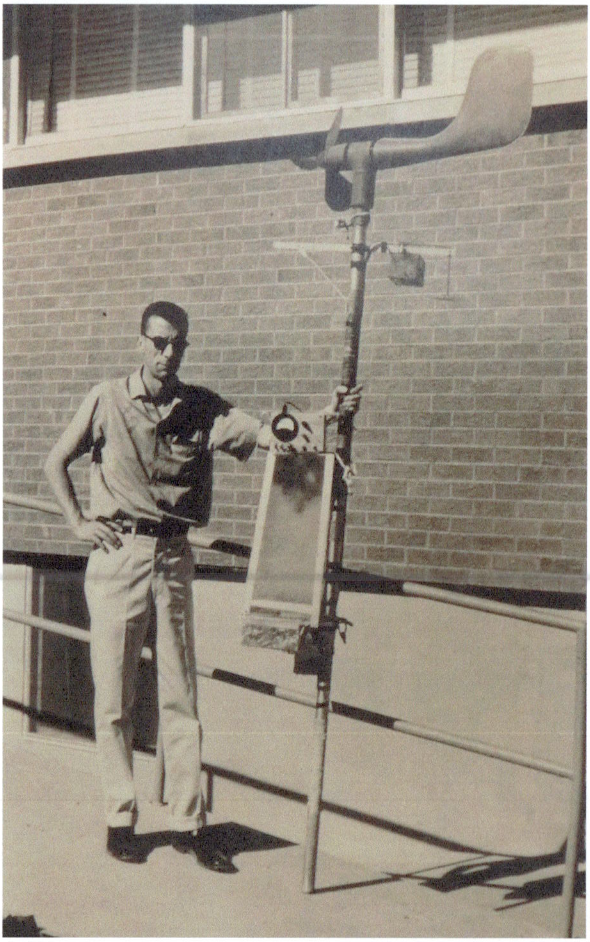

Fig. 8 Peter Sinclair with his portable measurement station with rather prominent anemometer and wind vane (from his Ph.D. thesis). The *box with electrical dial* is a sensitive microbarometer designed by Sinclair

Lambeth, observing 80 dust devils in just a few weeks, but they helped make their own luck by 'grooming' the surface around their mast to ensure dust availability.

After the early 1970s, dust devil research declined somewhat—a notable exception being Snow and McLelland's survey at White Sands (New Mexico) in the late 1980s, funded by the Defense Nuclear Agency—until the resurge stimulated by Mars Pathfinder in the late 1990s (Metzger et al. 1999). Several field campaigns took place, in Eloy (between Phoenix and Tucson, Arizona) and at ElDorado Playa near Boulder City (just outside Las Vegas, Nevada), in many instances in order to trial instruments for the aborted Mars 2001 lander in a campaign called MATADOR (e.g. Tratt et al. 2003), for Beagle 2, the Mars Science Laboratory and other projects.

2.8 Australia

As a large, arid country, Australia sees frequent dust devil activity, and a local Aboriginal term is frequently adopted—"willy-willy". Some notable researches have been stimulated there. The Australian Civil Aviation Authority (Bell 1969) noted that dust devils can present an appreciable hazard to light aircraft (which are an important means of transport and com-

Fig. 9 Arizona desert west of Tucson, with Peter Sinclair's mobile meteorology tower mounted on the surplus jeep he used to penetrate dust devils (from Sinclair's Ph.D. thesis)

munication, given Australia's large distances) and Spillane and Hess (1988) reviewed dry convective activity and its role in air accidents. Hess and Spillane (1990) reviewed the characteristics of dust devils in Australia, drawing on aviation reports, and noted that devils tend to ascend to a fraction (0.1 to 0.5) of the boundary layer height. They also suggested a condition of dust devil formation—$h/L > 50$, where h is the boundary layer height and L the Obukhov length (this type of condition may have originated with Deardorff in the USA). More recently, Oke et al. (2007a, 2007b) conducted a field study of dust devils, the most extensive reported since Snow and McLelland. Beyond dust devils proper, it may be noted that Australia's partly vegetated drylands (the "Bush") are subject to periodic wildfires, and these often produce quite prominent fire devils. Also related are observations by Hess et al. (1988), who identified convection-driven updrafts that they proposed are similar to dust devils, although they occurred over Port Phillip Bay in Victoria, Australia when cool air was advected over warm water.

Recently, the signatures of dust devil vortices were identified in infrasound (i.e. high-pass-filtered pressure) signals at a Comprehensive Test Ban Treaty Organization (CTBTO) monitoring site at Wurramunga in Australia's Northern Territory (Lorenz and Christie 2015).

2.9 Japan

Although there are no deserts or arid regions in the Japanese islands, dust devils are sometimes reported on fine days especially in warm seasons. They commonly occur over sand-paved playgrounds in schools and in sports fields, because strong surface heating under fine-weather conditions generates intense thermals locally. An example was reported in Tachikawa and Deguchi (1987) who showed that a whirlwind blew away a tent over the playground in an elementary school during a sports festival at the school in October 1986.

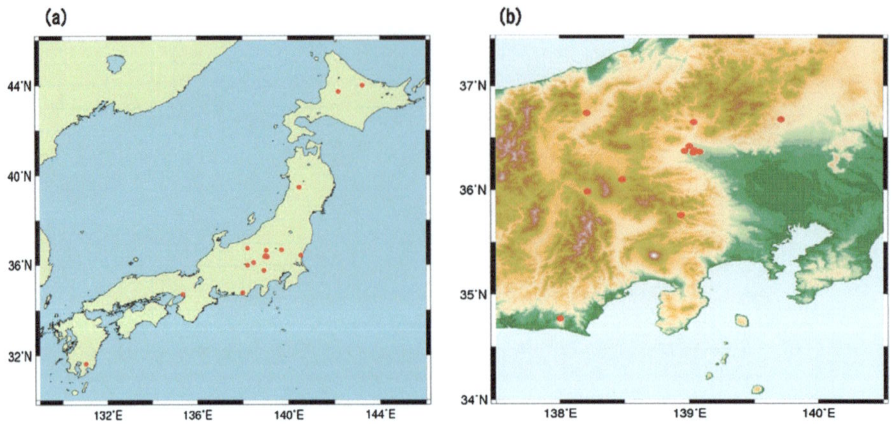

Fig. 10 The locations of dust devil occurrence in Japan from 1961 to 2009, based on JMA database. (**a**) The locations over the Japanese island, and (**b**) those in the central part of Japan

The Japan Meteorological Agency (JMA) archives the information on tornados and other gust events that occur in Japan on the web (http://www.data.jma.go.jp/obd/stats/data/bosai/tornado/index.html). Based on an inspection of this database for the present paper, it is found that there were 18 recorded occurrences of dust devils from 1961 to 2009 in Japan. Figure 10a exhibits the locations of dust devil occurrence in Japan during this time period. More frequent occurrence is seen in the central part of Japan. Figure 10b shows an enlarged map corresponding to a central part of Japan. Note that this JMA database is not comprehensive, because the information is obtained by field investigations by JMA officials after the events. For example, the event reported by Tachikawa and Deguchi (1987) is not included in the JMA's database. Although the JMA's database does not provide full coverage for the dust-devil occurrence in Japan, the geographical features shown in Fig. 10 seem to be consistent with the notion that dust devils favorably occur in the inland regions in Japan.

Since dust devils have typically a micro-scale (typically on the order of 1 m) with a very short lifetime (on the order of 10 seconds to a few minutes), direct meteorological observations of such events are very rare. The impacts of dust devils in Japan are not significant; however, damages such as shown in Tachikawa and Deguchi (1987) are notable. There are sometimes newspaper reports typically in the warm season every year on the occurrence of damage due to gusty winds produced by dust devils.

Recently remote-sensing techniques have become a powerful tool for measuring micro-scale vortices including dust devils in Japan. Inoue et al. (2011) successfully detected vortices such as tornados and mesocyclones, although not dust devils, during a winter storm event with the use of a low-level scanning X-band Doppler radar. Fujiwara et al. (2011, 2012) used a three-dimensional scanning Doppler lidar to detect dust-devil-like vortices that develop in the daytime under fair-weather convective conditions over an urban area. They revealed that invisible dust devils frequently occur from long-term observations. Doppler lidar was also used to detect an otherwise-invisible small-scale vortex over the sea i.e. a waterspout (Fujiwara and Fujiyoshi 2014). Ito and Niino (2014) applied a Particle Image Velocimetry (PIV) method to estimate the wind field in a dust devil which they encountered in Arizona, USA. They estimated the maximum tangential wind in the dust devil as $16\ \mathrm{m\,s^{-1}}$.

2.10 Russia, Canada, Argentina, Chile, Denmark and Other Countries

In Russia dust (sand) devils regularly occur over its southern arid and semi-arid parts spanning Kalmykia, Dagestan and Astrakhan regions (over the lowland north-west from the Kaspian Sea), and also over steppe territories in Voronezh and Rostov regions, and others. In particular, pilots flying over Kalmykia on light airplanes often observe high vertical dust columns associated with dust devils. Dust devils are observed in Kazakhstan, now an independent state, especially near the semi-dried Aral Sea. The observations by Przevalsky in the 1800s (reported in the 'France' section) were likely made in central Asia. Kennan (1886) describes 'great whirling columns of sand, 100–150 ft in height swept slowly and majestically across the sun-scorched plain' in the Kirgis steppes in Western Siberia.

A dust devil was observed (Roland Neuber, AWI Potsdam, personal communication to MK) in Svalbard in the late spring after snow and ice melting and forming a clear spot of dry sandy land, over which the dust devil developed: at 78°N, this observation is more poleward than that of Wegener in Iceland, and we have recorded it in Fig. 2.

In Chile dust devils occur in the north in the Atacama Desert and likely in the extreme south in Patagonia. In the Atacama Desert the dust devil activity was regularly reported by local people and also by travelers across the desert but the first regular field campaign to study dust devils was organized only in 2008–2009 (Kurgansky et al. 2011). A remarkable site was chosen near the settlement of Huara (about 50 km from the city of Iquique) with its unprecedented level of dust devil activity (up to more than 60 dust devils visible simultaneously in a 360° panorama). A site near Huara is also famous for its petroglyphs, the largest in South America. Etymologically the naming Huara is usually attributed to the star (with the reference to *aimara* language), but by an alternative version, it might be related to the dust devil, which is so frequent there. The Huara dust-devil field campaign was the second one on the South America sub-continent (historically, the first dust-devil field study in the South America of which we are aware is by Metzger et al. (2010)).

Recently, Lorenz and Radebaugh (2016) report observations of abundant dust devils at very high elevations (3600 m) in the Argentine Puna (a plateau somewhat southeast of, but higher than, the Atacama and Altiplano). The Campo Pedro Pomez site at which they measured several intense dust devils is the boundary between a plain of giant gravel ripples, and a field of yardangs carved in volcanic ash (ignimbrite)—a magnificent interplay of aeolian processes. Their report provides some circumstantial evidence that this site has quantitatively (2.5–10x) higher frequencies of dust devil occurrence than low-altitude sites like Eldorado. It may be noted, however, that over a century ago the Scottish astronomer Charles Piazzi-Smyth documented (Piazzi-Smyth 1858) several dust devil encounters, 'being peppered with little pebbles... almost blinded by dust' at an elevation of about 3000 m in Tenerife, while exploring its volcanic mountains as possible sites for astronomical observatories.

The development of meteorological instrumentation, which is generally rather small and affordable, for Mars landers opened the way for some countries with modest budgets to contribute to dust devil studies: in particular, Finland, Denmark and Canada contributed pressure sensors, a wind indicator and a lidar respectively to the Phoenix mars lander mission, stimulating several papers. More recently Finland and Spain have contributed to the instrumentation on the Curiosity rover, although so far rather few dust devils have been seen. One report of a dust devil in Canada exists, in sub-arctic Labrador at 65°N (Grant 1949).

Finally, although we are unaware of detailed reports on dust devils in Peru and Niger, there have been studies of dust devil tracks in those countries (e.g. Rossi and Marinangeli 2004; Hesse 2012 and Reiss et al. 2013). Additionally, the authors of this paper have seen dust devils in Ethiopia and the United Arab Emirates, and have heard of them in Botswana.

Fig. 11 Viking Orbiter 2 image (ID 038B25) from September 1976 at 72 m/pixel showing several dust devils with their shadows, the illumination coming from the bottom of the image. The crater is about 19 km in diameter. Image center is Latitude 40.4°N Longitude 215.0°E. The solar elevation was about 45 degrees, so the shadow length indicates many plumes were several km high. Viking images including this one were studied by Thomas and Gierasch (1985). The regular-spaced black dots are reseau marks to correct for geometric distortion in the vidicon camera

3 History of Dust Devil Studies—Mars

The existence of dust devils on Mars was proposed well before any direct evidence for their presence was acquired, both Neubauer (1966) and Gierasch and Goody (1973) predicting that Mars' arid surface, lack of cloud cover and large diurnal temperature variations should provide excellent opportunities for the presence of dust loaded convective vortices. Even before those works, Ryan (1964) had made relevant speculation about the formation of 'yellow clouds'.

Indirect confirmation of these predictions came first from meteorology data sent back by the Viking 2 lander. In their analysis of the Viking 2 wind speed, wind direction and temperature data Ryan and Lucich (1983) showed that excursions in these parameters could be fit by models of convective vortices passing over or near the lander. More than a hundred such encounters were noted, although only a few were thought to be sufficiently intense that they could have lifted sediments and become visible dust devils. Visual confirmation of the presence of dust devils on Mars would await the results of analysis of images returned by the Viking Orbiter spacecraft, more than five years after the missions themselves had come to an end—Thomas and Gierasch (1985) finally presented observations (see Fig. 11) of columnar dust clouds on the martian surface that provided convincing evidence for the presence of dust devils on Mars.

Observations of active martian dust devils were also made from the surface, first by the Mars Pathfinder mission (Metzger et al. 1999) and then in large numbers by the Mars Exploration Rover (MER) "Spirit" mission (Greeley et al. 2006, 2010). (Note that the Viking lander 'camera' was a facsimile imager which took many minutes to scan the scene, and so was not suitable for acquiring pictures of fast-moving objects like dust devils: famously the image of a tortoise in ground tests was smeared by its motion—e.g. Mutch 1978.) The Pathfinder images of dust devils (e.g. Fig. 12) are noisy and required significant processing to allow the dust-loaded vortices to be picked out against the dusty sky (note that unlike most subsequent Mars missions, Pathfinder data was transmitted directly to Earth, there being no relay orbiters yet in place, so the image bandwidth was somewhat more limited). The MER observations, however, were made from an elevated vantage point atop the Columbia

Fig. 12 Dust devil on Mars observed on the horizon by the Mars Pathfinder imager in 1997. Despite the modest image quality due to JPEG compression artifacts the detection of these features stimulated much modern dust devil research

Fig. 13 Dust devils seen on Sol 543 (July 13, 2005) of the Spirit rover mission at Gusev crater. Improvements due to the experience of scientist operators, to the better cameras, and the higher bandwidth, lead to much higher quality images. These are just four frames of a movie sequence, showing the dust devils evolving and migrating from right to left. Note that the dust devils in the foreground have a faster angular motion and are seen only in two frames, while the distant tall devil is seen in all four

Hills, and, in many images, dust devils on the valley floor are clearly visible against the background terrain beneath. In fact, the images of lines of dust devils marching across the valley are so similar to those of dust devils in arid regions on Earth that it is sometimes hard to tell which planet is being observed.

Further in-situ measurements (see also Chap. 2 by Murphy et al. 2016, and Chap. 10 by Neakrase et al. 2016) and surface observations of dust devils were made by the 2008 Phoenix Mission (Ellehoj et al. 2010), showing that dust devils were active in the martian

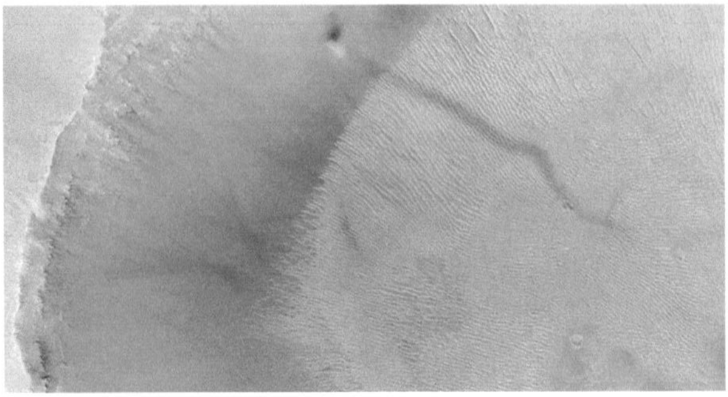

Fig. 14 A dust devil on Mars, photographed by Mars Global Surveyor, climbing the wall of a crater. The *long dark, slightly cycloidal streak* is the track left by the dust devil, seen as a bright column at the top of the image. A short shadow extends from the base of the devil to the top-right. MSSS MGS image E16-01487

arctic. A disappointment for martian dust devil research, though, was that the 2004 MER missions were not equipped with in-situ meteorology instruments. Hence while Spirit acquired glorious images of dust devils (e.g. Fig. 13), no correlations with vortex wind speeds, pressure and temperature excursions, nor contextual weather data could be made. Although this was rectified for the 2012 (and ongoing at time of writing) Mars Science Laboratory 'Curiosity' Rover, its landing site—Gale Crater—does not appear to be an area with much dust devil activity—only a single visible devil has been reported after several years of observations (Moores et al. 2015).

One of the major differences between dust devil activity on Earth and Mars is that martian dust devils often leave clear traces of their passage across the surfaces they move over—which is almost never the case on Earth—in the form of linear, curvilinear and even curlicue-shaped albedo markings. Dust devil tracks (see review Chapter by Reiss et al. 2016) were first recognized in Viking Orbiter images and interpreted to be 'tornado' tracks (Grant and Schultz 1987—their transient nature was recognized by comparison with Mariner 9 images; they were not suggesting moist tornados of course, but rather "baroclinic wave passage through an area of instability can trigger convective uplift and strong shear leading to tornadic-intensity vortex formation. On Mars in the late summer, the occurrence of atmospheric instability, a deep convective layer, and onset of baroclinic wave passage represent a similar situation"). However, with the arrival of the Mars Global Surveyor mission, and with it the Mars Orbiter Camera (MOC) providing narrow-angle images of $\sim$2 m/pixel, it soon became clear that these dark (and occasionally bright) lineaments were formed by dust devils: one noted MOC image showing a dust devil 'caught in the act' of forming a track (Fig. 14; Edgett and Malin 2000). It was recently noted (Lorenz and Zimbelman 2014—see also chapter in this volume by Reiss et al.) that in fact dust devil tracks can be clearly seen—but were not recognized at the time—in Mariner 9 images of Proctor crater in the Hellespontus region of Mars, images in which dunes on Mars were first reported. It was assumed (J. Cutts, personal communication to RL) that these dark markings were simply fissures on the ground filled with dark material.

The MOC dataset also provided the first opportunity to study dust devils as a population—the extensive coverage in time and space of both active dust devils and their tracks leading to several key conclusions, including (i) martian dust devils are, like on Earth, most active in spring and summer (e.g. Whelley and Greeley 2008), (ii) dust devils do not appear to trigger global dust storms on Mars (e.g., Balme et al. 2003a, 2003b) but do contribute to the persisting haziness of Mars' atmosphere (Cantor et al. 2006), (iii) dust devils are more common in some parts of Mars than in others, especially in the Amazonis Planitia region (Cantor et al. 2006). Further discussion on orbital observations of dust devils at Mars is given in the Chapter by Fenton et al. (2016).

In the last 15 years, martian dust devil data collection has focused on three areas: observations of dust devils tracks from orbit, observations of active dust devils from orbit, and observations and measurements of dust devils from the surface. Recent advances include the first measurements from orbit of horizontal forward velocity of martian dust devils (tens of meters per second, based on multi-image acquisitions from the ESA Mars Express mission; Stanzel et al. 2008; Reiss et al. 2011b and from the NASA Mars Reconnaissance Orbiter mission; Reiss et al. 2014) and, again from orbit, of wind speeds in martian dust devils, using comparison of the overlapping (but temporally slightly separated) RGB image strips in colour "HiRISE" images (wind speeds of 20–30 ms^{-1} were measured; Choi and Dundas 2011).

As for in-situ measurements on landers, while wind speed data have been limited (see chapter by Murphy et al.) the detection of vortices—at least some of which are dust-laden—in the long time-series of pressure measurements from the Pathfinder (1997—Murphy and Nelli 2002) and Phoenix (2008—Ellehoj et al. 2010) landers allowed for the first time the systematic evaluation of the vortex population (Lorenz 2012) and its comparison with visual observations of dust devils (Lorenz and Reiss 2015). For many years—indeed since Ryan and Lucich (1983) and until only recently—the high-time resolution meteorological data needed to detect vortices was better at Mars than at Earth. Mars data have stimulated much numerical modeling work, applying LES methods (e.g. see Chap. 7 by Spiga et al. 2016) at high spatial resolutions. An outstanding question remains—much more so at Mars than Earth—how the dust is actually lifted, since (as recognized by Ryan 1964) the thin Martian atmosphere requires very high wind speeds to overcome weight and cohesion of dust. Various solutions have been proposed, including 'sandblasting' by larger particles (Greeley 2002), and the 'Delta-P' (suction) effect (Balme and Greeley 2006).

4 Applications and Relevance of Dust Devil Research

The motivations for Mars and Earth dust devil research are several. Most obviously, 'because they are there'—dust devils are one of the most dynamic observable features of the Martian environment. Their transient nature makes them 'interesting' objects to study and chase on Earth, an appeal which has motivated much unfunded casual research. In this section we list some other motivations and applications of dust devil studies

4.1 Dust Devils as Dust Injectors into the Atmosphere

Dust in the terrestrial and Martian atmospheres is important for several reasons (e.g. Knippertz and Stuut 2014; Read and Lewis 2004). Dust aerosols interact with radiation through scattering and absorption of shortwave radiation and through absorption and re-emission of longwave radiation. In the terrestrial atmosphere, dust particles act as ice and cloud condensation nuclei and can therefore influence cloud properties and precipitation. Moreover, the removal of dust from the soil, and its redeposition elsewhere, can be important in transporting toxins (see later), nutrients, minerals and carbon, thereby influencing continental and marine ecosystems as well as animals and humans. For example, Saharan dust is known to be an important nutrient source for the Amazon forest (e.g. Bristow et al. 2010). Not least, dust presents an air quality concern, thereby directly affecting human life.

Although the importance of dust in the terrestrial and martian climates is evident, the contribution of dust devils to the Earth and Mars dust budgets is uncertain. On Earth, the first estimate by Koch and Renno (2005) was that dust devils may contribute by about 35 %

to the total dust burden. More recent estimates suggest a contribution of about a factor of 10 less ($\sim$3.4 %, Jemmett-Smith et al. 2015).

On Mars, a stronger focus has been on the contribution of dust devils to the dust cycle, leading to the development of dust devil parameterizations in Mars global climate models (e.g. Newman et al. 2002). While dust devils are no longer implicated in the formation of large-scale dust storms (and indeed their occurrence is anticorrelated, since dust-loading attenuates the convective instability that drives dust devils), dust lofting by dust devils is believed to be a significant contributor to Mars atmospheric dust loading. However, the need for model calibration against observed dust loadings, leads to uncertainties in the relative contributions of dust devils in comparison to other dust generating mechanisms.

Dust devil occurrence frequencies and associated dust transport can be assessed through a variety of methods, including in situ observations, remote sensing, and modeling. Results based on the different perspectives are synthesized in Chap. 11 of this volume (Klose et al. 2016) to provide a comprehensive picture on the role of dust devils in the terrestrial and martian climates, and the dust-lifting process is reviewed in Chap. 10 by Neakrase et al. (2016).

4.2 Formation of Dust Devil Tracks—The Martian Climate 'Memory'

The martian atmosphere is rather thin, and responds quickly to changes in solar heating (e.g. Goody and Belton 1967). It is thus in fact comparatively easy to model, more so than the Earth. All else being equal, one Martian year should look very much like the next, since the atmosphere is too thin to convey much information over long periods. This is in sharp contrast to the Earth, where many multiannual cycles exist due to heat stored in the oceans, a prominent example being the Pacific Decadal Oscillation.

Yet there are substantial interannual variations in Martian weather, notably in the occurrence or not of large dust storms. It seems the distribution of dust on the surface may be a principal controlling factor, and it has been observed that large-scale albedo changes on the Martian surface (e.g. Geissler 2005—although see also Szwast et al. 2006) are often the result of accumulated dust devil tracks (see Chapter by Reiss et al. 2016). Furthermore, these albedo changes are enough to cause variations in modeled temperatures and winds (Fenton et al. 2007). Thus, dust devils may be responsible for much of the variability in the Martian climate—they are a cause as well as an effect.

4.3 Atmospheric Electricity and Chemistry

The generation of strong electric fields by dust devils on Earth is well-documented, and electric field measurements are planned at Mars—see the Chap. 11 by Harrison et al. (2016).

A possible consequence of triboelectric charging on Mars, and the resulting corona discharge, is the generation of oxidizing chemicals, notably peroxides and ozone. These are important in that the nondetection of organic compounds by Viking (setting the stage for a sterile, dead world) may have been due to the presence of oxidants in the soil, and indeed the Phoenix lander discovered oxidizing perchlorates (e.g. Hecht et al. 2009). The association of the smell of ozone with electrical discharge is well-known, so this idea (Atreya et al. 2006) seems plausible. It is worth recalling Baddaley's observations of century and a half ago, where evidence of ozone was noted (see earlier).

Table 3 Dust devil and Tornado statistics compiled from US National Weather Service Hazards reports

	Dust devil death	Dust devil injury	Dust devil damage ($M)	Tornado death	Tornado injury	Tornado damage ($M)
2001		1		40	743	637
2002				55	968	1
2003	1	3	0.1	54	1087	1281
2004		5		35	396	549
2005				38	537	503
2006		1		67	990	759
2007				81	659	1407
2008	1	14	0.06	126	1714	1865
2009		3	0.23	31	351	584
2010			0.06	45	699	1135
2011		8	0.03	553	5483	9492
2012		2	0.27	70	822	1649
Total	2	37	0.75	1195	14449	19862
Average	0.17	3.08	0.06	100	1204	1655

4.4 Dust Devil Hazard

Damage and discomfort due to dust devils has been noted since Greek and Roman times earlier in this paper. In the modern world, dust devils, and dry convection generally, can present a hazard to aviation. Typically this is due to sudden changes in wind speed and direction, although sometimes reduced visibility is a factor too. As discussed earlier, some studies in this direction were undertaken in Australia. Lorenz and Myers (2005) review US air accident reports, finding around 100 incidents in the NTSB Air Accident database 1995–2005 pointing to dust devils as a contributing factor, and four fatal accidents.

An informal survey of news media and the Internet reveals a number of reports of injury due to dust devils. Usually these are associated with lightweight structures—barns, tents, children's bouncy castles and parachutes are often the mechanism by which a sudden wind can cause injury. In one instance, a vehicle in South Africa lost control on encountering a dust devil and an astronomer was killed. In another, a US airline pilot was out jogging, and apparently took shelter from a dust devil by a wooden barn, which collapsed, killing her. Table 3 summarizes injuries, deaths and damage from US National Weather Service (NWS) statistics.

It is seen that the damage and casualties due to dust devils are between two and three orders of magnitude less than those associated with tornados. It is interesting that the area swept by tornados of a given pressure (the intensity on the Fujita scale of a tornado is associated with a speed range, which can be equated with a dynamic pressure, since these vertical systems are in roughly cyclostrophic balance) can be plotted (Fig. 15) on a diagram together with pressure drop encounter statistics from dust devil surveys. The weight loading of various structures (which can be equated with the dynamic pressure required to lift them) is also shown. Clearly, tornados are associated with much stronger pressures, but occur much more seldom. The death rate by dust devil (including parachuting and air accidents) is crudely $\sim$10 in $\sim$30 years, or adopting a mean population for the period of $\sim$300 million, about 0.001 death per year per million.

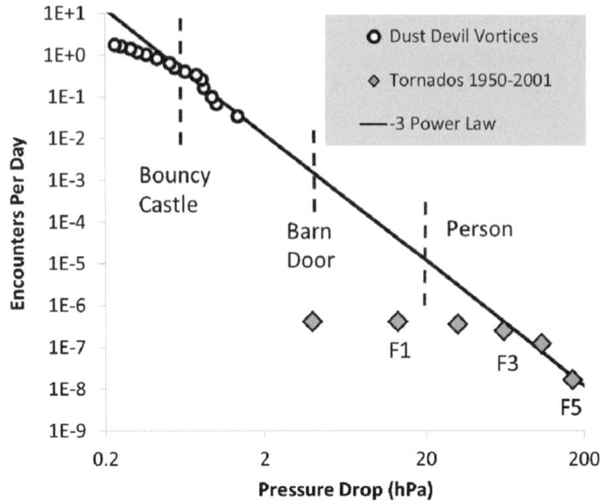

Fig. 15 The cumulative frequency that a spot will encounter a vortex of a given intensity, defined by a pressure drop. Dust devil vortex encounter frequencies in summer on Eldorado Playa in Nevada determined by a high-resolution pressure record in Lorenz and Lanagan (2014) are plotted as *circles*; NWS statistics of tornado mean swept area and intensity as defined in the text are plotted as *diamonds*. Weight loadings for various objects are noted by *dashed lines*

In addition to prompt injury, dust devils may adversely affect human health via injection of pathogens, toxins or irritants into the air (see Morman and Plumlee 2014). A particular concern in the USA is 'Valley Fever' wherein airborne fungal spores from the soil may be inhaled and lead to infection by Coccidioidomycosis, usually with minor symptoms but occasionally pneumonia, joint pain and fatige (e.g. Hector and Laniado-Laborin 2005). It may be noted that the dust at Eldorado Playa, near Boulder City, Nevada, USA—a popular field site for dust devil studies—has been documented (Buck et al. 2013) to contain large amounts of amphibolite, a carcinogenic natural asbestos.

On Mars, in principle, dust devils might be associated with wind extremes that could be problematic for landers or rovers during their descent and landing phases. This is a particular concern for parachute-borne vehicles, which would respond quickly to the wind shear associated with dust devil vortices.

4.5 Mars Solar Power and Planetary Protection

There is a direct impact on the operation of spacecraft on the surface of Mars, in that dust progressively accumulates on up-facing surfaces, such as solar arrays. Indeed, the Pathfinder mission observed a steady 0.28 % per day decline in solar power due to dust deposition. This trend guided expectations that later missions (notably the Mars Exploration Rovers) might survive only ~90 days. However, periodic sudden increases in solar power occurred due to 'clearing events' which swept the arrays clean. Lorenz and Reiss (2015) found that not only did these events seasonally coincide with the start of observed dust devils, but they occurred at a rate consistent with the vortex encounter rates measured by pressure drops on Pathfinder and Phoenix, extrapolated to a dust-lifting threshold intensity. Thus dust devils appear to significantly influence the capabilities of Mars surface assets.

Sand and dust transport on vehicles on the surface of Mars also has a possibly surprising relevance, in that stringent requirements (ultimately mandated by the UN Outer Space Treaty of 1967) on cleanliness and contamination control in order to comply with Planetary Protection policies. These ensure that bacteria or other contaminants are not introduced into Martian environments where they might flourish and disrupt efforts to detect extant or fossil life on Mars. The possibility exists that abrasive particles brought and/or removed by dust devils, might scour spacecraft surfaces on the upper part of the vehicle. Since these upper surfaces might not normally be thought to have contact with the Martian ground, these transport processes may demand additional cleaning or bioassay during mission preparation.

4.6 Dust Devils as Dynamical Objects

Like more general convective structures, such as the cells in Rayleigh-Benard convection, dust devils are interesting self-organized thermally-driven systems which deserve study as emergent entities. While they might be seen merely as parts of a convecting boundary layer, they can be usefully and accurately considered as discrete structures.

The dynamical structure of dust devils is of interest from a purely fluid-mechanical standpoint, and many studies of analytic models, numerical models and laboratory-scale vortices have been made. There are furthermore useful analogies between dust devils and tornado and other vortices.

A heat-engine formalism by Renno et al. (1998), analogous to similar thermodynamic studies of hurricanes, has been useful in reconciling the temperature perturbations seen in dust devils with the resultant pressure and velocity fields.

Statistical studies of dust devil populations have found that their observed diameters and heights, observed pressure drops, and the widths of dust devil tracks, have highly-skewed distributions, that can be fit by power-laws or other functions (e.g. Lorenz 2011; Kurgansky 2006). There exists an association of power law statistics with self-organized criticality and other forced physical systems. Thus dust devils are of interest from a more general statistical/physical standpoint.

5 Dust Devils in Culture

The very name ‘dust devil’ suggests an association with the occult, shared with the Arabic term ‘djinn’, appealing to the notion that dust devils conjure themselves out of thin air.

Because dust devils may have both clockwise and anticlockwise sense of rotation there was a belief among native people (Indians) that dust devils with one predefined sense of rotation correspond to a good genius and of opposite one to an evil spirit. Amazingly, both the native people in Chile, and the Navajo in the USA, about 10000 km distant from the Atacama Desert, have essentially the same belief in this respect.

The occurrence of dust devils in India has a long history, and they are a regular phenomenon in arid regions such as those found in the Hyderabad-Karnatak region. Whilst they are accepted nowadays as commonplace atmospheric phenomenon, in rural Indian farming areas old superstitions relating to dust devils still persist; some farmers working in fields still fear dust devils and will hold their children and belongings flat on the ground until it passes.

Like many evocative phrases, ‘dust devil’ has been appropriated by other fields of endeavour. ‘Dust Devils’ is a common name for school sports teams in desert areas of the USA. ‘Dust Devil’ is the title of a thriller novel by Roger Smith, set in South Africa, and is also that of a violent 1992 horror movie set in Namibia, but aside from scenes at the real-world ghost town of Kolmanskop (notable for being overrun by sand dunes) has no meteorological significance. Dust storms on Mars are a common feature in science fiction movies set on that planet, and dust devils are seen prominently in the popular 2015 movie ‘The Martian’. It may be noted that scientific publication of tests of dust penetration into

Fig. 16 Aerial view of dust devils at the 2014 Burning Man festival in the USA. This counterculture event is held on a playa which frequently sees dust devils (a web search will find many excellent videos and photos). This aerial image is notable in showing four dust devils apparently triggered by the updraft of the smoke plume from the burning icon. Note the vehicles for scale. Image by Duncan Rawlinson (duncan.co CC Attribution-NC license)

space suit designs for Martian surface operation was made possible by the entertainment industry (Gaier et al. 2010), being tested with simulated dust devils during the production of a documentary 'Storm Worlds'.

A prominent counterculture festival 'Burning Man' is held annually on a remote playa in the Nevada desert, culminating in its eponymous bonfire. Dust devils are frequently encountered at this location, with participants sometimes chasing them on bicycles or on foot. On some occasions, dust devils are actually triggered (see Fig. 16) by the smoke plume from the fire. Another cultural influence of dust devils is the film "Tornado" (2010) by the Belgian-born artist Francis Alys in Mexico, who possibly suggests the chaotic airflow in a dust devil as a metaphor for politics in that country. As reviewed by Rooney (2011), "the sight of his lean frame racing towards the twisters is at once ridiculous and hysterical—blithe qualities that quickly give way to gravitas as the artist physically enters the eye of the storm. Inside, chaos reigns and Alys, unprotected except for his handheld camera, is enveloped and pummeled by flying bits of sand, dust and dirt."

On a lighter note, 'Dust Devil' is the title of a US children's book (and should not be confused with the aforementioned crime novel). Finally, towards the end of a popular children's book (written in Swedish by Finnish author Tove Jansson and published in 1946) "Comet in Moominland", the heroes are lifted in the air by a "cyclone". The illustrations of the book reveal that this "cyclone" is only small in scale and there is no mention of rain. Also several other similar whirlwinds are seen in the illustration of the book. One of the characters, Snufkin, explains that they are caused by heating of the ground by the radiation of an approaching comet. Because of this formation mechanism, the whirlwind mentioned in "Comet in Moominland" should perhaps be classified as a gigantic dust devil instead of a cyclone or tornado. Comets are not traditionally associated with aeolian phenomena, but the discovery of ripples on the surface of a comet just prior to the workshop (Thomas et al. 2015) that stimulated this review paper reminds us that in science we must, as children do naturally, keep an open mind.

6 Conclusions

We have reviewed the history of dust devil research. Dust devils have been studied in a range of countries since the mid-19th century, initially somewhat informally in connection with other rotating wind systems. Instrumented studies of dust devils in their own right begin in earnest in the 1960s, perhaps in connection with nuclear weapons programs. Mars exploration emerged as a major stimulus of dust devil research, initially in the late 1960s, but again in the early 21st century.

Dust devils present an occasional terrestrial hazard, roughly 2–3 orders of magnitude less severe than tornados. Their role in terrestrial dust-lifting is probably not dominant, but may be significant at Mars. Dust devils appear to be a significant factor in surface albedo changes on Mars, and the removal of dust from lander or rover solar panels by dust devils is an important issue in spacecraft power management and operations. Dust devils engage the public and have interesting properties, both in their individual dynamics and as a population, which are appealing topics of scientific investigation.

Acknowledgements The work of R. Lorenz was supported by NASA Mars Fundamental Research Program grant NNX12AI04G "Quantitative Study of Dust Devils". Several reviewers and reviewing editors provided copious improvements to the paper.

References

Anonymous, Dancing dervishes or dust whirls. Mon. Weather Rev. **30**, 316 (1902)

A. Ansmann, M. Tesche, P. Knippertz, E. Bierwirth, D. Althausen, D. Müller, O. Schulz, Vertical profiling of convective dust plumes in southern Morocco during SAMUM. Tellus B **61**(1), 340–353 (2009)

B. Atalay, K. Wamsley, *Leonardo's Universe: The Renaissance World of Leonardo Da Vinci* (National Geographic, Washington, 2009)

S.K. Atreya, A.-S. Wong, N.O. Renno, W.M. Farrell, G.T. Delory, D.D. Sentman, S.A. Cummer, J.R. Marshall, S. Rafkin, D. Catling, Oxidant enhancement in martian dust devils and storms: implications for life and habitability. Astrobiology **6**, 439–450 (2006)

P.F.H. Baddeley, *Whirlwinds and Dust Storms of India* (Bell, London, 1860)

M. Balme, R. Greeley, Dust devils on Earth and Mars. Rev. Geophys. **44**, RG3003 (2006). doi:10.1029/2005RG000188

M. Balme, A. Hagermann, Particle lifting at the soil-air interface by atmospheric pressure excursions in dust devils. Geophys. Res. Lett. **33**(19) (2006). doi:10.1029/2006GL026819

M. Balme, S. Metzger, M. Towner, T. Ringrose, R. Greeley, J. Iversen, Friction wind speeds in dust devils: a field study. Geophys. Res. Lett. **30**(16), 1830 (2003a). doi:10.1029/2003GL017493

M.R. Balme, P.L. Whelley, R. Greeley, Mars: dust devil track survey in Argyre Planitia and Hellas Basin. J. Geophys. Res. **108**(E8), 5086 (2003b). doi:10.1029/2003JE002096

F. Bell, *Dust Devils and Aviation*. Report. Meteorol. Note, vol. 27 (Aust. Bur. of Meteorol, Melbourne, 1969)

D. Bowker, Meteorology and the ancient Greeks. Weather **66**(9), 249–251 (2011)

C.S. Bristow, K.A. Hudson-Edwards, A. Chappell, Fertilizing the Amazon and equatorial Atlantic with West African dust. Geophys. Res. Lett. **37**(14) (2010)

B.J. Buck, D. Goossens, R.V. Metcalf, B. McLaurin, M. Ren, F. Freudenberger, Naturally occurring asbestos: potential for human exposure, Southern Nevada, USA. Soil Sci. Soc. Am. J. **77**(6), 2192–2204 (2013)

E. Budde, Eine Beobachtung kleiner Tromben. Z. Österr. Ges. Meteorol. **18**, 462–465 (1883)

B.A. Cantor, K.M. Kanak, K.S. Edgett, MOC observations of Martian dust devils and their tracks (September 1997 to January 2006) and evaluation of theoretical vortex models. J. Geophys. Res. **111**, E12002 (2006)

D.S. Choi, C.M. Dundas, Measurements of Martian dust devil winds with HiRISE. Geophys. Res. Lett. **38**(24) (2011). doi:10.1029/2011GL049806

K.S. Edgett, M.C. Malin, Martian dust raising and surface albedo controls: thin, dark (and sometimes bright) streaks and dust devils in MGS high-resolution images. Lunar Planet. Sci. [CD-ROM], XXXI, Abstract 1073

M.D. Ellehoj, H.P. Gunnlaugsson, P.A. Taylor, H. Kahanpa, K.M. Bean, B.A. Cantor, B.T. Gheynani, L. Drube, D. Fisher, A.-M. Harri, C. Holstein-Rathlou, M.T. Lemmon, M.B. Madsen, M.C. Malin, J. Polkko, P.H. Smith, L.K. Tamppari, W. Weng, J. Whiteway, Convective vortices and dust devils at the Phoenix Mars mission landing site. J. Geophys. Res. **115**, E00E16 (2010)

L.K. Fenton, P.E. Geissler, R.M. Haberle, Global warming and climate forcing by recent albedo changes on Mars. Nature **446**(7136), 646–649 (2007)

L. Fenton, D. Reiss, M. Lemmon, B. Marticorena, S. Lewis, B. Cantor, Orbital observations of dust lofted by daytime convective turbulence. Space Sci. Rev. (2016, this issue). doi:10.1007/s11214-016-0243-6

W. Ferrel, Meteorological researches, part II. Cyclones, tornados and waterspouts. Am. J. Sci. **22**(127), 33–48 (1881)

W.D. Flower, Sand devils. London Meteorol. Off. Prof. Notes **5**(71), 1–16 (1936)

B. Franklin, Letter to Peter Collinson, dated Philadelphia, August 25, 1755 (see e.g. http://www.islandnet.com/~see/weather/history/bfrank1.htm, downloaded 1/21/2016)

C. Fujiwara, Y. Fujiyoshi, Detection of "invisible waterspout" using 3D scanning Doppler lidar. SOLA **10**, 127–130 (2014). doi:10.2151/sola.2014-026

C. Fujiwara, K. Yamashita, M. Nakanishi, Y. Fujiyoshi, Dust devil—like vortices in an urban area detected by a 3D scanning Doppler lidar. J. Appl. Meteorol. Climatol. **50**, 534–547 (2011). doi:10.1175/2010JAMC2481.1

C. Fujiwara, K. Yamashita, Y. Fujiyoshi, Observed effect of mesoscale vertical vorticity on rotation sense of dust devil-like vortices in an urban area. SOLA **8**, 25–28 (2012). doi:10.2151/sola.2012-007

J.R. Gaier, P.G. de Leon, P. Lee, T. McCue, E. Hodgson, J. Thrasher, Preliminary testing of a pressurized space suit and candidate fabrics under simulated Mars dust storm and dust devil conditions, in *40th International Conference on Environmental Systems, International Conference on Environmental Systems* ICES (2010), AIAA 2010-6247

P.E. Geissler, Three decades of Martian surface changes. J. Geophys. Res. **110**, E02001 (2005)

P. Gierasch, R.M. Goody, A model of a Martian great dust storm. J. Atmos. Sci. **30**, 749–762 (1973)

R.M. Goody, M.J.S. Belton, Radiative relaxation times for Mars: a discussion of Martian atmospheric dynamics. Planet. Space Sci. **15**, 247–256 (1967)

C.G. Grant, Dust devils in the sub-Arctic. Weather **4**, 402–403 (1949)

J.A. Grant, P.H. Schultz, Possible tornado-like tracks on Mars. Science **237**, 883–885 (1987)

R. Greeley, Saltation impact as a means for raising dust on Mars. Planet. Space Sci. **50**, 151–155 (2002)

R. Greeley et al., Active dust devils in Gusev crater, Mars: observations from the Mars exploration rover spirit. J. Geophys. Res. **111**, E12S09 (2006)

R. Greeley et al., Gusev crater, Mars: observations of three dust devil seasons. J. Geophys. Res. **115**, E00F02 (2010)

Z.L. Gu, *Wind-Blown Dust: Near Surface Gas-Solid Two-Phase Turbulent Flow* (Science Press, Beijing, 2010) (in Chinese)

Z.L. Gu, Y.Z. Zhao, Y. Li et al., Numerical simulation of dust lifting within dust devils—Simulation of an intense vortex[J]. J. Atmos. Sci. **63**(10), 2630–2641 (2006)

Z.L. Gu, W. Wei, Y.Z. Zhao, An overview of surface conditions in numerical simulations of dust devils and the consequent near-surface air flow fields. Aerosol Air Qual. Res. **10**, 272–281 (2010)

Z.L. Gu, W. Wei, J.W. Su et al., The role of water content in triboelectric charging of wind-blown sand[J]. Sci. Rep. **3**, 1337 (2013)

R.G. Harrison, E. Barth, F. Esposito, J. Merrison, F. Montmessin, K.L. Aplin, C. Borlina, J.J. Berthelier, G. Déprez, W.M. Farrell, I.M.P. Houghton, N.O. Renno, K.A. Nicoll, S.N. Tripathi, M. Zimmerman, Applications of electrified dust and dust devil electrodynamics to Martian atmospheric electricity. Space Sci. Rev. (2016, this issue). doi:10.1007/s11214-016-0241-8

M.H. Hecht et al., Detection of perchlorate and the soluble chemistry of martian soil at the Phoenix lander site. Science **325**(5936), 64–67 (2009)

R. Hector, R. Laniado-Laborin, Coccidioidomycosis—a fungal disease of the Americas (2005). PLoS Med. **2**(1). doi:10.1371/journal.pmed.0020002

J. Herschel (ed.), *A Manual of Scientific Enquiry* (Cambridge University Press, New York, 1851). Prepared for the Use of Her Majesty's Navy and Adapted for Travellers in General, reprinted in 1976

G.D. Hess, K.T. Spillane, Characteristics of dust devils in Australia. J. Appl. Meteorol. **29**, 498–507 (1990)

G.D. Hess, K.T. Spillane, R.S. Lourensz, Atmospheric vortices in shallow convection. J. Appl. Meteorol. Climatol. **27**, 305–317 (1988)

R. Hesse, Short-lived and long-lived dust devil tracks in the coastal desert of southern Peru. Aeolian Res. **5**, 101–106 (2012)

H.Y. Inoue, K. Kusunoki, W. Kato, H. Suzuki, T. Imai, T. Takemi, K. Bessho, M. Nakazato, S. Hoshino, W. Mashiko, S. Hayashi, T. Fukuhara, T. Shibata, H. Yamauchi, O. Suzuki, Finescale Doppler radar observation of a tornado and low-level misocyclones within a winter storm in the Japan Sea coastal region. Mon. Weather Rev. **139**, 351–369 (2011). doi:10.1175/2010MWR3247.1

J. Ito, H. Niino, Particle image velocimetry of a dust devil observed in a desert. SOLA **10**, 108–111 (2014). doi:10.2151/sola.2014-022

R.L. Ives, Behavior of dust devils. Bull. Am. Meteorol. Soc. **28**, 168–174 (1947)

T. Jefferson, Notes on the State of Virginia, Monticello, VA (1787)

B.C. Jemmett-Smith, J.H. Marsham, P. Knippertz, C.A. Gilkeson, Quantifying global dust devil occurrence from meteorological analyses. Geophys. Res. Lett. **42**(4), 1275–1282 (2015)

G. Kennan, A trip to the Altai mountains. Science **7**, 18–22 (1886)

M. Klose, B. Jemmett-Smith, H. Kahanpää, M. Kahre, P. Knippertz, M. Lemmon, S. Lewis, R. Lorenz, L. Neakrase, C. Newman, M. Patel, D. Reiss, A. Spiga, P. Whelley, Dust devil sediment transport: From lab to field to global impact. Space Sci. Rev. (2016, this issue)

P. Knippertz, J.-B. Stuut, *Mineral Dust: A Key Player in the Earth System* (Springer, New York, 2014)

J. Koch, N.O. Renno, The role of convective plumes and vortices on the global aerosol budget. Geophys. Res. Lett. **32**(18) (2005)

H. Koschmieder, Zur Trombenbildung. Arch. Meteorol. Geophys. Bioklimatol., Ser. A **4**, 203–219 (1951)

M.V. Kurgansky, Steady-state properties and statistical distribution of atmospheric dust devils. Geophys. Res. Lett. **33**, L19S06 (2006)

M. Kurgansky, R. Lorenz, N. Renno, T. Takemi, W. Wei, Z. Gu, Dust devil steady-state structure from a fluid dynamics perspective. Space Sci. Rev. (2016, this issue)

M.V. Kurgansky, A. Montecinos, V. Villagran, S.M. Metzger, Micrometeorological conditions for dust-devil occurrence in the Atacama Desert. Bound.-Layer Meteorol. **138**(2), 285–298 (2011)

R.L. Lambeth, On the measurement of dust devil parameters. Bull. Am. Meteorol. Soc. **47**, 522–526 (1966)

W. Lammert, Wirbel-Staubsturm. Meteorol. Z. **15**, 279 (1947)

Z. Le, Dust devil in Gobi Desert of Southern Xinjiang province[J]. Meteorol. Mon. **04**, 33 (1983) (in Chinese)

X.J. Lei, Y.L. Li, J.W. Du et al., Analysis on the spatio-temporal distribution and daily change rule of cyclone and dust devil in Shaanxi. J. Catastrophology **20**(02), 99–101 (2005) (in Chinese)

E. Less, Sandhose in der Nähe von Berlin. Meteorol. Z. **26**, 274–275 (1891)

J.P. Letzmann, Richtlinien zur Erforschung von Tromben, Tornados, Wasserhosen und Kleintromben (Guidelines for research on tornados, waterspouts, and whirlwinds), *Klimatologische Kommission, IMO Publ.*, vol. 38, ed. by E. Ijdo (1939), pp. 91–110. Secretariat de l'Organisation Meteorologique Internationale. Available online at: http://www.tordach.org/pdf/letzwww.pdf

R. Lorenz, On the statistical distribution of dust devil diameters. Icarus **215**(1), 381–390 (2011)

R.D. Lorenz, Pressure drops in dust devils: Earth and Mars. Planet. Space Sci. **60**, 370–375 (2012)

R. Lorenz, The longevity and aspect ratio of dust devils: effects on detection efficiencies and comparison of landed and orbital imaging at Mars. Icarus **226**(1), 964–970 (2013)

R.D. Lorenz, Vortex encounter rates with fixed barometer stations: Comparison with visual dust devil counts and large eddy simulations. J. Atmos. Sci. **71**, 4461–4472 (2014)

R.D. Lorenz, D. Christie, Dust devil signatures in infrasound records of the international monitoring system. Geophys. Res. Lett. **42**(6), 2009–2014 (2015)

R.D. Lorenz, P.D. Lanagan, A barometric survey of dust devil vortices on a desert playa. Bound.-Layer Meteorol. **53**, 555–568 (2014). doi:10.1007/s10546-014-9954-y

R.D. Lorenz, M.J. Myers, Dust devil hazard to aviation: a review of United States air accident reports. J. Meteorol. **30**(298), 178–184 (2005)

R. Lorenz, J. Radebaugh, Dust devils in thin air: vortex observations at a high elevation Mars analog site in the Argentinian puna. Geophys. Res. Lett. **43** (2016). doi:10.1002/2015GL067412

R.D. Lorenz, D. Reiss, Solar panel clearing events, dust devil tracks, and in-situ vortex detections on Mars. Icarus **248**, 162–164 (2015)

R. Lorenz, J. Zimbelman, *Dune Worlds: How Wind-Blown Sand Shapes Planetary Landscapes* (Springer, New York, 2014)

M.A. Lucanus, *The Civil War I–X (Pharsalia)* (Heinemann, London, 1926) (English Translation by J. Duff; the original was in 65 AD)

J.P. Mason, M.R. Patel, S.R. Lewis, The retrieval of optical properties from terrestrial dust devil vortices. Icarus **231**, 385–393 (2014)

H. Maurer, Entstehung einer Sandhose. Meteorol. Z. **51**, 275 (1934)

J.B. McGinnigle, Dust whirls in North-West Libya. Weather **21**, 272–276 (1966)

J.R. McGinnigle, A note on observed dust-whirl damage at Nicosia, Cyprus. Meteorol. Mag. **99**, 118–122 (1970)

S.M. Metzger, J.R. Carr, J.R. Johnson, T.J. Parker, M.T. Lemmon, Dust devil vortices seen by the Mars Pathfinder camera. Geophys. Res. Lett. **26**, 2781–2784 (1999)

S. Metzger, M. Kurgansky, A. Montecinos, V. Villagran, H. Verdejo, Chasing dust devils in Chile's Atacama Desert, in *41st Lunar and Planetary Science Conference*, Houston, TX, March 2010 (2010), Abstract #2564

A. Miethe, Das Entstehen der Windhosen. Prometheus **10**, 795–796 (1899)

J.E. Moores, M.T. Lemmon, H. Kahanpää, S.C. Rafkin, R. Francis, J. Pla-Garcia, A.R. Vasavada, Observational evidence of a suppressed planetary boundary layer in northern Gale Crater, Mars as seen by the Navcam instrument onboard the Mars Science Laboratory rover. Icarus **249**, 129–142 (2015)

S. Morman, G. Plumlee, Dust and human health, in *Mineral Dust: A Key Player in the Earth System*, ed. by P. Knippertz, J.-B. Stuut (Springer, New York, 2014), pp. 385–410

J.R. Murphy, S. Nelli, Mars Pathfinder convective vortices: frequency of occurrence. Geophys. Res. Lett. **29**(23), 18-1–18-4 (2002)

J. Murphy, K. Steakley, F. Esposito, G. Deprez, H. Kahanpaa, M. Lemmon, M. Patel, M. Balme, N. Murdoch, L. Neakrase, P. Whelley, R. Lorenz, Field measurements of terrestrial and Mars dust devils. Space Sci. Rev. (2016, this issue)

T.A. Mutch, The Martian Landscape, NASA Special Publication SP-425 (1978)

L.D.V. Neakrase, M.R. Balme, F. Esposito, T. Kelling, M. Klose, J.F. Kok, B. Marticorena, J. Merrison, M. Patel, G. Wurm, Particle lifting processes in dust devils. Space Sci. Rev. (2016, this issue)

F.M. Neubauer, Thermal convection in the Martian atmosphere. J. Geophys. Res. **71**, 2419–2426 (1966)

C.E. Newman, S.R. Lewis, P.L. Read, F. Forget, Modeling the Martian dust cycle: 1. Representations of dust transport processes. J. Geophys. Res. **107**(E12), 5123 (2002). doi:10.1029/2002JE001910

A.M.C. Oke, D. Dunkerley, N.J. Tapper, Willy-willies in the Australian landscape: sediment transport characteristics. J. Arid Environ. **71**(2), 216–228 (2007a)

A.M.C. Oke, N.J. Tapper, D. Dunkerley, Willy-willies in the Australian landscape: the role of key meteorological variables and surface conditions in defining frequency and spatial characteristics. J. Arid Environ. **71**(2), 201–215 (2007b)

C. Piazzi-Smyth, *Tenerife: An Astronomer's Experiment, or, Specialities of a Residence Above the Clouds* (Lovell Reeve, London, 1858), p. 180

S. Raasch, T. Franke, Structure and formation of dust devil-like vortices in the atmospheric boundary layer: a high resolution numerical study. J. Geophys. Res. **116**, D16120 (2011). doi:10.1029/2011JD016010

S. Rafkin, B. Jemmet-Smith, L. Fenton, J. Ito, D. Tyler, R. Lorenz, Dust devil formation. Space Sci. Rev. (2016, this issue)

P. Read, S. Lewis, *The Martian Climate Revisited: Atmosphere and Environment of a Desert Planet* (Springer, New York, 2004)

D. Reiss, J. Raack, A.P. Rossi, G. Di Achille, H. Hiesinger, First in-situ analysis of dust devil tracks on Earth and their comparison with tracks on Mars. Geophys. Res. Lett. **37**, L14203 (2010). doi:10.1029/2010GL044016

D. Reiss, J. Raack, H. Hiesinger, Bright dust devil tracks on Earth: implications for their formation on Mars. Icarus **211**, 917–920 (2011a)

D. Reiss, M. Zanetti, G. Neukum, Multitemporal observations of identical active dust devils on Mars with the High Resolution Stereo Camera (HRSC) and Mars Orbiter Camera (MOC). Icarus **215**, 358–369 (2011b)

D. Reiss, M.I. Zimmerman, D.C. Lewellen, Formation of cycloidal dust devil tracks by redeposition of coarse sands in southern Peru: implications for Mars. Earth Planet. Sci. Lett. **383**, 7–15 (2013)

D. Reiss, A. Spiga, G. Erkeling, The horizontal motion of dust devils on Mars derived from CRISM and CTX/HiRISE observations. Icarus **227**, 8–20 (2014)

D. Reiss, L. Neakrase, L. Fenton, P. Whelley, T. Statella, M. Zimmerman, A.P. Rossi, M. Balme, Dust devil trails. Space Sci. Rev. (2016, this issue)

N.O. Renno, M.L. Burkett, M.P. Larkin, A simple thermodynamical theory for dust devils. J. Atmos. Sci. **55**, 3244–3252 (1998)

T.J. Ringrose, M.C. Towner, J.C. Zarnecki, Convective vortices on Mars: a reanalysis of Viking Lander 2 meteorological data, sols 1–60. Icarus **163**, 78–87 (2003)

K. Rooney Francis alÿs: A story of deception, the Brooklyn Rail, July-August (2011). http://www.brooklynrail.org/2011/07/artseen/francis-als-a-story-of-deception-july11, downloaded 1/4/2016

A.P. Rossi, L. Marinangeli, The first terrestrial analogue to Martian dust devil tracks found in Tenere Desert, Niger. Geophys. Res. Lett. **31**(6) (2004)

J.A. Ryan, Notes on the Martian yellow clouds. J. Geophys. Res. **69**(18), 3759–3770 (1964)

J.A. Ryan, J.J. Carroll, Dust devil wind velocities: mature state. J. Geophys. Res. **75**, 531–541 (1970)

J.A. Ryan, R.D. Lucich, Possible dust devils, vortices on Mars. J. Geophys. Res. **88**, 11005–11011 (1983)

A. Schläfli, Über Staubtromben und den "Samum" in Unter-Mesopotamien. Z. Österr. Ges. Meteorol. **5**, 469–472 (1870)

P.C. Sinclair, Some preliminary dust devil measurements. Mon. Weather Rev. **22**(8), 363–367 (1964)

P.C. Sinclair, On the rotation of dust devils. Bull. Am. Meteorol. Soc. **46**, 388–391 (1965)

P.C. Sinclair, General characteristics of dust devils. J. Appl. Meteorol. **8**, 32–45 (1969)

P.C. Sinclair, The lower structure of dust devils. J. Atmos. Sci. **30**, 1599–1619 (1973)

J.T. Snow, T.M. McClelland, Dust devils at white sands missile range, New Mexico: 1. temporal and spatial distributions. J. Geophys. Res., Atmos. **95**(D9), 13707–13721 (1990)

A. Spiga, F. Forget, A new model to simulate the Martian mesoscale and microscale atmospheric circulation: validation and first results. J. Geophys. Res., Planets **114**(E2) (2009). doi:10.1029/2008JE003242

A. Spiga, E. Barth, Z. Gu, F. Homann, J. Ito, B. Jemmett-Smith, M. Klose, S. Nishizawa, S. Raasch, S. Rafkin, T. Takemi, D. Tyler, W. Wei, Large-eddy simulations of dust devils and convective vortices (2016)

K.T. Spillane, G.D. Hess, Fair weather convection and light aircraft, helicopter, and glider accidents. J. Aircr. **25**, 55–61 (1988)

C. Stanzel, M. Pätzold, D.A. Williams, P.L. Whelley, R. Greeley, G. Neukum (HRSC Co-Investigator Team), Dust devil speeds, directions of motion and general characteristics observed by the Mars express high resolution stereo camera. Icarus **197**(1), 39–51 (2008)

M.A. Szwast, M.I. Richardson, A.R. Vasavada, Surface dust redistribution on Mars as observed by the Mars global surveyor and Viking orbiters. J. Geophys. Res. **111**, E11008 (2006)

M. Tachikawa, K. Deguchi, A whirlwind attacked an athletic meeting. Wind Eng., JAWE **1987**(31), 13–16 (1987). Available at: https://www.jstage.jst.go.jp/article/jawe1982/1987/31/1987_31_13/_article (in Japanese with English figure captions)

P.C. Thomas, P.J. Gierasch, Dust devils on Mars. Science **230**, 175–177 (1985)

N. Thomas et al., The morphological diversity of comet 67P/Churyumov-Gerasimenko. Science **347**, 6220 (2015)

M.C. Towner, Characteristics of large Martian dust devils using Mars Odyssey thermal emission imaging system visual and infrared images. J. Geophys. Res., Planets **114**(E2) (2009). doi:10.1029/2008JE003220

D.M. Tratt, M.H. Hecht, D.C. Catling, E.C. Samulon, P.H. Smith, In situ measurements of dust devil dynamics: toward a strategy for Mars. J. Geophys. Res. **108**(E11), 5116 (2003). doi:10.1029/2003JE002161

A. von Humboldt, *Ansichten der Natur*, 3rd edn. (J.G. Cotta, Stuttgart, 1849), p. 26

A. Wegener, Staubwirbel auf Island. Meteorol. Z. **31**, 199–200 (1914)

P.L. Whelley, R. Greeley, The distribution of dust devil activity on Mars. J. Geophys. Res., Planets **113**(E7) (2008). doi:10.1029/2007JE002966

R.E. Wyett, Pressure drop in a dust devil. Mon. Weather Rev. **82**, 7 (1954)

Space Sci Rev (2016) 203:39–87
DOI 10.1007/s11214-016-0283-y

Field Measurements of Terrestrial and Martian Dust Devils

Jim Murphy[1] · Kathryn Steakley[1] · Matt Balme[2] · Gregoire Deprez[3] · Francesca Esposito[4] · Henrik Kahanpää[5,6] · Mark Lemmon[7] · Ralph Lorenz[8] · Naomi Murdoch[9] · Lynn Neakrase[1] · Manish Patel[2] · Patrick Whelley[10]

Received: 10 May 2016 / Accepted: 24 August 2016 / Published online: 29 September 2016

Abstract Surface-based measurements of terrestrial and martian dust devils/convective vortices provided from mobile and stationary platforms are discussed. Imaging of terrestrial dust devils has quantified their rotational and vertical wind speeds, translation speeds, dimensions, dust load, and frequency of occurrence. Imaging of martian dust devils has provided translation speeds and constraints on dimensions, but only limited constraints on vertical motion within a vortex. The longer mission durations on Mars afforded by long operating robotic landers and rovers have provided statistical quantification of vortex occurrence (time-of-sol, and recently seasonal) that has until recently not been a primary outcome of more temporally limited terrestrial dust devil measurement campaigns. Terrestrial measurement campaigns have included a more extensive range of measured vortex parameters (pressure, wind, morphology, etc.) than have martian opportunities, with electric field and direct measure of dust abundance not yet obtained on Mars. No martian robotic mission has yet provided contemporaneous high frequency wind and pressure measurements. Comparison of measured terrestrial and martian dust devil characteristics suggests that martian dust devils are larger and possess faster maximum rotational wind speeds, that the absolute magnitude of the pressure deficit within a terrestrial dust devil is an order of magnitude greater

✉ J. Murphy
murphy@nmsu.edu

1 New Mexico State University, Las Cruces, NM, USA

2 Open University, Milton Keynes, UK

3 Laboratoire Atmosphères, Guyancourt, France

4 INAF, Osservatorio Astronomico di Capodimonte, Naples, Italy

5 Finnish Meteorological Institute, Helsinki, Finland

6 Aalto University/School of Electrical Engineering, Espoo, Finland

7 Texas A&M University, College Station, TX, USA

8 Johns Hopkins University Applied Physics Lab, Laurel, MD, USA

9 ISAE-SUPAERO, Toulouse University, Toulouse, France

10 NASA Goddard Space Flight Center, Greenbelt, MD, USA

than a martian dust devil, and that the time-of-day variation in vortex frequency is similar. Recent terrestrial investigations have demonstrated the presence of diagnostic dust devil signals within seismic and infrasound measurements; an upcoming Mars robotic mission will obtain similar measurement types.

Keywords Dust devils

1 Introduction

Dust devils are small diameter, surface bordering, vertically aligned atmospheric convective vortices made visible by their entrainment of dust lifted from the surface (Fig. 1). Historic anecdotal reports (Lorenz et al. 2016) indicate qualitative awareness of these phenomena extending for millennia into the past. Quantitative awareness is more recent (Ives 1947; Wyett 1954). Dust devils are one mechanism that emplaces dust into atmospheric suspension, affecting air quality, atmospheric clarity, possible hazardous conditions to low flying aircraft, etc. Dust devils have also been identified on Mars, where they play a possibly substantial role in maintaining that planet's persistent atmospheric dust load, especially during the orbital aphelion season (Kahre et al. 2006). It is only during the modern era of scientific investigation and measurement that the true physical understanding of these phenomena on both planets have been achieved. In this presentation we focus upon the quantitative characterization of dust devils, on both Earth and Mars, which have resulted in our current understanding of their physical attributes.

Terrestrial dust devil attributes measured to date include winds (radial, azimuthal, vertical), translation speed, shape/height/width (visualized by suspended dust), central pressure deficit, dust load and its radial and vertical structure and particle size distribution, electric field, and surface dust lifting rate. Measurements have been provided in situ from mobile and stationary instrumented platforms (either individually or in a network) and/or remotely imaged at visible wavelengths, with some infrared wavelength measurements also available. Measurements obtained from Mars' surface include both in situ (pressure, wind) and remotely imaged (visible imaging) characterization, but remain deficient in other areas (electric field, dust lifting rate). While terrestrial field campaigns have usually been, until recently, of short time extent (days, weeks), martian 'campaigns' (robotic exploration missions) have been more time extended, limited by the lifetime of the mission or its most applicable instruments. However, no martian mission has provided continuous sampling for all forms of observations, though the Mars Phoenix Lander did provide almost continuous 0.5 Hz sampling of its meteorology measurements.

Balme and Greeley (2006) provided an extensive dust devil review. Here we emphasize subsequent gained knowledge in addition to reiterating their primary foci.

In Sect. 2 below we describe primarily surface based imaging characterization of terrestrial dust devils (Sect. 2.1) from a variety of field campaigns, followed by surface based imaging characterization of martian dust devils (Sect. 2.2) provided by the seven spacecraft that have operated/are operating there. [Note that Fenton et al. (2016) in this collection of papers addresses dust devil remote sensing from non-surface based measurements such as those provided by orbiting spacecraft.] This is followed in Sect. 3 by presentation of in situ measurements of terrestrial dust devils (Sect. 3.1) and martian dust devils (Sect. 3.2).

Fig. 1 Image of a terrestrial dust devil in Eldorado Valley, Nevada, USA, 2009. Notice, for scale, the 'chase vehicle' positioned at the lower left of the dust devil. Image provided by M. Balme

Emphasis is placed upon the types of measurements obtained and their indication of mean characteristics and identified extrema. [Assessment of dust devil population statistics is provided in the accompanying paper by Lorenz and Jackson (2016).] The terrestrial measurements span a large quantity of literature covering a large number of measurements opportunities. The martian literature is more limited and the measurement opportunities much more discrete than their terrestrial counterparts. The martian instrumentation is discussed in some detail. In Sect. 4 we provide a brief discussion of terrestrial versus martian dust devils. In Sect. 5 we address the topic of future measurements desired for a more complete characterization of terrestrial and martian dust devils. Concluding remarks are presented in Sect. 6.

2 Surface Obtained Imaging of Dust Devils

Most studies of dust devils on Earth have used either (i) an in-situ sampling methodology (i.e. by 'chasing' a dust devil with an instrumented vehicle, or by waiting for a dust devil to approach one or more immobile instrument stations), or (ii) a local remote sensing methodology, in which one or more observers document the local time, size, dust load, morphology, etc. of dust devils as they occur within some kind of fixed study area. Some studies have used a combination of these approaches by deploying both observers and in-situ instruments.

The main challenge for any field study is that dust devils are inherently changeable and somewhat unpredictable. Hence, measurements of size, lifetime and dustiness of a given dust devil are difficult to make—some dust devils start small and grow larger and dustier and persist for many minutes, whereas similar starting examples can disappear without a trace after just a few seconds. This presents a challenge for any qualitative study as "summarising" any dust devil with a few simple parameters is sometimes impossible. Another problem is that dust devils can travel quickly, and can move into and out of a defined study area (or even the entire field of view of an observer) during their lifetime. Similarly, studies

Table 1 Terrestrial dust devil observed characteristics

Parameter	Best observations or measurements	Key literature
Size (diameter)	1 to > 100 m; Strongly skewed size frequency distribution: small dust devils much more common than large; examples > 300 m are very uncommon	Carroll and Ryan (1970); Sinclair (1965); Lorenz (2011)
Size (height)	5 m to > 1000 m; only about 10 % are > 300 m	Sinclair (1965); Bell (1967); Flower (1936)
Lifetime	Seconds to minutes; larger dust devils have longer lifetimes; some reports of large dust devils lasting several hours	Flower (1936); Snow and McClelland (1990); Pathare et al. (2010); Lorenz (2013)
Morphology	Columnar, disordered or v-shaped; columnar vortices often include a v-shaped 'sand skirt'; larger dust devils often include sub-vortices	Metzger (1999)
Rotation sense	Equally clockwise and counter-clockwise	Flower (1936); Carroll and Ryan (1970)
Wind speeds (peak horizontal) at 2 m height	Usually peak at 5–10 $m\,s^{-1}$; recorded peak winds of up to 25 $m\,s^{-1}$ are not unusual	Ryan and Carroll (1970); Balme et al. (2003)
Wind speeds (peak vertical) at 1.0–4.5 m height	Usually ~ 25 % of peak horizontal winds; most peak measurements < 5 $m\,s^{-1}$; rare values of ~ 15 $m\,s^{-1}$ measured	Sinclair (1973); Fitzjarrald (1973); Metzger (1999); Metzger et al. (2011)
Horizontal translation speed	10–20 % greater than ambient wind speeds measured at 10 m height; values of 0 – –10 $m\,s^{-1}$ common; rare observations of values > 20 $m\,s^{-1}$	Balme et al. (2012)
Dust loading at 2 m height	Mean values (i.e., averaged across a profile within each dust devil rather than peak measurement) of 0.8–42 $mg\,m^{-3}$ measured for fine particles (0–10 mm diameter); mean values of up to 6–875 $mg\,m^{-3}$ for total suspended load; average values across > 20 dust devils reveal particle load of ~ 44 $mg\,m^{-3}$ for fine particles, ~ 300 $mg\,m^{-3}$ for total suspended load	Metzger et al. (2011)
Core temperature excursion	Wide range of temperature excursions measured—probably due to variations in sensor type; excursions of 1–5 °C common; larger excursions of > 20 °C reported	Metzger (1999); Tratt et al. (2003); Sinclair (1964); Sinclair (1973)
Core pressure excursion (DP)	Mobile sampling systems appear to give larger values: ~ 1–10 mbar; fixed monitoring stations give values < 1.5 mbar	Sinclair (1973); Metzger (1999); Lorenz and Lanagan (2014)
Electric fields	Field of ~ 10–100 kV/m measured; early measurements hampered by field reaching measurement limit of instrument	Jackson and Farrell (2006); Renno et al. (2004); Esposito et al. (2016)

relying on untended instruments can find it hard to distinguish between a population of long-lived dust devils and a population of more frequently occurring, but short lived dust devils.

A summary of observed characteristics is provided in Table 1.

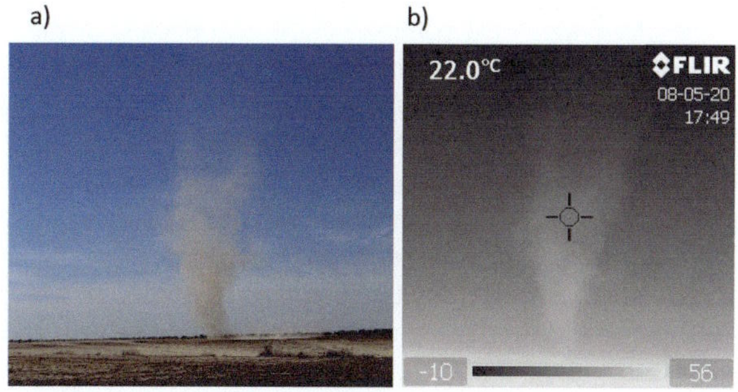

Fig. 2 (**a**) A digital camera image of a dust devil in Eloy, AZ in summer 2008. (**b**) A thermal image of the same devil a few seconds later with a FLIR Infracam hand-held thermal imager (240 × 240 pixels). Note that the image scale is not the same as in the optical image. The bar at the base of the thermal image indicates the brightness temperature grey scale in Celsius

2.1 Remotely Observed Terrestrial Dust Devil Characteristics

2.1.1 Morphology

Dust devil morphology has been measured almost exclusively by surface based remote imaging observation (Figs. 1 and 2). As described by Lorenz et al. (2016), dust devils have been observed for many centuries and many authors describe them as "dusty-columns" or "upright whirlwinds". However, as shown by Metzger (1999), many dust devils do not have this 'classic' form; many are simply disordered clouds of barely-spinning dust, others are more-rapidly spinning, v-shaped cones of entrained dust and sand, and others do have the well-known, rapidly-spinning columnar shape. In a study in Nevada, USA, Metzger (1999) found that only about 4 % of observed dust devils have a columnar shape. Even for those with columnar vortices, their morphology can differ: some have a clear core, others do not; some include a v-shaped "skirt" at their base, but others do not. Finally, some dust devils include sub-vortices that have their own centre of rotation but orbit the main circulation or which trail in their wake (e.g., Williams 1948; Ryan and Carroll 1970; Hallett and Hoffer 1971; Sinclair 1973; Metzger 1999). In many cases, the main circulation is barely dust-charged at all, so in these cases it is difficult to judge whether this is one large dust devil with subsidiary rotational elements, or a group of interacting smaller dust devils.

Most dust devils are higher than their widths. Hess and Spillane (1990) suggest that most are at least five times higher than their width, but again, many opposing observations exist. For those dust devils tall enough that a vertical structure can be observed, Sinclair (1966) provides a still-relevant summary: Region 1 is the zone nearest the ground, is heavily particle-loaded and often has a v-shaped form (Metzger 1999, refers to this as a "sand-skirt"). Region 2, at intermediate height, is the near- vertical column of rotating dust. Region 3, aloft, is where the rotation decays and where the dust devils "fades" into the ambient atmosphere. These regions have been associated with different flow regimes (see, for example, Balme and Greeley 2006, Fig. 9): region 1 is where the majority of the radial inflow occurs, while Region 2 is characterised by rotation and uplift, and Region 3 has poorly characterised flows as here the structure dissipates.

2.1.2 Size

Dust devils on Earth range in size from a few metres in diameter and height to 100 s of metres in diameter and perhaps more than a kilometre in height. While estimates or measurements of diameter are relatively easy to obtain, determining dust devil height is more difficult, especially from surface observations. Also, it must be recognised that dust devil height refers to the observable dust column height, as opposed to the height of any circulation associated with the dust devil. For the visible column, estimates suggest that most dust devils are less than 50 m high, with only about 8 % extending higher than 300 m (Sinclair 1965; Flower 1936; Williams 1948). However, observations made from the air identify taller dust devils that are 1–2.5 km in height (Bell 1967). Finally, Sinclair (1966) notes that measurable vertical wind speed and temperature excursions occur above large dust devils at heights of 2–4 km. However, it is not clear that these represent the upper parts of a dust devil per-se, or instead are associated with a broader circulation in which the dust devil is embedded.

The diameter-frequency of dust devils has been the subject of much study, with methodologies including simple "by-eye" observer surveys to more sophisticated studies that use time-lapse cameras or arrays of meteorology data. As noted by Lorenz (2011), most visual surveys (e.g., Sinclair 1965, 1969; Ryan and Carroll 1970; Snow and McClelland 1990) report that small dust devils are underreported by observers. Possible exceptions to this are the studies of Carroll and Ryan (1970), Pathare et al. (2010) and Balme et al. (2012), which both used small (1 km by 1 km or smaller) study areas. Several recent studies have focused on power law and other functional forms that best describe the dust devil size-frequency population (e.g., Kurgansky 2006; Lorenz 2009, 2011; Pathare et al. 2010)—a topic which is described in more detail by Lorenz and Jackson (2016) in this issue.

What is clear from all these studies is that the diameter-frequency distribution of dust devils is significantly skewed, with far more small dust devils occurring than large. Hence, the concept of an "average" dust devil diameter is not necessarily a useful one. What is clear, though, is that dust devils narrower than 5–6 meters in diameter are far more common than those wider than 10–12 m, and that dust devils wider than 50 m are actually rather rare (e.g. Balme and Greeley 2006, Fig. 3; Lorenz 2011, Table 1).

2.1.3 *Translation Speeds*

Compared with Mars, few measurements of translational speed of terrestrial dust devils have been made. This is mainly due to the lack of top-down remote sensing of dust devils available for the Earth. Some ad-hoc measurements were made in the twentieth century (e.g. Crozier 1970) but the only focussed study, aimed solely at measuring the forward motion of dust devils, is that of Balme et al. (2012), who used stereo photography of dust devils to locate them in time and space. Multiple observations were made of each dust devil, thus allowing a path and thus a velocity to be calculated for each dust devil. In addition, as Balme et al. (2012) employed two 10 m high meteorology masts within their study area, they were able to correlate dust devil motion with ambient wind speed and direction. During the 10 days of field sampling, covering two calendar years and two study sites, translation speeds of between 1 and 15 $m\,s^{-1}$ were measured for more than 100 dust devils.

Balme et al. (2012) found that dust devils translated in the same direction as ambient wind, and that the dust devil forward speed correlated well with ambient wind speed. In fact, they found that dust devils travel at about the same speed as the boundary layer winds a few tens of metres above ground. Interestingly, no correlation of translation speed with dust devil diameter was found. Therefore, Balme et al. (2012) conclude that dust devil forward motion is a good proxy for the wind field, and that dust devil motions is governed almost solely by local wind patterns—a finding that could prove to be important for Mars, where few meteorology data are available.

2.1.4 *Rotational and Vertical Speeds Within Dust Devils*

Measurement of the swirling winds within dust devils has generally been accomplished using in-situ sampling, though particle imaging velocimetry (Ito and Niino 2014) and remote measurements using LIght Detection And Ranging, or "LIDAR" (e.g., Schwiesow et al. 1977; Bluestein and Pazmany 2000) have also been employed. Obtaining statistically significant quantities of data is challenging for two main reasons. Firstly, even the most efficient sampling methodology (arguably, the mobile sampling platform approach) will only be able to sample a few dust devils per day, due to the short-lived and random nature of the phenomenon. Fixed sampling positions mean an even lower number of samples are likely, although this problem can be ameliorated by using many sensors in large arrays, or by sampling for very long periods of time. It is possible that dedicated LIDAR studies could also gather large quantities of data but to date no such attempts have been made. To date, the largest published study is that of Ryan and Carroll (1970), who sampled 80 dust devils. Another more recent large dataset exists (see preliminary report in Metzger et al. 2011) and includes more than 50 measurements from mobile in-situ sampling. However, these data are yet to be formally published. Aside from these two examples, individual studies usually report fewer than twenty encounters (Balme and Greeley 2006).

Both in-situ and remote sampling of wind speeds share the second problem—smaller and/or shorter-lived dust devils are harder to sample, and therefore are likely to be under represented in the data. This problem is likely to affect mobile "chase" strategies most significantly, as only the larger and longer-lived dust devils can be caught and penetrated to acquire data. Similarly, fixed-position remote-sensing studies are liable to target the most easily seen dust devils and could easily miss small, less dusty examples. The remaining methodology—that of fixed long-term meteorology stations—should be able to remove this sample bias, but again there is the issue of detectability, although this time the problem is how to tease out "detections" from the data. Significant progress has been made in this area recently; by using a single meteorology station and a theoretical dust devil 'signature', Lorenz (2016) was able to reconstruct peak wind speeds (and other signature parameters) and miss distance (i.e. the distance from the dust devil core to the sensor) for 27 dust devil events in 16 days of field time. However, the calculated diameters for these dust devils are all larger than 10 m—suggesting that these too are "exceptional" events and again the more typical, smaller examples have not been detected.

Despite these caveats, the wind speeds within dust devils have been measured. Some authors have measured only the magnitude of the horizontal wind speeds within the dust devils, whereas others have provided all three components (i.e. inflow, tangential and vertical wind speed). Speeds are usually quoted at a height of 2 m above the ground, but measurements both nearer the ground (e.g. Balme et al. 2003) and higher into the dust devil (e.g. Kaimal and Bussinger 1970) have been reported. The following key points have emerged: (i) the horizontal winds within dust devils can often reach $10\ \mathrm{m\,s^{-1}}$ (e.g., Ryan and Carroll 1970), can peak at $> 25\ \mathrm{m\,s^{-1}}$, but rarely, if ever, exceed $30\ \mathrm{m\,s^{-1}}$ (see Balme and Greeley 2006, Table 4, and Lorenz 2016, Table 1), (ii) vertical wind speeds are usually a factor of several less than the horizontal winds (e.g., Balme and Greeley 2006, Table 4), (iii) larger dust devils appear to contain stronger swirling winds, but vertical wind speeds do not correlate with diameter (e.g., Ryan and Carroll 1970), and (iv) the surface shear stress provided by the wind speeds within terrestrial dust devils appears sufficient to lift almost all sizes of naturally occurring loose sediments up to about granule-sized material (Balme et al. 2003).

2.1.5 *Dust Load*

Due to the fast-changing environment within dust devils, the concentrations of airborne dust and larger particles within dust devils are difficult to measure. The most complete study of particle loading in dust devils is that of Metzger et al. (2011), who used in-situ sampling on a mobile platform. Few other studies exist for terrestrial dust devils: a preliminary LIDAR observation at 100 m height (Renno et al. 2004) provided an estimate of dust load of $\sim 100\ \mathrm{mg\,m^{-3}}$ and there are reports of aircraft in-situ sampling of dust devils at 140 and 300 m height (Gillette and Sinclair 1990), but only flux data are given, not particle load. Other datasets exist for martian dust devils (e.g. Greeley et al. 2006, 2010), but Metzger et al. (2011) provide the main source of dust load data for Earth.

Metzger et al. (2011) present data from more than 30 encounters at two field sites and over four field seasons. They used both PM-10 sensors (sensitive to dust grade materials, 0.1–10 μm diameter) and total suspended load sensors (dust- to sand-grade materials). All measurements were made at the base of the dust devil (sensors were generally at 2 m height, but some measurements at 1, 2.8 and 4.5 m height are reported). They found that PM10 dust load had high intra- and inter-dust devil variability. In many cases they report both maximum and mean dust load per dust devil (rather than just reporting the peak dust load) and report a peak range of 6–162 $\mathrm{mg\,m^{-3}}$ and mean range of 0.8–42 $\mathrm{mg\,m^{-3}}$. The measured total suspended particle load (i.e. including larger sediments) was much higher: ranging from 6–875 $\mathrm{mg\,m^{-3}}$. Metzger et al. (2011) conclude that mean peak dust load (i.e. the amount of dust likely to lofted to height by a dust devil) is about three times less than the peak load measured, and that the total suspended particle load near the base of the dust devil is about ten time greater than the PM-10 fraction. They note, however, that the larger size fraction material is unlikely to be transported to great height, and is probably redeposited locally. This measurement is in agreement with observations of a "sand skirt" at the base of many dust devils. Oke et al. (2007) measured particle size within the bottom $\sim$ 1.5 meters of willy-willies, finding that sand sized particles were confined below $\sim$ 20 centimeters. Raack et al. (2014) find similar results in Morocco.

2.1.6 *Seasonal and Diurnal Frequency of Occurrence*

As they are convective vortices, driven primarily insolation, dust devils occur most frequently when there is strong, continuous sunshine. This is usually in the summer, but they can occur at any time of year when there is a significant thermal contrast between the ground and the atmosphere (for example, dust devils have been seen in the Canadian sub-arctic; Grant 1949). To our knowledge, no season-to-season monitoring of dust devil activity has been performed to further refine the seasonal frequency, though.

In terms of diurnal rate of occurrence, many authors have noted that dust devils form most frequently in the late morning and the early afternoon (see summary by Balme and Greeley 2006 and recent work by Kurgansky et al. 2011). However, many of these reports are based on observer surveys which are both qualitative and which are unlikely to have run throughout the day—so there is always a possibility of bias. Recent work by Lorenz and Lanagan (2014) using a continuous month-long survey of pressure excursions to detect dust devils showed that most dust devils occur between 10:00 and 16:00 local time. They do note that about 10 % of the day's dust devil events occur after 16:00 and there is a measurable tail of activity even after 18:00.

Several authors note that dust devil events are 'clustered' in time, with periods of more intense activity separated by periods of less activity. Carroll and Ryan (1970) suggest a periodicity of around 45 minutes and Renno et al. (2004) a periodicity of about 20 minutes. Lorenz and Lanagan's (2014) pressure-excursion data hint at a similar result.

In addition to surveys by human observers making either continuous records or recording at intervals (e.g. how many are seen at 15-minute intervals), the availability of time-lapse cameras, webcams etc. now allow new surveys with superior temporal coverage, and with quantitative detection criteria (e.g. optical contrast of 1 %). It is important in all such surveys that the detection criterion (size, contrast), and measurement cadence be documented—for instance images acquired at a given cadence or observing distance may preferentially detect a particular size of dust devils (Lorenz 2011, 2014; Kurgansky et al. 2011).

2.1.7 Thermal Imaging

The warm core of a dust devil and/or its suspended dust warmed by insolation absorption would be expected to provide a radiative thermal signature. Lorenz (2004) provided the apparent first scientific report of thermal infrared imaging of a dust devil (although Metzger et al. 2010, have since reported thermal imaging of Atacama dust devils, and Towner 2008, report orbital thermal imaging of dust devils at Mars). Thermal imagers have reduced significantly in cost in recent years, in part due to application in home improvement (to detect damp or poor insulation in walls). An example image is shown in Fig. 2.

The Lorenz (2004) observation reported a single dust devil as it moved away from an initial distance of 10 meters. The vortex temperature of 38–40 °C derived from the 8–14 μm emission was approximately the same as the measured ambient air temperature but greater than the background atmospheric 'brightness' temperature of 12–20 °C obtained for lines of sight that did not intersect the dust devil. Lorenz (2004) noted that the dust might even be physically warmer than the surrounding air due to its interception of sunlight, a factor that might enhance the intensity of a vortex when dust is lifted (as later discussed by Fuerstenau 2006). Thermal imaging might provide a higher-sensitivity means of detecting marginally visible dust devils under some circumstances (notably, low-light levels) but this has not been robustly demonstrated. It may be that thermal imaging could also help visualize the near-surface wind stress field around the devil (e.g. showing 'spiral' arms of the inflow, as can sometimes be seen on the ruffled surface of the sea around waterspouts) due to the wind-dependence of surface heat transfer.

2.2 Remotely Imaged Martian Dust Devil Characteristics

The opportunities for martian surface-based acquisition of visible imaging and subsequent characterization of dust devils has been limited to the two Viking Landers (1976–1982), Mars Pathfinder (1997), Mars Exploration Rovers Spirit (2004–2010) and Opportunity (2004–present), Phoenix Lander (2008), and Mars Science Laboratory Curiosity Rover (2012–present). Some of these missions provided no visible evidence of dust devil occurrence (Viking), while others provided the current best martian dust devil climatology (*Spirit*). Because imaging observations are discretely separated between robotic missions, we follow a chronological mission order presentation below.

A summary of observations is presented in Table 2.

2.2.1 Viking Lander & Mars Pathfinder Imaging

The two Viking Landers safely settled onto Mars' surface during 1976, at subtropical (VL1) and middle (VL2) northern latitudes. VL1 returned measurements spanning 2245 sols covering portions of four martian years, while VL2 returned measurements for 1050 sols. Despite the substantial temporal extent of the Viking lander imaging data sets, these landers were unable to image dust devils because their cameras were facsimile-type imagers not well-suited to detecting moving objects (Lorenz et al. 2016). As a consequence of this characteristic of the Viking camera system, the first opportunity to visually detect martian dust devils from a

Table 2 Martian dust devil observed characteristics

Parameter	Best observations or measurements	Key literature
Size (diameter)	Inferred core diameters 10–700 meters from Viking wind measurements (not visually confirmed); 15–550 m (most frequent 100–200 m) from Pathfinder IMP (for assumed 10 $m\,s^{-1}$ translation speed); median diameter 20–40 m from Spirit Nav Cam; MSL pressures and winds (16 m median; 21 meter mean)	Ryan and Lucich (1983); Ringrose et al. (2003); Ferri et al. (2003); Greeley et al. (2010); Kahanpää et al. (2016)
Size (height)	$\sim$ 10–400 m (though many images truncate vortex top); maximum verified lower limit $\sim$ 800 m	Greeley et al. (2006)
Lifetime/Detection Duration	120–180 seconds from imaging; wind effect determination duration (60–1000 seconds); FWHM from wind pressure measurements (5–20 seconds)	Ringrose et al. (2003); Ferri et al. (2003); Greeley et al. (2010); Ellehoj et al. (2010); Kahanpää et al. (2016); Steakley and Murphy (2016)
Morphology	Columnar, disordered or v-shaped; few columnar vortices include a v-shaped 'sand skirt'	Greeley et al. (2006); Ferri et al. (2003)
Rotation sense	Equally clockwise and counter-clockwise	Ryan and Lucich (1983)
Wind speeds (peak horizontal) at $\sim$ 2 m height	Maximum measured speeds are $\sim$ 30–40 $m\,s^{-1}$; maximum inferred core boundary speeds approach 100 $m\,s^{-1}$ for two instances which correspond with the greatest spatial extrapolation to that core boundary position	Ryan and Lucich (1983); Ringrose et al. (2003); Ellehoj et al. (2010)
Wind speeds (peak vertical) from imaging	Maximum $\sim$ 17 $m\,s^{-1}$, with median 1.0–1.6 $m\,s^{-1}$	Greeley et al. (2010)
Horizontal translation speed	A few to $\sim$ 25 $m\,s^{-1}$ (median $\sim$ 2 $m\,s^{-1}$) (Note: ambient wind speeds were not correspondingly available)	Greeley et al. (2010)
Dust loading	700 $mg\,m^{-2}$ for a horizontal path through a dust devil, implying 20 $mg\,m^{-3}$ for a 35-m diameter dust devil, for Pathfinder; 0.002–250 $mg\,m^{-3}$ for Spirit; coarse particles (> 63 micrometers) confined to < 30 cm above the surface	Metzger et al. (1999); Greeley et al. (2010); Oke et al. (2007)
Temperature excursion	Measured excursions of 1–6 °C for Viking & Pathfinder & Phoenix; MSL/REMS	Ryan and Lucich (1983); Murphy and Nelli (2002); Ringrose et al. (2003); Ellehoj et al. 2010); Kahanpää et al. (2016)
Core pressure excursion (ΔP)	0.3 to $\sim$ 5 Pa; lower limit arises from limit imposed upon the analyses, with smaller magnitudes more frequent; no measurement regarding verified vortex maximum ΔP excursion is available	Murphy and Nelli (2002); Ellehoj et al. (2010); Kahanpää et al. (2016); Steakley and Murphy (2016)
Electric fields	No measurements available	

surface vantage point occurred with the Mars Pathfinder Lander (MPF) mission. It is tempting to speculate that the facsimile-type imager on the Viking landers may have a few dark lines in parts of a few images caused by dust devil passages during their slow scans, but these would be challenging to detect and attribute.

MPF began operation on Mars' surface on 04 July, 1997 in the northern subtropical (19.3 N, 33.4 W; Golombek et al. 1997) Ares Vallis region located $\sim$ 1000 kilometers east-southeast of Viking Lander 1. The mission provided measurements spanning 83 sols covering the latter third of northern summer through very early northern autumn (L_s 142–183). There was some expectation that Pathfinder would have the opportunity to detect dust devil signatures within its meteorology (MET) measurements (Seiff et al. 1997).

From analysis of multi-color images provided by the Imager for Mars Pathfinder (IMP) (Smith et al. 1997), Metzger et al. (1999) reported the first identification of martian dust plumes within surface-obtained images. Image enhancement via band subtraction was employed to isolate the low-contrast signature of the dust plume. Dust plumes were most evident as occultation features at blue (430 nm) wavelengths among the Imager for Mars Pathfinder (IMP) wavelength filters, which also included 530 and 670 nm, due to the reduced dust scattering at that shorter wavelength. Five dust plumes were identified within 16 analyzed IMP images obtained near mid-sol on sols 10 and 11 of the 83-sol mission when landing site panoramas were being acquired. Dust devil diameter (14–79 meter) and height (46–450 meters) and translation speed (0.5–4.6 $\mathrm{m\,s^{-1}}$) were estimated from the angular width and motion derived from multiple images and the inferred distance from the lander obtained from identification of foreground and background features. The plume/vortex dust load ($\sim 7 \times 10^{-5}\ \mathrm{kg\,m^{-3}}$) was estimated to be $\sim$ 10,000 times greater than the background dust load.

Ferri et al. (2003) applied the Metzger et al. (1999) band subtraction technique to a more extensive set of horizon-containing IMP images and identified 14 dust plumes/devils that included the 5 from Metzger et al. (1999). Only one of these identifications occurred for an image that was not part of the panorama captured during mission sols 10–11. To quantify vortex size, a constant translation speed of 10 $\mathrm{m\,s^{-1}}$ was assumed, with the observed angular rate of motion from consecutive images being employed to estimate vortex distance and subsequently vortex size. The 14 identified vortices were estimated to span the size range of 10–570 meters and to have been positioned 1–25 km from the lander. A surface vortex surface area coverage of 2×10^{-4} (0.02 %) was estimated for the 0900–1500 local true solar time (LTST) time interval. This fractional coverage, coupled with a derived vortex vertical dust flux estimate of $7 \times 10^{-5}\ \mathrm{kg\,m^{-2}\,s} - 1$, resulted in an estimated total vortex-induced vertical dust flux of $3.6 \times 10^{-9}\ \mathrm{kg\,m^{-2}\,s^{-1}}$, which exceeded by an order of magnitude the estimated local dust deposition rate (Landis and Jenkins 2000).

In addition to images, one IMP-provided opacity measurement from direct solar imaging on sol 14 resulted in larger values at all wavelengths compared to more than 10 additional opacity observations obtained that same sol (Smith and Lemmon 1999). This increased opacity event could possibly have been the result of a dust-laden vortex occulting the Sun but no additional supporting measurements, including contemporaneous MET measurements, are available.

Thus, the Pathfinder mission did verify that dust devils, or at least dust plumes (since motion within a plume was not identified), are visible from Mars' surface, but these imaging results did not provide a rigorous quantitative characterization of martian dust devils.

2.2.2 *Mars Exploration Rovers (Spirit & Opportunity)*

After the experience gained with the martian dust devils identified in the MPF IMP images, the Mars Exploration Rover (MER-A & B—*Spirit & Opportunity*) science teams prepared for the possibility of observing dust devils at the two landing sites: Gusev Crater (*Spirit*) and Meridiani Planum (*Opportunity*). MER science teams planned on using similar band-subtraction techniques for viewing dust devils as was originally used by Metzger

Fig. 3 Three images from a typical *Spirit*—Navigation camera sequence used to create martian dust devil "movies". *Numbers* in the *lower left* of *each image* indicate the number of seconds since the image sequence began. This sol 456 dust devil viewed from the west flank of the Columbia Hills translates from left to right, passing over a small bright toned depression as it moves

et al. (1999) to enhance the contrast to make these difficult phenomena easier to study. The Panorama Cameras (Pancams) on the MERs were higher resolution than the Imager for Mars Pathfinder (IMP) and capable of better images, but it was not known if either of the two locations would be capable of producing better, more visible dust devils than those observed in Ares Vallis with MPF. Orbital images suggested that Gusev Crater (*Spirit*) would have a good chance of seeing dust devils because of a swath of features across the crater with many dust devil tracks (Greeley et al. 2003). However, prior to Spirit's landing, active dust devils had not been observed from orbit in Gusev crater, casting doubt that the swath of dark features were in fact dust devil tracks. Spirit landed near the end of Southern Hemisphere Summer ($\sim L_s$ 330) [Squyres et al. 2004]. The first part of the mission produced no imaging-detected dust devils at either location for the rovers, and many searches through images using the band subtraction technique yielded no results (Lemmon et al. 2004), although one new dust devil track was observed from orbit, having formed between L_s 12 and 22 (Lemmon et al. 2015). On sol 421 (L_s 173), while *Spirit* was perched on the Columbia Hills near the center of the Gusev, the first dust devil was observed, differing greatly from the nebulous wisps from the MPF images. The MER image was crisp and detailed and as with subsequent observations, the dust devils were clearly visible as distinct from the background (Fig. 3). Observations at Meridiani by *Opportunity* have been limited to a few sightings of individual dust devils that could be a result of poorer viewing geometry and/or limited liftable dust (Lemmon et al. 2015).

After the first year of the mission, an elevated vantage point on Husband Hill offered several advantages over the previous lander geometry (Greeley et al. 2006). Sitting above the plains of the crater floor, the dust devils appeared bright against the ground, and darker against the sky above the horizon. The elevated viewing angle also allowed each dust devil to be more precisely located against surface features such as smaller craters and hollows, and rock patterns. More precise locations allowed better distances to be known, allowing for better estimations of sizes. Once the first dust devil season officially started, it became clear that dust devils could be seen easily in all of the rover's camera systems including not only the Pancam, but also the lower resolution monochrome Navigation Camera (Navcam) and both the forward and rear Hazard Cameras (Hazcams). Specific imaging campaigns were designed to make use of what was known about dust devil statistics. Initial dedicated imaging occurred during $\sim$ 0900–1700 LST. Subsequently, Navcam and Pancam images were subframed upon the ground/sky boundary and the most common locales for dust devil occurrence. Subframing reduced the amount of storage space for each image and allowed multiple image, "movie", sequences to be obtained (Fig. 3). The ~ 20 second frame rate on the movie sequences was limited by the refresh rate of the cameras' CCDs and the image buffer, but the image acquisition times refresh cycle was well known, allowing time between frames to be accurately established. Due to the favorable viewing geometry, on many sols not only sizes of and distances to the dust devils could be determined, but also translation speeds (horizontal) of the dust devils suggesting background ambient wind speeds in Gusev. In some movie sequences, detail was high enough that pockets of dust from some of the hollows could be followed from frame to frame allowing rough estimates of vertical velocities to be determined.

The longevity of the MER campaigns allowed for repeat seasonal studies of dust devil activity, which was particularly useful in Gusev Crater. *Spirit* had observed ~ 533 dust devils in the first documented dust devil season (Greeley et al. 2006). This high number of individual vortices was due in part to excellent viewing geometry from atop the Columbia Hills, from where much more of the crater floor of Gusev was visible. As *Spirit*'s traverse led it further south into the saddle and eventually to the "Home Plate" feature, *Spirit*'s view of the crater floor was obstructed by the hills. Greeley et al. (2010) describes the three total observed seasons in detail. The second dust devil season began on about sol 1101 (L_s 181°), which was comparable to the first season's start around sol 421 (L_s 173.2°). The second dust devil season was truncated by the onset of a set of planet-encircling dust storms that restricted insolation at the surface, presumably prohibiting the formation of dust devils while the background atmospheric dust opacity rose to a peak tau of 4.31 (Greeley et al. 2010; Lemmon et al. 2015). While the dust opacity was so high, solar power for rover operations was limited and fewer images were taken, but of the images that were acquired, no dust devils were observed during the dust storm and the upper limit for dust devil frequency was an order of magnitude below pre-storm levels (Lemmon et al. 2015). The second season, with limited viewing geometry and the presence of strong regional dust storms, produced an observed 101 individually identified dust devils. The following martian year, 127 more vortices were observed when the third dust devil season began around sol 1785 (L_s 189°). Still located near Home Plate, *Spirit*'s view of the plains was still restricted similarly to the second season (Fig. 4). Over three Mars years, Lemmon et al. (2015) found that changes in dust devil frequency correlated with changes in surface insolation, whether the changes were seasonal or from dust storms, and that dust devil frequency fell to $1/e$ with each 18 $\mathrm{W\,m^{-2}}$ reduction in mean insolation (roughly 10 % of the peak insolation) (Fig. 5).

Spirit and *Opportunity* were not equipped with meteorological instrument packages for temperature and pressure measurements, but over the three dust devil seasons recorded at Gusev, several key measurements were made using the camera systems. For the first time on Mars there was repeat coverage at the same site for dust devil seasonal information. Correcting for sample bias, Greeley et al. (2010) estimated that the diurnal distribution of dust devils at Gusev began after 0900 and tailed off before 1700 LTST. This time range was consistent for all three seasons at Gusev (Fig. 6). Peak activity occurred around 1300 LTST with some indication of possible burst of activity near the end of the day between 1400 and 1600 LTST. Tracking dust devils within movie sequences yielded estimates of both translational and vertical velocities. Translational speeds, which could serve as a rough surrogate to the background boundary layer winds across Gusev Crater were estimated to be between a few meters per second up to $\sim 27\ \mathrm{m\,s^{-1}}$ and maximum velocities tended to occur near the end of local springtime. Minimum vertical wind speeds

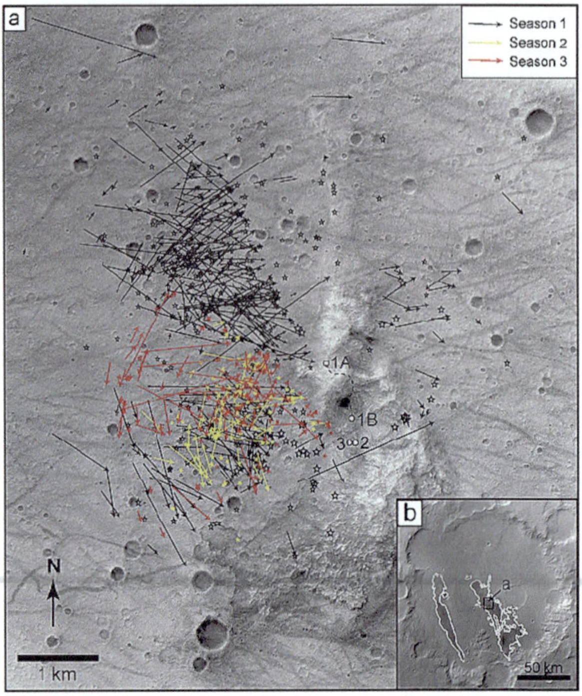

Fig. 4 (After Greeley et al. 2010) Three seasons of dust devil occurrence locations within Gusev Crater. (**a**) Mosaic of HiRISE images of Spirit's operation area, with vectors indicating dust devil tracks from Spirit Navigation camera 'movies', *color* coded for each season (year); stars indicate locations of active dust devils from single frames. (**b**) Mars Orbiter Camera Wide Angle red image R21-00168 inset of Gusev Crater showing the Gusev Low Albdeo Zone (GLAZ, outlined) where dust devil tracks are observed, and the location of Fig. 4a

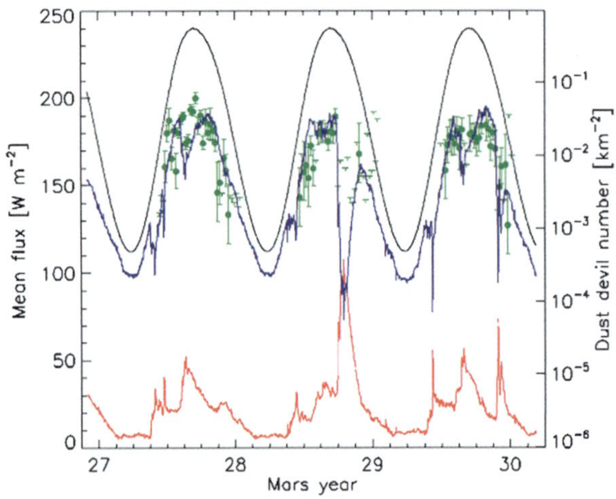

Fig. 5 Insolation at the Spirit rover site. The *continuous curve* shows the modeled top-of-atmosphere, sol average insolation (*upper*, *black*), direct plus diffuse surface insolation (*middle*, *blue*) and atmospheric absorption of sunlight (*lower*, *red*). *Symbols* (*green*) show dust devil number density (*right axis*) reported by Greeley et al. (2010) with the T symbols indicating upper limits). (For the interpretation of the references to color in this figure legend, the reader is referred to the web version of this article)

within a few dust devil columns were estimated based on visual identification of small clumps of dusty material present from frame to frame in the movie sequences. Vertical wind speeds were estimated to be 0.04 up to $\sim 17\ \mathrm{m\,s^{-1}}$ with medians being between 1.0 and $1.6\ \mathrm{m\,s^{-1}}$. The data from the MER rovers, and Spirit in particular provide the most complete observation of a dust devil season on Mars and rival any campaign attempted on Earth.

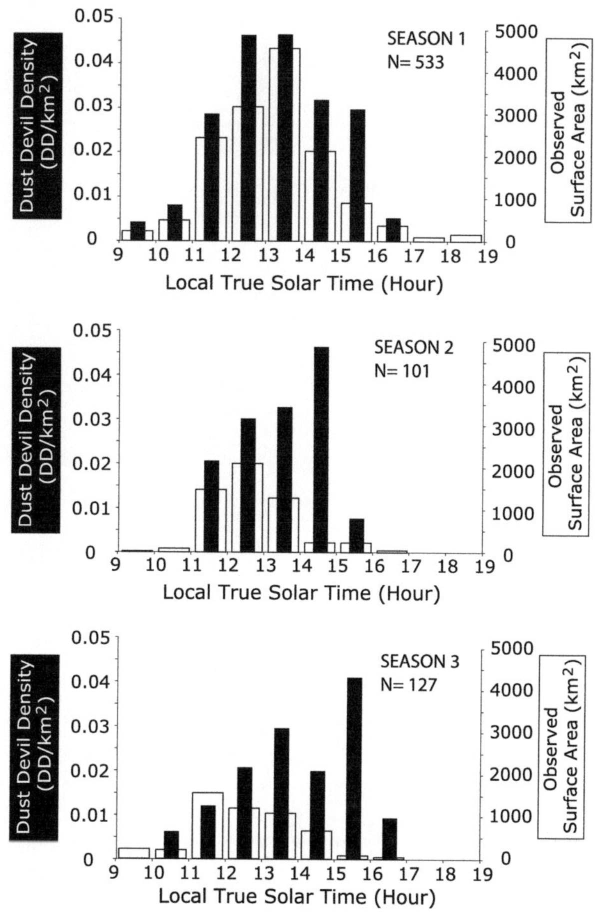

Fig. 6 (After Greeley et al. 2010) Number density of Spirit observed dust devils versus Local True Solar Time (LTST) for three seasons (years). *N* is the total number of dust devils observed each season

2.2.3 *Mars Phoenix Lander*

NASA's Mars Phoenix spacecraft (Smith et al. 2008, 2009) landed at an arctic location (68.2 °N, 234.3 °E) in the Martian Northern Plains on 25 May 2008, in early northern hemisphere summer (L_s 77°). The primary mission lasted for 90 sols; contact was lost after 151 sols (circa 5 months), in late summer (L_s 148°). Equipped with the Surface Stereo Imager (SSI), with adequate resolution to image dust devils, and a high-resolution pressure sensor, Phoenix became the second Mars lander, after Pathfinder, that had the capacity to detect dust devils both visually and by meteorological measurements. It was, however, not until sol 104 ($L_s = 125°$) that the first dust devil was spotted by SSI (Ellehoj et al. 2010). However, there had not been many opportunities to detect dust devils before this. Besides imaging geological targets close to the lander, this camera was used to monitor the Telltale wind indicator (Holstein-Rathlou et al. 2010). The first dust devil was detected serendipitously while imaging a panorama. After this first detection, Phoenix was commanded to take image sets aimed to search for dust devils. These sets consisted of 13 to 50 sequential images of the horizon.

Ellehoj et al. (2010) investigated the images taken by Phoenix of dust devils. Image contrast was enhanced, as had been previously done with the Pathfinder (Metzger et al. 1999;

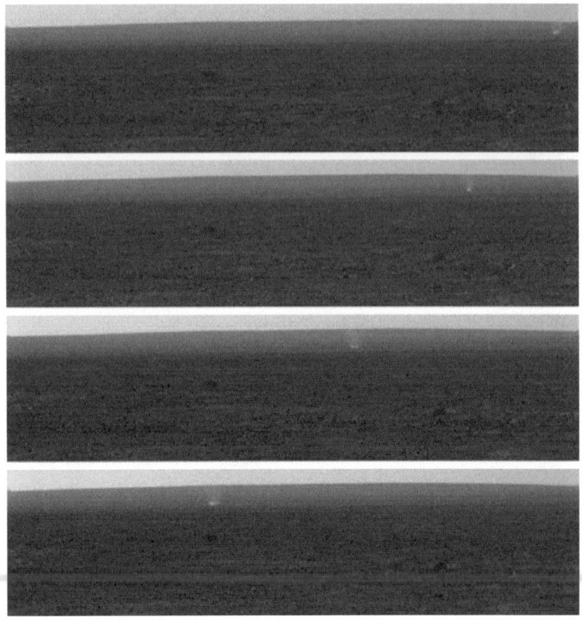

Fig. 7 This two-minute and 37-second time sequence (*top*-to-*bottom*) of contrast-enhanced Phoenix SSI images shows the translation of a mid-afternoon dust devil seen southwest of the lander on the mission's 109th sol ($L_s = 127$)

Ferri et al. 2003) and MER images (Greeley et al. 2006, 2010), allowing the detection of features with only an approximately 3 % difference compared to background albedo. 37 individual dust devils were identified in the SSI images obtained between sol 104 and sol 138 ($L_s = 125$–142) (Fig. 7). The beginning and end of the Phoenix "dust devil season" could not be determined because the dust devil imaging campaign lasted only for a small fraction of the Martian year. However, the first Phoenix observations of dust devils were shortly (14 sols) after the Sun set for the first time in the mission—still near peak northern summer insolation, as at Gusev crater for southern summer, but when temperature contrasts could develop through diurnal cycles. The diurnal time range of SSI dust devil observations spanned 11:00 to 16:00 LTST, matching results of MER Spirit and the timing of vortices detected by the Phoenix pressure sensor (Sect. 3), although no dust devil search imaging was performed in the morning hours before 11:00 due to operational constraints on available energy to heat and aim the camera for use.

All dust devils were seen in the southwest direction, despite two-thirds of the horizon images having been obtained from other directions, with unobstructed views in all azimuths. The distances of the dust devils from the lander could not be determined in the featureless landscape so their physical size could not be evaluated. In most cases the dust devil was visible in several sequential images and was thus seen to move across the frame. As the distances were unknown, translation speeds could not be assessed. In most cases the direction of motion had a an eastward component, agreeing with the wind directions measured by the Telltale wind indicator and orbital observations of dust devils and their tracks in the same area (Holstein-Rathlou et al. 2010; Reiss et al. 2014). The dust devils detected by Phoenix were too faint to enable determination of vertical wind speeds as had been done for some of the dust devils imaged by MER Spirit (Ellehoj et al. 2010).

2.2.4 *Mars Science Laboratory/Curiosity Rover*

The US/NASA Mars Science Laboratory (MSL) Curiosity rover has performed meteorological measurements since August 2012 in 154 kilometer diameter Gale crater, centered just south of the Martian equator (4.6 °S, 137.4 °E). Studying the modern Martian environment is one of the science goals of NASA's Mars Science Laboratory (MSL) mission (Grotzinger et al. 2012). MSL is, unlike the MERs, equipped with a meteorological station, a video-capable color science camera, and a MER-like high signal-to-noise navigation camera system with a reasonably high frame rate. With this payload and a planned operational lifetime of more than one Martian year, MSL could have become an almost perfect lander for the study of dust devils. However, the landing site in Gale crater turned out to be less than ideal for this purpose.

Prior to MSL's landing it was expected that very few, if any, dust devils would occur at this site. No dust devil tracks had been seen in orbital images of the crater floor (Fred Calef, Jet Propulsion Laboratory/Caltech, personal communication, 2012). Further, atmospheric modeling (Tyler and Barnes 2013; 2015) suggested that the depth of the daytime boundary layer inside the crater is suppressed. This suppressed, shallow depth was expected to reduce vortex activity, or at least vortex intensity, since the thermodynamic efficiency of vortices depends upon the boundary layer depth according to the so-called heat engine model (Renno et al. 1998, 2000). Nevertheless, an extensive campaign of imaging Dust Devil Search Movies was initiated after MSL's landing (Moores et al. 2015). The Dust Devil Search Movies are taken using MSL's monochrome Navigation Cameras (Navcam) and consist of four to eight frames with the central elevation on the horizon. All together 91 Dust Devil Search Movies were imaged during the first 360 sols of the mission. The result matched the expectations: only one very faint dust devil was detected on sol 41 ($L_s = 173$). This virtual non-detection, combined with MSL's meteorological measurements, can be used to constrain the conditions where dust devils can form on Mars (Kahanpää et al. 2016; Klose et al. 2016).

Collectively, five spacecraft equipped with cameras suitable for imaging moving objects have landed on Mars: Mars Pathfinder, MER Spirit, MER Opportunity, Mars Phoenix and MSL. Despite the different landing sites, spanning latitudes from 14.6 °S (Spirit) to 68.2 °N (Phoenix), all of these landers have succeeded in imaging dust devils (or at least dust plumes), indicating that dust devils occur on all latitudes on Mars. While the other landers have imaged from tens to hundreds of dust devils, MER Opportunity and MSL have observed only a few, in spite of the long durations of these missions and active search for dust devils, showing that there are strong variations in local dust devil occurrence rates.

While martian surface obtained imaging has provided direct evidence for dust devils, only the Spirit rover within Gusev Crater has provided measurements from which internal vortex characteristics have been quantified and then only minimally. The Spirit observations did provide valuable seasonal and time-of-sol occurrence characterization; Phoenix provided some seasonal indication of occurrence variation but the number of images available for analysis were more limited than the number and seasonal extent available from Spirit.

3 Surface Obtained Meteorology Measurements of Dust Devils

In addition to their visual manifestation, dust devils can also be characterized via in situ measurements of their thermodynamic conditions, including the central pressure drop, wind speed, temperature, suspended dust load, surface dust lifting magnitude, electrostatic state, etc. For such measurements to be of statistical characterization value, high sampling frequency and long duration (seasonal, annual) measurements are best, but until recently such long lived measurement opportunities were not the norm. Previous in situ measurements have provided much of the knowledge currently available regarding dust devil thermodynamics. Despite their often incomplete coverage and potential biases these measurements are the foundation upon which current and future measurements of terrestrial dust devils are constructed.

A compilation of measured characteristics is provided in Table 1.

3.1 Meteorological Measurements of Terrestrial Dust Devils

Sporadic serendipitous meteorological encounters with dust devils were reported in the early literature, such as an encounter with the barograph at a small airport (Wyett 1954) but the first systematic measurements began in the 1960s. A particular challenge is that dust devil phenomena generally occur on timescales (seconds) that require fast instrument response and data acquisition. Sinclair (1966) made pressure, temperature and wind measurements with a hand-carried recording instrument station, and later a more elaborate station mounted on a jeep. The mobile sensor platform allowed penetration of dust devils within a reasonably short 'hunting' season, and data acquisition arrangements included a cine camera recording instrument readings.

Also in the 1960s, two sets of fixed-station investigations were performed. Lambeth (1966) set up an array of 6 meteorology stations at White Sands Missile Range, and recorded (with chart recorders) 19 encounters in a several month period. This rather low encounter rate proved discouraging, leading the author to recommend vehicle-borne measurements. On the other hand, Ryan and Carroll (1970) made temperature and wind measurements at a single fixed mast in the Mojave desert, but groomed the ground around the mast to ensure dust availability.

Field studies of dust devils saw a renaissance in the late 1990s and early 2000s with the observation of dust devils on Mars by Mars Pathfinder (Sects. 2.2.1, 3.2.2), and the prospects for observing more there with a lander planned for 2001 (later cancelled) and by Beagle 2. These impending Mars missions lead to several field campaigns in Arizona and Nevada (e.g. Tratt et al. 2003; Ringrose et al. 2003; Renno et al. 2004; etc.) during which, again, the measurements were principally vehicle-borne.

The presentation below of terrestrial dust devil measurements follows a measured parameter structure, with emphasis upon more recent measurements.

3.1.1 Pressure Measurements

While providing a time-efficient means of acquiring measurements in dust devils, vehicle-borne chase measurements do not reproduce how measurements are acquired on Mars, where a single fixed station records data over an extended time, but not continuously. In addition to vehicle disturbance of measurements (e.g. visible in Tratt et al. 2003) and often rather poorly-documented distance histories, the tendency to chase the biggest, slowest devils leads to strong selection biases, which make it difficult to estimate the characteristics of the dust devil population. Technological developments in flash memory in the late 2000s allowed inexpensive data acquisition (Lorenz 2012b) with compact and low-power systems that could be deployed for months, sampling at >1 Hz without requiring operator visits to download data or replace power supplies. Furthermore, when only pressure and light levels are recorded, the logging package can be placed in

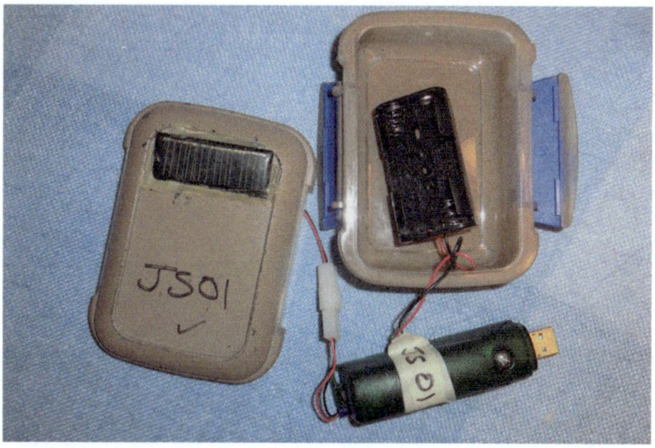

Fig. 8 Pressure and solar-flux logger used by Lorenz and Jackson (2015) and elsewhere. The commercial logger itself is essentially a USB memory stick (*green cylindrical object* at *lower right*) which can accommodate a single AA battery to operate for several days. Here a 2 × AA battery holder (which due to peculiarities of the power supply system yields about a month of operation at 2 Hz) is included, as well as a solar cell to record the shadow of dust devils. The whole unit can fit in a pocket. Note the logger identifier, necessary as these loggers have been installed in arrays

a box with a volume of less than 0.5 liters (Fig. 8) and simply set on the ground: this can be done in such a way (with a camouflaged housing) such that attrition by theft or vandalism at open sites is minimal. The systems are inexpensive enough (~ $150) compared with the cost of deployment and retrieval that they can be considered somewhat expendable. These systems have allowed large numbers (hundreds to thousands) of unperturbed vortex encounters to be obtained without chase biases, finally yielding robust statistics on vortex populations from pressure drops (Lorenz and Lanagan 2014; Jackson and Lorenz 2015) and on dust loading (Lorenz and Jackson 2015). Furthermore, the small measurement stations can be deployed in spatial arrays to make simultaneous measurements that resolve the two-dimensional horizontal structure of dust devils (Lorenz et al. 2015a, 2015b, 2015c).

The pressure loggers developed by Lorenz (2012a, 2012b) are compact enough that they can be easily carried in a pocket and operate without attention for many days, allowing opportunistic pressure measurements during other field activities without extensive preparation. For example, Lorenz and Radebaugh (2016) report the first in-situ measurements of vortex activity at a high-elevation site (a yardang and gravel-ripple field at 3800 m in the Argentinian Andes high plain—Puna) using such methods, indicating higher levels of dust devil activity than reported previously elsewhere (see also Chap. 8 of this volume, Lorenz et al. 2016).

Pressure drops associated with dust devils have been recorded opportunistically at a few sites (notably by chart-recording barographs as early as the 1950s—see Chap. 1 of this volume, Lorenz et al. 2016). A number of pressure traces were obtained in vehicle encounters by Sinclair (1969) and subsequent studies, but until recently, the pressure signatures of vortex encounters were better-documented at Mars (Lorenz 2012a) than on Earth. The 'expendable' small pressure loggers advanced by Lorenz (2012b) were deployed at Eldorado playa near Boulder City, NV, and showed promise. Three such loggers were operated for a one-month period in June 2012 (Lorenz and Lanagan 2014), generating the first statistically-robust terrestrial dataset of fixed-station encounters (and allowing comparison of their power-law statistics with those at Mars—see Lorenz et al. 2016, in this issue). Note

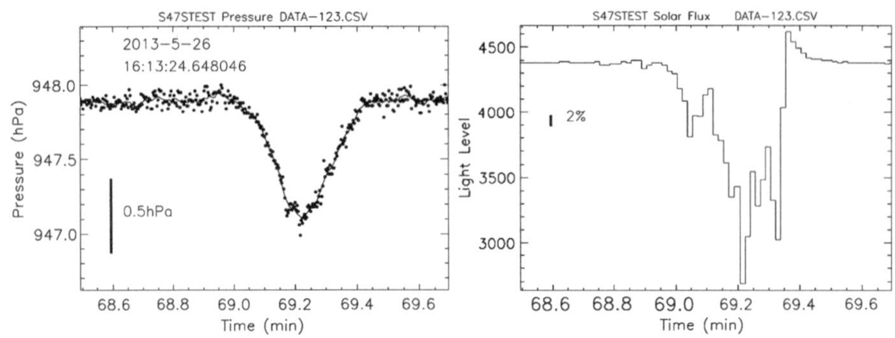

Fig. 9 A solar logger encounter with a large dust devil on Eldorado in summer 2013 (from Lorenz and Jackson 2015). A broad pressure drop lasting about 20 s is seen; this pressure drop of about 0.7 mbar is typically encountered not more than once in a few weeks. The solar flux measurement at *right* shows that the devil was heavily dust-laden, blocking about 30 % of the light: the two prominent troughs are probably the wall of the dust column. Note that the light level *rises* after the encounter—part of the normally dark sky is occupied by dust which scatters light onto the solar cell in addition to the direct solar beam which is now no longer shadowed. This indicates the devil was moving away from the solar azimuth

that the recorded pressure drop at a point is a function of the vortex pressure field (related to diameter and intensity—see the companion chapter by Kurgansky et al. 2016) and the trajectory of the dust devil relative to the measurement station. Importantly, the data were made available for the use of other workers.

Jackson and Lorenz (2015) extended the Eldorado study with observations from several sites over summers 2012 and 2013, and the intervening winter, giving insights into the seasonal variation of vortex occurrence, and interannual variability (e.g. due to dust availability—flooding of the playa altering the surface texture and thus the lifting threshold—see Neakrase et al. 2016 in this issue). That study also explored automatic detection methodologies (finding over 1000 events), since these measurement approaches develop many Gigabytes of data, for which the manual vortex detection employed by Lorenz and Lanagan (2014) would have been prohibitive: the $\sim$1200 station-days of data comprise some 120 million measurements.

Lorenz and Jackson (2015) performed another study at 4 locations at Eldorado in summer 2013, using loggers with a solar cell to record dust devil shadows (Fig. 8). This study found that about half of pressure encounters were accompanied by measurable light level drops (Fig. 9). Some dustless vortices occurred, and in others, the devil's shadow missed the logger.

The Eldorado site is an open area, used for various recreational purposes, at which large unattended installations may encounter human interference (although artfully-concealed small loggers have been generally unaffected, although occasionally damaged by flooding). Lorenz et al. (2015b) employed another field site, La Jornada Experimental Range in New Mexico (operated by the US Department of Agriculture) where access is restricted and so larger installations can be left safely unattended.

This facility was used by Lorenz et al. (2015b) to deploy a line array of pressure/solar loggers in summer 2013. In addition to providing population statistics at this site, the survey noted that the number of vortices encountered varied quite substantially over a distance of a few tens of meters, due presumably to the influence of topography on dust devil migration and/or the effects of different scrub bushes on dust availability and surface roughness. This array study generated for the first time (since a pioneering chart-recorded 6-station study by

Lambeth 1966) simultaneous measurements exposing the horizontal pressure structure of dust devils. An example dataset is shown in Fig. 10, illustrating the radial variation of the measured pressure drop detected as the vortex passed across the pressure sensor array.

3.1.2 Wind

In a supplemental investigation to pressure logging, Lorenz (2016) obtained a high-quality set of wind speed and direction data at the Jornada Experimental Range in summer 2014, using the same logger technology as the Lorenz and Jackson (2015) effort. These data, uncontaminated by vehicle motion effects, allowed rather accurate vortex model fits, e.g. Fig. 11, wherein the pressure, wind speed and direction histories are simultaneously fit. The superposition of the circumferential vortex winds and the ambient wind field result in quite distinctive wind direction and speed histories, which resolve most of the geometric ambiguities intrinsic to fitting a pressure time series alone.

3.1.3 Other Terrestrial Dust Devil Observations

Electric Fields/Saltation—Dust Flux Dust on Earth is mainly lifted from the surface through the process of 'saltation' (Bagnold 1941; Shao 2008): when wind friction velocity/surface stress overcomes a threshold, its drag force causes larger particles with size around 100 micrometers to be the first to move. They jump over the surface, where they reimpact and initiate the motion of particles of a wide range of sizes, including dust. Indeed, due to dust sized particles protruding minimally upward into the wind affected near-surface atmosphere and also strong interparticle forces, dust grains are difficult to be lifted directly by the wind force (Gillette et al. 1974; Greeley and Iversen 1985; Shao et al. 1993).

The collisions among particles during saltation are also responsible for electric charge transfer between grains. Even if the exact mechanism for this process is still not clearly understood, some experiments and observations show that this process is size dependent (Freier 1960; Inculet et al. 2006; Duff and Lacks 2008; Lowell and Truscott 1986; Kok and Renno 2008; Desch and Cuzzi 2000; Forward et al. 2009; Gill 1948; Latham 1964; Latham and Stow 1968; Harper 1967; Horn et al. 1993; Lacks and Levandovsky 2007). Considering that, in general, the smallest particles are transported higher into the atmosphere by local turbulence while larger particles remain closer to the surface, this translates in a charge separation and consequently in an enhancement of the atmospheric electric field.

So generally, a variation of the electrical properties of the atmosphere is observed during dust events including dust storm and dust devils with electric fields up to 150 kV/m being measured (Esposito et al. 2016; Schmidt et al. 1998; Renno et al. 2004; Kok and Renno 2006, 2008; Harper 1967). Fig. 12 shows an example of electric field observed during a dust devil (Esposito et al. 2016).

Esposito et al. (2016) undertook field test campaigns in the West Sahara desert to study dust lifting process by monitoring simultaneously weather parameters (pressure, wind, relative humidity, temperature, solar irradiance), soil properties (temperature, moisture), sand and dust dynamics (dust size distribution and abundance, sand saltation rate and flux), and the atmospheric electric field (with a field mill). They monitored several dust storms and devils. They found that there is a very strong correlation between the concentration of dust lifted during a dust storm and the atmospheric electric field intensity. The same behavior was observed also during dust devils, indicating that a similar dust electrification process was in action (Fig. 13).

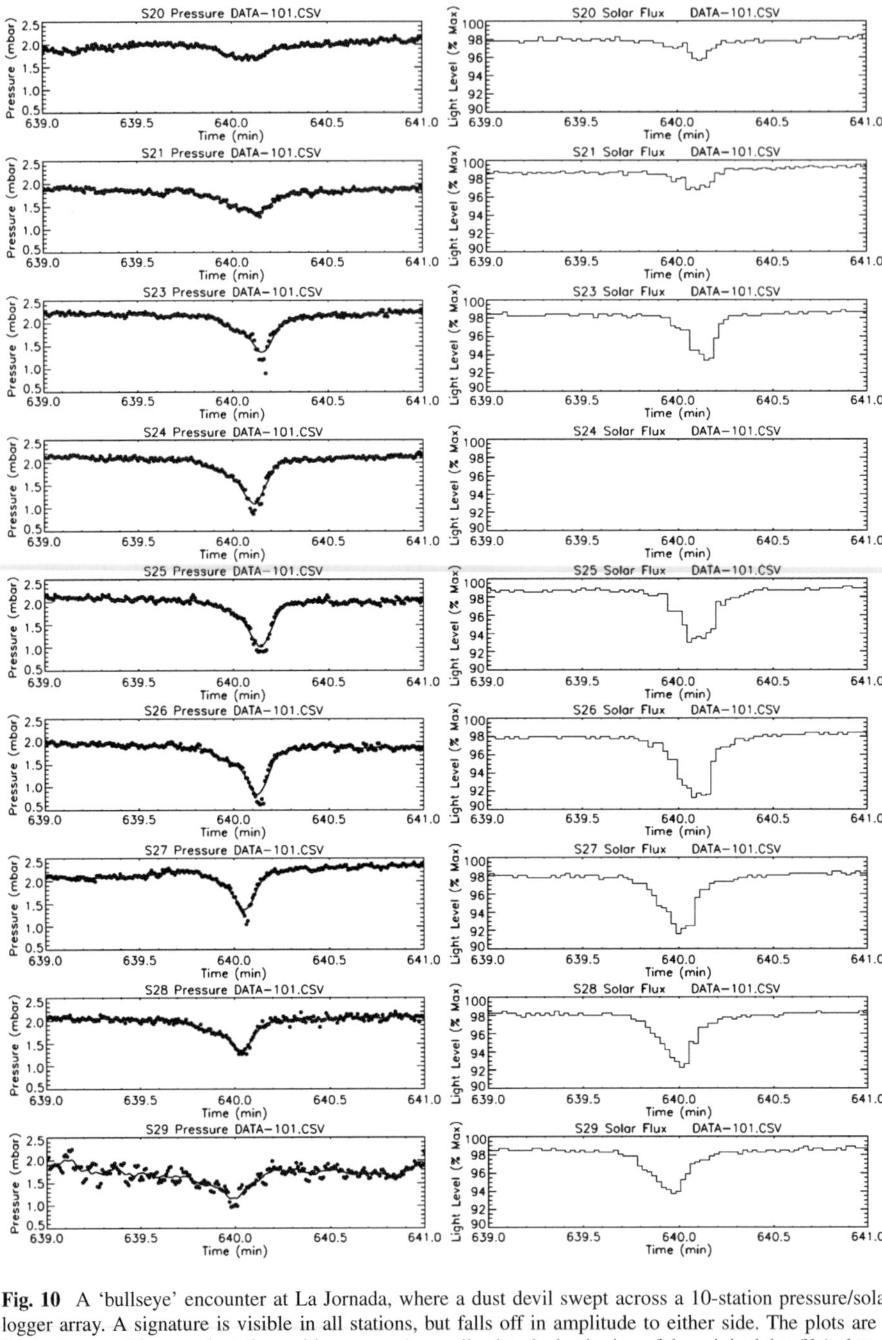

Fig. 10 A 'bullseye' encounter at La Jornada, where a dust devil swept across a 10-station pressure/solar logger array. A signature is visible in all stations, but falls off in amplitude to either side. The plots are a 2-minute record from each station, with pressure (normalized to the beginning of the original datafile) plotted on the *left*, and solar flux (normalized to maximum) on the *right*. The devil is dust-laden, as evidenced by the sunlight drop that is simultaneous with the pressure signal. Note that no solar data is available on station S24 (which had been flipped over, perhaps by animal action). Note that the profile is asymmetric—the onset of the pressure drop is shallower and longer than the decay, a feature often seen and perhaps associated with the advection of the devil in the ambient wind field. [After Lorenz et al. 2015b]

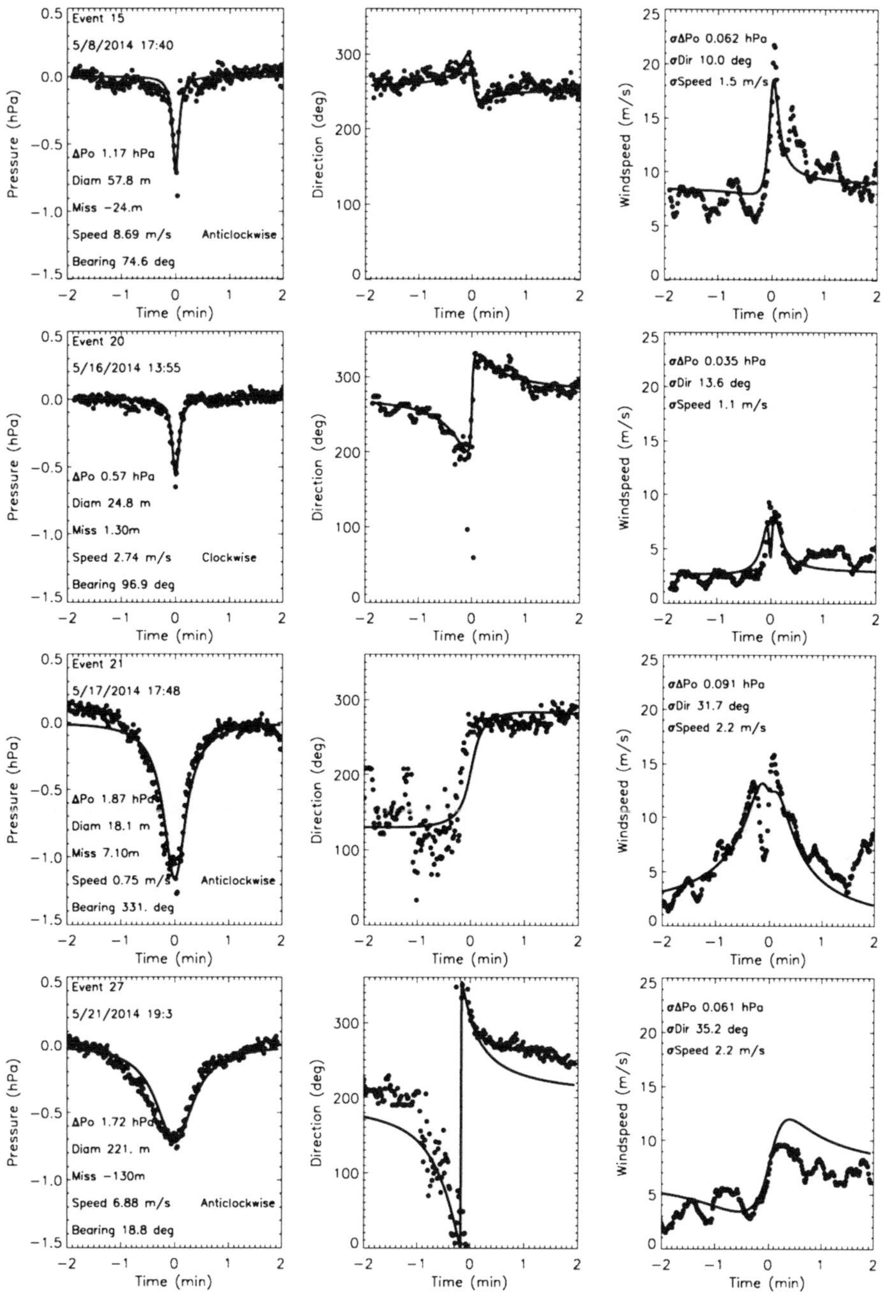

Fig. 11 Three encounters with 1-Hz pressure, wind direction and speed (left to right) data (*grey points*) of dust devils at La Jornada (Lorenz 2016) with vortex model fits (*black lines*). The wind direction histories are particularly distinctive—speed histories tend to be somewhat noisy. It may be noted that the 'eye' of the vortex, where near-solid-body rotation within the wall results in very reduced windspeed at the *center* of the devil, clearly resolved in the *lower-right plot* (although not well captured by the model fit)

Fig. 12 An example of the electric field accompanying a dust devil as measured by Esposito et al. (2016) in the West Sahara desert: atmospheric electric field enhancement

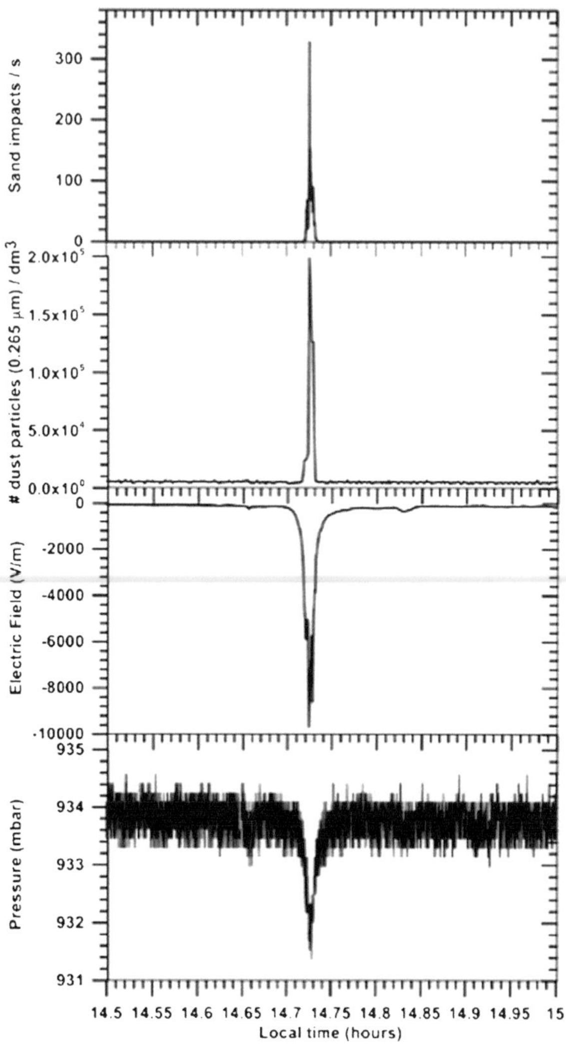

The 2014 West Sahara Campaign depicted in Esposito et al. (2016) has also been the opportunity to test the Micro-ARES electric field sensor of the DREAMS (Dust Characterisation, Risk Assessment, and Environment Analyser on the Martian Surface) science package for ExoMars 2016 (Bettanini et al. 2014). The instrument is based on the Relaxation Probe principle (Berthelier 2000, Molina-Cuberos et al. 2010) which requires a more complex post-processing than a classic field-mill sensor. One of the capabilities that had to be tested was the ability for the instrument to detect the electric field variations during the nearby passage of a dust devil, the passage being confirmed by the classical weather measurements. It appears that such events have been observed (Fig. 14) during the 4 days campaign and properly detected by the instrument. A more detailed study of the results will show if the instrument is able to detect single particle collisions with the electrode, thus giving access to the particle electric charge during such events. A more detailed overview of the

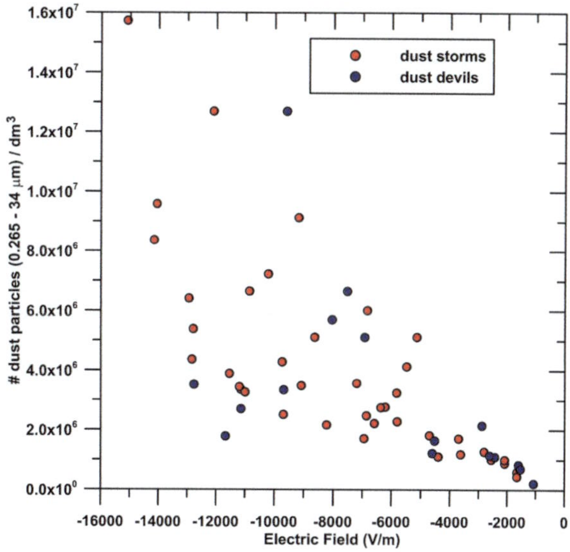

Fig. 13 Correlation between the abundance of lifted dust and the intensity of atmospheric electric field during dust storms (*red marks*) and dust devils (*blue marks*)

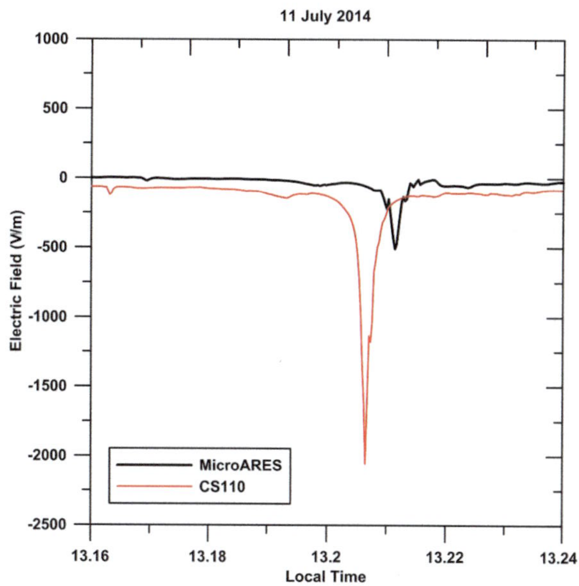

Fig. 14 Comparison of Micro-ARES and Commercial Field-mill sensor DC measurements during a dust-devil event confirmed by pressure and wind parameters measurements. The amplitude and time differences are explained by the installation height (respectively 0.8 and 2 m) of the instruments and their distance (approx. 30 meters)

Micro-ARES and the results gathered during the 2014 West Sahara Campaign is presented in Harrison et al. (2016) in this issue.

Seismic Signals Pressure fluctuations in the atmosphere induce an elastic response in the ground that can be detected as a ground tilt by seismic stations installed on, or close to, the Earth's surface. This effect has been known since the 1970s (Sorrells 1971; Sorrells et al. 1971) and is one of the reasons that terrestrial seismic stations are typically installed deep underground in vaults.

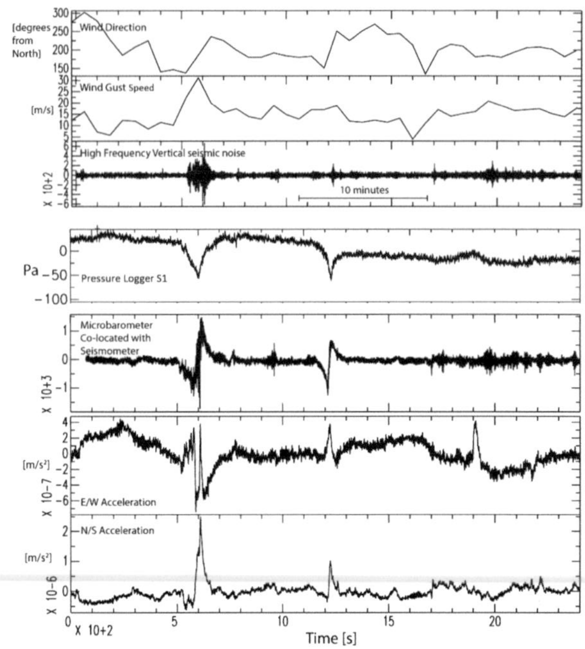

Fig. 15 Two distinct seismic signals recorded during the passage of two dust devils during mid-afternoon in late spring (2014) on a playa in the Mojave Desert, USA. The seismic measurements were accompanied by pressure and wind measurements; time was not precisely synchronized between the two sets of measurements; the meteorology measurements have been time shifted to coincide with the seismic signals. A third seismic signal (at ~ 19 minutes) is not accompanied by a meteorologically detected dust devil. [After Lorenz et al. 2015b]

The InSight lander mission, selected under the NASA Discovery programme and now scheduled for a May 2018 launch, will perform the first comprehensive surface-based geophysical investigation of Mars. The seismic instrument SEIS (Seismic Experiment for Internal Structures) is the critical instrument for delineating the deep interior structure of Mars (Lognonné et al. 2012). SEIS will be deployed directly onto the surface of the planet and will, therefore, be sensitive to the atmospheric seismic signals. In fact, due to the lack of microseism-producing oceans, such atmospheric seismic signals are likely to be the dominating background seismic noise on Mars (Lognonné and Mosser 1993).

In preparation for the InSight mission, and to understand the effects of a surface deployment of a seismometer, representative field experiments were carried out in California close to the Goldstone Deep Space Communications Complex (Lorenz et al. 2015a). The experiment included a seismometer buried at very shallow depth, together with a suite of meteorological instruments. During this field campaign, ground tilt was measured by the seismometer at the same time as vortex encounters were documented by an array of pressure loggers (Lorenz et al. 2015a). The negative load of a dust devil vortex pulls up the ground as it passes, causing the ground—and seismometer—to tilt away from the dust devil center. This first identification of the isolated seismic signature of a dust devil has shown that a seismometer appears to be capable of tracking close encounters with dust devils and, in addition, that seismometers may be more effective than in-situ meteorological instruments at detecting dust devils at long-range (Lorenz et al. 2015a).

Peak surface accelerations of 1–2 $\times 10^{-6}$ m^2 s^{-1} were measured during the passage of a dust devil presenting a measured pressure drop of 80 Pa (0.8 millibars) accompanied by a measured 2 % decrease in the short-circuit current of a solar cell mounted with a pressure logger located 30 meters from the seismometer (Fig. 15). The measured acceleration magnitudes and knowledge of the local surface material structure provided by a seismic survey enabled estimation of the dust devil's decreased surface mass loading/negative point load on an elastic half-space. The measured acceleration values and inferred vortex center miss

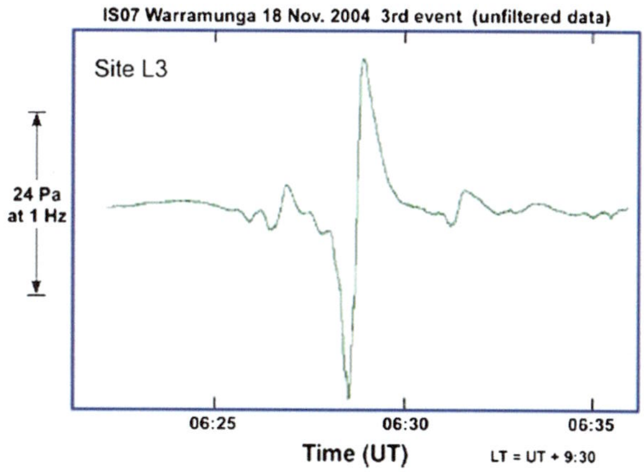

Fig. 16 Microbaragraph measurement of a dust devil pressure signature, showing the up-down heartbeat signature of the vortex. Two additional smaller amplitude subsidiary events are present ~ 1 minute before and 2 minutes after the primary event. [After Lorenz and Christie 2015]

distances imply a mass load of ~ 8000 Newtons and tilt accelerations of $7 \times 10^{-7}\ \mathrm{m\,s^{-2}}$ for a 5-meter diameter vortex with a central pressure deficit of 200 Pa passing 10 meters from the seismometer. A larger diameter (100 meter) dust devil also with a 200 Pa central pressure deficit could provide a total mass load of 300 metric tons and accelerations of 10^{-6}–$10^{-5}\ \mathrm{m\,s^{-2}}$ for miss distances of 50–200 meters, though the size of such a dust devil violates consideration of it as a point-source. The seismometer system employed by Lorenz et al. (2015a, 2015b, 2015c) included a microbaragraph that coincident with the seismic signals registered the tell-tale infrasound 'heartbeat' dust devil signature identified by Lorenz and Christie (2015).

Infrasound and Acoustic Measurements It was noted by Lorenz (2012a, 2012b) that while meteorological stations tend to record data at only 15-minute intervals, continuous pressure measurements, made with sufficient sensitivity and sample rate to detect dust devil vortices are made for other applications, notably for monitoring compliance with international treaties on nuclear testing. Some of these stations, operated by the Comprehensive Test Ban Treaty Organization (CTBTO) are located in desert areas, and dust devil vortex signatures at a CTBTO station in Australia are reported by Lorenz and Christie (2015). Since atmospheric effects are a major perturbation to seismic signals, many seismic stations (such as those in the USARRAY) now also record pressure and other meteorological parameters. Data-mining of such records may be a fruitful avenue of research. A subtlety to be noted is that microbarographs used for infrasound studies (and those e.g. at CTBTO stations) are high-pass-filtered pressure records, such that the principal component in the signal is the derivative of the pressure signal. Thus the typical dip seen in pressure time series in fact appears as a down-up-down 'heartbeat' signature (Fig. 16).

Lorenz and Christie (2015) investigated dust devil pressure signatures within microbaragraph measurements obtained as part of CTBT monitoring from a station located in Australia. The high pass filtered signal of a dust devil pressure measurement appears as a 'heartbeat' signal that resembles the temporal derivative of measured pressure provide by a pressure sensor with a sampling of ~ 1 Hz. This heartbeat signal exhibits declining signal magnitude as the pressure minimum is approached and an abrupt transition in the sign of the signal to a maximum magnitude that subsequently declines in concert with the measured pressure increase as the dust devil's influence wanes. Contemporaneous microbaragraph and pressure sensor measurements verify the dust devil production of the microbaragraph's heartbeat signal. While this is not a direct measure of the infrasound generation produced by a dust devil, it is a distinct signal from which dust devil occurrence can be derived.

Edmonds (2014) attempted to detect the infrasound emission that a dust devil, or that interacting dust devil vortices, might produce. Theory (Powell 1964; Howe 2003) indicates that two vortices 'orbiting' around each other are capable of generating infrasound emission, as is a single non-circular vortex (Howe 2003). Williams (2001) addressed the attenuation of acoustic signals within Mars' tenuous atmosphere and concluded that infrasound frequencies (< 20 Hz) at which dust devils are effective emitters (Bedard 2005) experience smaller dissipation than do higher audible frequencies. That work was motivated by the inclusion on the ill-fated 1999 US/NASA Mars Polar Lander of a microphone intended to listen for martian sounds. Edmonds' (2014) work was motivated by the inclusion on the NASA InSight lander of a high-frequency pressure sensor (Banfield 2014). Edmonds (2014) conducted a field exercise in the desert of southern New Mexico, USA in Spring 2014 attempting to detect the infrasound and audible frequency emission from dust devils. Using a microphone/recording acoustic system and a microbaragraph, three dust devils recorded from distances of a few to 10's of meters coincided with measured audible (> 20 Hz) signals. The recordings indicate amplitude 'ridges' within distinct acoustic frequency ranges, distinct from sounds attributed to wind-induced movement of vegetation objects (vegetation, sand, etc.). However, no infrasound detection was identified.

3.2 Mars Surface Obtained Meteorological Measurements of Dust Devils

While there has been until recently a general dearth of seasonal or longer temporal coverage of terrestrial dust devil in situ measurements arising from limited duration measurement campaigns, but a plethora of measured dust devil parameters/characteristics from these same studies, in situ measurement of martian dust devils has suffered from opposite conditions. Martian lander and rover missions have provided durations extending from 83 sols (Pathfinder) to multiple martian years (Viking Lander 1 spanned more than 3 Mars years, Spirit Rover 3 Mars years, Opportunity Rover 6+ Mars years and continuing, and MSL almost two Mars years and counting). However, some missions have suffered from meteorological instrument failures (Viking Lander 1, MSL) or calibration issues (Pathfinder) while others did not carry any direct meteorological instruments at all (Opportunity, Spirit). The robotic vehicles that did carry meteorological instrumentation generally have provided high frequency, one to a few Hz, sampling sufficient to characterize dust devil signatures but did not do so continuously, except for the Phoenix Lander. It is from the measurements provided by the five meteorology-instrumented vehicles (Viking Landers 1 & 2, Pathfinder, Phoenix, MSL) that our current thermodynamic understanding of martian dust devils has been obtained.

A compilation of measured characteristics is provided in Table 2.

3.2.1 Viking Landers

The two US/NASA Viking Lander spacecraft which arrived at Mars in 1976 provided the first opportunity for in situ sensing of the meteorological signatures of passing martian dust devils. Viking Lander 1's landing location was 23 N, 48 W, while Viking Lander 2's was 48 N, 226 W.

As previously described, a characteristic transient drop in atmospheric pressure is detected when a convective vortex passes over/near a deployed pressure sensor, as has been amply demonstrated for terrestrial dust devils. The digital quantization of the long lived

Viking lander pressure sensors was 8.8 Pa (Hess et al. 1977; Tillman et al. 1993), which subsequent missions (Murphy and Nelli 2002; Ellehoj et al. 2010) indicated was too large to unambiguously detect a martian dust devil pressure signature of several Pascals or less magnitude. Additionally, the Viking Lander pressure measuring strategy was not generally focused upon detection of short duration events but rather upon characterization of diurnal and seasonal variations. A typical time interval between pressure measurements was 17 minutes (Ryan and Lucich 1983). Pressure sampling did include some brief time periods early in the mission during which measurements were obtained at a rate of ~ 1 per second, but much more often sampling rates were once each 16 or 32 seconds extending to once per 65 to 105 minutes. There is no publication that addresses assessment of the complete Viking lander pressure record for identification of dust devil/convective vortex signatures. The VL1 pressure record spans from L_s 97.1 (MY 12) through L_s226.7 (MY 15), while VL2 spans L_s 117 (MY12) through L_s 57.1 (MY 14).

While Viking Lander pressure measurements were not amenable to dust devil studies, Viking's measured winds were. Ryan and Lucich (1983) investigated wind vector measurements provided by Viking Meteorology Instrument System (VMIS) at both landers, with measurement sampling intervals ranging from 2–112 seconds. Vortices were identified by temporal rotation of the measured wind direction accompanied by a concurrent wind speed variation illustrative of an imbedded Rankine-type vortex, and a concurrent temperature maximum. A total of 118 vortices were identified during the mission's first year, 40 vortices at VL1 spanning summer through winter and 78 at VL2 spanning summer through spring. The greatest likelihood of vortex detection occurred almost equally (~ 65 % of the sols investigated) at VL1 during summer and VL2 during spring.

Identified vortex disturbance influence persisted for several to ~ 10 minutes, with the most pronounced vortex effects present for tens of seconds to several minutes. Inferred vortex core diameters, the distance from vortex center at which tangential wind speed maximized, were generally tens to several hundreds of meters, with radii of disturbance effects extending out ten times the core radius. Several inferred core diameters extended to ~ 500–1000 meters, implying radii of vortex influence extending to ~ 5–10 km. Inferred vortex rotation was equally divided between cyclonic ('counter-clockwise') and anticyclonic ('clockwise'), consistent with terrestrial experience (Sinclair 1973) and suggestive that the vorticity is generated locally at small scales. Several of the vortices at both Viking lander sites were sufficiently intense to generate winds ($> 35\ \mathrm{m\,s^{-1}}$) deemed capable of lifting surface dust (Greeley and Iversen 1985). There were no concurrent imaging observations invoked to address the presence or lack of dust in the detected vortices.

It is unclear how unambiguous VL1 wind directions employed in this investigation were derived subsequent to sol 45 and the failure of the wind direction quadrant sensor (Murphy et al. 1990). VL2's wind instrument did not suffer from such a failure.

Ringrose et al. (2003) readdressed the dust devil/convective vortex signatures present in Viking Lander 2 wind measurements. Using a phase picker detection technique comparing a running mean value to a threshold value (also used terrestrially in Hecht et al. 2001), instances where a short term average wind speed or wind direction varied by more than $6\ \mathrm{m\,s^{-1}}$ or 40 degrees azimuth, respectively, from longer term averages were flagged as possible vortex signatures. For verified vortex signatures, minimum distance from the vortex center and vortex diameter were derived from a Rankine vortex fit to the measured wind. Applying this technique to VL2's first 60 sols resulted in 38 identified vortex occurrences, nine of which suffered from lander interference which makes them suspect. Maximum measured wind speeds were 12–15 $\mathrm{m\,s^{-1}}$ and maximum wind vector rotation was 300 degrees. Inferred vortex core diameters arising from the Rankine vortex fits ranged from a few 10's to a few hundred meters. Detected vortex durations ranged from 60–1080 seconds. Maximum inferred vortex core diameter tangential wind speeds were $\sim 100\ \mathrm{m\,s^{-1}}$ for two events. These two events coincided with the largest inferred 'miss distance from core center' ($\sim$ 1700 m) and largest vortex core diameters ($>$ 350 m), while maximum inferred speeds for smaller ($<$ 250 m) miss distances and smaller core diameters ($<$ 100 m) ranged from 3 to $70\ \mathrm{m\,s^{-1}}$. Daytime hour-of-occurrence of detected vortices exhibited late morning and early afternoon maxima. A secondary minimum was also evident at 0930 local time. There was no description of the completeness of coverage of measurements during the 30-minute time intervals into which the vortex occurrences were binned for this time-of-sol evaluation, so the time-of-sol distribution might not be representative of conditions.

Measured atmospheric temperature increases of several degrees Centigrade accompanied some, but not all, of the wind-detected vortices identified by both Ryan and Lucich (1983) and Ringrose et al. (2003).

The Viking Landers did provide for the first time in situ evidence of martian convective vortices, in the form of wind vector temporal variation consistent with the passage of a vortex. Vortex occurrences were frequent, with at least one-half of the mission sols investigated possessing a detected vortex occurrence. This detection rate is certainly a lower limit since VMIS was not in continuous operation during any of the sols investigated.

3.2.2 Mars Pathfinder

From its northern subtropical landing location, Pathfinder provided pressure measurements that spanned the mission's 83 late summer-early autumn sols. Pathfinder's deflecting diaphragm, variable reluctance pressure sensor (Seiff et al. 1997) provided 14-bit, $\sim$ 0.25 Pa resolution in its 600–1000 Pa surface operating mode. While Pathfinder's Atmospheric Structure Investigation/Meteorology (ASI/MET) system (pressure, temperature, wind) generally provided greater temporal sampling resolution (0.25 to 2 Hz) than did the Viking Lander's, ASI/MET like Viking VMIS was not continuously operated. During the mission's first $\sim$ 30 sols, 3-minute, 0.25 Hz measuring sessions initiated at the start of most LTST hours were augmented by 15-minute and 60-minute continuous measurement sessions at 1 Hz sampling cadence. On five occasions during the mission, the first starting at 0600 LTST of Sol 25, the ASI/MET system was continuously operated for a complete sol at a sampling rate of 0.25 Hz. These 'Presidential MET' sessions were initiated on Sols 25, 32, 38, 55, and 68. Subsequent to Sol 30, the ASI/MET system was operated during daylight hours only, $\sim$ 0900–1600 LTST, except during the Presidential MET sessions.

Schofield et al. (1997) presented the first in situ contemporaneous measurement of pressure, wind and temperature within a martian dust devil/convective vortex, from measurements obtained during early afternoon, 1353 LST, during the Sol 25-initiated Presidential MET session (Fig. 17). The quantified vortex pressure drop magnitude was $\sim$ 2.5 Pa. Subsequently, Murphy and Nelli (2002) assessed the entire Pathfinder pressure data archive and identified the occurrence of 79 vortices, $\sim$ 1 per sol, with pressure drop magnitudes equal to or exceeding 0.5 Pa. Vortex identification was based upon a pressure drop magnitude determined from the difference between a 3rd order polynomial fit to measured pressures during a 15-minute interval and the unaltered pressure measurements. The maximum pressure drop magnitude identified was 4.8 Pa at 1132 LST on Sol 34. The LST time of detection of the 79 vortices spanned 0930–1700.

The discontinuous temporal coverage provided by ASI/MET indicated that the true number of detectable vortices would exceed 79. Normalizing the number of vortices detected throughout the mission during each 15-minute LST time window by the percentage of time

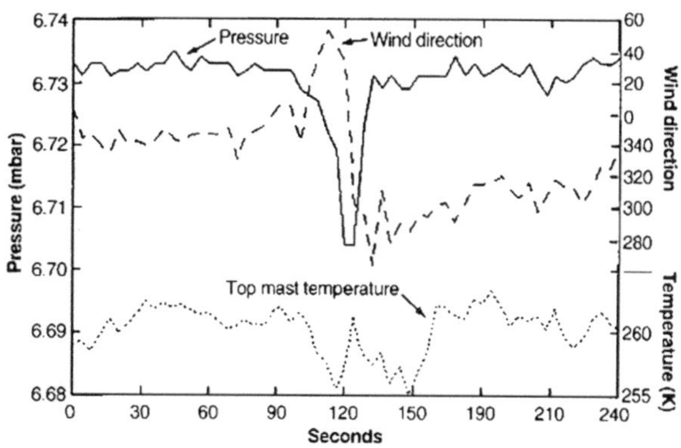

Fig. 17 Mars Pathfinder ASI/MET measured Time series of 0.25 Hz pressure, temperature, and wind direction during a 4-minute time period during the early afternoon of Sol 25 of the mission. After Schofield et al. (1997)

the ASI/MET system was operating during that 15-minute window resulted in an estimate of 210 detectable vortices occurring during the 83 sol mission, equating to ~ 2.5 vortices per sol. This 2.5 per sol estimate is less than the ~ 4 vortices per sol detected during the five Presidential MET sols of almost continuous pressure sensor operation (Murphy and Nelli 2002).

Ferri et al. (2003) determined the duration of 19 of the larger magnitude (> 1 Pa) pressure signature events identified by Murphy and Nelli (2002). Durations ranged from 14–51 seconds, with a mean value of 28 seconds and median value of 25 seconds.

Both the directly detected and normalized inferred Pathfinder vortex occurrences exhibited a maximum hourly occurrence during very early afternoon (Fig. 18). Vortex activity after the early afternoon peak declined more gradually than through the afternoon than its increase during the morning.

Applying Metzger et al.'s (1999) visible plume dust mass load estimates and the frequency of occurrence derived from the Pathfinder pressure measurements, Murphy and Nelli (2002) estimated that local dust devil activity could provide dust to the atmosphere at the rate of an optical depth of 0.01 per sol over a 1.5 km diameter area centered upon the lander. This estimate is approximately equal to the opacity decay rate measured during the decline of the second global scale martian dust storm during 1977 (Pollack et al. 1979).

Pathfinder's ASI/MET system included a wind sensor (Seiff et al. 1997). Additionally, a 'wind sock' was mounted on the ASI/MET mast with IMP-provided images enabling wind vector derivation from observed windsock orientation (Sullivan et al. 2000). The ASI/MET wind sensor was a thermal/mechanical instrument using the measured overheat of resistively heated sensor elements (wires) from which wind speed and direction were intended to be derived. The sensor consisted of a vertically oriented 2.7 cm diameter, 3 cm tall solid exterior cylinder around which were arrayed six 'segments' of azimuthally confined vertically aligned 8-wire length windings. A pulsed 20 milliamp current provided unheated segment wire temperature from measured voltage drops across each segment. When the sensor was operated in a continuous 52 milliamp high current mode the segment wire temperatures were resistively heated. Segments positioned in the upwind direction experienced ventilative cooling, with overheat magnitudes and their azimuthal structure providing the signal indicative of the wind vector. The sensor was not designed with its own reference temperature against which segment temperature could be assessed for 'overheat' magnitude. It was anticipated that thermocouples mounted on the ASI/MET would provide the reference temperature. However, large magnitude 'turbulent' temperature variations (~ 10–15 K; Schofield et al. 1997) inhibited unambiguous determination of a reference temperature for wind sensor segment overheat determination purposes.

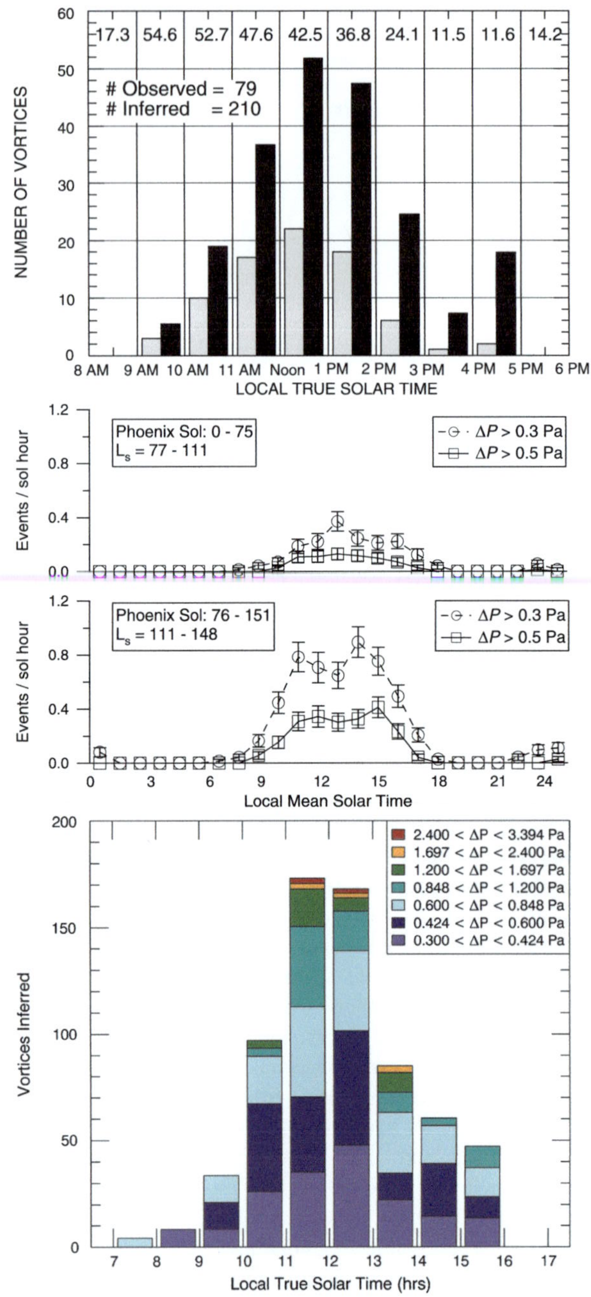

Fig. 18 Martian dust devil occurrence frequency versus time-of-sol from pressure signatures from Mars Pathfinder (*top*; Murphy and Nelli 2002), Mars Phoenix (*middle*; Ellehoj et al. 2010) and Mars Science Lab (*bottom*; Steakley and Murphy 2016)

The wind sensor was generally operated for ∼ 12 seconds of low-current pulsed mode to establish unheated segment temperatures, after which continuous high current was imposed for ∼ 150 seconds. It would be during these 150 second intervals that wind speed and direction could be quantified.

Despite these unresolved operational/calibration issues for the Pathfinder wind sensor, the sensor did provide signals that were qualitatively correlated to dust devil occurrences identified in Pathfinder's pressure measurements (Murphy and Nelli 2002; Schofield et al. 1997). Wind sensor signals did suggest substantial and abrupt changes in wind direction and speed in conjunction with measured dust devil pressure signatures (see, for instance, Fig. 8 in Schofield et al. 1997), however no systematic study of vortex winds has been published.

In addition to the ASI/MET's thermal wind sensor, Pathfinder also included three mechanical 'wind socks' mounted at three heights along the ASI/MET mast (Sullivan et al. 2000). IMP provided images of the solid, metallic, inverted-cone shaped socks and their tilt orientation from which wind speed and direction were derived using pre-flight calibration information. Twelve image sequences obtained over a time period of ~ 100 seconds several times per sol provided the dataset from which wind vector derivation was attempted. Data from only four of the sampled 42 sols provided imaging indicative of minimum wind speed (few meters per second) necessary to overcome windsock inertia and induce detectable windsock deflection (Sullivan et al. 2000). Thus, Pathfinder's windsock experiment did not aid in advancing understanding of the dynamic components of martian dust devils.

Schofield et al. (1997) noted that one pressure-indicated vortex passage detected by MPF was accompanied by a transient drop in output from the lander's solar panels, suggesting that the vortex was dust-laden and the dust column created at least a partial shadow that crossed the lander. This observation underscores the utility of making available engineering data such as solar array current data from lander missions—in effect the arrays serve as a 'free' instrument.

3.2.3 *Mars Exploration Rovers (Spirit and Opportunity)*

Neither Spirit nor Opportunity were outfitted with meteorological measuring instrumentation. The MERs' mini-TES instrument was used to diagnose the vertical atmospheric structure within the bottom few hundred meters to 1–2 km (Smith et al. 2004). While those measurements illustrated the superadiabatic conditions present during mid-sol, there was no identification of such a measurement probing a dust devil.

It had been noted during the MPF mission that a calibrated solar cell on the Sojourner rover recorded a progressive decline in cell current of ~ 0.25 % per day (Landis and Jenkins 2000), due to the accumulation of airfall dust on the cell. This obscuration set expectations of operational lifetime of solar-powered landers and rovers on Mars (e.g. the Mars Exploration Rovers Spirit and Opportunity had a nominal mission duration of 90 days).

In practice, it was observed that while the MER solar power per day declined due to dust accumulation, sudden reversals of the decline were seen, and camera images of the arrays showed that dust had been removed. While dust devils were suspected, the likelihood remains that straight-line gusts may have been responsible. However, Lorenz and Reiss (2015) showed that not only did the seasonal onset of dust-clearing events coincide with the appearance of dust devils (Greeley et al. 2006; 2010), but also the rate at which dust-clearings occurred was coincident with the rate of vortex encounters seen in pressure drops by Phoenix and Pathfinder, extrapolated to a dust lifting threshold of a few Pa.

3.2.4 *Mars Phoenix Lander*

The US/NASA Phoenix lander meteorologically detected, at its northern arctic latitude landing site, dust devil/convective vortex signatures that complement its SSI imaging dust devil detections (Sect. 2.2.3). Phoenix was equipped with a meteorological package (MET), including pressure, air temperature and wind sensors and a Light Detection And Ranging (LIDAR) instrument for measuring dust and ice particles in the atmosphere (Taylor et al. 2008; Whiteway et al. 2008). The pressure sensor was based on Barocap® silicon diaphragm sensor heads manufactured by Vaisala Inc. and had a very high resolution of 0.1 Pa (Taylor et al. 2010), limited by the noise level. In contrast to the Vikings, Pathfinder and MSL, Phoenix logged atmospheric pressure and air temperature almost continuously, with an invariant sampling rate (0.5 Hz) (Taylor et al. 2010; Davy et al. 2010). The mission spanned 151 sols, almost twice the duration of the Pathfinder mission, extending from early spring through mid northern hemisphere summer (L_s 77° to L_s 148°). The continuous Phoenix pressure record enabled the detection of a greater number of convective vortices than had Pathfinder.

Ellehoj et al. (2010) surveyed the entirety of pressure measurements provide by Phoenix and identified 502 (~ 3.3 per sol) transient pressure drops similar to the vortex signatures that Murphy and Nelli (2002) had detected in the Pathfinder data. The magnitudes of these Phoenix pressure drops ranged from 0.3 Pa (an imposed lower limit) to 3.6 Pa, with 197 occurrences possessing a magnitude greater than the 0.5 Pa detection threshold used by Murphy and Nelli (2002). Vortex identification was based on the search for 20 s long time intervals that fulfilled the following two criteria: 1) mean pressure more than 0.1 Pa lower than mean of the previous and next 20 s intervals, and 2) minimum pressure more than 0.3 Pa lower than mean of the previous and next 20 s intervals.

The Full Width at Half Maximum (FWHM) durations of events with magnitude > 0.5 Pa ranged from less than 1 s to circa 35 s, the mean being circa 9 s (Table 1 in Ellehoj et al. 2010). [Note that there is a typographical error in the caption of Fig. 7 in the Ellehoj et al. (2010); the shown quantity is actually full duration, i.e. $2 \times$ FWHM (H. P. Gunnlaugsson, Aarhus University, personal communication, 2015)]. It is unclear if these durations are comparable to the durations reported by Ferri et al. (2003) for the Pathfinder pressure drops, as Ferri et al. (2003) did not explain how their durations are defined.

Most transient pressure drop events identified in the Phoenix data occurred between 06:00 and 18:00 Local Mean Solar Time (LMST). However, unlike Pathfinder, Phoenix also detected also 29 events between 21:00 and 01:00 LMST (mostly with the Sun low in the arctic sky), interpreted as being caused by turbulence induced by air passing over Heimdal crater, the only major topographic feature in the vicinity of the lander (Ellehoj et al. 2010). The general shape of the diurnal distribution of vortex activity resembled that detected by Pathfinder, but at the Phoenix site the vortex activity stayed high until circa 15:00 in the afternoon while at the Pathfinder sites the activity started to fall already at circa 13:00.

Phoenix operated long enough to detect some seasonal variation in vortex activity. The number of identified transient pressure drops generally increased around Phoenix sol 75 ($L_s = 111$, about 40 sols after summer solstice and 15 sols before the Sun set for the first time) and the proportion of events with large pressure drops became higher at the same time (Ellehoj et al. 2010). Before this, an average of 0.6 events with magnitude > 0.5 Pa were observed per sol, but after this 2.0 events per sol, a number comparable to the estimated number of vortices that passed by Pathfinder per sol during the same season. Phoenix also detected more frequent vortex activity coinciding with passing cold fronts associated to low-pressure baroclinic systems. The clearest example of this was Phoenix sol 95, when 36 pressure drops larger than 0.3 Pa were identified in contrast to 6 and 7 vortices on the preceding and following sols, respectively. A concurrent cloud feature suggestive of a cold front was seen to cross the Phoenix landing site in images taken by the Mars Color Imager (MARCI) onboard the Mars Reconnaissance Orbiter on that sol (Ellehoj et al. 2010).

The payload on the Mars Phoenix lander included a mechanical wind sensor, the so-called Telltale (Gunnlaugsson et al. 2008; Holstein-Rathlou et al. 2010), consisting of a lightweight cylindrical mass dangling on a thread attached to a crossbar at the top of the meteorological mast. The Telltale was designed to be deflected by wind and the deflection was observed by imaging the Telltale with the SSI. SSI did not monitor the Telltale continuously. Imaging sequences were implemented during only a limited number of hours per sol, and during these sequences readings were acquired with time intervals longer than 50 seconds (Holstein-Rathlou et al. 2010).

Telltale imaging on nine occasions occurred within 10 seconds of a pressure minimum associated to a passing vortex (Ellehoj et al. 2010). In these events the wind vector was observed to differ by 1.4 to 9.3 $\mathrm{m\,s^{-1}}$ from background wind speeds, magnitudes greater than typical changes between consecutive Telltale images. This observed magnitude range of the wind vector perturbations is in agreement with the range of the observed pressure drops assuming cyclostrophic balance (Ellehoj et al. 2010). Even the strongest vortex-related wind perturbation detected from Telltale measurements, however, is below any estimate of the dust lifting threshold on Mars (Neakrase and Greeley 2010), which is not surprising considering that the derived wind speeds are 'snapshots of the wind in random points inside a vortex, not maximum wind speeds, and the great majority of the vortices detected by pressure measurements on Mars are actually too weak to lift dust (Moores et al. 2015; Kahanpää et al. 2016; Steakley and Murphy 2016).

3.2.5 MSL

MSL's Rover Environmental Monitoring Station (REMS) includes sensors for pressure, air and ground temperature, wind speed and direction, humidity and UV radiation measurements (Gómez-Elvira et al. 2012). REMS executes five-minute 1 Hz data acquisition sessions at the start of each LMST hour, with 15 minute and one hour 1 Hz "extended measurement blocks" implemented at variable times of the sol (Gómez-Elvira et al. 2014). The REMS pressure sensor is, as was MET Phoenix, comprised of Vaisala Barocap® silicon diaphragm sensor heads. REMS' pressure sensor noise is slightly higher (0.2 Pa peak-to-peak) as a result of the faster sampling rate and thus shorter integration time (Harri et al. 2014).

Two major studies have characterized the meteorological signatures of convective vortices identified in the MSL/REMS data (Kahanpää et al. 2016; Steakley and Murphy 2016), identifying dust devils from their temporary pressure declines using slightly different detection criteria. The Kahanpää et al. (2016) criteria are more consistent with the Phoenix dust devil detections (Ellehoj et al. 2010) while the Steakley and Murphy (2016) criteria are more consistent with the Pathfinder dust devil detections (Murphy and Nelli 2002). Despite the different identification criteria both studies come to similar conclusions about vortex activity at Gale Crater.

Few dust devils were anticipated within Gale Crater based upon a lack of observed dust devil tracks and a suggestion of a suppressed atmospheric boundary layer depth (Tyler and Barnes 2013; Haberle et al. 2014). However, approximately 250 pressure vortex signatures were identified during the first Martian year of the mission: Kahanpää et al. (2016) report 252 transient pressure drops with magnitude exceeding 0.5 Pa during the first 681 sols of the mission, and Steakley and Murphy (2016) report 245 pressure drops with magnitude exceeding 0.3 Pa during the first 707 sols of the mission. The largest reported pressure drop, 2.97 Pa and 2.86 a by Kahanpää et al. (2016) and Steakley and Murphy (2016), respectively, was detected on MSL sol 403 at 13:02 Local Mean Solar Time (LMST). Steakley and Murphy's (2016) identification of fewer vortices than Kahanpää et al. (2016) despite their lower detection threshold and a 5 % longer study time is partly explained by their criterion that a pressure drop must have a magnitude clearly above background noise. This criterion probably deselects many pressure drops with magnitudes below 0.5 Pa. Also, some larger pressure drops reported by Kahanpää et al. are apparently deselected when local turbulence raises the noise level of the pressure signal. Moreover, Kahanpää et al. (2016) are less strict with the criterion that a pressure drop must have a "dust devil like" shape and they report many pressure events with several minima or otherwise irregular shape.

The Full Width at Half Maximum (FWHM) durations of the pressure drops reported by Kahanpää et al. (2016) follow a distribution almost identical to that reported by Ellehoj et al. (2010) for the Phoenix data, ranging from less than 1 s to circa 54 s, the mean being circa 9 s. Steakley and Murphy's (2016) FWHM durations range from 1–20 seconds with a median value of 5.3 seconds.

The time-of-sol distribution of MSL's daytime transient pressure drops resembles that detected by Pathfinder and Phoenix, ranging from around 9:00 to 16:00 Local True Solar Time (LTST) with maximum occurrence around noon (Kahanpää et al. 2016; Steakley and Murphy 2016) (Fig. 18). Several night-time pressure 'wiggles' were also identified (Kahanpää et al. 2016). These night-time events exhibit wave-like fluctuations rather than isolated pressure drops and are interpreted as the result of gravity waves initiated by topographic winds (Haberle et al. 2014).

The extended duration of REMS operation provides the first opportunity to assess seasonal vortex occurrence from measured martian pressures. Kahanpää et al. (2016) estimated the mean number of vortices per sol causing pressure drops larger than 0.5 Pa separately for 8 "seasons" during MSL's first martian year. These estimates ranged from 0.5 per sol (L_s s 67.5–112.5) to 1.8 per sol (L_s 202.5–247.5). Steakley and Murphy (2016) find a spring (L_s 180–270) occurrence maximum of 1.5 per sol that is twice the minimum occurrence rate (L_s 90–180). This continuous vortex occurrence through the year is in contrast to the visual MER Spirit imaging observations of detected dust devils only during the "dust devil season" (Greeley et al. 2010, referred in Sect. 2.2.1), but is in rough accordance with the vortex observations performed using the Viking wind data (Ryan and Lucich 1983).

During MSL's first 681 sols there was only one case when more than 3 pressure drops larger than 0.5 Pa were detected within the same LMST hour (Kahanpää et al. 2016). This exception occurred on sol 664 when 16 pressure drops were identified by Kahanpää et al. (2016) between 11:00 and 12:00 LMST. Steakley and Murphy (2016) also identified sol 664 as experiencing the greatest number, four, of verified vortices. This "sol 664 vortex burst" resembles the peaks in vortex activity detected by Phoenix coincident with a passing dust storm front seen in MARCI images.

REMS includes a hot-film anemometer (Domínguez et al. 2008; Gómez-Elvira et al. 2012) designed to distinguish the 3-dimensional wind field. The sensor is mounted on two horizontally aligned booms attached to MSL's Remote Sensing Mast (RSM) which are separated by 120 degrees in azimuth. On each boom there are three identical hot-film anemometer boards sensing wind speed in different directions. Unfortunately, three of the altogether six boards were damaged during MSL's landing (Gómez-Elvira et al. 2014), initiating an ongoing re-calibration effort. To date only 2-dimensional wind measurements have been retrieved (Sara Navarro, Centro de Astrobiología/CSIC-INTA, personal communication, 2015). The compromised wind sensor raw data do qualitatively reveal wind variations concurrent with the transient vortex pressure drops (Kahanpää et al. 2016), with 87 % of the pressure drops being accompanied by wind sensor events. Magnitudes of these wind perturbations could not be determined.

Available quantified REMS wind measurements consist of derived 5-minute median values. Kahanpää et al. (2016) used these median winds to derive vortex diameters from measured pressure drop durations, assuming that the vortices moved with the velocity of the background (median) wind. Resulting vortex diameters, encompassing pressure perturbation exceeding 0.5 Pa, range from 2.3 m to 755 m, with a mean of 21 meters and median of 16 meters. The distribution of these is similar to the distribution of dust devil diameters observed by MER Spirit (Greeley et al. 2010).

Solar irradiance loggers have been used terrestrially to detect obscuration of the Sun resulting from dust within a dust devil (Lorenz and Jackson 2015). Compared to camera observations, solar irradiance measurements are better suited for long measurement campaigns with fast sampling rate because of the reduced data volume and wider field of view. REMS' ultraviolet (UV) sensor has been used for a similar survey on Mars (Zorzano et al. 2013). REMS' six UV diodes measure downwelling solar flux in different spectral bands within an upward facing 30-degree half-angle cone field-of-view (Gómez-Elvira et al. 2012). Observations of UV dips coincident with pressure dips are rare. Kahanpää et al. (2016) reported one very weak UV obscuration among its 252 pressure events. Steakley and Murphy (2016) identified 2 pressure events that appear to correspond to UV flux drops, but these pressure events cannot be confirmed as dust devils due to an instrumental error described in Harri et al. (2014) as the shadow effect. Although these events were eliminated from the sample, they may show signs of a double trough signature (Steakley and Murphy 2016) which could be produced by repeated passage into and out of the core of a cylindrical dust devil (Mason et al. 2013). Zorzano et al. (2013) searched for UV obscurations not concurrent with pressure dips, but found no clear signs of dust devils during MSL's first 100 sols. This lack of UV obscurations suggests that most or all vortices detected in the REMS pressure data are dustless, a result that agrees with the detection of only one dust devil by the MSL cameras (Moores et al. 2015).

4 Discussion

Despite the differing range of dust devils parameters measured by individual terrestrial campaigns, and their more extensive characterization than has been provided for martian dust devils, comparison between terrestrial and martian dust devils does indicate some differences and some similarities. Martian dust devils exhibit larger maximum widths and heights (Fig. 19) and faster maximum rotational wind speeds than their terrestrial counterparts (see Tables 1 and 2 and references therein). The more extreme inferred martian dust devil winds correspond to what is perceived to be the largest dust devils, which also correspond to the measurement location being most distant from the presumed core location.

Measured terrestrial maximum pressure drops are one order of magnitude larger than the measured maximum martian pressure drops. Renno et al. (1998, 2000) suggests that the potential pressure drop magnitude of a dust devil is proportional to the surface pressure. Since surface pressure on Earth is two orders of magnitude greater than on Mars, other aspects affecting potential vortex intensity, including the pressure thickness of the convective layer ('boundary layer height') and the near surface fraction of the mechanical dissipation of energy and the thermodynamic efficiency, must play a role in martian vortices only being one order of magnitude smaller in their depression than terrestrial dust devils. The larger observed size of martian dust devils (Fig. 1) could be one manifestation of these environmental characteristics.

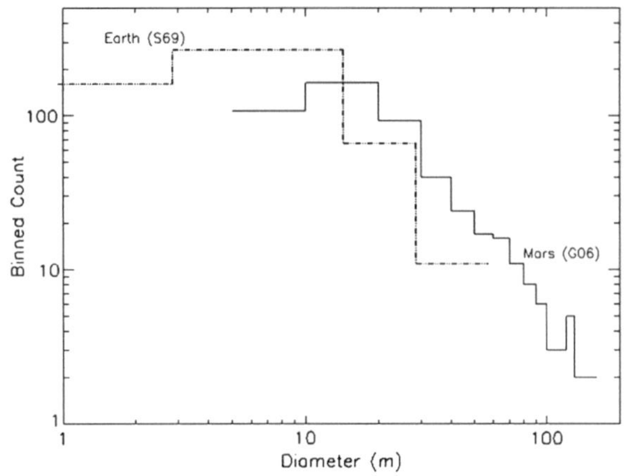

Fig. 19 Differential diameter counts of dust devils from Sinclair's (1969) study at Tucson, and the martian Spirit observations of Greeley et al. (2006) involve roughly comparable survey areas. The roughly linear fall-off towards larger diameters on these logarithmic axes corresponds to a power-law diameter distribution; the turn-over at small diameters may indicate a minimum intrinsic size, or a poor detection efficiency for small devils. In this instance the modal diameter on Earth is more like 5 m, only a factor 3 smaller than on Mars. It is notable that the largest devil seen in the Mars set is larger than that at Earth, but perhaps a longer survey on Earth (pushing the *curve upwards*) would allow the expected number of detections of larger devils on Earth to rise above unity. As discussed in Lorenz and Jackson (2016) in this issue care must be exercised in drawing size population conclusions from different types of data due to observational biases, especially at the small size end of the distribution

The relative frequency of occurrence of both terrestrial and martian devils with a pressure drop magnitude, ΔP, normalized by the local mean surface pressure exhibits a power law functional form (Fig. 20) represented by, ΔP^{-x}, with $x \sim 2$ (Lorenz 2012a, 2012b). A similar functional form for observed vortex diameter has also been proposed (Lorenz 2009), but an exponential fit has also been suggested, especially at smaller diameters (Fig. 19). These issues are more thoroughly explored in the accompanying papers by Lorenz and Jackson (2016) and Kurgansky et al. (2016).

Terrestrial field measurements have been obtained from numerous geographic locations and continents (Africa, Australia, North America, South America). There are some sites (Eldorado Valley, Nevada USA; Eloy, Arizona USA) that have served as the locales for a number of field campaigns. Martian measurements have also been obtained from discrete locations with a preference (5 of 7 successful missions) for equatorial/subtropical locations, though northern middle and polar latitudes have also been investigated. While martian robotic missions do not offer the opportunity to upgrade or augment instrumentation after launch (and often a substantial time period prior to launch), repeated terrestrial field campaigns to the same location provide the possibility of building upon previous measurements. A number of campaigns over 10–15 years at the Eldorado Valley site have used previously employed and accompanying new instrumentation (Metzger 1999; Balme et al. 2003; Ringrose et al. 2007; Metzger et al. 2010, 2011; Balme et al. 2012; Mason et al. 2014; Jackson and Lorenz 2015). These campaigns have employed mobile and deployable stationary measurement platforms, vertical masts with 10 meter heights to probe the vertical structure of encountered dust devils, stereo camera imaging to improve quantification of dust devil position and size, and recently networks of pressure loggers with partial coverage by solar flux loggers. None of the martian robotic missions to date have 'returned' to a previous lander/rove location, and there is no pronounced motivation to do so for dust

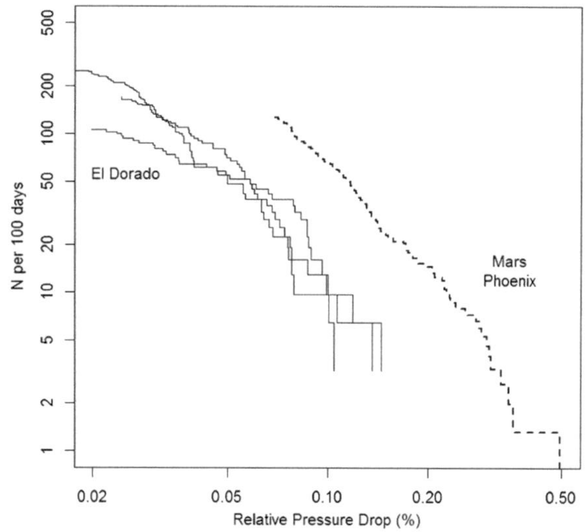

Fig. 20 The number of times a vortex-induced pressure drop will be encountered by a fixed station on Earth and Mars is presented as a cumulative count. Because the absolute pressures on Mars are > 100 times smaller than on Earth, it is more illuminating to use a relative pressure drop. Thus the 'once a week' vortex on Earth corresponds to about 0.1 % or 1 mbar pressure drop, while that on Mars is about 0.3 % or ~ 0.02 mbar (2 Pa). Note that despite the differences in temperature and gas compositions between the two planets, the densities and pressures are approximately in the same proportion. Since the pressure drop of a vortex in cyclostrophic balance is $\sim rV^2$, then a fixed relative pressure drop corresponds roughly to a fixed maximum tangential speed V on both bodies: 0.1 % for a 10 m/s wind speed

devil measurements. One opportunity to do so could be realized with a Mars sample return set of missions which could have a second spacecraft retrieve a sample previously cached by a preliminary mission. If the preliminary mission's measurements identified unique dust devil characteristics at the cache site, possibly the cache retrieving mission could carry instrumentation to enhance characterization of the dust devils at the cache site, but mission lead time development needs and other mission priorities would likely preclude such a an instrumentation selection.

5 Future Dust Devil Measurements

5.1 Earth

The basic features and correlations of properties of terrestrial dust devils have been measured, and somewhat robust statistics on the population are emerging. What is presently lacking are data to support an evaluation of the dependence of dust devil morphology on environmental parameters. For example, there is a somewhat anecdotal (see Chap. 6, Kurgansky et al. 2016) correlation of columnar vortices with very smooth land surfaces such as playa, whereas rougher (e.g. scrubby) terrain tends to have more conical devils.

Present profiles of dust devils do not allow discrimination of the various analytic functions used as axisymmetric idealizations of vortex structure (Kurgansky et al. 2016). On the other hand, it is not clear that better measurements will help, as it may be that nonidealities are more significant than the difference between models. A prominent area that merits further study is the frequency and nature of multi-cored vortices and what controls their formation. Another deviation from axisymmetry is evident in simulations (e.g. Toigo et al. 2003)—that winds spiral into the vortex core in azimuthally-concentrated bands, like those seen in hurricane clouds or in the arms of galaxies. Such azimuth variations have not yet been formally documented in field observations. Finally, field measurements from Sinclair (1973) and Lorenz (2012b) appear to show an asymmetry in the pressure profile of migrating dust devils, the 'attack' slope being shallower than the 'decay'. This may be a generic feature of a migrating vortex, tilted by wind shear, but has not been quantified.

Another question is the possible role in dust devil energy balance of sunlight intercepted by lofted dust. It is sometimes noticed that devils become more intense when they become dustier—but is this cause, or effect? If effect, is it sensible heat in the dust and air pulled off the ground, or is the solar heat introduced by the absorbing dust and introduced into the vortex directly a significant factor? Field measurements, perhaps including thermal as well as video imaging (to relate flow speeds to dust temperature) may illuminate this question.

Related to this question is that of dust-lifting per-se: do larger, more vigorous, but less frequently occurring, dust devils lift more dust than smaller, more numerous ones? Can the dust lifting of a dust devil ever be disentangled from the properties of the surface (i.e. is the dustiness of a dust devil controlled mainly be surface properties or by its internal properties?). Only by coordinated measurements of dust content and wind speed (likely in-situ measurement), and dust devil diameter (likely remote measurements) from controlled field-sites can these questions be addressed. Such measurements would help provide better understanding of the environmental effects of dust devils and provide a model for comparison with Mars.

Circumstantial observations, dating as far back as Flower (1936) and a correlation of satellite-observed dust devil heights and estimates of planetary boundary layer thickness from model predictions and occultation measurements (Fenton and Lorenz 2015) suggest dust devil heights and spacings may be controlled by the planetary boundary layer (PBL) thickness. In-situ measurements that establish this correlation more securely (perhaps also validating the anecdotal suggestion that small devils are more common early in the day and larger ones in the late afternoon, consistent with the diurnal growth of the PBL) would similarly be useful.

In summary, terrestrial dust devil characterization would benefit from dedicated campaigns of long-lived networks of uniformly instrumented stations measuring pressure, wind (3-D), temperature, dust load, electric field, etc. Measurements at a variety of heights above the surface would be valuable. High cadence (1 Hz) measurements in conjunction with imaging of the network would aid in correlating visible dust devil manifestation with interior conditions. The use of several networks in dissimilar field sites would enable the influence of local surface properties to be disentangled from the properties of individual dust devils.

5.2 Mars

Measurement characterization of martian dust devils is limited by the scarcity of observations, especially those that provide simultaneous measurements of a variety of geophysical parameters. Since dust devils play a substantial role in maintaining Mars' persistent dust haze, especially during the aphelion season, better characterizing their attributes will enable improved understanding of the their role in providing/maintaining the thermodynamically important atmospheric dust load.

Robotic spacecraft sent to Mars' surface should carry imaging systems capable of providing high resolution and fast frame rates covering a wide field of view. Monochrome imaging is adequate (Greeley et al. 2010) and would not interfere with other onboard color imagers dedicated to geologic studies. An imaging system that provides full azimuthal coverage and onboard software to identify and retain images that display dust devil diagnostic temporal variability (thus reducing downlink volume of dust-devil deficient images) would enable greater temporal coverage than has been provided previously. The dust-devil imaging strategy employed by Spirit (and Opportunity) Navigation Cameras did provide substantial numbers of dust devil observations, targeted to the times of sol when they were most frequent.

No robotic spacecraft has yet provided unambiguous contemporaneous measurement of surface pressure and wind. Thus, the dynamic connection between vortex magnitude and velocity fields has not been established. Hot wire / hot film anemometers have been the only type of high-frequency sampling wind instruments to operate on Mars' surface, and all have suffered from instrument failure and/or calibration difficulties and placement issues (Lenoir et al. 2011; Lorenz and Sotzen 2014). These instruments have generally been low mass, which are desirable, but their poor performance record suggests other sensor types might be warranted. An enhanced REMS wind sensor is selected for NASA's InSight lander (scheduled for a 2018 launch). NASA's 2020 Mars Rover will carry a meteorology package (Rodriguez-Manfredi et al. 2014). Acoustic and laser Doppler anemometers have been suggested as possible technologies applicable for Mars (Merrison et al. 2006; Wilson et al. 2008; Esposito 2011; Banfield 2012; Leonard-Pugh et al. 2012; Rafkin et al. 2013). Both technologies enable high frequency sampling, 10 Hz or more, of the three-dimensional wind field, which provides opportunity for determination of turbulent characteristics of the flow. Acoustic anemometry has the added advantage of providing simultaneous atmospheric temperature measurement which when coupled to the winds affords quantification of the turbulent thermal flux. To date neither technology has been selected for flight. Wind socks provide wind sampling at very low frequencies (Sullivan et al. 2000; Holstein-Rathlou et al. 2010), but their data bandwidth requirement is large compared to their low response time and inability to unambiguously provide the vertical wind component.

Pressure measurements have provided the most valuable and plentiful in situ dust devil/convective vortex measurements to date. The high resolution (0.1–0.2 Pa) and high sampling frequencies (2–0.25 Hz) implemented have proved adequate to capture pressure signatures. Care must be taken for future opportunities to avoid sensor thermal variation which complicates pressure value determination, as well as inlet orientation which can provide unfavorable dynamic pressure contamination, and measurement tube length/diameter which affects sensor response time (Ellehoj et al. 2010). The low power and mass requirements of the necessary equipment as well as the tremendous value of martian surface pressure measurements for both local (dust devil) and larger scale atmospheric analyses warrant pressure sensors being included on every robotic spacecraft landed upon Mars' surface.

Contemporaneous measurement of received solar flux and pressure/wind signatures of vortex passage are necessary to assess the 'dustiness' of vortices and thus the frequency of occurrence of dust devils vs. 'clear' convective vortices. Such discrimination would aid the assessment of vortex thermal vs. dust vertical fluxes. Dedicated solar flux instruments can provide appropriate measurements (Gómez-Elvira et al. 2012; Kahanpää et al. 2016) with a cadence coupled to that of other science instruments. Engineering systems can also provide solar flux measurements at no additional 'cost (Schofield et al. 1997), though likely at a lower cadence than desired for science purposes.

To date no martian robotic lander/rover has provided electric field or dust concentration or dust particle flux/size distribution measurements. ESA's Schiaparelli lander's MicroARES system will provide initial electric field measurements during that missions short duration operations in late 2016 (Esposito et al. 2016). A laser Doppler anemometer could provide particle concentration in addition to velocity, but as previously stated no such instrument is currently selected for a Mars robotic mission. A vertically oriented LIDAR system onboard the NASA Phoenix lander provided vertical resolution of aerosol concentration within the bottom few kilometers of the atmosphere (Whiteway et al. 2008). A horizontally scanning LIDAR system could provide signals from which radial variations in near-surface aerosol abundance within several kilometers of a lander/rover could be obtained.

Dust devil infrasound/acoustic and seismic measurements are still minimally available for terrestrial studies, and their full diagnostic value remains immature yet promising; a seismometer, for example, appears to be capable of tracking close encounters with dust devils, recovering an estimate of the azimuth history and constraining the integral of the pressure field (relating to diameter and core pressure drop). In combination with wind and pressure measurements, the dust devil parameters and miss distance may be reconstructed or at least constrained (Lorenz et al. 2015a). The NASA InSight mission will provide meteorological and seismic measurements that will improve understanding of dust devils for both martian and terrestrial applications. In addition, the seismic signals of dust devils may also provide information about the surface and subsurface properties at the InSight landing site by acting as calibration loads (Lorenz et al. 2015a, 2015b). A Mars microphone is going to be part of the payload of the 2018 Exomars landing platform (Zelenyi et al. 2015). The microphone, with a bandwidth of 100 Hz to 10 kHz, will record, for the first time, audio signals at the surface of Mars and is expected to contribute to atmospheric science investigations including dust devil studies (Maurice et al. 2016).

Ideally a network of meteorologically outfitted stations designed to characterize dust devils will be deployed upon Mars' surface within a very limited spatial footprint, possibly as one component of a more globally extensive network. The PASCAL network mission developed and proposed to from NASA Ames in the late 1990's/early 2000's developed some network concepts but was not selected to launch.

6 Conclusions

Dust devils are inherently difficult to measure. Their occurrence is somewhat unpredictable in both space and time and they are fast-moving, short-lived phenomena. Observational 'survey' data have provided information about size-frequency distributions, sense of rotation and other parameters, but to assess the wind and pressure structure, and the dust-loading behaviour of a dust devil has required in-situ measurement.

To obtain these data, terrestrial dust devils have either been "chased" or "monitored". The former method can be subject to observation bias, but the latter can require long field campaigns and multiple sets of instruments if statistically significant numbers of data are to be obtained. Recently, advances in both instrumentation and data recording technology have enabled new methodologies to be developed to solve these problems. Long-term, high cadence monitoring using multiple sensors, coupled with automatically controlled imaging provides a way to get the 'best of both worlds'. In the next ten years, new field studies should be able to provide much better data about population-statistics of dust devils and their environmental role.

On Mars, dust devils studies have generally 'piggy-backed' on other camera or meteorological studies. Nevertheless, significant progress in measuring population statistics and pressure-structure have been possible. These data have been complemented by remote-sensing observations from orbit—a method not widely applied on Earth.

Future studies for Mars should focus on in situ, high cadence sampling of pressure, wind speed and dust load. Along with the continuing operation of the US/NASA MSL Curiosity rover, the next opportunity for martian dust devil measurements will be provided by the ESA/Russian ExoMars Schiaparelli Lander scheduled for ~ 1 week's operation on Mars surface during late 2016. The longer lived US/NASA InSight lander is scheduled to arrive in late 2018, followed by the ESA ExoMars Rover and the US/NASA 2020 Rover in early 2021.

Acknowledgements All authors thank the International Space Science Institute for organizing and hosting the February 2015 Bern, Switzerland Workshop and for providing lodging accommodations. J. Murphy thanks the New Mexico State University College of Arts & Sciences Faculty Travel Award program for travel support. K. Steakley thanks the New Mexico State University (NMSU) Aggies Go Global organization, NMSU Graduate School, Associated Students of NMSU, New Mexico Space Grant Consortium, and NMSU Astronomy Department for funding support. M. Balme acknowledges funding from the UK Space Agency (grant ST/L00643X/1) and from STFC, the UK Science and Technology Facilities Council (grant ST/L000776/1). R. Lorenz acknowledges the support of NASA Mars Fundamental Research Program Grant NNX12AI04G. N. Murdoch was supported by a CNES-provided Post-doctoral award. M. Patel acknowledges support from the UK STFC and UK Space Agency under grants ST/I003061/1 and ST/P001262/1 and as part of the project UPWARDS-633127, funded by the European Union's Horizon 2020 Programme (H2020-Compet-08-2014). P. Whelley was supported by the NASA Post Doctoral Program and NASA's *Remote, In Situ, and Synchrotron Studies for Science and Exploration* project.

The authors thank two anonymous reviewers for their thorough evaluations which have resulted in an improved final paper.

References

R.A. Bagnold, *The Physics of Blown Sand and Desert Dunes* (Dover, Mineola, 1941)

M.R. Balme, A. Pathare, S.M. Metzger, M.C. Towner, S.R. Lewis, A. Spiga, L.K. Fenton, N.O. Renno, H.M. Elliott, F.A. Saca, T.I. Michaels, P. Russell, J. Verdasca, Field measurements of horizontal forward motion velocities of terrestrial dust devils: towards a proxy for ambient winds on Mars and Earth. Icarus **221**(2), 632–645 (2012)

M. Balme, R. Greeley, Dust devils on Earth and Mars. Rev. Geophys. **44**, RG3003 (2006)

M. Balme, S. Metzger, M. Towner, T. Ringrose, R. Greeley, J. Iversen, Friction wind speeds in dust devils: a field study. Geophys. Res. Lett. **30**(16), 1830 (2003). doi:10.1029/2003GL017493

D. Banfield, Mars acoustic anemometer, *AGU Fall Meeting 2012*, San Francisco, California, USA. Poster presentation (2012), http://abstractsearch.agu.org/meetings/2012/FM/P23A-1916.html

D. Banfield, Atmospheric observations from the Mars insight mission, in *Fifth International Workshop on the Mars Atmosphere: Modelling and Observations*, Oxford, UK (2014). Oral presentation, http://www-mars.lmd.jussieu.fr/oxford2014/abstracts/banfield_oxford2014.pdf

A.J. Bedard, Low frequency atmospheric acoustic energy associated with vortices produced by thunderstorms. Mon. Weather Rev. 241–243 (2005)

F. Bell, Dust devils and aviation, report, Meteorol. Note 27, Aust. Bur. of Meteorol., Melbourne, Victoria, 1967

J.-J. Berthelier, ARES, atmospheric relaxation and electric field sensor, the electric field experiment on NETLANDER. Planet. Space Sci. **48**(12–14), 1193–2000 (2000)

C. Bettanini et al., The DREAMS experiment on the ExoMars 2016 mission for the study of martian environment during the dust storm season. MetroAeroSpace **167**(173), 29–30 (2014). doi:10.1109/MetroAeroSpace.2014.6865914

H.B. Bluestein, A.L. Pazmany, Observations of tornadoes and other convective phenomena with a mobile, 3-mm wavelength, Doppler radar: the spring 1999 field experiment. Bull. Am. Meteorol. Soc. **81**, 2939–2951 (2000)

J.J. Carroll, J.A. Ryan, Atmospheric vorticity and dust devil rotation. J. Geophys. Res. **75**, 5179–5184 (1970)

W.D. Crozier, Dust devil properties. J. Geophys. Res. **75**, 4583–4585 (1970)

R. Davy, J.A. Davies, P.A. Taylor, C. Lange, W. Weng, J. Whiteway, H.P. Gunnlaugson, Initial analysis of air temperature and related data from the Phoenix MET station and their use in estimating turbulent heat fluxes. J. Geophys. Res. **115**, E00E13 (2010). doi:10.1029/2009JE003444

S.J. Desch, J.N. Cuzzi, The generation of lightning in the solar nebula. Icarus **143**, 87–105 (2000). doi:10.1006/icar.1999.6245

M. Domínguez, V. Jiménez, J. Ricart, L. Kowalski, J. Torres, S. Navarro, J. Romeral, L. Castañer, A hot film anemometer for the martian atmosphere. Planet. Space Sci. **56**(8), 1169–1179 (2008). doi:10.1016/j.pss.2008.02.013

N. Duff, D.J. Lacks, Particle dynamics simulations of triboelectric charging in granular insulator systems. J. Electrost. **66**, 51 (2008). doi:10.1016/j.elstat.2007.08.005

R. Edmonds, Examination of two martian atmosphere phenomena: dust devil acoustics and gravity wave forcing of dust storm development. PhD dissertation, New Mexico State University, 2014, p. 223

M.D. Ellehoj, H.P. Gunnlaugsson, P.A. Taylor, H. Kahanpää, K.M. Bean, B.A. Cantor, B.T. Gheynani, L. Drube, D. Fisher, A.-M. Harri, C. Holstein-Rathlou, M.T. Lemmon, M.B. Madsen, M.C. Malin, J. Polkko, P.H. Smith, L.K. Tamppari, W. Weng, J. Whiteway, Convective vortices and dust devils as the Phoenix Mars mission landing site. J. Geophys. Res. **115**, E00E16 (2010). doi:10.1029/2009JE003413

F. Esposito, MEDUSA: observation of atmospheric dust and water vapor close to the surface of Mars. Mars **6**, 1–12 (2011). http://www.marsjournal.org/contents/2011/0001/. doi:10.1555/mars.2011.0001

F. Esposito, R. Molinaro, C.I. Popa, C. Molfese, F. Cozzolino, L. Marty, K. Taj-Eddine, G. Di Achille, G. Franzese, S. Silvestro, G.C. Ori, The role of atmospheric electric field in the dust lifting process. Geophys. Res. Lett. (2016, in press)

L.K. Fenton, R. Lorenz, Dust devil height and spacing with relation to the martian planetary boundary layer thickness. Icarus **260**, 246–262 (2015). doi:10.1016/j.icarus2015.07.028

L. Fenton, D. Reiss, M. Lemmon, B. Marticorena, S. Lewis, B. Cantor, Orbital observations of dust lofted by daytime convective turbulence. Space Sci. Rev. (2016). doi:10.1007/s11214-016-0243-6

F. Ferri, P.H. Smith, M.T. Lemmon, N. Renno, Dust devils as observed by Mars Pathfinder. J. Geophys. Res. **108**(E12), 5133 (2003). doi:10.1029/2000JE001421

D.E. Fitzjarrald, A field investigation of dust devils. J. Appl. Meteorol. **12**, 808–813 (1973)

W.D. Flower, Sand devils. Lond. Meteorol. Off. Prof. Notes **5**(71), 1–16 (1936)

K.M. Forward, D.J. Lacks, R.M. Sankaran, Particle-size dependent bipolar charging of martian regolith simulant. Geophys. Res. Lett. **36**(13), L13201 (2009)

G.D. Freier, The electric field of a large dust devil. J. Geophys. Res. **65**, 3504 (1960). doi:10.1029/JZ065i010p03504

S.D. Fuerstenau, Solar heating of suspended particles and the dynamics of martian dust devils. Geophys. Res. Lett. **33**, 19 (2006). doi:10.1029/2006GL026798

E.W.B. Gill, Friction electrification of sand. Nature **162**, 568 (1948)

D.A. Gillette, P. Sinclair, Estimation of suspension of alkaline material by dust devils in the United States. Atmos. Environ. **24**(A), 1135–1142 (1990)

D.A. Gillette, I.H. Blifford, D.W. Fryrear, Influence of wind velocity on size distributions of aerosols generated by wind erosion of soils. J. Geophys. Res. **79**, 4068–4075 (1974)

M. Golombek et al., Overview of the Mars Pathfinder mission and assessment of landing site predictions. Science **278**, 5344 (1997). doi:10.1126/science.278.5344.1743

J. Gómez-Elvira et al., REMS: the environmental sensor suite for the Mars science laboratory rover. Space Sci. Rev. **170**, 583–640 (2012). doi:10.1007/s11214-012-9921-1

J. Gómez-Elvira et al., Curiosity's rover environmental monitoring station: overview of the first 100 sols. J. Geophys. Res., Planets **119**, 1680–1688 (2014). doi:10.1002/2013JE004576

C.G. Grant, Dust devils in the sub-arctic. Weather **4**, 402–403 (1949)

R. Greeley, J.D. Iversen, *Wind as a Geological Process*. Cambridge Planetary Science Series (Cambridge University Press, Cambridge, 1985). ISBN 0521356927

R. Greeley et al., Wind related features in Gusev crater, Mars. J. Geophys. Res. **108**(E12), 8077 (2003)

R. Greeley, P.L. Whelley, R.E. Arvidson, N.A. Cabrol, D.J. Foley, B.J. Franklin, P.G. Geissler, M.P. Golombek, R.O. Kuzmin, G.A. Landis, M.T. Lemmon, L.D.V. Neakrase, S.W. Squyres, S.D. Thompson, Active dust devils in Gusev Crater, Mars: observations from the Mars exploration rover spirit. J. Geophys. Res. **111**(E12) (2006). doi:10.1029/2006JE002743

R. Greeley, D.A. Waller, N.A. Cabrol, G.A. Landis, M.T. Lemmon, L.V. Nekarase, M. Pendelton Hoffer, S.D. Thompson, Gusev Crater, Mars: observations of three dust devil seasons. J. Geophys. Res. **115** (2010)

J. Grotzinger et al., Mars science laboratory mission and science investigation. Space Sci. Rev. **170**(1), 5–56 (2012). doi:10.1007/s11214-012-9892-2

H.P. Gunnlaugsson et al., Telltale wind indicator for the Mars Phoenix lander. J. Geophys. Res. **113**, E00A04 (2008). doi:10.1029/2007JE003008

R.M. Haberle, J. Gómez-Elvira, M. de la Torre Juárez, A.-M. Harri, J.L. Hollingsworth, H. Kahanpää, M.A. Kahre, M. Lemmon, F.J. Martín-Torres, M. Mischna, J.E. Moores, C. Newman, S.C.R. Rafkin, N. Rennó, M.I. Richardson, J.A. Rodríguez-Manfredi, A.R. Vasavada, M.-P. Zorzano-Mier, REMS/MSL Science Teams, Preliminary interpretation of the REMS pressure data from the first 100 sols of the MSL mission. J. Geophys. Res., Planets **119**, 440–453 (2014). doi:10.1002/2013JE004488

J. Hallett, T. Hoffer, Dust devil systems. Weather **26**, 247–250 (1971)

W.R. Harper, *Contact and Frictional Dissipation* (Clarendon Press, Oxford, 1967)

A.-M. Harri, M. Genzer, O. Kemppinen, H. Kahanpää, J. Gomez-Elvira, J.A. Rodriguez-Manfredi, R. Haberle, J. Polkko, W. Schmidt, H. Savijärvi, J. Kauhanen, E. Atlaskin, M. Richardson, T. Siili, M. Paton, M. de la Torre Juarez, C. Newman, S. Rafkin, M.T. Lemmon, M. Mischna, S. Merikallio, H. Haukka, J. Martin-Torres, M.-P. Zorzano, V. Peinado, R. Urqui, A. Lapinette, A. Scodary, T. Mäkinen, L. Vazquez, N. Rennó, The REMS/MSL Science Team, Pressure observations by the curiosity rover: initial results. J. Geophys. Res., Planets **119**, 82–92 (2014). doi:10.1002/2013JE004423

R.G. Harrison, E. Barth, F. Esposito, J. Merrison, F. Montmessin, K.L. Aplin, C. Borlina, J.J. Berthelier, G. Déprez, W.M. Farrell, I.M.P. Houghton, N.O. Renno, K.A. Nicoll, S.N. Tripathi, M. Zimmerman, Applications of electrified dust and dust devil electrodynamics to martian atmospheric electricity. Space Sci. Rev. (2016). doi:10.1007/s11214-016-0241-8

M. Hecht, M.D. Tratt, D. Catling, S. Samulon, MATADOR dust devil campaign (2001)

G.D. Hess, K.T. Spillane, Characteristics of dust devils in Australia. J. Appl. Meteorol. **29**, 498–507 (1990)

S.L. Hess, R.M. Henry, C.B. Leovy, J.A. Ryan, J.E. Tillman, Meteorological results from the surface of Mars: viking 1 and 2. J. Geophys. Res. **82**(28), 4559–4574 (1977). doi:10.1029/JS082i028p04559

C. Holstein-Rathlou et al., Winds at the Phoenix landing site. J. Geophys. Res. **115**, E00E18 (2010). doi:10.1029/2009JE003411

R.G. Horn, D.T. Smith, A. Grabbe, Contact electrification induced by monolayer modification of a surface and relation to acid-base interactions. Nature **366**, 442–443 (1993). doi:10.1038/366442a0

M.S. Howe, *Theory of Vortex Sound* (Cambridge University Press, Cambridge, 2003)

I.I. Inculet, G.S.P. Castle, G. Aartsen, Generation of bipolar electric fields during industrial handling of powders. Chem. Eng. Sci. **61**, 2249–2253 (2006). doi:10.1016/j.ces.2005.05.005

J. Ito, H. Niino, Particle image velocimetry of a dust devil observed in a desert. Sci. Online Lett. Atmos. **10**(1), 108–111 (2014)

R.L. Ives, Behavior of dust devils. Bull. Am. Meteorol. Soc. **28**, 168–174 (1947)

D.J. Lacks, A. Levandovsky, Effect of particle size distribution on the polarity of triboelectric charging in granular insulator systems. J. Electrost. **65**, 107 (2007)

T.L. Jackson, W.M. Farrell, IEEE Trans. Geosci. Remote Sens. **44**, 2942 (2006). doi:10.1109/TGRS.2006.875785

B. Jackson, R. Lorenz, A multiyear dust devil vortex survey using an automated search of pressure time series. J. Geophys. Res. **120**(2), 401–412 (2015). doi:10.1092/2014JE004712

H. Kahanpää, C. Newman, J. Moores, M-P. Zorzano, J. Martín-Torres, S. Navarro, A. Lepinette, M.T. Lemmon, B. Cantor, P. Valentín-Serrano, A. Ullán, W. Schmidt, Convective vortices and dust devils at the MSL landing site: annual variability. J. Geophys. Res. Planets **121** (2016). doi:10.1002/2016JE005027

M. Kahre, J. Murphy, R. Haberle, Modeling the martian dust cycle and surface dust reservoirs with the NASA Ames general circulation model. J. Geophys. Res., Planets **111**, E6 (2006)

J.C. Kaimal, J.A. Bussinger, Case studies of a convective plume and a dust devil. J. Appl. Meteorol. **9**, 612–620 (1970)

M. Klose, B.C. Jemmet-Smith, H. Kahanpää, M. Kahre, P. Knippertz, M. Lemmon, S. Lewis, R. Lorenz, L. Neakrase, C. Newman, M. Patel, D. Reiss, A. Spioga, P. Whelley, Space Sci. Rev. (2016). doi:10.1007/s11214-016-0261-4

J.F. Kok, N.O. Renno, Enhancement of the emission of mineral dust aerosols by electric forces. Geophys. Res. Lett. **33**, L19S10 (2006)

J.F. Kok, N.O. Renno, Electrostatics in wind-blown sand. Phys. Rev. Lett. **100**, 014501 (2008)

M.V. Kurgansky, Size distribution of dust devils in the atmosphere. Izv., Atmos. Ocean. Phys. **42**(3), 319–325 (2006)

M.V. Kurgansky et al., Dust devil steady-state structure from a fluid dynamic perspective. Space Sci. Rev. (2016). doi:10.1007/s11214-016-0281-0

M.V. Kurgansky, A. Montecinos, V. Villagran, S.M. Metzger, Micrometeorological conditions for dust-devil occurrence in the Atacama Desert. Bound.-Layer Meteorol. **138**, 285–298 (2011). doi:10.1007/s10546-010-9549-1

R.I. Lambeth, On the measurement of dust devil parameters. Bull. Am. Meteorol. Soc. **47**, 522–526 (1966)

G.A. Landis, P.P. Jenkins, Measurement of the settling rate of atmospheric dust on Mars by the MAE instrument on Mars Pathfinder. J. Geophys. Res. **105**(E1), 1855–1857 (2000). doi:10.1029/1999JE001029

J. Latham, The electrification of snowstorms and sandstorms. Q. J. R. Meteorol. Soc. **90**, 383 (1964). doi:10.1002/qj.49709038310

J. Latham, C.D. Stow, A laboratory investigation of the electrification of snowstorms. Q. J. R. Meteorol. Soc. **94**, 415 (1968)

M.T. Lemmon, M.J. Wolff, M.D. Smith, R.T. Clancy, D. Banfield, G.A. Landis, A. Ghosh, P.H. Smith, N. Spanovich, B. Whitney, P. Whelley, R. Greeley, S. Thompson, J.F. Bell III, S.W. Squyres, Atmospheric imaging results from the Mars exploration rovers: spirit and opportunity. Science **306**, 1753–1756 (2004)

M.T. Lemmon, M.J. Wolff, J.F. Bell III., M.D. Smith, B.A. Cantor, P.H. Smith, Dust aerosol, clouds and the atmospheric optical depth record over 5 Mars years of the Mars exploration rover mission. Icarus **251**, 96–111 (2015)

B. Lenoir, D. Banfield, D.A. Caughey, Accommodation study for an anemometer on a martian lander. J. Atmos. Ocean. Technol. **28**, 210–218 (2011). doi:10.1175/2010JTECHA1490.1

E. Leonard-Pugh, C. Wilson, S. Calcutt, L. Davis, Capacitive ultrasonic transducer development for acoustic anemometry on Mars in *44th Annual Meeting of the Division for Planetary Sciences of the American Astronomical Society*, Reno, NV, USA (2012). Poster presentation, http://adsabs.harvard.edu/abs/2012DPS....4421523L

P. Lognonné, B. Mosser, Planetary seismology. Surv. Geophys. **14**, 239–302 (1993)

P. Lognonné, W.B. Banerdt, K. Hurst, D. Mimoun, R. Garcia, M. Lefeuvre, J. Gagnepain-Beyneix, M. Wieczorek, A. Mocquet, M. Panning, E. Beucler, S. Deraucourt, D. Giardini, L. Boschi, U. Christensen, W. Goetz, T. Pike, C. Johnson, R. Weber, K. Larmat, N. Kobayashi, J. Tromp, Insight and single-station broadband seismology: from signal and noise to interior structure determination, in *43rd Lunar and Planetary Science Conference*, Houston, TX, March 2012 (2012). Abstract #1983

R. Lorenz, Thermal imaging of a desert dust devil. J. Meteorol. **29**(292), 275–276 (2004)

R.D. Lorenz, Power law of dust devil diameters on Mars and Earth. Icarus **203**(2), 683–684 (2009)

R. Lorenz, On the statistical distribution of dust devil diameter. Icarus **215**(1), 381–390 (2011)

R.D. Lorenz, Power law distribution of pressure drops in dust devils: observation techniques and Earth–Mars comparison. Planet. Space Sci. **60**, 370–375 (2012a). doi:10.1016/j.pss.2011.11.003

R.D. Lorenz, Observing desert dust devils with a pressure logger. Geosci. Instrum. Method. Data Syst. **1**, 209–220 (2012b)

R. Lorenz, The longevity and aspect ratio of dust devils: effects on detection efficiencies and comparison of landed and orbital imaging at Mars. Icarus **226**, 964–970 (2013)

R.D. Lorenz, Vortex encounter rates with fixed barometer stations: comparison with visual dust devil counts and large eddy simulations. J. Atmos. Sci. **71**, 4461–4472 (2014)

R.D. Lorenz, Heuristic estimation of dust devil vortex parameters and trajectories from single-station meteorological observations: application to InSight at Mars. Icarus **271**(2–16), 326–337 (2016)

R.D. Lorenz, D. Christie, Dust devil signatures in infrasound records of the international monitoring system. Geophys. Res. Lett. **42**(6), 2009–2014 (2015)

R.D. Lorenz, B.K. Jackson, Dust devils and dustless vortices on a desert playa observed with surface pressure and solar flux logging. GeoResJ **5**, 1–11 (2015). doi:10.1016/j.grj.2014.11.002

R.D. Lorenz, B. Jackson, Dust devil populations and statistics. Space Sci. Rev. (2016). doi:10.1007/s11214-016-0277-9

R.D. Lorenz, P.D. Lanagan, A barometric survey of dust devil vortices on a desert playa. Bound.-Layer Meteorol. **53**, 555–568 (2014). doi:10.1007/s10546-014-9954-y

R. Lorenz, J. Radebaugh, Dust devils in thin air: vortex observations at a high elevation Mars analog site in the Argentinian Puna. Geophys. Res. Lett. **43** (2016). doi:10.1002/2015GL067412

R.D. Lorenz, D. Reiss, Solar panel clearing events, dust devil tracks, and in-situ vortex detections on Mars. Icarus **248**, 162–164 (2015)

R.D. Lorenz, K.S. Sotzen, Buoyant thermal plumes from planetary landers and rovers: application to sizing of meteorological masts. Planet. Space Sci. **90**, 81–89 (2014). doi:10.1016/j.pss.2013.10.011

R.D. Lorenz, S. Kedar, N. Murdoch, P. Lognonné, T. Kawamurak, D. Mimoun, W.B. Banerdt, Seismometer signature of dust devils: implication for InSight, in *2015 European and Planetary Science Conference*, Nantes (2015a)

R.D. Lorenz, S. Kedar, N. Murdoch, P. Lognonné, T. Kawamurak, D. Mimoun, W.B. Banerdt, Seismometer detection of dust devil vortices by ground tilt. Bull. Seismol. Soc. Am. BSSA-S-15-00169 (2015b)

R.D. Lorenz, L.D. Neakrase, J.D. Anderson, In-situ measurement of dust devil activity at La Jornada Experimental Range, New Mexico, USA. Aeolian Res. **19**, 183–194 (2015c)

R.D. Lorenz, M. Balme, Z. Gu et al., History and application of dust devil studies. Space Sci. Rev. (2016). doi:10.1007/s11214-016-0239-2

J. Lowell, W.S. Truscott, Triboelectrification of identical insulators. II. Theory and further experiments. J. Phys. D, Appl. Phys. **19**, 1281–1298 (1986). doi:10.1088/0022-3727/19/7/018

J.P. Mason, M.R. Patel, S.R. Lewis, Radiative transfer modelling of dust devils. Icarus **223**, 1–10 (2013)

J.P. Mason, M.R. Patel, S.W. Lewis, The retrieval of optical properties from terrestrial dust devil vortices. Icarus **231**, 385–393 (2014)

S. Maurice, R.C. Wiens, W. Rapin, D. Mimoun, X. Jacob, B. Betts, S. Clegg, A. Cousin, O. Gasnault, O. Forni, J. Lasue, P.-Y. Meslin, J.F. Bell, G. Delory, A microphone supporting LIBS investigation on Mars. Lunar Planet. Sci. Conf. **47**, 3044 (2016)

J.P. Merrison, H.P. Gunnlaugsson, K. Kinch, T.L. Jacobsen, A.E. Jensen, P. Nørnberg, H. Wahlgreen, An integrated laser anemometer and dust accumulator for studying wind-induced dust transport on Mars. Planet. Space Sci. **54**(11), 1065–1072 (2006). doi:10.1016/j.pss.2006.05.026

S.M. Metzger, Dust devils as aeolian transport mechanisms in southern Nevada and in the Mars Pathfinder landing site. Ph.D. thesis, Univ. of Nev., Reno (1999)

S.M. Metzger, J.R. Carr, J.R. Johnson, T.J. Parker, M.T. Lemmon, Dust devil vortices seen by the Mars Pathfinder camera. Geophys. Res. Lett. **26**(18), 2781–2784 (1999)

S.M. Metzger, M. Kurgansky, A. Montecinos, V. Villagram, H. Verdejo, Chasing dust devils in Chile's Atacama Desert. LPSC Abstract 2564 (2010)

S. Metzger, M. Balme, A. Pathare, N. Renno, M. Towner, A. Spiga, H. Elliott, High-resolution dust devil sampling for sediment loads, wind speeds, temperature and pressure excursions, in *42nd Lunar and Planetary Science Conference*. Texas (2011). Abstr. #1608

G.J. Molina-Cuberos et al., A new approach for estimating Titan's electrical conductivity based on data from relaxation probe sensors on the Huygens experiment. Planet. Space Sci. **58**(14–15) (2010). doi:10.1016/j.pss.2010.09.014

J.E. Moores et al., Observational evidence of a suppressed planetary boundary layer in northern Gale Crater, Mars as seen by the Navcam instrument onboard the Mars Science Laboratory rover. Icarus **249**(15), 129–142 (2015). doi:10.1016/j.icarus.2014.09.02

J.R. Murphy, S. Nelli, Mars Pathfinder convective vortices: frequency of occurrence. Geophys. Res. Letters **29**(23) (2002). doi:10.1029/2002GL015214

J. Murphy, C.B. Leovy, J. Tillman, Observations of martian surface winds at the Viking Lander 1 site. J. Geophys. Res. **95**, B9 (1990). doi:10.1029/JB095iB09o14555

L. Neakrase et al., Particle lifting processes in dust devils. Space Sci. Rev. (2016, this issue)

L.D.V. Neakrase, R. Greeley, Dust devil sediment flux on Earth and Mars: laboratory simulations. Icarus **206**, 306–318 (2010). doi:10.1016/j.icarus.2009.08.028

A.M.C. Oke, D. Dunkerley, N.J. Tapper, Willy-willies in the Australian landscape: sediment transport characteristics. J. Arid Environ. **71**, 216–228 (2007)

A.V. Pathare, M.R. Balme, S.M. Metzger, A. Spiga, M.C. Towner, N.O. Renno, F. Saca, Assessing the power law hypothesis for the size-frequency distribution of terrestrial and martian dust devils. Icarus **209**(2), 851–853 (2010)

J.B. Pollack, D.S. Coburn, F.M. Flasar, R. Kahn, C.E. Carslton, D. Pidek, Properties and effects of dust particles suspended in the martian atmosphere. J. Geophys. Res. **84**(B6), 2929–2945 (1979)

A. Powell, Theory of vortex sound. J. Acoust. Soc. Am. **36**(1), 177–195 (1964)

J. Raack, D. Reiss, G.G. Ori, K. Taj-Eddine, Vertical grain size distributions in dust devils: analyses of in situ sampled from southern Morocco. EPSC Abstract, EPSC2012-427-1, 2014

S. Rafkin, D. Banfield, J. Silver, K. Nowicki, R. Dissly, A. Stanton, in *An Instrument to Measure Turbulent Fluxes in the Atmosphere of Mars and Other Planets, European Planetary Science Congress 2013*. EPSC Abstracts, vol. 8, London, UK (2013), EPSC2013-575, Poster presentation, http://meetingorganizer.copernicus.org/EPSC2013/EPSC2013-575.pdf

D. Reiss, A. Spiga, G. Erkeling, The horizontal motion of dust devils on Mars derived from CRISM and CTX/HiRISE observations. Icarus **277**, 8–20 (2014)

N.O. Renno, M.L. Burkett, M.O. Larkin, A simple thermodynamical theory for dust devils. J. Atmos. Sci. **55**, 3244–3252 (1998)

N.O. Renno, A.A. Nash, J. Lunine, J. Murphy, Martian and terrestrial dust devils: test of a scaling theory using Pathfinder data. J. Geophys. Res. **105**(E1), 1859–1865 (2000)

N.O. Renno, V.J. Abreu, J. Koch, P.H. Smith, o.K. Hartogensis, H.A.R. De Bruin, D. Burose, G.T. Delory, W.M. Farrell, C.J. Watts, J. Garatuza, M. Parker, A. Carswell, MATADOR 2002: a pilot field experiment on convective plumes and dust devils. J. Geophys. Res. **109**, E07001 (2004). doi:10.1029/2003JE002219

T.J. Ringrose, M.C. Towner, J.C. Zarnecki, Convective vortices on Mars: a reanalysis of Viking Lander 2 meteorological data, sols 1–60. Icarus **163**(1), 78–87 (2003). doi:10.1016/S0019-1035(03)00073-3

T.J. Ringrose, M. Patel, M. Towner, M. Balme, S. Metzger, J. Zarnecki, The meteorological signatures of dust devils on Mars. Planet. Space Sci. **55**, 14 (2007)

J.A. Rodriguez-Manfredi et al., MEDA: an environmental and meteorological package for Mars 2020, in *45th Lunar and Planetary Science Conference*, The Woodlands, Texas, USA (2014). Poster presentation, http://ssed.gsfc.nasa.gov/IPM/PDF/1125.pdf

J.A. Ryan, J.J. Carroll, Dust devil wind velocities: mature state. J. Geophys. Res. **75**, 531–541 (1970)

J.A. Ryan, R.D. Lucic, Possible dust devils, vortices on Mars. J. Geophys. Res. **88**(C15), 11005–11011 (1983)

D.S. Schmidt, R.A. Schmidt, J.D. Dent, Electrostatic force on saltating sand. J. Geophys. Res. **103**(D8), 8997–9001 (1998)

J.T. Schofield, J.R. Barnes, D. Crisp, R.M. Haberle, S. Larsen, J.A. Magalhaes, J.R. Murphy, A. Seiff, G. Wilson, The Mars Pathfinder atmospheric structure investigation meteorology (ASI/MET) experiment. Science **278**(5344), 1752–1758 (1997). doi:10.1126/science.278.5344.1752

R.L. Schwiesow, R.E. Cupp, M.J. Post, P.C. Sinclair, R.F. Abbey, Velocity structures of waterspouts and dust devils as revealed by Doppler lidar measurements. Bull. Am. Meteorol. Soc. **58**, 677 (1977)

A. Seiff, J.E. Tillman, J.R. Murphy, J.T. Schofield, D. Crisp, J.R. Barnes, C. LaBaw, C. Mahoney, J.D. Mihalov, G.R. Wilson, R. Haberle, The atmosphere structure and meteorology instrument on the Mars Pathfinder lander. J. Geophys. Res. **103**(E2), 4045–4056 (1997)

Y.P. Shao, *Physics and Modelling of Wind Erosion*, 2nd edn. (Springer, Heidelberg, 2008)

Y. Shao, M.R. Raupach, P.A. Findlater, Effect of saltation bombardment on the entrainment of dust by wind. J. Geophys. Res. **98**, 12719–12726 (1993)

P.C. Sinclair, On the rotation of dust devils. Bull. Am. Meteorol. Soc. **46**, 388–391 (1965)

P.C. Sinclair, A quantitative analysis of the dust devil. Ph.D. Dissertation, University of Arizona, 1966, p. 292

P. Sinclair, General characteristics of dust devils. J. Appl. Meteorol. **8**, 32–45 (1969)

P. Sinclair, The lower structure of dust devils. J. Atmos. Sci. **30**, 1599–1619 (1973)

P.H. Smith et al., Introduction to special section on the Phoenix Mission: landing site characterization experiments, mission overviews, and expected science. J. Geophys. Res. **113**(E3) (2008)

P.H. Smith, M.T. Lemmon, Opacity of the Mars atmosphere measured by the imager for Mars Pathfinder. J. Geophys. Res. **104**, 8975–8985 (1999)

P.H. Smith, M.G. Tomasko, D. Britt, D.G. Crowe, R. Reid, H.U. Keller, N. Thomas, F. Gliem, P. Rueffer, R. Sullivan, R. Greeley, J.M. Knudsen, M.B. Madsen, H.P. Gunnlaugsson, S.F. Hviid, W. Goetz, L.A. Soderblom, L. Gaddis, R. Kirk, The imager for Mars Pathfinder experiment. J. Geophys. Res. **102**(E2), 4003–4025 (1997)

M.D. Smith, M.J. Wolff, M.T. Lemmon, N. Spanovich, D. Banfield, C.J. Budney, R.T. Clancy, A. Ghosh, G.A. Landis, P. Smith, B. Whitney, P.R. Christensen, S.W. Squyres, First atmospheric sciences results from the Mars Exploration Rovers mini-TES. Science **306**(5702), 1750–1753 (2004). doi:10.1126/science.1104527

P.H. Smith, L.K. Tamppari, R.E. Arvidson, D. Bass, D. Blaney, W.V. Boynton, A. Carswell, D.C. Catling, B.C. Clark, T. Duck, E. DeJong, D. Fisher, W. Goetz, H.P. Gunnlaugsson, M.H. Hecht, V. Hipkin, J. Hoffman, S.F. Hviid, H.U. Keller, S.P. Kounaves, C.F. Lange, M.T. Lemmon, M.B. Madsen, M. Malin, W.J. Markiewicz, J. Marshall, C.P. McKay, M.T. Mellon, D.W. Ming, R.V. Morris, N. Renno, W.T. Pike, U. Staufer, C. Stoker, P. Taylor, J. Whiteway, A.P. Zent, Water at the Phoenix landing site. Science **325**, 58–61 (2009)

J.T. Snow, T.M. McClelland, Dust devils at white sands missile range, New Mexico: 1. Temporal and spatial distributions. J. Geophys. Res. **95**, 13,707–13,721 (1990)

G.G. Sorrells, A preliminary investigation into the relationship between long-period noise and local fluctuations in the atmospheric pressure field. Geophys. J. **26**, 71–82 (1971)

G.G. Sorrells, J.A. McDonald, Z.A. Der, E. Herrin, Earth motion caused by local atmospheric pressure changes. Geophys. J. **26**, 83–98 (1971)

S.W. Squyres et al., The Spirit rover's Athena science investigation at Gusev Crater, Mars. Science **305**, 794–799 (2004)

K. Steakley, J. Murphy, A year of convective vortex activity at Gale Crater. Icarus **278**, 180–193 (2016)

R. Sullivan, R. Greeley, M. Kraft, G. Wilson, M. Golombek, K. Herkenhoff, J. Murphy, P. Smith, Results of the imager for Mars Pathfinder windsock experiment. J. Geophys. Res. **105**(E10), 24547–24562 (2000). doi:10.1029/1999JE001234

P.A. Taylor, D.C. Catling, M. Daly, C.S. Dickinson, H.P. Gunnlaugsson, A.M. Harri, C.F. Lange, Temperature, pressure, and wind instrumentation in the Phoenix meteorological package. J. Geophys. Res. **113**, E00A10 (2008). doi:10.1029/2007JE003015

P.A. Taylor et al., On pressure measurement and seasonal pressure variations during the Phoenix mission. J. Geophys. Res. **115**, E00E15 (2010). doi:10.1029/2009JE003422

J.E. Tillman, N.C. Johnson, P. Guttorp, D.B. Percival, The martian annual atmospheric pressure cycle: years without great dust storms. J. Geophys. Res. **98**(E6), 10963–10971 (1993). doi:10.1029/93JE01084

A.D. Toigo, M.I. Richardson, S.P. Ewals, P.J. Gierasch, Numerical simulation of martian dust devils. J. Geophys. Res., Planets **108**(E6) (2003)

M.C. Towner, Characteristics of large martian dust devils using Mars odyssey thermal emission imaging system visual and infrared images. J. Geophys. Res. **114**, E02010 (2008)

D.M. Tratt, M.H. Hecht, D. Catling, E.C. Samulon, P.H. Smith, In situ measurement of dust devil dynamics: toward a strategy for Mars. J. Geophys. Res. **108**(E11), 5116 (2003)

D.T. Tyler, J.R. Barnes, Mesoscale modeling of the circulation in the Gale Crater region: an investigation into the complex forcing of convective boundary layer depths. Mars **8**, 58–77 (2013). doi:10.1555/mars.2013.0003

D.T. Tyler, J.R. Barnes, Convergent crater circulations on Mars: influence on the surface pressure cycle and the depth of the convective boundary layer. Geophys. Res. Lett. **42**, 7343–7350 (2015)

J. Whiteway, M. Daly, A. Carswell, T. Duck, C. Dickinson, L. Komguem, C. Cook, Lidar on the Phoenix mission to Mars. J. Geophys. Res. **113**, E00A08 (2008). doi:10.1029/2007JE003002

N.R. Williams, Development of dust whirls and similar small-scale vortices. Bull. Am. Meteorol. Soc. **29**, 106–117 (1948)

J.P. Williams, Acoustic environment of the martian atmosphere. J. Geophys. Res. **106**(E3), 5033–5041 (2001)

C.L. Wilson, A. Davis, D. Hutchins, M.C. Towner, An ultrasonic anemometer for Mars. J. Acoust. Soc. Am. **123**(5), 3401 (2008). doi:10.1121/1.2934100

R.E. Wyett, Pressure drop in a dust devil. Mon. Weather Rev., 7–8 January 1954

L.M. Zelenyi, O.I. Korablev, D.S. Rodionov, B.S. Novikov, K.I. Marchenkov, O.N. Andreev, E.V. Larionov, Scientific objectives of the scientific equipment of the landing platform of the ExoMars-2018 mission. Sol. Syst. Res. **49**(7), 509–517 (2015)

M-P. Zorzano, F.J. Martín-Torres, H. Kahanpää, J. Moores, S. Navarro, A. Lepinette, E. Sebastian, J. Gómez-Elvira, REMS Team, The MSL Science Team), Radiation obscuration by dust devils at Gale as observed by the REMS UV sensor, in *EGU General Assembly 2013*. Geophysical Research Abstracts, vol. 15 (2013), EGU2013-11155, http://meetingorganizer.copernicus.org/EGU2013/EGU2013-11155.pdf

Space Sci Rev (2016) 203:89–142
DOI 10.1007/s11214-016-0243-6

Orbital Observations of Dust Lofted by Daytime Convective Turbulence

Lori Fenton[1] · Dennis Reiss[2] · Mark Lemmon[3] · Béatrice Marticorena[4] · Stephen Lewis[5] · Bruce Cantor[6]

Received: 26 October 2015 / Accepted: 29 February 2016 / Published online: 31 March 2016

Abstract Over the past several decades, orbital observations of lofted dust have revealed the importance of mineral aerosols as a climate forcing mechanism on both Earth and Mars. Increasingly detailed and diverse data sets have provided an ever-improving understanding of dust sources, transport pathways, and sinks on both planets, but the role of dust in modulating atmospheric processes is complex and not always well understood. We present a review of orbital observations of entrained dust on Earth and Mars, particularly that produced by the dust-laden structures produced by daytime convective turbulence called "dust devils". On Earth, dust devils are thought to contribute only a small fraction of the atmospheric dust budget; accordingly, there are not yet any published accounts of their occurrence from orbit. In contrast, dust devils on Mars are thought to account for several tens of percent of the planet's atmospheric dust budget; the literature regarding martian dust devils is quite rich.

✉ L. Fenton
lfenton@seti.org

D. Reiss
dennis.reiss@uni-muenster.de

M. Lemmon
lemmon@tamu.edu

B. Marticorena
Beatrice.Marticorena@lisa.u-pec.fr

S. Lewis
stephen.lewis@open.ac.uk

B. Cantor
cantor@msss.com

1 SETI Institute, 189 Bernardo Ave. Suite 100, Mountain View, CA, 94043, USA

2 Institut für Planetologie, WWU Münster, Münster, Germany

3 Texas A&M University, College Station, USA

4 Laboratoire Interuniversitaire des Systèmes Atmosphériques (LISA), Créteil, France

5 The Open University, Milton Keynes, UK

6 Malin Space Science Systems, San Diego, USA

Because terrestrial dust devils may temporarily contribute significantly to local dust loading and lowered air quality, we suggest that martian dust devil studies may inform future studies of convectively-lofted dust on Earth.

As on Earth, martian dust devils form most commonly when the insolation reaches its daily and seasonal peak and where a source of loose dust is plentiful. However this pattern is modulated by variations in weather, albedo, or topography, which produce turbulence that can either enhance or suppress dust devil formation. For reasons not well understood, when measured from orbit, martian dust devil characteristics (dimensions, and translational and rotational speeds) are often much larger than those measured from the ground on both Earth and Mars. Studies connecting orbital observations to those from the surface are needed to bridge this gap in understanding. Martian dust devils have been used to remotely probe conditions in the PBL (e.g., CBL depth, wind velocity); the same could be done in remote locations on Earth. Finally, martian dust devils appear to play a major role in the dust cycle, waxing and waning in relative importance and spatial patterns of occurrence with the planet's orbital state. Orbital studies of terrestrial dust devils would provide a basis for comparative planetology that would broaden the understanding of these dusty vortices on both planets.

Keywords Atmospheric dust · Dust devil · Mars · Dust storm · Boundary layer

Abbreviations

ADEOS	Advanced Earth Observing Satellite
AOT	aerosol optical thickness
ASTER	Advanced Spaceborne Thermal Emission Spectrometer
AVHRR	Advanced Very High Resolution Radiometer
CALIOP	Cloud-Aerosol Lidar with Orthogonal Polarization
CALIPSO	Cloud-Aerosol Lidar and Infrared Pathfinder Satellite Observations
CBL	convective boundary layer
CTX	Context Camera
DD	dust devil
DOD	dust optical depth
DOT	dust optical thickness
EPF	emission phase function
EY	Earth year
FOV	field of view
GCM	global circulation model
GLAS	Geoscience Laser Altimeter System
GLI	Global Imager
GOCART	Goddard Chemistry Aerosol Radiation and Transport
HiRISE	High Resolution Imaging Science Experiment
HRSC	High Resolution Stereo Camera
ICESat	Ice, Cloud and land Elevation Satellite
IR	infrared
IRIS	Infrared Interferometric Spectrometer
IRTM	Infrared Thermal Mapper
LITE	Lidar In-Space Technology Experiment
LW	long wave
MARCI	Mars Reconnaissance Orbiter Mars Color Imager
MCS	Mars Climate Sounder

MEX	Mars Express
MGS	Mars Global Surveyor
MOC NA	Mars Orbiter Camera Narrow Angle
MOC WA	Mars Orbiter Camera Wide Angle
MODIS	Moderate-resolution Imaging Spectro-radiometer
MOLA	Mars Orbiter Laser Altimeter
MRO	Mars Reconnaissance Orbiter
MSG	Meteosat Second Generation
MSL	Mars Science Laboratory Multi-angle Imaging Spectro-Radiometer
MY	Mars Year
MVIRI	Meteosat Visible Infra-Red Imager
NOAA	National Oceanic and Atmospheric Administration
ODY	Mars Odyssey
PBL	planetary boundary layer
POLDER	Polarization and Directionality of the Earth's Reflectances
PSD	particle size distribution
SeaWiFS	Sea-viewing Wide Field of view Sensor
SRC	Super-resolution Camera
SEVIRI	Spinning Enhanced Visible Infra-Red Imager
SPICAM	Spectroscopy for Investigation of Characteristics of the Atmosphere of Mars
SW	short wave
TES	Thermal Emission Spectrometer
THEMIS IR	Thermal Emission Imaging System, Infrared camera
THEMIS VIS	THEMIS Visible camera
TIROS	Television and Infra-Red Observation Satellite
UV	ultraviolet
VCS-MA	Vidicon Camera System—Medium Angle
VIS: TOMS	Total Ozone Mapping Spectrometer Visual Imaging Subsystems
VL1	Viking Lander 1
VL2	Viking Lander 2
VO1	Viking Orbiter 1
VO2	Viking Orbiter 2

1 Introduction

Spaceborne observations of lofted dust began with the first weather satellites in the 1960s (see Fig. 1a). Until recently, most such phenomena have been associated with dust hazes and smog that have been transported far from their source regions (e.g., United States 1964), but new data have highlighted a complex interplay between dust emission and daytime dry convective turbulence in the planetary boundary layer. The CBL is composed of structured turbulent eddies, containing tens-of-meter-scale vortices that form most commonly in narrow updrafts at the intersections of three or more kilometer-scale convection cells (e.g., Willis and Deardorff 1979; Hess and Spillane 1990; Kanak et al. 2000). When dust-laden, these vortices become visible to the eye as dust devils (DDs). Non-rotating gusts may also entrain dust, likely occurring most commonly along upwelling sheets where two convection cells meet. For a detailed review of the meteorological context of convectively-lofted dust, we refer the reader to Chaps. 5 and 7 of this volume.

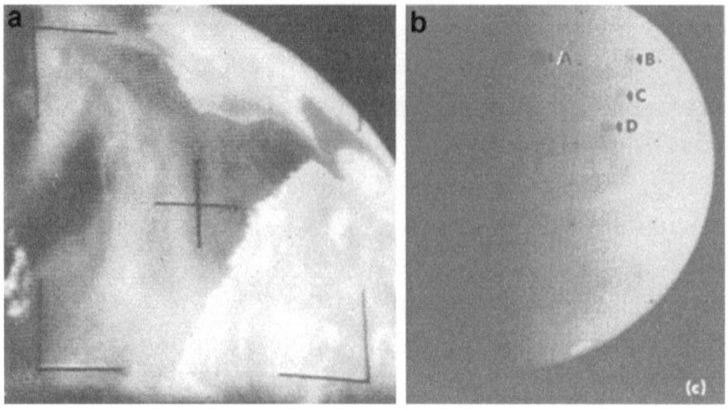

Fig. 1 Early spacecraft images of dust storms. (**a**) An early image of dust over the Arabian Sea and Gulf of Oman; the view is towards the southwest, with Iran and Pakistan on the right and the Arabian Peninsula on the left. Image from Schnapf (1964). (**b**) In 1971, Mariner 9 entered orbit around Mars during an intense, global dust storm, that obscured all surface features except for four tall volcano peaks (labeled A through D) and the south polar cap (white spot at lower right). Image from Masursky et al. (1972)

On Earth, the amount of dust lofted by daytime convective turbulence, mainly by DDs, was first estimated by Koch and Renno (2005) to be ~0.7 Tg/year, or 34 ± 19 % of the global terrestrial mineral dust budget. Jemmett-Smith et al. (2015) revised this estimate down to only ~3.4 % of the global terrestrial mineral dust budget by refining estimates of the temporal and spatial occurrence of dust entrainment. Although small, this input could be significant on a regional scale (e.g., Gillette and Sinclair 1990; Jemmett-Smith et al. 2015), with potentially significant environmental consequences and hazards (Goudie and Middleton 2006). DDs have not yet been identified in images obtained from terrestrial satellites, although the tracks they sweep out on the surface have been studied (see Sect. 3.1 and Chap. 4). The continued monitoring of such features from orbital platforms opens new avenues of research that, informed by the extensive studies performed on Mars, could prove to be of use to the field of terrestrial climate science.

Dust on Mars has long been known from Earth-based telescopic observations; Martin and Zurek (1993) summarized observations of "yellow clouds" dating as far back as 1873. These phenomena were correctly attributed to lofted dust, but the determining details of their formation, development, and dissipation required a closer inspection from orbiting spacecraft. The best first look at martian dust from space is that from Mariner 9, which entered orbit around Mars in November 1971, in the midst of one of the most intense planet-encircling dust storms on record (see Fig. 1b). Features related to dust entrained by sub-kilometer-scale daytime convective turbulence, such as DDs, were not expected to be resolved in images from 1970s-era spacecraft; as a result they were only identified many years later after careful inspection of these data sets (Thomas and Gierasch 1985).

We present a review of atmospheric dust research from orbital spacecraft, both on Earth and Mars, with a particular emphasis on DDs. This chapter is complementary to a review of field measurements, which can be found in Chap. 2. We first summarize the orbital platforms used and the observed spatial and temporal patterns of dust lifting and transport, in part to provide a context for the DD studies, but also because instrument capabilities dictate what may be learned from the data sets they produce. We then describe the current body of literature on orbital observations of martian DDs, including regional and global surveys, spatial and temporal patterns of their occurrence, physical characteristics, their relation to the martian dust cycle, and the potential role of convectively-lofted dust as a climate forcing mechanism (a more detailed discussion of how dust lofted by DDs relates to the climate system can be found in Chap. 11). The conclusions begin with a comparison of global mean estimates of dust load, the contribution contributed by DDs, and the DOT, providing a high-level comparison of the quantity of atmospheric dust on Earth and Mars. This is followed by a summary of the major knowledge gaps that could be addressed with use of orbital data. In particular, we emphasize that, although observations of terrestrial DDs from space have yet to be reported, the extensive surveys and detailed investigations from Mars suggest that similar work on Earth could be quite informative.

The martian year is 668.6 sols (martian days) long, nearly twice as long as that of the Earth. An annual "calendar" for Mars is denoted by the solar longitude, or L_s, in which the year begins at the northern vernal equinox ($L_s = 0°$) and circumscribes its orbit around the sun over the following 360° (i.e., northern summer solstice occurs at $L_s = 90°$, autumnal equinox occurs at $L_s = 180°$, and northern winter solstice occurs at $L_s = 270°$). Mars years (MY) are numbered beginning with MY 1 on 11 April, 1955, following the convention of Clancy et al. (2000); a convenient tool for converting Earth dates to Martian dates can be found at http://www-mars.lmd.jussieu.fr/mars/time/martian_time.html. Because the martian sol is 24 hours, 39 minutes, and ~35.2 seconds long, the convention is to divide the day into 24 "hours" that are ~3698.7 seconds in duration. Lacking oceans, Mars has no equivalent elevation for sea level, so the topographic datum is defined as the equipotential surface at the mean equatorial radius (3396.2 km), which is located at a pressure level of ~520 Pa at $L_s = 0°$ (Smith et al. 2001b).

2 Orbital Measurements of Lofted Dust

2.1 Earth

2.1.1 Aerosols on Earth

Compared to Mars, sources of atmospheric particulate matter (aerosols) are numerous on the Earth. Primary aerosols, directly emitted as particles, are distinguished from secondary aerosols, resulting from chemical or physical transformation of gaseous precursors (e.g., Boucher et al. 2013). Primary aerosols are produced by the mechanical action of the atmosphere on the surface: over the continents, it produces mineral dust (see Fig. 2); over the sea, it produces sea salts. Secondary aerosols are produced by the combustion of biomass and all types of fuels, as well as by chemical reactions of gases naturally emitted by vegetation, oceanic surfaces, and volcanic activity. In terms of radiative forcing and health impacts, most of the scientific attention is focused on secondary aerosols resulting from human activities, because of the aerosols' size (mostly $<\sim 1$ μm) and composition. In terms of annual emissions, mineral dust (emitted in arid and semi-arid regions) and sea-salts are the most abundant aerosols on the Earth, with annual emissions estimated respectively to 1000–4000 Tg and 1400–6800 Tg (Boucher et al. 2013). Total anthropogenic annual emissions are of the order of 400 Tg, whereas the atmospheric input of cosmic dust in the terrestrial atmosphere is estimated to be 0.0018 to 0.1 $\mathrm{Tg\,yr^{-1}}$, based on daily input reported by Plane (2012). Aerosols emitted in the terrestrial troposphere have lifetimes of a few days, indicating that the concentrations of the different aerosols are not homogeneously distributed. Aerosols are often subjected to transport ranging from tens to thousands of kilometers, but the highest concentrations are observed within and immediately downwind of their source regions. In addition, different aerosol sources are located in different geographic areas, producing regions with different dominant aerosol types: mineral dust is prevalent downwind of the main desert areas (North Africa, Asia, etc.), sea salts are common over sea-surfaces and

Fig. 2 Visible MODIS images of dust plumes over (**a**) the Yellow Sea and (**b**) downwind of North Africa. Adapted from Marticorena and Formenti (2013)

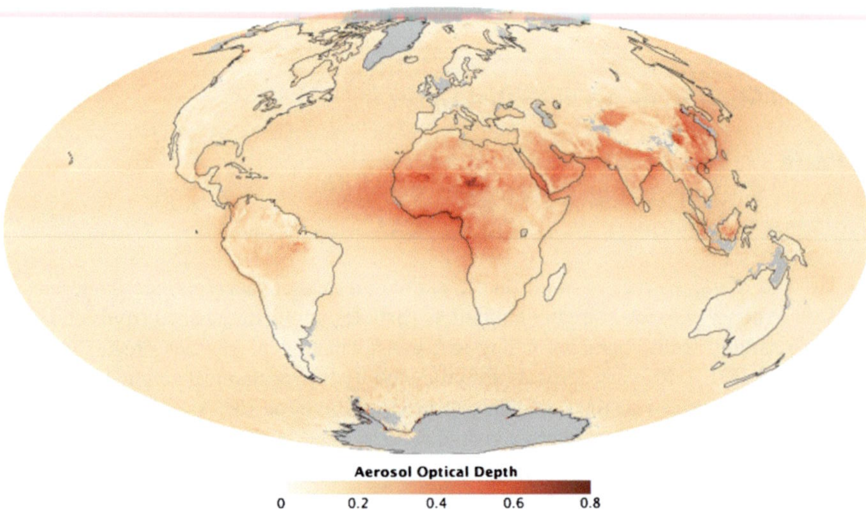

Fig. 3 Mean Aerosol Optical Depth (AOD, which is used interchangeably with AOT) from June 2000 through May 2010 from MISR. (NASA map by Robert Simmon, based on MISR data)

remote coastal areas, and anthropogenic aerosols form over the developed and developing countries of the northern hemisphere. Figure 3 shows a global map of AOT retrieved from MISR data over a 5 year period, showing regions of strong aerosol emission on continents and its transport over both land and water.

Different aerosols interact with radiation in different ways: many of them (e.g., sulfates, nitrates, organic aerosols) primarily backscatter solar radiation, but a few types have strong absorption properties (e.g., carbonaceous aerosols). Mineral dust has the capacity not only to backscatter radiation in the visible range, but also to absorb radiation in the IR and UV range (e.g., Redmond et al. 2010). Regarding the variety of different aerosols and the geographic distribution of their sources, detection from orbital platforms is a powerful tool for investigating the spatial and temporal variability of the atmospheric content (Lenoble et al. 2013; Chiapello 2014).

2.1.2 *Orbital Platforms*

Quantitative measurements of the atmospheric aerosol load, and of particular interest to this review, mineral dust, can be derived from satellite measurements of backscattered radiation; many commonly-used instruments are listed in Table 1. The columnar extinction of solar radiation by atmospheric aerosols is quantified by the AOT (or equivalently, the AOD). For a given aerosol type, and assuming homogeneous properties exist along the atmospheric column, the AOT is proportional to the vertically-integrated atmospheric concentration weighted by the extinction efficiency. The extinction efficiency is itself largely controlled by the aerosol size distribution and composition.

The first operational algorithms to detect and quantify the aerosol atmospheric load from instruments onboard satellite platforms were developed in the 1990s and applied to observations from weather satellites: the European Meteosat (Jankowiak and Tanré 1992) and the American NOAA AVHRR (Swap et al. 1996; Husar et al. 1997). The AOT derived at the global scale revealed that desert dust is responsible for the largest and most persistent aerosol loads over the world's oceans (Herman et al. 1997; Husar et al. 1997).

For many years, the retrieval of AOT was restricted to surfaces with low albedo, and in particular, to oceanic surfaces. As an alternative, indicators of the presence of an absorbing aerosol (carbonaceous aerosols and mineral dust) have been developed based on measurements both in the UV (Herman et al. 1997; Torres et al. 1998) and in the IR (Legrand et al. 1994; 2001). These aerosol indices have been used widely, in particular for mineral dust source identification (i.e., Prospero et al. 2002). Sensors of novel generation, with spectral capabilities or additional types of measurements (i.e. polarization, several view angles), have significantly increased the capacity to characterize the different aerosol types and to retrieve AOT over land surfaces (e.g., Martonchik et al. 2004). As an example, the MODIS data subsets from Collection 5 includes the retrieval of aerosols over "dark targets", i.e., surfaces dark enough to enable the separation of the surface and aerosol signals, such as vegetated areas. For mineral dust studies, the recent developments for the retrieval of AOT over bright desert surfaces (e.g., the "Deep Blue" algorithm by Hsu et al. 2004; 2013) offers new perspectives on the investigation of dust emission and dust storms inside and close to source regions. AOT can also be retrieved from IR observations of the SEVIRI instruments onboard the Meteosat Second Generation (Banks and Brindley 2013; Carrer et al. 2014).

Retrieved AOTs may differ depending on the instrument and the algorithm. In addition, temporally-averaged AOTs can vary from one sensor to another depending on the temporal sampling. Typically, satellites in geostationary orbits provide higher sampling rates than polar-orbiting satellites, but for limited regions. Another limitation is that AOT retrieval from satellites is still generally constrained to clear-sky conditions.

Most of the algorithms applied to satellite sensors allow the retrieval of the AOT with an uncertainty that has been significantly reduced from that of early sensors (Meteosat, AVHRR) relative to that of the current satellite missions, which are dedicated to aerosol research (e.g., sensors from the A-Train: MODIS, MISR, POLDER). These new sensors allow the retrieval of additional parameters that provide information on aerosol size, shape or optical properties. The spectral dependence of the AOT, known as the Angström coefficient, can be used to discriminate aerosols of different size distributions. Aerosols with a significant coarse mode, such as mineral dust or sea salt, have Angström coefficient values close

Table 1 Instruments on spacecraft orbiting Earth, discussed in the text, that are or were commonly used to study atmospheric dust

Mission: Instrument	Spectral range (μm)	Spatial resolution (m/px)	Footprint area (km^2) or coverage	Local time
TIROS-6 (1962–1963):				
VCS-MA	1 band: visible	2500	518×10^3	Various
Meteosat-1 to Meteosat-7:				
MVIRI	0.5–0.9 μm 5.7–7.1 μm 10.5–12.5 μm	2500 5000 5000	Full disk	Every 30 min
Meteosat-8 to Meteosat-11:				
SEVIRI	1 band: 0.6–0.9 μm 12 bands: 0.56–14.4 μm	1000 3000	Full disk	Every 15 min
NOAA-6, NOAA-8, NOAA-10, TIROS-N:				
AVHRR	1 band: 0.55–0.68 μm 3 bands: 0.725–11.5 μm	1100	Global: 1x/day Global: 2x/day	07:30, 14:30
NOAA-7, NOAA-9, NOAA-11 to NOAA-14:				
AVHRR/2	1 band: 0.55–0.68 μm 4 bands: 0.725–12.5 μm	1100	Global: 1x/day Global: 2x/day	Various
NOAA-15 to NOAA-19, Metop-A to Metop-C:				
AVHRR/3	1 band: 0.58–0.68 μm 5 bands: 0.725–12.5 μm	1100	Global: 1x/day Global: 2x/day	Various
ADEOS, Meteor-3 to Meteor-5, Nimbus-7, TOMS Earth Probe:				
TOMS	6 bands: 312.5–380 nm	50 km	Global: 1x/day	Various
Terra, Aqua:				
MODIS	2 bands: 645, 858 nm 5 bands: 0.469–2.130 μm 29 bands: 0.412–14.235 μm	250 500 1000	Global: 1x/day Global: 1x/day Global: 2x/day	10:30, 13:30
MISR (terra)	4 bands: 0.4464–0.8664 μm	250–275	Global: 1x/9 days	10:30
OrbView-2:				
SeaWiFS	8 bands: 0.412–0.865 μm	1100, 4500	Global: 1x/day	12:00
ADEOS-2:				
GLI	29 bands: 0.38–2.21 μm 6 bands: 3.715–12.0 μm	250 1000	Global: 0.5x/day Global: 1x/day	10:30
ADEOS, ADEOS-2, PARASOL:				
POLDER	9 bands: 0.444–10.2 μm 3 polarizations	6500	~Global: 1x/day	Various
ICESat:				
GLAS	532, 1064 nm	Horiz.: 66 Vert.: 76.8	–	Various
Space Shuttle Discovery:				
LITE	355, 532, 1064 nm	Horiz.: 300 Vert.: 15	–	Various
CALIPSO:				
CALIOP	532, 1064 nm 2 polarizations	Horiz: 70 Vert: 30	–	13:30

to 0, whereas aerosols dominated by fine-mode particles, such as particles from fossil fuel combustion and biomass burning, have Angström coefficients higher than 1. Simultaneous information on AOT and Angström coefficients can thus be used to estimate the atmospheric load of mineral dust containing large amounts of coarse particles. Some of the recent sensors enable discrimination of the contribution of fine and coarse modes to the total AOT (e.g., MODIS), which, for example, helps to distinguish fine pollution aerosols from coarse mineral dust. Aerosol retrieval from the POLDER instrument can be used to distinguish spherical and non-spherical aerosols within the coarse mode (Herman et al. 2005), and thus further refine the detection of mineral dust (Tanré et al. 2011; Peyridieu et al. 2010). Several parameters (Angström exponent, size fraction, absorption), especially those derived from MODIS data, can be combined to separate the different types of aerosols (Kaufman et al. 2005; Ginoux et al. 2012). However, the unambiguous identification of dust is still challenging, particularly in regions where dust can be mixed with other optically active species (e.g., biomass burning aerosols). It must also be noted that most of these aerosol products are available from instruments onboard polar-orbiting platforms, providing one observation per day at one time of the day. This orbital configuration is relevant for documenting medium to long-range transport of mineral dust, identifying source regions, and for climatological studies, but it is not well-suited to monitoring specific dust storms at regional and local scales. Spatial resolution is also an issue, with the AOT collection most frequently binned to a spatial resolution on of the order of 1°. The spatial and temporal resolution provided by geostationary platforms is much higher. Color composites of observations from SEVIRI instruments onboard the MSG satellites have been widely used to monitor dust events over the Sahara and the Sahel either for large continental dust storms (e.g. Slingo et al. 2006) or to identify dust plumes associated with mesoscale convective systems (e.g. Marticorena et al. 2010). The AOT over land, including desert surfaces, can be retrieved from SEVIRI measurements with a nadir spatial resolution of 3 km, as well as an extra high-resolution visible channel at a nadir resolution of 1 km (Banks and Brindley 2013).

In parallel with the development of new, but vertically-integrated, aerosol products, lidar techniques have been developed both from ground-based and airborne platforms. They provide the unique opportunity to document the vertical structure of mineral dust distribution. The first lidar observations of dust from space were provided by LITE, which flew on the Space Shuttle Discovery in 1994 (e.g, Berthier et al. 2006), and GLAS, on ICEsat (Spinhirne et al. 2005), although it is interesting to note that these sensors were preceded by the original MOLA on the failed Mars Orbiter mission (Zuber et al. 1992). Since 2006, lidar aerosol and cloud observations have been available from the CALIPSO mission (Winker et al. 2010). The CALIOP lidar onboard CALIPSO measures aerosol profiles with a 30 m vertical resolution and 70 m horizontal resolution, and thanks to depolarization measurements, it enables aerosol classification, including identification of nonspherical particles typified by mineral dust (Omar et al. 2009).

2.1.3 *Spatial and Temporal Patterns of Dust Lifting*

From satellite observations, significant progress has been made regarding specific dust transport events, including determining their regional transport patterns, seasonal and interannual variability, and long-term trends in relation to climatic conditions or change in anthropogenic pressure. Climatology of aerosols and/or mineral dust has been extensively investigated with different aerosol products, from the "historical" weather sensors (Meteosat, AVHRR) and from dedicated sensors (e.g., Remer et al. 2008). The transport of mineral dust from the Sahara, the most intense dust source on Earth, has been the most frequently studied, allowing, for example, identification of the synoptic conditions associated with its transport pattern over the Atlantic Ocean (Huang et al. 2010). Lidar observations from CALIOP have been used to estimate the flux of the Saharan dust exported along this pathway (Yu et al. 2015). A re-analysis of the SeaWiFS decadal data set has been performed with the "Deep Blue" algorithm to analyze the trends in the variation of aerosols, including mineral dust, both over oceanic and continental surfaces of the world (Hsu et al. 2012). Since AOT retrieval can be performed only in clear-sky conditions, dust transport from other source regions can be difficult to detect. This is the case, for example, for dust coming from sources at high latitudes in the southern hemisphere, where the viewing geometry is not always ideal and bright sea ice interferes with AOT retrieval (e.g., Gassó and Stein 2007). However, dust has also been detected by remote sensing at high latitudes, such as in Alaska, New Zealand or Iceland (Prospero et al. 2012). The distribution of the main sources of desert dust on the Earth has been investigated using the TOMS aerosol index with a spatial resolution of $1^\circ \times 1.25^\circ$ (Prospero et al. 2002). To investigate the link between dust emission occurrence and meteorological processes, Schepanski et al. (2009) used a combination of aerosol index and AOT retrieval in the IR with a 15-min timescale at $1^\circ \times 1^\circ$ resolution. Over time, there has been a trend toward the production of ever-higher resolution products, largely motivated by the need for aerosol retrievals over continental urban areas for air quality applications. As an illustration, the Collection 6 algorithm that produces MODIS aerosol products includes, in addition to the standard retrieval at a resolution of 10 km × 10 km, a new product with a resolution of 3 km × 3 km (Levy et al. 2013). The AOTs derived from SEVIRI are also available in the native geographical projection of MSG/SEVIRI (AERUS-GEO product, Carrer et al. 2014).

Figure 4 illustrates continental regions prone to high aerosol concentrations that are either entrained locally or transported from elsewhere (Ginoux et al. 2012). Areas of high optical depth include the Sahara and Sahel deserts, the Middle East, northern India, the coasts of the Aral and Caspian seas, basins in central Asia, the Lake Eyre Basin in Australia, and deserts in Namibia, Chile, and Peru. Prospero et al. (2002) found that most sources of mineral dust are located in topographic depressions that contain deep alluvial and lacustrine deposits subject to (and often built by) episodes of flooding. Because of the higher proportion of land coverage, the northern hemisphere is dustier than the southern hemisphere. In most areas, the dusty season occurs in local spring and/or summer, with the season of minimum dust occurring in local autumn (e.g., Ginoux et al. 2012).

Monitoring interannual variations in dust loading has shed some light on the forcing mechanisms that control dust entrainment and transport. Most multi-year studies have focused on dust transport from northern Africa over the north Atlantic (Chiapello 2014). Years with increased dust in the north Atlantic correlate with lower rainfall rates in the Sahel during the preceding year (Prospero and Lamb 2003; Moulin and Chiapello 2004; Chiapello et al. 2005). For example, long-term measurements indicate a peak in dustiness over the north Atlantic in the 1980s that has declined in the years since (Foltz and McPhaden 2008; Evan and Mukhopadhyay 2010). This decline in DOD/AOT corresponds with an increase in rainfall in the Sahel since the mid-1980s (Chiapello et al. 2005), which may have increased vegetation cover, thus reducing dust emission (Cowie et al. 2013). However, Chin et al. (2014) proposed that dust emission decreased instead as a result of reduced wind speeds in north Africa, driven by increased sea surface temperatures in the North Atlantic. Establishing a causal relationship between observed trends is critical for understanding the role of aerosols in the Earth's climate system.

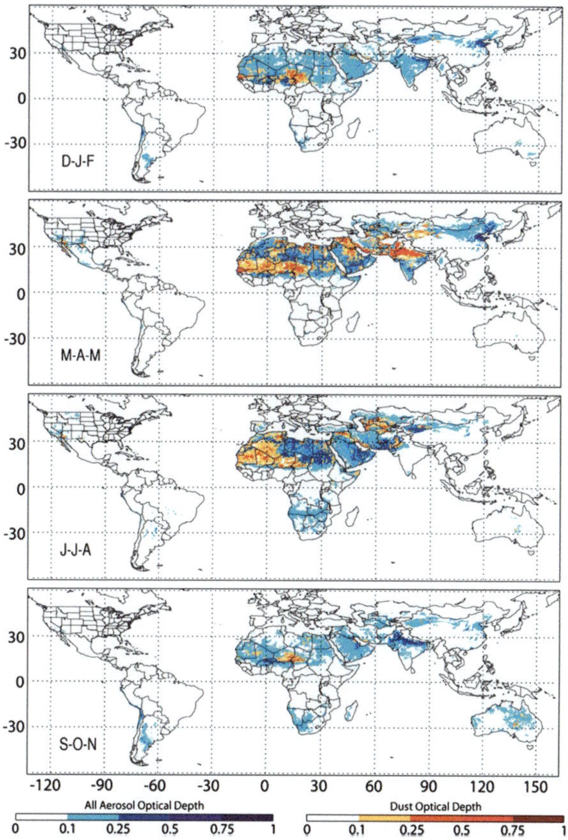

Fig. 4 Seasonal distribution of AOT (*blue*) and DOD (*yellow* to *red*) measured from MODIS Deep Blue aerosol products, from Ginoux et al. (2012). AOT is sensitive to all particulates, whereas DOD is derived from AOT and considered representative of mineral dust from natural sources

2.1.4 *Changed Perspectives on the Role of Dust in the Terrestrial Climate System?*

Figure 3 and Fig. 4 show the spectacular extent of mineral dust plumes and the fact that mineral dust is responsible for the highest measured AOT on Earth. The analysis of a long time-series from orbital observations highlights the interannual variability of the dust content and multidecadal trends in relation with climatic parameters. In addition, satellite observations have been used to estimate the impact of aerosol forcing. For example, based on multiple satellite data sets and a radiative model, Zhu et al. (2007) estimated the shortwave (visible) and longwave (IR) radiative impact of dust downwind of the three largest mineral dust source regions: eastern Asia, the Arabian Peninsula and the Sahara Desert. The mean seasonal and regionally-averaged reduction of radiative flux (visible + IR) at the surface has been estimated in clear-sky conditions to be 5.9 W m^{-2}, 17.8 W m^{-2}, and 14.2 W m^{-2}, over the Yellow Sea, the Arabian Sea and the west African coasts, respectively. The relative contributions of shortwave and longwave heating both at the surface and at the top of the atmosphere have been estimated and found to be very different for these three regions. The dust plume over the Arabian Sea was found to produce the largest effect on atmospheric heating, mainly due to shortwave heating. The maximum longwave effect on heating rates occurred over the western African coast downwind of the Sahara, resulting in strong cooling throughout the dust layer that offset up to 80 % of the shortwave heating, with moderate heating below. Finally, the net radiative heating rate over the Yellow Sea is the smallest among these three regions. This heating or cooling effect can impact atmospheric dynamics at both local and synoptic scales. In the eastern Atlantic, Wong et al. (2009) suggested that Saharan dust contributed to about 50 % of the detected heating rate anomalies and thus has a substantial impact on atmospheric stability. This effect is also suspected to influence the development of cyclones and may explain a possible inhibition on the formation of tropical cyclones, revealed by an anti-correlation between north tropical cyclone activity and Saharan dust cover (Evan et al. 2006).

Mineral dust, being of natural origin, is not accounted for in the estimation of the radiative forcing as defined by the IPCC, except for the fraction of the global dust load attributed to "anthropogenic" dust (i.e., that which is emitted from disturbed land). However, the large radiative effect of mineral dust and its variability in space and time must be accounted for in the estimation of the change in the aerosol load and its impact on climate. An analysis of global simulations and observations of multi-decadal (1980 to 2009) aerosol variations suggests that the strong variability of mineral dust emission and transport has partly dampened the changes in anthropogenic aerosol loads and highlights the fact that natural aerosols, such as mineral dust, play an important role in determining the regional and global aerosol budget, even over major pollution source regions (Chin et al. 2014). A comparison of the simulations of the mineral dust cycle by 15 global models, mainly driven by meteorological re-analysis, has been performed in the frame of the AeroCom project (Huneeus et al. 2011), and it shows large discrepancies between models. On average, the climatology of AOT due to mineral dust was reproduced to within a factor of two, and the surface concentrations and deposition were reproduced to within a factor of ten. The capacity of climatic models to reproduce the present-day AOT climatology is even lower (Evan et al. 2014). Further work is needed to understand the response of aerosols to, and their interaction with, Earth's changing climate system (Boucher et al. 2013).

Paleoarchives from ice cores (e.g., Vallelonga and Svensson 2014), deep-sea sediments (e.g., Winckler et al. 2008) and loess sequences (e.g., Muhs et al. 2014) have revealed variations of the atmospheric dust load variations between glacial and interglacial periods (Maher et al. 2010), with dust concentrations at the Last Glacial Maximum (25 ka BP) of 80–100 times that of interglacials and the present-day. Understanding the link between the mineral dust cycle and climate for past periods on Earth is still challenging. Uncertainties mainly arise from the need for climate models to properly represent dust emission processes and their link with surface properties (e.g., soils, vegetation, moisture, etc.) and the feedback that dust radiative effects can have on the climate, which are equally as complex as modeling the mineral dust cycle on Mars.

2.2 Mars

In this subsection we discuss our knowledge of aerosols in the martian atmosphere, how lifting has been observed from orbital platforms and the spatial and temporal variability of dust in the atmosphere that results. We briefly review, by comparison with the Earth, what orbital observations tell us about the role of convectively-lofted dust, such as that from DDs, in the martian climate system.

2.2.1 *Aerosols on Mars*

Many of the processes that create aerosols on Earth, such as sulfate emissions from volcanic eruptions and fossil fuel combustion, sea salt from spray, and smoke and soot from fires, do not occur on Mars (although volcanic eruptions and possibly sea salt spray contributed to atmospheric aerosols in the distant past). Failing to account for light scattering by atmospheric dust led early 20th century efforts to overestimate the surface air pressure of Mars by more than a factor of five (Kieffer et al. 1992). Mineral dust is a major forcing mechanism in the martian atmosphere, and it is intimately linked with interannual variability in the martian climate (e.g., Zurek et al. 1992; Read and Lewis 2004). Lifting may be accomplished by winds linked to large-scale weather systems or atmospheric tides (Wang et al. 2003; Hinson and Wang 2010; Hinson et al. 2012; Wang and Richardson 2015), local mesoscale gusts or topographic flows (Spiga and Lewis 2010; Mulholland et al. 2015), or on much smaller scales by convective motions (Spiga and Forget 2009), such as the DDs (Balme and Greeley 2006; Greeley et al. 2010) that are the primary focus of this review.

The primary effect of martian dust is to provide local heating to the atmosphere through absorption of solar shortwave radiation. Dust in the atmosphere also absorbs, scatters and re-radiates radiation at longer wavelengths, such as thermal infrared emission originating from the surface (Smith 2004). The net effect is to warm the atmosphere where it is most dusty and in daylight, and to cool the surface below regions with very high dust opacity. If atmospheric dust loading varies from place-to-place, this may introduce or steepen horizontal temperature gradients within the atmosphere that are, in turn, linked to winds. Winds in the atmosphere both advect dust and may lift more from the surface. In a heavily dust-laden atmosphere, the effect is to warm the atmosphere relative to the surface, which increases its static stability and tends to ultimately reduce both vertical convection during the day and large-scale wave-like instabilities, both reducing the likelihood of dust lifting from the surface and ultimately leading to the slow decay of planet-encircling dust events (e.g., Cantor 2007). In this way, atmospheric dust provides complex positive and negative feedbacks to the martian climate system.

Dust aerosols have an additional potential feedback as nuclei for cloud ice particles, which in turn impact atmospheric radiative heating and cooling (Montmessin et al. 2004; Wilson et al. 2008; Madeleine et al. 2012a; Hinson et al. 2014; Navarro et al. 2014; Steele et al. 2014a; 2014b), although, unlike on Earth, it seems likely that there will always be a sufficient supply of small dust particles on Mars to nucleate the relatively thin water ice clouds that have been observed (Heavens et al. 2010; Madeleine et al. 2012b). Clouds, in turn, may further increase the complexity of the climate feedbacks, by accelerating the removal of dust from the atmosphere by scavenging smaller particles, thereby enhancing the sedimentation rate (Madeleine et al. 2012a; Navarro et al. 2014).

2.2.2 *Orbital Platforms*

Spacecraft orbiting Mars with instruments useful for monitoring the dust aerosol distribution are listed in Table 2. The reader is referred to Snyder and Moroz (1992) for a review of the early reconnaissance of Mars including additional television camera and other instrumentation.

Most orbital observations of martian atmospheric dust have been made with nadir-viewing cameras that sense reflected solar light, or with spectrometers that sense thermal emissions of dust and gasses. The first quantitative observations came from the Mariner 9 Infrared Interferometric Spectrometer (IRIS), which provided 5–50 μm spectral coverage showing CO_2 and dust emission/absorption features that were used to monitor the decay of the 1971 planet-encircling dust storm (Hanel et al. 1972). CO_2 has a strong and distinctive 15-μm band that allows temperature sounding. Silicate dust has a 9-μm band that, with atmospheric temperature, can be used to derive column opacity. Orbital instruments tend to

Table 2 Instruments on spacecraft orbiting Mars, discussed in the text, that have been used to study atmospheric dust

Mission: Instrument	# Bands/Channels: spectral range[a] (μm)	Spatial resolution[b] (m/px)	Footprint area (km^2)	Local time
Mariner 9 (14 November 1971–27 October 1972):				
IRIS	750: 5–50	110 km/px	~ 9500	various
VO1 (19 June 1976–17 August 1980), VO2 (7 August 1976–25 July 1978):				
VIS	5: 0.35–0.7	60–80	~7×10^3	various
IRTM	broadband: 0.3–3.5 28: 6–30	> 1500	> 2	
Phobos 2 (29 January 1989–27 March 1989):				
Auguste	2: 1.9, 3.7	occultation	occultation	various
MGS (12 September 1997–2 November 2006):				
MOC WA	Red: 0.575–0.625 Blue: 0.4–0.45	230–7000	~$21–240 \times 10^3$	~13:00–15:00
MOC NA	0.5–0.9	1.4–13	~10–600	
TES	VNIR: 0.3–2.9 Thermal IR: 5.1–150 300: 5.8–50	~6×3 km/px	~18	
MOLA	1.064	160	0.02	
ODY (24 October 2001–present):				
THEMIS VIS	5: 0.42–0.86	18–70	~$300–20 \times 10^3$	~14:00–18:00
THEMIS IR	9: 6.8–14.9	100	~$840–47 \times 10^3$	
MEX (25 December 2003–present):				
HRSC	4: 0.53–0.97	> 12.5	~$60–400 \times 10^3$	Various
PFS	1.2–45	10, 20 km/px	100, 400	
OMEGA	0.38–1.05 0.93–5.1	350–12000	various	
SPICAM-UV	118–320 nm	various	various	
SPICAM-IR	1–1.7	various	various	
MRO (10 March 2006–present):				
CTX	1: 0.5–0.8	4.9–10.6	~8×10^3	~14:00–16:00
HiRISE	3: 0.57–0.83	0.25–1	~10–600	
CRISM	545: 0.362–3.920	18–36	various	
MARCI	7: 0.258–0.718	1–10 km/px	various	
MCS	broadband: 0.3–3.0 8: 11.8–42.1	sounder	various	

[a] Units are in μm unless otherwise stated

[b] Units are in m/px unless otherwise stated

be least sensitive to boundary layer dust, as a strong temperature contrast with the surface is helpful in thermal emission sounding. The Viking Orbiters made similar measurements with the IRTM. Martin (1986) modeled the data using albedo from the solar channel, thermal inertia from 20-μm emission, and gas and dust emissions from 7- to 15-μm channels, and derived a time-series of AOT that compared well with contemporaneous Viking Lander data. IRTM data was used to characterize the onset and decay of both planet-encircling dust storms (Martin and Richardson 1993). Fenton et al. (1997) applied the Martin (1986) technique to IRIS data to quantify the decay of the 1971 storm, showing that dust spatial distributions varied less as the storm decayed, with exponential timescales ranging from 42 to 67 sols.

Contemporaneously, imaging of the surface was used for geologic mapping and meteorology. The Viking Orbiters mapped the planet with two vidicon visible-light cameras (VIS). Solar reflectance imaging of dust is hampered by the extensive dust coverage of the surface, so that a low dust load in the atmosphere is difficult to distinguish. However, dust storms have been distinguished due to color, morphology, and temporal changes. Baroclinic storms were identified using imaging and thermal data from VO (Hunt and James 1979). Local and regional dust storms were identified and confirmed as dust—rather than water ice—using color data, with most occurring near perihelion and southern summer, including many at the receding cap edge (Briggs et al. 1979). Orbital monitoring of martian atmospheric dust has been essentially continuous since the arrival of MGS in 1997, with each orbiter carrying at least one imager (see Table 2).

On polar orbiting spacecraft, cameras with limb-to-limb fields of view build up a daily map over the course of ~12 2-hour orbits. MOC WA obtained daily global maps that have been used to track the evolution of local, regional, and planet encircling dust storms (e.g., Cantor et al. 2001; Cantor 2007). Cantor et al. (2001) characterized the source regions of local and regional dust storms, showing differences compared to the Viking era. Frontal storms frequently follow the receding (springtime) polar cap (Wang and Fisher 2009). Regional storms tend to originate in the low-lying Acidalia, Utopia, Arcadia, and Hellas Planitiae (Wang and Richardson 2015). Storms exhibiting visible structures on the cloud tops have been interpreted to indicate regions of active dust lifting; these features are most common in the low-lying planitiae that produce regional storms (Guzewich et al. 2015). MARCI data have shown north polar region dust storms at all times of year, but especially in early northern spring and mid-summer (Cantor et al. 2010). Mars Daily Global Maps (MDGMs) allow the tracking of individual storms (see Fig. 5) and the aggregation of data over time. Figure 6 shows typical storm tracks, including the north-to-south Acidalia, Arcadia, and Utopia cross-equatorial storms; and the Hellas and Solis sources of southern east-west storm tracks.

In addition to imaging, thermal emission has been used to monitor atmospheric dust on each orbiter. Interferometric sounders like IRIS comprise MGS TES and Mars Express PFS. Multi-channel sounders like IRTM comprise ODY THEMIS IR and MRO MCS. MGS TES observed thermal emission of dust and gas, and has been used to track the zonal and seasonal development of the 9-μm AOT (Smith 2004). TES data show 3 Mars years of dust variations (MYs 24–26), in context with temperature and water vapor and ice variations, and include one planet-encircling dust storm (in MY 25). THEMIS has been used to obtain maps with 5 reflectance bands and 10 thermal emission bands (Christensen et al. 2004). The thermal bands have allowed the cross-calibration with TES results and the temporal extension of the 9 μm optical depth maps past the end of the MGS mission (Smith 2009). Unlike MGS, Odyssey, and MRO, Mars Express is not in a circular or polar orbit and it samples varied local times each periapse. However, the effects of dust storms on the thermal environment can still be studied; for example, Määt-

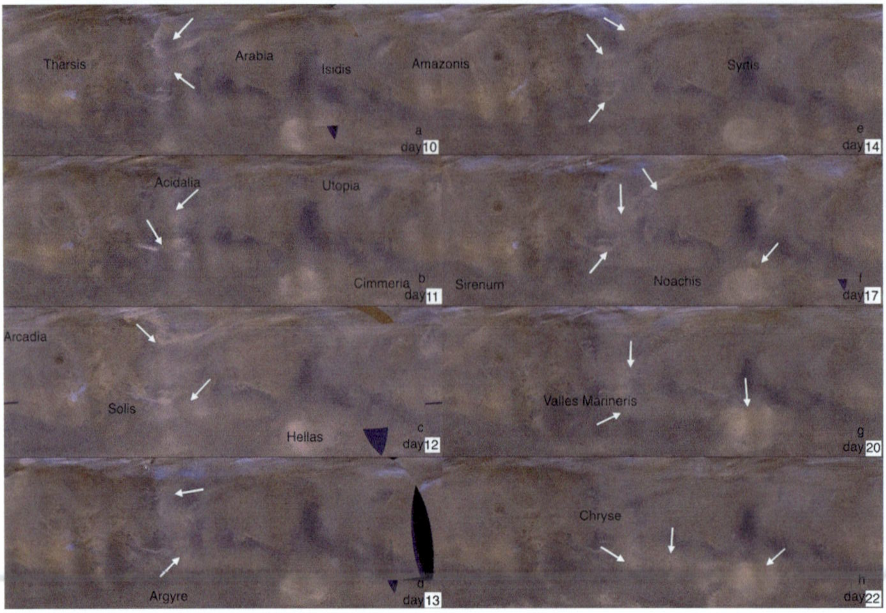

Fig. 5 Characteristic evolution of dust storms (from Wang and Richardson 2015) showing 60°S to 60°N in a cylindrical projection for $L_s = 214–228°$ in MY 27 (2005). Numbers indicate the relative sol

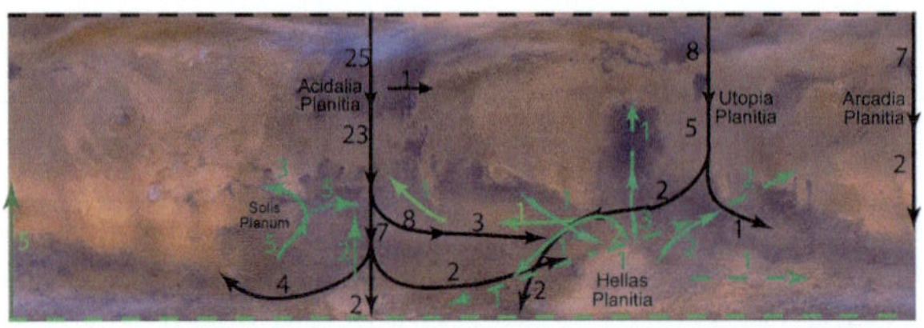

Fig. 6 Routes for storm sequences as described (modified from Wang and Richardson 2015). *Black* indicates storms of northern origin; *green* indicates those of southern origin; *dashed lines* indicate polar storms. *Numbers* indicate how many storms were observed in each trajectory segment

tänen et al. (2009) found that the thermal impact of one local dust storm was confined to the lowest two scale heights. PFS data suggest the dust is well mixed with the gas even far from dust storms, with a mean AOT of 0.25 at 0-km elevation (Zasova et al. 2005; Grassi et al. 2007). In addition to nadir-looking measurements, MCS acquires multi-channel radiometry in a limb-scanning geometry. It thus retrieves AOT as well as vertical profiles of dust (Kleinböhl et al. 2009). Detached layers have been found with the vertical profiling, helping to diagnose heating and circulation (Heavens et al. 2011b). For comparison across bands, the ratio of visible AOT to infrared AOT was determined by contemporaneous rover-based observations to vary from about 3 in northern summer to about 1.3 during southern summer dust storms, with the differences likely coming from different particle sizes (Lemmon et al. 2015).

Mars Express and MRO carry infrared reflectance mapping spectrometers, OMEGA and CRISM, respectively. For the purpose of monitoring dust aerosols, these can function as context imagers (Reiss et al. 2014a, 2014b). However, they have the advantage of additional gas absorption bands, which enable vertical sounding. Further, they aim in differing directions, and can image the same area in multiple viewing geometries (an emission phase function, or EPF, sequence) or even image in a limb-viewing geometry. These capabilities are used to study physical properties of the dust (e.g., Wolff et al. 2009) and the vertical distribution (Smith et al. 2013).

The first occultation spectrometer to orbit Mars, capable of vertical sounding of AOT from photometry of the Sun as it rose (or set) above the limb of the planet, was flown on the Soviet Phobos mission. During its short operational lifetime, the Phobos mission obtained 9 solar-occultation profiles, in 2 infrared channels, of the dust extinction at altitudes of 12–35 km (Korablev et al. 1993). These data suggested 1–2 μm dust existed in radiatively significant quantities at altitudes of 15–25 km. The Mars Express SPICAM has UV and IR channels, and operates in nadir mode and limb mode, as well as stellar (UV only) and solar occultation modes. Solar occultations have been used for a climatology of vertical distribution of the dust over 4 Mars years including the MY 28 (2007) planet encircling dust storm (Määttänen et al. 2013). Such observations traced the summer to winter dust transport pathway at high altitudes. Combining UV and IR occultation data allowed the particle size distribution (PSD) to be inferred. A bimodal size distribution was found, with a small mode unstable to coagulation and too large to be supplied by meteoric inflow, suggesting a continual supply of fine surface dust from DDs and other winds (Fedorova et al. 2014). An additional vertical sounder, MOLA, used nadir-looking LIDAR for altimetry. As a by-product, scattering from aerosols was also seen, showing column abundance, vertical extent, and relationship to clouds (Smith et al. 2001a). MOLA tracked dust storms for 1.25 Mars years, including the MY 25 (2001) planet encircling dust storm, and identified dust-ice fogs and possible DDs (Neumann et al. 2003).

2.2.3 *Spatial and Temporal Patterns of Dust Lifting*

Having summarized observations made from spacecraft in the previous subsection, we now turn to a description of the dust loading of the martian atmosphere as observed from space. It is important to note first that although lifting processes are sometimes observed directly (e.g. Cantor et al. 2006) or inferred from the sudden growth of dust loading in a region (Cantor et al. 2001; Wang et al. 2003; Strausberg et al. 2005; Wang 2007; Wang and Richardson 2015) or by the texture of dust clouds (Guzewich et al. 2015), lifting is rarely observed directly and the observations are really of dust once it is airborne, and potentially after advection over large distances from its original source. Similarly, landed spacecraft may see passing DDs (Ferri et al. 2003; Greeley et al. 2010) or monitor changing background dust loading (Colburn et al. 1989; Smith and Lemmon 1999; Smith et al. 2004; Lemmon et al. 2015), without observing the lifting process. The sparse nature of the coverage in both space and time for Mars relative to that of the Earth means that it is difficult to quantify the relative sizes and distributions of dust lifting sources based on observations alone.

Martian dust is generally observed at either visible or infrared wavelengths. Earlier observations of Mars dust, with particular emphasis on large and planet-encircling events were documented by Martin and Zurek (1993). This record was later extended to include smaller and regional storms using more detailed visible wavelength imaging from polar orbiters (e.g., Cantor et al. 2001; Cantor 2007; Wang 2007; Wang and Fisher 2009; Wang and Richardson 2015). The emphasis in all these studies is, however, on discrete dust storms. These occur principally from northern hemisphere autumn equinox to spring equinox, i.e., throughout the winter period. Dust loading is much lower and discrete storms are rarely observed throughout the northern hemisphere summer half of the year ($L_s = 0°–180°$), which is the period when Mars is presently furthest from the Sun (aphelion is close to $L_s = 70°$), and the circulation and winds are at their least intense. It is during this period, when the atmosphere is relatively clear and the surface-atmosphere thermal contrast is at its greatest, that convection (Petrosyan et al. 2011), and so DDs (Rennó et al. 1998; Newman et al. 2002a; 2002b), are likely to be most active and to contribute to the background dust levels that are observed.

Polar orbiters have also provided an almost continuous record of near-infrared dust optical depths ranging from MGS/TES (Smith 2004), through ODY/THEMIS (Christensen et al. 2004), to more recent MRO/MCS observations (McCleese et al. 2010; Heavens et al. 2011a). The MRO/MCS observations are limb soundings that include information on the vertical distribution of the dust (Heavens et al. 2011b) in contrast to previous infrared soundings that were mostly nadir observations of total dust opacity. MRO/MCS has revealed dust to be more complex in its vertical distribution than previously suspected, with layering possibly related to deep convective motions (Spiga et al. 2010; Rafkin 2012).

The observations described above are neatly summarized into an eight-year climatology of martian dust by Montabone et al. (2015). Figure 7 shows the infrared column dust absorption optical depth from this dataset, averaged over all longitudes, as a function of latitude and time of year (L_s). The dust absorption optical depth was measured at 9.3 μm, and then normalized to a consistent reference pressure of 610 Pa. This optical depth should be multiplied by about 2.6 to get an equivalent broadband visible dust total extinction. Two features are of particular note: the variability in the timing and occurrence of the periods of high dust loading in northern hemisphere winter ($L_s = 180°–360°$), in particular the planet-encircling dust events in MY 25 (2001) and MY 28 (2007); and the remarkable repeatability and much lower levels of dust loading in northern hemisphere summer ($L_s = 0°–180°$).

2.2.4 *Changed Perspectives on the Role of Dust in the Martian Climate System?*

In contrast to those performed for Earth, the majority of recent martian dust observations have been made from space (mostly from orbiting spacecraft, but including images from the Hubble Space Telescope), with the exception of a handful of discrete surface landers and earlier telescopic observations from Earth (see Chap. 2 for further detail on dust observations from the martian surface). Hence, understanding of the role of dust on a global scale has been gradually accumulated primarily through orbital monitoring rather than any other source of observations, in direct contrast to the Earth. The major qualitative change in our understanding of dust on Mars has come from the increased temporal and spatial resolution of observations. Before the late 1970s, observations naturally tended to select the largest dust events only, since these could be most readily observed either by a ground-based telescope or from a spacecraft that was either on a fly-past or in a high orbit with limited resolution (Martin and Zurek 1993). Dust variability has now been observed over a much wider range of spatial and temporal scales with the advent of polar orbiting spacecraft in relatively low orbits carrying optical cameras, such as MGS/MOC and MRO/HiRISE and MARCI (Cantor et al. 2001; Cantor 2007; Bell et al. 2009; McEwen et al. 2010), and infrared sounders, such as MGS/TES, ODY/THEMIS and MRO/MCS (Smith et al. 2000; 2001b; Christensen et al. 2004; Smith 2004; McCleese et al. 2010; Heavens et al. 2011a), and of landers with increasing spectroscopic capability (Smith et al. 2004). Despite this progress, many questions about the relative role of DDs within the martian dust cycle remain unanswered.

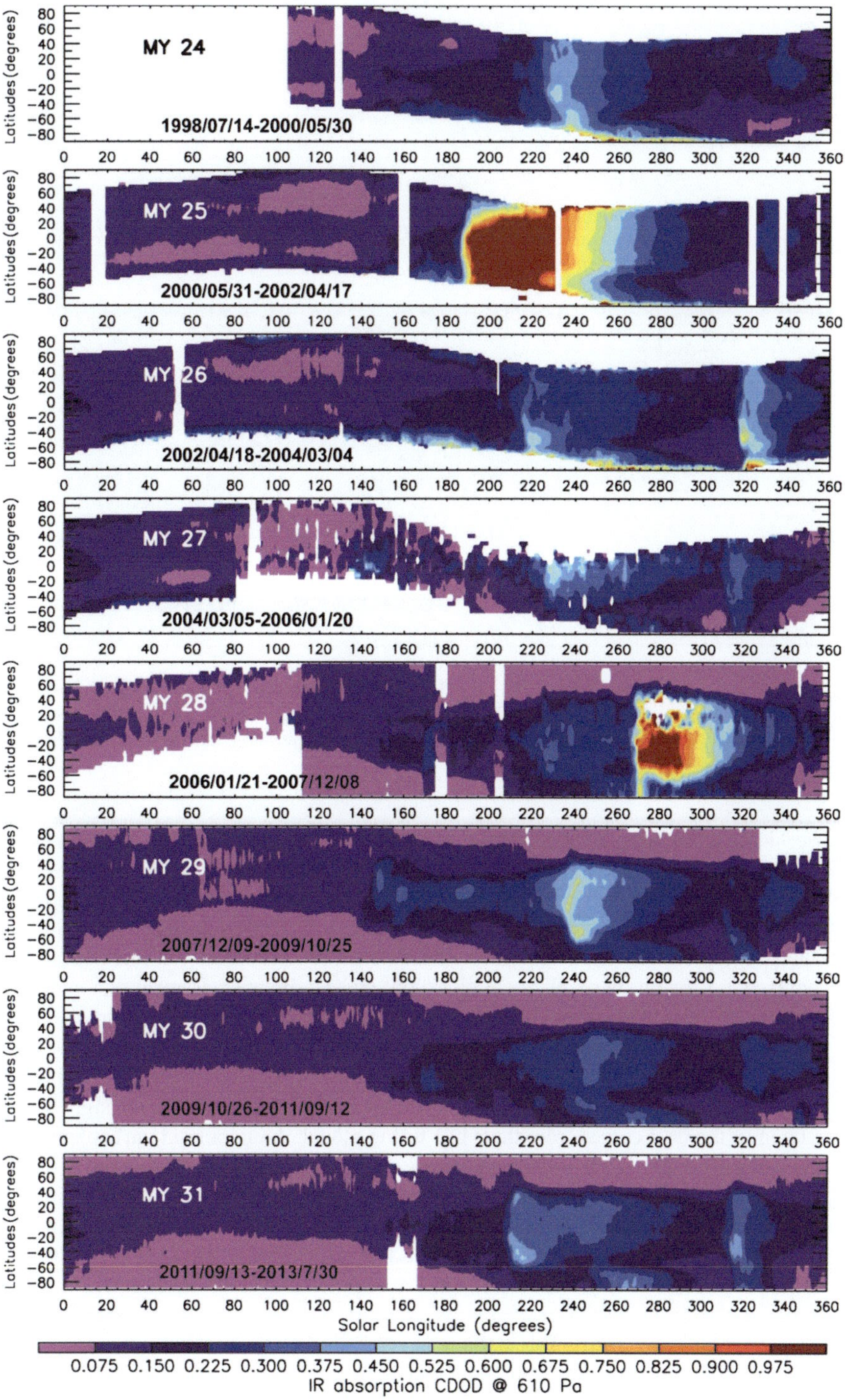

Fig. 7 Climatology of column dust optical depth (CDOD) at 9.3 µm spanning MYs 24–31 (April 1999–July 2013), from Montabone et al. (2015)

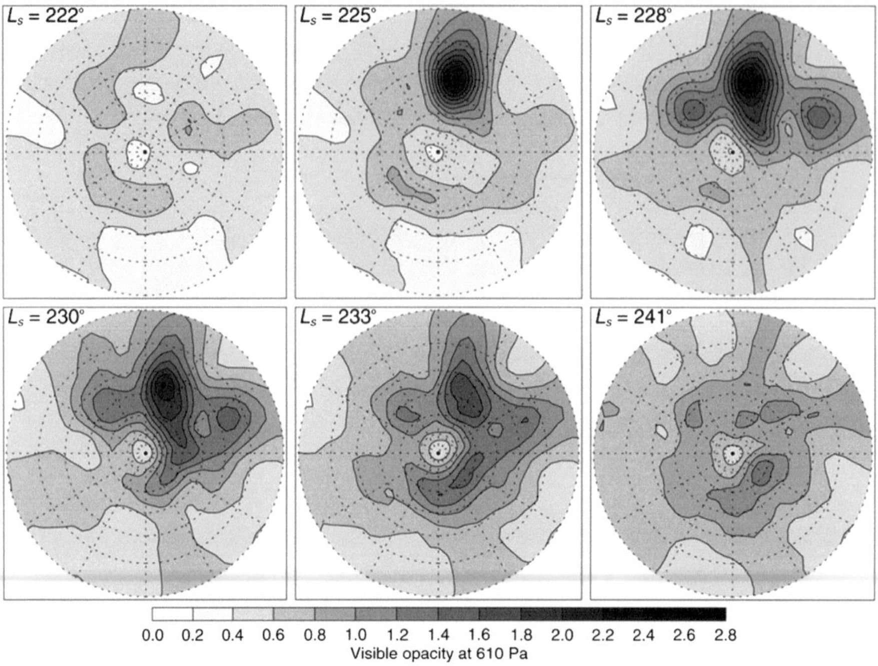

Fig. 8 Visible optical depth from the model assimilation at six intervals during the MY 23 (1997) Noachis dust storm, normalized to 610 Pa to remove topographic effects (from Lewis et al. 2007). The plots are in polar stereographic projection with the south pole at the center, the equator at the edge, and with the prime meridian pointing upward

Considerable progress can be made through modeling the martian dust cycle (Newman et al. 2002a; 2002b; Basu et al. 2004, 2006; Newman et al. 2005; Kahre et al. 2006; Mulholland et al. 2013; 2015). These are discussed further in Chap. 11 of this volume. It is worth noting at this stage that different combinations of dust lifting schemes can be tuned to produce results that are very broadly in accordance with the observed dust opacity on Mars throughout the year (see Fig. 8). Of these, Kahre et al. (2006) make a direct assessment that about one half of the total dust lifted through the year comes from their DD lifting sub-model, with the other half coming from near-surface wind stress lifting, including saltation processes. Newman et al. (2005) do not provide a comparable number, but an estimate from figures included for their present day simulation suggests that a smaller, but perhaps not significantly so, fraction of the dust comes from their DD lifting model. An important caveat is that the total dust opacity has been broadly tuned to observations, but there is no guarantee that all the near-surface wind stress processes are properly accounted for in these moderate resolution models. The DD lifting sub-model may, in practice, also be accounting for small-scale winds, such as at the polar cap edges, that are missed by the models. The estimate of roughly half the total dust lifted being from DDs is therefore potentially an overestimate.

Finally, it can be noted that dust opacity measurements have begun to be assimilated into Mars GCMs. Early efforts (Montabone et al. 2005; Lewis et al. 2007) permit dust to be tracked in three dimensions even from sparse spacecraft observations, but are not yet sufficiently developed to isolate the various sources of dust at the surface unambiguously, although work is ongoing. Figure 8 demonstrates this by showing dust broadband visible optical depth, normalized to 610 Pa, at intervals of 4, 4, 4, 4, and 12 sols made during a

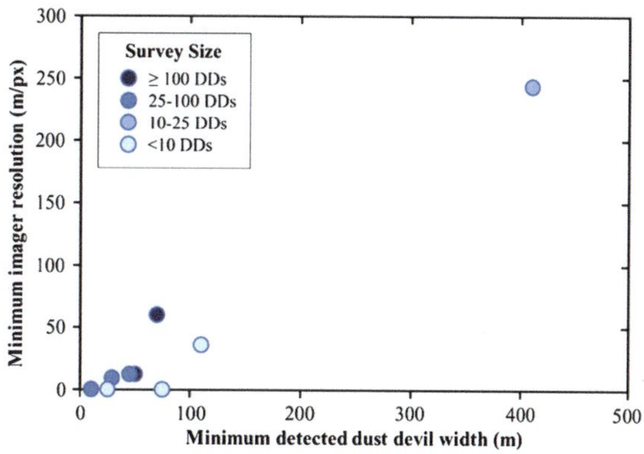

Fig. 9 The minimum detected dust devil diameter on Mars from different orbital image surveys, showing a trend towards smaller dust devils with increased sensor resolution

regional dust storm in Noachis Terra in MY 23 (1997), at a time when MGS/TES was taking limited data (about one orbit per day) during its aerobraking hiatus observing phase. The plots are polar stereographic with the south pole at the center, the equator at the edge, the prime meridian pointing upward and a grid spacing of 15° in latitude and 30° in longitude. The evolution of the dust in longitude as well as latitude can be tracked by the data assimilation technique.

3 Dust Devils

In this section we describe the discovery of prevalent DD formation as observed from orbiting spacecraft, primarily that on Mars. As with dust storms, the physical characteristics and spatio-temporal patterns of convectively-lofted dust tell a story about atmospheric conditions in the PBL, filling a gap in knowledge on Mars that is yet largely unaddressed on Earth.

3.1 Earth

Terrestrial DDs have not yet been observed directly with orbital data. Given their prevalence in orbital images of Mars, it is a mystery as to why none have yet been reported on Earth. Ground-based measurements suggest that they are wide enough to be visible in high resolution images (Balme and Greeley 2006; Lorenz 2011). A plot of the minimum martian dust devil diameter detected from imagers of varying spatial resolution shows that smaller dust devils are detected with finer resolution cameras (see Fig. 9); the same trend is likely present on Earth, implying that image resolution is not the problem. Note from Fig. 9 that the more extensive surveys are more likely to identify smaller (and more plentiful) DDs, indicating that sample size is important for orbital DD detection (see Chap. 8 for further discussion of DD size distributions). There may be other factors at play in the failure to detect terrestrial DDs from orbit. For example, DDs form most frequently when insolation is at a maximum, whereas most high resolution imagers orbit sun-synchronously with equatorial crossing times ~10:30, before much DD activity gets underway. In addition, Earth is much more cloudy than Mars, which would inhibit DD detection from above (also, clouds are also likely to suppress DD formation). Despite these limiting factors, we propose that DDs should be visible in orbital images of arid landscapes on Earth, and that a search for their presence in images should be considered.

In contrast to direct DD detection, tracks of DD passages have been identified in orbital images in several desert regions on Earth. The first DD tracks on Earth were identified in ASTER images by Rossi and Marinangeli (2004) in the Ténéré Desert, Niger. Further DD tracks on Earth were found in publicly-available high resolution satellite images such as GeoEye, Quickbird, and WorldView through the web interfaces of Google Earth and Bing Maps. Neakrase et al. (2008, 2012) found DD tracks in the east-central Sahara, including southwestern Libya, southern Libya, northeastern Chad, and the Egypt–Libya border. Reiss et al. (2010) identified DD tracks in the Turpan depression desert (north-west China) and analyzed them in-situ (Reiss et al. 2010, 2011a). Hesse (2012) and Reiss et al. (2013) observed DD tracks in the coastal desert of southern Peru in orbital imagery, which were also analyzed in situ by Reiss et al. (2013). For further details on DD tracks, we refer the reader to Chap. 4.

3.2 Mars

The first orbital detections of DDs on Mars were found in VO images by Thomas and Gierasch (1985), although convective vortices had been previously identified on the surface, in temperature and wind data from VL1 and VL2 (Ryan and Lucich 1983). These discoveries were no surprise, with their presence having been previously predicted both from a theoretical standpoint and from observations of dust clouds (Ryan 1964; Neubauer 1966; Gierasch and Goody 1973). These early investigations from 1970s-era Viking data verified the suspected prevalence of insolation-driven free convection in the martian atmosphere, fueled by a superadiabatic layer near the surface.

Since those first observations, DDs have been observed by nearly every camera in orbit around Mars; Fig. 10 shows visible images of DDs from each instrument used for their study. As on Earth, martian DDs stand out as distinctive nearly-vertical, tapered dust columns, with shadows extending to the base of the dust column at the surface. With every successive mission to Mars, new cameras have imaged the surface with generally increasing spatial and spectral resolution (see Table 2), allowing for both DD monitoring and detailed study of DD morphology and dynamics. Although a broad view from above may appear the best way to monitor DD occurrence and physical characteristics, it is worth emphasizing that there are biases in orbital data sets. For example, DD densities measured from orbit appear to be higher than those measured from the surface. Lorenz (2013) proposed that this enhancement is caused by the (typically nearly nadir) viewing angle being partly aligned with the (typically nearly vertical) DDs, such that the bright dust columns are foreshortened and thus appear brighter than they would from an oblique or horizontal perspective (e.g., from the ground). A further bias in orbital imagery is towards larger DDs. This is caused in part by camera resolution, which cannot reliably detect any features smaller than a few pixels. However, as Lorenz (2013) discussed, this is exacerbated by DD longevity, which is correlated with DD size (Sinclair 1969).

DDs have not been observed in every location on Mars. However, the lack of detection from orbit does not imply that vortices have not formed or cannot form at a given location or time. Areas that lack sufficient amounts of loose dust may experience vigorous daytime convective turbulence, but fail to entrain enough dust to render vortices visible in orbital images. Surface lineations interpreted to be DD tracks have been identified in many locations where DDs have not been directly observed (for further discussion see Chap. 4). Another factor influencing DD detection is the local time of image acquisition, as DD formation is

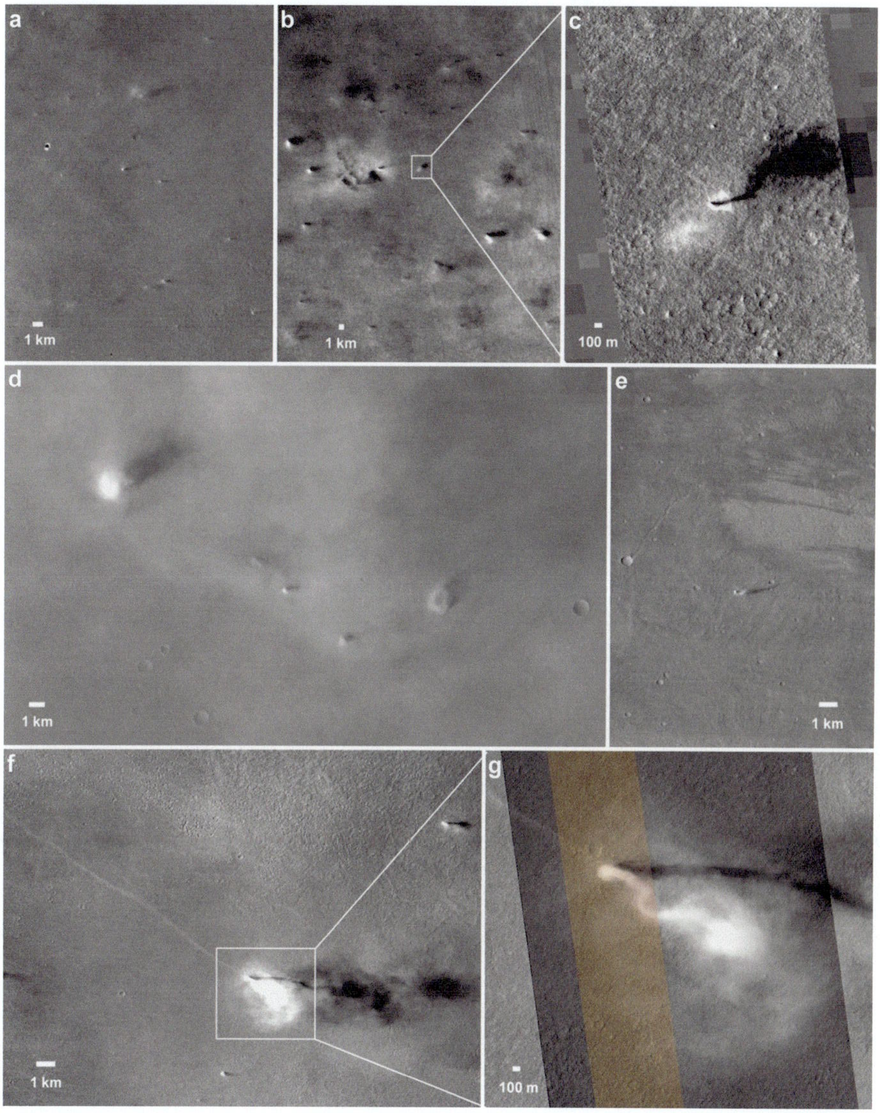

Fig. 10 Examples of DDs imaged by cameras orbiting Mars. (**a**) VO2 image 038B25, (**b**) MOC WA image E23/01275, with box showing the location of the concurrent (**c**) MOC NA image E23/01274, (**d**) HRSC image H2054_0000_ND3, (**e**) THEMIS VIS image V02326010, (**f**) CTX image G21_026394_2155, with a box showing the location of the concurrent (**g**) HiRISE image ESP_026394_2160 (both color and red images are shown)

highly sensitive to the thermal contrast between the ground and lower PBL (e.g., Deardorff 1978). Image data sets routinely obtained during peak DD formation hours are more likely to capture them in action (e.g., MOC, CTX, and HiRISE), whereas DDs are less likely to be seen in images obtained either at a different routine local time or at varying local times (e.g., VO, THEMIS, and HRSC).

Table 3 Global-scale DD surveys on Mars

Global survey	Earth Date: YYYY/MM/DD	Mars Date: MY/L_s	Mission: Camera
Thomas and Gierasch (1985)	1976/06/22–1980/07/30	12/84°–14/142°	VO1: VIS VO2: VIS
Fisher et al. (2005)	1999/03/9–2001/07/31	24/107°–25/206°	MGS: MOC WA MGS: MOC NA MO: THEMIS VIS
Cantor et al. (2006)	1997/09/15–2006/01/21	23/181°–27/359°	MGS: MOC WA MGS: MOC NA
Stanzel et al. (2008)	2004/01–2006/07	26/325°–28/86°	MEX: HRSC
Towner (2009)	2004/04/18–2006/03/03	27/20°–28/19°	MO: THEMIS IR MO: THEMIS VIS
Reiss et al. (2014a, 2014b)	2008/01–2011/12	29/11°–31/50°	MRO: CRISM MRO: CTX MRO: HiRISE

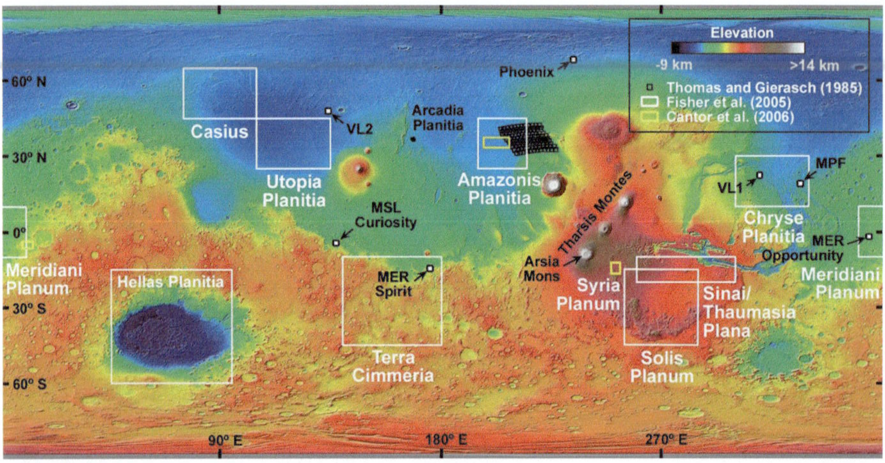

Fig. 11 A MOLA shaded relief map of Mars with colorized elevation, showing locations referred to in this work. DD monitoring sites for three global-scale surveys are noted. Note that Cantor et al.'s (2006) study included nearly the entire planet and that the regions indicated here are those with dedicated image targeting for DD activity

3.2.1 DD Inventories

There have been several global-scale surveys of martian DDs, each using image datasets of vastly different spatial resolution and temporal coverage (see Table 3). The first survey was conducted by Thomas and Gierasch (1985) using VIS images from VO during MYs 12–14, corresponding to the late 1970s on Earth (an example is shown in Fig. 10a). Their study required that DDs be transient phenomena for positive detections (i.e., identified in only one of two overlapping images), to ensure they were not mistaken for landforms. Although DDs typically have a distinctive morphology, small hills can replicate their shape, and so the requirement of transience is often still in use today. Because of the limited amount of overlapping high resolution VO image coverage, Thomas and Gierasch (1985) identified only 99 DDs, found in two limited regions in the northern hemisphere lowlands (see Fig. 11).

Although the VO images did not fully sample the martian surface, Thomas and Gierasch (1985) did examine the entire data set, so that their survey was as "global" as the 1970s-era mission permitted. A similar but unpublished survey of DDs in VO images by Wennmacher et al. (1996) focused on these areas and the Viking Lander sites, finding more than 30 DDs, many of which were previously unidentified. An automated pattern recognition algorithm detected 313 individual DDs in the high resolution VO images of Amazonis Planitia as a first step towards applying the algorithm to HRSC images, although the rest of the VO data set has not yet been investigated (Stanzel 2007).

The advent of the MGS mission in 1997 permitted the first truly global-scale inventories and interannual monitoring surveys of geomorphic features and atmospheric phenomena, including DDs (Edgett and Malin 2000; Malin and Edgett 2001). MOC WA images typically have a lower spatial resolution than the best VO images (see Fig. 10b), but their dramatically broader spatial and temporal coverage produced an unprecedented record of the martian surface and atmosphere. These images have been used to identify DDs $\geq$ 230 m in apparent height or width (here referred to as "large DDs"). In contrast, the MOC NA images sampled the surface at much higher resolution, revealing surface details to complement the context provided by MOC WA images (see Fig. 10c). Although they occasionally captured "small DDs" ($\geq\sim$ 28 m in width, $\geq\sim$ 170 m in height, as measured by Fisher et al. 2005), the small footprint of MOC NA images precluded sampling the likely DD population in a statistically significant way (see Table 2). DD studies using MOC images benefited from the spacecraft's nearly polar orbit, which obtained images in the early afternoon when DDs commonly form (typically 13:00–15:00), albeit at a cost of diurnal time coverage.

Fisher et al. (2005) searched MOC WA and NA images in nine broad regions on Mars (see Fig. 11), greatly expanding on the Thomas and Gierasch (1985) survey. Their study areas spanned a broad range of elevation, latitude, and topographic relief, specifically targeting areas known to be sources of atmospheric dust storms (Hellas Planitia, Solis and Sinai Plana) and spacecraft landing sites (Meridiani Planum, Chryse Planitia). In MOC NA and red MOC WA images spanning $\sim$1.25 MY from MYs 24–25 (1999–2001) in each study area, Fisher et al. (2005) tallied the number of images containing DDs into seasonal bins centered on each solstice and equinox. Depending on season and location, DDs were identified in 0 % to 54 % of the MOC WA images and 0 % to 18 % of the MOC NA images. The following year, Cantor et al. (2006) published a significantly more extensive survey of martian DDs across the entire planet, including all MOC NA and MOC WA images with spatial resolutions $<$500 m/px, spanning $\sim$4.5 MY from MYs 23–27 (1997–2006). To investigate the likely seasonal dependence of DD formation rate, three monitoring sites were established in locations where DDs had frequently been observed, targeting red MOC WA images on a weekly to biweekly basis over the last $\sim$3 MY of the survey (see Fig. 11). Cantor et al. (2006) found 11,456 DDs in 0.4 % of the inspected images; this work remains the most exhaustive DD imaging survey ever published. Attempts have been made at automating DD detection in MOC WA images (Gibbons et al. 2005; Yang et al. 2006), but full surveys of the image data set have yet to be completed.

The MEX mission arrived at Mars in 2003, bringing the HRSC. This camera provides stereo color images with an image footprint area on the order of that of MOC WA, but with a spatial resolution only slightly lower than that of MOC NA (see Fig. 10d). These images effectively span the gap in resolution between MOC WA and MOC NA images, although they are not obtained as regularly and thus do not provide as much detail on seasonal trends in DD activity. However, MEX's orbit permits acquisition at a wider range of local times than MGS, providing more information on the diurnal pattern of DD activity. Stanzel et al. (2008) conducted a DD survey with HRSC images spanning $\sim$1.3 MY from MYs 26–28 (2004–2006), partly overlapping in time with the MOC survey of Cantor et al. (2006). They focused on three targeted areas: Amazonis Planitia, Chryse Planitia, and Syria Planum, which were selected for their previous DD detections in orbiter and lander data. Stanzel et al. (2008) identified 205 DDs, measuring their location as well as translational velocity, diameter, and height. The average measured DD diameter and height were 230 and 660 m, respectively, indicating that the previous studies using MOC WA images were unable to capture DDs smaller than the mean size, highlighting the limitations of image resolution.

THEMIS obtains images of the surface in both visible and thermal IR wavelengths. This data set is not optimal for a large-scale DD survey because (1) these images have a relatively low spatial resolution relative to other recent data sets (although it is significantly better than that of MOC WA), and (2) the sun-synchronous orbit typically passes over the surface with local times later than 16:00 h, when diurnal DD activity has typically waned. As a result, there are only a few DD studies involving THEMIS images (Fisher et al. 2005; Cushing et al. 2005; Towner 2009). However, the thermal signal of a DD on either Mars or Earth is rarely measured, and thus THEMIS IR images can provide valuable information regarding heat transfer between DDs and the surrounding air. Towner (2009) searched THEMIS VIS images from 20°S–50°N and from $L_s = 0–270°$ of MYs 27–28 for DDs, specifically targeting the northern hemisphere summer season (see Fig. 10e). They found only 8 DDs in THEMIS VIS images with simultaneous THEMIS IR coverage; the small number of detections is likely a result of imaging times during the late afternoon (15:42–16:30).

Recent studies of martian DDs have mainly used images from CTX, HRSC, HiRISE, and CRISM, often in combination. Each of these data sets has its own advantages, reflecting technological advancements over the last few decades (see Table 2). CTX images combine a relatively high spatial resolution similar to that of MOC NA but with an image footprint typically $\sim 100\times$ larger. HRSC images have a moderate spatial resolution, but they cover broad swath areas and produce multiple images that can be used to track DD motion. HiRISE images span only a small area, but they provide astonishing detail in color with image resolutions reaching to 25 cm/px; however, they are unsuitable for broad surveys. Some DD inventories have focused on specific regions, such as Arsia Mons (Reiss et al. 2009) and Amazonis Planitia (Fenton and Lorenz 2015); others specifically searched through temporally-overlapping image data sets to estimate DD velocities and lifetimes (Reiss et al. 2011b, 2014a). For example, Reiss et al. (2014a) compiled a global inventory of DDs using CRISM VNIR images in order to determine DD velocities through comparison with CTX and HiRISE images, which are typically obtained with temporal offsets within ~1 minute (see Sect. 3.2.5). These detailed studies step beyond simple descriptions of DD morphology and behavior, delving into DD dynamics and their relation to the CBL.

3.2.2 *Spatial Patterns of DD Formation*

To first order, DD occurrences are widespread across the surface of Mars. Cantor et al. (2006) found DDs at nearly all latitudes outside the polar regions, with detections ranging from 71.9°S to 62.2°N (see Fig. 12). Similarly, Reiss et al. (2014a) identified DDs ranging from 68.4°S to 68.3°N. Martian DDs appear to form over different kinds of terrain, as well as a wide range of surface Minnaert albedos (~0.11 to ~0.22; Cantor 2007 reported a global mean of ~0.18).

Despite their occurrence over a wide range of terrains, martian DDs are much more prevalent in some areas than in others. The most DD-prone region on the planet is northern Amazonis Planitia, a 2809 km diameter, low-lying basin centered on 197.09°E, 25.75°N. Of all surveyed DDs on Mars, Cantor et al. (2006) found that 7 out of 8 (87.5 %) were

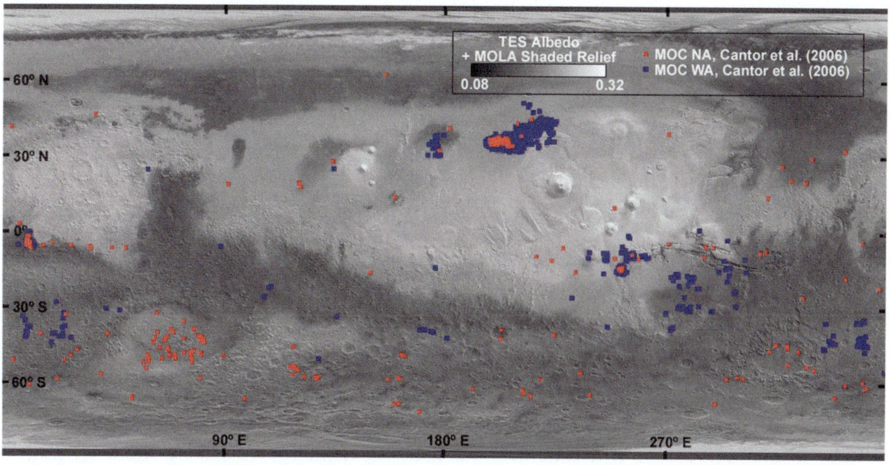

Fig. 12 Locations of MOC WA and NA images with DDs identified in the Cantor et al. (2006) survey, superposed on a composite of TES surface albedo and MOLA shaded relief. Note the high density of DDs in Amazonis Planitia, their widespread distribution in the southern midlatitudes, and the different spatial patterns of DDs detected by the two different cameras

located on this plain (see Fig. 12). Fisher et al. (2005) found DDs in 32 % of MOC WA and 12 % of MOC NA images in Amazonis Planitia, whereas their other study sites averaged <1 %. The high detection rate in Amazonis Planitia is likely biased because this area was an observational target for MOC WA; however, CRISM images also showed that DDs are numerous there, with 53 % of all identified DDs located on this plain (Reiss et al. 2014a). It is likely a fortuitous combination of this exceptional abundance and VO high resolution image targeting that led to their initial discovery by Thomas and Gierasch (1985). In contrast, Stanzel et al. (2008) found only six Amazonis Planitia DDs in HRSC images. However, the HRSC images used in that study are not likely representative of the region: many of the images do not extend into the area most densely populated by DDs. Because of their relatively large size and frequent occurrence, DDs in Amazonis Planitia have been the target of much scrutiny (see Sect. 3.2.4).

West of the Amazonis Planitia site, there is a minor center of DD generation in Arcadia Planitia, where Cantor et al. (2006) found 0.98 % of the DDs in their survey and Thomas and Gierasch (1985) identified the only two DDs in their survey that were not located in Amazonis Planitia. Otherwise, Fisher et al. (2005) and Cantor et al. (2006) found that both large and small DDs are fairly uncommon in both MOC WA and NA images in the northern hemisphere (see Fig. 12). However, approximately half of the DDs identified by Stanzel et al. (2008) were located in southern Chryse Planitia, although Fisher et al. (2005) and Cantor et al. (2006) found few MOC images with DDs in this area. Stanzel et al. (2008) attributed this high DD density to the low elevation (and thus high air pressure) of Chryse Planitia, coupled with channeled air flows down Simud Vallis (an ancient outflow channel that flowed northward into Chryse Planitia). The spatial patterns of DD production zones on Mars (most notably those of Amazonis and Chryse Planitiae) vary more from one spacecraft mission to another than they do from interannual observations. These disparities are likely created by differences in the sampled locations, local times, seasons, and imaging spatial resolutions of each mission, indicating that no single instrument is perfectly suited to monitoring all degrees of DD activity over all regions on Mars.

The MOC and HRSC surveys found that in the southern hemisphere, large DDs were concentrated in the high-standing Solis, Syria, Thaumasia, and Sinai Plana (with elevations $>\sim 3$ km), as well as in Noachis Terra (located west of Hellas Planitia, with elevations $\sim$1–2 km). In contrast to the clustered production zones of the north, the MOC surveys found that small southern hemisphere DDs are more widely distributed, with a slight concentration in the low-lying Hellas Planitia (elevation $<\sim -4$ km). The difference in small vs. large southern DDs also appears in their latitudinal distribution, in which large DDs are uncommon south of ~50°S, but smaller DDs are well-distributed down to ~70°S. The HRSC survey found a relatively high density of DDs from 50–60°S, matching the high density of DD tracks in the same area found by Whelley and Greeley (2008). Showing consistency with the MOC surveys, these HRSC DDs were typical in size compared to others found elsewhere in that study (100–500 m in diameter), but none were as large as the towering dust columns of the high plateaus and Amazonis Planitia (i.e., few exceeded heights of 1 km). Stanzel et al. (2008) attributed this southern midlatitude activity to the enhanced southern summer insolation relative to that in the northern hemisphere, caused by Mars' eccentric orbit. However, it is not clear what process would create large DDs in some areas but not in others.

Fisher et al. (2005) and Cantor et al. (2006) observed DDs at elevations on Mars ranging from the deepest basin (< -6 km) to the highest plateaus (>7 km). However, the DD distribution with elevation does not follow a simple trend: larger DDs are more prevalent at higher elevations and (with the exception of Amazonis Planitia) smaller DDs are more prevalent at lower elevations; this is the result of the clustering on the southern tropical Plana and in Hellas Planitia discussed above. Cantor et al. (2006) attributed the abundance of small DD observations at low elevation to a correspondingly higher air pressure, in which the threshold conditions for particle entrainment are reduced relative to those elsewhere on Mars. Thus, vortices need not be so intense to entrain dust at lower elevations, so that a higher proportion of the more common, weaker vortices become visible. In turn, the large DDs at high elevation could be created by a relatively deep CBL (Hinson et al. 2008; Spiga et al. 2010) accompanied by strong updrafts (Spiga and Lewis 2010), which could in turn produce taller DDs (Fenton and Lorenz 2015).

There are many locations on Mars where DD production appears to be enhanced by local conditions. Cantor et al. (2006) noted that DDs are more common near the boundaries of contrasting albedo features, which are known to generate vortices in terrestrial field studies (Snow and McClelland 1990; Rennó et al. 2004). In some cases, DDs cluster in groups or align in rows (Biener et al. 2002; Fisher et al. 2005; Stanzel et al. 2006; Fenton and Lorenz 2015). It is possible that some surface factor, such as a roughness element or albedo contrast, locally enhances DD production in these locations. Stanzel et al. (2006) proposed that DDs might form along an air mass boundary, similar to those that been observed on Earth (e.g., Markowski and Hannon 2006). Alternatively, Fenton and Lorenz (2015) proposed that these regularly-spaced DDs could instead denote vertices of intersecting convection cells, where DDs are most likely to form. Many such scenarios are possible.

Despite many years of continued monitoring from orbit and landed missions, there have been no simultaneous observations of DDs both from above and on the ground. Despite a dedicated imaging campaign by MGS MOC, no DDs were identified from orbit that were within the same fields of view as the MER Spirit and Opportunity cameras (Cantor et al. 2006). This result is consistent with the lack of DD detection in Opportunity images (Greeley et al. 2010). However, Greeley et al. (2006) found 533 DDs in Spirit images during the most complete monitoring season in MY 27 (2005). These DDs were typically 10–20 m in diameter, with only a few large enough to have been resolved in MOC WA images. A similar attempt was made to coordinate observations from HRSC with Spirit rover images (Stanzel 2007). However, once again, no DDs were captured from orbit, as a result of mismatching fields of view with the rover and differences in the image acquisition times. The smallest DDs imaged by HRSC were ~50 m in diameter, so that the most abundant size range observed from Spirit is still not resolved by the newer camera in orbit. This disconnect between orbital and landed surveys has made it difficult to use orbital observations of DDs to make quantitative assessments of conditions at the surface.

3.2.3 Temporal Patterns of DD Formation

All martian DD surveys indicate that, like their terrestrial counterparts, they most frequently form when insolation is near its seasonal and diurnal maximum. HRSC images, obtained at a range of local times, show that DDs across the martian surface form from noon through the afternoon, with peak hours between ~14:00–15:00 (Stanzel et al. 2008). DDs were rarely seen in THEMIS images, which were obtained from ~15:00–17:30, indicating that the circulations that produce these vortices typically shut down by midafternoon (Cushing et al. 2005; Fisher et al. 2005; Towner 2009). These observations are consistent with diurnal DD activity observed from the surface (e.g., Greeley et al. 2006).

On a seasonal timescale, Cantor et al. (2006) found that DD numbers reached a maximum during $L_s = 145$–$150°$ in the northern hemisphere and at $L_s = 305$–$310°$ in the southern hemisphere (i.e., local summer), although some studies have identified DDs during other seasons—even during local winter (e.g., Fisher et al. 2005; Cantor et al. 2006; Stanzel et al. 2006, 2008; Fenton and Lorenz 2015). All of the global-scale surveys found that the seasonal DD distribution is sensitive to latitude, such that terrains at higher latitudes are less likely to produce DDs during local winter. For example, Cantor et al. (2006) found no DDs within half of a season surrounding the local winter solstice poleward of 50° in either hemisphere (particularly in the north). In contrast, they found that lower latitudes have two seasonal DD peaks, centered near either solstice. In the latitudinal zone 0–30°S, Cantor et al. (2006) found that DDs formed year-round, mainly as large DDs on the high plateaus of Solis, Syria, Thaumasia, and Sinai Plana.

Interannual variability superposed on the seasonal trend of DD activity is often linked to the occurrence of dust storms. Cantor et al. (2006) found that dust storms can influence DD production; most notably, that DDs have not been observed from orbit in the midst of dust storms. This is generally thought to be caused by a less strongly superadiabatic temperature profile near the ground, as a result of increased atmospheric heating aloft (from suspended dust absorbing solar radiation) and decreased atmospheric heating of the surface (from less solar radiation reaching the ground). However, the increase in atmospheric optical depth during dust storms also reduces image contrast, making DDs more difficult to identify: if they were present under the dust haze, they would be harder to spot. Because of these factors, DD observations are generally anti-correlated with dust events. For example, Cantor et al. (2006) noted that the onset of summertime DDs in the southern midlatitudes (40–70°S) appears to be delayed until southern spring dust storms abate.

Despite the general anti-correlation with dust storms, the detailed interaction of DD activity with weather fronts is often more complex. DDs are often seen in front of the leading edge of dust storms, indicating that mechanical turbulence from storm fronts produces vortices that are vigorous enough to entrain dust (Cantor et al. 2006; Stanzel et al. 2008). In some cases, Cantor et al. (2006) showed that the same storm may trigger DD production in one stage of its evolution, and suppress DD activity in another stage. For example, the MY 25 (2001) global dust event during southern spring and summer (see Fig. 7) may have caused

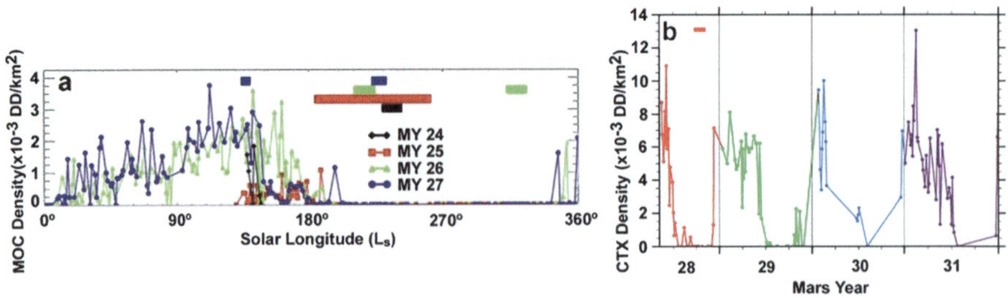

Fig. 13 Multiple years of DD densities measured in Amazonis Planitia from (**a**) MOC WA images, modified from Cantor et al. (2006) and (**b**) CTX images, modified from Fenton and Lorenz (2015). *Colored bars* correspond to the duration of global-scale dust events

a short spike in DD activity during northern autumn in Amazonis Planitia as a dust haze layer associated with the expanding storm passed overhead. However, as the same storm slowly decayed a few months later, the dust-laden atmosphere appears to have delayed the start of the southern summer DD season in Syria Planum by 40° of L_s relative to the following two years.

Loose dust on the surface (i.e., that which has been transported and then deposited in the wake of dust storms) can also modulate DD production rates by changing the surface albedo. For example, Cantor et al. (2006) investigated the passage of a series of late winter dust storms over MER Spirit in 2005 (MY 27, $L_s = 170$–173°). These storms produced strong winds that removed surface fines, reducing the albedo of dark surface areas by 10–14 % and forming new dark surface features that contrasted with surrounding bright terrain by 12–18 %. Despite a dedicated MOC WA imaging campaign to observe weather patterns in the vicinity of Spirit, Cantor et al. (2006) found that only one such image captured DDs (in MY 27, $L_s = 264.7°$): they were located near the margins of the newly-darkened ground a few months after the storms, during the peak of the DD season (unfortunately they were too far from Spirit to be visible from the ground).

Both planet-encircling dust storm activity and surface albedo patterns vary in degree and location from one year to the next, influencing the timing of and possibly the density of DD production. It is likely that DD activity could be predicted based on the occurrence of interannually-varying storms and their resulting albedo patterns. It is also possible that longer-term changes in the martian climate system affect DD formation rates, as well as their efficiency in lifting dust. For example, Cantor et al. (2006) predicted that the subtle change in insolation during the 11 year solar cycle could also impact DD formation rates, although no studies have yet sought to test this hypothesis.

3.2.4 *DD Observations at Specific Locations*

Northern Amazonis Planitia Northern Amazonis Planitia is far and away the dominant producer of both large and small DDs on Mars. Figure 13 shows that there is a regularly-occurring active ("on") season that typically begins with sporadic activity just prior to the northern spring equinox and extends until just past the northern autumn equinox (Cantor et al. 2002; Fisher et al. 2005; Cantor et al. 2006; Fenton and Lorenz 2015). DDs generally do not form during the rest of the year (the "off" season), although Fenton and Lorenz (2015) reported a few isolated cases during late fall and early winter. Superposed on the seasonal pattern is a higher frequency variation in DD density on the order of 5–10° L_s, possibly related to regional weather patterns (e.g., dust storms and fronts) that might briefly enhance or suppress DD formation.

Figure 13 shows that the DD "on" seasons during Mars Years 26 (2002) and 27 (2004) in Amazonis Planitia experienced a gradual growth in activity, peaking at MOC WA densities of $\sim 2 \times 10^{-3}$ DD/km^2 at $L_s \sim 140°$ (Fisher et al. 2005; Cantor et al. 2006). A survey of the same region using CTX images from MYs 28–31 (2006–2013) produced density peaks of $\sim 8 \times 10^{-3}$ DD/km^2 (Fenton and Lorenz 2015); this relatively high density reflects the increase in spatial resolution of CTX images versus that of MOC WA images rather than any interannual variation (see Table 2). Fenton and Lorenz (2015) observed no similar growth in activity during northern spring and summer in MYs 29–31 (2007–2011), contrasting with the trend in MYs 26–27 (2002–2004). It is not yet clear whether this behavior is only apparent in MOC WA images or interannual variations have affected the DD production rate.

The most significant departures from the "on vs. off" seasonal trend in Amazonis Planitia appear in MYs 25 (2001) and 27 (2004), when DD production appears low late in summer (MOC WA densities $<1 \times 10^{-3}$ km^2). The low DD occurrence in MY 25 occurs long before the onset of the MY 25 global dust event (just after northern autumn solstice), and therefore this lull in activity is unlikely to be related to this storm. Fisher et al. (2005) attributed the weak DD activity in MY 25 to low image exposures early in the MGS mission, making features less crisp (and thus DDs more difficult to identify). Unfortunately no MOC WA images of the area earlier during the "on" season in MY 25 are available for comparison, leaving this problem unresolved. The early drop in DD production during the late summer of MY 27 ($L_s \sim 155°$) occurred as dust haze from a global-scale dust event passed over Amazonis Planitia, following the general trend that high altitude dustiness suppresses DD production by increasing the stability of the lower CBL.

It is not immediately apparent why DD production rates in northern Amazonis Planitia are so high relative to the rest of the planet. By treating the lower atmosphere as a heat engine, the intensity of DD production can be represented as a function of the pressure thickness of the convective boundary layer (Rennó et al. 1998). When using this relation to create a parameterization of dust lifting by DD activity in martian atmospheric models (such as GCMs), Amazonis Planitia stands out as a major source of atmospheric dust (Newman et al. 2002a; Basu et al. 2004; Kahre et al. 2006). This appears to be caused by the low thermal inertia of the surface, which allows the daytime surface temperature in Amazonis Planitia to soar (e.g., Putzig et al. 2005). However, data from TES indicate that there are other regions on Mars with equally high summertime surface temperatures, similarly deep CBLs, and comparable sensible heat fluxes, but these regions do not produce correspondingly high densities of DDs (Fisher et al. 2005). It is possible that yet unidentified local and regional factors, such as dust availability or albedo/thermal inertia contrasts in the vicinity of the high DD density region, constructively interfere to enhance DD production in Amazonis Planitia.

Syria and Eastern Meridiani Plana Being located not far south of the equator, Cantor et al.'s (2006) Syria Planum monitoring site (see Fig. 11) experienced two seasonal peaks in DD production, one at each solstice. The southern winter peak is the stronger of the two, lasting from $L_s = 109$–$132°$, with typical MOC WA densities exceeding 1.5×10^{-3} DD/km^2. The slightly weaker southern summer peak ranged from $L_s = 265$–$315°$, with MOC WA densities less than 1.5×10^{-3} DD/km^2. It is not clear why the winter peak would be more intense than the summer peak, as the insolation during winter would be less than that during summer. Perhaps some dynamical effect from the nearby Tharsis volcanoes somehow plays a role in producing conditions conducive to DD formation. Further investigation with data from subsequent years might provide clues (or at least better statistics) to address this question.

Like its northern counterpart in Amazonis Planitia, Syria Planum experienced some inter-annual variability in DD activity (Cantor et al. 2006). In the two full years of monitoring in Syria Planum during MY 26–27 (2002–2005), the second year's winter season was slightly more active, although the reason for this is not known. In both years, the winter DDs stopped abruptly after $L_s \sim 135°$; in the case of MY 27 this shutdown may be related to the occurrence of a planet-encircling dust storm. The summer DD season onset in MY 25 was delayed by $\sim$40° of L_s relative to those in MYs 26 and 27, likely as a result of the extensive MY 25 global dust event.

Cantor et al.'s (2006) monitoring site in Meridiani Planum experienced a lower rate of DD formation than the other two sites, although (perhaps because of this) it was not monitored as frequently. DDs at this nearly equatorial site were found nearly year-round, with a short southern winter break from $L_s \sim$ 130–190°. A slight peak in southern summer activity occurred from $L_s =$ 272.5–302.5°, with maximum MOC WA densities less than $\sim 1.4 \times 10^{-3}$ DD/km^2. An enhancement during late southern summer in MY 27 (2005) relative to that in MY 26 (2004) was attributed to local dust storm activity that may have deposited or abraded a source of dust for entrainment.

Arsia Mons Surprisingly, DDs have been identified at elevations exceeding 16 km above the datum. Cushing et al. (2005) was the first to report the occurrence of 3 DDs in the Arsia Mons caldera in THEMIS VIS and IR images, and Cantor et al. (2006) identified 3 in MOC WA images. In a search of MOC NA, MOC WA, THEMIS VIS, HRSC, and CTX images, Reiss et al. (2009) found 28 DDs, 11 of which formed at elevations >16 km; most of these formed in the caldera. The DDs on Arsia Mons appear to form year-round, although there is a peak in activity during late local summer from $L_s =$ 340–360°. The air pressure at this altitude is $\sim$1 mbar, six times lower than the global annual average surface air pressure, and wind stresses required to directly lift dust are 2–3 times higher than those found at the surface. Although it may seem unlikely that dust-laden vortices could form in such a sparse atmosphere (indeed, many Earth researchers might have supposed the same of any location on Mars prior to their discovery), the direct observations and many bright tracks attest to their profusion on the caldera floor.

It is not clear how DDs form and entrain dust at the top of this tall volcano, but it is notable that they have not been reported at the tops of the other, equally tall Tharsis volcanoes. Reiss et al. (2009) proposed that the illumination of a dusty surface at low air pressure enhances entrainment through a combination of a thermophoretic force and solid state greenhouse effect. Comparing mesoscale and large eddy simulations at different elevations on Mars, Spiga and Lewis (2010) suggested that winds produced by convective activity are relatively stronger at high elevation, enhancing the likelihood of DD formation on Arsia Mons (Lorenz and Myers (2005) proposed that a similar effect on Earth may contribute to more DD-related aviation incidents at high elevation in the United States). In addition, their occurrence late in the afternoon ($\sim$15:00–16:00) in THEMIS VIS images is rare on Mars (e.g., Fisher et al. 2005; Stanzel et al. 2008), so that mechanical turbulence from some local factor (e.g., the 110 km wide, smooth caldera floor surrounded by steep cliffs $\sim$1 km high) may be required to trigger DD generation at this altitude and local time.

In addition to DDs, Arsia Mons is also host to a 50 km wide "spiral dust storm" that regularly forms in late southern winter, shortly before the southern spring equinox (see Fig. 14; Malin et al. 2010). Mesoscale model simulations by Rafkin et al. (2002) suggest that this storm is part of a larger thermally-driven circulation up the slopes of Arsia Mons that reaches

Fig. 14 A seasonally-repeating spiral cloud 50 km wide on the caldera floor of Arsia Mons, ~16.3 km above the datum

~30 km above the caldera; the dust is derived from the lower slopes of the volcano and carried aloft by daytime upslope winds that can exceed 25 m/s. Recent observations from MCS indicate that detached dust layers are common at this season over the Tharsis volcanoes, at altitudes higher (~65 km) and with mass mixing ratios higher (~150 ppm) than previously modeled (Heavens et al. 2015). This study suggests that daytime slope winds racing up the high-relief Tharsis volcanoes force the terrain to serve as a chimney, transporting dust to high altitudes. Although this mechanism is driven by mesoscale circulations, which are not a major topic of this review, it is possible that these slope winds bring the dust, and possibly the mechanical turbulence, that is responsible for DD generation in the Arsia Mons caldera.

3.2.5 Physical Characteristics of DDs

Dimensions DD dimensions have been measured from orbital image data from several missions (Thomas and Gierasch 1985; Fisher et al. 2005; Stanzel et al. 2006, 2008; Towner 2009; Choi and Dundas 2011; Reiss et al. 2011b, 2014a, 2014b; Fenton and Lorenz 2015; see also Table 4). The diameter is usually determined by measuring the visible extent of the dust-laden columnar vortex, although in some cases the extent of the vortex shadow width was used (e.g., Fisher et al. 2005). The uncertainty in measuring DD diameter from orbital imagery depends on the spatial ground resolution of the image data. For example, Stanzel et al. (2008) observed a DD in a HRSC nadir image with a spatial ground resolution of 25 m/px which was simultaneously imaged by the HRSC SRC camera with a spatial ground resolution of 5 m/px, obtaining diameters of 90 m versus 42 m, respectively. This suggests that errors in measuring DD diameters from orbit can be quite large (up to ~50 %) due to dust clouds surrounding the vortex, causing uncertainties in accurately determining the dust-laden vortex diameter. However, DD diameters measured with higher resolution image data are more accurate.

DD heights are calculated by measuring the shadow length and dividing it by the tangent of the solar incidence angle. There are several uncertainties for accurate measurements of shadow lengths, such as detached dust clouds and diffuse shadows. In addition, DDs are not necessarily vertical; they are typically tilted towards the direction of motion (i.e.,

Table 4 The range of DD diameters and heights measured from various orbital platforms on Mars, including the number of DDs measured (N)

Reference	Instrument	Minimum resolution (m/px)	N	Diameter range (m)	Height range (km)
Thomas and Gierasch (1985)	VO1/VO2 VIS	60	$\sim$100	70–1000	1–2.5
Neumann et al. (2003)	MOLA	$\sim$160	7	N/A	8–10
Fisher et al. (2005)	MOC NA	9.28	$\sim$25	30–510	0.03–0.8
Fisher et al. (2005)	MOC WA	230	14	N/A	3.5–8.5
Stanzel et al. (2008)	HRSC	12.5	$\sim$200	45–1650	0.1–2.2
Towner (2009)	THEMIS VIS	36	8	110–335	0.25–1.9
Choi and Dundas (2011)	HiRISE	0.25	4	25–250	0.13–0.65
Reiss et al. (2011b)	MOC WA	244	13	410–1180	0.3–1.35
Reiss et al. (2011b)	HRSC	12.5	26	45–860	0.06–1.7
Reiss et al. (2014a)	CTX or HiRISE	0.25	44	10–280	0.04–4.4
Reiss et al. (2014b)	HiRISE	0.25	3	75–175	0.37–0.87
Fenton and Lorenz (2015)	CTX	5.77	2038	N/A	< 0.1–16.5
Mean/median of the above data sets:				220/160	0.63/0.45

downwind) with height (e.g., McGinnigle 1966; Maxworthy 1973). Thus, the error in measuring DD heights is estimated to be between $\sim$20 to $\sim$30 % (e.g., Reiss et al. 2014a; Fenton and Lorenz 2015). The accuracy in measuring heights is probably higher for relatively short DD, which exhibit sharper shadows (Fenton and Lorenz 2015). In 2012, the HiRISE team announced the tallest recorded DD, with an estimated height of 20 km (http://www.uahirise.org/ESP_026394_2160; this DD is also shown in Figs. 10f and 10g). Fenton and Lorenz (2015) proposed that this height is an overestimate, produced by a combination of DD lean and shear by winds aloft, both of which would artificially stretch the dust column and make it appear taller. Imagers are not the only sensor that can estimate DD height: some DDs may have been dense enough to trigger returns in MOLA data (Neumann et al. 2003), raising the question of whether terrestrial lidars such as CALIOP have been able to detect convectively-lofted dust.

Table 4 summarizes orbital measurements of DD dimensions obtained by different instruments. DD diameters range between 10 and 1650 m and heights between 0.03 and 16.5 km. Fisher et al. (2005), Stanzel et al. (2006, 2008), Towner (2009), Choi and Dundas (2011), and Reiss et al. (2011b, 2014a, 2014b) provided detailed information about simultaneously measured DD diameters and heights. Based on this data set of about 300 measurements, average diameters and heights are 220 m and 0.63 km (median = 160 m and 0.45 km), respectively. Figure 15 shows a plot of measured DD dimensions, showing a general correlation of height and diameter. However, the wide scattering of the measurements also indicates that there is no clear analytical relationship between these two dimensions. This is not surprising, considering the large variety in DD morphologies (Balme and Greeley 2006; see also Chaps. 5 and 6). DD height-to-diameter ratios, based on the data shown in Fig. 15, range from $\sim$0.5 to 28 with a mean of 3.8 (median = 2.5). This ratio is in broad agreement with the terrestrial and martian median height-to-diameter ratio of $\sim$5 reported by Lorenz (2013), based on the studies by Flower (1936) and Fisher et al. (2005).

The minimum detectable DD diameter is dependent on the spatial ground resolution of each the orbital camera. At least two to three pixels are needed to resolve a DD, including any surrounding dust clouds. Thus DDs with diameters $<\sim$ 500 m are not identifiable in

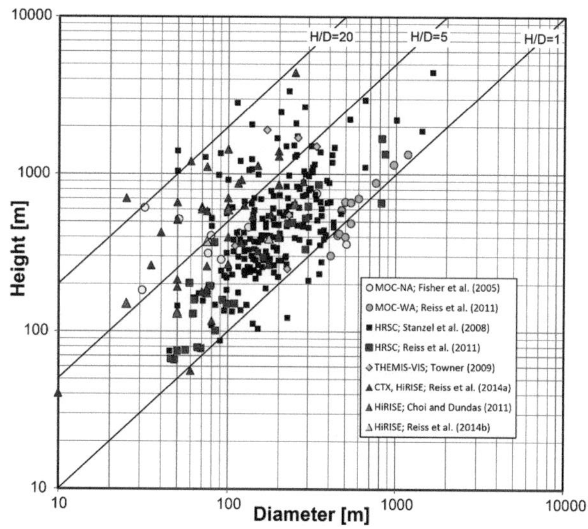

Fig. 15 DD diameter vs. height, based on measurements from orbital imagers described in the literature (see Table 4). *Black lines* show height-to-diameter ratios of 1, 5, and 20. Minimum detectable DD diameter depends on, among other factors, image resolution (see also Fig. 9 and Table 4)

MOC WA images with spatial ground resolutions of ~230 m/px, whereas detection of DDs with diameters as small as ~1 m is theoretically possible in HiRISE imagery, with spatial ground resolutions of up to 0.25 m/px (although the smallest observed diameter in HiRISE to date is ~10 m). In general, orbital image data reveals larger DDs in comparison to those identified in lander observations. Greeley et al. (2010) observed ~760 DDs in images taken from MER Spirit on the martian surface, with diameters ranging between 2 and 276 m; the median diameter was 30 m, more than 5× smaller than that obtained from orbital camera images. Full DD heights from lander images could only be determined for 44 DDs (most extended beyond the top of the images), which ranged from 0.01 to 0.36 km (Greeley et al. 2006), compared to 0.03 to 16.5 km from orbital data.

DDs on Mars can be orders of magnitude larger than terrestrial ones. On Earth, in situ measurements of DD diameters and heights are typically in the range of ~1 to <100 m and ~0.01 to <1.2 km, respectively (Flower 1936; Williams 1948; Sinclair 1965; Snow and McClelland 1990; Mattsson et al. 1993). However, Bell (1967) and Sinclair (1964) reported DDs as high as 2.5 km and 3.8 km, respectively. Although the typical size ranges of terrestrial DDs are comparable to those observed at martian landing sites, the diameters and heights obtained with orbital instruments imply that DDs on Mars can be several orders of magnitude larger. The lack of DD detections in terrestrial satellite imagery also suggests that DDs on Earth are smaller in size. This discrepancy might be caused by differences in the PBL on both planets, which limits the height extent of convective vortices. The CBL depth on Earth over land is usually below 2–3 km (e.g., Garratt 1994), in contrast to Mars, which typically has CBL depths between 4 and 10 km (Hinson et al. 2008). Fenton and Lorenz (2015) analyzed seasonal DD height and PBL variations in Amazonis Planitia on Mars. They found that median DD heights are generally about 0.2× PBL depth, in agreement with terrestrial studies (Willis and Deardorff 1979; Hess and Spillane 1990; Ansmann et al. 2009) comparing DD heights with that of the CBL. The difference in DD heights on Mars and Earth is likely caused by "a greater solar energy (absorbed by the surface) to atmospheric mass ratio on Mars (tens of times greater than that on Earth), providing martian DDs with a more effective power source" (Kok et al. 2012). It is not yet clear whether DD dimensions on Earth correlate with length scales in the terrestrial PBL.

Translational Ground Speeds Dynamic processes on Mars can be captured by orbital instruments or platforms acquiring time-delayed imaging data of the same surface area. Hence, the horizontal ground speed and direction of motion of DDs translating across the surface can be directly measured when imaged by such datasets. For example, Cantor et al. (2006) used two VO images taken 5 s apart from each other, which showed the same DD in the overlapping area of the two images. The DD moved ~880 m towards the northeast across the surface in the time between the two image observations, resulting in a horizontal ground speed of $17.6 \pm 2.6\ \mathrm{m\,s^{-1}}$ (Cantor et al. 2006). However, such observations in acquired image pairs of frame cameras are rare, because the overlapping areas required to contain the same DD are relatively small. The unique imaging capabilities of the HRSC pushbroom instrument (Jaumann et al. 2007) allow systematic measurements of DD translational ground speeds. The HRSC consists of nine line sensors simultaneously acquiring superimposed image swaths of the martian surface. The different emission angles (+18.9° to −18.9°) of each of the nine image channels result in a time delay at which the same surface area is covered. The maximum time delay between the outermost forward- and backward-looking image channels is about one minute. Stanzel et al. (2006, 2008) measured translational ground speeds of 205 DDs using HRSC images, with results ranging between 1 and 59 $\mathrm{m\,s^{-1}}$, and a mean of 13 $\mathrm{m\,s^{-1}}$ (median: 10.4 $\mathrm{m\,s^{-1}}$). Reiss et al. (2011b) measured translational ground speeds of an additional 26 DDs in HRSC images, which ranged between 3 and 22 $\mathrm{m\,s^{-1}}$, with a mean of 12 $\mathrm{m\,s^{-1}}$ (median: 11.1 $\mathrm{m\,s^{-1}}$) in the Syria-Claritas region on Mars.

Another method of systematically obtaining DD translational ground speeds was recently introduced by Reiss et al. (2014a) using the CRISM imaging spectrometer in combination with the CTX and/or HiRISE cameras, which are all located on the same instrument platform onboard MRO (see Table 2). The CRISM instrument uses an active pointing system for tracking targets with long exposure times (Murchie et al. 2007). The center of a surface target is imaged by forward- and backward-looking angles in the flight direction, in contrast to the exclusively nadir-looking angles of the CTX and HiRISE instruments, resulting in positive and negative time offsets of $\sim \pm 1$ minute between the CRISM and CTX/HiRISE surface observations. Using this method, Reiss et al. (2014a) measured translational ground speeds of 44 DDs, which ranged from 4–25 $\mathrm{m\,s^{-1}}$ with a mean of 12 $\mathrm{m\,s^{-1}}$ (median: ~11.5 $\mathrm{m\,s^{-1}}$). A further method measuring DD translational ground speeds is the usage of the HiRISE blue-green, red, and infrared color swaths, which cover the same surface area with a time delay of ~0.1 seconds. However, such measured horizontal ground speeds should be seen as an estimate due to the very short time interval between the color swaths, which introduces significant uncertainties. Choi and Dundas (2011) and Reiss et al. (2014b) estimated DD translational ground speeds in the range of ~5–20 $\mathrm{m\,s^{-1}}$ using HiRISE color swaths.

The comparison of some DD translational ground speeds and directions of motion with predicted wind speeds and directions from the Mars Climate Database (MCD; Forget et al. 1999) indicated that DDs move with speeds and in directions of the ambient wind field (Stanzel et al. 2006, 2008). A systematic study comparing measured DD translational ground speeds and directions of motion with MCD-predicted wind speeds and directions at various heights by Reiss et al. (2014a) showed that DDs on Mars move with ambient wind fields and with ground speeds commensurate with those at height within the PBL, hence faster than near-surface winds. This is in agreement with terrestrial in situ studies by Balme et al. (2012), who compared measured DD translational ground speeds and directions of motion with wind and direction measurements at 10 m height. They found that DDs move at speeds and in directions reflecting the ambient wind field within the PBL at about 20–30 m above the surface (Balme et al. 2012). As pointed out by Balme et al. (2012) and shown by Reiss

Table 5 Translational speeds of martian DDs, including the number of DDs considered (N), the observed range, mean and median values

Method	Reference	N	Range (m/s)	Mean (m/s)	Median (m/s)
Orbiter images	Stanzel et al. (2008)	194	1–59	13	10.4
	Reiss et al. (2011b)	26	3–22	12	11.1
	Reiss et al. (2014a)	44	4–25	12	11.5
Lander images	Greeley et al. (2010)	498	0.1–27	N/A	1.5–2.5[a]

[a]Range of medians from three analyzed seasons

et al. (2014a), orbital measurements of DD translational ground speeds and directions of motion can therefore be used as a proxy for local regional wind regimes within the PBL on Mars.

Table 5 summarizes horizontal ground speed measurements from both orbiting and landed missions on Mars. Greeley et al. (2010) measured horizontal ground speeds of nearly 500 DDs at the MER Spirit landing site in Gusev crater based on time sequential lander images. Although the range of translational ground speeds of the large dataset obtained by Greeley et al. (2010) is in agreement with orbitally-derived ranges of DD translational ground speeds, the median speeds are lower than those measured from orbit. Because orbital data can only resolve larger DDs, in contrast to lander missions, this might suggest that larger DDs move faster. However, neither the orbital nor landing site nor even terrestrial data sets have found a relationship between translational ground speeds and DD diameter (Stanzel et al. 2008; Greeley et al. 2010; Balme et al. 2012; Reiss et al. 2014a). This implies that the differences between the globally-derived orbiter and locally-obtained lander results are reflecting differences in global and local wind regimes on Mars.

Tangential Velocities The threshold friction velocity required to directly entrain surface dust into the thin martian atmosphere is estimated to be $\sim 30\ \mathrm{m\,s^{-1}}$ (Greeley and Iversen 1985). Thus measurements of tangential velocities provide a first order estimate of the intensity of DDs and their capacity to lift dust from the martian surface, although many other mechanisms are suggested to lower the threshold friction velocity for dust lifting by DDs on Mars and Earth (see Chaps. 10 and 11). Measurements of tangential velocities from orbit are rare, because multiple satellite images taken of the same DD within a short time period are needed. In addition, contrast features in the DD cloud must be visible for measuring the displacement between the image observations. Cantor et al. (2006) measured a tangential velocity of $14.1 \pm 0.3\ \mathrm{m\,s^{-1}}$ in the outermost visible part of a DD, with a diameter of ~ 370 m and a height of ~ 900 m, from two VO images taken 5 seconds apart. Choi and Dundas (2011) used the blue-green, red, and infrared central color swaths of the HiRISE camera, taken 0.1 seconds apart, to automatically track contrast features in four DDs with diameters between ~ 25 and 250 m and heights between ~ 150 and 650 m. Their wind vector measurements yielded typical tangential velocities approaching ~ 20 and $30\ \mathrm{m\,s^{-1}}$, with maximum velocities reaching $\sim 45\ \mathrm{m\,s^{-1}}$. The strongest tangential velocities in all DDs occurred along the outer edge of the visible dust columns (Choi and Dundas 2011).

Terrestrial in situ measurements by Sinclair (1973) and laboratory studies by Greeley et al. (2003) of DD velocities are in good agreement with a Rankine vortex model, in which the core exhibits solid body rotation and tangential velocities linearly increase with radius r,

Table 6 Tangential velocities of DDs on Mars and Earth, including the number of DDs considered (N), the mean velocity (V_{mean}), and the maximum velocity (V_{max})

Method	Reference	N	V_{mean} ($m\,s^{-1}$)	V_{max} ($m\,s^{-1}$)
Orbiter images (Mars)	Cantor et al. (2006)	1	14.3	
	Choi and Dundas (2011)	4	25	45
In situ (Earth)	Ryan and Carroll (1970)	80	4.2	9.5
	Fitzjarrald (1973)	11	7.3	11.5
	Sinclair (1973)	3	10.8	11.5
	Metzger (1999)	5	13.6	22
	Metzger et al. (2011)	12	11.7	16
Theoretical calculation (Mars)	Rennó et al. (2000)	19	11.5	17.7
	Ringrose et al. (2003)	7	16.6	46

reaching a peak outside the radius, and then decreasing as a function of r^{-1}. A radial profile of one DD measured by Choi and Dundas (2011) was in good agreement with the Rankine vortex model, but another one decreased as a function of $r^{-1/2}$ instead of r^{-1} outside the solidly rotating central region (Choi and Dundas 2011). A velocity profile closer to the $r^{-1/2}$ distribution was also measured in situ on Earth by Tratt et al. (2003) and attributed to non-conservation of angular momentum caused by frictional losses near the surface (Tratt et al. 2003; Balme and Greeley 2006). For further discussion of vortex models used to describe DDs, we refer the reader to Chap. 6.

Table 6 summarizes tangential velocity measurements of DDs by orbital instruments on Mars, by in situ measurements on Earth, and by modeling approaches based on in situ measurements for Mars. The tangential velocities of DDs on Mars measured from orbit by Choi and Dundas (2011) are about two to three times higher than terrestrial peak tangential velocities measured in situ by Ryan and Carroll (1970), Fitzjarrald (1973), Sinclair (1973), Metzger (1999), Tratt et al. (2003), and Metzger et al. (2011). One reason for this discrepancy might be that the martian orbital measurements are from large and relatively intense DDs (>50 m in diameter), whereas terrestrial measurements mainly sample smaller DDs (<10 m in diameter). The largest data set provided by Ryan and Carroll (1970) indicated that larger DDs are more intense, exhibiting greater tangential velocities (Balme and Greeley 2006). The disparity between orbital measurements of martian DD tangential velocities and those measured *in situ* on Earth is yet another example of the disconnect in understanding between orbital and surface observations of DDs.

Tangential velocities are related to the pressure drop across the vortex by assuming cyclostrophic balance (Rennó et al. 1998). For further details we refer the reader to Rennó et al. (1998) and Chaps. 6 and 10. Tratt et al. (2003) applied this method to two terrestrial DDs passing directly over the instruments and observed a good agreement between measured and calculated tangential velocities. Theoretically-calculated minimum tangential velocities for DDs on Mars by Rennó et al. (2000) using meteorological data (pressure and temperature pertubations) from the Mars Pathfinder landing site ranged between 8.6 and 17.7 $m\,s^{-1}$. These predicted minimum values for convective vortices are uncertain because it is unclear if the DDs passed directly over the lander instruments.

Ringrose et al. (2003) analyzed meteorological data from VL2 and found that seven convective vortices passed directly over the instruments. They calculated tangential velocities

Table 7 DD durations on Mars, including the number of DDs measured (N), mean and minimum duration, and mean DD diameter

Method	Reference	N	Mean duration (min)	Minimum duration (min)	Mean diameter (m)
Orbiter images	Stanzel et al. (2008)	12	13	3.7–32.5	185
	Reiss et al. (2011a, 2011b)	2	N/A	26	700
	Reiss et al. (2011a, 2011b)	1	N/A	74	820
Lander images	Greeley et al. (2006)	533	2.83	0.35–32.27	19
	Greeley et al. (2010)	101	2.15	N/A	24
	Greeley et al. (2010)	127	2.6	N/A	39

ranging between 2.8 and 46 $m\,s^{-1}$ for inferred core diameters between ~22 and ~313 m, using the Rankine vortex approximation. Their derived peak velocity of 46 $m\,s^{-1}$ is in agreement with peak velocities derived from orbital measurements by Choi and Dundas (2011).

DD Duration Terrestrial observations and duration measurements of DDs imply that larger DDs are active longer than smaller ones (Flower 1936; Ives 1947; Sinclair 1969; Snow and McClelland 1990; Metzger 1999; Pathare et al. 2010). On Mars, Stanzel et al. (2008) and Reiss et al. (2011b) constrained minimum DD durations with orbital data sets. Stanzel et al. (2008) measured translational ground speeds of 12 DDs in HRSC images and divided the length of their adjacent tracks (for a detailed review about DD tracks we refer the reader to Chap. 4) by their speed. The derived durations are minimum values, because the DDs used as the end point were still active during observation and likely continued. In addition, the starting point of the DD tracks are uncertain because their formation depends on surface substrate properties (see Chap. 4), hence the observed DDs might have been active before the track formation started. The calculated minimum durations by Stanzel et al. (2008) range between 3.7 and 32.5 minutes with a mean value of 13 minutes for DDs characterized by a mean diameter of about 185 m. Reiss et al. (2011b) observed DDs in HRSC data, some of which some could be retraced to DDs in a MOC WA image acquired 26 minutes earlier, using the measured translational ground speeds and directions of motion from the HRSC image data. Two DDs with mean diameters of ~700 m had a minimum lifetime of 26 minutes. The inferred minimum duration for one DD with a diameter of ~820 m is 74 minutes based on its adjacent track length and translational ground speed. Greeley et al. (2006, 2010) derived mean minimum durations of ~2.5 minutes for DDs with a mean diameter of ~30 m at the MER Spirit landing site in Gusev crater on Mars. Table 7 summarizes minimum and mean durations of DDs on Mars. There is a diameter-duration relationship for DDs on Mars in which larger DDs last longer than smaller ones; a similar relationship has been determined from *in situ* measurements on both Mars and Earth (Lorenz 2013), although the underlying reasons for it have yet to be determined.

3.2.6 Relation of DDs to Regional/Global Circulation

The significance of the role of DDs in the martian climatic system has long been debated. Entrained dust affects both the thermal structure and the circulation of the atmosphere (e.g., Kahn et al. 1992). Therefore, if DDs loft a sufficient quantity of dust, they could contribute significantly to the seasonal and interannual dust cycles that dominate Mars' climate system. Years ago, it was postulated that DDs could trigger dust storms (Kahn et al. 1992), but orbital observations of DDs and dust storms clearly show that this is not the case (e.g., Balme et al. 2003; Cantor et al. 2006; Stanzel et al. 2008). Rather, DDs appear to be responsible for producing a global background haze (with optical depths in the visible of ~0.1–0.3) that persists year-round, even during northern spring when dust storm occurrence is at a minimum (e.g., Clancy et al. 2000). Evidence for this is inferred from dust flux estimates from both lander and orbiter observations of DD activity (Ryan and Lucich 1983; Murphy and Nelli 2002; Balme et al. 2003; Ferri et al. 2003; Fisher et al. 2005; Balme and Greeley 2006; Cantor et al. 2006; Whelley and Greeley 2008; Reiss et al. 2011b). Regional to global-scale dust storms entrain much of the dust that contributes to the measured AOT, but it is clear that they do not account for all of it; DDs have been considered a major source of the remaining component (Guzewich et al. 2015). In addition, recent aerosol measurements indicate the presence of a fine particle population with a mean radius of ~50 nm (Fedorova et al. 2014). Without a continuous source from the surface, these aerosols would quickly coagulate; Fedorova et al. (2014) attributed their production to continuous lofting from both wind stress and DDs. Finally, many studies using GCM simulations agree that DDs are required to maintain the observed low-level background dust loading during the aphelion season, when dust storms are not prevalent and thus cannot be a reliable source of entrained dust (Newman et al. 2002a; Basu et al. 2004; Kahre et al. 2006).

A major question in martian science is to understand how long-term changes in dust loading affect the climate. Mars has undergone dramatic changes in orbital variations in the past 10 Ma, with the eccentricity ranging from 0 to 0.115 (currently at a relatively high value of 0.0935), obliquity ranging between 15°–45° (currently at a moderate value of 25.19°), and continual precession of the aerocentric longitude of perihelion (perihelion currently occurs at $L_s = 251°$, nearly coinciding with southern summer solstice at $L_s = 270°$); each of these orbital parameters oscillates with periods ranging from ~50 ka–2.4 Ma (Laskar et al. 2004). Atmospheric modeling under different orbital states has provided some insight on how DD activity impacts erosion and deposition of dust on the surface. At a high obliquity (45°), DDs are predicted to be more active than they are today, but they would loft far less dust than that entrained by saltation impact driven by high wind stresses; conversely, at low obliquity (<15°), DDs may be the dominant source of atmospheric dust (Newman et al. 2005). In the current perihelion state, dust lifting is thought to be caused both by DDs and saltation impact; however, DDs are predicted to dominate dust entrainment in the opposing perihelion position, which last occurred 22.5 ka (Haberle et al. 2006).

Predictions of dust erosion and deposition rates by both wind stress and DDs can be used to place constraints on the age of dust deposits on the martian surface, connecting climate cycles to geological processes on the surface. Vast swaths of dust-covered terrain in the low and mid northern latitudes of Mars, often termed the "low thermal inertia continents", are regions of net dust deposition (this includes Amazonis Planitia). Estimating that they could be as much as 2 m thick, Christensen (1986) proposed that these regions are currently accumulating new deposits, but that in opposing perihelion states, these deposits would erode, perhaps as a result of DD activity, and reform elsewhere. Haberle et al. (2006) tested this hypothesis with GCM simulations, comparing the present day to that at 22.5 and 72.5 ka, when Mars was at the opposite longitude of perihelion. They found that over the last ~100 ka, DDs have caused net erosion of these deposits. Newman et al. (2005) looked at longer timescales when the obliquity was higher, finding that some of the dusty areas undergo accumulation at high obliquity (>35°), and that as a result these deposits may have formed ~> 500 ka. However, their model results predict regions of net deposition that only partially correspond with the pattern of observed dust deposits, indicating that other processes that have not yet been considered must also impact the evolution of dust reservoirs

Table 8 Comparison of dust loading measurements on Earth and Mars

Earth			Mars	
Global mean AOT and DOT:				
Data/Model	EY	550 nm AOT	MY (EY)	DOT
MODIS	2002	0.188[a]	25 (00/05–02/04)	0.67*
	2001	0.159[b]	26 (02/04–04/03)	0.43
SeaWiFS	2001	0.124[b]	27 (04/03–06/01)	0.40
	1997–2010	0.13[c]	28 (06/01–07/12)	0.47*
TOMS	2001	0.153[b]	29 (07/12–09/10)	0.38
MISR	2001	0.167[b]	30 (09/10–11/09)	0.33
GOCART	2001	0.118[b]	31 (11/09/13/07)	0.36
	2000–2007	0.14[d]	7 MY mean	0.43
Model	EY	550 nm DOT	MY w/o GDE	0.39
GOCART	2000–2007	0.042[d]		
AeroCom	N/A	0.023[e]		
Mineral dust emission (Tg/Earth yr):				
1000–4000[f]			7 MY mean	400–1300
			MY w/o GDE	400–1200
			MY 25	600–2000
Percentage lofted by daytime dry convective turbulence:				
~3.4 %[g]			~25–75 %[h,i,j]	

*MYs with a global-scale dust event

Table values from [a]Bellouin et al. (2008), [b]Chin et al. (2014), [c]Hsu et al. (2012), [d]Chin et al. (2009), [e]Huneeus et al. (2011), [f]Boucher et al. (2013), [g]Jemmett Smith et al. (2015), [h]Cantor et al. (2006), [i]Kahre et al. (2006), [j]Newman et al. (2002b)

on Mars. The full impact of DDs on dust erosion, transport, and deposition is not yet understood, but atmospheric modeling suggests that these processes can play a significant role in altering both the climate and surface geology on Mars; the reader is referred to Chap. 11 for further discussion.

4 Conclusions

4.1 High-Level Comparison of Atmospheric Dust on Earth and Mars

Although data sets and methods differ, it is possible to perform a semi-quantitative comparison of global mean dust measurements on Earth and Mars, placing these estimates in a planetary context. Table 8 lists the global mean AOT and DOT, dust emission rate, and the percentage estimated to be contributed by convective turbulence (e.g., by DDs).

Global mean AOT values (at a wavelength of 550 nm) for Earth have been measured by a number of sensors and simulated with at least one aerosol model, with estimates ranging from 0.118 to 0.188. However, there are not yet any estimates of the global mean DOT from orbital data sets. Chin et al. (2009) used GOCART to estimate that mineral dust consisted of ~30 % of the AOT, with a 550 nm DOT of 0.042, second in magnitude only to sulfate aerosols. The AeroCom project simulated the Earth's atmospheric dust budget with a median model that uses monthly output from twelve aerosol models. Global DOT estimates from these models ranged from 0.01 to 0.053, with a median model estimate at 0.023.

A similar estimate can be made for Mars, using the climatology of Montabone et al. (2015). The 9.3 μm equivalent column DOTs were converted to broadband visible DOTs by multiplying by a factor of 2.0, which is consistent with that expected for 1.5–2.0 μm dust particles (e.g., Lemmon et al. 2015). Further, full-extinction DOTs were estimated from absorption-only values by multiplying by an additional suggested factor of 1.3. The resulting area-weighted and annually-averaged DOTs for each of seven MYs are listed in Table 8, as is the mean DOT for all seven years (0.43). Also listed is the mean DOT excluding MYs 25 and 28 (0.39), which experienced global-scale dust events (see Fig. 7, and note that the global mean DOT for MY 25 is twice that of MY 30). Although the martian DOTs are broadband equivalent values and the terrestrial DOTs were estimated at 550 nm, the low Angstrom coefficient values found for mineral dust imply that the terrestrial broadband AOT and DOT values are similar to those estimated at 550 nm, allowing for approximate comparisons of the two planetary atmospheres. The global mean martian optical depths are ∼2.5–3 times that of terrestrial aerosols, and ∼15–20 times that contributed by dust on Earth. Mars is, indeed, a dusty planet.

As stated in Sect. 2.1.1, annual mineral dust emission rates on Earth range from 1000–4000 Tg/EY (Boucher et al. 2013). To estimate a comparable mean flux of dust into the Martian atmosphere, we consider several observations. Optical depth is proportional to the column number density, area, and extinction efficiency. Column mass (CM) is proportional to column number density, volume, and particle density. Thus column mass may be expressed as $CM = (AOT/Q) \times (V/A) \times \rho$; where we estimate the particle volume to area ratio, V/A, to be $4/3 \times 1.5 \times 10^{-6}$ m (Lemmon et al. 2004). This estimate assumes the CM is close to that for mono-disperse spheres; the extinction efficiency, Q, is 2–4; and the particle density, ρ, is 2.5 $\mathrm{g\,cm^{-3}}$. Thus for the 7 MY mean DOT, we find a column mass of $0.5-1.1 \times 10^{-6}\ \mathrm{g\,m^{-2}}$ and an atmospheric aerosol mass of 80–150 Tg. During the aftermath of the 1971, 1977a, 1977b, and 2007 global dust storms, when lifting was inhibited, the sedimentation timescale was 42–67 sols (Pollack et al. 1979; Fenton et al. 1997; Lemmon et al. 2015). Thus the characteristic supply (and sedimentation) rate must be ∼1–4 $\mathrm{Tg\,sol^{-1}}$, or ∼400–1300 Tg per Earth year. Years without global-scale dust events have a slightly lower mean emission rate of ∼400–1200 $\mathrm{Tg\,yr^{-1}}$; the dustiest year with an available annual mean DOT is MY 25, which had a dust emission rate of ∼600–2000 $\mathrm{Tg\,yr^{-1}}$. The martian dust emission estimate is comparable, if on the low end, to that of Earth. Given that the continental surface area of Earth is close to the global surface area of Mars, this result suggests that the mean dust lifting rate per unit area is similar on both planets. This result is somewhat surprising, given the quite variable nature of dust emission in space and time on both Earth and Mars (e.g., little dust lifting is expected over terrestrial rain forests or on the martian polar caps). For comparison, meteoric input of dust to the Martian atmosphere is relatively insignificant, at 0.002–0.003 $\mathrm{Tg\,yr^{-1}}$ (Flynn 1992), similar to that estimated for Earth, 0.0018–0.1 $\mathrm{Tg\,yr^{-1}}$ (Plane 2012).

Jemmett-Smith et al. (2015) estimated that only ∼3.4 % of the mineral dust emission rate is caused by convective turbulence. Without better measurements from the surface, it is difficult to make a similarly precise estimate for Mars. Atmospheric models that include parameterizations of dust flux from convective turbulence suggest that daytime convective activity contributes to as much as 30–50 % of the global martian dust budget (Newman et al. 2002b; Kahre et al. 2006). A second approach makes use of the global-scale DD inventory of Cantor et al. (2006), who estimated a global dust flux $> 4 \times 10^{-3}\ \mathrm{kg\,m^{-2}\,MY^{-1}}$, which amounts to >600 Tg/MY (or >300 Tg yr^{-1}). Given the range of dust emission rates in a year without a global-scale dust event, DDs would contribute to >25–75 % of the estimated dust flux, a percentage similar to that predicted by the atmospheric models.

From these rough estimates, we may conclude that, although mean dust emission rates are similar on Earth and Mars, Mars' dust budget is more dependent on small-scale convective turbulence, whereas Earth's dust budget is generated mainly by frontal storms. Such storms occur on Mars as well (e.g. Wang and Richardson 2015), but the current understanding suggests that, unlike on Earth, dust lofting by convective activity is comparable in magnitude to these storms on Mars.

4.2 Summary and Knowledge Gaps

This review demonstrates the breadth of work being done with orbiting spacecraft data to understand lofted dust, particularly that entrained by daytime convective turbulence (most notably by DDs), on both Earth and Mars. In the past two decades, the availability of several high quality, high resolution spacecraft instruments orbiting Earth has led to an improved understanding of the nature of entrained mineral dust. This includes not just dust sources, transport pathways, and sinks, but also characteristics of aerosol particle size, shape, composition, and optical properties. With this knowledge base, connections are being made between mineral dust patterns and weather patterns, both on daily and interannual time scales. The same is true of Mars, but to a much more limited extent, with progress being constrained by the type and extent of the available data sets. Despite this impediment, ongoing work into convectively-lofted dust on Mars is in many ways more advanced than that on Earth.

It is clear that on both Earth and Mars, mineral dust aerosols play a significant and yet largely unconstrained role in modulating the climate, by acting as an internal forcing mechanism in the climate system. Atmospheric dust appears to play a relatively more significant role on Mars, where dust is plentiful, sinks are more limited (i.e., there are no oceans), and there are fewer processes competing with dust (e.g., clouds and precipitation are limited). Similarly, convectively-lofted dust is much more prominent on Mars, where the thin atmosphere and typical daytime superadiabatic conditions produce vigorous turbulent activity that encourages the formation of dusty plumes and DDs.

Several years of monitoring dust and DD activity on Mars has led to a general understanding about the spatial and temporal patterns of dust lifting. DDs are widespread phenomena on Mars, reaching to nearly all latitudes and elevations. Like those on Earth, they occur most often when insolation is at its seasonal and diurnal peak. This correlation is modulated by weather patterns, most notably those that generate dust storms, which can either suppress or enhance DD production. In addition, local factors such as albedo contrast on the surface, nearby topographic relief, and changes in dust source availability also appear to influence DD formation. It is likely that remote monitoring of DD occurrence and density would be of value in studying similar processes on Earth.

Physical characteristics of martian DDs observed from orbit differ from those measured from landed missions, which are in some ways similar to many of the field surveys performed on Earth. From orbit, DDs are generally wider, taller, translate along the ground faster, and rotate faster than those measured from the surface. To some degree, these differences are a factor of image resolution: for example, only the largest DDs are visible from orbit, whereas measurements of much smaller and more plentiful DDs dominate the data sets from the surface (see also Fig. 9). Despite this measurement bias, some true physical variations are clear. Examples include the towering DDs of Amazonis Planitia, which are not common in most other locations on Mars, and the fact that many DD translational ground speeds are much higher than those measured on Earth. The failure to obtain simultaneous observations of individual DDs from orbit and the surface, despite valiant attempts to do so, leaves a gap in understanding the differences between the observed DD characteristics. Given that DD dimensions and translational speeds could be indicative of conditions within the PBL, it may be of value to resolve the discrepancy between orbital and surface measurements, both for Mars and Earth atmospheric studies. This could be accomplished through coordinated orbital and surface observations on both Earth and Mars, or with LES that resolve individual DDs, thus allowing virtual viewing both from above and from the ground.

Despite decades of studying martian DDs, many questions remain: why are DDs so numerous and large in Amazonis Planitia, and less so at other similarly low-lying plains with comparable albedos, thermal inertias, and latitudes? why do DDs form at elevation on Arsia Mons, but not at the tops of other high-standing volcanoes on Mars? how do DDs contribute to surface erosion and the deposition of long-term dust reservoirs on the surface? how can DDs be used to characterize the present-day atmosphere of Mars, and how does interannual variation in DD activity interact with and reflect other interannual changes (e.g., albedo and thermal inertia patterns, annual H_2O and CO_2 cycles)? what locally-dependent factors control DD size and production rates? Further research on the expanding data sets will surely address many of these questions, as will continued attempts to model DD behavior under increasingly realistic conditions. Similar questions could be posed for Earth, using the available literature on Mars as a starting point. The first question to address is why terrestrial DDs have not yet been observed from orbit. We suggest that the numerous high resolution cameras, and perhaps lidars as well, continually monitoring the Earth's surface are an untapped resource for DD studies, both for comparative planetology with Mars and to better understand aerosol entrainment mechanisms on Earth.

In summary, dust entrained by daytime convective turbulence, particularly that of DDs, appears to impact (at least) local atmospheric conditions on Earth and the global climate on Mars. Much has been learned about atmospheric dust from orbital sensor data over the last few decades, but varying methods, data sets, and scientific goals have led to different insights on Earth and Mars. Future investigations on each planet have the potential to inform further research on the other planet. For example, scouring the extensive terrestrial image data sets for DDs, as has been done for Mars, could prove to be quite fruitful. Likewise, observations and validated simulations of terrestrial DDs from both the surface and from above could resolve the current discrepancy between landed and orbital spacecraft observations of martian DDs. Studies on both planets indicate that DDs are an integral part of the dust cycle, which likely varies in relative importance with temporal and spatial changes in climate. We propose that further progress can be accelerated when the terrestrial and planetary communities work together to determine how convectively-lofted dust impacts planetary atmospheres.

Acknowledgements The authors would like to thank two anonymous reviewers and an editor for many suggestions that greatly improved the manuscript.

References

A. Ansmann, M. Tesche, P. Knippertz, E. Bierwirth, D. Althausen, D. Müller, O. Schulz, Vertical profiling of convective dust plumes in southern Morocco during SAMUM. Tellus B **61**, 340–353 (2009). doi:10.1111/j.1600-0889.2008.00384.x

M.R. Balme, P.L. Whelley, R. Greeley, Mars: dust devil track survey in Argyre Planitia and Hellas Basin. J. Geophys. Res. **108**, 5086 (2003). doi:10.1029/2003JE002096

M. Balme, R. Greeley, Dust devils on Earth and Mars. Rev. Geophys. **44**, RG3003 (2006). doi:10.1029/2005RG000188

M.R. Balme, A. Pathare, S.M. Metzger, M.C. Towner, S.R. Lewis, A. Spiga, L. Fenton, N.O. Renno, H.M. Elliott, F.A. Saca, T. Michaels, P. Russell, J. Verdasca, Field measurements of horizontal forward motion velocities of terrestrial dust devils: towards a proxy for ambient winds on Mars and Earth. Icarus **221**, 632–645 (2012). doi:10.1016/j.Icarus2012.08.021

J.R. Banks, H.E. Brindley, Evaluation of MSG-SEVIRI mineral dust retrieval products over North Africa and the Middle East. Remote Sens. Environ. **128**, 58–73 (2013). doi:10.1016/j.rse.2012.07.017

S. Basu, M.I. Richardson, R.J. Wilson, Simulation of the Martian dust cycle with the GFDL Mars GCM. J. Geophys. Res. **109**, E11006 (2004). doi:10.1029/2004JE002243

S. Basu, J. Wilson, M.I. Richardson, A. Ingersoll, Simulation of spontaneous and variable global dust storms with the GFDL Mars GCM. J. Geophys. Res.-Planets **111** (2006). doi:10.1029/2005je002660

F. Bell, Dust devils and aviation, Meteorol. Note 27, Commonwealth of Australia, Bureau of Meteorology (1967) p. 10

J.F. Bell, M.J. Wolff, M.C. Malin, W.M. Calvin, B.A. Cantor, M.A. Caplinger, R.T. Clancy, K.S. Edgett, L.J. Edwards, J. Fahle, F. Ghaemi, R.M. Haberle, A. Hale, P.B. James, S.W. Lee, T. McConnochie, E. Noe Dobrea, M.A. Ravine, D. Schaeffer, K.D. Supulver, P.C. Thomas, Mars Reconnaissance Orbiter Mars Color Imager (MARCI): Instrument description, calibration, and performance. J. Geophys. Res. **114** (2009). doi:10.1029/2008JE003315

N. Bellouin, A. Jones, J. Haywood, S.A. Christopher, Updated estimate of aerosol direct radiative forcing from satellite observations and comparison against the Hadley Centre climate model. J. Geophys. Res. **113**, D10205 (2008). doi:10.1029/2007JD009385

S. Berthier, P. Chazette, P. Couvert, J. Pelon, F. Dulac, F. Thieuleux, C. Moulin, T. Pain, Desert dust aerosol columnar properties over ocean and continental Africa from Lidar In-Space Technology Experiment (LITE) and meteosat synergy. J. Geophys. Res. **111**, D21202 (2006). doi:10.1029/2005JD006999

K.K. Biener, P.E. Geissler, A.S. McEwen, C. Leovy, Observations of martian dust devils in MOC wide angle camera images, in *Lunar Planet. Sci. Conf. XXXIII, League City, TX* (2002) Abst. #2004

O. Boucher, D. Randall, P. Artaxo, C. Bretherton, G. Feingold, P. Forster, V.-M. Kerminen, Y. Kondo, H. Liao, U. Lohmann, P. Rasch, S.K. Satheesh, S. Sherwood, B. Stevens, X.Y. Zhang, in *Climate Change 2013: The Physical Science Basis*, ed. by T.F. Stocker, D. Qin, G.-K. Plattner, M. Tignor, S.K. Allen, J. Boschung, A. Nauels, Y. Xia, V. Bex, P.M. Midgley (Cambridge University Press, New York, 2013), p. 571. doi:10.1017/CBO9781107415324.016

G.A. Briggs, W.A. Baum, J. Barnes, Viking Orbiter imaging observations of dust in the Martian atmosphere. J. Geophys. Res. **84**, 2795–2820 (1979). doi:10.1029/JB084iB06p02795

B.A. Cantor, MOC observations of the 2001 Mars planet-encircling dust storm. Icarus **186**(1), 60–96 (2007). doi:10.1016/j.icarus.2006.08.019

B.A. Cantor, P.B. James, M. Caplinger, M.J. Wolff, Martian dust storms: 1999 Mars Orbiter Camera observations. J. Geophys. Res. **106**, 23653 (2001). doi:10.1029/2000JE001310

B. Cantor, M. Malin, K.S. Edgett, Multiyear Mars Orbiter Camera (MOC) observations of repeated Martian weather phenomena during the northern summer season. J. Geophys. Res. **107**(E3), 5014 (2002). doi:10.1029/2001JE001588

B.A. Cantor, K.M. Kanak, K.S. Edgett, Mars Orbiter camera observations of Martian dust devils and their tracks (September 1997 to January 2006) and evaluation of theoretical vortex models. J. Geophys. Res. **111**, E12002 (2006). doi:10.1029/2006JE002700

B.A. Cantor, P.B. James, W.M. Calvin, MARCI and MOC observations of the atmosphere and surface cap in the north polar region of Mars. Icarus **208**, 61–81 (2010). doi:10.1016/j.icarus.2010.01.032

D. Carrer, X. Ceamanos, B. Six, J.-L. Roujean, AERUS-GEO: a newly available satellite-derived aerosol optical depth product over Europe and Africa. Geophys. Res. Lett. **41**, 7731–7738 (2014). doi:10.1002/2014GL061707

I. Chiapello, C. Moulin, J.M. Prospero, Understanding the long-term variability of African dust transport across the Atlantic as recorded in both Barbados surface concentrations and large-scale TOMS optical thickness. J. Geophys. Res. **110**, D18S10 (2005). doi:10.1029/2004JD005132

I. Chiapello, Dust observations and climatology, in *Mineral Dust: A Key Player in the Earth System*, ed. by P. Knippertz, J.-B.W. Stuut (Springer, Dordrecht, 2014), p. 149. doi:10.1007/978-94-017-8978-3

M. Chin, T. Diehl, O. Dubovik, T.F. Eck, B.N. Holben, A. Sinyuk, D.G. Streets, Light absorption by pollution, dust, and biomass burning aerosols: a global model study and evaluation with AERONET measurements. Ann. Geophys. **27**, 3439–3464 (2009). doi:10.5194/angeo-27-3439-2009

M. Chin, T. Diehl, Q. Tan, J.M. Prospero, R.A. Kahn, L.A. Remer, H. Yu, A.M. Sayer, H. Bian, I.V. Geogdzhayev, B.N. Holben, S.G. Howell, B.J. Huebert, N.C. Hsu, D. Kim, T.L. Kucsera, R.C.L.M.I. Mishchenko, X. Pan, P.K. Quinn, G.L. Schuster, D.G. Streets, S.A. Strode, O. Torres, X.-P. Zhao, Multidecadal aerosol variations from 1980 to 2009: a perspective from observations and a global model. Atmos. Chem. Phys. **14**, 3657–3690 (2014). doi:10.5194/acp-14-3657-2014

D.S. Choi, C.M. Dundas, Measurements of martian dust devil winds with HiRISE. Geophys. Res. Lett. **38** (2011). doi:10.1029/2011GL049806

P.R. Christensen, Regional dust deposits on Mars: physical properties, age, and history. J. Geophys. Res. **91**(B3), 3533–3545 (1986). doi:10.1029/JB091iB03p03533

P.R. Christensen, B.M. Jakosky, H.H. Kieffer, M.C. Malin, H.Y. McSween Jr., K. Nealson, G.L. Mehall, S.H. Silverman, S. Ferry, M. Caplinger, M. Ravine, The thermal emission imaging system (THEMIS) for the Mars 2001 Odyssey mission. Space Sci. Rev. **110**, 85 (2004). doi:10.1007/978-0-306-48600-5_3

R.T. Clancy, B.J. Sandor, M.J. Wolff, P.R. Christensen, M.D. Smith, J.C. Pearl, B.J. Conrath, R.J. Wilson, An intercomparison of ground-based millimeter, MGS TES, and Viking atmospheric temperature measurements: seasonal and interannual variability of temperatures and dust loading in the global Mars atmosphere. J. Geophys. Res. **105**, 9553–9571 (2000). doi:10.1029/1999JE001089

D.S. Colburn, J.B. Pollack, R.M. Haberle, Diurnal variations in optical depth at Mars. Icarus **79**, 159–189 (1989). doi:10.1016/0019-1035(89)90114-0

S.M. Cowie, P. Knippertz, J.H. Marsham, Are vegetation-related roughness changes the cause of the recent decrease in dust emission from the Sahel? Geophys. Res. Lett. **40**, 1868–1872 (2013). doi:10.1002/grl.50273

G.E. Cushing, T.N. Titus, P.R. Christensen, THEMIS VIS and IR observations of a high-altitude martian dust devil. Geophys. Res. Lett. **32**, 23 (2005). doi:10.1029/2005GL024478

J.W. Deardorff, Observed characteristics of the outer layer, in *Short Course on the Planetary Boundary Layer*, ed. by A.K. Blackadar (Am. Meteorol. Soc., Boston, 1978), 101 pp.

K.S. Edgett, M.C. Malin, Martian dust raising and surface albedo controls: thin, dark (and sometimes bright) streaks and dust devils in MGS MOC high resolution images, in *Lunar Planet. Sci. Conf. XXXI, Houston, TX* (2000) Abst. #1073

A.T. Evan, J. Dunion, J.A. Foley, A.K. Heidinger, C.S. Velden, New evidence for a relationship between Atlantic tropical cyclone activity and African dust outbreaks. Geophys. Res. Lett. **33**, L19813 (2006). doi:10.1029/2006GL026408

A.T. Evan, S. Mukhopadhyay, African dust over the northern tropical Atlantic: 1955–2008. J. Appl. Meteorol. Climatol. **49**, 2213–2229 (2010). doi:10.1175/2010JAMC2485.1

A.T. Evan, C. Flamant, S. Fiedler, O. Doherty, An analysis of aeolian dust in climate models. Geophys. Res. Lett. **41**, 5996–6001 (2014). doi:10.1002/2014GL060545

A.A. Fedorova, F. Montmessin, A. Rodin, O.I. Korablev, A. Määttänen, L. Maltagliati, J.-L. Bertaux, Evidence for a bimodal size distribution for the suspended aerosol particles on Mars. Icarus **231**, 239–260 (2014). doi:10.1016/j.icarus.2013.12.015

L.K. Fenton, J.C. Pearl, T.Z. Martin, Mapping Mariner 9 dust opacities. Icarus **130**, 115–124 (1997). doi:10.1006/icar.1997.5810

L.K. Fenton, R. Lorenz, Dust devil height and spacing with relation to the martian planetary boundary layer thickness. Icarus **260**, 246–262 (2015). doi:10.1016/j.icarus2015.07.028

F. Ferri, P.H. Smith, M. Lemmon, N.O. Rennó, Dust devils as observed by Mars Pathfinder. J. Geophys. Res. **108** (2003). doi:10.1029/2000JE001421

J.A. Fisher, M.I. Richardson, C.E. Newman, M.A. Szwast, C. Graf, S. Basu, S.P. Ewald, A.D. Toigo, R.J. Wilson, A survey of martian dust devil activity using Mars global surveyor Mars orbiter camera images. J. Geophys. Res. **110**, E03004 (2005). doi:10.1029/2003JE002165

D.E. Fitzjarrald, A field investigation of dust devils. J. Appl. Meteorol. **12**, 808–813 (1973). doi:10.1175/1520-0450(1973)012<0808:AFIODD>2.0.CO;2

W.D. Flower, Sand devils. London Meteorol. Off. Prof. Notes. 5, No. 71 (1936)

G.J. Flynn, The meteoritic contribution to dust and aerosols in the atmosphere of Mars, in Workshop on the Martian Surface and Atmosphere Through Time, Boulder, CO, USA, Report No. NASA-CR-190418, pp. 51–52 (1992)

G.R. Foltz, M.J. McPhaden, Trends in Saharan dust and tropical Atlantic climate during 1980–2006. Geophys. Res. Lett. **35**, L20706 (2008). doi:10.1029/2008GL035042

F. Forget, F. Hourdin, R. Fournier, C. Hourdin, O. Talagrand, M. Collins, S.R. Lewis, P.L. Read, J.-P. Huot, Improved general circulation models of the martian atmosphere from the surface to above 80 km. J. Geophys. Res. **104**(E10), 24155–24175 (1999). doi:10.1029/1999JE001025

J.R. Garratt, *The Atmospheric Boundary Layer* (Cambridge Univ. Press, New York, 1994)

S. Gassó, A.F. Stein, Does dust from Patagonia reach the sub-Antarctic Atlantic Ocean? Geophys. Res. Lett. **34**, L01801 (2007). doi:10.1029/2006GL027693

A. Gibbons, F. Yang, P. Mlsna, P. Geissler, Automated procedures for detecting martian dust devils, in *Lunar Planet. Sci. Con. XXXVI, League City, TX* (2005) Abst. #2005

P.J. Gierasch, R.M. Goody, A model of a Martian Great dust storm. J. Atmos. Sci. **30**(2), 169–179 (1973). doi:10.1175/1520-0469(1973)030<0169:AMOAMG>2.0.CO;2

D.A. Gillette, P.C. Sinclair, Estimation of suspension of alkaline material by dust in the United States. Atmos. Environ. **24A**(5), 1135–1142 (1990). doi:10.1016/0960-1686(90)90078-2

P. Ginoux, J.M. Prospero, T.E. Gill, N.C. Hsu, M. Zhao, global-scale attribution of anthropogenic and natural dust sources and their emission rates based on MODIS Deep Blue aerosol products. Rev. Geophys. **50**, RG3005 (2012). doi:10.1029/2012RG000388

A. Goudie, N. Middleton, *Desert Dust in the Global System* (Springer, Berlin, 2006)

D. Grassi, V. Formisano, F. Forget, C. Fiorenza, N.I. Ignatiev, A. Maturilli, L.V. Zasova, The martian atmosphere in the region of Hellas basin as observed by the planetary Fourier spectrometer (PFS-MEX). Planet. Space Sci. **55**, 1346–1357 (2007). doi:10.1016/j.pss.2006.12.006

R. Greeley, J. Iversen, *Wind as Geologic Process on Earth, Mars, Venus, and Titan* (Cambridge Univ. Press, New York, 1985)

R. Greeley, M.R. Balme, J.D. Iversen, S.M. Metzger, R. Mickelson, J. Phoreman, B. White, Martian dust devils: laboratory simulations of particle threshold. J. Geophys. Res. **108** (2003). doi:10.1029/2002JE001987

R. Greeley, P.L. Whelley, R.E. Arvidson, N.A. Cabrol, D.J. Foley, B.J. Franklin, P.G. Geissler, M.P. Golombek, R.O. Kuzmin, G.A. Landis, M.T. Lemmon, L.D.V. Neakrase, S.W. Squyres, S.D. Thompson, Active dust devils in Gusev crater, Mars: observations from the Mars Exploration Rover Spirit. J. Geophys. Res. **111**, E12S09 (2006). doi:10.1029/2006JE002743

R. Greeley, D.A. Waller, N.A. Cabrol, G.A. Landis, M.T. Lemmon, L.D.V. Neakrase, M. Pendeleton Hoffer, S.D. Thompson, P.L. Whelley, Gusev crater, Mars: observations of three dust devil seasons. J. Geophys. Res. **115**, E00F02 (2010). doi:10.1029/2010JE003608

S.D. Guzewich, A.D. Toigo, L. Kulowski, H. Wang, Mars Orbiter Camera climatology of textured dust storms. Icarus **258**, 1–13 (2015). doi:10.1016/j.icarus.2015.06.023

R.M. Haberle, M.A. Kahre, J.R. Murphy, Role of dust devils and orbital precession in closing the Martian dust cycle. Geophys. Res. Lett. **33**, L19504 (2006). doi:10.1029/2006GL026188

R. Hanel, B. Conrath, W. Hovis, V. Kunde, P. Lowman, W. Maguire, J. Pearl, J. Pirraglia, C. Prabhakara, B. Schlachman, G. Levin, P. Straat, T. Burke, Investigation of the Martian environment by infrared spectroscopy on Mariner 9. Icarus **17**, 423–442 (1972). doi:10.1016/0019-1035(72)90009-7

N.G. Heavens, J.L. Benson, D.M. Kass, A. Kleinbohl, W.A. Abdou, D.J. McCleese, M.I. Richardson, J.T. Schofield, J.H. Shirley, P.M. Wolkenberg, Water ice clouds over the Martian tropics during northern summer. Geophys. Res. Lett. **37** (2010). doi:10.1029/2010gl044610

N.G. Heavens, D.J. McCleese, M.I. Richardson, D.M. Kass, A. Kleinböhl, J.T. Schofield, Structure and dynamics of the Martian lower and middle atmosphere as observed by the Mars Climate Sounder: 2. Implications of the thermal structure and aerosol distributions for the mean meridional circulation. J. Geophys. Res., Planets **116**, E01010 (2011a). doi:10.1029/2010JE003713

N.G. Heavens, M.I. Richardson, A. Kleinbohl, D.M. Kass, D.J. McCleese, W. Abdou, J.L. Benson, J.T. Schofield, J.H. Shirley, P.M. Wolkenberg, Vertical distribution of dust in the Martian atmosphere during northern spring and summer: high-altitude tropical dust maximum at northern summer solstice. J. Geophys. Res., Planets **116** (2011b). doi:10.1029/2010JE003692

N.G. Heavens, B.A. Cantor, P.O. Hayne, D.M. Kass, A. Kleinböhl, D.J. McCleese, S. Piqueux, J.T. Schofield, J.H. Shirley, Extreme detached dust layers near Martian volcanoes: evidence for dust transport by mesoscale circulations forced by high topography. Geophys. Res. Lett. **42**, 3730–3738 (2015). doi:10.1002/2015GL064004

J.R. Herman, P.K. Bhartia, O. Torres, C. Hsu, C. Seftor, E. Celarier, Global distribution of UV-absorbing aerosols from Nimbus 7/TOMS data. J. Geophys. Res. **102**, 16911 (1997). doi:10.1029/96JD03680

M. Herman, J.-L. Deuzé, A. Marchand, B. Roger, P. Lallart, Aerosol remote sensing from POLDER/ADEOS over the ocean: improved retrieval using a nonspherical particle model. J. Geophys. Res. **110**, D10S02 (2005). doi:10.1029/2004JD004798

G.D. Hess, K.T. Spillane, Characteristics of dust devils in Australia. J. Appl. Meteorol. **29**, 498–506 (1990). doi:10.1175/1520-0450(1990)029<0498:CODDIA>2.0.CO;2

R. Hesse, Short-lived and long-lived dust devil tracks in the coastal desert of southern Peru. Aeolian Res. **5**, 101–106 (2012). doi:10.1016/j.aeolia.2011.10.003

D.P. Hinson, M. Pätzold, S. Tellmann, B. Häusler, G.L. Tyler, The depth of the convective boundary layer on Mars. Icarus **198**, 57–66 (2008). doi:10.1016/j.icarus.2008.07.003

D.P. Hinson, H. Wang, Further observations of regional dust storms and baroclinic eddies in the northern hemisphere of Mars. Icarus **206**, 290–305 (2010). doi:10.1016/j.icarus.2009.08.019

D.P. Hinson, H. Wang, M.D. Smith, A multi-year survey of dynamics near the surface in the northern hemisphere of Mars: short-period baroclinic waves and dust storms. Icarus **219**, 307–320 (2012). doi:10.1016/j.icarus.2012.03.001

D.P. Hinson, S.W. Asmar, D.S. Kahan, V. Akopian, R.M. Haberle, A. Spiga, J.T. Schofield, A. Kleinböhl, W.A. Abdou, S.R. Lewis, M. Paik, S.G. Maalouf, Initial results from radio occultation measurements with the Mars Reconnaissance Orbiter: a nocturnal mixed layer in the tropics and comparisons with polar profiles from the Mars Climate Sounder. Icarus **243**, 91–103 (2014). doi:10.1016/j.icarus.2014.09.019

N.C. Hsu, S.C. Tsay, M.D. King, J.R. Herman, Aerosol properties over bright-reflecting source regions. IEEE Trans. Geosci. Remote Sens. **42**(3), 557–569 (2004). doi:10.1109/TGRS.2004.824067

N.C. Hsu, R. Gautam, A.M. Sayer, C. Bettenhausen, C. Li, M.J. Jeong, S.-C. Tsay, B.N. Holben, Global and regional trends of aerosol optical depth over land and ocean using SeaWiFS measurements from 1997 to 2010. Atmos. Chem. Phys. **12**, 8037–8053 (2012). doi:10.5194/acp-12-8037-2012

N.C. Hsu, M.-J. Jeong, C. Bettenhausen, A.M. Sayer, R. Hansell, C.S. Seftor, J. Huang, S.-C. Tsay, Enhanced Deep Blue aerosol retrieval algorithm: the second generation. J. Geophys. Res., Atmos. **118**, 9296–9315 (2013). doi:10.1002/jgrd.50712

J. Huang, C. Zhang, J.M. Prospero, African dust outbreaks: a satellite perspective of temporal and spatial variability over the tropical Atlantic Ocean. J. Geophys. Res. **115**, D05202 (2010). doi:10.1029/2009JD012516

N. Huneeus, M. Schulz, Y. Balkanski, J. Griesfeller, J.A. Prospero, S. Kinne, S. Bauer, O. Boucher, M. Chin, F. Dentener, T. Diehl, R. Easter, D. Fillmore, S. Ghan, P. Ginoux, A. Grini, L. Horowitz, D. Koch, M. Krol, W. Landing, X. Liu, N. Mahowald, R. Miller, J.-J. Morcrette, G. Myhre, J.E. Penner, J. Perlwitz, P. Stier, T. Takemura, C. Zender, Global dust model intercomparison in AeroCom phase I. Atmos. Chem. Phys. **11**, 7781–7816 (2011). doi:10.5194/acp-11-7781-2011

G.E. Hunt, P.B. James, Martian extratropical cyclones. Nature **278**, 531–532 (1979). doi:10.1038/278531a0

R.B. Husar, J.M. Prospero, L.L. Stowe, Characterization of tropospheric aerosols over the oceans with the NOAA AVHRR optical thickness product. J. Geophys. Res. **102**, 16889–16909 (1997). doi:10.1029/96JD04009

R.L. Ives, Behavior of dust devils. Bull. Am. Meteorol. Soc. **28**, 168–174 (1947)

I. Jankowiak, D. Tanré, Satellite climatology of Saharan dust outbreaks: method and preliminary results. J. Climate **5**(6), 646–656 (1992). doi:10.1175/1520-0442(1992)005<0646:SCOSDO>2.0.CO;2

R. Jaumann, G. Neukum, T. Behnke, T.C. Duxbury, K. Eichentopf, J. Flohrer, S. van Gasselt, B. Giese, K. Gwinner, E. Hauber, H. Hoffmann, A. Hoffmeister, U. Köhler, K.-D. Matz, T.B. McCord, V. Mertens, J. Oberst, R. Pischel, D. Reiss, E. Ress, T. Roatsch, P. Saiger, F. Scholten, G. Schwarz, K. Stephan, M. Wählisch, the HRSC Co-Investigator Team, The high-resolution stereo camera (HRSC) experiment on Mars Express: instrument aspects and experiment conduct from interplanetary cruise through the nominal mission. Planet. Space Sci. **55**, 928–952 (2007). doi:10.1016/j.pss.2006.12.003

B.C. Jemmett-Smith, J.H. Marsham, P. Knippertz, C.A. Gilkeson, Quantifying global dust devil occurrence from meteorological analyses. Geophys. Res. Lett. **42**, 1275–1282 (2015). doi:10.1002/2015GL063078

R.A. Kahn, T.Z. Martin, R.W. Zurek, S.W. Lee, The Martian dust cycle, in *Mars*, ed. by H.H. Kieffer, B.M. Jakosky, C.W. Snyder, M.S. Matthews (Univ. of Ariz. Press, Tucson, 1992), pp. 1017–1053

M.A. Kahre, J.R. Murphy, R.M. Haberle, Modeling the Martian dust cycle and surface dust reservoirs with the NASA Ames general circulation model. J. Geophys. Res. **111** (2006). doi:10.1029/2005JE002588

K.M. Kanak, D.K. Lilly, J.T. Snow, The formation of vertical vortices in the convective boundary layer. Q. J. R. Meteorol. Soc. **126**, 2789–2810 (2000). doi:10.1002/qj.49712656910

Y.J. Kaufman, I. Koren, L.A. Remer, D. Tanré, P. Ginoux, S. Fan, Dust transport and deposition observed from the Terra-Moderate Resolution Imaging Spectrometer (MODIS) spacecraft over the Atlantic Ocean. J. Geophys. Res. **110**, D10S12 (2005). doi:10.1029/2003JD004436

H.H. Kieffer, B.M. Jakosky, C.W. Snyder, The planet Mars: from antiquity to the present, in *Mars*, ed. by H.H. Kieffer, B.M. Jakosky, C.W. Snyder, M.S. Matthews (University of Arizona Press, Tucson, 1992), pp. 12–13

A. Kleinböhl, J.T. Schofield, D.M. Kass, W.A. Abdou, C.R. Backus, B. Sen, J.H. Shirley, W.G. Lawson, M.I. Richardson, F.W. Taylor, N.A. Teanby, D.J. McCleese, Mars Climate Sounder limb profile retrieval of atmospheric temperature, pressure, and dust and water ice opacity. J. Geophys. Res. **114**, E10006 (2009). doi:10.1029/2009JE003358

J. Koch, N.O. Renno, The role of convective plumes and vortices on the global aerosol budget. Geophys. Res. Lett. **32**, L18806 (2005). doi:10.1029/2005GL02342

J.F. Kok, E.J.R. Parteli, T.I. Michaels, D.B. Karam, The physics of wind-blown sand and dust. Rep. Prog. Phys. **75**, 10 (2012). doi:10.1088/0034-4885/75/10/106901

O. Korablev, V.A. Krasnopolsky, A.V. Rodin, E. Chassefière, Vertical structure of Martian dust measured by solar infrared occultations from the Phobos spacecraft. Icarus **102**, 76–87 (1993). doi:10.1006/icar.1993.1033

J. Laskar, A.C.M. Correia, M. Gastineau, F. Joutel, B. Levrard, P. Robutel, Long term evolution and chaotic diffusion of the insolation quantities of Mars. Icarus **170**, 343–364 (2004). doi:10.1016/j.icarus.2004.04.005

M. Legrand, C. N'doumé, I. Jankowiak, Satellite-derived climatology of the Saharan aerosol, in *Proc. SPIE 2309, Passive Infrared Remote Sensing of Clouds and the Atmosphere II* (1994), pp. 127–135. doi:10.1117/12.196669

M. Legrand, A. Plana-Fattori, C. N'doumé, Satellite detection of dust using the IR imagery of Meteosat: 1. Infrared difference dust index. J. Geophys. Res. **106**, 18251–18274 (2001). doi:10.1029/2000JD900749

M.T. Lemmon, M.J. Wolff, M.D. Smith, R.T. Clancy, D. Banfield, G.A. Landis, A. Ghosh, P.H. Smith, N. Spanovich, B. Whitney, P. Whelley, R. Greeley, S. Thompson, J.F. Bell, S.W. Squyres, Atmospheric imaging results from the Mars exploration rovers: Spirit and Opportunity. Science **306**, 1753–1756 (2004). doi:10.1126/science.1104474

M.T. Lemmon, M.J. Wolff, J.F. Bell, M.D. Smith, B.A. Cantor, P.H. Smith, Dust aerosol, clouds, and the atmospheric optical depth record over 5 Mars years of the Mars Exploration Rover mission. Icarus **251**, 96–111 (2015). doi:10.1016/j.icarus.2014.03.029

J. Lenoble, L. Remer, D. Tanré, *Aerosol Remote Sensing* (Springer, Heidelberg, 2013). doi:10.1007/978-3-642-17725-5

R.C. Levy, S. Mattoo, L.A. Munchak, L.A. Remer, A.M. Sayer, F. Patadia, N.C. Hsu, The Collection 6 MODIS aerosol products over land and ocean. Atmos. Meas. Tech. **6**, 2989–3034 (2013). doi:10.5194/amt-6-2989-2013

S.R. Lewis, P.L. Read, B.J. Conrath, J.C. Pearl, M.D. Smith, Assimilation of thermal emission spectrometer atmospheric data during the Mars Global Surveyor aerobraking period. Icarus **192**, 327–347 (2007). doi:10.1016/j.icarus.2007.08.009

R. Lorenz, On the statistical distribution of dust devil diameters. Icarus **215**, 381–390 (2011). doi:10.1016/j.icarus.2011.06.005

R. Lorenz, The longevity and aspect ratio of dust devils: effects on detection efficiencies and comparison of landed and orbital imaging at Mars. Icarus **226**, 964–970 (2013). doi:10.1016/j.icarus.2013.06.031

R.D. Lorenz, M.J. Myers, Dust devil hazard to aviation: a review of United States air accident reports. J. Meteorol. **30**(298), 179–184 (2005)

A. Määttänen, T. Fouchet, O. Forni, F. Forget, H. Savijärvi, B. Gondet, R. Melchiorri, Y. Langevin, V. Formisano, M. Giuranna, J.-P. Bibring, A study of the properties of a local dust storm with Mars Express OMEGA and PFS data. Icarus **201**, 504–516 (2009). doi:10.1016/j.icarus.2009.01.024

A. Määttänen, C. Listowski, F. Montmessin, L. Maltagliati, A. Reberac, L. Joly, J.-L. Bertaux, A complete climatology of the aerosol vertical distribution on Mars from MEx/SPICAM UV solar occultations. Icarus **223**, 892–941 (2013). doi:10.1016/j.icarus.2012.12.001

J.-B. Madeleine, F. Forget, E. Millour, T. Navarro, A. Spiga, The influence of radiatively active water ice clouds on the Martian climate. Geophys. Res. Lett. **39**, L23202 (2012a). doi:10.1029/2012GL053564

J.-B. Madeleine, F. Forget, A. Spiga, M.J. Wolff, F. Montmessin, M. Vincendon, D. Jouglet, B. Gondet, J.-P. Bibring, Y. Langevin, B. Schmitt, Aphelion water-ice cloud mapping and property retrieval using the OMEGA imaging spectrometer onboard Mars Express. J. Geophys. Res., Planets **117**, E00J07 (2012b). doi:10.1029/2011JE003940

B.A. Maher, J.M. Prospero, D. Mackie, D. Gaiero, P.P. Hesse, Y. Balkanski, Global connections between aeolian dust, climate and ocean biogeochemistry at the present day and at the last glacial maximum. Earth-Sci. Rev. **99**, 61–97 (2010). doi:10.1016/j.earscirev.2009.12.001

M.C. Malin, K.S. Edgett, Mars Global Surveyor Mars Orbiter Camera: interplanetary cruise through primary mission. J. Geophys. Res. **106**(E10), 23429–23570 (2001). doi:10.1029/2000JE001455

M.C. Malin, K.S. Edgett, B.A. Cantor, M.A. Caplinger, G.E. Danielson, E.H. Jensen, M.A. Ravine, J.L. Sandoval, K.D. Supulver, An overview of the 1985–2006 Mars Orbiter Camera science investigation. Mars **5** (2010). doi:10.1555/mars.2010.0001

P. Markowski, C. Hannon, Multiple-Doppler radar observations of the evolution of vorticity extrema in a convective boundary layer. Mon. Weather Rev. **134**, 355–374 (2006). doi:10.1175/MWR3060.1

B. Marticorena, B. Chatenet, J.L. Rajot, Temporal variability of mineral dust concentrations over West Africa: analyses of a pluriannual monitoring from the AMMA Sahelian Dust Transect. Atmos. Chem. Phys. **10**, 8899–8915 (2010). doi:10.5194/acp-10-8899-2010

B. Marticorena, P. Formenti, Fundamentals of aeolian sediment transport: long-range transport of dust, in *Treatise Geomorphol*, vol. 11, ed. by J. Shroder, N. Lancaster, D.J. Sherman, A.C.W. Bass (Academic Press, San Diego, 2013), pp. 64–84

L.J. Martin, R.W. Zurek, An analysis of the history of dust activity on Mars. J. Geophys. Res. **98**, 3221 (1993). doi:10.1029/92JE02937

T.Z. Martin, Thermal infrared opacity of the Mars atmosphere. Icarus **66**, 2–21 (1986). doi:10.1016/0019-1035(86)90003-5

T.Z. Martin, M.I. Richardson, New dust opacity mapping from Viking Infrared Thermal Mapper Data. J. Geophys. Res. **98**, 10941–10949 (1993). doi:10.1029/93JE01044

J.V. Martonchik, D.J. Diner, R. Kahn, B. Gaitley, B.N. Holben, Comparison of MISR and AERONET aerosol optical depths over desert sites. Geophys. Res. Lett. **31**, L16102 (2004). doi:10.1029/2004GL019807

H. Masursky, R.M. Batson, J.F. McCauley, L.A. Soderblom, R.L. Wildey, M.H. Carr, D.J. Milton, D.E. Wilhelms, B.A. Smith, T.B. Kirby, J.C. Robinson, C.B. Leovy, G.A. Briggs, T.C. Duxbury, C.H. Acton

Jr., B.C. Murray, J.A. Cutts, R.P. Sharp, S. Smith, R.B. Leighton, C. Sagan, J. Veverka, M. Noland, J. Lederberg, E. Levinthal, J.B. Pollack, J.T. Moore Jr., W.K. Hartmann, E.N. Shipley, G. de Vaucouleurs, M.E. Davies, Mariner 9 television reconnaissance of Mars and its satellites: preliminary results. Science **175**, 294–305 (1972). doi:10.1126/science.175.4019.294

J.O. Mattsson, T. Nihlén, W. Yue, Observations of dust devils in a semi-arid district of southern Tunisia. Weather **48**, 359–363 (1993). doi:10.1002/j.1477-8696.1993.tb05814.x

T. Maxworthy, A vorticity source for large-scale dust devils and other comments on naturally occurring columnar vortices. J. Atmos. Sci. **30**, 1717–1722 (1973). doi:10.1175/1520-0469(1973)030<1717:AVSFLS>2.0.CO;2

D.J. McCleese, N.G. Heavens, J.T. Schofield, W.A. Abdou, J.L. Bandfield, S.B. Calcutt, P.G.J. Irwin, D.M. Kass, A. Kleinböhl, S.R. Lewis, D.A. Paige, P.L. Read, M.I. Richardson, J.H. Shirley, F.W. Taylor, N. Teanby, R.W. Zurek, Structure and dynamics of the Martian lower and middle atmosphere as observed by the Mars Climate Sounder: seasonal variations in zonal mean temperature, dust, and water ice aerosols. J. Geophys. Res., Planets **115** (2010). doi:10.1029/2010JE003677

A.S. McEwen, M.E. Banks, N. Baugh, K. Becker, A. Boyd, J.W. Bergstrom, R.A. Beyer, E. Bortolini, N.T. Bridges, S. Byrne, B. Castalia, F.C. Chuang, L.S. Crumpler, I. Daubar, A.K. Davatzes, D.G. Deardorff, A. DeJong, W.A. Delamere, E. Noe Dobrea, C.M. Dundas, E.M. Eliason, Y. Espinoza, A. Fennema, K.E. Fishbaugh, T. Forrester, P.E. Geissler, J.A. Grant, J.L. Griffes, J.P. Grotzinger, V.C. Gulick, C.J. Hansen, K.E. Herkenhoff, R. Heyd, W.L. Jaeger, D. Jones, B. Kanefsky, L. Keszthelyi, R. King, R.L. Kirk, E.J. Kolb, J. Lasco, A. Lefort, R. Leis, K.W. Lewis, S. Martinez-Alonso, S. Mattson, G. McArthur, M.T. Mellon, J.M. Metz, M.P. Milazzo, R.E. Milliken, T. Motazedian, C.H. Okubo, A. Ortiz, A.J. Philippoff, J. Plassmann, A. Polit, P.S. Russell, C. Schaller, M.L. Searls, T. Spriggs, S.W. Squyres, S. Tarr, N. Thomas, B.J. Thomson, L.L. Tornabene, C. Van Houten, C. Verba, C. Weitz, J.J. Wray, The high resolution imaging science experiment (HiRISE) during MRO's primary science phase (PSP). Icarus **205**, 2–37 (2010). doi:10.1016/j.icarus.2009.04.023

J.B. McGinnigle, Dust whirls in north-west Libya. Weather **21**, 272–276 (1966). doi:10.1002/j.1477-8696.1966.tb05204.x

S.M. Metzger, Dust devils as aeolian transport mechanisms in southern Nevada and in the Mars Pathfinder landing site, PhD thesis, Univ. Nevada, Reno (1999), 208 pp.

S.M. Metzger, M.R. Balme, M.C. Towner, B.J. Bos, T.J. Ringrose, M.R. Patel, In situ measurements of particle load and transport in dust devils. Icarus **214**, 766–772 (2011). doi:10.1016/j.icarus.2011.03.013

L. Montabone, F. Forget, E. Millour, R.J. Wilson, S.R. Lewis, B. Cantor, D. Kass, A. Kleinböhl, M.T. Lemmon, M.D. Smith, M.J. Wolff, Eight-year climatology of dust optical depth on Mars. Icarus **251**, 65–95 (2015). doi:10.1016/j.icarus.2014.12.034

L. Montabone, S.R. Lewis, P.L. Read, Interannual variability of Martian dust storms in assimilation of several years of Mars global surveyor observations. Adv. Space Res. **36**, 2146–2155 (2005). doi:10.1016/j.asr.2005.07.047

F. Montmessin, F. Forget, P. Rannou, M. Cabane, R.M. Haberle, Origin and role of water ice clouds in the Martian water cycle as inferred from a general circulation model. J. Geophys. Res., Planets **109**, E10004 (2004). doi:10.1029/2004JE002284

C. Moulin, I. Chiapello, Evidence of the control of summer atmospheric transport of African dust over the Atlantic by Sahel sources from TOMS satellites (1979–2000). Geophys. Res. Lett. **31**, L02107 (2004). doi:10.1029/2003GL018931

D.R. Muhs, S.R. Cattle, O. Crouvi, D.-D. Rousseau, J. Sun, M.A. Zárate, Loess records, in *Mineral Dust: A Key Player in the Earth System*, ed. by P. Knippertz, J.-B.W. Stuut (Springer, Dordrecht, 2014), p. 411. doi:10.1007/978-94-017-8978-3D

D.P. Mulholland, P.L. Read, S.R. Lewis, Simulating the interannual variability of major dust storms on Mars using variable lifting thresholds. Icarus **223**, 344–358 (2013). doi:10.1016/j.icarus.2012.12.003

D.P. Mulholland, A. Spiga, C. Listowski, P.L. Read, An assessment of the impact of local processes on dust lifting in martian climate models. Icarus **252**, 212–227 (2015). doi:10.1016/j.icarus.2015.01.017

S. Murchie et al., R. Arvidson, P. Bedini, K. Beisser, J.-P. Bibring, J. Bishop, J. Boldt, P. Cavender, T. Choo, R.T. Clancy, E.H. Darlington, D. Des Marais, R. Espiritu, D. Fort, R. Green, E. Guinness, J. Hayes, C. Hash, K. Heffernan, J. Hemmler, G. Heyler, D. Humm, J. Hutcheson, N. Izenberg, R. Lee, J. Lees, D. Lohr, E. Malaret, T. Martin, J.A. McGovern, P. McGuire, R. Morris, J. Mustard, S. Pelkey, E. Rhodes, M. Robinson, T. Roush, E. Schaefer, G. Seagrave, G. Seelos, P. Silverglate, S. Slavney, M. Smith, W.J. Shyong, K. Strohbehn, H. Taylor, P. Thompson, B. Tossman, M. Wirzburger, M. Wolff, Compact reconnaissance imaging spectrometer for Mars (CRISM) on Mars reconnaissance orbiter (MRO). J. Geophys. Res. **112** (2007). doi:10.1029/2006JE002682

J.R. Murphy, S. Nelli, Mars Pathfinder convective vortices: frequency of occurrence. Geophys. Res. Lett. **29**, 11–14 (2002). doi:10.1029/2002GL015214

T. Navarro, J.-B. Madeleine, F. Forget, A. Spiga, E. Millour, F. Montmessin, A. Määttänen, Global climate modeling of the Martian water cycle with improved microphysics and radiatively active water ice clouds. J. Geophys. Res., Planets **119**, 1479–1495 (2014). doi:10.1002/2013JE004550

L.D.V. Neakrase, J. McHone, P.L. Whelley, R. Greeley, Saharan dust devil tracks: Mars analog field study areas, in *Abstracts with Programs* (Geol. Soc. Am., Houston, 2008), p. 262, Abst. #40

L.D.V. Neakrase, J. McHone, P.L. Whelley, R. Greeley, Terrestrial analogs to Mars: east-central Saharan dust devil tracks, in *Lunar Planet. Sci. Conf. XLIII, League City, TX* (2012) Abst. #2009

F.M. Neubauer, Thermal convection in the martian atmosphere. J. Geophys. Res. **71**(10), 2419–2426 (1966). doi:10.1029/JZ071i010p02419

G.A. Neumann, D.E. Smith, M.T. Zuber, Two Mars years of clouds detected by the Mars Orbiter Laser Altimeter. J. Geophys. Res. **108**, 5023 (2003). doi:10.1029/2002JE001849

C.E. Newman, S.R. Lewis, P.L. Read, F. Forget, Modeling the Martian dust cycle, 1. Representations of dust transport processes. J. Geophys. Res. **107**(E12), 5123 (2002a). doi:10.1029/2002JE001910

C.E. Newman, S.R. Lewis, P.L. Read, F. Forget, Modeling the Martian dust cycle 2. Multiannual radiatively active dust transport simulations. J. Geophys. Res. **107**(E12), 5124 (2002b). doi:10.1029/2002JE001920

C.E. Newman, S.R. Lewis, P.L. Read, The atmospheric circulation and dust activity in different orbital epochs on Mars. Icarus **174**, 135–160 (2005). doi:10.1016/j.icarus.2004.10.023

A.H. Omar, D.M. Winker, M.A. Vaughan, Y. Hu, C.R. Trepte, R.A. Ferrare, K.-P. Lee, C.A. Hostetler, C. Kittaka, R. Rogers, R.E. Kuehn, Z. Liu, The CALIPSO automated aerosol classification and lidar ratio selection algorithm. J. Atmos. Ocean. Technol. **26**, 1994–2014 (2009). doi:10.1175/2009JTECHA1231.1

A.V. Pathare, M.R. Balme, S.M. Metzger, A. Spiga, M.C. Towner, N.O. Renno, F. Saca, Assessing the power law hypothesis for the size–frequency distribution of terrestrial and martian dust devils. Icarus **209**, 851–853 (2010). doi:10.1016/j.icarus2010.06.027

S. Peyridieu, A. Chédin, D. Tanré, V. Capelle, C. Pierangelo, N. Lamquin, R. Armante, Saharan dust infrared optical depth and altitude retrieved from AIRS: a focus over North Atlantic—comparison to MODIS and CALIPSO. Atmos. Chem. Phys. **10**(4), 1953–1967 (2010). doi:10.5194/acp-10-1953-2010

A. Petrosyan, B. Galperin, S.E. Larsen, S.R. Lewis, A. Määttänen, P.L. Read, N. Renno, L.P.H.T. Rogberg, H. Savijrvi, T. Siili, A. Spiga, A. Toigo, L. Vázquez, The Martian atmospheric boundary layer. Rev. Geophys. **49** (2011). doi:10.1029/2010RG000351

J.M.C. Plane, Cosmic dust in the Earth's atmosphere. Chem. Soc. Rev. **41**(19), 6507–6518 (2012). doi:10.1039/C2CS35132C

J.B. Pollack, D.S. Colburn, F.M. Flasar, R. Kahn, C.E. Carlston, D. Pidek, Properties and effects of dust particles suspended in the Martian atmosphere. J. Geophys. Res. **84** (1979). doi:10.1029/JB084iB06p02929

J.M. Prospero, P.J. Lamb, African droughts and dust transport to the Caribbean: climate change implications. Science **302**, 1024–1027 (2003). doi:10.1126/science.1089915

J.M. Prospero, P. Ginoux, O. Torres, S.E. Nicholson, T.E. Gill, Environmental characterization of global sources of atmospheric soil dust identified with the Nimbus 7 total ozone mapping spectrometer (TOMS) absorbing aerosol product. Rev. Geophys. **40**(1), 1002 (2002). doi:10.1029/2000RG000095

J.M. Prospero, J.E. Bullard, R. Hodgkins, High-latitude dust over the North Atlantic: inputs from Icelandic proglacial dust storms. Science **335**, 1078–1082 (2012). doi:10.1126/science.1217447

N. Putzig, M. Mellon, K. Kretke, R. Arvidson, Global thermal inertia and surface properties of Mars from the MGS mapping mission. Icarus **173**(2), 325–341 (2005). doi:10.1016/j.icarus.2004.08.017

S.C.R. Rafkin, The potential importance of non-local, deep transport on the energetics, momentum, chemistry, and aerosol distributions in the atmospheres of Earth, Mars, and Titan. Planet. Space Sci. **60**, 147–154 (2012). doi:10.1016/j.pss.2011.07.015

S.C.R. Rafkin, M.R.V.S. Maria, T.I. Michaels, Simulation of the atmospheric thermal circulation of a martian volcano using a mesoscale numerical model. Nature **419**, 697–699 (2002). doi:10.1038/nature01206

P.L. Read, S.R. Lewis, *The Martian Climate Revisited: Atmosphere and Environment of a Desert Planet* (Springer/Praxis Books, Berlin/New York, 2004)

H.E. Redmond, K.D. Dial, J.E. Thompson, Light scattering and absorption by wind blown dust: theory, measurement, and recent data. Aeolian Res. **2**(1), 5–26 (2010). doi:10.1016/j.aeolia.2009.09.002

D. Reiss, D. Lüsebrink, H. Hiesinger, T. Kelling, G. Wurm, J. Teiser, High altitude dust devils on Arsia Mons: testing the greenhouse and thermophoresis hypothesis of dust lifting, in *Lunar and Planetary Science Conference XL, the Woodlands, TX* (2009) Abst. #1961

D. Reiss, J. Raack, A.P. Rossi, G. Di Achille, H. Hiesinger, First in-situ analysis of dust devil tracks on Earth and their comparison with tracks on Mars. Geophys. Res. Lett. **37**, L14203 (2010). doi:10.1029/2010GL044016

D. Reiss, J. Raack, H. Hiesinger, Bright dust devil tracks on Earth: implications for their formation on Mars. Icarus **211**, 917–920 (2011a). doi:10.1016/j.icarus.2010.09.009

D. Reiss, M. Zanetti, G. Neukum, Multitemporal observations of identical active dust devils on Mars with the High Resolution Stereo Camera (HRSC) and Mars Orbiter Camera (MOC). Icarus **215**, 358–369 (2011b). doi:10.1016/j.icarus.2011.06.011

D. Reiss, M.I. Zimmerman, D.C. Lewellen, Formation of cycloidal dust devil tracks by redeposition of coarse sands in southern Peru: implications for Mars. Earth Planet. Sci. Lett. **383**, 7–15 (2013). doi:10.1016/j.epsl.2013.09.033

D. Reiss, A. Spiga, G. Erkeling, The horizontal motion of dust devils on Mars derived from CRISM and CTX/HiRISE observations. Icarus **227**, 8–20 (2014a). doi:10.1016/j.icarus.2013.08.028

D. Reiss, N.M. Hoekzema, O.J. Stenzel, Dust deflation by dust devils on Mars derived from optical depth measurements using the shadow method in HiRISE images. Planet. Space Sci. **93–94**, 54–64 (2014b). doi:10.1016/j.pss.2014.01.016

L.A. Remer, R.G. Kleidman, R.C. Levy, Y.J. Kaufman, D. Tanré, S. Mattoo, J.V. Martins, C. Ichoku, I. Koren, H. Yu, B.N. Holben, Global aerosol climatology from the MODIS satellite sensors. J. Geophys. Res. **113**, D14S07 (2008). doi:10.1029/2007JD009661

N.O. Rennó, M.L. Burkett, M.P. Larkin, A simple thermodynamical theory for dust devils. J. Atmos. Sci. **55**, 3244–3252 (1998). doi:10.1175/1520-0469(1998)055<3244:ASTTFD>2.0.CO;2

N.O. Rennó, A.A. Nash, J. Lunine, J. Murphy, Martian and terrestrial dust devils: test of a scaling theory using Pathfinder data. J. Geophys. Res. **105**, 1859–1865 (2000). doi:10.1029/1999JE001037

N.O. Rennó, V.J. Abreu, J. Koch, P.H. Smith, O.K. Hartogensis, H.A.R. De Bruin, D. Burose, G.T. Delory, W.M. Farrell, C.J. Watts, J. Garatuza, M. Parker, A. Carswell, MATADOR 2002: a pilot field experiment on convective plumes and dust devils. J. Geophys. Res. **109** (2004). doi:10.1029/2003JE002219

T.J. Ringrose, M.C. Towner, J.C. Zarnecki, Convective vortices on Mars: a reanalysis of Viking Lander 2 meteorological data, sols 1–60. Icarus **163**, 78–87 (2003). doi:10.1016/S0019-1035(03)00073-3

A.P. Rossi, L. Marinangeli, The first terrestrial analogue to Martian dust devil tracks found in Ténéré Desert, Niger. Geophys. Res. Lett. **31**, L06702 (2004). doi:10.1029/2004GL019428

J.A. Ryan, Notes on the martian yellow clouds. J. Geophys. Res. **69**(18), 3759–3770 (1964). doi:10.1029/JZ069i018p03759

J.A. Ryan, J.J. Carroll, Dust devil wind velocities: mature state. J. Geophys. Res. **75**, 531–541 (1970). doi:10.1029/JC075i003p00531

J.A. Ryan, R.D. Lucich, Possible dust devils, vortices on Mars. J. Geophys. Res. **88**(C15), 11005–11011 (1983). doi:10.1029/JC088iC15p11005

K. Schepanski, I. Tegen, M.C. Todd, B. Heinold, G. Bonisch, B. Laurent, A. Macke, Meteorological processes forcing Saharan dust emission inferred from MSG-SEVIRI observations of subdaily dust source activation and numerical models. J. Geophys. Res. **114**, D10201 (2009). doi:10.1029/2008JD010325

A. Schnapf, *TIROS: A Story of Achievement, AED P-5167A* (Radio Corporation of America, Princeton, 1964), p. 37

P.C. Sinclair, Some preliminary dust devil measurements. Mon. Weather Rev. **92**, 363–367 (1964). doi:10.1126/science.27.693.594

P.C. Sinclair, On the rotation of dust devils. Bull. Am. Meteorol. Soc. **46**, 388–391 (1965)

P.C. Sinclair, General characteristics of dust devils. J. Appl. Meteorol. **8**, 32–45 (1969). doi:10.1175/1520-0450(1969)008<0032:GCODD>2.0.CO;2

P.C. Sinclair, The lower structure of dust devils. J. Atmos. Sci. **30**, 1599–1619 (1973). doi:10.1175/1520-0469(1973)030<1599:TLSODD>2.0.CO;2

A. Slingo, T.P. Ackerman, R.P. Allan, E.I. Kassianov, S.A. McFarlane, G.J. Robinson, J.C. Barnard, M.A. Miller, J.E. Harries, J.E. Russell, S. Dewitte, Observations of the impact of a major Saharan dust storm on the atmospheric radiation balance. Geophys. Res. Lett. **33**, L24817 (2006). doi:10.1029/2006GL027869

M.D. Smith, Interannual variability in TES atmospheric observations of Mars during 1999–2003. Icarus **167**(1), 148–165 (2004). doi:10.1016/j.icarus.2003.09.010

M.D. Smith, THEMIS observations of Mars aerosol optical depth from 2002–2008. Icarus **202**, 444–452 (2009). doi:10.1016/j.icarus.2009.03.027

P.H. Smith, M. Lemmon, Opacity of the Martian atmosphere measured by the Imager for Mars Pathfinder. J. Geophys. Res. **104**, 8975 (1999). doi:10.1029/1998JE900017

M.D. Smith, J.C. Pearl, B.J. Conrath, P.R. Christensen, Mars global surveyor thermal emission spectrometer (TES) observations of dust opacity during aerobraking and science phasing. J. Geophys. Res. **105**, 9539–9552 (2000). doi:10.1029/1999JE0010

D.E. Smith, M.T. Zuber, H.V. Frey, J.B. Garvin, J.W. Head, D.O. Muhleman, G.H. Pettengill, R.J. Phillips, S.C. Solomon, H.J. Zwally, W.B. Bannerdt, T.C. Duxbury, M.P. Golombek, F.g. Lemoine, G.A. Neumann, D.D. Rowlands, O. Aharonson, P.G. Ford, A.B. Ivanov, C.L. Johnson, P.J. McGovern, J.B. Abshire, R.S. Afzal, X. Sun, Mars orbiter laser altimeter: experiment summary after the first year of global mapping of Mars. J. Geophys. Res. **106**, 23689–23722 (2001a). doi:10.1029/2000JE001364

M.D. Smith, J.C. Pearl, B.J. Conrath, P.R. Christensen, One Martian year of atmospheric observations by the thermal emission spectrometer. Geophys. Res. Lett. **28**, 4263–4266 (2001b). doi:10.1029/2001GL013608

M.D. Smith, M.J. Wolff, M.T. Lemmon, N. Spanovich, D. Banfield, C.J. Budney, R.T. Clancy, A. Ghosh, G.A. Landis, P. Smith, B. Whitney, P.R. Christensen, S.W. Squyres, First atmospheric science results from the Mars exploration rovers Mini-TES. Science **306**, 1750–1753 (2004). doi:10.1126/science.1104257

M.D. Smith, M.J. Wolff, R.T. Clancy, A. Kleinböhl, S.L. Murchie, Vertical distribution of dust and water ice aerosols from CRISM limb-geometry observations. J. Geophys. Res. **118**, 321–334 (2013). doi:10.1002/jgre.20047

J.T. Snow, T.M. McClelland, Dust devils at White Sands Missile Range, New Mexico 1. Temporal and spatial distributions. J. Geophys. Res. **95**, 13707–13721 (1990). doi:10.1029/JD095iD09p13707

C.W. Snyder, V.I. Moroz, Spacecraft exploration of Mars, in *Mars*, ed. by H.H. Kieffer, B.M. Jakosky, C.W. Snyder, M.S. Matthews (University of Arizona Press, Tucson, 1992), pp. 71–119

A. Spiga, F. Forget, A new model to simulate the Martian mesoscale and microscale atmospheric circulation: validation and first results. J. Geophys. Res. **114** (2009). doi:10.1029/2008je003242

A. Spiga, F. Forget, S.R. Lewis, D.P. Hinson, Structure and dynamics of the convective boundary layer on mars as inferred from large-eddy simulations and remote-sensing measurements. Q. J. R. Meteorol. Soc. **136**, 414–428 (2010). doi:10.1002/qj.563

A. Spiga, S.R. Lewis, Martian mesoscale and microscale wind variability of relevance for dust lifting. Mars **5**, 146–158 (2010). doi:10.1555/mars.2010.0006

J.D. Spinhirne, S.P. Palm, W.D. Hart, D.L. Hlavka, E.J. Welton, Cloud and aerosol measurements from GLAS: overview an initial results. Geophys. Res. Lett. **32** (2005). doi:10.1029/2005GL023507

C. Stanzel, M. Pätzold, R. Greeley, E. Hauber, G. Neukum, Dust devils on Mars observed by the high resolution stereo camera. Geophys. Res. Lett. **33**, L11202 (2006). doi:10.1029/2006GL025816

C. Stanzel, Studying martian dust devils by applying pattern recognition algorithms to multi-mission camera images. PhD thesis, University of Cologne (2007). 158 pp.

C. Stanzel, M. Pätzold, D.A. Williams, P.L. Whelley, R. Greeley, G. Neukum, The HRSC Co-Investigator Team, Dust devil speeds, directions of motion and general characteristics observed by the Mars express high resolution stereo camera. Icarus **197**, 39–51 (2008). doi:10.1016/j.icarus2008.04.017

L.J. Steele, S.R. Lewis, M.R. Patel, The radiative impact of water ice clouds from a reanalysis of Mars climate sounder data. Geophys. Res. Lett. **41**, 4471–4478 (2014a). doi:10.1002/2014GL060235

L.J. Steele, S.R. Lewis, M.R. Patel, F. Montmessin, F. Forget, M.D. Smith, The seasonal cycle of water vapour on Mars from assimilation of thermal emission spectrometer data. Icarus **237**, 97–115 (2014b). doi:10.1016/j.icarus.2014.04.017

M.J. Strausberg, H. Wang, M.I. Richardson, S.P. Ewald, A.D. Toigo, Observations of the initiation and evolution of the 2001 Mars global dust storm. J. Geophys. Res. **110**, E02006 (2005). doi:10.1029/2004JE002361

R. Swap, S. Ulanski, M. Cobbett, M. Garstang, Temporal and spatial characteristics of Saharan dust outbreaks. J. Geophys. Res. **101**(D2), 4205–4220 (1996). doi:10.1029/95JD03236

D. Tanré, F.M. Bréon, J.L. Deuzé, O. Dubovik, F. Ducos, P. François, P. Goloub, M. Herman, A. Lifermann, F. Waquet, Remote sensing of aerosols by using polarized, directional and spectral measurements within the A-Train: the PARASOL mission. Atmos. Meas. Tech. **4**, 1383–1395 (2011). doi:10.5194/amt-4-1383-2011

P. Thomas, P.J. Gierasch, Dust devils on Mars. Science **230**, 175–177 (1985). doi:10.1126/science.230.4722.175

O. Torres, P.K. Bhartia, J.R. Herman, Z. Ahmad, Derivation of aerosol properties from satellite measurements of backscattered ultraviolet radiation: theoretical basis. J. Geophys. Res. **103**, 17099–17110 (1998). doi:10.1029/98JD00900

M.C. Towner, Characteristics of large Martian dust devils using Mars Odyssey Thermal Emission Imaging System visual and infrared images. J. Geophys. Res. **114**, E02010 (2009). doi:10.1029/2008JE003220

D.M. Tratt, M.H. Hecht, D.C. Catling, E.C. Samulon, P.H. Smith, In situ measurement of dust devil dynamics: toward a strategy for Mars. J. Geophys. Res. **108** (2003). doi:10.1029/2003JE002161

United States, Catalogue of meteorological satellite data – TIROS VI television cloud photography. Key to Meteorological Records Documentation No. **5**(36), (1964) 400 pp.

P. Vallelonga, A. Svensson, Ice core archives of mineral dust, in *Mineral Dust: A Key Player in the Earth System*, ed. by P. Knippertz, J.-B.W. Stuut (Springer, Dordrecht, 2014), p. 463. doi:10.1007/978-94-017-8978-3

H. Wang, M.I. Richardson, R.J. Wilson, A.P. Ingersoll, A.D. Toigo, R.W. Zurek, Cyclones, tides, and the origin of a cross-equatorial dust storm on Mars. Geophys. Res. Lett. **30** (2003). doi:10.1029/2002GL016828

H. Wang, Dust storms originating in the northern hemisphere during the third mapping year of Mars Global Surveyor. Icarus **189**, 325–343 (2007). doi:10.1016/j.icarus.2007.01.014

H. Wang, J.A. Fisher, North polar frontal clouds and dust storms on Mars during spring and summer. Icarus **204**, 103–113 (2009). doi:10.1016/j.icarus.2009.05.028

H. Wang, M.I. Richardson, The origin, evolution, and trajectory of large dust storms on Mars during Mars years 24–30 (1999–2011). Icarus **251**, 112–127 (2015). doi:10.1016/j.icarus.2013.10.033

A. Wennmacher, F.M. Neubauer, M. Pätzold, J. Schmitt, K. Schulte, A search for dust devils on Mars, in *Lunar Planet. Sci. Conf. XXVII, League City, TX* (1996). Abst. #1417

P.L. Whelley, R. Greeley, The distribution of dust devil activity on Mars. J. Geophys. Res. **113**, E7 (2008). doi:10.1029/2007JE002966

N.R. Williams, Development of dust whirls and similar small-scale vortices. Bull. Am. Meteorol. Soc. **29**, 106–117 (1948)

G.E. Willis, J.W. Deardorff, Laboratory observations of turbulent penetrative convection planforms. J. Geophys. Res. **84**, 295–302 (1979). doi:10.1029/JC084iC01p00295

R.J. Wilson, S.R. Lewis, L. Montabone, M.D. Smith, Influence of water ice clouds on Martian tropical atmospheric temperatures. Geophys. Res. Lett. **35** (2008). doi:10.1029/2007GL032405

G. Winckler, R.F. Anderson, D. McGee, M.Q. Fleisher, N. Mahowald, Half a million years of coherent dust flux variations in the tropical Pacific and Antarctica, in *8th Annual V.M. Goldschmidt Conference* (Pergamon–Elsevier, Vancouver, 2008) pp. A1026

D.M. Winker, J. Pelon, J.A. Coakley Jr., S.A. Ackerman, R.J. Charlson, P.R. Colarco, P. Flamant, Q. Fu, R.M. Hoff, C. Kittaka, T.L. Kubar, H. Le Treut, M.P. McCormick, G. Mégie, L. Poole, K. Powell, C. Trepte, M.A. Vaughan, B.A. Wielicki, The CALIPSO mission: a global 3D view of aerosols and clouds. Bull. Am. Meteorol. Soc. **91**, 1211–1229 (2010). doi:10.1175/2010BAMS3009.1

M.J. Wolff, M.D. Smith, R.T. Clancy, R. Arvidson, M. Kahre, F. Seelos, S. Murchie, H. Savijärvi, Wavelength dependence of dust aerosol single scattering albedo as observed by the Compact Reconnaissance Imaging Spectrometer. J. Geophys. Res. **114**, E00D04 (2009). doi:10.1029/2009JE003350

S. Wong, A.E. Dessler, N.M. Mahowald, P. Yang, Q. Feng, Maintenance of lower tropospheric temperature inversion in the Saharan Air Layer by dust and dry anomaly. J. Climate **22**(19), 5149–5162 (2009). doi:10.1175/2009JCLI2847.1

F. Yang, P.A. Mlsna, P. Geissler, Gaussian-based filters for detecting martian dust devils, in *2006 IEEE Southwest Symp. Image Anal. Interpret.*, pp. 46–50 (2006). doi:10.1109/SSIAI.2006.1633719

H. Yu, M. Chin, T. Yuan, H. Bian, L.A. Remer, J.M. Prospero, A. Omar, D. Winker, Y. Yang, Y. Zhang, Z. Zhang, C. Zhao, The fertilizing role of African dust in the Amazon rainforest: a first multiyear assessment based on data from Cloud-Aerosol Lidar and Infrared Pathfinder Satellite Observations. Geophys. Res. Lett. **42** (2015). doi:10.1002/2015GL063040

M.T. Zuber, D.E. Smith, S.C. Solomon, D.O. Muhleman, J.W. Head, J.B. Garvin, J.B. Abshire, J.L. Bufton, The Mars observer laser altimeter investigation. J. Geophys. Res. **97**(E5), 7781–7797 (1992). doi:10.1029/92JE00341

L. Zasova, V. Formisano, N. Moroz, D. Grassi, N. Ignatiev, M. Giuranna, G. Hansen, M. Blecka, A. Ekonomov, E. Lellouch, S. Fonti, A. Grigoriev, H. Hirsch, I. Khatuntsev, A. Mattana, A. Maturilli, A. Moshkin, D. Patsaev, G. Piccioni, M. Rataj, B. Saggin, Water clouds and dust aerosols observations with PFS MEX at Mars. Planet. Space Sci. **53**, 1065–1077 (2005). doi:10.1016/j.pss.2004.12.010

A. Zhu, V. Ramanathan, F. Li, D. Kim, Dust plumes over the Pacific, Indian, and Atlantic oceans: climatology and radiative impact. J. Geophys. Res. **112**, D16208 (2007). doi:10.1029/2007JD008427

R.W. Zurek, J.R. Barnes, R.M. Haberle, J.B. Pollack, J.E. Tillman, C.B. Leovy, Dynamics of the atmosphere of Mars, in *Mars*, ed. by H.H. Kieffer, B.M. Jakosky, C.W. Snyder, M.S. Matthews (University of Arizona Press, Tucson, 1992), pp. 835–933

Space Sci Rev (2016) 203:143–181
DOI 10.1007/s11214-016-0308-6

Dust Devil Tracks

Dennis Reiss[1] · Lori Fenton[2] · Lynn Neakrase[3] · Michael Zimmerman[4] · Thiago Statella[5] · Patrick Whelley[6] · Angelo Pio Rossi[7] · Matthew Balme[8]

Received: 1 July 2016 / Accepted: 31 October 2016 / Published online: 24 November 2016

Abstract Dust devils that leave dark- or light-toned tracks are common on Mars and they can also be found on the Earth's surface. Dust devil tracks (hereinafter DDTs) are ephemeral surface features with mostly sub-annual lifetimes. Regarding their size, DDT widths can range between ∼1 m and ∼1 km, depending on the diameter of dust devil that created the track, and DDT lengths range from a few tens of meters to several kilometers, limited by the duration and horizontal ground speed of dust devils. DDTs can be classified into three main types based on their morphology and albedo in contrast to their surroundings; all are found

✉ D. Reiss
dennis.reiss@uni-muenster.de

L. Fenton
lfenton@seti.org

L. Neakrase
lneakras@nmsu.edu

M. Zimmerman
Michael.Zimmerman@jhuapl.edu

T. Statella
thiago.statella@cba.ifmt.edu.br

P. Whelley
patrick.l.whelley@nasa.gov

A.P. Rossi
an.rossi@jacobs-university.de

M. Balme
Matt.Balme@open.ac.uk

1 Institut für Planetologie, Westfälische Wilhelms-Universität, Wilhelm-Klemm-Str. 10, 48149 Münster, Germany

2 SETI Institute, 189 Bernardo Ave. Suite 100, Mountain View, CA, 94043, USA

3 Department of Astronomy, New Mexico State University, Las Cruces, NM, USA

4 The Johns Hopkins University Applied Physics Laboratory, Laurel, MD, USA

5 Instituto Federal de Educação, Ciência e Tecnologia de Mato Grosso – IFMT, 95 Zulmira Canavarro, 780025-200 Cuiabá, Brazil

on both planets: (a) dark continuous DDTs, (b) dark cycloidal DDTs, and (c) bright DDTs. Dark continuous DDTs are the most common type on Mars. They are characterized by their relatively homogenous and continuous low albedo surface tracks. Based on terrestrial and martian in situ studies, these DDTs most likely form when surficial dust layers are removed to expose larger-grained substrate material (coarse sands of $\geq$500 μm in diameter). The exposure of larger-grained materials changes the photometric properties of the surface; hence leading to lower albedo tracks because grain size is photometrically inversely proportional to the surface reflectance. However, although not observed so far, compositional differences (i.e., color differences) might also lead to albedo contrasts when dust is removed to expose substrate materials with mineralogical differences. For dark continuous DDTs, albedo drop measurements are around 2.5 % in the wavelength range of 550–850 nm on Mars and around 0.5 % in the wavelength range from 300–1100 nm on Earth. The removal of an equivalent layer thickness around 1 μm is sufficient for the formation of visible dark continuous DDTs on Mars and Earth. The next type of DDTs, dark cycloidal DDTs, are characterized by their low albedo pattern of overlapping scallops. Terrestrial in situ studies imply that they are formed when sand-sized material that is eroded from the outer vortex area of a dust devil is redeposited in annular patterns in the central vortex region. This type of DDT can also be found in on Mars in orbital image data, and although in situ studies are lacking, terrestrial analog studies, laboratory work, and numerical modeling suggest they have the same formation mechanism as those on Earth. Finally, bright DDTs are characterized by their continuous track pattern and high albedo compared to their undisturbed surroundings. They are found on both planets, but to date they have only been analyzed in situ on Earth. Here, the destruction of aggregates of dust, silt and sand by dust devils leads to smooth surfaces in contrast to the undisturbed rough surfaces surrounding the track. The resulting change in photometric properties occurs because the smoother surfaces have a higher reflectance compared to the surrounding rough surface, leading to bright DDTs. On Mars, the destruction of surficial dust-aggregates may also lead to bright DDTs. However, higher reflective surfaces may be produced by other formation mechanisms, such as dust compaction by passing dust devils, as this may also cause changes in photometric properties. On Mars, DDTs in general are found at all elevations and on a global scale, except on the permanent polar caps. DDT maximum areal densities occur during spring and summer in both hemispheres produced by an increase in dust devil activity caused by maximum insolation. Regionally, dust devil densities vary spatially likely controlled by changes in dust cover thicknesses and substrate materials. This variability makes it difficult to infer dust devil activity from DDT frequencies. Furthermore, only a fraction of dust devils leave tracks. However, DDTs can be used as proxies for dust devil lifetimes and wind directions and speeds, and they can also be used to predict lander or rover solar panel clearing events. Overall, the high DDT frequency in many areas on Mars leads to drastic albedo changes that affect large-scale weather patterns.

Keywords Dust devil · Dust devil track · Vertical convective vortices · Mars · Earth · Deflation

[6] NASA Goddard Space Flight Center, Greenbelt, MD, USA

[7] Department of Earth and Space Sciences, Jacobs University Bremen, Bremen, Germany

[8] Department of Physical Science, The Open University, Milton Keynes, UK

1 Introduction

Dust devil tracks (DDTs) are dark or bright surface lineaments left by passages of dust devils. Before dust devils had ever been detected on Mars, they were hypothesized to exist (Ryan 1964; Neubauer 1966; Sagan et al. 1971; Gierasch and Goody 1973). For a complete historical review about dust devils we refer here to Lorenz et al. (2016). The first direct observations of dust devils in Viking Orbiter images were made by Thomas and Gierasch (1985). For general information about orbital observations of dust devils we refer here to Fenton et al. (2016). Indirectly, Ryan and Lucich (1983) detected dust devils by their analysis of meteorological data acquired by Viking Lander 2 (see also Lorenz and Jackson 2016). Dark filamentary lineaments on the martian surface—which would nowadays be referred to as dust devil tracks (DDTs)—have already been visible in Mariner 9 images (Sagan et al. 1972, 1973; Cutts and Smith 1973; Veverka 1975, 1976). These observed albedo patterns were termed variable features (Sagan et al. 1972) because their size, shape and position were observed to change, which was correctly attributed to active aeolian processes on Mars. Most of these albedo features correspond with either albedo changes caused by a global-scale dust storm in 1971, or wind streaks with lifetimes lasting from seasons to decades (Malin and Edgett 2001). For further detailed classifications and summaries of variable albedo features on Mars, we refer the reader to Thomas et al. (1981) and Greeley and Iversen (1985). However, dark lineaments found on the floor of Proctor Crater (Fig. 1A) were not described (Sagan et al. 1972) or these and other patterns of dark filamentary lineaments found elsewhere on Mars were either interpreted as linear seif dunes (Cutts and Smith 1973) or interpreted to be formed due to topographically controlled joints where strong winds are able to erode or deposit surface material (Veverka 1976).

Grant and Schultz (1987) first interpreted the dark, ephemeral, filamentary lineaments on Mars as features formed during the passage of 'tornado-like vortices'. They linked the surface lineaments to local, ephemeral, and intense atmospheric phenomena because of the absence of structural or topographic control, gaps in lineaments, their non-destructive nature, seasonal occurrence, ephemeral nature, and year-to-year variation in distribution (Fig. 1). They favored tornadic-intensity vortices caused by baroclinic wave passage through an area of atmospheric instability triggering convective uplift and strong shear, although they discussed and did not preclude their formation by dust devils. Based on terrestrial studies, Grant and Schultz (1987) found it unlikely that dust devils could cross large topographic obstacles and can leave such long and wide surface lineaments (up to 75 km long and 1 km wide), which would imply very large dust devils with very long durations. The development of the albedo difference between the dark surface lineaments to their surroundings was attributed by Grant and Schultz (1987) to the redistribution of coarser material into a narrow band (appearing darker due to coarse grained material versus finer grainer material in the surroundings), as an analogue to observed tornado tracks on Earth.

About two decades after the end of the two Viking Orbiter missions, the Mars Orbiter Camera (MOC) onboard the Mars Global Surveyor (MGS) orbiter revealed that some of the dark lineaments (especially the larger ones) in Proctor crater interpreted by Grant and Schultz (1987) as DDTs are permanent features caused by boundaries between differing surface textures (Malin and Edgett 2001). However, the camera provided the first direct evidence of dust devils creating surface tracks (Edgett and Malin 2000). High-resolution orbital images acquired by the Mars Orbiter Camera–Near Angle (MOC-NA) instrument showed numerous active dust devils in the process of creating tracks (Edgett and Malin 2000; Malin and Edgett 2001; Cantor et al. 2006) (Fig. 2). Additionally, the high-resolution capability of the MOC-NA (as fine as 1.4 m/pixel) revealed the existence of not only dark

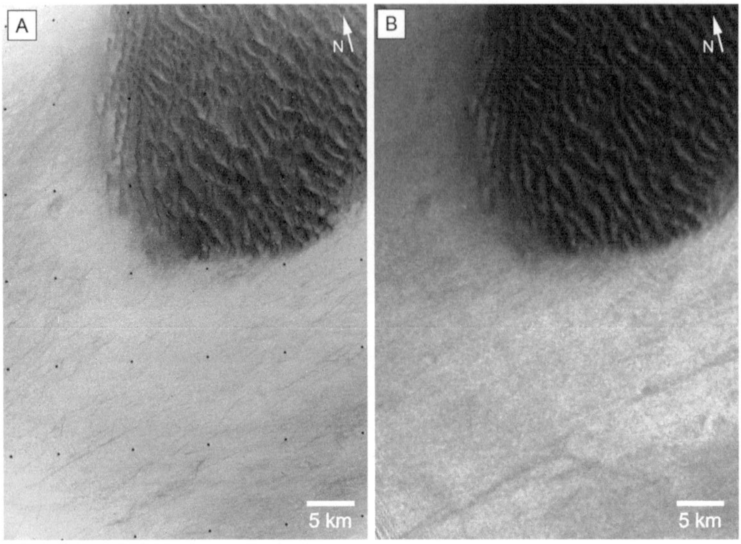

Fig. 1 First observations of dark, ephemeral tracks (later found to be dust devil tracks; DDT's) on Mars in Proctor crater at 29.5°E and 48°S. (**A**) Example of dark filamentary lineaments in Mariner 9 images described by Veverka (1976). Mariner 9 image R229/09807499 acquired on 1972-03-07 with an image resolution of ~65 m/pxl. (**B**) Viking image F510A46 acquired on 1977-11-09 with an image resolution of ~175 m/pxl. The changes of dark lineaments to the Mariner 9 image (**A**) were mapped by Grant and Schultz (1987) (see Fig. 3 in Grant and Schultz 1987) and interpreted as ephemeral tornado-like tracks (Grant and Schultz 1987)

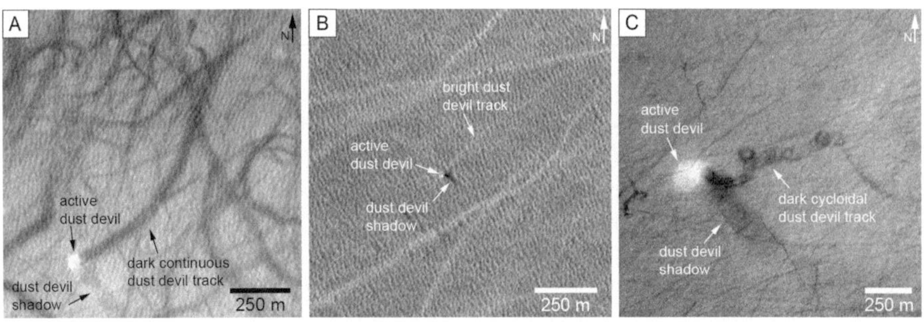

Fig. 2 First direct observations that surface lineaments on Mars are created by passing dust devils. (**A**) Active dust devil with a dark continuous track resembling some of the dark filamentary tracks observed in Mariner 9 and Viking Orbiter images (MOC-NA image M1103289 in Hellas basin at 59.2°S and 22.1°E). (**B**) MOC-NA images also revealed that bright dust devil tracks (MOC-NA image S0501277 in Syria Planum at 14.9°S and 250.9°E) and (**C**) cycloidal patterns of dust devil tracks exist on Mars (MOC-NA image M1001267 in Promethei Terra at 54.1°S and 117.2°E)

(Fig. 2A) but also bright DDTs (Fig. 2B) on Mars (Edgett and Malin 2000; Malin and Edgett 2001). Furthermore, variations in dark DDT morphologies were detected, showing both dark continuous DDTs (Fig. 2A) and dark cycloidal DDTs (Fig. 2C) (Edgett and Malin 2000; Malin and Edgett 2001).

Dust devil tracks on Earth were discovered relatively recently, largely due to the limited general availability of high spatial (and temporal) resolution imagery. The first known observation of DDTs on Earth was reported by Louis Maher who imaged bright DDTs from a

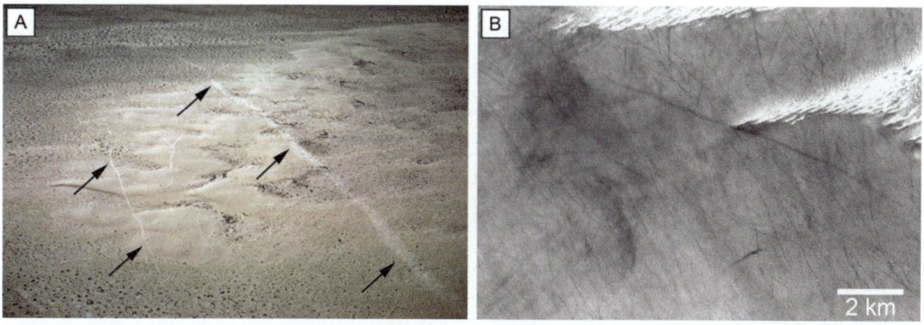

Fig. 3 (**A**) Aerial image from a lightplane of bright dust devil tracks on dunes east of Chaco River, about 25 km northeast of Sheep Springs, NM, USA. View to SSW. (Image #052-21, 23 August 1959, Image credit: Louis J. Maher, Jr., http://geoscience.wisc.edu/~maher/air/air00.htm; ftp.geology.wisc.edu/maher/air.) (**B**) Dark DDTs in the Ténéré desert (Niger) at 10.5°W and 18.85°N observed in an ASTER satellite image acquired on 26 May 2001 (Rossi and Marinangeli 2004)

lightplane occurring on dunes from in Sheep Springs, New Mexico, USA on 23 August 1959 (Fig. 3A). This image was not published and is only available in the Internet, but a general summary of the geology by lightplane roundtrip is given in Maher (1968). The first report of DDTs detected on satellite data dates back to 2002 (Rossi and Marinangeli 2004). These dark DDTs occur in the Ténéré desert (Niger) (Fig. 3B) and have been observed over several years in ASTER data (Rossi and Marinangeli 2004). Mostly low-sinuosity tracks occur on diverse terrains: transverse dune fields, sand sheets and interdune seif zones (Rossi and Marinangeli 2004). The highest concentration of tracks could be observed on smooth interdunes and sand sheets. Their formation seems to have some seasonal dependency, with the largest number and surface density of tracks observed during spring (Rossi and Marinangeli 2004; Reiss and Rossi 2011).

During the following decade several more occurrences elsewhere were documented (Neakrase et al. 2008, 2012; Reiss et al. 2010, 2011b, 2013; Hesse 2012). Overall, the number of detections is extremely low compared to those on Mars, where DDTs are rather ubiquitous. The physiographic, geologic, and geomorphologic setting of the areas where DDTs have been discovered and studied is somewhat variable, but all documented DDT discoveries share a climatic setting characterized by semi-arid to hyper-arid conditions (Fig. 4).

In this review paper, we summarize the current knowledge about DDTs on Mars and Earth. The paper is structured as follows: In Sect. 2. we classify DDTs by their morphology and albedo contrast to their surroundings, and for each DDT type we present possible formation mechanisms for both planets. In addition, in this section we give further information about each DDT type, such as lifetimes, if available. In Sect. 3, we present the spatial and temporal DDT distribution on Mars. In Sect. 4, we present how DDTs can be used as proxies for dust devil activity, dust devil minimum durations, ambient wind directions, surface wind conditions, and predictions for landed spacecraft solar panel clearing events. Then, Sect. 5 highlights global albedo changes caused by DDTs and how these changes impact the climate system of Mars. In Sect. 6, we present recent advances in detecting DDTs automatically in satellite imagery from Mars. Finally, in Sect. 7 we summarize the paper and concurrently identify knowledge gaps and point out future directions in DDT research.

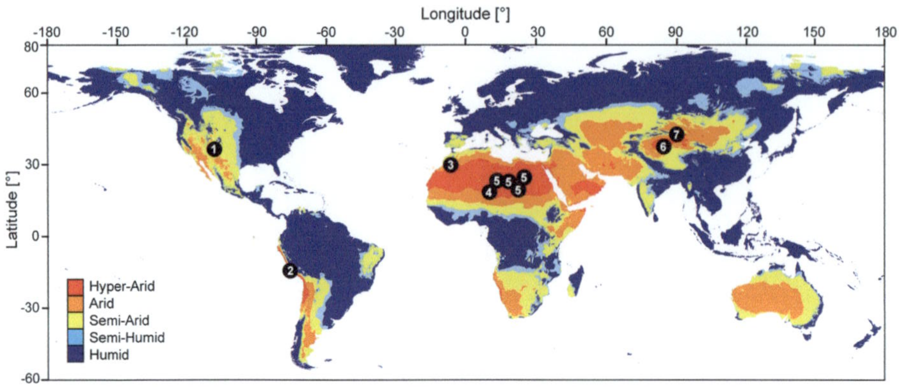

Fig. 4 Global distribution of reported DDTs on Earth. (1) Maher (1968); (2) Hesse (2012), Reiss et al. (2013); (3) Reiss et al. (2012a, 2012b); (4) Rossi and Marinangeli (2004); (5) Neakrase et al. (2008, 2012); (6) Reiss (2016); (7) Reiss et al. (2010, 2011a). Background: Humid to Hyper-Arid climate zones from CGIAR-CSI Global-Aridity database (http://www.cgiar-csi.org; Zomer et al. 2007, 2008)

2 Morphology, Classification and Formation

In plan view, DDTs generally show linear, curvilinear, curved, meandering or looping streak morphologies (Fig. 5A and B). They might be confused with other low albedo features such as wind streaks (Balme et al. 2003). However, DDTs are typically singular, not completely straight and crisscross other tracks, whereas streaks formed by wind gusts exhibit multiple straight parallel to subparallel lineaments (Fig. 5C) (Cantor et al. 2006). Regarding their size, most DDTs have relatively constant widths. On Mars, they generally range from 10 m to 200 m in width and up to a few kilometers in length (e.g., Balme and Greeley 2006; Fisher et al. 2005; Verba et al. 2010). One might expect DDT width distribution would resemble that of dust devils, because small dust devils occur more often than larger ones (Sinclair 1969; Carroll and Ryan 1970; Snow and McClelland 1990; Pathare et al. 2010; Greeley et al. 2010; Lorenz 2009, 2011). However, reported observations of DDT widths seem to be biased towards wider dust devils, which might be because DDTs mainly form as a result of more intense, large dust devils, because certain surface properties only enable DDT formation from the larger dust devils, and/or because the DDT images that have been analyzed have low spatial resolution and thus do not capture the smaller DDTs. Interestingly, in some regions DDT widths observed in high resolution HiRISE images seem to be exclusively dominated by small DDTs between 1 and 10 m in width and 100 meters to few kilometers in length, which is possibly related to a low thickness of the Planetary Boundary Layer (PBL) in these regions suppressing larger dust devil size populations (Reiss and Lorenz 2016). In general, observed DDT sizes on Earth tend to be narrower than martian DDTs, with dimensions ranging generally from 1 to a few 10s of meters in width and between a few 100s of meters to several kilometers in length (Rossi and Marinangeli 2004; Hesse 2012; Reiss et al. 2010, 2011b, 2013). The observation that terrestrial DDTs are narrower than their martian counterparts may reflect the smaller dust devil size population on Earth relative to Mars (Fenton et al. 2016).

DDTs on Earth and Mars can broadly be classified into three categories based on their morphology and the albedo contrast relative to their surroundings. Most DDTs are dark compared to their surroundings, but they can differ in morphology, exhibiting either continuous dark lineaments or discontinuous dark cycloidal patterned surface streaks. More rarely,

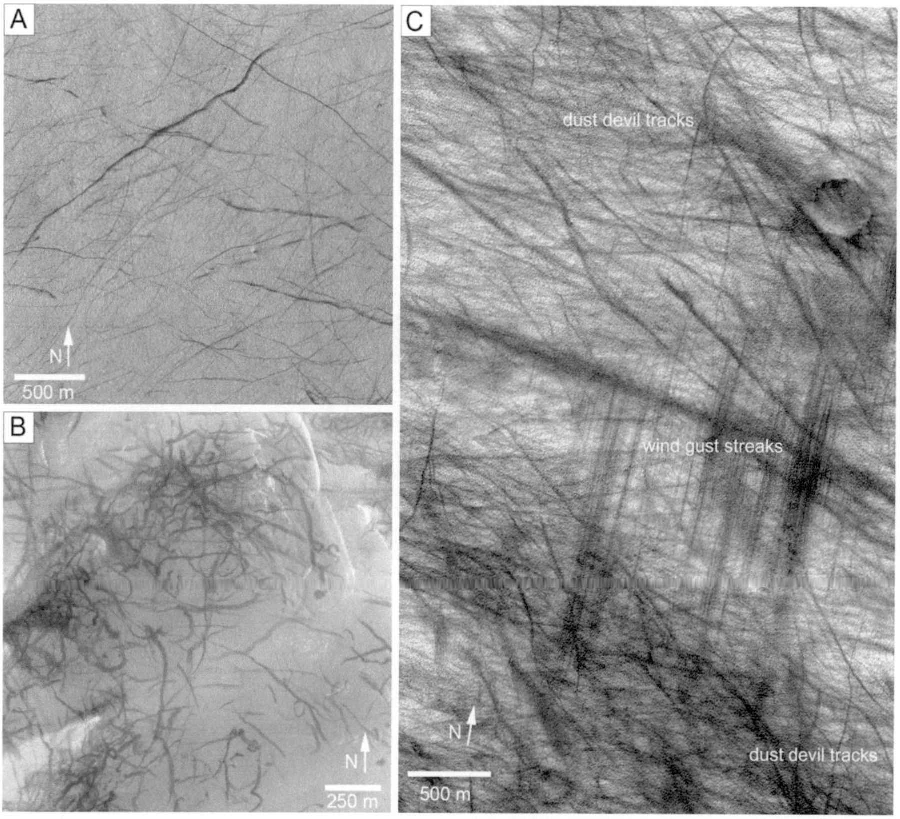

Fig. 5 (**A**) Linear to curvilinear DDTs in the Thyles Rupes region at 68.53°S and 145.02°E (HiRISE image ESP_013751_1115). (**B**) Curved, meandering and looping DDTs on the Russell crater dune field at 54.27°S and 12.95°E (HiRISE image PSP_005383_1255). (**C**) Wind gust streaks and DDTs in Malea Planum at 67.15°S and 43.9°E (MOC-NA image R1103946). The wind gust streaks shown are linear, aligned in a parallel pattern in the north-south direction, whereas the DDTs are linear to curvilinear, aligned mostly in a west-east direction. See also Cantor et al. (2006)

relatively bright lineated tracks occur. Martian examples of each of these three DDT categories are shown in Fig. 2 and each of these categories have also been found on Earth. In the following section, we classify and present the current knowledge about the individual DDT types.

2.1 Dark Continuous DDTs

One of the most common low albedo lineaments on the martian surface are dark continuous DDTs. They are characterized by their relatively homogenous and continuous low albedo surface track pattern in contrast to dark cycloidal patterned DDTs (see Sect. 2.3). Figure 6 shows two examples from Gusev crater, one obtained from orbit by the HiRISE camera and one from the martian surface by the Navcam camera onboard MER-A (Spirit) rover.

The formation of dark continuous DDTs is suggested to be linked to the removal of a surficial dust layer by passing dust devils and their fading with time due to subsequent airfall dust deposition (e.g., Malin and Edgett 2001; Balme et al. 2003). The albedo contrast between dark continuous DDTs and their surroundings might be explained by compositional

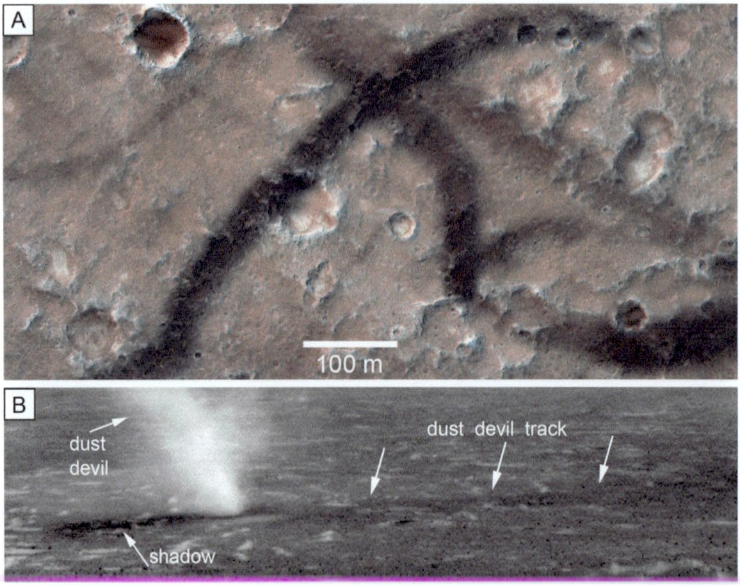

Fig. 6 (**A**) High-resolution orbital view of dark continuous DDTs in Gusev crater at 14.6°S and 175.5°E (HiRISE IRB-image PSP_006524_1650 with a spatial resolution of 25 cm/pxl). (**B**) Active dust devil (diameter of ~100 m) in Gusev crater moving from right to left leaving a dark continuous DDT in its wake (MER-A (Spirit) Navcam image 2n176788730radadaep1560l0c1). For the full image sequence see also Greeley et al. (2010)

differences or photometric differences between the eroded dust and exposed surface material. Airfall dust on the surface of Mars is similar in composition to the global soil and basaltic crust, but it is enriched in S, Cl, and Fe (e.g., Rieder et al. 1997; Goetz et al. 2005; Yen et al. 2005; Berger et al. 2016). The removal of surface dust by dust devils likely exposes soils with a different composition, which might result in the formation of DDTs. However, photometric effects due to changes of grain size might be a more common cause of albedo differences because the reflectance depends on particle size. Larger grains have greater internal photon path lengths which increases absorption, whereas smaller grains have proportionally more surface reflections that shorten internal photon path lengths (e.g., Hapke 1981, 1993; Clark and Roush 1984). The surface-to-volume ratio is a function of grain size and as a consequence the reflectance decreases with increasing grain size in the visible and near-infrared wavelengths. The effect that changing a surface's grain size has on its photometric properties was studied by Wells et al. (1984), who used Mars-analog materials of Mauna Kea volcanic soil to conduct reflectance measurements of varying amounts of deposited dust particles (1–5 μm in diameter) on a larger-grained substrate (<44–250 μm in diameter) at visible and near-infrared wavelengths (0.4–1.2 μm). Their laboratory experiments showed that the spectral and photometric properties of the substrate material are significantly affected even after deposition of very small amounts of dust particles (Fig. 7). Conversely, the erosion of fine dust particles from a coarser grained substrate such as sand would lead to a decreased reflective surface, hence leading to a darker surface area in the wake of passing dust devils.

The MER-A (Spirit) rover landed in Gusev crater within an area exhibiting various aeolian features, including seasonal dust devil activity and DDT formation (Greeley et al. 2005, 2006a; Fig. 6). The Spirit rover crossed a dark continuous DDT (Greeley et al. 2005),

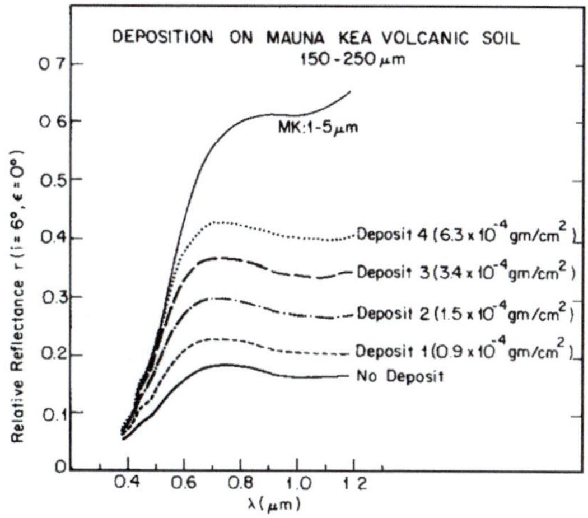

Fig. 7 Reflectance changes of 150–250 μm particles (Mauna Kea volcanic soil) after progressive deposition of 1–5 μm particle sizes (from Wells et al. 1984)

making it possible to investigate the surface substrate inside and outside the track with the Microscopic Imager (MI) (Herkenhoff et al. 2003). These in situ studies revealed that the surface substrate within the dark DDT area, consisting of coarse sand (500–1000 μm), was relatively free of the fine grained dust (Fig. 8A) compared to the brighter regions outside the track (Fig. 8B) (Greeley et al. 2005). This implies dust devils surficially remove dust, leading to photometric changes within the DDT area (Fig. 8A). Specifically, the albedo difference between the DDT and its surroundings can be explained by the removal of a thin dust layer leading to the exposure of coarse sand grains within the track area; hence, the reflectance in the visible and near-infrared wavelengths within the track area decreases in comparison to the surroundings, which are still dust covered, because brightness is photometrically inversely proportional to grain size (Greeley et al. 2005). In Fig. 8A, the texture of the surface consisting of coarse sands within the track area is still indicative of a thin layer of dust material coating the coarse sand, even after surficial dust removal exposing the coarse sand by the passing dust devil. This indicates that for this DDT, compositional effects did not or only minorly contribute to its formation. However, this is the only in situ study of a martian DDT to date, such that compositional differences producing dark continuous DDTs might be relevant elsewhere on Mars.

How much dust needs to be eroded for DDTs to form? In most cases, the amount of eroded dust from the surface is given in an equivalent layer thickness not including pore space between the dust grains. This dust deflation produced by dust devils creating DDTs was estimated to be in the range of a few to several tenths of microns (Malin and Edgett 2001; Balme et al. 2003). Direct measurements from rover imaging instruments such as the MI are difficult because the instrument is not able to spatially resolve dust grains (see also Fig. 8). However, the removed thickness of a dust layer can be inferred indirectly from active dust devils leaving tracks using observed characteristics of their behavior, although there exist relatively large uncertainties associated with such obtained dust fluxes. In the following dust deflation estimates from lander, orbital and large-eddy simulations (LES) are presented.

Figure 6B shows an active dust devil in Gusev crater, leaving a track in its wake, captured in time-sequential images acquired by the Navcam onboard the MER rover Spirit (Greeley et al. 2010). From the sequential images, Greeley et al. (2010) were able to estimate the vertical speed within the dust devil and its dust load, calculating a dust flux of $1.6 \times 10^{-5}\ \mathrm{kg\,m^{-2}\,s^{-1}}$. They also approximated the diameter of the dust devil to be ~100 m

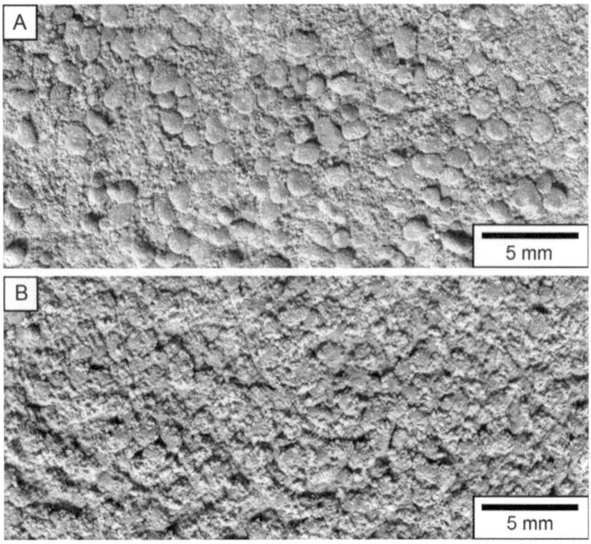

Fig. 8 Microscopic images (**A**) inside (MI image 2m129820106cfd0400p2943m2f1) and (**B**) outside of a DDT (MI image 2m132840805cfd2000p2937m2f1) in Gusev crater (see also Greeley et al. 2005). (**A**) Exposed coarse sand within the DDT, which is relatively free of fine dust compared to (**B**) outside the DDT. Note that the spatial image resolution of the MI is not able to resolve single dust grains, but the texture is indicative of fine dust coating coarse sand grains

and the horizontal ground speed to be 4.4 $m\,s^{-1}$ (Greeley et al. 2010). Using the estimated value of the dust flux, and the measured diameter and horizontal ground speed, it is possible to calculate the eroded dust layer because it is known how long the dust devil stayed above a specific point on the surface (Metzger 1999). In the case of the dust devil in Fig. 6B, which is leaving a track in Gusev crater, the dust devil needed about 23 s to cross a given spot on the surface; this translates into an eroded dust equivalent thickness layer of $\sim$1.5 μm using a dust grain density of 3000 $kg\,m^{-3}$.

A similar method was used by Reiss et al. (2014b) but estimating the eroded dust thickness from orbital images and assuming typical vertical speeds within vortices. The diameters, dust loads, and horizontal ground speeds of two active dust devils leaving tracks were measured from orbital imagery. Using the range of typical vertical speeds found in dust devil cores (0.1–10 $m\,s^{-1}$) as measured on both Earth and Mars (Ryan and Carroll 1970; Fitzjarrald 1973; Sinclair 1966, 1973; Metzger 1999; Metzger et al. 2011; Greeley et al. 2006b, 2010) the minimum and maximum dust fluxes can be calculated. In combination with the measured dust devil diameter and the horizontal ground speed, the dust deflation can then be calculated. The maximum eroded dust equivalent thickness layer of both dust devils leaving tracks was $<\sim$2 μm using a dust grain density of 3000 $kg\,m^{-3}$ (Reiss et al. 2014b).

Along with direct observations, dust deflation can also be calculated using large-eddy simulations (LES). Numerical calculation of DDT formation by Michaels (2006) using the Mars Regional Atmospheric Modeling System (MRAMS) produced an eroded dust equivalent thickness layer ranging from $\sim$1–8 μm within the track area (Fig. 9). However, in the majority of the track area the eroded equivalent thickness layer was less than 1.5 μm (Michaels 2006) which is consistent with calculated values from direct observations on Mars (e.g., Reiss et al. 2014b). The average diameter of airborne dust particles on Mars is around 3 μm (Pollack et al. 1995; Tomasko et al. 1999; Markiewicz et al. 1999; Lemmon et al. 2004; Wolff et al. 2006), hence the erosion of less or about one monolayer of surficial dust can be sufficient to form dark continuous DDTs.

Albedo measurements between DDTs and their surroundings are rare. Statella et al. (2015) calculated albedo contrasts between DDTs and their surroundings in 5 regions on

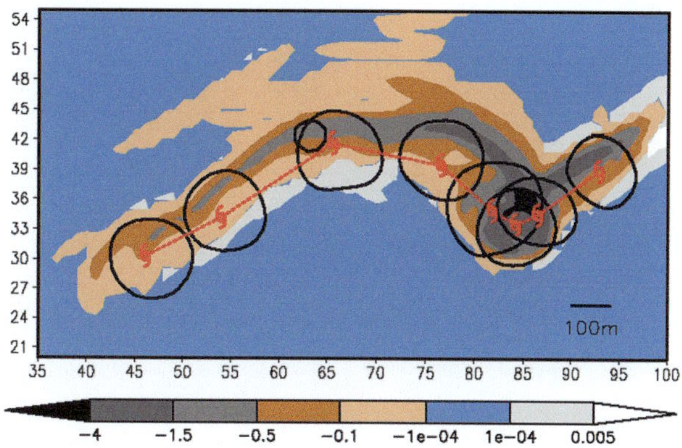

Fig. 9 Numerical simulation of DDT formation on Mars (from Michaels 2006). *Colors* indicate the depth of net surface dust reservoir change in μm (negative values = dust removal). The *red line* indicates the position of the vortex center, and *black contours* indicate the extent of the vortex core walls (Michaels 2006)

Mars (Argyre, Aeolis, Eridania, Noachis, and Hellas) using HiRISE images with band passes between 550 and 850 nm. The mean albedo contrasts are between ∼2–3 % ± ∼1.5 %.

Observed lifetimes based on multi-temporal image coverage of dark continuous DDTs are relatively short, ranging from a few weeks to less than one martian year (e.g., Balme et al. 2003; Cantor et al. 2006; Greeley et al. 2010; Verba et al. 2010; Reiss and Lorenz 2016). DDTs are erased by the steady or seasonal settling of atmospheric dust (e.g., Balme et al. 2003; Cantor et al. 2006) or an increased dust deposition after dust storm events (Greeley et al. 2010; Verba et al. 2010; Reiss 2016). At higher latitudes, seasonal frost deposits can also lead to the erasure of DDTs (Cantor et al. 2006). These different short- to long-term erasure mechanisms likely explain the relatively broad range of DDT lifetimes.

Terrestrial in situ studies of dark continuous DDTs are rare. First observations of dark continuous DDTs were made by Rossi and Marinangeli (2004) on medium-resolution satellite images in the Ténéré desert (Niger). In the following years, several more detections of dark continuous DDTs in satellite images have been reported in the Saharan desert (Neakrase et al. 2008, 2012) and in the Turpan desert in northwestern China (Reiss et al. 2010) (Fig. 10). Reiss et al. (2010) analyzed dark continuous DDTs in the Turpan desert in northwestern China that were previously detected on high-resolution satellite imagery. Figure 10B shows an example of an active dust devil leaving a dark continuous track in its wake. The occurrence of dark continuous DDTs is limited to rippled surfaces consisting of coarse to very coarse sand grains (500–2000 μm) (Reiss et al. 2010, 2012a). Microscopic images taken with a handheld device revealed that the sand substrates on the ripple surfaces are relatively free of fine-grained dust (<63 μm) compared to the areas outside of the track area (Fig. 11) (Reiss et al. 2010). This suggests that passing dust devils erode dust deposits located on top of the sandy surfaces, leading to photometric changes, hence darker surface areas (dark continuous DDTs) where the dust is removed. Dark continuous DDTs only occur on surfaces consisting of sand with grain sizes >500 μm, although most of the Turpan desert area is covered by a large dune field consisting of fine sand (∼125 μm), on which dark continuous DDTs were not observed to form (Reiss et al. 2010). This suggests that for photometric changes (rather than compositional changes) to result in DDTs, the grain size differences between the substrate material and the overlying dust needs to be sufficiently large. This terrestrial formation mechanism is in agreement with the formation of dark continuous DDTs on Mars.

For the dark continuous DDTs analyzed from the Turpan desert, the removed equivalent dust layer thickness was estimated to be about 2 μm (Reiss et al. 2010), based on size measurements of dust particles in the obtained microscopic images (Fig. 11). A subsequent

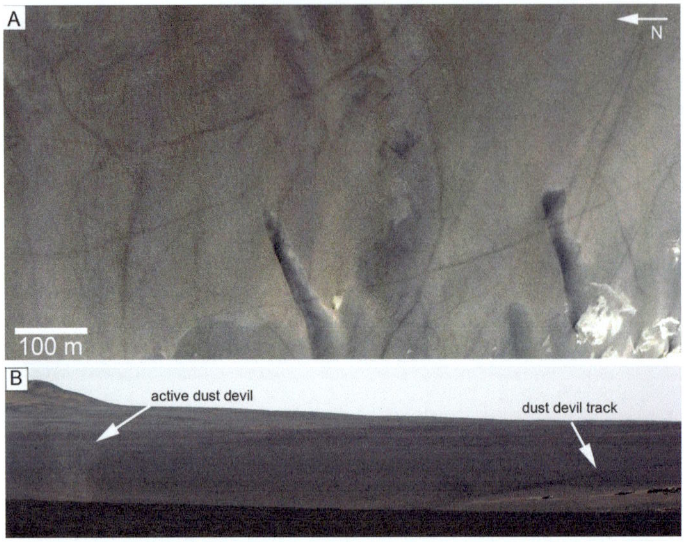

Fig. 10 (**A**) High-resolution satellite image showing several dark continuous DDTs in the Turpan desert acquired on 03 April 2005 (Quickbird image with a resolution of 0.6 m/pxl accessed through Google Earth). (**B**) Example of an active dust devil (diameter ∼3 m) in the Turpan desert leaving a dark continuous DDT on 20 April 2010. See also Reiss et al. (2010)

Fig. 11 Microscopic imagery (magnification factor 20×) within (**A** and **C**) and outside (**B** and **D**) of dark continuous DDT areas taken in the Turpan desert. **A** and **B** were taken on ripple surfaces, and **C** and **D** on large ripple surfaces. All photographs were taken on ripple crests. See also Reiss et al. (2010)

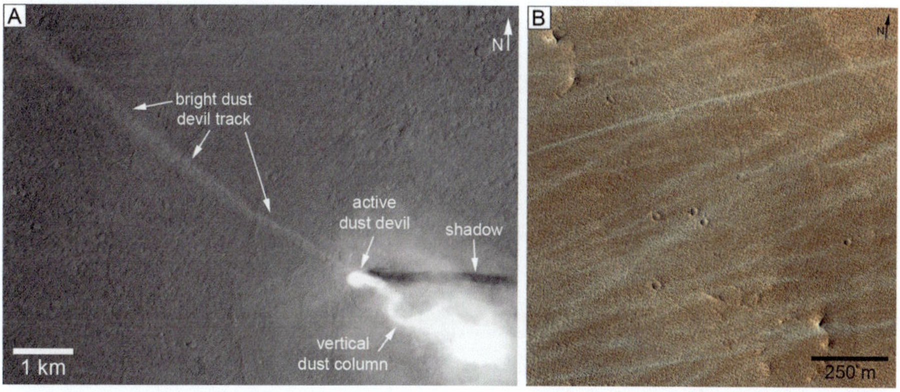

Fig. 12 (**A**) Active dust devil in Amazonis Planitia leaving a bright DDT in its wake (CTX image G21_026394_2155_XN_35N158W at 34.9°N and 201.7°E). (**B**) Numerous bright DDTs in Syria Planum (HiRISE image PSP_005453_1680 at 11.7°S and 258°E)

study using a larger number of microscopic image data and refined techniques estimated the removed equivalent dust layer thickness to be about 1.2 μm (Reiss et al. 2012a). These dust removal thicknesses are in the same range as those observed on Mars, indicating that the deflation of relatively thin dust layers are sufficient for the formation of dark continuous DDTs on both planets.

Reiss et al. (2012a) measured albedo contrasts of two DDTs and their surroundings in the Turpan desert using a pyranometer in the visible wavelength range from 300 to 1100 nm. The dark continuous DDTs were ∼0.5 and ∼0.6 % darker than the adjacent terrain.

Based on multi-temporal image coverage the lifetime of dark continuous DDTs in the Ténéré desert (Niger) is sub-annual (Rossi and Marinangeli 2004).

2.2 Bright DDTs

Bright DDTs resemble dark continuous DDTs but exhibit a higher albedo than the surrounding areas (Fig. 12). On Mars, they are less common than dark ones (Edgett and Malin 2000; Malin and Edgett 2001; Cantor et al. 2006), have lifetimes of less than five terrestrial months (Cantor et al. 2006), and seem to be confined to specific regions such as Amazonis Planitia, Syria Planum (Cantor et al. 2006) and Arsia Mons (Cushing et al. 2005). These regions are known to exhibit a relatively thick dust cover (Ruff and Christensen 2002) indicating that a sufficient amount of fine dust on the surface is required for their formation (Reiss 2014). However, comprehensive studies about the distribution or formation mechanisms of bright DDTs on Mars are lacking.

Whelley and Greeley (2008) suggested that bright DDTs on Mars might be formed by (1) removal of dark dust, (2) exposure of a bright underlying substrate, or (3) a compaction mechanism by the downdraft of dust devils. Hoffer and Greeley (2010) proposed a compaction mechanism based on laboratory experiments in which a reorientation of individual dust grains resulted in a closer packing and produced higher reflective surfaces due to changes in photometric properties. Based on terrestrial in situ studies (see the following section), Reiss et al. (2011b) proposed that bright DDTs on Mars might be formed due to the destruction of dust aggregates as passing dust devils leave a smoother, higher reflective surface within the track area in contrast to the rougher (at mm-scale), lower reflective track surrounding areas. Interestingly, MER rovers have observed weakly bound dust aggregates

Fig. 13 (**A**): Bright DDT (∼1 m width) observed on a sand dune in the Turpan desert. (**B**) Rough surface texture due to soil aggregates caused by raindrop impacts outside the DDT. (**C**) Smooth surface texture within the DDT area, occurring after the dust devil destroyed the aggregates. (**D**) Direct observation of an active dust devil leaving a bright track in its wake. Note that the dust column of the dust devil is faint, but the sand skirt near the ground is clearly visible. See also Reiss et al. (2011b)

≥100 μm to several mm in size (Vaughan et al. 2010)—held together by van der Waals, electrostatic, or other inter-particle forces—on the martian surface (e.g., Herkenhoff et al. 2004, 2006; Sullivan et al. 2008; Bridges et al. 2010).

On Earth, Maher (1968) suggested that bright DDTs (see also Fig. 3A) are caused by the disturbance of desert varnish by passing dust devils. However, this mechanism is highly unlikely because active dust devils are not strong enough to disturb hard crusts, which normally need to be scratched. Reiss et al. (2011b) observed bright DDTs in the Turpan desert and made the first in situ studies. Usually, dark continuous DDTs occur in this area (see Sect. 2.1), but after a rainfall event only bright DDTs were observed in the field (Reiss et al. 2011b) (Fig. 13). Raindrop impacts onto the surface caused the formation of aggregates of sand, silt, and clay, leading to rough surface textures (Fig. 13B). Due to their weak cohesion, passing dust devils easily destroyed the aggregates, leading to smooth surface textures at the millimeter scale (Fig. 13C). The albedo difference between the track area and the surroundings can be explained simply by photometric effects between the smoother, higher reflective track area and the rougher (at mm-scale), lower reflective surroundings (Reiss et al. 2011b). Based on field observations, bright DDTs in the Turpan desert remained only for a few days likely due to the destruction of surficial aggregates by strong winds (Reiss et al. 2011b).

2.3 Dark Cycloidal DDTs

Dark cycloidal DDTs were first observed in satellite imagery by Hesse (2012) in southern Peru. In plan view, they are characterized by a low albedo cycloidal pattern of overlap-

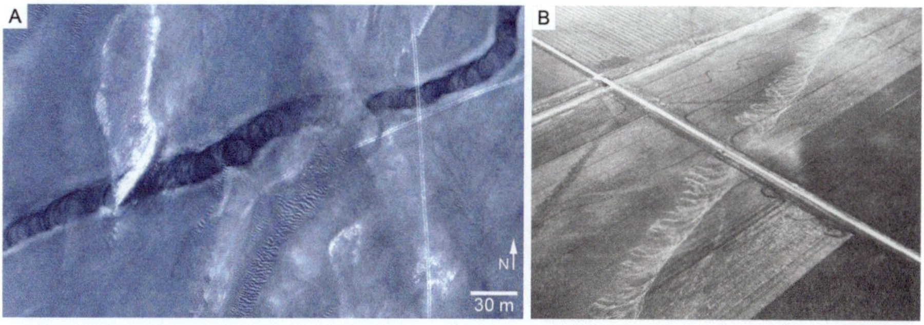

Fig. 14 (**A**) Dark cycloidal DDT in southern Peru at 14.2°S and 75.9°W (DigitalGlobe satellite image accessed through Google Earth). (**B**) Cycloidal tornado track left by Anchor (No. 2) tornado of 3 April 1974 (Image credit: Fujita 1974. Weatherwise from Davies-Jones 1986). Note that the tornado track in (**B**) is bright due to the deposition of relatively bright corn stubbles

ping scallops, sometimes accompanied by lateral bright margins on one or both track sides (Fig. 14A). In comparison to dark continuous DDTs, the tracks of dark cycloidal DDTs are not continuously dark, but rather only the overlapping scallops show a low albedo, which forms the cycloidal track pattern. In morphology, the tracks resemble ground marks left by tornadoes (Fig. 14B), which form their cycloidal pattern by depositing debris (including corn stubbles on farmland, etc.) gathered by suction vortices (e.g., Van Tassel 1955; Prosser 1964; Fujita et al. 1970; Fujita 1971, 1974).

Multi-temporal satellite images show that the lifetime of cycloidal DDTs in southern Peru can vary drastically. In some regions, dark cycloidal DDTs can still be identified after several years (long-lived) whereas in other regions they disappear within one year (short-lived) (see examples in Hesse 2012 and Reiss et al. 2013). In situ studies of short-lived and long-lived dark cycloidal DDTs in southern Peru revealed that cycloidal track patterns are formed by redeposition of sand-sized material, which is eroded by dust devils from the outer track margins and is subsequently deposited in annular patterns in the vortex cores (Reiss et al. 2013). Figures 15A and B show an example of long-lived cycloidal DDT in southern Peru with accompanying bright margins as seen from orbit and in the field, respectively. The long-lived cycloidal DDTs are located on a desert pavement surface consisting of a layer of crusted fine-material (<250 μm) overlain by very coarse sand grains (1–2 mm) (Fig. 15E). The bright marginal area of the track shows a reduced fraction of very coarse sand grains (Fig. 15D) compared to the undisturbed desert pavement surface (Fig. 15E). In contrast, there is an increased fraction of very coarse sand grains within the dark cycloidal DDT area (Fig. 15C) relative to the undisturbed desert pavement surface (Fig. 15E). This implies erosion of the surficial very coarse sand at the outer margins of vortices (leaving deflated bright margins) and their subsequent deposition in annular patterns within the vortices (forming the deposited cycloidal pattern) (Reiss et al. 2013). The albedo differences between the dark track, bright margins, and the surrounding area can be explained by changes in photometric properties caused by different amounts of surficial very coarse sand. The long lifetime of dark cycloidal DDTs in this area can simply be explained by their occurrence on desert pavements that do not experience much aeolian activity. Field experiments on desert pavements in other region on Earth showed that the recovery of relatively small areas cleared from stones and granules occurs at very low rates of about 1 % per year, which indicates full surface recovery times of up to 80 years (Haff and Werner 1996).

Compared to long-lived dark cycloidal DDTs, such as those shown in Fig. 15, short-lived dark cycloidal DDTs rarely have lateral bright margins. Examples of short-lived dark

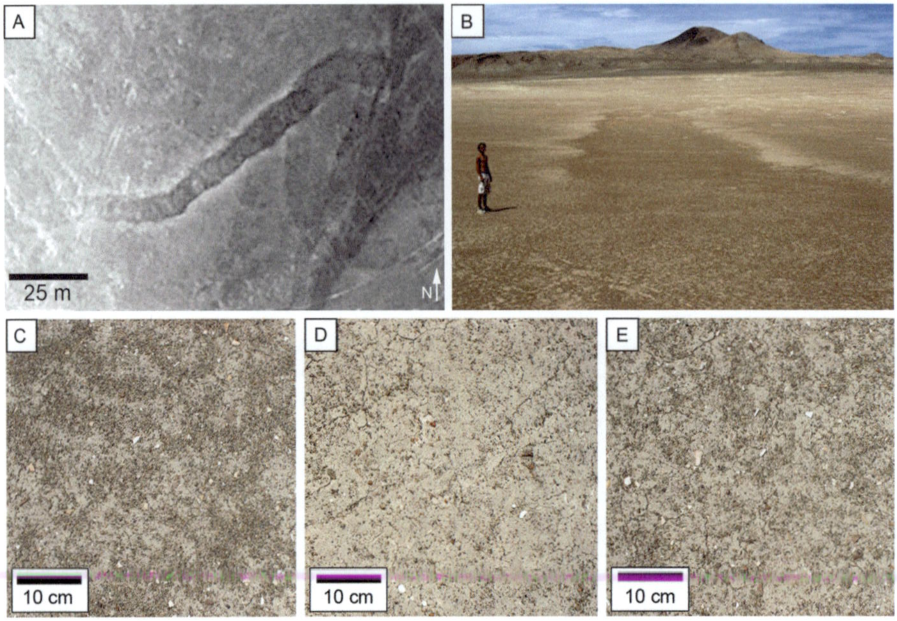

Fig. 15 (**A**) Long-lived dark cycloidal DDT in southern Peru at 14.4°S and 75.8°W (DigitalGlobe satellite image accessed through Google Earth). (**B**) Field photograph of the same DDT as in A, imaged facing southwest. Note the bright margins of the dark track area. (**C**) Top-view image of the dark cycloidal track area. (**D**) Top-view image of the bright margin area. (**E**) Top-view image of an area outside the dark and bright margin area of the cycloidal track area. For further information see also Reiss et al. (2013)

cycloidal DDTs in southern Peru are shown in Fig. 16A, which occur on sand sheets consisting of granule ripples dominated by coarse to very coarse sand (0.5–2 mm). Ripple troughs are relatively free of coarse sand, and bright patches of the underlying fine grained material (<250 μm) are visible (Fig. 16B and C). In situ studies of the dark cycloidal DDT shown in Fig. 16B by Reiss et al. (2013) revealed that the bright patches (Fig. 16C) within the track area are covered by coarse sand (Fig. 16D), indicating that active dust devils redistribute sand material in annular patterns within the track area leading to dark cycloidal DDTs. The rarer occurrence of bright marginal areas along the dark track of short-lived dark cycloidal DDTs can be explained by the large amount and thickness of coarse sand (in contrast to the desert pavement region where long-lived DDTs usually exhibit bright lateral areas) impeding complete exposure of the bright underlying surface by erosion. However, some dust devils seem to be strong enough to remove enough sand to create bright marginal areas along one (Fig. 16A, arrow 2) or both sides (Fig. 14A) of the dark cycloidal tracks. The relatively short lifetime of the DDTs on these surfaces can be explained by their occurrence on active aeolian sand sheets, which mobilize the coarse grains and destroy the tracks.

Large-eddy simulations (LES) have revealed critical details of large-scale vertical convective vortices, including how interactions with the surface can lead to their considerable intensification (Lewellen et al. 2000; Lewellen and Lewellen 2007a, 2007b), how massive loadings of debris can reorganize the momentum distribution and damage potential of a vortex (Lewellen et al. 2008), and how debris transport can leave behind visible deposits of debris, or surface marks (Michaels 2006; Lewellen and Zimmerman 2008). Most of these simulations have focused on the physics of tornadoes and their debris clouds, but a simple dimensionless scaling (Lewellen et al. 2008; Zimmerman and Lewellen 2010) allows the

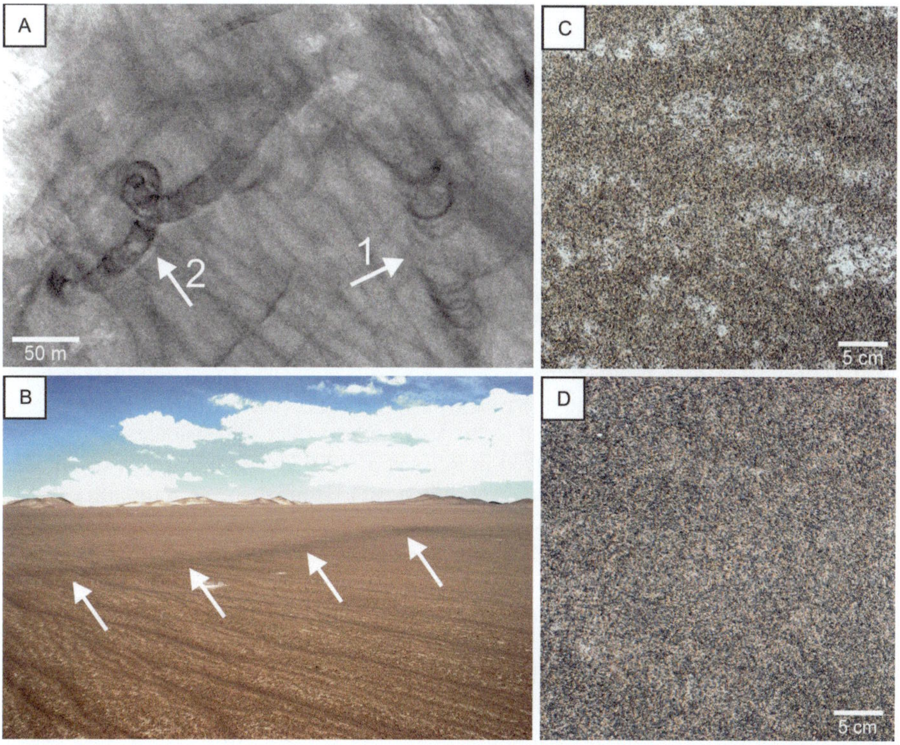

Fig. 16 (**A**) Short-lived dark cycloidal DDT (*arrow 1* and *2*) in southern Peru at 14.2°S and 75.9°W (DigitalGlobe satellite image accessed through Google Earth). (**B**) Dark cycloidal DDT (*white arrows*) observed near the DDTs shown in **A**. (**C**) Top-view image of undisturbed ripple surface next to the DDT shown in **B**. (**D**) Top view image of the dark cycloidal DDT area shown in **B**. For further information see also Reiss et al. (2013)

results to be scaled to terrestrial dust devils, though the latter are typically smaller and not as intense as tornadoes. This approach has allowed existing simulations of tornado tracks to be repurposed for studying how surface marks are generated by dust devils in the field (Reiss et al. 2013).

The dimensionless scaling between tornadoes and dust devils is enabled by several similar physical characteristics. Both classes of vortex are fed by a wide but shallow inflow layer that is relatively low in angular momentum compared to the embedding flow aloft. They also exhibit a corner flow region—where the near-surface flow intensifies while also turning rapidly upward—and a swirling, rising annular core aloft (Lewellen et al. 2000). Tornadoes and dust devils can accumulate large debris clouds that leave contrasting patterns of deposition on the ground (Lewellen et al. 2008; Zimmerman and Lewellen 2010; Reiss et al. 2013). A dimensionless parameterization applicable to tornadoes and dust devils has been discussed at length in Lewellen et al. (2000, 2008), Zimmerman and Lewellen (2010), Reiss et al. (2013). Given a radius R_c and peak swirl velocity V_c in the upper core, a background angular momentum level Γ_∞ in which the vortex is embedded, a ground-relative translation speed U_t, the depleted flux of angular momentum Υ flowing into the corner from the surface layer, a surface roughness length z_0, a debris particle's terminal velocity w_t, and gravitational acceleration g, at least four dimensionless parameters can be formed:

1. Corner flow swirl ratio, $S_c = R_c \Gamma_\infty^2 / \Upsilon$, which determines whether the vortex is dominated by swirling flow or radial/vertical flow-through.
2. Translation ratio, $A_t = U_t / V_c$, defining the surface-relative translation speed as a fraction of the characteristic swirl velocity.
3. Acceleration ratio, $A_a = V_c^2 / g R_c$, scaling the characteristic centripetal acceleration to planetary gravity.
4. Debris type, $A_v = V_c / w_t$, which is a measure of how effectively debris can be entrained and lofted.

Simulations have revealed that leading aspects of the vortex are largely encoded within (1–4) (Lewellen et al. 2000; Lewellen and Lewellen 2007a, 2007b; Reiss et al. 2013).

The particles comprising surface marks at the site of the Peruvian dust devil study of Reiss et al. (2013) were nearly monodisperse, with a size of 0.5–2 mm and mass density 2000 $\mathrm{kg\,m^{-3}}$, making a direct comparison with large-eddy simulations (with a similarly monodisperse debris population) possible. To derive approximate dimensionless parameters associated with the dust devils observed in Reiss et al. (2013), and thus select the closest possible tornado simulation for comparison, the debris terminal speed, translation speed, core radius aloft, peak swirl velocity aloft, and near-surface depleted angular momentum flux must be estimated. Under terrestrial gravity, the terminal speed for a spherical grain in the size and density range of Reiss et al. (2013) is $w_t = 3.2\text{–}9.2\ \mathrm{m\,s^{-1}}$. The observed dust devil tracks give an approximate width of the vortex (although in some simulated cases, debris can be thrown a considerable distance inward), setting $R_c \sim 25$ m. Meteorological measurements of the Peruvian dust devils are not available, but a core velocity aloft $V_c = 25\ \mathrm{m\,s^{-1}}$ and translation speed $U_t \sim 4\ \mathrm{m\,s^{-1}}$ are tentatively assumed based on dust devil climatology (Balme and Greeley 2006). Together with $g = 9.81\ \mathrm{m\,s^{-1}}$, these parameters set the ratios $A_t = 0.15$, $A_a = 2.1\text{–}2.5$ and $A_v = 2.7\text{–}2.8$. It is impossible to estimate the near-surface flow properties required to derive S_c for the dust devils in Reiss et al. (2013); however, with set ranges for the other three dimensionless ratios, an attempt has been made to select the "best" swirl ratio that reproduces major characteristics of the surface tracks observed in the field.

Figure 17 shows a surface track from a large eddy tornado simulation with $A_t = 0.15$, $A_a = 2.2$, $A_v = 4.3$, and $S_c = 9.3$. This best matches the qualitative appearance of the surface track in Fig. 13A. Sand is removed from the margins of the simulated track, carried inward, and deposited in cycloidal marks, which are preferentially laid down to the back and right of the vortex (with respect to an observer on the ground looking along the direction of travel of the vortex). This is consistent with laboratory experiments on dust devil track formation (Greeley et al. 2004). In the simulation that produced Fig. 17 (and many others, cf. Zimmerman and Lewellen 2010), bands of alternating debris concentration swirl inward from a wide area across the surface. In the corner flow, some debris is unable to follow the upward accelerating flow and slips back to the surface, where it is deposited in sharp cycloidal bands. The remainder is lofted and centrifuged outward into the tornado's debris cloud, or in the case of a dust devil, the sand skirt. In Fig. 17, some of this debris falls back into the right margin of the track, where it forms diffuse bands of deposition. The left-right asymmetry in the far-field deposition is due to a rightward and forward tilt of the vortex induced by surface-relative translation (Lewellen and Zimmerman 2008). The simulated debris field is infinitely deep in the case of Fig. 17; that is, removal is limited only by the negative feedback between debris removal and air momentum. In the Peruvian case study by Reiss et al. 2013, the effective debris field of millimeter-size grains is probably much more limited. However, the comparison between Figs. 17 and 14A demonstrates that cycloidal mark formation is consistent with inward transport of sand from the margins of the track.

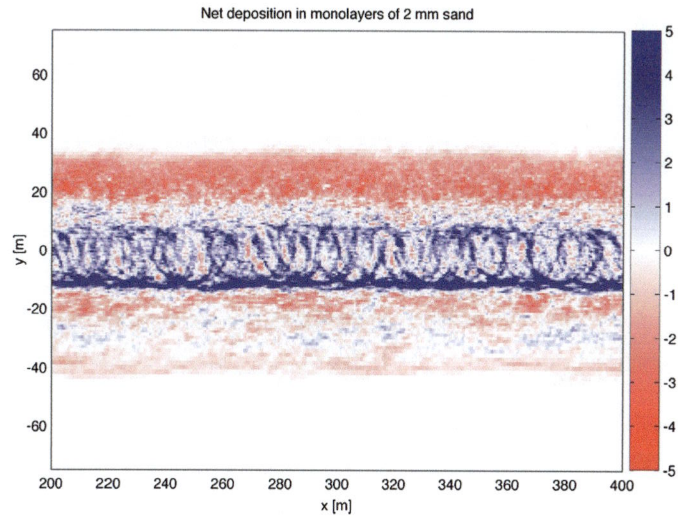

Fig. 17 Simulated dust devil track (scaled from a LES tornado simulation). Positive and negative values correspond to net deposition and removal, respectively. Reproduced from Reiss et al. (2013)

Cycloidal dust devil tracks have been studied in the laboratory (Greeley et al. 2004) with the Arizona State University Vortex Generator (ASUVG). The ASUVG is a 2.5-m by 2.5-m moveable table assembly with an independently mounted motor and fan blade mounted in a cylindrical housing (see Fig. 2 in Greeley et al. 2003). The table can translate both vertically and horizontally for simulating surface movement. The cylinder/fan-blade assembly can also translate vertically, which is used to control vortex parameters such as tangential wind speed and vortex diameter (Greeley et al. 2003). Previous experiments (Greeley et al. 2003) demonstrated that the ASUVG could replicate fundamental vortex morphology at lab scales at both Earth ambient and Mars analog conditions, including the characteristic 'wobble' of natural dust devils. The dust devil track experiments consisted of depositing a uniform, thin (400–800 μm thick) layer of ∼125 to 200-μm silica sand on the test surface. In some experiments there was a 2-μm coating of red dust as well. As the ASUVG generated the vortex, the test surface was pushed beneath the rotating column of air, where the laboratory dust devil was allowed to interact with the sediment. Results showed a reorganization of the sand portions of the sediment, or the resulting 'track' left behind after passage of the vortex (Fig. 18A). In many of the experiments, the resulting track was cycloidal in morphology and provided a sense of the movement direction of the vortex (Fig. 18).

Analogous to the martian conditions, cycloidal tracks can be shown to be depositional features, primarily due to sand grains being reorganized by the dust devil. Larger particles that are initially lifted by the dust devil are too heavy to be lofted up into the column and are subsequently thrown out of the dust devil with angular momentum. The resulting pattern is a cycloidal track that opens in the direction of dust devil movement. This interpretation assumes that dust devils act both as an erosional and depositional agent, removing sand from the front of the dust devil and leaving the heavier particles in its wake.

Dark cycloidal DDTs are also observed on Mars (Fig. 19), although they seem to be much rarer than dark continuous DDTs. Morphologically they resemble terrestrial dark cycloidal DDTs observed in southern Peru (Figs. 14A, 15A and 16A), indicating that they might be formed by the same mechanism process. Entrainment of sand on Mars is much more

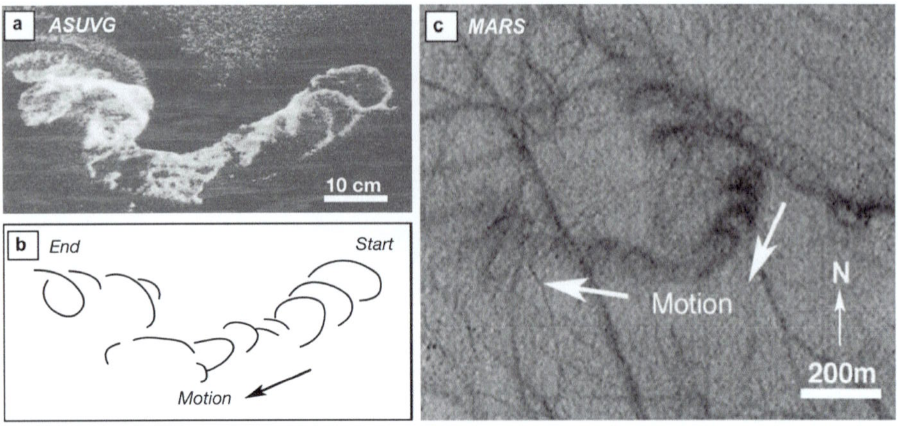

Fig. 18 (After Greeley et al. 2004 and Neakrase 2009) (**a**) Photo showing the experiment run with the ASUVG with 125–200 μm sand demonstrating the creation of a cycloidal track. (**b**) Schematic cartoon of the ASUVG track. (**c**) MOC-NA image (M10-03516) showing cycloidal tracks east of Hellas basin on Mars

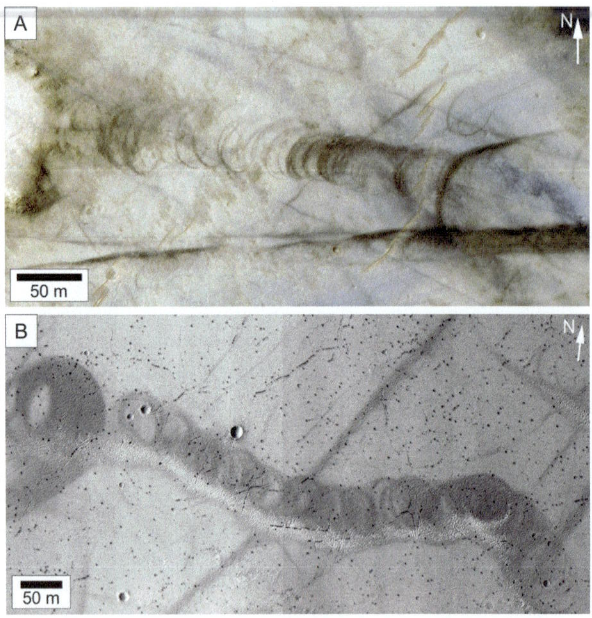

Fig. 19 Dark cycloidal DDTs on Mars. (**A**) Dark cycloidal DDT in Schiaparelli crater at 5.2°S and 17.7°E (HiRISE image PSP_006477_1745). (**B**) Dark cycloidal DDT in Brazos crater at 5.6°S and 18.9°E (HiRISE image PSP_006477_1745). For comparison with terrestrial dark cycloidal DDTs, see tracks white arrows 1 and 2 in Fig. 15A

difficult than on Earth due to the low atmospheric pressure. The threshold shear velocity required to move sand grains with a diameter of ∼200 μm in diameter by saltation under martian atmospheric conditions is ∼1.5 m s^{-1} compared to 0.2 m s^{-1} for Earth (Iversen and White 1982; Kok et al. 2012). Wind speed measurements from martian landed spacecraft suggest that such shear velocities are rarely exceeded (e.g., Zurek et al. 1992; Holstein-Rathlou et al. 2010). However, there is much direct evidence that at least fine sand material is actively moved under present-day martian atmospheric conditions (e.g., Sullivan et al. 2008; Geissler et al. 2010; Silvestro et al. 2010; Bridges et al. 2011). Furthermore, dust devil tangential speeds can reach higher values than average martian wind speeds (e.g., Cantor et al. 2006; Choi and Dundas 2011), hence the redeposition of sand creating observed dark cycloidal DDTs on Mars seems plausible. However, direct evidence by in situ observations with rovers for this formation mechanism of dark cycloidal DDTs on Mars is lacking.

3 Spatial and Temporal DDT Distribution on Mars

Figure 20 shows a global map of Mars with the location and extent of DDT studies. Most studies did not discriminate between the different types of dark continuous, dark cycloidal and bright DDTs. In many cases only darker-toned DDTs were analyzed, because bright DDTs seem to be limited to dusty regions on Mars. Many studies also did not distinguish dark continuous from dark cycloidal DDTs, hence they are defined only as dark (dark toned or low albedo) DDTs.

3.1 Elevations

DDTs on Mars have been found to occur at all elevations, from the top of the tallest volcano (Olympus Mons) to the bottom of the deepest basin (Hellas Planitia) (Malin and Edgett 2001; Cantor et al. 2006). This is surprising because at high elevations the annual mean atmospheric pressure is only at around 1 mbar (avg. martian atmospheric pressure is around 7 mbar), but active dust devils and DDTs were observed in these low pressure environments (Cushing et al. 2005; Reiss et al. 2009), implying that dust devils can form even at very low ambient atmospheric pressures. Interestingly, Balme et al. (2003) found no correlation of DDT abundance with elevation in the study regions of Argyre and Hellas basin (elevations ranging from −6440 to +6130 m and −8208 to +4886 m, respectively), but this would be expected because increased atmospheric pressures at low elevations favor dust devil occurrence (Greeley et al. 2003). In addition, in a global study Whelley and Greeley (2008) did

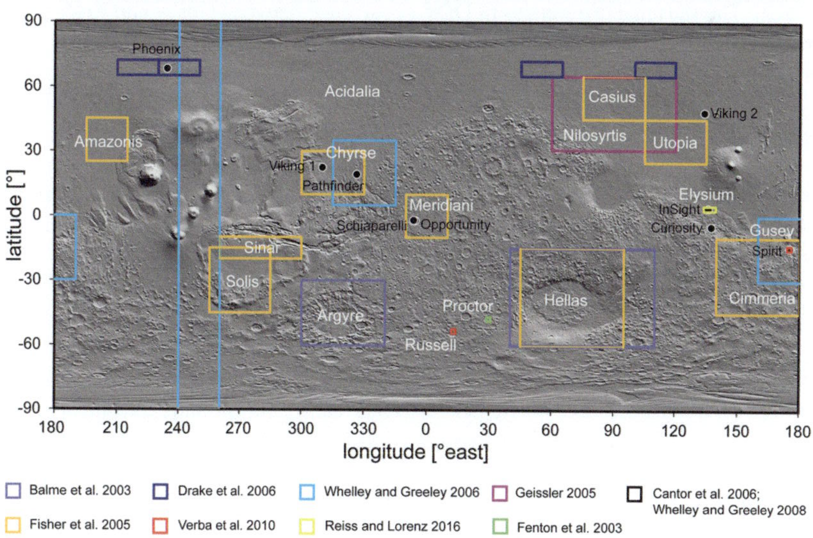

Fig. 20 Location and extent of DDT study regions. *Color extent* and *white text* correspond to study regions given in the references. *Black dots* and *black letters* show the location and name of landed missions or future landing sites on Mars. Not included are the study regions of Ormö and Komatsu (2003) and Calef and Sharpton (2005) because they do not give specific statistics about DDTs

not find a correlation of DDTs with elevation. The reason for the independence of dust devil and DDT occurrence with elevation is unclear.

3.2 Seasonal Occurrence

Dust devils form most frequently when insolation is around its seasonal maximum (e.g., Fisher et al. 2005; Cantor et al. 2006; Stanzel et al. 2008; Greeley et al. 2010; for a summary see also Fenton et al., 2016, this issue), hence DDT occurrence should exhibit a seasonal maximum during spring and summer. While most studies record the time when the image data containing DDTs was acquired, this does not provide the exact time of DDT formation (e.g., Balme et al. 2003). DDTs are likely formed prior to the image acquisition; hence the recorded seasons are probably actually later in the season. Balme et al. (2003) observed an increase of DDTs in the Argyre and Hellas Planitiae during early spring, followed by a more distinctive increase in late spring and reaching its maximum during summer, whereas in fall and winter only a few DDTs formed because insolation is then near its seasonal minimum. The same trend in seasonal DDT occurrence was observed by Fisher et al. (2005) and Whelley and Greeley (2006, 2008) in the northern and southern hemisphere, although higher frequencies in fall were recorded.

Local-scale studies have used repeat coverage of the same region, recording newly formed DDTs (Verba et al. 2010; Reiss 2016). This method gives a more precise determination of seasonal DDT formation. Using MOC-NA images of the Proctor crater floor (47°S and 30°E), Fenton et al. (2003, 2005) observed DDT formation only from mid spring through late summer ($Ls = 223\text{–}354°$). Verba et al. (2010) found that the DDT activity in Gusev Crater (14.6°S and 175.4°E) was mostly confined to a period between $Ls = 235°$ and 12° (mid spring to early fall) with an peak activity at $Ls = 245°$ (mid spring). In Russel Crater (53.3°S and 12.9°E) the active dust devil season ranged from $Ls = 172°$ to 40° (around the start of spring to early fall) with peak activity at $Ls = 316°$ (mid summer) (Verba et al. 2010). The seasonal differences in DDT occurrence between Gusev and Russell Crater are probably caused by the latitudinal difference in location, in which Russell Crater receives more insolation at perihelion during the southern summer, whereas DDT formation starts earlier in Gusev Crater due to its location within the tropical zone (Verba et al. 2010). At the proposed InSight landing site in Elysium Planitia (3.9°N and 136.7°E), the seasonally limited observations did not allow a conclusion about seasonal DDT activity, but the analysis points toward activity throughout the year with peak activities in northern spring and southern summer (Reiss 2016), both probably related to the equatorial location of this region.

3.3 Distribution

Based on a global MOC-NA image survey, Cantor et al. (2006) observed DDTs in the latitude range between 80°S and 80°N; hence, DDTs can occur globally except on the permanent polar caps. This hemispheric dichotomy was also detected by Whelley and Greeley (2006) latitudinal analysis (pole-to-pole survey). On a global scale, Whelley and Greeley (2008) mapped around 55,000 DDTs in 1238 MOC-NA images, which were selected with a seasonally stratified random sampling technique. They measured the percentage of the surface area covered by DDTs per image. Their results show a seasonal peak in the percentage of DDT cover near 60° in both hemispheres (Fig. 21). In the northern hemisphere, the peak areal coverage is 10 % between 40°N and 65°N during northern spring and summer, whereas in the southern hemisphere the peak coverage is 92 % during southern spring and summer in a latitude band between 45°S and 75°S (Whelley and Greeley 2008). Whelley and Greeley (2006, 2008) attributed this to the asymmetric solar heating on Mars because of the planet's orbital eccentricity, where the southern hemisphere receives about 40 % more solar insolation during the southern summer.

On a regional and local scale, the spatial DDT distribution can vary drastically. Drake et al. (2006) surveyed DDTs and wind streaks in four different regions, all located in the same latitudinal band between 65°N and 72°N. Although they did not distinguish between DDTs and wind streaks or give additional information about the seasonal coverage, the percentage of images containing these features varied from 3.5 %–20.9 % (Drake et al. 2006). In another study, Fisher et al. (2005) analyzed active dust devils and DDTs in nine different regions located across Mars using MOC-NA imagery. They showed that all DDT detections were seasonally separated, although individual tracks were not counted. The maximum percentage of images containing DDTs in one season between the regions varied from 0.79 %–52.17 % (Fisher et al. 2005). In some regions such as Casius, more than 50 % of the images contained DDTs in two of four seasons, whereas in other regions such as Utopia, only 1 % of the images contained DDTs in one season and 0 % in the other three seasons (Fisher et al. 2005). The same strong variations of DDT occurrences can also be found in other study regions. For example, Geissler (2005) analyzed MOC-NA images in the Nilosyrtis study area (60–120°E and 30–65°N) and observed a strong zonation of DDT occurrence confined to latitudes between 45°N and 65°N. DDT areal densities can vary even on a scale of 10s of kilometers: Fenton et al. (2003) found that dark continuous DDTs were plentiful in all but the northern area of Proctor crater floor, where none were visible.

3.4 Areal Density

The hemispheric asymmetry and strong regional spatial variations in DDT occurrence on Mars are also reinforced by statistically studies that normalize the number of DDTs by surface area, expressed as the number of DDTs per square kilometer. Whelley and Greeley (2006) mapped DDTs in MOC-NA images in a pole-to-pole survey and in some other regions in both hemispheres (Fig. 20). They calculated average densities of 0.6 DDTs/km^2 for the southern and 0.06 DDTs/km^2 for the northern hemisphere. In addition, they measured

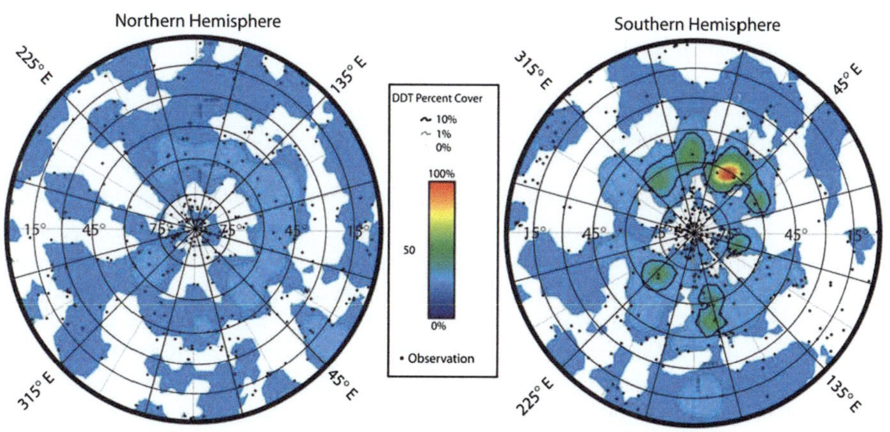

Fig. 21 Interpolation maps showing the percentage of DDT image cover in spring and summer for the northern and southern hemispheres. *Black dots* indicate the location of analyzed MOC-NA images, *isochrones* and *colors* show the percentage of DDT coverage. From Whelley and Greeley (2008)

regional differences with peak seasonal densities of about 0.02 DDTs/km^2 in Ares Vallis and about 0.1 DDTs/km^2 in Gusev Crater (Whelley and Greeley 2006). In the Hellas and Argyre basins, DDT densities showed a latitudinal zonation, often reaching 50–100 DDTs/km^2 in the southern areas of both study regions compared to values of less than 1–5 DDTs/km^2 in the northern areas (see also Fig. 4 in Balme et al. 2003). The average density from one continuous year was 0.81 DDTs/km^2 for Argyre and 0.47 DDTs/km^2 for Hellas although densities in both study regions reached seasonal peaks of about 2.5 DDTs/km^2 (Balme et al. 2003).

On a local scale, repeat imaging of the same surface areas with high-resolution image data such as HiRISE (limited to interesting areas or landing sites) provides information of DDT formation rates expressed as DDTs/km^2/sol (in which one sol is one martian day). Using this technique, Verba et al. (2010) mapped newly formed DDTs between image observations in Russell and Gusev Craters over the course of a martian year, calculating seasonal DDT formation rates. Formation rates ranged from 0.0011 to 0.103 DDTs/km^2/sol in Gusev Crater and from 0.04 to 0.95 DDTs/km^2/sol in Russell Crater (Verba et al. 2010). Reiss and Lorenz (2016) used the same technique to derive DDT formation rates at the proposed InSight landing site. They calculated formation rates from 0.002 to 0.08 DDTs/km^2/sol (Reiss and Lorenz 2016). The much higher formation rates in Russell Crater relative to those in Gusev Crater and the InSight landing site shows large variability in DDT frequencies on a local scale.

As discussed above, local, regional, and global variations in DDT densities can partly be explained by their latitudinal location, which controls the receivable insolation, controlling dust devil generation, hence also DDT densities. However, the large regional variations in DDT distributions and densities are not so easily explained as DDT distributions and densities vary strongly between study regions within the same latitudinal band or within individual regions (Balme et al. 2003; Drake et al. 2006; Geissler 2005). Furthermore, one study analyzed the frequency of active dust devils versus DDT abundance within several study regions, revealing a large discrepancy (Fisher et al. 2005). For example, in the Casius study region, no active dust devils were observed but many DDTs were, whereas in the Amazonis Planitia, many active dust devils were observed but only a few DDTs were detected (Fisher et al. 2005). The probable reason causing these large local, regional and global discrepancies in DDT densities and dust devil occurrences are discussed in the following section.

3.5 Thermophysical Surface Properties

Because DDT distributions and densities are not always consistent with dust devil activity, researchers suggested relatively early on that DDTs are not solely controlled by dust devil activity but by differences in surface materials, such as dust cover thickness or underlying substrate properties (Balme et al. 2003; Fisher et al. 2005; Whelley and Greeley 2006, 2008). Most DDTs on Mars are dark continuous tracks suggested to be formed by dust erosion, which exposes a substrate of relatively coarse-grained material (see Sect. 2.1). If the dust layer is too thick, a passing dust devil may not be strong enough to expose the underlying substrate, hence no DDT would be formed. On the other hand, thin dust layers may enable even weak dust devils to expose the substrate, leading to areas with many DDTs. Another factor contributing to the formation of dark continuous DDTs is that on Earth the distribution of DDTs seems to be controlled by the particle-size difference between the substrate material to the lofted material (Reiss et al. 2010). The substrate material needs to be sufficiently large enough relative to the lofted grains to create albedo differences by photometric changes (see Sect. 2.1). For example, terrestrial studies have shown that dark continuous DDTs occur

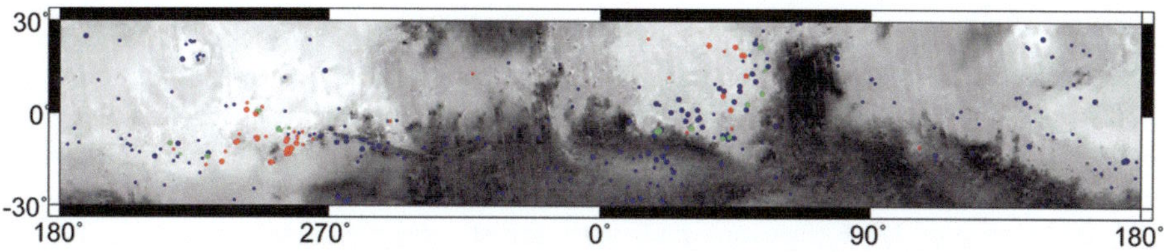

Fig. 22 Distribution of dark and bright DDTs from a MOC-NA survey (Reiss 2014). *Blue* = dark DDT; *Red* = bright DDT; *Green* = dark and bright. *Small circles* = few DDTs; *Large circles* = many DDTs. *Background*: TES Albedo (Christensen et al. 2001)

only in areas where the substrate consists of coarse to very coarse sands, and no DDTs are observed in areas consisting of fine sand grains, implying that when particle sizes of the substrate materials are too small, the photometric effects are not large enough to create visible albedo contrasts between the track and the undisturbed area (Reiss et al. 2010, 2012a).

The distribution of identified DDTs on Mars can be compared to derived global thermophysical data sets, such as Thermal Emission Spectrometer (TES) albedo maps (Christensen et al. 2001), thermal inertia maps (Mellon et al. 2002; Putzig et al. 2005), and dust cover index (DCI) maps (Ruff and Christensen 2002). For example, Balme et al. (2003) found that the distribution of DDTs in the Argyre and Hellas study were correlated with surfaces indicative of a thin dust layer, suggesting that dust availability controls the abundance of DDTs. Whelley and Greeley (2006, 2008) found that DDTs correlate with high thermal inertia (interpreted to be rocky) in the southern hemisphere, but with low thermal inertia (interpreted to be dusty) in the northern hemisphere. This suggests that a high thermal inertia surface with a thin dust cover is not a prerequisite for DDT formation and a low thermal inertia surface with a relatively thick dust cover does not preclude DDT formation (Whelley and Greeley 2008). Reiss (2014) surveyed DDTs in MOC-NA images with a resolution of <3 m/pxl in the equatorial region (latitude range from 30°S to 30°N) and compared the geographic location of images containing dark or bright DDTs with these areas' themophysical surface properties (Fig. 22). Results showed that bright DDTs occur in areas of high albedo ($\sim$0.28), low thermal inertia ($\sim$75 $\mathrm{J\,m^{-2}\,s^{-1/2}\,K^{-1}}$), and high dust cover (DCI$\sim$0.94), whereas dark DDTs occur in areas of moderate albedo ($\sim$0.2), moderate thermal inertia ($\sim$250 $\mathrm{J\,m^{-2}\,s^{-1/2}\,K^{-1}}$), and moderate dust cover (DCI$\sim$0.96), indicating that bright DDTs occur in regions with a relatively thick dust cover and dark DDTs in regions with a relatively thin dust cover (Reiss 2014).

4 DDTs as proxies

4.1 Dust Devil Activity

Inferring dust devil activity from DDT frequencies is difficult because DDT formation is, amongst others factors, controlled by surface properties (see Sect. 3.5). This is probably one of the main reasons why there are large regional variations in DDT densities (see Sect. 3.4). Another reason why it is difficult to infer dust devil activity from DDTs is that only intense dust devils (e.g., high tangential wind speed, large pressure deficit or high vertical wind speed within the core) are able to create tracks if the dust layer is several microns thick, meaning that using DDTs as a proxy for DD frequency underestimates the actual number of dust devils. Furthermore, the horizontal ground speeds of dust devils are controlled by ambient wind speeds (e.g., Balme et al. 2012; Reiss et al. 2014a), which define how long a dust devil crosses a specific point at the surface; hence, local or regional variations in wind speeds directly influences the formation of DDTs. Direct observations of active dust devils on Mars show that on a global scale, only ~14 % leave tracks (Cantor et al. 2006). However, on a regional or local scale, large variations can be expected. Verba et al. (2010) compared measured DDT formation rates with directly observed active dust devils by the MER Spirit (Greeley et al. 2006b, 2010) in Gusev Crater, implying that only every 100th to 500th dust devil is able to leave a track. Therefore, deducing dust devil activity from DDTs is problematic. However, global calculations using DDT frequencies as a proxy for dust devil activity and derived calculations of the contribution to dust entrainment on Mars (Whelley and Greeley 2006, 2008) may give first-order estimates on a global scale (see also Klose et al., 2016).

4.2 Dust Devil Durations

The duration of dust devils on Mars can be calculated by measuring their translational ground speeds (see also Fenton et al. 2016, this issue) and their adjacent DDT lengths. Such derived dust devil durations are only minimum values, however, because the observed dust devils used as end points are still active. Other uncertainties surround the DDT starting points, because the corresponding dust devils might have been active before they started forming tracks. Nevertheless, these calculation methods are powerful tools for determining dust devil durations. They were used by Stanzel et al. (2008), who observed four dust devils in HRSC images and measured their translational ground speeds and adjacent DDT lengths, thereby calculating minimum dust devil durations in the range of 3.7–32.5 minutes. In addition, they used eight additional DDT length measurements in the vicinity of the four dust devils to derive a mean minimum durations of 13 minutes (Stanzel et al. 2008). In a later study, Reiss et al. (2011a) examined one large dust devil (diameter of ~820 m) on HRSC images. After measuring the translational ground speed and an adjacent DDT length, they calculated the dust devil minimum duration to be 74 minutes. Dust devil durations can be additionally estimated from measured DDT lengths and widths using a dust devil longevity-diameter relationship (Lorenz 2013a) in combination with assumed translational ground speeds (Lorenz and Jackson 2016). These studies show that, compared with direct duration measurements of smaller dust devils in Gusev crater by Greeley et al. (2006b, 2010), larger dust devils are active longer than smaller ones (Reiss et al., 2011a, see also Fenton et al. 2016). For a summary of terrestrial and martian dust devil durations, we refer here to Lorenz (2013a). Additional information about the influence of dust devil longevity and DDT lengths can be found in Lorenz and Jackson (2016).

4.3 Ambient Wind Directions

The orientation of DDTs can be used to infer wind directions because dust devils move approximately parallel to the ambient wind field (e.g., Wegener 1914; Flower 1936; Crozier 1970; Balme et al. 2012; Reiss 2014). Precise wind directions without the need of climate models can be inferred directly from imaged active dust devils leaving tracks in their wakes or from the morphology of cycloidal DDTs (Greeley et al. 2004; see also Sect. 2.3). However, most DDTs are dark continuous tracks in which the wind direction is ambiguous, e.g., west-east oriented DDTs would indicate winds from the west or east. On Earth, DDT orientations in the Ténéré Desert were compared with regional winds inferred from wind streaks, but no correlation was found (Rossi and Marinangeli 2004). However, wind directions during the formation of the DDTs from (for example) meteorological stations were unknown and might have been consistent with observed DDT orientations. On Mars, Fenton et al. (2005) used the PSU/NCAR 5th generation mesoscale model (MM5) to show that dust devil tracks that form on the floor of Proctor crater during spring and summer are oriented to light mid-latitude westerly winds that blow during the early afternoon in these seasons. Drake et al. (2006) provided an estimate of wind directions from the orientation of DDTs for the Phoenix landing site candidates. Later, DDT orientations in Mars's Gusev and Russell craters were measured and compared with wind direction predictions from the NASA Ames general circulation model (GCM); for active dust devil seasons, estimates made using the DDT orientations offered good agreement with the predicted wind directions (Verba et al. 2010). More recently, DDT orientations at the InSight landing site region were compared with the Mars Climate Database (MCD) wind direction predictions (Reiss and Lorenz 2016). The resulting seasonal comparison between the DDT alignment and the MCD-predicted wind directions showed that dust devils moved from SE to NW until early northern autumn, and then they reverse directions and move from NW to SE at around $Ls = 200°$, consistent with the seasonal reversal in direction of the Hadley circulation (Reiss and Lorenz 2016). In general, seasonal ambient wind fields can be directly inferred from DDT orientations in combination with wind direction predictions from atmospheric models. Such direct deductions of wind directions from DDT measurements reflect the real ambient wind fields during the DDT formation times and are more accurate than climate model predictions, especially during times when regional weather fronts are active (e.g., Reiss et al. 2014a), which are not resolved in climate models.

4.4 Surface Wind Conditions

The shape of DDTs can probably be used to infer surface wind conditions. DDTs exhibit large variations in plan-view of linear, curvilinear, curved, meandering or looping track patterns (see Sect. 2.1). Terrestrial field measurements showed that dust devils have a greater variability in the ground track direction at lower ambient wind speeds (Balme et al. 2012).

Lorenz (2016) showed that dust devil migration directions are consistent with a 'random' component of motion associated with convection vector-added to the ambient wind. Hence, at lower ambient wind speeds and/or high random component curved or looping track patterns should be formed, whereas at larger ambient wind speeds and/or small random component linear DDT morphologies should be expected. This relationship and its effect on in-situ measurements (pressure drop) are discussed in detail in Lorenz (2013b). One parameter for the classification of DDT shapes is the sinuosity (ratio between the DDT length measured (a) along the total path length and (b) as a straight line from the start and end point). Verba et al. (2010) measured a DDT mean sinuosity of $\sim$1.3 and $\sim$1.08 in Russell and Gusev Crater, respectively, and Reiss and Lorenz (2016) measured a mean sinuosity of $\sim$1.03 at the InSight landing site region in Elysium Planitia. This would suggest higher ambient wind speeds in Elysium Planum and Gusev Crater than in Russell Crater. Such a relationship between the sinuosity of DDTs and ambient wind speeds could ideally be tested in future LES simulations (see Spiga et al. 2016, this issue).

4.5 Solar Panel Clearing Predictions

The power supply of many landers or rovers on Mars is provided by solar arrays. Atmospheric dust deposition on the solar panels causes a decline in electrical power output with time (e.g., Landis and Jenkins 2000; Crisp et al. 2004). Without solar panel clearing events by wind gusts or dust devils, lander or rover science operations can be limited as it has been the case for the Spirit (MER–A) rover at Gusev crater (e.g., Greeley et al. 2010). Lorenz and Reiss (2015) showed that intense dust devils were primarily responsible for recurrent solar panel clearing events (recurrence interval of 100–700 sols) at Gusev crater. DDTs can serve as a proxy for solar panel clearing recurrence interval estimates, because the formation of dark continuous DDTs by the removal of a thin layer of dust is in principle the same process a clearing of dust from solar panels. Reiss (2016) mapped newly formed DDTs in repeat HiRISE image observations covering the same surface area at the InSight landing site region. Calculated seasonal DDT formation rates (in DDT km^{-2} sol^{-1}) and DDT area formation rates (in DDT km^{-2} km^{-2}) give estimates how often DDTs are formed and how often a specific point on the surface is crossed by an intense dust devil (able to leave a DDT), respectively (Reiss and Lorenz 2016). Measured DDT formation rates were used to find a scaling factor to the calculated seasonal dust devil activity (DDA) index (Rennó et al. 1998), which is defined as the flux of energy available to drive dust devils estimate seasonal dust devil activity, and then integrated over the year to estimate a mean annual DDT formation rate for the InSight landing site region (Reiss and Lorenz 2016). As a result, a maximum solar panel clearing recurrence interval of $\sim$11 Mars years was estimated (Reiss and Lorenz 2016). Applying the same technique for Gusev Crater using measured DDTs of Verba et al. (2010) gives average solar panel clearing recurrence interval of $\sim$160 sols (Reiss and Lorenz 2016). This is in relatively good agreement with solar panel dust clearing events (Vaughan et al. 2010) and with the calculated recurrence interval of 100–700 sols for intense dust devils (pressure drop $>$6 Pa) (Lorenz and Reiss 2015) at the MER Spirit rover in Gusev crater, indicating that DDTs can be used effectively as a proxy to predict the rate of solar panel clearing events. However, such estimated solar panel clearing recurrence intervals by dust devils from DDTs should be seen as an upper limit, because less intense dust devils not able to leave tracks likely contribute in clearing dust on solar panels (Reiss and Lorenz 2016).

5 Impacts on Mars Climate

It has long been recognized that minute amounts of dust can significantly alter the spectral and photometric properties of the martian surface; for example, the reflectance at 0.56 μm of a typical dark region on Mars will increase by 35 % after dust deposition of only 9×10^{-5} g/cm^2 (Wells et al. 1984). As a result, fallout from dust storms, particularly that from global-scale events, can brighten the surface, often increasing the albedo of large regions by several percent or more (e.g., Christensen 1988; Smith 2004; Geissler 2005; Szwast et al. 2006). Eventually, and often episodically, these dust deposits are swept away during local dust storms, along storm fronts, or by regional winds enhanced by topography or albedo contrasts (Geissler 2005; Szwast et al. 2006). These changes in the albedo impact more than just the surface, however. For example, surface brightening caused by fallout from the 2001 global-scale dust event decreased daytime surface and atmospheric temperatures that lasted long after the storm itself abated (Smith 2004). Atmospheric modeling indicates that the observed albedo changes on Mars influence annual mean near-surface air temperatures and wind stresses, which in turn affect dust lifting (Fenton et al. 2007). As a result, the albedo changes caused by dust erosion and deposition may be regarded as a dynamic climatological variable that is critical to understanding the martian climate system (Szwast et al. 2006).

Within dark DDTs, the surface albedo can be as much as $\sim$3 % lower than that adjacent to the tracks (Statella et al. 2015, see also Sect. 2.1). Geissler (2005) observed dark DDTs in some of the regions where the surface albedo had decreased, and attributed part of the observed regional darkening to their occurrence. However, Szwast et al. (2006) argued that dust devils and other small-scale lifting processes are not capable of significantly changing global patterns of dust cover (a recent study by Geissler et al. (2016) concedes this point). In the areas debated by Geissler (2005) and Szwast et al. (2006), the seasonal (local spring and summer) peak percentage covered by dust devil tracks is $\sim 10\ \%$ (Whelley and Greeley 2008), suggesting that dust devils are less effective at removing dust than other processes. However, the regions on Mars most densely covered by seasonal dust devil tracks are in the southern mid- to high-latitudes, where the seasonal peak coverage percentages reach up to 92 % (Whelley and Greeley 2008). It is more plausible that dust devils significantly contribute to surface erosion and albedo darkening in the southern hemisphere, although their impact has not yet been investigated. In addition, DDTs may more significantly alter surface albedos in specific, smaller-scale regions, such as on dark dunes (Bennett et al. 2016, in press).

Haberle et al. (2006) ran simulations from a general circulation model (GCM) based on the NASA/Ames GCM (Kahre et al. 2006), both for present-day Mars and under orbital conditions of the past, when Mars' longitude of perihelion was opposite that of today (i.e., perihelion occurred near northern summer solstice rather than the current position just prior to southern summer solstice at $Ls = 251°$). Using a parameterization of dust lifting by dust devils based on the thermodynamic model of Rennó et al. (1998), Haberle et al. (2006) modeled the seasonal pattern of dust entrainment for the different martian orbital states. The present-day case predicts a significant amount of dust lifting during the southern summer; the other two cases predict a lower but more seasonally balanced rate of dust lifting. It is likely that as Mars' orbit forces climatic shifts to occur, the rate of dust devil (and dust devil track) production will vary in frequency and with location on the planet. If dust devil tracks do indeed contribute significantly to changes in surface albedo in some regions, then they are one of many factors influencing climate change on Mars.

6 Automatic Detection of DDTs on Mars

As more successful missions are launched to study Mars, the amount of remotely sensed data we have for Mars also rises. Previous missions have produced thousands of orbital images depicting the martian surface, which requires automated analysis approaches in order to quickly and accurately extract relevant information from the data.

Once such automated method for detecting martian DDTs in digital images was presented and evaluated by Statella et al. (2011, 2012). This method, mainly based on Mathematical Morphology, is based on the following steps: filtering, track candidate selection, track candidate enhancement and track detection. The method was tested on MOC-NA and HiRISE images.

The first step is filtering using the morphological surface area opening and surface area closing operators in order to attenuate the high reflectance of boulders, ripples and the dark spots caused by their shadows. In Fig. 23B and C the bright faces of boulders and their shadows have been suppressed from the original (cropped) HiRISE image (Fig. 23A).

In the next step, a morphological path closing (Hendriks 2010) with a specified length and oriented in the four directions (0°, 45°, 90° and 135°) is applied for the selection of all possible dark paths. This is performed in an indirect way, since all the desired long dark structures are suppressed (Fig. 24A). The considered paths are not strictly straight lines, as they are allowed to deflect inside a 90° aperture cone centered in each of the four adopted directions of search. The maximum lengths of the paths are defined by the dimensions of

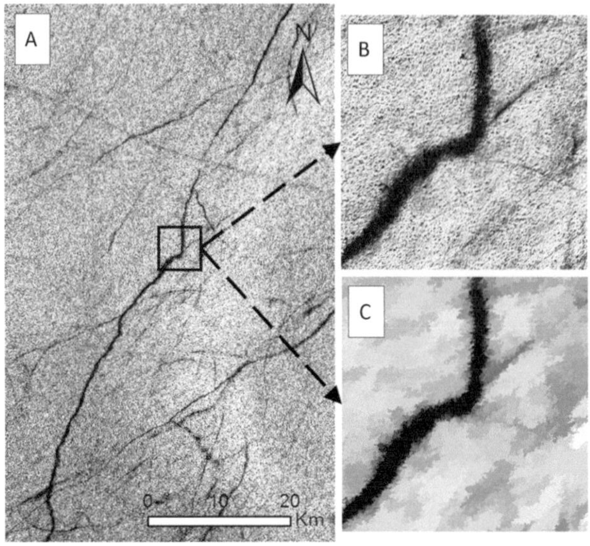

Fig. 23 Filtering step: (**A**) original HiRISE image ESP_013557_1245; (**B**) detail of (**A**) and (**C**) the result after filtering. The rectangle annotated in (**A**) indicates the region selected for the details shown in (**B**, **C**)

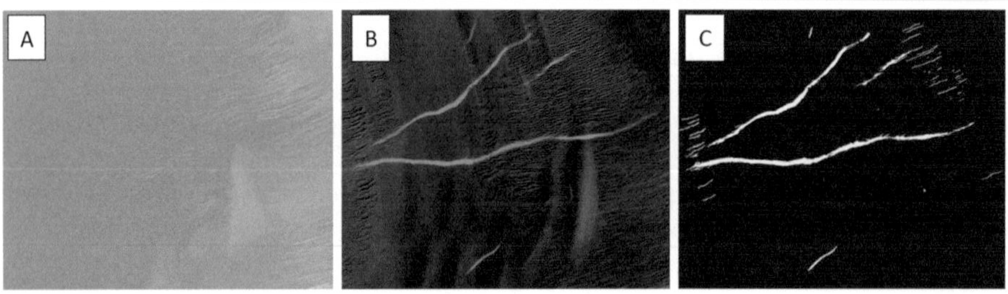

Fig. 24 (After Statella et al. 2012) Selection of DDT candidate, enhancement and binarization: (**A**) Path closing transformation suppresses the candidates; (**B**) Top-hat by closing recovers and enhances the candidates; (**C**) A constrained Otsu's method is applied for the segmentation of the tracks

each image and calculated as $\sqrt{m^2+n^2} \times 2$, where m and n are the number of columns and rows of the image, respectively.

In the third step, the dark paths are then recovered and enhanced by a morphological top-hat (difference between the closed image, in which the tracks were removed by the path closing, and the original image) transformation (Fig. 24B). This grey level output must go through a thresholding to become a binary image that shows only the dust devil tracks. For that purpose, Statella et al. (2012) applied an Otsu's algorithm (Otsu 1979) constrained to look for a threshold in the interval $[k_{mean}, k_{\max}]$ of the histogram. Such intervals were defined based on a preliminary analysis of the binarization performance; the binarization by the constrained Otsu's algorithm is shown in Fig. 24C. The method was applied to 90 MOC NA and 110 HiRISE images, from which 2 original images and detection results are shown in Fig. 25.

The results were analyzed according to the procedure proposed by Bandeira et al. (2011).

For each of the 200 processed images, a ground truth image was constructed manually in order to estimate the global accuracy of the method using pixel based comparison (Statella et al. 2012). The mean global accuracy after processing the 200 images was 92.02 % ± 4.87 %. The lowest accuracy was 69.15 % and the highest was 99.34 %. The accuracy

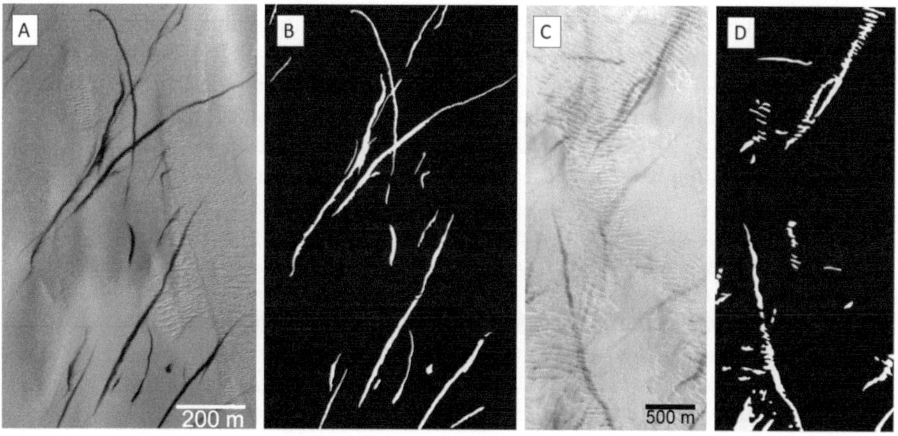

Fig. 25 (**A** and **C**) Original HiRISE (PSP_006163_1345) and MOC (M10-01206) images. (**B** and **D**) Corresponding images that have been processed by the proposed method. Images **B** and **D** are binary and show *pixel tracks in white* and *background in black*

was not affected by the variation in spatial resolution of the images and is not sensitive to variations in latitude and solar longitude. The results show that the measure of ~92 % ± 5 % for the detection process allows the features to be estimated accurately. Therefore, the method can be a very useful tool for intensively mapping dust devil tracks.

Then, the availability of black and white images with the detected tracks allows extracting their characteristics in an exhaustive way. Automated approaches have been developed to obtain those features in an extensive way using information from of the whole detected tracks including average widths, lengths and orientations. These methods permit to better describe the dust devil tracks, namely, by inferring their direction of movement (Statella et al. 2014), estimating the albedo contrast with their neighborhoods (Statella et al. 2015) or computing the distribution of widths (Statella et al. 2016).

7 Summary and Future Directions

DDTs exhibit a linear to curvilinear, curved, meandering or looping but not entirely straight morphology in plan-view, which makes it relatively easy to differentiate them from aeolian wind gust streaks. On both Earth and Mars, their sizes typically reflect the width and duration of dust devils (see also Fenton et al. 2016, this issue), and they have widths that range from one to hundreds of meters and lengths of up to several kilometers. They can be classified into three different types based on their morphology and albedo: (a) dark continuous, (b) dark cycloidal, and (c) bright DDTs. All three types can be found on both planets, although terrestrial DDTs are rarely observed.

The most common type of DDT on Mars is dark continuous DDTs, which have been observed on both planets in satellite imagery and analyzed in situ by MER Spirit on and on Earth. Both studies revealed that dust erosion by dust devils exposes coarser grained substrate consisting of coarse sand particles (0.5 < particle diameter < 2 mm), which changes the photometric properties of the surface and leads to low albedo track areas in contrast to the undisturbed surroundings. The formation mechanism of dark continuous DDTs is only based on one in situ study on each planet, so we can not exclude alternative effects, such as compositional differences between the eroded dust layer and the exposed substrate or additional effects such as redistribution. Further in situ investigations of dark continuous DDTs on both planets would help verify the proposed formation mechanism. Dust deflation calculations as well as numerical simulations indicate that removing an equivalent dust layer thickness of around 1 μm would be sufficient to cause albedo differences that lead to visible tracks on Mars. On Earth, measured values are similar, requiring the removal of an equivalent dust layer having a thickness of 1–2 μm. Future measurements that can quantify the dust deflation thickness within dark continuous DDTs would be very helpful for improving estimates about how much dust devils contribute to dust entrainment in the atmosphere, especially on Mars (see also Klose et al. 2016, this issue). DDTs are visible due to the lower albedo track relative to the adjacent terrain. Measurements of albedo differences between the track area and the undisturbed surrounding terrain have rarely been conducted. Mean albedo contrasts are in the range of ∼2.5 % wavelength range 550–850 nm) on Mars and around 0.5 % (wavelength range 300 to 1100 nm) on Earth. Further albedo contrast measurements would help quantify larger-scale albedo changes that influence the recent climate on Mars. The lifetime of dark continuous DDTs is sub-annual on both planets and depends on atmospheric dust deposition rates, as deposited dust obscures tracks. This track obscuration can happen relatively quickly after dust storm events because of increased atmospheric opacity and dust settling rates.

The second type of DDT, dark cycloidal DDTs, can be observed on Mars and Earth in satellite imagery. In plan-view morphology, they resemble tornado tracks on Earth, and terrestrial in situ studies revealed that they formed when sand-sized grains that has eroded from the outer vortex area is redeposited in annular patterns in the central vortex region. Terrestrial dark cycloidal DDTs can be either short- or long-lived; they exhibit sub-annual lifetimes on aeolian active surfaces, but they can be visible for several years on aeolian inactive surfaces. On Mars, neither in situ studies nor specific studies have been conducted to date. The proposed formation mechanism for these DDTs on Mars (re-deposition of sand) has only been suggested from terrestrial analog, laboratory, and modelling studies. Thus, to confirm the proposed formation mechanism, in situ studies on Mars carried out by future rover missions would be necessary. However, this would require that dark cycloidal DDTs occur close to the rover landing site, which is unlikely. But, in the near future, specific studies on martian dark cycloidal DDTs, based on satellite imagery and combined with numerical modelling seem to be most promising method to verify the proposed formation mechanism and to determine their lifetimes.

The third type of DDT, bright DDTs, are brighter than their surroundings and occur on both planets. However, in situ studies have so far only been conducted on Earth, and they indicate that surficial aggregates of fine particles are destroyed by passing dust devils, which causes changes in the photometric properties inside the track area (i.e., smooth surfaces inside the track versus rough surface texture of the undisturbed surroundings). Based on this terrestrial analog, the same formation mechanism might also be valid for the formation of the bright martian DDTs, because in situ studies provided evidence of surficial dust aggregates that can easily be lofted and destroyed by passing dust devils. Furthermore, the distribution of bright DDTs on Mars seems to be predominantly in regions with a relatively thick dust cover. However, the lack of in situ studies on Mars leaves the formation mechanism of the martian bright DDTs unclear. Other proposed formation mechanisms, such as removal of dark dust, exposure of a bright underlying substrate, a compaction mechanism by the downdraft of dust devils, might also lead to tracks having higher reflective surfaces. The lifetimes of bright DDTs seem to be sub-annual, similar to the dark continuous DDTs on Mars and Earth. Future in situ and orbital martian studies are needed to understand the formation mechanism of bright DDTs on Mars.

In general, the suggested formation mechanisms of different DDT types are all based on only a few in situ studies. Therefore, other formation mechanisms leading to the same morphologies can not be excluded. In addition, it is also plausible that two different processes are simultaneously responsible for creating tracks, e.g., removal of dust and redistribution of sand-sized material.

On Earth, observed DDTs are limited to semi-arid to arid regions and have been reported in the Sahara desert, arid regions in northwestern China, and the coastal Peruvian desert. On Mars, they occur globally (except on the permanent polar caps) and can be found at all topographic elevations. They are formed most frequently during spring and summer, when insolation is around its maximum, triggering dust devil formation (see also Rafkin et al. 2016, this issue). Mapped DDT densities in the southern hemisphere (0.6 DDTs/km^2) are about 10 times higher than those in the northern hemisphere (0.06 DDTs/km^2), which is probably mainly related to Mars's orbital eccentricity in which the southern hemisphere receives about 40 % stronger insolation during southern summer than the northern hemisphere during northern summer. Regionally, large variations in DDT distributions and densities occur on Mars and there exists no relationship between observed active dust devils and DDT frequencies, implying that DDT formation is strongly controlled by surface properties, e.g. dust cover thickness. Currently, there exists no distribution study in which all three DDT types have been separately mapped on Mars. However, future studies mapping the three different types separately and relating their location to thermophysical surface properties may give insights into the formation mechanism of DDTs on Mars, especially for the less common dark cycloidal and bright DDTs. In general, the automatic detection of DDTs from satellite imagery could help to minimize time consuming mapping procedures and include larger data sets.

DDTs can indirectly be used as indicators for several processes. Obviously, their frequent occurrence and relatively long visibility on the surface, in contrast to direct observations of active dust devils trough satellite imagery snapshots (see Fenton et al. 2016, this issue), have been used to infer global dust devil activity and quantify how much dust devils contribute to the global dust entrainment into Mars's atmosphere. Although the derived calculations can be used as a first-order estimate, several uncertainties remain to be resolved in the future to constrain the contribution of dust devils in replenishing the background dust opacity on Mars from DDTs. Using DDTs as a proxy for dust devil frequency is problematic because the formation of DDTs and therefore the DDT spatial density in an area is strongly controlled by surface properties, such as dust availability and albedo contrast with the substrate, the intensity of dust devils, and the horizontal ground speeds of dust devils, which are related to the ambient wind fields. These parameters vary on local and regional scales and likely represent the main reasons for the large regional differences in DDT frequencies and proportions of how many active dust devils are able to leave tracks. However, future studies could decompose the individual parameters that influence DDT frequencies on local to global scales, hence providing better knowledge about dust devil activity as well as their contribution to dust entrainment on Mars (see also Klose et al. 2016, this issue). However, while DDTs can not yet be used as an accurate proxy for dust devil frequency, DDTs can be used as good proxies for inferring minimum dust devil durations (lifetimes), wind directions (from the morphology of dark cycloidal DDTs, ambient wind directions from DDT orientations with the help of climate models, and predicting solar panel dust clearing recurrence intervals, which is important for future Mars landing missions that depend on power from solar arrays.

Finally, DDT formation mechanisms may reveal new insights about other active processes on Mars related to albedo changes. Mass wasting and aeolian processes with dark or bright features, such as slope streaks (e.g., Sullivan et al. 2001; Schorghofer and King 2011), wind streaks (e.g., Thomas et al. 1981; Geissler et al. 2008), or Recurring Slope Lineae (RSL) (e.g., McEwen et al. 2011; Ojha et al. 2015), may share the same formation mechanisms. Currently, there are no comprehensive studies comparing such features with each other or quantifying their similarities or differences, such as albedo contrasts. However, future comprehensive analog studies between these different albedo contrast features on Mars may help improve our knowledge about the different formation processes and constrain their formation mechanisms.

References

M. Balme, R. Greeley, Dust devils on Earth and Mars. Rev. Geophys. **44**, RG3003 (2006). doi:10.1029/2005RG000188

M.R. Balme, P.L. Whelley, R. Greeley, Mars: dust devil track survey in Argyre Planitia and Hellas Basin. J. Geophys. Res. **108**, 5086 (2003). doi:10.1029/2003JE002096

M.R. Balme, A. Pathare, S.M. Metzger, M.C. Towner, S.R. Lewis, A. Spiga, L. Fenton, N.O. Renno, H.M. Elliott, F.A. Saca, T. Michaels, P. Russell, J. Verdasca, Field measurements of horizontal forward motion velocities of terrestrial dust devils: towards a proxy for ambient winds on Mars and Earth. Icarus **221**, 632–645 (2012)

L. Bandeira, J.S. Marques, J. Saraiva, P. Pina, Automated detection of martian dune fields. IEEE Geosci. Remote Sens. Lett. **8**, 626–630 (2011)

K.A. Bennett, L. Fenton, J.F. Bell III, The albedo of martian dunes: insights into aeolian activity and dust devil formation. Aeolian Res. (2016). doi:10.1016/j.aeolia.2016.08.009, in press

J.A. Berger, M.E. Schmidt, R. Gellert, J.L. Campbell, P.L. King, R.L. Flemming, D.W. Ming, B.C. Clark, I. Pradler, S.J.V. Vanbommel, M.E. Minitti, A.G. Fairén, N.I. Boyd, L.M. Thompson, G.M. Perrett, B.E. Elliott, E. Desouza, A global Mars dust composition refined by the Alpha-Particle X-ray Spectrometer in Gale Crater. Geophys. Res. Lett. **43**, 67–75 (2016)

N.T. Bridges, M.E. Banks, R.A. Beyer, F.C. Chuang, E.Z. Noe Dobrea, K.E. Herkenhoff, L.P. Keszthelyi, K.E. Fishbaugh, A.S. McEwen, T.I. Michaels, Aeolian bedforms, yardangs, and indurated surfaces in the Tharsis Montes as seen by the HiRISE Camera: evidence for dust aggregates. Icarus **205**, 165–182 (2010)

N.T. Bridges, M.C. Bourke, P.E. Geissler, M.E. Banks, C. Colon, S. Diniega, M.P. Golombek, C.J. Hansen, S. Mattson, A.S. Mcewen, M.T. Mellon, N. Stantzos, B.J. Thomson, Planet-wide sand motion on Mars. Geology (2011). doi:10.1130/G32373.1

F.J. Calef, V.L. Sharpton, Enigmatic linear features in the Northern Hemisphere of Mars: their formation process. Geophys. Res. Lett. **32**, L24202 (2005). doi:10.1029/2005GL023868

B.A. Cantor, K.M. Kanak, K.S. Edgett, Mars Orbiter Camera observations of Martian dust devils and their tracks (September 1997 to January 2006) and evaluation of theoretical vortex models. J. Geophys. Res. **111**, E12002 (2006). doi:10.1029/2006JE002700

J.J. Carroll, J.A. Ryan, Atmospheric vorticity and dust devil rotation. J. Geophys. Res. **75**, 5179–5184 (1970)

D.S. Choi, C.M. Dundas, Measurements of Martian Dust Devil Winds with HiRISE. Geophys. Res. Lett. **38**, L24206 (2011). doi:10.1029/2011GL049806

P.R. Christensen, Global albedo variations on Mars–Implications for active aeolian transport, deposition, and erosion. J. Geophys. Res. **93**, 7611–7624 (1988). doi:10.1029/JB093iB07p07611

P.R. Christensen, J.L. Bandfield, V.E. Hamilton, S.W. Ruff, H.H. Kieffer, T.N. Titus, M.C. Malin, R.V. Morris, M.D. Lane, R.L. Clark, B.M. Jakosky, M.T. Mellon, J.C. Pearl, B.J. Conrath, M.D. Smith, R.T. Clancy, R.O. Kuzmin, T. Roush, G.L. Mehall, N. Gorelick, K. Bender, K. Murray, S. Dason, E. Greene, S. Silverman, M. Greenfield, Mars Global Surveyor Thermal Emission Spectrometer experiment: investigation description and surface science results. J. Geophys. Res. **106**, 23823–23871 (2001)

R.N. Clark, T.L. Roush, Reflectance spectroscopy: quantitative analysis techniques for remote sensing applications. J. Geophys. Res. **89**, 6329–6340 (1984)

D. Crisp, A. Pathare, R.C. Ewell, The performance of gallium arsenide/germanium solar cells at the Martian surface. Acta Astronaut. **54**, 83–101 (2004)

W.D. Crozier, Dust devil properties. J. Geophys. Res. **75**, 4583–4585 (1970)

G.E. Cushing, T.N. Titus, P.R. Christensen, THEMIS VIS and IR observations of a high-altitude Martian dust devil. Geophys. Res. Lett. **32**, L23202 (2005). doi:10.1029/2005GL024478

J.A. Cutts, R.S.U. Smith, Eolian deposits and dunes on Mars. J. Geophys. Res. **78**, 4139–4154 (1973)

R.P. Davies-Jones, Tornado dynamics, in *Thunderstorm Morphology and Dynamics* (Univ. of Okla. Press, Norman, 1986), pp. 197–388, Chap. 10

N.B. Drake, L.K. Tamppari, R.D. Baker, B.A. Cantor, A.S. Hale, Dust devil tracks and wind streaks in the North Polar Region of Mars: a study of the 2007 Phoenix Mars Lander Sites. Geophys. Res. Lett. **33**, L19S02 (2006). doi:10.1029/2006GL026270

K.S. Edgett, M.C. Malin, Martian dust raising and surface albedo controls: thin, dark (and sometimes bright) streaks and dust devils in MGS MOC high resolution images, in *Lunar and Planetary Science Conference XXXI* (2000). Abstract #1073

L.K. Fenton, J.L. Bandfield, A.W. Ward, Aeolian processes in Proctor Crater on Mars: sedimentary history as analyzed from multiple data sets. J. Geophys. Res. **108(E12**, 5129 (2003). doi:10.1029/2002JE002015

L.K. Fenton, A.D. Toigo, M.I. Richardson, Aeolian processes in Proctor Crater on Mars: mesoscale modeling of dune-forming winds. J. Geophys. Res. **110**, E06005 (2005). doi:10.1029/2004JE002309

L.K. Fenton, P.E. Geissler, R.M. Haberle, Global warming and climate forcing by recent albedo changes on Mars. Nature **446**, 646–649 (2007). doi:10.1038/nature05718

L. Fenton, D. Reiss, M. Lemmon, B. Marticorena, S. Lewis, B. Cantor, Orbital observations of dust lofted by daytime convective turbulence. Space Sci. Rev. (2016). doi:10.1007/s11214-016-0243-6

J.A. Fisher, M.I. Richardson, C.E. Newman, M.A. Szwast, C. Graf, S. Basu, S.P. Ewald, A.D. Toigo, R.J. Wilson, A survey of Martian dust devil activity using Mars Global Surveyor Mars Orbiter Camera images. J. Geophys. Res. **110**, E03004 (2005). doi:10.1029/2003JE002165

D.E. Fitzjarrald, A Field investigation of Dust Devils. J. Appl. Meteorol. **12**, 808–813 (1973)

W.D. Flower, Sand devils. London Meteorol. Off. Prof. Notes. **5**, 1–16 (1936)

T.T. Fujita, Proposed Mechanism of Suction Spots accompanied by Tornadoes. Preprints of papers presented at the Seventh Conference on Severe Local Storms, October 5–7, 1971, Kansas City, Mo., 208–213 (1971)

T.T. Fujita, Jumbo tornado outbreak of 3 April 1974. Weatherwise **27**, 116–126 (1974)

T.T. Fujita, D.L. Bradbury, C.F. van Thullenar, Palm Sunday tornadoes of April 11, 1965. Mon. Weather Rev. **98**, 29–69 (1970)

P.E. Geissler, Three decades of Martian surface changes. J. Geophys. Res. **110**, E02001 (2005). doi:10.1029/2004JE002345

P.E. Geissler, J.R. Johnson, R. Sullivan, K. Herkenhoff, D. Mittlefehldt, R. Fergason, D. Ming, R. Morris, S. Squyres, L. Soderblom, M. Golombek, First in situ investigation of a dark wind streak on Mars. J. Geophys. Res. **113**, 1–16 (2008)

P.E. Geissler, R. Sullivan, M. Golombek, J.R. Johnson, K. Herkenhoff, N. Bridges, A. Vaughan, J. Maki, T. Parker, J. Bell, Gone with the wind: eolian erasure of the Mars Rover tracks. J. Geophys. Res. **115**, 1–17 (2010)

P.E. Geissler, L.K. Fenton, M.-T. Enga, P. Mukherjee, Orbital monitoring of martian surface changes. Icarus **278**, 279–300 (2016). doi:10.1016/j.icarus.2016.05.023

P.J. Gierasch, R.M. Goody, A model of a martian great dust storm. J. Atmos. Sci. **30**, 169–179 (1973)

W. Goetz, P. Bertelsen, C.S. Binau, H.P. Gunnlaugsson, S.F. Hviid, K.M. Kinch, D.E. Madsen, M.B. Madsen, M. Olsen, R. Gellert, G. Klingelhöfer, D.W. Ming, R.V. Morris, R. Rieder, D.S. Rodionov, P.A. de Souza, C. Schröder, S.W. Squyres, T. Wdowiak, A. Yen, Indication of drier periods on Mars from the chemistry and mineralogy of atmospheric dust. Nature **436**, 62–65 (2005)

J.A. Grant, P.H. Schultz, Possible tornado-like tracks on Mars. Science **237**, 883–885 (1987)

R. Greeley, J.D. Iversen, *Wind as a geological process on Earth, Mars, Venus, and Titan* (Cambridge University Press, New York, 1985), p. 333

R. Greeley, M.R. Balme, J.D. Iversen, S.M. Metzger, R. Mickelson, J. Phoreman, B. White, Martian dust devils: laboratory simulations of particle threshold. J. Geophys. Res. **108**, 5041 (2003). doi:10.1029/2002JE001987

R. Greeley, P.L. Whelley, L.D.V. Neakrase, Martian dust devils: directions of movement inferred from their tracks. Geophys. Res. Lett. **31**, L24702 (2004). doi:10.1029/2004GL021599

R. Greeley, R. Arvidson, J.F. Bell III., P. Christensen, D. Foley, A. Haldemann, R.O. Kuzmin, G.A. Landis, L.D.V. Neakrase, G. Neukum, S.W. Squyres, R. Sullivan, S.D. Thompson, P.L. Whelley, D. Williams, Martian variable features: new insight from the Mars Express Orbiter and the Mars Exploration Rover Spirit. J. Geophys. Res. **110**, E06002 (2005). doi:10.1029/2005JE002403

R. Greeley, R.E. Arvidson, P.W. Barlett, D. Blaney, N.A. Cabrol, P.R. Christensen, R.L. Fergason, M.P. Golombek, G.A. Landis, M.T. Lemmon, S.M. McLennan, J.N. Maki, T. Michaels, J.E. Moersch, L.D.V. Neakrase, S.C.R. Rafkin, L. Richter, S.W. Squyres, P.A. de Souza, R.J. Sullivan, S.D. Thompson, P.L. Whelley, Gusev crater: wind-related features and processes observed by the Mars Exploration Rover Spirit. J. Geophys. Res. **111**, E02S09 (2006a). doi:10.1029/2005JE002491

R. Greeley, P.L. Whelley, R.E. Arvidson, N.A. Cabrol, D.J. Foley, B.J. Franklin, P.G. Geissler, M.P. Golombek, R.O. Kuzmin, G.A. Landis, M.T. Lemmon, L.D.V. Neakrase, S.W. Squyres, S.D. Thompson, Active dust devils in Gusev crater, Mars: observations from the Mars Exploration Rover Spirit. J. Geophys. Res. **111**, E12S09 (2006b). doi:10.1029/2006JE002743

R. Greeley, D.A. Waller, N.A. Cabrol, G.A. Landis, M.T. Lemmon, L.D.V. Neakrase, M. Pendeleton Hoffer, S.D. Thompson, P.L. Whelley, Gusev crater, Mars: observations of three dust devil seasons. J. Geophys. Res. **115**, E00F02 (2010). doi:10.1029/2010JE003608

R.M. Haberle, M.A. Kahre, J.R. Murphy et al., Role of dust devils and orbital precession in closing the Martian dust cycle. Geophys. Res. Lett. **33**, 1–4 (2006). doi:10.1029/2006GL026188

P.K. Haff, B.T. Werner, Dynamical processes on desert pavements and the healing of surfical disturbances. Quat. Res. **45**, 38–46 (1996)

B. Hapke, Bidirectional reflectance spectroscopy 1. Theory. J. Geophys. Res. **86**, 3039–3054 (1981)

B. Hapke, *Introduction to the Theory of Reflectance and Emittance Spectroscopy* (Cambridge University Press, New York, 1993)

C.L.L. Hendriks, Constrained and dimensionality-independent Path Openings. IEEE Trans. Image Process. **19**, 1587–1595 (2010)

K.E. Herkenhoff, S.W. Squyres, J.F. Bell, J.N. Maki, H.M. Arneson, P. Bertelsen, D.I. Brown, S.A. Collins, A. Dingizian, S.T. Elliott, W. Goetz, E.C. Hagerott, A.G. Hayes, M.J. Johnson, R.L. Kirk, S. McLennan, R.V. Morris, L.M. Scherr, M.A. Schwochert, L.R. Shiraishi, G.H. Smith, L.A. Soderblom, J.N. Sohl-Dickstein, M.V. Wadsworth, Athena Microscopic Imager investigation. J. Geophys. Res. **108**, 8065 (2003). doi:10.1029/2003JE002076

K.E. Herkenhoff et al., Textures of the soils and rocks at Gusev crater from Spirit's Microscopic Imager. Science **305**, 824–826 (2004)

K.E. Herkenhoff et al., Overview of the Microscopic Imager investigation during Spirit's first 450 sols in Gusev crater. J. Geophys. Res. **110**, E02S04 (2006). doi:10.1029/2005JE002574

R. Hesse, Short-lived and long-lived dust devil tracks in the coastal desert of southern Peru. Aeolian Res. **5**, 101–106 (2012). doi:10.1016/j.aeolia.2011.10.003

M.P. Hoffer, R. Greeley, Bright tracks on Mars: alternate dust devil tracks, in *41th Lunar Planet. Sci.* (2010). Abstract 2713

C. Holstein-Rathlou, H.P. Gunnlaugsson, J.P. Merrison, K.M. Bean, B.A. Cantor, J.A. Davis, R. Davy, N.B. Drake, M.D. Ellehoj, W. Goetz, S.F. Hviid, C.F. Lange, S.E. Larsen, M.T. Lemmon, M.B. Madsen, M. Malin, J.E. Moores, P. Nørnberg, P. Smith, L.K. Tamppari, P.A. Taylor, Winds at the Phoenix landing site. J. Geophys. Res. **115**, E00E18 (2010)

J.D. Iversen, B.R. White, Saltation threshold on Earth, Mars and Venus. Sedimentology **29**, 111–119 (1982)

M.A. Kahre, J.R. Murphy, R.M. Haberle, Modeling the Martian dust cycle and surface dust reservoirs with the NASA Ames general circulation model. J. Geophys. Res. **111**, 1–25 (2006). doi:10.1029/2005JE002588

M. Klose, B.C. Jemmett-Smith, H. Kahanpää, M. Kahre, P. Knippertz, M.T. Lemmon, S.R. Lewis, R.D. Lorenz, L.D.V. Neakrase, C. Newman, M.R. Patel, D. Reiss, Space Sci. Rev. (2016). doi:10.1007/s11214-016-0261-4, this issue

J.F. Kok, E.J.R. Parteli, T.I. Michaels, D.B. Karam, The physics of wind-blown sand and dust. Rep. Prog. Phys. **75**, 72 (2012)

G.A. Landis, P.P. Jenkins, Measurement of the settling rate of atmospheric dust on Mars by the MAE instrument on Mars Pathfinder. J. Geophys. Res. **105**, 1855–1857 (2000)

M.T. Lemmon, M.J. Wolff, M.D. Smith, R.T. Clancy, D. Banfield, G.A. Landis, A. Ghosh, P.H. Smith, N. Spanovich, B. Whitney, P. Whelley, R. Greeley, S. Thompson, J.F. Bell, S.W. Squyres, Atmospheric imaging results from the Mars Exploration Rovers: spirit and opportunity. Science **306**, 1753–1756 (2004)

D.C. Lewellen, W.S. Lewellen, Near-surface intensification of tornado vortices. J. Atmos. Sci. **64**, 2176–2194 (2007a)

D.C. Lewellen, W.S. Lewellen, Near-surface vortex intensification through corner flow collapse. J. Atmos. Sci. **64**, 2195–2209 (2007b)

D.C. Lewellen, M.I. Zimmerman, Using tornado surface marks to help decipher near-ground wind fields, in *24th Conf. Severe Local Storms* (2008)

D.C. Lewellen, W.S. Lewellen, J. Xia, The influence of a local swirl ratio on tornado intensification near the surface. J. Atmos. Sci. **57**, 527–544 (2000)

D.C. Lewellen, B. Gong, W.S. Lewellen, Effects of finescale debris on near-surface tornado dynamics. J. Atmos. Sci. **65**, 3247–3262 (2008)

R.D. Lorenz, Power law of dust devil diameters on Mars and Earth. Icarus **203**, 683–684 (2009). doi:10.1016/j.icarus.2009.06.029

R.D. Lorenz, On the statistical distribution of dust devil diameters. Icarus **215**, 381–390 (2011). doi:10.1016/j.icarus.2011.06.005

R.D. Lorenz, The longevity and aspect ratio of dust devils: effects on detection efficiencies and comparison of landed and orbital imaging at mars. Icarus **226**, 964–970 (2013a). doi:10.1016/j.icarus.2013.06.031

R.D. Lorenz, Irregular dust devil pressure drops on Earth and Mars: effect of cycloidal tracks. Planet. Space Sci. **76**, 96–103 (2013b)

R.D. Lorenz, Heuristic estimation of dust devil vortex parameters and trajectories from single-station meteorological data: application to InSight at Mars. Icarus **271**, 326–337 (2016)

R.D. Lorenz, B. Jackson, Dust devil populations and statistics. Space Sci. Rev. (2016). doi:10.1007/s11214-016-0277-9, this issue

R.D. Lorenz, D. Reiss, Solar panel clearing events, dust devil tracks, and in-situ vortex detections on Mars. Icarus **248**, 162–164 (2015). doi:10.1016/j.icarus.2014.10.034

R.D. Lorenz, M.R. Balme, G. Zhaolin, H. Kahanpää, M. Klose, M.V. Kurgansky, M.R. Patel, D. Reiss, A.P. Rossi, A. Spiga, T. Takemi, W. Wei, History and applications of dust devil studies. Space Sci. Rev. (2016). doi:10.1007/s11214-016-0239-2, this issue

L.J. Maher, Geology education by lightplane. AOPA Pilot **11**(1), 60–65 (1968)

M.C. Malin, K.S. Edgett, Mars global surveyor Mars orbiter camera: interplanetary cruise through primary mission. J. Geophys. Res. **106**, 23429–23570 (2001)

W.J. Markiewicz, R.M. Sablotny, H.U. Keller, N. Thomas, D. Titov, P.H. Smith, Optical properties of the Martian aerosols as derived from Imager for Mars Pathfinder midday sky brightness data. J. Geophys. Res. **104**, 9009–9017 (1999)

A.S. McEwen, L. Ojha, C.M. Dundas, S.S. Mattson, S. Byrne, J.J. Wray, S.C. Cull, S.L. Murchie, N. Thomas, V.C. Gulick, Seasonal flows on warm Martian slopes. Science **80**(333), 740–743 (2011)

M.T. Mellon, K.A. Kretke, M.D. Smith, S.M. Pelkey, A global map of thermal inertia from Mars Global Surveyor Mapping Mission, in *Lunar Planet. Sci. [CDROM] XXXIII* (2002). Abstract 1416

S.M. Metzger, Dust devils as aeolian transport mechanisms in southern Nevada and in the Mars Pathfinder landing site. PhD thesis, Univ. of Nev, Reno, p. 208 (1999)

S.M. Metzger, M.R. Balme, M.C. Towner, B.J. Bos, T.J. Ringrose, M.R. Patel, In situ measurements of particle load and transport in dust devils. Icarus **214**, 766–772 (2011)

T.I. Michaels, Numerical modeling of Mars dust devils: albedo track generation. Geophys. Res. Lett. **33**, L19S08 (2006). doi:10.1029/2006GL026268

L.D.V. Neakrase, A laboratory study of sediment flux within dust devils on Earth and Mars. ProQuest (2009), 184 pages

L.D.V. Neakrase, J. McHone, P.L. Whelley, R. Greeley, Saharan dust devil tracks: Mars analog field study areas. Geol. Soc. Am. Abstr. **40**, 262 (2008)

L.D.V. Neakrase, J. McHone, P.L. Whelley, R. Greeley, Terrestrial analogs to Mars: East-central Saharan dust devil tracks, in *43rd Lunar and Planetary Science Conference* (2012), Abstract #2009

F.M. Neubauer, Thermal convection in the Martian atmosphere. J. Geophys. Res. **71**, 2419–2426 (1966)

L. Ojha, M.B. Wilhelm, S.L. Murchie, A.S. Mcewen, J.J. Wray, J. Hanley, M. Massé, M. Chojnacki, Spectral evidence for hydrated salts in recurring slope lineae on Mars. Nat. Geosci. **8**, 829–832 (2015)

J. Ormö, G. Komatsu, Mars Orbiter Camera observation of linear and curvilinear features in the Hellas basin: indications for multiple processes of formation. J. Geophys. Res. **108**, 5059 (2003). doi:10.1029/2002JE001980

N. Otsu, A threshold selection method from gray-level histograms. IEEE Trans. Syst. Man Cybern. **9**, 62–66 (1979)

A.V. Pathare, M.R. Balme, S.M. Metzger, A. Spiga, M.C. Towner, N.O. Renno, F. Saca, Assessing the power law hypothesis for the size–frequency distribution of terrestrial and martian dust devils. Icarus **209**, 851–853 (2010)

J.B. Pollack, M.E. Ockert-Bell, M.K. Shepard, Viking Lander image analysis of Martian atmospheric dust. J. Geophys. Res. **100**, 5235–5250 (1995)

N.E. Prosser, Aerial photographs of a tornado path in Nebraska, May 5, 1964. Mon. Weather Rev. **92**, 593–598 (1964)

N.E. Putzig, M.T. Mellon, K.A. Kretke, R.E. Arvidson, Global thermal inertia and surface properties of Mars from the MGS mapping mission. Icarus **173**, 325–341 (2005)

S. Rafkin, B. Jemmett-Smith, L. Fenton, R. Lorenz, T. Takemi, J. Ito, D. Tyler, Dust devil formation. Space Sci. Rev. (2016). doi:10.1007/s11214-016-0307-7

D. Reiss, Morphology, formation and distribution of dust devil tracks on Mars: insights from terrestrial analogs, in: *45th Lunar Planet. Sci. Conf.* (2014). #2011

D. Reiss, First observations of terrestrial dust devils in orbital image data: comparison with dust devils in Amazonis Planitia, Mars, in: *47th Lunar Planet Sci. Conf.* (2016). #2912

D. Reiss, R.D. Lorenz, Dust devil track survey at Elysium Planitia, Mars: implications for the InSight landing sites. Icarus **266**, 315–330 (2016)

D. Reiss, A.P. Rossi, Seasonal dust devil track observations on Earth and Mars: relationships to atmospheric dust opacity, in: *42nd Lunar Planet. Sci. Conf.* (2011). #2186

D. Reiss, D. Lüsebrink, H. Hiesinger, T. Kelling, G. Wurm, J. Teiser, High Altitude Dust Devils on Arsia Mons, Mars: Testing the Greenhouse and Thermophoresis Hypothesis of Dust Lifting, in *40th Lunar Planet. Science Conf.* (2009). #1961

D. Reiss, J. Raack, A.P. Rossi, G. Di Achille, H. Hiesinger, First in-situ analysis of dust devil tracks on Earth and their comparison with tracks on mars. Geophys. Res. Lett. **37**, L14203 (2010). doi:10.1029/2010GL044016

D. Reiss, J. Raack, H. Hiesinger, Bright dust devil tracks on Earth: implications for their formation on Mars. Icarus **211**, 917–920 (2011b). doi:10.1016/j.icarus.2010.09.009

D. Reiss, M. Zanetti, G. Neukum, Multitemporal observations of identical active dust devils on Mars with the High Resolution Stereo Camera (HRSC) and Mars Orbiter Camera (MOC). Icarus **215**, 358–369 (2011a)

D. Reiss, J. Raack, A. Maturilli, A.P. Rossi, G. Erkeling, Dust devil tracks in the Turpan Depression desert (China): implications for their formation on Mars, in *43rd Lunar and Planetary Science Conference* (2012a). Abstract #2227

D. Reiss, J. Raack, G.G. Ori, K. Taj-Eddine, In situ analysis of dust devil tracks in southern Morocco: comparison with other terrestrial and martian tracks, in *EPSC 2012* (2012b). Abstract #EPSC2012-389

D. Reiss, M.I. Zimmerman, D.C. Lewellen, Formation of cycloidal dust devil tracks by redeposition of coarse sands in southern Peru: implications for Mars. Earth Planet. Sci. Lett. **383**, 7–15 (2013). doi:10.1016/j.epsl.2013.09.033

D. Reiss, N.M. Hoekzema, O.J. Stenzel, Dust deflation by dust devils on Mars derived from optical depth measurements using the shadow method in HiRISE images. Planet. Space Sci. **93–94**, 54–64. (2014b). doi:10.1016/j.pss.2014.01.016

D. Reiss, A. Spiga, G. Erkeling, The horizontal motion of dust devils on Mars derived from CRISM and CTX/HiRISE observations. Icarus **227**, 8–20. (2014a). doi:10.1016/j.icarus.2013.08.028

N.O. Rennó, M.L. Burkett, M.P. Larkin, A simple thermodynamical theory for dust devils. J. Atmos. Sci. **55**, 3244–3252 (1998)

R. Rieder, T. Economou, H. Wänke, A. Turkevich, J. Crisp, J. Brückner, G. Dreibus, H.J. McSween Jr., The chemical composition of Martian soil and rocks returned by the mobile alpha proton X-ray spectrometer: preliminary results from the X-ray mode. Science **278**, 1771–1774 (1997)

A.P. Rossi, L. Marinangeli, The first terrestrial analogue to Martian dust devil tracks found in Ténéré Desert. Niger. Geophys. Res. Lett. **31**, L06702 (2004). doi:10.1029/2004GL019428

S.W. Ruff, P.R. Christensen, Bright and dark regions on Mars: particle size and mineralogical characteristics based on Thermal Emission Spectrometer data. J. Geophys. Res. **107**, 5217 (2002)

J.A. Ryan, Notes on the Martian yellow clouds. J. Geophys. Res. **69**, 3759–3770 (1964)

J.A. Ryan, J.J. Carroll, Dust devil wind velocities: mature state. J. Geophys. Res. **75**, 531–541 (1970)

J.A. Ryan, R.D. Lucich, Possible dust devils, vortices on Mars. J. Geophys. Res. **88**, 11005–11011 (1983)

C. Sagan, J. Veverka, P. Gierasch, Observational consequences of Martian wind regimes. Icarus **15**, 253–278 (1971)

C. Sagan, J. Veverka, P. Fox, R. Dubisch, J. Lederberg, E. Levinthal, L. Quam, R. Tucker, J.B. Pollack, B.A. Smith, Variable features on Mars: preliminary mariner 9 television results. Icarus **17**, 346–372 (1972)

C. Sagan et al., Variable features in Mars: 2. mariner 9 global results. J. Geophys. Res. **78**, 4163–4196 (1973)

N. Schorghofer, C.M. King, Sporadic formation of slope streaks on Mars. Icarus **216**, 159–168 (2011)

S. Silvestro, L.K. Fenton, D.A. Vaz, N.T. Bridges, G.G. Ori, Ripple migration and dune activity on Mars: evidence for dynamic wind processes. Geophys. Res. Lett. **37**, L20203 (2010). doi:10.1029/2010GL044743

P.C. Sinclair, A quantitative analysis of the dust devil. PhD thesis, University of Arizona, 292 p (1966)

P.C. Sinclair, General characteristics of dust devils. J. Appl. Meteorol. **8**, 32–45 (1969)

P.C. Sinclair, The lower structure of dust devils. J. Atmos. Sci. **30**, 1599–1619 (1973)

M.D. Smith, Interannual variability in TES atmospheric observations of Mars during 1999–2003. Icarus **167**, 148–165 (2004). doi:10.1016/j.icarus.2003.09.010

J.T. Snow, T.M. McClelland, Dust devils at white sands missile range, New Mexico 1. Temporal and spatial distributions. J. Geophys. Res. **95**, 13707–13721 (1990)

A. Spiga, E. Barth, Z. Gu, F. Hoffmann, J. Ito, B. Jemmett-Smith, M. Klose, S. Nishizawa, S. Raasch, S. Rafkin, T. Takemi, D. Tyler, W. Wei, Large-eddy simulations of dust devils and convective vortices. Space Sci. Rev. (2016). doi:10.1007/s11214-016-0284-x, this issue

C. Stanzel, M. Pätzold, D.A. Williams, P.L. Whelley, R. Greeley, G. Neukum, HRSC Co-Investigator Team, Dust devil speeds, directions of motion and general characteristics observed by the Mars Express High Resolution Stereo Camera. Icarus **197**, 39–51 (2008). doi:10.1016/j.icarus.2008.04.017

T. Statella, P. Pina, E.A. da Silva, A study on automatic methods based on mathematical morphology for martian dust devil tracks detection. Lect. Notes Comput. Sci. **7042**, 533–540 (2011)

T. Statella, P. Pina, E.A. da Silva, Image processing algorithm for the identification of martian dust devil tracks in MOC and HiRISE images. Planet. Space Sci. **70**, 46–58 (2012). doi:10.1016/j.pss.2012.06.003

T. Statella, P. Pina, E.A. da Silva, Automated determination of the orientation of dust devil tracks in mars orbiter images. Adv. Space Res. **53**, 1822–1833 (2014). doi:10.1016/j.asr.2013.05.012

T. Statella, P. Pina, E.A. Da Silva, Extensive computation of albedo contrast between martian dust devil tracks and their neighboring regions. Icarus **250**, 43–52 (2015). doi:10.1016/j.icarus.2014.11.023

T. Statella, P. Pina, E.A. da Silva, Automated width measurements of Martian dust devil tracks. Aeolian Res. **20**, 1–6 (2016). doi:10.1016/j.aeolia.2015.11.001

R. Sullivan, P. Thomas, J. Veverka, M. Malin, K.S. Edgett, Mass movement slope streaks imaged by the Mars Orbiter Camera. J. Geophys. Res. **106**, 23607–23633 (2001)

R. Sullivan, R. Arvidson, J.F. Bell, R. Gellert, M. Golombek, R. Greeley, K. Herkenhoff, J. Johnson, S. Thompson, P. Whelley, J. Wray, Wind-driven particle mobility on Mars: insights from Mars Exploration Rover observations at "El Dorado" and surroundings at Gusev Crater. J. Geophys. Res. **113**, E06S07 (2008)

M.A. Szwast, M.I. Richardson, A.R. Vasavada, Surface dust redistribution on Mars as observed by the Mars Global Surveyor and Viking orbiters. J. Geophys. Res. **111**, E11008 (2006). doi:10.1029/2005JE002485

P. Thomas, P.J. Gierasch, Dust devils on Mars. Science **230**, 175–177 (1985)

P. Thomas, J. Veverka, S. Lee, A. Bloom, Classification of wind streaks on Mars. Icarus **45**, 124–153 (1981)

M. Tomasko, L. Doose, M. Lemmon, P. Smith, E. Wegryn, Properties of dust in the Martian atmosphere from the Imager on Mars Pathfinder. J. Geophys. Res. **104**, 8987–9007 (1999)

E.L. Van Tassel, The North Platte Valley tornado outbreak of June 27, 1955. Mon. Weather Rev. **83**, 255–264 (1955)

A.F. Vaughan, J.R. Johnson, K.E. Herkenhoff, R. Sullivan, G.A. Landis, W. Goetz, M.B. Madsen, Pancam and Microscopic Imager observations of dust on the Spirit Rover: cleaning events, spectral properties, and aggregates. Mars J. **5**, 129–145 (2010)

C.A. Verba, P.E. Geissler, T.N. Titus, D.A. Waller, Observations from High Resolution Imaging Science Experiment (HiRISE): Martian dust devils in Gusev and Russell craters. J. Geophys. Res. **115**, E09002 (2010). doi:10.1029/2009JE003498

J. Veverka, Variable features on Mars V: evidence for crater streaks produced by wind erosion. Icarus **25**, 595–601 (1975)

J. Veverka, Variable features on Mars. VII. Dark filamentary markings on Mars. Icarus **27**, 495–502 (1976)

A. Wegener, Staubwirbel auf Island. Meteorol. Zeitschrift. **31**, 199–200 (1914)

E. Wells, J. Veverka, P. Thomas, Mars: experimental study of albedo changes caused by dust fallout. Icarus **58**, 331–338 (1984)

P.L. Whelley, R. Greeley, Latitudinal dependency in dust devil activity on Mars. J. Geophys. Res. **111**, E10003 (2006). doi:10.1029/2006JE002677

P.L. Whelley, R. Greeley, The distribution of dust devil activity on Mars. J. Geophys. Res. **113**, E07002 (2008). doi:10.1029/2007JE002966

M.J. Wolff et al., Constraints on dust aerosols from the Mars Exploration Rovers using MGS overflights and Mini-TES. J. Geophys. Res. **111**, E12S17 (2006). doi:10.1029/2006JE002786

A.S. Yen, R. Gellert, C. Schröder, R.V. Morris, J.F. Bell, A.T. Knudson, B.C. Clark, D.W. Ming, J.A. Crisp, R.E. Arvidson, D. Blaney, J. Brückner, P.R. Christensen, D.J. DesMarais, P.A. de Souza, T.E. Economou, A. Ghosh, B.C. Hahn, K.E. Herkenhoff, L.A. Haskin, J.A. Hurowitz, B.L. Joliff, J.R. Johnson, G. Klingelhöfer, M.B. Madsen, S.M. McLennan, H.Y. McSween, L. Richter, R. Rieder, D. Rodionov, L. Soderblom, S.W. Squyres, N.J. Tosca, A. Wang, M. Wyatt, J. Zipfel, An integrated view of the chemistry and mineralogy of martian soils. Nature **436**, 49–54 (2005)

M.I. Zimmerman, D.C. Lewellen, Taxonomy and analysis of tornado surface marks, in *25th Conf. Severe Local Storms* (2010)

R.J. Zomer, D.A. Bossio, A. Trabucco, L. Yuanjie, D.C. Gupta, V.P. Singh, Trees and water: smallholder agroforestry on irrigated lands in Northern India. IWMI Research Report 122, International Water Management Institute, Colombo, Sri Lanka, pp 45. (2007)

R.J. Zomer, A. Trabucco, D.A. Bossio, O. van Straaten, L.V. Verchot, Climate change mitigation: a spatial analysis of global land suitability for clean development mechanism afforestation and reforestation. Agric. Ecosyst. Environ. **126**, 67–80 (2008)

R.W. Zurek, J.R. Barnes, R.M. Haberle, J.B. Pollack, J.E. Tillman, C.B. Leovy, Dynamics of the atmosphere of Mars, in *Mars*, ed. by H.H. Kieffer et al. (Univ. of Ariz. Press, Tucson, 1992), pp. 799–817

Space Sci Rev (2016) 203:183–207
DOI 10.1007/s11214-016-0307-7

Dust Devil Formation

S. Rafkin[1] · B. Jemmett-Smith[2] · L. Fenton[3] · R. Lorenz[4] · T. Takemi[5] · J. Ito[6] · D. Tyler[7]

Received: 16 May 2016 / Accepted: 27 October 2016 / Published online: 21 November 2016

Abstract The essential dynamical characteristic of convective vortices, including dust devils, is a highly localized vorticity tube that extends into the vertical. This chapter is concerned with both the generation of vorticity and the subsequent focusing of that vorticity into a tight vortex, and with the environmental conditions that are conducive to the formation of convective vortices in general and dust devils in particular. A review of observations, theory, and modeling of dust devil formation is provided.

Keywords Dust devils · Mars atmosphere · Earth atmosphere · Atmospheric dynamics

1 Introduction

The essential dynamical characteristic of convective vortices, including dust devils, is a highly localized vorticity tube that extends into the vertical. This chapter is concerned with both the generation of vorticity and the subsequent focusing of that vorticity into a tight vortex, and with the environmental conditions that are conducive to the formation of convective vortices in general and dust devils in particular.

For a dust devil to form, three major criteria must be met. First, a convective vortex must be produced. Second, the vortex must have sufficient strength for the associated winds to loft sand, dust or other particulates into air. And third, there must be a sufficient source of mobile

✉ S. Rafkin
rafkin@boulder.swri.edu

1 Southwest Research Institute, Boulder, CO, USA

2 University of Leeds, Leeds, UK

3 SETI Institute, Mountain View, CA, USA

4 Johns Hopkins Applied Physics Laboratory, Laurel, MD, USA

5 Kyoto University, Gokasho, Uji, Kyoto, Japan

6 The University of Tokyo, Tokyo, Japan

7 Oregon State University, Corvallis, OR, USA

surface material to produce visible opacity once that material is lifted. Each of these three criteria is explored through a review of observational studies, numerical modelling results, and theoretical consideration for both Earth and Mars. A taxonomy of dustless devils, trail-forming and non-trail-forming dust devils, etc. is discussed in Chap. 8 of this issue (Lorenz et al. 2016a)

While tremendous progress has been made toward understanding dust devil formation, the story is by no means complete. Open questions and issues that remain are discussed in Sect. 4, as well as measurements and experiments that are likely to lead to continued progress in dust devil formation theory.

2 Observations

Observations of dust devils on both Earth and Mars can provide direct clues to the environmental conditions that are conducive to dust devil formation. From a qualitative perspective, locations on Earth with a propensity for dust devils have long been known (Chap. 1 of this issue, Lorenz et al. 2016b presents a historical review), and it is of no surprise that these locations often serve as natural laboratories for dust devil field campaigns. Identifying common characteristics of land surface, surface properties and atmospheric state at locations favorable for dust devil formation provides quantitative information that can be used to refine theoretical formation models.

On Earth, arid locations with limited vegetation, especially deserts, are often associated with dust devils (Chap. 11 of this issue presents some global evaluations, Klose et al. 2016, this issue). Limited precipitation usually implies limited vegetation, which allows for sand and soil to be directly exposed to the wind. Further, the root systems of plants can anchor and stabilize soils, thereby limiting sediment mobility. The cohesion of surface particles and soils increases as soil moisture or atmospheric humidity increases (e.g., Azizov 1977; Jackson and Nordstrom 1997; Fecan et al. 1999; Neuman 2003) so that dry conditions also tend to favor grain mobility. Finally, limited soil moisture and vegetation implies that insolation is directed toward the heating of the ground rather than toward evaporation or transpiration; i.e., the Bowen Ratio (Bowen 1926) is large. All other things being equal, warm ground temperatures lead to warmer near-surface air conditions and the potential for energetic convection in a thermally unstable atmosphere.

Surface-based observations of Martian dust devils are limited. Data volume constraints preclude continuous observational campaigns, but targeted campaigns and serendipitous imaging during the Martian afternoon have confirmed that dust devils are present at some time at almost every location visited (e.g., Metzger et al. 1999; Waller 2011; Moores et al. 2015). Even in the absence of imaging, pressure signals of convective vortices are present, although it is unknown whether dust devils are present (e.g., Moores et al. 2015). Large dust devils and dust devil tracks have been observed from Mars orbit, and these provide the only observational data to evaluate global dust devil activity on Mars.

2.1 Terrestrial studies and field programs

There were few studies focused on dust devils prior to the mid-20th century (see Chap. 1 of this issue, Lorenz et al. 2016b). Flower (1936), who referred to dust devils as sand devils, and Williams (1948) noted that dust devils were strictly afternoon phenomena with peak occurrences one to two hours after noon. A short review by Battan (1958) provides insight into the accepted knowledge of the time. Of the few dust devil measurements that had been made, the slow response of instruments and potentially large instrument errors call into question the fidelity and veracity of the measurements. With regard to formation mechanisms, Battan (1958) refers to Ives (1947), who suggested that a superadiabatic lapse rate is required for dust devil initiation. This connection to atmospheric state is one of the first mentions in the literature. Battan (1958) further states that the kinetic energy of a dust devil occurs by virtue of reductions in potential and internal energy, a hypothesis that was described by Michelson (1951). This is necessarily required by conservation of energy in the absence of diabatic forcing, but it is an indirect assertion that direct radiative heating is not a primary source of energy for dust devils.

The field work of Sinclair (1964, 1969) in the Avra Valley and the Tuscon basin, Arizona between 1960 and 1962 is widely regarded as the first series of focused investigations of dust devils. The output of Sinclair's research is covered in depth in Chaps. 1 and 2 of this book; those observations and findings relevant to dust devil formation are noted here. Sinclair considered the dust devil as a special type of thermal that forms when a convective updraft later develops into a vortex. In this conceptual model, no doubt influenced by his sailplane experiences, the radial velocity component near the ground (i.e., the air swept towards the center) pulls in warm boundary layer air, which leads to the elevated temperature in the core. As with previous studies (e.g., Flower 1936 and Williams 1948) and the many that would follow, the characteristic diurnal variation of dust devil activity is noted with peak frequencies one to two hours after noon.

Another important observation made by Sinclair (1969) was that dust devil frequency was suppressed when the wind exceeded a certain value. There are at least two possible reasons to explain this. First, increasing wind speeds may enhance mechanical mixing, thereby reducing the magnitude of the near-surface lapse rate and reducing the vigor of convective mixing associated with dust devils. Secondly, wind or wind shear could disrupt the coherence of the dynamical structures in which dust devils are embedded. Greatest activity was observed with wind speeds between 1 to 10 mph (0.5 to 4.5 m/s). It was also noted that upstream mountains and hills appeared to enhance dust devil formation, possibly through increased generation of turbulence and eddy shedding. This may be the first study to implicate direct topographical influence on dust devils. No preference for a given rotation sense was observed.

Observing dust devils in northwest Libya, McGinnigle (1966) found that low background winds were conducive to dust devil formation, consistent with the wind threshold finding of Sinclair (1969). It was further noted that surfaces with low thermal conductivity and high thermal capacity with spatially uneven heating also produced favorable dust devil conditions. Low thermal conductivity would allow the uppermost soil layers to heat in the afternoon without substantial energy loss to the cooler layers below. The tendency to produce a hot ground, however, would be tempered by high thermal capacity. The uneven heating of the surface suggests that larger-scale, thermally-driven circulations have some influence on dust devil formation, perhaps in a manner analogous to the topographical effect described by Sinclair.

Ryan and Carroll (1970) expanded on the results of Sinclair with more accurate measurements of horizontal and vertical wind. A 500 × 300 m area of the Mojave desert was scraped to produce a more uniform surface and to provide loose material suitable for lifting (approval of the destruction of fragile, native desert habitats would likely be difficult today). Little mixing was found to occur between the dust devil core and the surrounding air. Observations also indicated that the only point of air entry into the dust devil core was from the near-surface friction layer, which was consistent with the conceptual model of Sinclair.

In follow-up work, Ryan (1972) confirmed a relationship between lapse rate and dust devil frequency. A strong correlation was found between dust devil frequency and the lapse

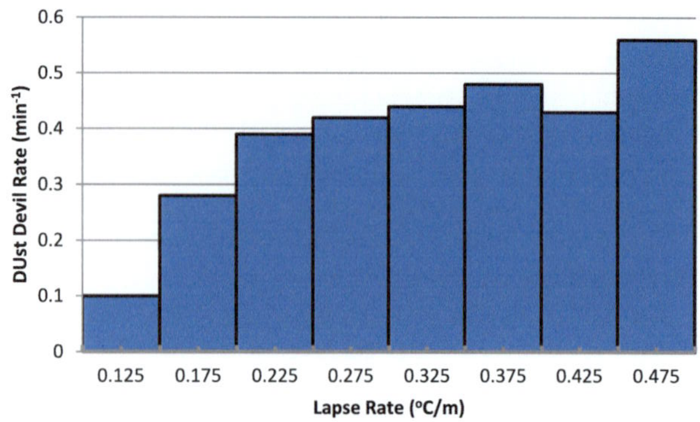

Fig. 1 Dust devil rate as a function of lapse rate from 0.3 m to 10 m (based on tabular data in Ryan 1972)

rate in the superadiabatic layer from 0.3 to 10 m above ground level (Fig. 1); however, little to no correlation was found between dust devils and higher level lapse rates in the superadiabatic layer.

Snow and McClelland (1990) conducted a field campaign in the New Mexico desert in 1986/1987 and found results broadly consistent with previous studies. Dust devils appeared to be drawn towards flatter, cleared areas of ground. They postulated this behavior was the result of these areas warming faster, producing a series of thermals which drew in unstable air from around the region. However, the location of peak dust devil activity did shift from 1986 to 1987, which could have been due to the higher precipitation that occurred in 1986, potentially changing the surface thermal properties. It is also possible that the soil moisture increased cohesion of normally mobile surface material, although this was not mentioned. The importance of dust availability was specifically noted. Numerous convective vortices may have been present but were not visually detected; material must be available and mobile. Frequency histograms showed a dip in dust devil frequency after a period of peak activity. Sinclair (1969) also noticed this tendency and attributed the temporary decline in activity to a temporary stabilization of the atmosphere by the previous thermals, although some care is needed in interpreting this assertion. If dust devils tend to be concentrated at the corners of convection cells in the planetary boundary layer (see Sect. 3.3, and also Chapter 7 by Spiga et al. 2016, this issue), then an immediate encounter with another devil at that same spot is probably less likely. Some vortex pseudoperiodicities have also been noted in meteorology time series data (e.g. Renno et al. 2004; Lorenz and Christie 2015), as well as regular spacing of dust devils in field observations on Earth (e.g. McGinnigle 1966) or in Mars orbital imaging (Fenton and Lorenz 2015). An autocorrelation of dust devil occurrences may show a modest peak at a period commensurate with cell advection timescale, i.e. the cell size (the convective boundary layer thickness, typically ∼2 km) divided by the ambient wind speed (a few m/s), typically 5–20 minutes on Earth, with low probabilities of encounter at shorter intervals.

On the other hand, the conditional probability of a dust devil encounter at a fixed point at very short intervals (within a minute or two) of another encounter may be rather high. Convection cells are often irregular, so corners with vortices can sometimes be close together, and field observers have often reported close pairs of dust devils (the so-called 'little brother'—see also Chap. 11 of this issue, Kurgansky et al. 2016). Depending on whether one is interested in the population of dust devil entities, or the frequency of dust devil action at a fixed point, the occasionally cycloidal migration path of dust devils in low-wind conditions can also lead to multiple encounters with the same dust devil (e.g. Lorenz 2013).

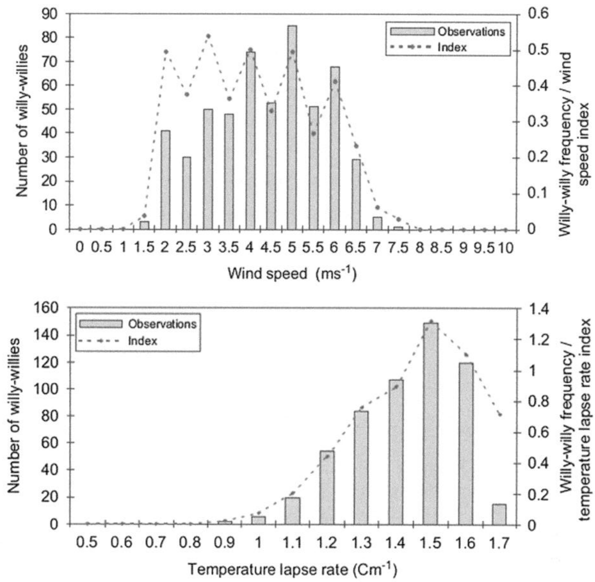

Fig. 2 Dust devil frequency as a function of ambient wind speed (*top*) and lapse rate (*bottom*) from Oke et al. (2007)

As an additional point, on timescales of the order of hours or more, dust devil occurrence is also correlated in that some days are favorable for formation, and other days are not (e.g. due to stronger ambient winds). Thus, as discussed in the chapter on dust devil populations (Lorenz et al. 2016a), statistics of number of dust devils observed per day, or number seen in individual orbit images, have fat-tailed distributions, with much larger numbers of high-incidence epochs than would be expected from a Poisson distribution with a single fixed probability of occurrence. Observations also suggested that dust devil activity was modulated by sensible heat flux and surface property variations, possibly including surface roughness. Snow and McClelland (1990) also present tabulated data of duration vs length of traverse. It is clear, dividing one by the other, that dust devils are more abundant for lower translation speeds (and thus, one assumes, lower ambient wind). Nonetheless, at least one devil was observed to migrate by 1800m in 2 minutes, implying a speed of 15 m/s.

Dust devil field experiments and observational campaigns have dramatically increased in sophistication since the early studies of Sinclair. The sheer number of these studies as well as the quantitative return from these experiments has also increased dramatically. The first study to successfully quantify wind shear in dust devils was performed by Balme et al. (2003). Based on field measurements of 10 dust devils that passed directly over an instrument cluster, the peak friction velocity, u_{*peak} was found to relate to the peak tangential wind speed V_{peak}:

$$u_{*peak} = (0.08 \pm 0.01) V_{peak} \quad (1)$$

Dust devils were found to occur under conditions of light winds (<5 m/s at 2 m height), generally clear skies and temperatures of 35 to 45 °C. A superadiabatic lapse rate was also observed during dust devil activity. During the measurement period, the ambient friction velocity was <0.3 m/s, but this increased to between 0.9 and 2.4 m/s in the dust devils.

Oke et al. (2007) observed 557 dust devils (“willi-willies” in Australian vernacular) in New South Wales, Australia, and found that dust devils were restricted to ambient winds of 1.5 to 7.5 m/s with peak activity at ∼3 m/s (Fig. 2). Ansmann et al. (2009) found very similar wind criteria in their lidar measurements of convective vortices, including dust devils.

A minimum lapse of ~0.9 K/m measured from 0.12 to 2.52 m AGL was necessary to initiate dust devils (Fig. 2), and the frequency of occurrence increased with lapse rate and with decreasing surface cover. Although the connection between lapse rate and activity is strong, the authors note that lapse rates cannot be compared to those from previous studies, because of different measurement heights and the lack of an agreed upon experimental protocol (this problem remains—see Sect. 4). Observations suggest that wind speeds >7.5 m/s are likely to weaken the superadiabatic lapse rate due to turbulent mixing in the lower boundary layer; however, Large Eddy Simulations also show that the organization of convection changes from cellular convection to more linear convection as wind speeds increase, and this can fundamentally alter the conditions in a way detrimental to dust devil formation (See LES section below). A hypothesis to explain the lack of dust devil production for wind speeds <1.5 m/s is that a small amount of wind may be required to unsettle the air pockets near the ground and to provide an initial source of rotary motion using the so called 'spin-up' theory (tilting and stretching of ambient shear vorticity—see Sect. 3.1). Changes in vegetation (i.e., rough to less-rough) could also initiate eddies due to changes in friction, which could also lead to shear in the airflow. The beneficial frictional effects of vegetation, if any, would tend to counter the soil stabilizing effect of the vegetation.

Based on Monin-Obukhov (MO) surface layer scaling theory, Deardorff (1978) proposed a stability parameter criteria for dust devil occurrence that was consistent with empirical evidence: $-h/L > 100$, where h is the convective boundary layer (CBL) height and L is the Obokhuv length. L may be regarded as the height above the ground where the production of turbulence by shear and buoyancy are equal, i.e., the Richardson number is unity. Hess and Spillane (1990) also relied on MO theory to generate a list of both necessary and probable conditions for the formation of dust devils. The proposed necessary conditions were: (i) relatively level and fairly smooth terrain; (ii) intense surface heating with temperature gradients greatly exceeding the dry adiabatic lapse rate; and, (iii) winds at the surface below a critical value. Conditions that were probably required were: (i) variability in wind direction; and (ii) vertical wind shear of the horizontal wind. The necessary conditions are all satisfied by small values of the Obukhov length, given by:

$$L = \frac{-u_*^3}{\frac{kgH_o}{\bar{\theta}_v \rho C_p}(1 + 0.61[\frac{C_p \bar{T}}{b\beta}])} \tag{2}$$

where $u_*^2 = \tau_0/\rho$, τ_0 is the surface stress, ρ the air density, k the von Kármán constant ($k \approx 0.41$), g the acceleration due to gravity, $\bar{\theta}_v$ the average boundary layer virtual potential temperature, H_0 the surface heat flux, C_p the specific heat at constant air pressure, $\bar{T}$ the average boundary layer temperature, b the heat of vaporization at temperature $\bar{T}$, β the Bowen ratio ($\equiv H_o/bE_o$) and E_0 the surface evaporation flux of water.

For $-\frac{h}{L} \gg 1$ the flow field is dominated by convection and thus the *mean* wind is not a significant factor in the formation of dust devils. Hess and Spillane (1990) found a critical threshold value of $-\frac{h}{L} \gg 50$, which is less than Deardorff (1978), but still solidly in the convective regime. From Eq. (2), it can be inferred that small values of friction velocity would tend to drive L to small values. This is consistent with large values of wind speed (and therefore, presumably large wind stress) suppressing dust devils, particularly given the cubed dependence. It is also consistent with the observations of Lyons et al. (2008) that indicated a reduction in frictional drag due to a decrease in surface roughness increases the likelihood of dust devils. Large values of sensible heat flux, H_o would seemingly decrease L and therefore increase the value of $\frac{h}{L}$, also consistent with observations. This is accomplished directly with H_o in the denominator of Eq. (2), but also by minimizing the virtual temperature term (the term with the 0.61 multiplicative factor) through a reduction of the Bowen ratio. As potential temperature increases, the denominator of Eq. (2), decreases, producing an increase in L and a commensurate decrease in $\frac{h}{L}$, also consistent with observations.

The correlation between dust devil occurrence and $\frac{h}{L}$ appears universal for Earth, although the critical threshold may be site-specific. Kurgansky et al. (2011) found the same general, necessary conditions in the Atacama Desert as Hess and Spillane (1990) and Lyons et al. (2008) did in Australia. Small dust devils were only able to form over darker soil (lower albedo), which should be significantly more heated than lighter soil. This would presumably lead to higher near-surface air temperatures and greater lapse rates. In contrast, the larger dust devils observed by Kurgansky et al. (2011) formed irrespective of the underlying soil characteristics. Mean wind speeds of 2 to 8 m/s enhanced activity, which is consistent with Oke et al. (2007), whereas values above 8 m/s quickly curtailed activity. The energy conversion hypothesis of Michelson (1951) is again invoked with the suggestion that background wind serves to inhibit convective instability in the first few meters above the ground.

A recent study by Broersen (2013) in Jordan found most dust devils formed in conditions that were strongly heated (the lapse rate, measured by the temperature difference between 1 m and 2.5 m above the ground, had to be above about 1 K/m), and winds were modest (most devils were encountered at ambient windspeeds above 2 m/s and below 7 m/s). He also reports a histogram of wind speed standard deviation, with most devils occurring for ranges of 0.4–1.0 m/s. The significance of this observation is not clear in that strong dust devil activity will in itself cause variation. h/L values were not reported, but the need for an unstable lapse rate and some wind is consistent with other studies.

2.2 Martian Dust Devils

There have been only a few meteorological measurements at the surface of Mars (e.g., Hess et al. 1977; Gómez-Elvira et al. 2014). None of the experiments made measurements sufficient to directly compute L or to explore the range of h/L conducive to dust devils. Of the data that has been collected, there are indications that MO theory is at least approximately valid (e.g., Sutton et al. 1978; Tillman et al. 1994). Although direct meteorological measurements on Mars are limited, Mars is known to exhibit a very steep superadiabatic afternoon lapse rate (e.g., Sorbjan et al. 2009). The Bowen ratio may be assumed to be large (nearly infinite), except perhaps in the very few places where water ice may be sublimating into the atmosphere. There are also very large expanses of nearly flat plains, and wind speeds are known to be sufficient to lift dust. Finally, the CBL depth, h, is generally 2 to 10 times higher than on Earth (e.g., Hinson et al. 2008). All of these conditions are factors that would favor dust devil formation, assuming that extrapolations from Earth to Mars are appropriate. The covariance of vertical velocity and temperature may be larger than on Earth due to the vigorous turbulence associated with extreme convective instability, but the heat flux may be quite small, because of the low atmospheric density (Haberle et al. 1993; Michaels and Rafkin 2004). This is at least suggestive that the magnitude of the heat flux is not a prime indicator of dust devil formation, but it is the covariance—an indicator of convective energy. On the other hand, Mars is heated most directly by radiation, not by turbulent fluxes. So, perhaps it is simply the energy input near the surface that is relevant, whether it be by turbulent fluxes or radiative heating.

Large dust devils—sometimes of diameter near 1 km—have also been observed from orbit (e.g., Fisher et al. 2005), and the tracks from dust devils (Reiss and Lorenz 2016;

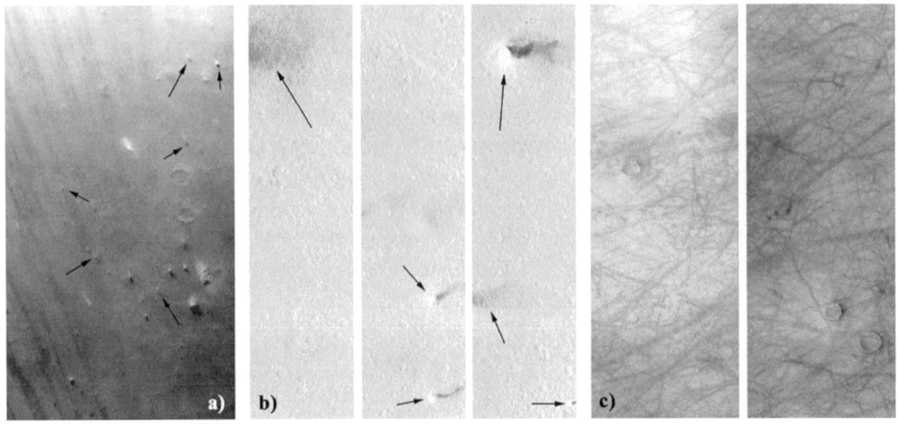

Fig. 3 Dust devils and dust devil tracks imaged from orbit. From Fig. 1 of Fisher et al. (2005). (**a**) Dust devils in the MOC WA images (some examples indicated by arrows), (**b**) dust devils and shadows in MOC NA images (indicated by *arrows*), and (**c**) dust devil tracks observed in MOC NA images

Fisher et al. 2005; Balme et al. 2003) provide additional clues about their preferred locations and density (Fig. 3; see also Chaps. 4 and 5). Most previous and current Mars orbiters with cameras sufficient to resolve at least the largest dust devils are in a sun synchronous orbit, meaning that only two local times of day, 12 Mars hours apart, are observed. Fortunately, the daytime imaging has occurred from roughly 1400–1600 local time, near and slightly after when peak dust devil activity might be expected. Global orbital surveys of dust devils (e.g., Fisher et al. 2005; Cantor et al. 2006) noted a seasonal peak in dust devil activity, which like Earth, corresponds to a time of maximal ground and low-level air temperatures.

While dust devils or their tracks were found globally, the northern Amazonis and Arcadia Planitia were found to have the highest number of dust devil activity indicators (e.g., Fisher et al. 2005; Cantor et al. 2006; Reiss et al. 2014). Interestingly, nearby regions, such as Utopia Planitia, had lower evidence of activity. The reason for this is not clear. There are more direct observations of dust devils in Amzonis Planitia, although there are not a great number of dust devil tracks. Clearly then, tracks are more efficiently created in some regions. Tracks could also be more efficiently covered by dust sedimentation in some regions. Nevertheless, there are likely real meteorological and physiographical differences that favor dust devil formation in some locations. For example, perhaps mean winds are too strong in some locations. Or, perhaps the surface thermal inertia and albedo produce favorable warm temperatures in some locations. Variations in the height of the CBL, which are tied to topographic elevation and very likely to albedo and thermal inertia, could also play a role (e.g., Hinson et al. 2008).

3 Models of Dust Devils

There are three basic types of models that may be applied to better understand the formation mechanisms responsible for dust devils. The first looks at dust devils from primarily a dynamical perspective. The second takes a thermodynamic approach. The final model type is numerical and directly simulates in a physically consistent way both the dynamic and

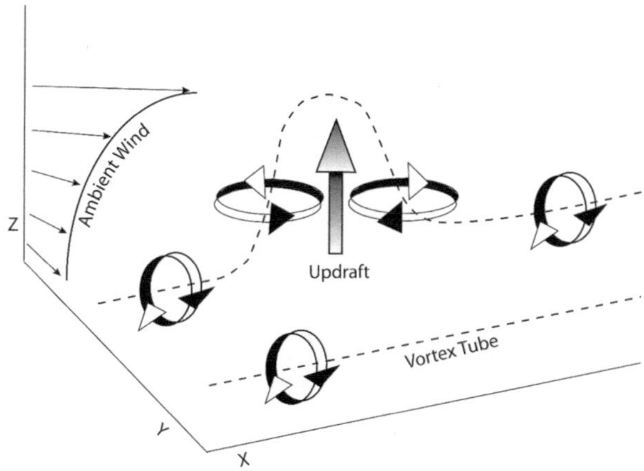

Fig. 4 Schematic of vorticity tilting. The ambient wind results in a horizontal tube in the X-Y plane. An updraft will tilt the vortex tube, creating counter-rotating vortices aligned with Z

thermodynamic behavior of the dust devil or its environment. Large Eddy Simulations are representative of this third model type.

3.1 Dynamical Models

Dynamical models of vortex dynamics are discussed in a companion chapter (Chapter 6; Kurgansky et al. 2016). The vorticity equation provides a basic place to approach dust devils from the dynamical perspective. Taking the curl of the Navier-Stokes equation results in the following expression:

$$\frac{D\vec{\omega}}{Dt} = \frac{\partial\vec{\omega}}{\partial t} + (\vec{u}\cdot\nabla)\vec{\omega} = \underbrace{(\vec{\omega}\cdot\nabla)\vec{u}}_{\text{Tilting}} - \underbrace{\vec{\omega}(\nabla\cdot\vec{u})}_{\text{Stretching}} + \frac{1}{\rho^2}\underbrace{\nabla\rho\times\nabla p}_{\text{Baroclinic}} + \underbrace{\nabla\times\left(\frac{\nabla\cdot\tau}{\rho}\right)}_{\text{Diffusion}} \tag{3}$$

where $\vec{\omega} = \nabla \times \vec{u}$, $\vec{u}$ is the velocity vector, τ is the stress tensor, ρ is density, and p is pressure. There first two terms on the right side of Eq. (3) describe how existing vorticity may be modified in magnitude due to stretching or in direction due to tilting. Tilting may be visualized as a non-uniform velocity field perpendicular to a vortex tube producing a change in the tube's orientation (Fig. 4). The stretching term represents conservation of angular momentum. A diverging flow will expand the radius of parcels circulating in a vortex tube, and this will necessarily decrease vorticity via the circulation theorem. The last two terms represent source/sink terms for vorticity. These are important because they provide the only means by which vorticity may arise from an initially irrotational flow. For the baroclinic term to be non-zero, the gradients of density and pressure must not be aligned. Through the ideal gas law, this can be envisioned as requiring a thermal gradient on a constant pressure surface. Under such conditions, hot air will tend to rise and cold air will tend to sink. The relative motion will generate vorticity. For example, a warm thermal will generate a circulation with rising motion in the core of the updraft, sinking air at the peripheral interface with the cooler environment, converging horizontal flow beneath the thermal that feeds the updraft, and diverging air aloft that feeds the downdrafts.

Finally, the diffusion term is exactly as the name implies. If $\frac{\nabla\cdot\tau}{\rho}$ is written as a viscous dissipation term $\upsilon\nabla^2\vec{u}$, where υ is viscosity, it then becomes clear that the curl of this term is, in fact, the diffusion of *existing* vorticity. In equating the divergence of the stress tensor with viscous dissipation, the term now depends on the presence of vorticity. It is no longer an irrotational source/sink. Typically, the diffusion term for vorticity acts to decrease the magnitude of vorticity, but countergradient diffusion could, in theory, result in an increase.

In the case of a dust devil, it is mostly the vertical component of the vorticity vector that is of interest, $\zeta = \boldsymbol{k} \cdot \vec{\omega}$. From Eq. (3), the vertical component of the vorticity is governed by

$$\frac{D\zeta}{Dt} = -\left(\frac{\partial w}{\partial x}\frac{\partial v}{\partial z} - \frac{\partial w}{\partial y}\frac{\partial u}{\partial z}\right) - \zeta \nabla_h \cdot \vec{v}_h - \frac{1}{\rho^2}\boldsymbol{k} \cdot (\nabla_h p \times \nabla_h \rho) + \boldsymbol{k} \cdot \left[\nabla \times \left(\frac{\nabla \cdot \tau}{\rho}\right)\right] \quad (4)$$

Looking only at the vertical component makes visualization of each term clear. The vertical shear of the horizontal wind ($\frac{\partial u}{\partial z}$ or $\frac{\partial v}{\partial z}$) produces a vortex tube parallel to the ground. If there is a horizontal gradient of vertical motion acting along that tube ($\frac{\partial w}{\partial x}$ or $\frac{\partial w}{\partial y}$), the initial horizontal orientation of the tube will be distorted (Fig. 4). Similarly, it is now clear that horizontal convergence will cause an increase in vertical vorticity, which can be visualized as contraction of the radius of the vortex tube. Since convergence implies compensating vertical motion in an incompressible fluid, an equally valid interpretation is that the ends of the vortex tube are being pulled apart so that the vortex tube stretches and thins like a piece of taffy.

Dust devils do not occur in isolation; they are embedded in a larger background environment with vorticity. For example, surface friction will generally result in a near surface horizontal vorticity sheet resulting from the increase of the horizontal wind with height. Scorer (1978) suggested that the rotation of dust devils is fed by friction induced drag from the surface. Necessarily, this vorticity would still need to be tilted and stretched into the vertical. A vorticity tilting and stretching model for dust devils is consistent with the general lack of preference for rotational direction; a tilted vortex tube should produce a pair of counter-rotating vortices.

Pressure will respond to a small scale circulation within a vortex, as governed by centrifugal adjustment theory; the circulation and pressure will tend toward a dynamically balanced steady-state system where the force due to the pressure gradient matches the centrifugal force. To achieve this balance from an unbalanced state, the system generates acoustic and mixed acoustic-gravity waves to redistribute mass and momentum (e.g., Kalashnik and Visheratin 2010; Lelong and McWilliams 1999). This is a process exactly analogous to geostropic adjustment, which occurs at larger scales where the Coriolis acceleration rather than centrifugal acceleration is a dominant term. As rotational speed increases, the pressure perturbation also intensifies.

Dynamical models do not fully explain the observational evidence that critical values of heat flux or lapse rate are necessary for the formation of dust devils, nor do they explain why strong winds would tend to suppress dust devils. In fact, strong winds should generally lead to a larger source of background vorticity.

3.2 Thermodynamic Models

Another approach to understanding the formation and maintenance of dust devils is to examine the energetics. Renno et al. (1998) modeled the dust devil as a Carnot-like engine, which followed from similar work that described hurricanes as heat engines (Emanuel 1986). The underlying concept is that dust devils convert heat to kinetic energy.

The Bernoulli principle, including the effect of friction may be written as

$$\frac{d}{dt}\left(\frac{1}{2}\boldsymbol{v} \cdot \boldsymbol{v} + gz\right) + \alpha dp - \boldsymbol{f} \cdot d\boldsymbol{l} = 0, \quad (5)$$

where $\boldsymbol{v}$ is velocity, g is gravitational acceleration, z is height above the ground, α is specific volume, p is pressure, $\boldsymbol{f}$ is the frictional acceleration and $d\boldsymbol{l}$ is a segment along a path. Equation (5) states the sum total of kinetic energy, potential energy, pressure work and frictional work must sum to zero; energy is conserved. Integrating Eq. (5) over the material volume of a circulation and demanding mass conservation produces the convenient result

$$\oint \alpha dp = \oint \boldsymbol{f} \cdot d\boldsymbol{l}, \tag{6}$$

which can be evaluated piecewise around a dust devil circulation. Equation (6) states that the pressure work done by the dust devil circulation must be balanced by the frictional loss of energy. From the 1st Law of Thermodynamics, $\oint \alpha dp = \oint T ds$, so that the work done may also be equated with the heat input into the system.

To first order, a parcel of air flowing along the surface toward the center of the dust devil undergoes an isothermal expansion. In order to maintain an isothermal state, heat must be added to the parcel in order to maintain the nearly constant input temperature. Within the dust devil, the air rises adiabatically and is then ejected horizontally at the top of the circulation. This forms one half of a Carnot engine cycle: an isothermal leg and an adiabatic leg. The temperature at which the parcel is ejected should be near the level of neutral buoyancy and will depend on the initial input temperature and the environmental lapse rate. The second half of the Carnot circuit is more nebulously defined, but is not terribly important to the overall theory.

Renno et al. (1998) defines two efficiency factors. The first is the efficiency, γ, of surface friction compared to the overall circulation in dissipating energy

$$\gamma = \int_{\infty}^{0} \boldsymbol{f} \cdot d\boldsymbol{l} / \oint \boldsymbol{f} \cdot d\boldsymbol{l}, \tag{7}$$

where the first integral represents the surface leg flowing from ∞ to the center of the dust devil. The second efficiency factor, η, is the standard definition of engine efficiency relating the fraction of input heat used to do work

$$\eta = \oint T ds / \int_{\infty}^{0} T ds. \tag{8}$$

For a Carnot engine, the input and output temperatures are directly related to efficiency.

Equations (7) and (8) are combined with the Ideal Gas Law with the following result:

$$-RT_s \ln\left(\frac{p_o}{p_\infty}\right) = \eta\gamma \int_{\infty}^{0} T ds \tag{9}$$

where T_s is the input temperature. The key point resulting from Eq. (9) is that the pressure deficit at the core of the vortex can be directly related to the heat input and the efficiencies of the system.

The heat engine theory is not specific to a vortex; no rotation is required. However, if a specific cyclostrophic dynamical balance is assumed (e.g., a Rankine vortex), the radial velocity associated with the pressure deficit may be calculated. If lifting thresholds for surface particulates are known, this can be further extended to determine the critical pressure deficit that will produce a tangential wind sufficient to lift dust. For given efficiencies, greater heat input will lead to a larger pressure drop and greater tangential winds at the core of the vortex. There is no dynamical explanation for how the vortex arises, only how a pressure drop is produced. The source of vorticity for the dust devil is not explained by this theory.

Lyons et al. (2008) makes reference to Renno et al. (1998), whereby the intensity of dust devils is a function of surface pressure, pressure thickness of the convective layer and the near-surface fraction of mechanical dissipation of energy and the horizontal thermodynamic efficiency. In their description, key ingredients for dust devil formation are the magnitude of convective buoyancy and rate of frictional dissipation at the surface. The ratio h/L (cf. Hess and Spillane 1990) can be expressed in terms of the convective velocity scale (w^*) and the friction velocity (u^*), such that $w^*/u^* > 5$ is a requirement for dust devil formation, and convection and friction at the surface need to be balanced for a dust devil to maintain a quasi-steady state.

3.3 Numerical Models

Numerical models of dust devils are discussed in the chapter by Spiga et al. (2016). Numerical models can implicitly include both dynamics and thermodynamics. This presents a clear advantage compared to models that have one or the other, but it can also be disadvantageous, because of the potential difficulties involved in deconvolving the importance of the multitude of processes that can be acting simultaneously. Consequently, all models should be viewed as necessary and complementary to one another for understanding dust devils.

Large Eddy Simulations (LESs) are the primary numerical model for simulating the environment of dust devils. The foundational assumption of the LES is that the domain be large enough in both the vertical and horizontal dimensions that it captures a statistically significant ensemble of dominant-scale eddies, and that the resolution is sufficient to fully resolve those eddies. As with all models, LESs have a subgrid scale below which the effect of unresolved motions must be parameterized, if important. Ideally, the LES will have sufficient resolution such that the dominant unresolved eddies fall into the inertial subrange where their main role is to dissipate energy via the turbulent energy cascade. The scale of the dominant eddies to be resolved should scale with the height of the boundary layer.

Dust devils are at a scale that is equal to, and more typically smaller than, the depth of the boundary layer. Fully resolving a dust devil in an LES can be challenging, since it requires very high resolution grid spacing while simultaneously demanding that there be a domain sufficient to capture an ensemble of eddies.

In most LES simulations, the land surface is taken to be homogeneous, while with few exceptions the real world has heterogeneity in albedo, thermal inertia, topography, and land cover (e.g., vegetation). As previously noted, observations suggest that dust devil activity may be modulated by this heterogeneity. In principle, many models could introduce heterogeneity to study impacts on dust devil formation. This has yet to be done, although Shao et al. (2013) have modeled the impact, which is noticeable, of heterogeneities on eddy and boundary layer structures.

A reasonable convective boundary layer depth on Earth may be 2 km, whereas on Mars it may be 8 km. If this is taken as the scale of the dominant eddy, and if one wishes to contain at least 10 eddy lengths in each dimension of the domain (this is probably close to the minimum that could be considered a statistically valid ensemble population), then that requires a horizontal domain of at least 20 km × 20 km on Earth (80 km × 80 km on Mars). A large dust devil on Earth might be 40 m across, and requiring at least four grid points to resolve such a feature would place the necessary grid spacing at ~ 10 m. Consequently, a minimally adequate domain would necessitate a horizontal grid of 2000 × 2000 points (8000 × 8000 points for Mars).

The vertical domain must be greater than the convective depth in order to keep the upper lid a reasonable distance from the circulations of interest. Furthermore, if there is any hope of resolving a dust devil with the most intense circulation within the lowest 100 m of the atmosphere, the vertical grid spacing must be fine near the surface. Assuming a depth twice the maximum convective boundary layer depth with a vertical spacing of 100 m, plus an additional 10 levels of higher resolution near the surface, a model would need a minimum of $\sim$50 levels. In total, a minimum domain of $2000 \times 2000 \times 50$ is required. This is extremely challenging for current computer systems, and in practice, most simulations will not provide such an extensive domain, although there are a few recent instances of massively parallel simulations with suflarge domains at high resolution (e.g., Raasch and Franke 2011). More typically, a sacrifice in the horizontal domain is often necessary, and the finer the grid spacing desired, the greater is the reduction in domain size that must be implemented (Table 1). Similar scaling arguments can be made for Mars. Dust devils are larger on Mars (e.g., Balme and Greeley 2006), so the grid spacing might, in principle, be made larger compared to Earth. However, the depth of the boundary layer is a factor of 4 to 5 times larger, so that gain is quickly offset by larger domain requirements.

The challenge of completely resolving a dust devil within an LES means that detailed information on the structure and dynamics of dust devils themselves may be limited. However, information about the dust devil environment and the convective structures that initiate dust devils can be more easily extracted.

Kanak (2006) provided a fairly comprehensive review of simulations of the convective boundary layer of dust devils for both Earth and Mars. Every study considered in this review exhibits large-scale cellular convective patterns in the horizontal plane, and the intersection of convective cells with one another is closely linked with the location of the most intense vertical vortices.

The first example of vertical vortices Kanak (2006) was able to identify within a LES was Mason (1989), although little attention was given to those results as they were not the focus of the work. Mason's results indicated that vortices occurred at the vertices of cellular convective pattern branches. Kanak et al. (2000) and Kanak (2005a) replicated this pattern for Earth, and simulations of the Mars convective boundary layer and dust devils yielded the same finding (Rafkin et al. 2001; Michaels and Rafkin 2004; Toigo et al. 2003; Fenton and Michaels 2010; Fujiwara et al. 2011; Gheynani and Taylor 2010, 2011; Ito et al. 2011,2013; Klose and Shao 2013; Ohno and Takemi 2010a, 2010b; Raasch and Franke 2011; Spiga 2012; Spiga and Lewis 2010).

Kanak (2005a) was the most advanced terrestrial LES study of dust devils at the time. It is of specific interest to the topic of dust devil formation, because one of the prime objectives of the work was to explore how updraft rotation was achieved from an initially irrotational environment. A specific source of vorticity was not identified, but it was noted that some small amount of randomness in initial conditions was necessary to produce the overall random convective structure from which dust devils emerged (Fig. 5). In addition to finding dust devils at the intersection of convective cell branches, vortices were also found along the convective cell branches. These "book end" vortices were highly suggestive of a tilting of a horizontal vortex tube by a local maximum in the updraft. The result of such a process is counter rotating vortices.

Analysis of one simulated dust devil from Kanak (2005a) showed that the temperature increase within the core was consistent with field experiments (Sinclair 1973), but the pressure drop of 43 Pa is over-predicted compared to field measurements of 2 to 3 Pa. Furthermore, tangential and radial velocities are much lower in the LES compared to experiments. Thus, although the pressure deficit was greater, the pressure gradient must be smaller in the LES than in those observed. A direct comparison of the LES to observations is not entirely analogous because the simulated dust devil has a core diameter of 30 m, 6 times greater than

Table 1 Summary of LES dust devil simulations from Kanak (2006)

Study	Equations	Grid points	Domain, km	Grid spacing, m	Boundary conditions	Ambient wind
Mason (1989)	B	$64 \times 64 \times 98$	$32 \times 32 \times 5$	$50 \times 50 \times 5$–150	NS	none
Cortese and Balachandar (1995)	B	$96 \times 96 \times 97$	Non-dimensional	Non-dimensional	NS	none
Kanak et al. (2000)	SC	$86 \times 86 \times 48$	$3 \times 3 \times 2.1$	$35 \times 35 \times 10$–80	NS	none
Fiedler and Kanak (2001)	B	$128 \times 128 \times 33$	Non-dimensional	Non-dimensional	FS	none
Kanak (2005a)	SC	$370 \times 370 \times 86$	$0.740 \times 0.740 \times 12$	$2 \times 2 \times 4.12$–24	NS	none
Rakfin et al. (2001)	FC	$180 \times 180 \times 36$	$18 \times 18 \times 9$	$100 \times 100 \times 12$–250	NS	Backing wind profile
Toigo et al. (2003)	FC	$100 \times 100 \times 57$	$10 \times 10 \times 7.5$	$100 \times 100 \times 10$–100	NS	Both wind and no wind
Michaels and Rafkin (2004)	FC	$200 \times 200 \times 61$	$6 \times 6 \times 23$	$30 \times 30 \times 4$–45	NS	5 ms^{-1} mean wind
Kanak (2005b)	SC	$150 \times 150 \times 60$	$6 \times 6 \times 10$	$40 \times 40 \times 40$–295	NS	none

For equations, B = Boussinesq, SC = supercompressible, and FC = fully compressible. For lower boundary conditions, FS = Free slip, and NS = no slip. Grid points, domain size and grid spacing are given in the x, y, and z directions, respectively. For the grid spacing, the z coordinate is often stretched with height, so a range of grid spacings is given. In these simulations, the smallest (largest) value is the vertical grid spacing near the bottom (top) domain boundary

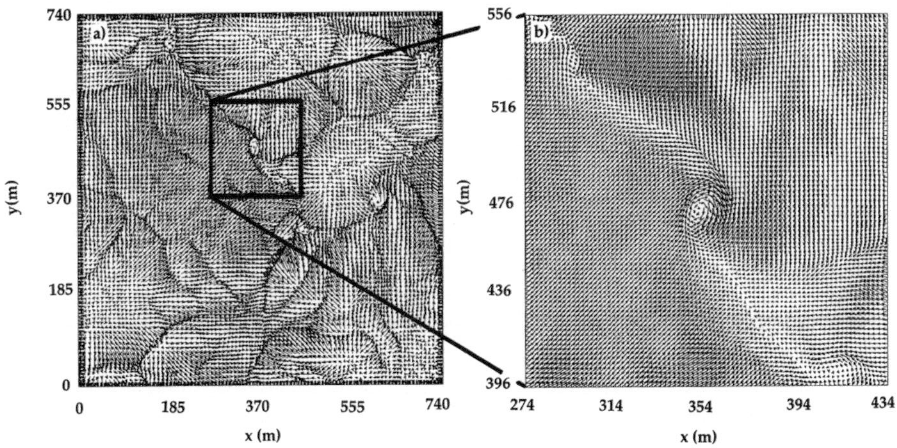

Fig. 5 An example of convective cell structures and the development of a vortex in the LES simulations of Kanak (2005a)

the observed case. This is at least suggestive of insufficient model resolution. Although the broad circulation of a dust devil was captured in the model, the full dynamical details were not. It is interesting to note that radar observations of dust devils do show that dust devils (the true, optically thick phenomena) are embedded in much broader but invisible circulations (Bluestein and Pazmany 2000; Bluestein et al. 2004; Fujiwara et al. 2011) much the same way that tornadoes are embedded in a larger mesocyclone. Like tornadoes, how much vorticity is derived from the parent circulation and how much is derived from baroclinic vorticity generated at the boundary of the warm updraft and cold downdraft is still an open discussion.

By conducting LESs of terrestrial dust devils under various mean wind conditions, Ohno and Takemi (2010a) found that the stability parameter $\frac{w_*}{u_*}$ of greater than 5 is favorable for the generation of dust devils, which agrees with the observational studies such as Hess et al. (1988), Hess and Spillane (1990), and Lyons et al. (2008). In Ohno and Takemi (2010a), mean winds of 0, 5, 10, and 15 $\mathrm{m\,s^{-1}}$ were examined. It was shown that with the increase in mean wind the number of vortices decreases and the lifetime of vortices becomes shorter; vertical vortices were preferentially generated and developed in the modest wind (i.e., 5 and 10 $\mathrm{m\,s^{-1}}$) conditions. On the other hand, vortices under no mean wind condition (which corresponds to the highest stability parameter) were not the strongest among the various mean wind cases. The results suggested that the stability parameter provides a lower limitation, that is, a necessary condition, for the occurrence of dust devils.

As for Earth, the first LES simulation of the Mars convective boundary layer (Rafkin et al. 2001) contained dust devil-like vortices, but these were largely not discussed, as they were not the focus. Later studies that specifically targeted the vortices (e.g., Toigo et al. 2003; Michaels and Rafkin 2004) were remarkably similar to those for Earth. Interestingly, the simulations of Martian dust devils pre-dates similar studies for Earth. Perhaps this is because dust devil tracks are so prominent in Mars imagery, or because dust is such an important radiative driver in the Martian climate so that the role of dust devils in the dust cycle and climate system has been given more attention. It does appear that the structures and dust devils on Earth and Mars are quite similar and are very likely driven by the very same processes.

On Mars, like Earth, the strongest vortices form at the intersection of convective cells. In the case of Michaels and Rafkin (2004), vortices did not appear in the morning until a surface heated by insolation forced the structure of the boundary layer to evolve from linear, roll-like convection to cellular convection (Fig. 6). The bulk Richardson Number is an indicator of the broad convective structure. When shear dominates over buoyancy forces, as might be expected in strong wind conditions, or when the boundary layer has not yet become substantially unstable, the magnitude of the Richardson Number is driven to very small values. In that case convection transitions increasingly from a cellular structure to a linear, roll structure.

Toigo et al. (2003) simulated Martian thermal vortices with a LES and looked specifically at the origin of vorticity as well as the energetics compared to the thermodynamic engine model. In rough agreement with the thermodynamic engine model, energy budgets suggested a balance between buoyancy generation and loss through dissipation and transport. Vertical vorticity was generated through the tilting of horizontal vorticity with the vorticity being generated through the baroclinic process. The production of vorticity from horizontal variations of buoyancy is in contrast to most terrestrial LES studies that derive buoyancy from the ambient shear.

Sykes et al. (1988) found that $\frac{u_*}{w_*} \geq 0.35$ is a critical condition for the appearance of roll structures on Earth. Thus, this might be roughly considered as a criterion in which dust devils become increasingly unlikely. The aspect ratio of the rolls (i.e., the width of the rolls divided by the depth of the rolls) decreases as the parameter increases. In other words, the rolls become increasingly narrow as the relative magnitude of the horizontal wind increases. Recall that the Lyons et al. (2008) found $\frac{w_*}{u_*} > 5 \rightarrow \frac{u_*}{w_*} < 0.20$ was the necessary condition for dust devils.

Jemmett-Smith et al. (2015) used meteorological analyses (output by the ECMWF) to quantify global dust devil occurrence on Earth. The previous studies of Deardorff (1978), Hess and Spillane (1990), Lyons et al. (2008), and Kurgansky et al. (2011), were analyzed to produce w^*/u^* criteria for dust devils, which yielded similar thresholds limits of between 5.0 and 6.3, above which dust devil activity is expected to occur. When applied to hourly global meteorological analyses, the $w^*/u^* > 5$ criterion was shown to be insufficient. Too many potential dust devil hours were predicted to occur during the nighttime. By applying the 8.5 K/m lapse rate criterion from Ansmann (2009) combined with the $w^*/u^* > 5$ criterion from Lyons et al. (2011), results produced a diurnal cycle of activity similar to that observed. It was suggested that the primary reason why the $w^*/u^* > 5$ proves insufficient, is that the criterion can be met during situations with weak convection (low w^*) and low values of u^* (hence many potential hours occur at night). Currently there is no known minimum $w*$ threshold. Kurgansky et al. (2011) suggested that dust devil activity peaks when $w^* \sim= 2.8$ m/s, but provide no lower threshold. Care should be taken in interpreting w^*/u^* values from Jemmett-Smith et al. (2015), as the values are derived from large-scale (global-scale) models where eddy properties are parameterized, and topography, land surface heterogeneity and other sub-grid scale features complicate the comparison to observations and modeling valid at the local or eddy scale.

The structure of convective boundary layer and the location of dust devil occurrence under various mean wind conditions as studied in Ohno and Takemi (2010a) can be seen in Fig. 7. Similar to the feature shown in Fig. 5, it is seen that with the increase in mean wind there is a transition in the convective structure from cellular to roll type. The locations of dust devils are seen at the intersections of convective cells in no wind condition, while those are seen along the linear updraft regions in higher wind conditions.

The morphing of cellular convection to linear rolls may very well explain the observed suppression of dust devils under strong wind conditions. If dust devils are favored at the

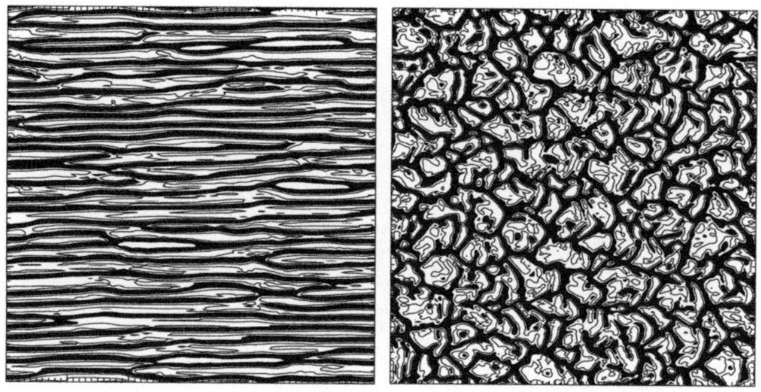

Fig. 6 Linear roll structures in the mixed layer for small Richardson Number (*left*), and cellular convection when buoyancy dominates over shear for larger magnitude Richardson Number (*right*). From Michaels and Rafkin (2004)

intersection of convective cells, then the number of those intersections will decrease as roll structures emerge. With a strong enough wind, intersections will be effectively absent and then the necessary condition for dust devils will also be absent. This hypothesis is in contrast to the idea that strong winds more effectively mix the boundary layer and reduce the superadiabatic lapse rate that feeds convection.

Moore's Law as applied to processor capacity, speed, memory, and storage capacity means that what was not computationally possible only a few years ago is possible now or may become possible in the very near future. The study of Ohno and Takemi (2010b) represented a new paradigm for LES modeling where both the number of grid points was increased by many factors over previous studies and the grid spacing was dramatically reduced. They performed a simulation with 20 million cell grids and a domain size of $1000 \times 1000 \times 1500$ m with horizontal grid spacing of 3 m and 185 vertical levels stretched with height. This can be compared to the experiments in Table 1.

Consistent with previous studies, Ohno and Takemi (2010a) found that the conditions under which dust devils form and are maintained are: (i) fairly flat terrain, (ii) intense surface heat flux and (iii) weak ambient winds. Similar to Kanak's work (Kanak et al. 2000; Kanak 2005a) many dust devils form due to tilting of horizontal vorticity towards the vertical, leading to a pair of vortices with opposite signs. Rayleigh-Bernard (cellular) convection became completely turbulent after approximately 5 convective turnover periods of ~ 500 s. In the highly unstable region close to the ground, the lapse rate was 0.47 °K/m, which agrees with that measured by Sinclair (1973). Many vortices were generated but the authors chose to focus on a small number of strong ones. Intense vortices were strengthened by the merging of multiple vortices, provided they all have the same rotation sense. Vortices merging with those of opposite rotation tended to cancel out and thus reduce the intensity. This is the first study to suggest that merging of vortices may be an important process. In later studies, Ohno and Takemi (2010b) demonstrated that strong vortices are intensified through the merger of multiple vortices with the same rotation sense.

The work of Raasch and Franke (2011) is the current state-of-the-art in numerical modelling of dust devils. A few of their simulations were configured with 2 billion cell grids and a domain size of $4096 \times 4096 \times 1704$ m (height = 1740 m). The domain was perhaps too small to capture an ensemble of eddies, but the grid spacing was 2m, which should be more than sufficient to resolve dust devil structures. If dust devil formation does require capturing the full interaction of eddies, then the small domain could be an issue. Their lowest simulated perturbation pressure was -72.4 Pa and the largest vorticity was 6.8 s^{-1}, which is just slightly lower than observed values of $7\ s^{-1} < \zeta < 14\ s^{-1}$ (Sinclair 1973). The longest vortex lifetime was around 660 seconds. Results show that the flow field exhibited the typical cellular structure of convergence lines Higher background wind appeared to drastically reduce the number of vortices, because the near-surface cellular flow regime cannot develop

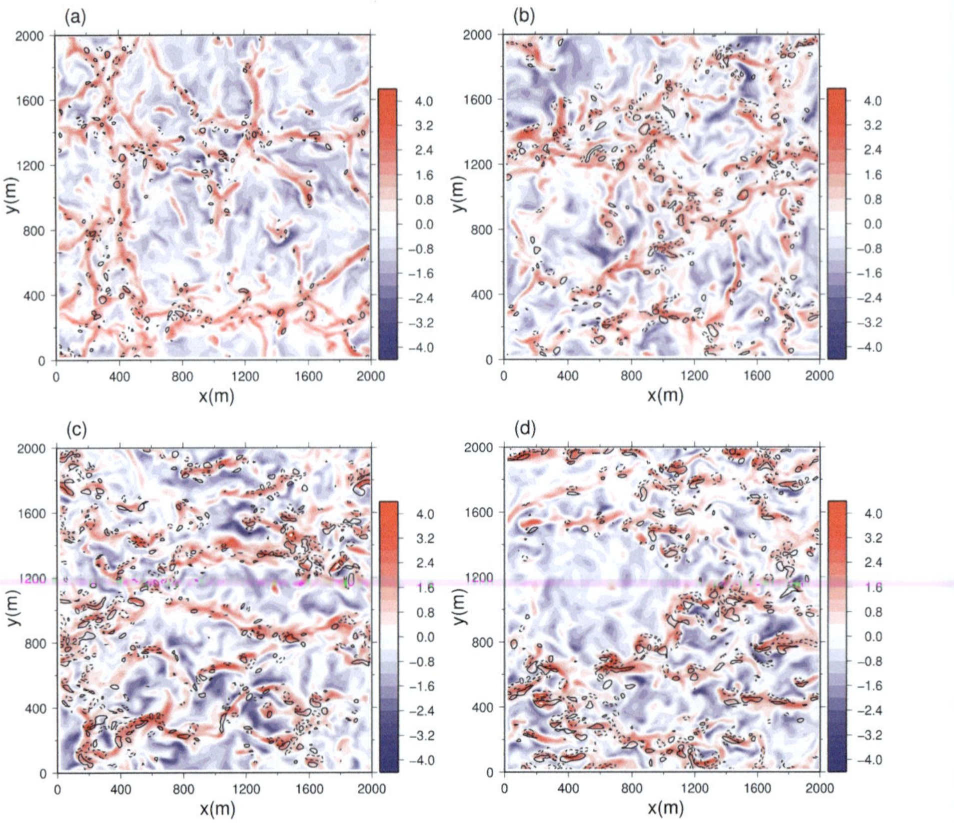

Fig. 7 The horizontal structure of convective boundary layer at the height of 50 m under the conditions with mean wind of (**a**) 0, (**b**) 5, (**c**) 10, and (**d**) 15 $m\,s^{-1}$. *Color shading* indicates the vertical velocity (the legend to the *right of each panel*; positive in *red*, and negative in *blue*), and the *solid (dotted) contour* denotes positive (negative) vorticity with the magnitude of 0.2 s^{-1}

under shear-dominated conditions. This supports the notion that as the boundary layer transitions to a linear structure the number of cell intersections and thus dust devils is reduced. They specifically noted, "The number and strength of vortices significantly decreases in a more shear dominated CBL, where the hexagonal pattern changes to a band-like pattern."

Raasch and Franke (2011) explicitly examined the vorticity budget to develop a theory for dust devil maintenance. They could not analyze the formation mechanism, because the analysis was only performed once a pressure and vortex was identified. Both the tilting term and stretching term were found to be important. Vertical shear of the horizontal wind did produce vorticity that flowed toward the updraft where it was ultimately tilted into the vertical. Stretching by the updraft then took over and further spun up the vortex (Fig. 7). Perhaps surprisingly, the baroclinic term was found to be an order of magnitude smaller than the tilting and stretching terms. What is not clear is how the vortex tubes terminate, as shown in Fig. 8. A horizontal vortex tube that is tilted into vertical should produce a pair of vortices, whereas only one vortex is shown.

Ito et al. (2013) looked specifically at the formation mechanisms of strong vortices in a convective mixed layer using a small domain (1.6 × 1.6 × 1.8 km) with a spacing of 5m. Their findings are similar to that of Raasch and Franke (2011): horizontal vorticity is tilted

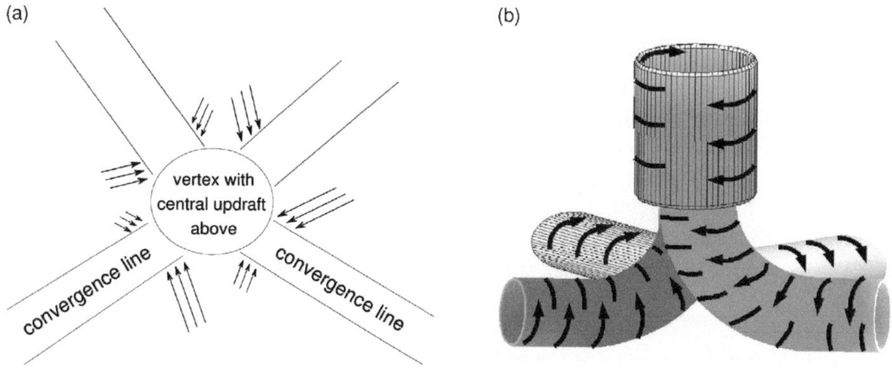

Fig. 8 Schematic of dust devil formation according to Raasch and Franke (2011). Horizontal vortex tubes due to vertical shear of the horizontal wind are concentrated along convergence line updrafts and are then tilted into the vertical where the convective cell boundaries intersect

and stretched into the vertical. However, the initial source of that vorticity is undetermined. The vorticity could be derived from the background wind profile, but the effect of topography and a heterogeneous surface could not be ruled out. Baseline simulations used a free-slip condition. A simulation in which surface friction was added tested the potential importance of friction explicitly, and it was found not to have a large effect. This finding is perhaps in conflict with the heat engine theory, because friction should increase the amount of work that the dust devil needs to do in order to remain in steady state. It also conflicts with Zhao et al. (2004), who indicated that surface friction is very important.

4 Summary, Open Questions, and Future Work

As noted in the introduction of this chapter, three criteria must be met: a convective vortex must be produced; the vortex must have sufficient strength for the associated winds to loft sand, dust or other particulates into air; and, there must be a sufficient source of mobile surface material to produce visible opacity once that material is lifted. Field observations, laboratory studies, and modeling results all provide insight into how each of these criteria are met for dust devils, but a definitive and comprehensive explanation for the formation of dust devils remains elusive.

With regard to both Earth and Mars, it seems clear that a convective environment is necessary for vortex formation. On Earth, suitable instability, as might be measured by lapse rate, and an appropriate convective regime, as might be characterized by w^*/u^* appear as necessary conditions, but are perhaps not sufficient. On Mars, an unstable lapse rate also appears necessary, but the universality of terrestrial-derived convective parameters is uncertain. For the second dust devil criteria, an optically visible vortex (i.e., a true dust devil) must also involve site specific dust lifting processes. That lifting (and/or saltation) must necessarily connect to surface wind stress, which is itself connected back to the overall convective environment.

All evidence points to wind stress as the dominant aerosol lifting mechanism for both Earth and Mars (e.g., Balme et al. 2003), but second order effects, which are far less understood than wind stress lifting, could also be important in dust devil formation in some instances. Balme and Hagermann (2006) examined the implication of the so-called "Δp effect" in dust entrainment. While probably not important for vortex generation, it could be an important ingredient to make an optically thick vortex. The hypothesis is based on the fact that dust devils have a low pressure in the center just above the ground and this, combined with higher (ambient) pressure air pockets trapped beneath the soil, is sufficient to give an upward resultant force and thus to release dust into the atmosphere. While this sounds reasonable in theory, they do not consider horizontal surface shear stresses, which result from swirling winds. Also, there is an assumption that the soil is impermeable and is essentially a solid plate, which is likely to be an oversimplification—in reality some soil degassing is expected to take place. Chapter 10 covers the "Δp effect" in greater detail. Another secondary effect could be perturbations to the fair weather electric field. Dust devils on Earth are known to produce substantial locally enhanced electric fields due to collisional charging and size/charge sorting due to differential transport and sedimentation of sand and dust. The same is likely true for Mars (Barth et al. 2016). An enhanced electric force could help lift particles off the surface.

Additional observational campaigns on Earth and Mars are needed, as are additional modeling studies that can help to interpret observational data. An important step toward determining the processes and conditions necessary for dust devil formation would be to first determine the necessary and sufficient conditions for the total inventory of thermal vortices (both dusty and dustless). Do such universal necessary and sufficient conditions even exist? From within the population of thermal vortices, what fraction form dust devils, and are there threshold kinematic, thermodynamic, or boundary layer conditions that mark the transition from dustless to dusty? Answering these questions implies the ability to inventory the total number of dustless (i.e., invisible) circulations while simultaneously making quantitative measurements of the vortices as well as the larger scale environment in which they are embedded. Remote sensing instrumentation of some combination of radar, lidar and thermal imaging would be helpful in this regard, but frequent vertical soundings via balloon, tethersondes or similar are also important to obtain boundary layer depth, lapse rate, and wind shear conditions. An ideal situation would be remote sensing scans capturing the complete evolution of a vortex emerging from the ambient environment. Since w^*/u^* appears to be an important parameter, and possibly heat flux, these quantities should be a measurement priority for future field experiments.

Of equal importance to the measurement of various atmospheric and surface properties, it is essential to establish standards for measurement. For example, because the lapse rate of the convective boundary layer varies from highly superadiabatic near the surface to more neutral at greater altitudes, it is not sufficient to report a single lapse rate without stating the altitude range over which the value is valid. Furthermore, if different experiments report the lapse rate over different altitudes, it is difficult (perhaps impossible) to quantify the environment in a meaningful, consistent way. Table 2 summarized key parameters reported in numerous studies, and illustrates how a lack of measurement standards hinders direct comparison. It is imperative that the community establish standards for measuring and reporting data. This should be done in collaboration with numerical modelers who requires these data for both model validation for providing boundary conditions and forcing. Consideration might be given to obtaining sufficient information to nondimensionalize measurements as a way to normalize disparate measurements that are likely to come from different future field programs.

On Mars, in situ observations are considerably lacking. The best meteorological data remain that of the Viking Landers (Hess et al. 1977). Only the short-lived Mars Pathfinder mission (Schofield et al. 1997) and the Mars Science Laboratory (Gómez-Elvira et al. 2014) have subsequently carried full meteorological stations. Pathfinder had temperature measurements at different heights above the ground on a 1 m mast, which provided lapse rate information. The wind sensor on the Mars Science Laboratory (MSL) was destroyed on landing

Table 2 Observed and modeled threshold conditions for dust devil formation

	Study	$-h/L$	w_*/u_*	[a]u_{min} (m/s)	[a]u_{max} (m/s)	[a]γ (K/m)
Field Study	Sinclair (1969)	–	–	–	~ 8.9	–
	Ryan and Carroll (1970)	–	–	1.1	> 5.8	–
	Ryan (1972)	–	–	–	–	> 0.1 (0.3–10)
	Deardorff (1978)	100	6.3	–	–	–
	Hess et al. (1988)	50	5	–	–	–
	Hess and Spillane (1990)	76	5.1	0.1	18	–
	Oke et al. (2007)	–	–	1.5	7.5	> 0.9 (0.01–0.25)
	Ansmann et al. (2009)	–	–	–	7	> 0.9 (1–2)
	Kurgansky et al. (2011)	66	5.5	–	8	–
LES	Ohno and Takemi (2010a)	29.5	4.2	–	–	–
	Ito et al. (2010)	9.3	2.9	–	–	–

[a]u_{min} and u_{max} are the minimum and maximum wind speeds at which DDs were observed, respectively; γ is the lapse rate measured between the two heights listed in parentheses (in meters)

and the overall accommodation of the meteorological sensors was poor. The 2020 rover will carry a similar meteorological package to MSL with similar accommodation problems. Data from a fully capable, properly accommodated weather station remains a high priority (MEPAG 2015), but that is unlikely to be realized until at least the mid 2020s, if not later. Until that time, only incremental advances from quantitative dust devil studies comparing Earth and Mars are anticipated.

One thing that is clear is that lapse rate alone is not sufficient to determine whether dust devils will form. Not all unstable atmospheres produce dust devils, even if dust is available to be lifted. Proper energetics are necessary, but not sufficient. The structure of the convection within the boundary layer matters, as determined from parameters like $\frac{u_*}{w_*}$ and h/L. It is the characteristic depth of the convective boundary layer, the relative strength of buoyancy to shear, and parameters normalized by the depth of the of convective boundary layer that are of importance. For the purposes of predicting dust devil activity on a day to day basis, it is necessary to predict the fully evolved state of the convective boundary layer; this cannot be done by simply by looking at a sounding. Nevertheless, some estimate of the bulk Richardson number in the boundary layer, as calculated by the magnitude of the expected lapse rate divided by the magnitude of the shear within the boundary layer, might be informative. Generally, some wind (and therefore some shear which is produced by friction) is advantageous. Likewise, the atmosphere must be neutral and ideally superadiabatic. Thus, large, negative bulk Richardson numbers would tend to be favorable.

In additional to standard measures of atmospheric properties, full documentation of vegetation, soil type, surface albedo and thermal inertia, soil moisture, and surface particle type(s) and distribution should be documented as part of field programs. Local and regional topography may play a key role in initiating vortices and in modulating the overall convective boundary layer and should also be fully documented.

From the modeling side, LES simulations with sufficiently large domains are needed. To date, even the most ambitious modeling studies have made sacrifices in either horizontal domain size, model depth, or resolution (or some combination). The continued advance in computational capabilities will fill this need over time.

One promising technique that could partial mitigate the trade problem between model domain and model resolution is the implementation of adaptive grids. Adaptive grids analyze the flow and automatically refine the computational mesh based on predetermined criteria—for example, strong gradients or regions of high vorticity. Adaptive grid refinement has largely been applied to applied engineering problems (e.g., flow around buildings), but the technique may be generalized to atmospheric LES and potentially dust devil simulations. The method is not without drawbacks, such as spurious energy, enstrophy and/or vorticity production where modeling grid structures are changing (e.g., Goodfriend et al. 2015)

Larger modeling domains with higher resolution are not a panacea. Even when computer resources allow for adequate domains and resolution on a regular basis, LES models will generally still be unable to properly represent mesoscale and large-scale forcing (e.g., circulations within basins and synoptic wave influences). They will also not be able to fully represent local or regional topographic influences.

In the context of available observational evidence and modeling work on Earth and Mars, dust devil formation requires conversion of potential energy to kinetic energy following, roughly, the thermodynamic Carnot engine model. In doing so, however, the dynamics are governed by the vorticity equation, which determines the origin and amplification of a vortex circulation. A significant amount of vorticity appears to be derived from the ambient environment. Horizontal vortex tubes generated by the vertical wind shear profile can be tilted and stretched by updrafts. This dynamical process may be sufficient for some dust devils, but baroclinic vorticity generated along warm updraft boundaries may also be an important term. Further, the structures in which convection are embedded can provide a source of vorticity, with near surface convergence/divergence boundaries providing vorticity already oriented in the vertical. The strongest source of this vorticity is generally found at the vertices of intersecting convective cells. Merging vortices may either combine to strengthen or weaken vorticity depending on the sign of the vortices. This entire description remains more qualitative rather than quantitative, and it will likely remain so until observations can adequately constrain the vorticity and energy budget of both the overall convective boundary layer and the convective vortices within.

References

A. Ansmann et al., Vertical profiling of convective dust plumes in southern Morocco during SAMUM. Tellus **61B**, 340–353 (2009)

A. Azizov, Influence of soil moisture on the resistance of soil to wind erosion. Sov. Soil Sci. **9**, 105–108 (1977)

M. Balme, R. Greeley, Dust devils on Earth and Mars. Rev. Geophys. **44**, RG3003 (2006). doi:10.1029/2005RG000188

M. Balme, A. Hagermann, Particle lifting at the soil-air interface by atmospheric pressure excursions in dust devils. Geophys. Res. Lett. (2006). doi:10.1029/2006GL026819

M. Balme, S. Metzger, M. Towner, T. Ringrose, R. Greeley, J. Iversen, Friction wind speeds in dust devils: a field study. Geophys. Res. Lett. **30**(16) (2003). doi:10.1029/2003JE002096

E.L. Barth, W.M. Farrell, S.C.R. Rafkin, Electric field generation in Martian dust devils. Icarus **268**, 253–265 (2016). doi:10.1016/j.icarus.2015.12.020

L.J. Battan, Energy of a dust devil. J. Meteorol. **15**, 235–237 (1958)

H.B. Bluestein, A.L. Pazmany, Observations of tornadoes and other convective phenomena with a mobile, 3 mm wavelength, Doppler radar: the spring 1999 field experiment. Bull. Am. Meteorol. Soc. **81**, 2939–2952 (2000)

H.B. Bluestein et al., Doppler radar observations of dust devils in Texas. Mon. Weather Rev. **132**(1), 209–224 (2004)

I.S. Bowen, The ratio of heat losses by conduction and by evaporation from any water surface. Phys. Rev. **27**, 779–787 (1926)

T. Broersen, Quantification of soil erosion by dust devil in the Jordan Badia. Master's Thesis, Utrecht Univ. (2013)

B.A. Cantor, K.M. Kanak, K.S. Edgett, Mars Orbiter Camera observations of Martian dust devils and their tracks (September 1997 to January 2006) and evaluation of theoretical vortex models. J. Geophys. Res. **111**, E12002 (2006). doi:10.1029/2006JE002700

T.A. Cortese, S. Balachandar, High performance spectral simulation of turbulent flows in massively parallel machines with distributed memory. Int. J. High Perform. Comput. Appl. **9**(3), 187–204 (1995)

J.W. Deardorff, Observed characteristics of the outer layer, in *Short Course on the Planetary Boundary Layer*, ed. by A.K. Blackadar (Am. Meteorol. Soc., Boston, 1978), 101 pp.

K.A. Emanuel, An air-sea interaction theory for tropical cyclones. Part I: Steady-state maintenance. J. Atmos. Sci. **43**(6), 585–605 (1986)

F. Fecan, B. Marticorena, G. Bergametti, Parametrization of the increase of the aeolian erosion threshold wind friction velocity due to soil moisture for arid and semi-arid areas. Ann. Geophys. **17**, 149–157 (1999)

L.K. Fenton, T.I. Michaels, Characterizing the sensitivity of daytime turbulent activity on Mars with the MRAMS LES: early results. Mars **5**, 159–171 (2010). doi:10.1555/mars.2010.0007

L.K. Fenton, R. Lorenz, Dust devil height and spacing with relation to the martian planetary boundary layer thickness. Icarus **260**, 246–262 (2015)

B.H. Fiedler, K.M. Kanak, Rayleigh-Benard convection as a tool for studying dust devil formation. Atmos. Sci. Lett. **2**, 86–93 (2001). doi:10.1006/asle.2001.0043

J.A. Fisher et al., A survey of Martian dust devil activity using Mars Global Surveyor Mars Orbiter Camera images. J. Geophys. Res., Planets **110**(E3) (2005)

W.D. Flower, *Sand Devils* (HM Stationery Office, London, 1936)

C. Fujiwara, K. Yamashita, M. Nakanishi, Y. Fujiyoshi, Dust devil-like vortices in an urban area detected by a 3D scanning Doppler lidar. J. Appl. Meteorol. Climatol. **50**(3), 534–547 (2011). doi:10.1175/2010JAMC2481.1

B.T. Gheynani, P.A. Taylor, Large-eddy simulations of vertical vortex formation in the terrestrial and martian convective boundary layers. Bound.-Layer Meteorol. **137**(2), 223–235 (2010). doi:10.1007/s10546-010-9530-z

B.T. Gheynani, P.A. Taylor, Large eddy simulation of typical dust devil-like vortices in highly convective martian boundary layers at the Phoenix lander site. Planet. Space Sci. **59**(1), 43–50 (2011). doi:10.1016/j.pss.2010.10.011

J. Gómez-Elvira et al., Curiosity's rover environmental monitoring station: overview of the first 100 sols. J. Geophys. Res., Planets **119**, 1680–1688 (2014). doi:10.1002/2013JE004576

E. Goodfriend, F.K. Chow, M. Vanella, E. Balaras, Improving large-eddy simulation of neutral boundary layer flow across grid interfaces. Mon. Weather Rev. **143**(8), 3310–3326 (2015)

R.M. Haberle et al., A boundary-layer model for Mars: Comparison with Viking lander and entry data. J. Atmos. Sci. **50**, 1544–1559 (1993)

G.D. Hess, K.T. Spillane, Characteristics of dust devils in Australia. J. Appl. Meteorol. **29**, 498–507 (1990)

S.L. Hess, R.M. Henry, C.B. Leovy, J.A. Ryan, J.E. Tillman, Meteorological results from the surface of Mars: Viking 1 and 2. J. Geophys. Res. **82**(28), 4559–4574 (1977). doi:10.1029/JS082i028p04559

G.D. Hess, K.T. Spillane, R.S. Lourensz, Atmospheric vortices in shallow convection. J. Appl. Meteorol. **27**, 305–317 (1988)

D.P. Hinson, M. Pätzold, S. Tellmann, B. Häusler, G.L. Tyler, The depth of the convective boundary layer on Mars. Icarus **198**(1), 57–66 (2008). doi:10.1016/j.icarus.2008.07.003

J. Ito, R. Tanaka, H. Niino, M. Nakanishi, Large Eddy simulation of dust devils in a diurnally-evolving convective mixed layer. J. Meteorol. Soc. Jpn. **88**(1), 63–77 (2010). doi:10.2151/jmsj.2010-105

J. Ito, H. Niino, M. Nakanishi, Effects of ambient rotation on dust devils. Sci. Online Lett. Atmos. **7**, 165–168 (2011). doi:10.2151/sola.2011-042

J. Ito, H. Niino, M. Nakanishi, Formation mechanism of dust devil-like vortices in idealized convective mixed layers. J. Atmos. Sci. **70**(4), 1173–1186 (2013). doi:10.1175/JAS-D-12-085.1

R.L. Ives, Behavior of dust devils. Bull. Am. Meteorol. Soc. **28**, 168–174 (1947)

N.L. Jackson, K.F. Nordstrom, Effects of time-dependent moisture content of surface sediments on aeolian transport rates across a beach, Wildwood, New Jersey, U.S.A.. Earth Surf. Process. Landf. **22**, 611–621 (1997)

B.C. Jemmett-Smith et al., Quantifying global dust devil occurrence from meteorological analyses (2015)

M.V. Kalashnik, K.N. Visheratin, Cyclostrophic adjustment and nonlinear oscillations in the core of an intense atmospheric vortex. Izv., Atmos. Ocean. Phys. **46**(5), 591–596 (2010)

K.M. Kanak, Numerical simulation of dust devil-scale vortices. Q. J. R. Meteorol. Soc. **131**, 1271–1292 (2005a)

K.M. Kanak, Numerical simulation of dust devil-like vortices in the terrestrial and Martian convective boundary layers. Paper presented at Dust Devils on Earth and Mars: Workshop, Ariz. State Univ. Planet. Geol. Group, Flagstaff (2005b)

K.M. Kanak, On the numerical simulation of dust devil-like vortices in terrestrial and martian convective boundary layers. Geophys. Res. Lett. **32**, L19S05 (2006)

K.M. Kanak, D.K. Lilly, J.T. Snow, The formation of vertical vortices in the convective boundary layer. Q. J. R. Meteorol. Soc. **126**(569), 2789–2810 (2000)

M. Klose, Y. Shao, Large-eddy simulation of turbulent dust emission. Aeolian Res. **8**, 49–58 (2013). doi:10.1016/j.aeolia.2012.10.010

Klose et al., Space Sci. Rev. (2016). Chapter 11, this issue. doi:10.1007/s11214-016-0261-4

M.V. Kurgansky et al., Space Sci. Rev. (2016, this issue). doi:10.1007/s11214-016-0281-0

M.V. Kurgansky et al., Micrometeorological conditions for dust-devil occurrence in the Atacama Desert. Bound.-Layer Meteorol. **138**(2), 285–298 (2011). doi:10.1007/s10546-010-9549-1

M.P. Lelong, J.C. McWilliams, Cyclostrophic adjustment in shallow-water and Boussinesq fluids. J. Fluid Mech. (1999)

R.D. Lorenz, Irregular dust devil pressure drops on Earth and Mars: effect of cycloidal tracks. Planet. Space Sci. **76**, 96–103 (2013). doi:10.1016/j.pss.2013.01.001

R.D. Lorenz, D. Christie, Dust devil signatures in infrasound records of the International Monitoring System. Geophys. Res. Lett. **42**, 2009–2014 (2015). doi:10.1002/2015GL063237

R. Lorenz et al. Chapter 8 (2016a, this issue)

R. Lorenz et al. Chapter 1 (2016b, this issue)

T.J. Lyons et al., Clearing enhances dust devil formation. J. Arid Environ. **72**, 1918–1928 (2008)

P.J. Mason, Large-eddy simulation of the convective atmospheric boundary layer. J. Atmos. Sci. **46**(11), 1492–1516 (1989)

J.B. McGinnigle, Dust whirls in north-west Libya. Weather **21**, 272–276 (1966)

S.M. Metzger, J.R. Carr, J.R. Johnson, T.J. Parker, M.T. Lemmon, Dust devil vortices seen by the Mars Pathfinder camera. Geophys. Res. Lett. **26**(18), 2781–2784 (1999). doi:10.1029/1999GL008341

MEPAG, Mars Scientific Goals, Objectives, Investigations, and Priorities: 2015. V (2015)

T.I. Michaels, S.C.R. Rafkin, Large-eddy simulation of atmospheric convection on Mars. Q. J. R. Meteorol. Soc. **130**(599), 1251–1274 (2004). doi:10.1256/qj.02.169

I. Michelson, On the dust devils. Diss., California Institute of Technology (1951)

J.E. Moores et al., Observational evidence of a suppressed planetary boundary layer in northern Gale Crater, Mars as seen by the Navcam instrument onboard the Mars Science Laboratory rover. Icarus **249**, 129–142 (2015). doi:10.1016/j.icarus.2014.09.020

M.C. Neuman, Effects of temperature and humidity upon the entrainment of sedimentary particles by wind. Bound.-Layer Meteorol. **108**(1), 61–89 (2003)

H. Ohno, T. Takemi, Mechanisms for intensification and maintenance of numerically simulated dust devils. Atmos. Sci. Lett. **32**, 27–32 (2010a). doi:10.1002/asl.249

H. Ohno, T. Takemi, Numerical study for the effects of mean wind on the intensity and evolution of dust devils. Sci. Online Lett. Atmos. **6A**, 5–8 (2010b). Special edition. doi:10.2151/sola.6A-002

A.M.C. Oke et al., Will-willies in the Australian landscape: the role of key meteorological variables and surface conditions in defining frequency and spatial characteristics. J. Arid Environ. **71**, 201–215 (2007)

S. Raasch, T. Franke, Structure and formation of dust devil-like vortices in the atmospheric boundary layer: a high-resolution numerical study. J. Geophys. Res. **116**(D16), 1–16 (2011). doi:10.1029/2011JD016010

S.C.R. Rafkin, R.M. Haberle, T.I. Michaels, Mars regional atmospheric modeling system: model description and selected simulations. Icarus **151**(2), 228–256 (2001). doi:10.1006/icar.2001.6605

D. Reiss, R.D. Lorenz, Dust devil track survey at Elysium Planitia, Mars: implications for the InSight landing sites. Icarus **266**, 315–330 (2016)

D. Reiss, A. Spiga, G. Erkeling, The horizontal motion of dust devils on Mars derived from CRISM and CTX/HiRISE observations. Icarus **227**, 8–20 (2014). doi:10.1016/j.icarus.2013.08.028

N.O. Renno et al., A simple thermodynamical theory for dust devils. J. Atmos. Sci. **55**, 3244–3252 (1998)

N.O. Renno et al., MATADOR 2002: a pilot field experiment on convective plumes and dust devils. J. Geophys. Res. **109**(E7) (2004). doi:10.1029/2003JE002219

J.A. Ryan, Relation of dust devil frequency and diameter to atmospheric temperature. J. Geophys. Res. **77**(36), 7133–7137 (1972)

J.A. Ryan, J.J. Carroll, Dust devil wind velocities: mature state. J. Geophys. Res. **75**(3), 531–541 (1970)

J.T. Schofield, J.R. Barnes, D. Crisp, R.M. Haberle, S. Larsen, J.A. Magalhaes, J.R. Murphy, A. Seiff, G. Wilson, The Mars Pathfinder atmospheric structure investigation/meteorology (ASI/MET) experiment. Science **278**(5344), 1752–1758 (1997)

R.S. Scorer, *Environmental Aerodynamics* (Halsted, New York, 1978)

Y. Shao, S. Liu, J. Schween, S. Crewell, Large-Eddy atmosphere—land surface modeling over heterogeneous surfaces: model development and comparison with measurements. Bound.-Layer Meteorol. **148**, 333–356 (2013). doi:10.1007/s10546-013-9823-0

P.C. Sinclair, Some preliminary dust devil measurements. Mon. Weather Rev. **22**(8), 363–367 (1964)

P.C. Sinclair, General characteristics of dust devils. J. Appl. Meteorol. **8**, 32–45 (1969)

P.C. Sinclair, The lower structure of dust devils. J. Atmos. Sci. **30**(8), 1599–1619 (1973)

J.T. Snow, T.M. McClelland, Dust devils at White Sands missile range, New Mexico. 1. Temporal and spatial distributions (1990)

Z. Sorbjan, M.J. Wolff, M.D. Smith, Thermal structure of the atmospheric boundary layer on Mars based on mini-TES observations. Q. J. R. Meteorol. Soc. **135**, 1776–1787 (2009). doi:10.1002/qj.510

A. Spiga, Comment on "Observing desert dust devils with a pressure logger" by Lorenz (2012). Geosci. Instrum. Method. Data Syst. Discuss. **2**(2), 593–601 (2012). doi:10.5194/gid-2-593-2012

A. Spiga et al., Space Sci. Rev. (2016). Chapter 7, this issue. doi:10.1007/s11214-016-0284-x

A. Spiga, S. Lewis, Martian mesoscale and microscale wind variability of relevance for dust lifting. Mars **5**, 146–158 (2010). doi:10.1555/mars.2010.0006

R.I. Sykes, W.S. Lewellen, D.S. Henn, A Numerical Study of the Development of Claud-Street Spacing. J. Atmos. Sci. **45**(18), 2556–2570 (1988)

J.L. Sutton, C.B. Leovy, J.E. Tillman, Diurnal variations of the martian surface layer meteorological parameters during the first 45 sols at two Viking Lander sites. J. Atmos. Sci. **35**, 2346–2355 (1978)

J.E. Tillman, L. Landberg, S.E. Larsen, The boundary layer of Mars: fluxes, stability, turbulent spectra, and growth of the mixed layer. J. Atmos. Sci. **51**, 1709–1727 (1994). doi:10.1175/1520-0469(1994)051<1709:TBLOMF>2.0.CO;2

A.D. Toigo, M.I. Richardson, S.P. Ewald, P.J. Gierasch, Numerical simulation of martian dust devils. J. Geophys. Res. **108**(E6), 1–14 (2003). doi:10.1029/2002JE002002

D.A. Waller, Active dust devils on Mars: a comparison of six spacecraft landing sites. M.S. Thesis, Arizona State Univ. (2011), 138 pp. https://repository.asu.edu/items/9212

N.R. Williams, Development of dust whirls and similar small-scale vortices. Bull. Am. Meteorol. Soc. **29**, 106–117 (1948)

Y. Zhao, Z. Gu, Y. Yu, Y. Ge, Y. Li, X. Feng, Mechanism and large eddy simulation of dust devils. Atmos.-Ocean **42**(1), 61–84 (2004). doi:10.3137/ao.420105

Space Sci Rev (2016) 203:209–244
DOI 10.1007/s11214-016-0281-0

Dust Devil Steady-State Structure from a Fluid Dynamics Perspective

Michael V. Kurgansky[1] · Ralph D. Lorenz[2] · Nilton O. Renno[3] · Tetsuya Takemi[4] · Zhaolin Gu[5] · Wei Wei[6]

Received: 15 September 2015 / Accepted: 22 August 2016 / Published online: 5 September 2016

Abstract Simple analytical models for the flow structure of dust devils in steady state, and a "thermophysical" scaling theory that explains how these flow structures are maintained are reviewed. Then, results from high-resolution numerical simulations are used to provide insights into the structure of dust-devil-like vortices and study the impact of surface roughness on them. The article concludes with an overview of the influence of lofted dust on the flow structure of dust devils and a discussion of open questions.

Keywords Mars · Atmosphere · Dust devils · Vortices

✉ M.V. Kurgansky
kurgansk@ifaran.ru

R.D. Lorenz
Ralph.Lorenz@jhuapl.edu

N.O. Renno
nrenno@umich.edu

T. Takemi
takemi@storm.dpri.kyoto-u.ac.jp

Z. Gu
guzhaoln@mail.xjtu.edu.cn

W. Wei
wei_wei@whut.edu.cn

1 A.M. Obukhov Institute of Atmospheric Physics, Russian Academy of Sciences, Pyzhevsky 3, 119017 Moscow, Russia

2 Applied Physics Laboratory, Johns Hopkins University, 11100 Johns Hopkins Road, Laurel, MD 20723, USA

3 Department of Climate and Space Sciences and Engineering, College of Engineering, University of Michigan, Space Research Building, 2455 Hayward St., Ann Arbor, MI 48109, USA

4 Disaster Prevention Research Institute, Kyoto University, Gokasho, Uji, Kyoto 611-0011, Japan

5 Xi'an Jiaotong University, Xi'an, Shaanxi 710049, China

6 Wuhan University of Technology, Wuhan, Hubei 430063, China

1 Introduction

Atmospheric vortices driven by thermal instability and made visible by the dust and debris they lift while moving with the wind[1,2] over dry desert and steppe terrains (e.g. Balme and Greeley 2006; Balme et al. 2012) are reviewed in this chapter. On Earth, dust devils are a potentially hazardous (e.g. the introductory chapter to this issue, Lorenz et al. 2016; Lorenz and Myers 2005) meteorological phenomenon that affects air quality, atmospheric circulations, and aerosol transport. On Mars, ubiquitous dust devils play an important role in dust lifting, strongly affecting the radiative properties and dynamics of the atmosphere (e.g. Ferri et al. 2003).

Dust devils are forced by buoyancy created by the solar heating of the surface, and their principal source of vorticity is the local wind shear. Dust devils form within rising plumes of warm air in the convective boundary layer (CBL) (e.g. Battan 1958; Ryan and Carroll 1970; Cortese and Balachandar 1993; Shapiro and Kogan 1994; Rennó et al. 1998; Michaels and Rafkin 2004), which are also referred to as "swirling rising plumes" (SRPs) (Zhao et al. 2004). These SRPs originate in areas where the air near the ground is characterized by a superadiabatic temperature lapse rate (Sinclair 1976; Oke et al. 2007—see also the chapter in this issue on dust devil formation by Rafkin et al.). Once formed, intense terrestrial dust devils can extend to about 1–2 km altitude, enabling dust particles to be carried aloft and transported laterally or vertically by the larger scale flow (Hess and Spillane 1990). On Mars, with CBL depths approximately three times larger than on Earth, dust devils can reach height of several kilometers (Fenton and Lorenz 2015; and references therein). According to an existing paradigm (e.g. Willis and Deardorff 1979; Hess and Spillane 1990; Kanak et al. 2000) the tallest dust-devil vortices form in narrow updrafts at the intersections of convective cells within the CBL; these updrafts can reach to the top of the CBL.

Since the sand and dust grains that make dust devils visible (Rennó et al. 1998) have different sizes and shapes, the airflow within dust devils is extremely complex (e.g. Gu et al. 2006; see also Sect. 6 and references therein). As described for a typical case in the literature (Balme and Greeley 2006), the flow in the upper portion of dust devils contains fine dust, while that in their bottom is surrounded by a "skirt" of coarse-grained sand particles.

The dust devil is one of the smaller-scale members of the broad spectrum of columnar atmospheric vortices that includes tornadoes, waterspouts, landspouts, cold air funnels, steam devils, and fire whirls. The dust devil is distinguished from most of these other vortices by the unimportance of wet thermodynamic processes. Whether dust devils are dynamically simpler than moist convective vortices remains uncertain, since dust devils are embedded in a strongly heated boundary layer, which provides unique complexity.

Intense columnar atmospheric vortices have been studied extensively in the fluid dynamical literature. Classical Rankine, Lamb–Oseen and Burgers–Rott analytical vortex models (e.g. Lugt 1983; Newton 2001) are popular among fluid dynamicists and dynamic meteorologists. Dust devils historically attracted less attention than the larger-scale and more violent tornadoes (e.g., Maxworthy 1973). This is evident when inspecting Bengtsson and Lighthill (1982) who provided a comprehensive (up to that date) review of the theoretical modeling and the results of laboratory modeling and field studies of intense atmospheric

[1] The individual dust devil path may deviate from the straight line parallel to the wind direction, as, e.g., evident by dust-devil cycloidal tracks (Reiss et al. 2013).

[2] According to Balme et al. (2012), the difference between the direction of the 2-h averaged dust devil ground velocity and the direction of the 10-m ambient wind is less than 90° in all instances, and only 9 out of 52 measurements show a difference of more than 30°.

vortices. The same is still true today as indicated by the historical review by Lorenz et al. (2016, the introductory chapter in this issue). An analysis of theoretical vortex models applied to terrestrial and Martian dust devils (Cantor et al. 2006) shows significant discrepancies between some model predictions and field measurements. Except for a general review by Balme and Greeley (2006) the work by Cantor et al. (2006) is the only existing notable recent attempt to review certain aspects of fluid dynamics of dust devils. Currently, emphasis is placed on large eddy simulation (LES) of dust devil-like vortices (e.g. Zhao et al. 2004; Kanak 2005, 2006; Ohno and Takemi 2010a, 2010b; Gheynani and Taylor 2010; Raasch and Franke 2011; see more in Sects. 4 and 5). On the other hand, the development of a reliable theoretical fluid dynamical model is imperative for the interpretation of images and measurements of dust devils (e.g. Metzger et al. 2010).

Rotunno (2013) and Davies-Jones (2015) reviewed the fluid dynamics of tornadoes. Many aspects of these reviews and of earlier reviews by Lewellen (1976, 1993) are directly applicable to dust devils, thus they are emphasized and used in this review. Despite the apparent dynamic similarity between tornadic and dust devil vortices, there is an essential difference between them associated with the altitudinal position of their main buoyant forcing features (see also Zhao et al. 2004, and Sect. 4). In tornadoes, the buoyant forcing is due to the release of energy by the condensation of water vapor occurring at a considerable elevation above the ground; whereas in dust devils this forcing is independent of moist processes and is located immediately adjacent to the surface. The convergence of radial air flows in dust devils is confined to a surface layer a few meters thick (e.g. Hess and Spillane 1990), and both the changes in enthalpy of the converging air parcels and the depth of the convective circulation constraint the vertical and tangential velocities of the vortex (Rennó et al. 1998; see also Sect. 3).

The steady-state properties of dust devils are reviewed in this chapter, recognizing the fact they also have stages of growth and decay. Very little information on such a life cycle is available in the literature because it requires following individual dust devils that may last hundreds of seconds and traverse many kilometers, which is nearly impossible in the field. Dust devil vortices have been studied by means of LES in Raasch and Franke (2011)—the evolution of their core pressure drop and vorticity over their lifetimes of ~ 600 s shows growth for about 30 % of their duration, and decay for about 20 %, with a steady state for about half of the dust devil lifetime. Consistently with the steady-state approximation, we do not consider the fluid dynamical work on dust devil formation, discussed by Michelson (1951) (see a reference to Michelson's work in Barcilon 1967b), Barcilon and Drazin (1972), and more recently by Onishchenko et al. (2014). These and other similar articles are more appropriate as part of the review by Rafkin et al. (this issue).

Two observationally based concepts can be laid into the foundation of a work aiming at an analytical treatment of steady-state dust devil-like vortices. First, Balme and Greeley (2006) indicate that the vertical velocity W in dust devils is generally about a quarter of the value of the maximum rotational velocity V_m, that is, the vortex Rossby number (e.g. Barcilon 1967a, 1967b) equals $\alpha \equiv W/V_m \sim 1/4$.[3] Using video of a dust devil observed from close distance in the desert, Ito and Niino (2014) applied particle image velocimetry (PIV) to obtain $V_m = 16\ \mathrm{m\,s^{-1}}$ and $W = 6\ \mathrm{m\,s^{-1}}$, i.e. $\alpha = 3/8$; in this case confirming that $\alpha < 1$.

[3]The vortex Rossby number (also named the helical parameter in Kurgansky (2005)) has an intrinsic (internal) character and essentially differs from the reciprocal of swirl ratio S in Davies-Jones (1973), the latter being an external non-dimensional parameter defined in terms of input variables defining the "vortex simulator" experiment (e.g. the imposed volumetric flow rate $\hat{Q}$ through the convection zone of the radius R in the vortex chamber, and the imposed ambient circulation $\hat{\Gamma}$; so that $S = R\hat{\Gamma}/2\hat{Q}$, see e.g. Rotunno (1984), Bluestein (2013), etc.).

(a)
(b)

Fig. 1 Two characteristic morphological forms of dust devils: (**a**) rope-type vortices and (**b**) vase-type vortices (cf. Schwiesow 1981). Whereas the vase-type vortices are often more intense in their lower part, the rope-type vortices extend to the maximum altitude and transport dust aloft most efficiently. Photos are taken near Huara in Chile's Atacama Desert in January 2009 (see Kurgansky et al. 2011)

Second, it is known from observations (Sinclair 1973) that terrestrial dust devils have a warm core with typical temperature perturbations from 4 to 8 K above ambient background values. In summary, warm-core baroclinic vortices with a less-than-unity vortex Rossby number fit dust-devil steady-state fluid dynamics quite well.

A major challenge to a theory is the fact that dust devils may have two characteristic morphological forms as shown in Fig. 1(a, b).[4] variety of the dust-devil corner flow appearances[5] in the dust devils is discussed in Sect. 5. The rope-like vortex similar to that shown in Fig. 1(a) bears many similarities to the narrowest dust devil (A) studied in Bluestein et al. (2004) and shown in Fig. 2, with the tangential wind profile similar to that of a Rankine combined vortex (see Sect. 2). Whereas, a more diffuse, vase-like vortex similar to that shown in Fig. 1(b) is analogous to the widest dust devil (B) investigated in Bluestein et al. (2004) and shown in Fig. 3, with the angular momentum within the vortex core being concentrated near the radius of maximum tangential wind.

[4]Figure 4 from Balme and Greeley (2006) (image credit S. Metzger, Planetary Science Institute, Tuscon, USA) shows "typical" dust devil morphology ranging from (a) a narrow, tightly defined column (~ 1 m in diameter) through (b) an inverted V-shaped dust cloud with a less well defined column (~ 10 m in diameter) to (c) a poorly defined inverted V shape with no visible internal column (~ 50 m in diameter at ground). In this paper, two morphological types of dust devils are considered, corresponding to (a) and (b) + (c) types in Balme and Greeley (2006); cf. Schwiesow (1981).

[5]The corner flow domain is the part of the vortex flow, which is immediately below the vortex core and is adjacent to the ground. In the corner flow domain, the main vortex updraft adjusts to the inward air motion within in the surface boundary layer beneath the peripheral vortex flow.

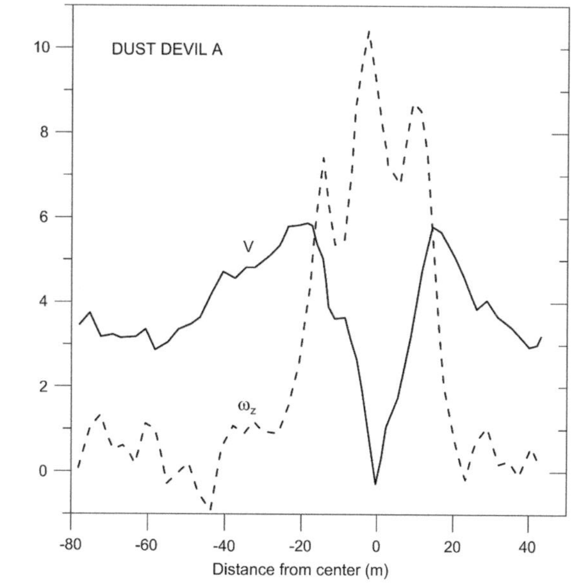

Fig. 2 Approximate vortex-relative azimuthal wind component V (*solid line*; $\mathrm{m\,s^{-1}}$) and vertical vorticity ω_z (*dashed line*; $\times 10\ \mathrm{s^{-1}}$) in dust devil A, at 1515:33 CDT, as a function of distance from the *center* of the vortex (m). Negative (positive) distances are measured to the *left* (*right*) of the vortex center, as viewed by the radar. The vorticity at the *center* is estimated as the average of the vorticities computed just to the *right* and *left* of the *center* (adapted from Bluestein et al. 2004)

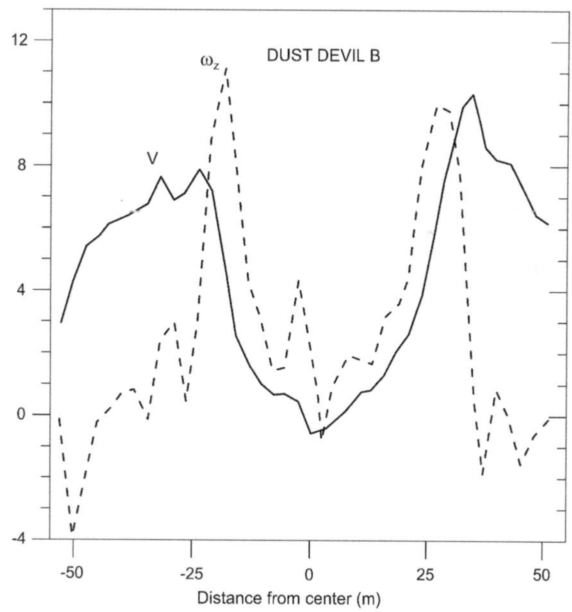

Fig. 3 As in Fig. 2, but for dust devil B at 1518:01 CDT, when it was most intense (adapted from Bluestein et al. 2004)

Here we focus more on tightly organized "vortex chimneys", or tube (rope) vortices of nearly cylindrical shape, which extend high aloft without significant radial expansion and that are efficient in transporting dust into the atmosphere (Fig. 1(a)). However, keeping in mind possible applications to dust devils of other morphological types, different distributions of angular momentum within the vortex core will be considered.

Zhao et al. (2004) argue that columnar, rope-type vortices occur over smooth ground, whereas inverse-cone vortices occur over rough, sandy surfaces such as those with bushes

Fig. 4 Photo taken on 1 December 2009 at 1600 h local time in Chile's Atacama Desert with a front-looking onboard camera on a chase vehicle. A large, poorly structured dust-laden vortex is accompanied with a tightly-organized, small "tube" column (Metzger et al. 2010; Kurgansky et al. 2011)

(see Sect. 5 of this review). Kurgansky et al. (2011) observed that vortices with quite different appearances might sometimes occur over virtually identical surfaces (see Fig. 4). Here, we add that for inverse-cone, diffuse vortices the impact of buoyancy is supposedly relatively minor.

In theoretical studies, tornadoes and dust devils are usually idealized as tall and thin columnar vortices. The vortex flow is assumed axisymmetric, vertical diffusion of momentum is neglected because it is small compared to the radial diffusion, and the radial component of momentum balance equation is approximated by the cyclostrophic balance equation. The similarity approach is often used to solve the problem, and the vertical velocity is, as a rule, assumed to be a linearly increasing function of height to compensate the radial converging flow at the lowermost portions of the vortex. These classical vortex models were constructed not only for homogeneous fluids with constant density (Long 1958, 1961) but also for baroclinic (buoyant) fluids which allow for the treatment of the convective forcing, albeit in a schematic way (Gutman 1957; Kuo 1966, 1967); see also Barcilon (1967b) and Nikulin (1980). Paradoxically, radial turbulent momentum exchange can result not in the smoothing out but, instead, in the sharpening of the tangential velocity profile, similar to the action of "negative viscosity" (Lilly 1969). This effect is discussed in more detail below.

For the modeling of dust devils, the assumption of linear dependence of the vertical velocity on height may be restrictive and can realistically be applied only to a shallow, few meters thick, corner flow region. Beyond this region, an assumption of complete similarity of the wind profile at different altitudes is possible, but it eliminates the possibility to reproduce the downdraft, which reportedly develops in the central part of vortex core starting from a certain altitude, though some controversy concerning this issue remains. Certain observational evidence for such a downdraft comes from field studies by Sinclair (1973) and by Kaimal and Businger (1970) and the downdraft motion finds support in some recent LES studies of dust-devil vortices (see also Sects. 4, 5) and laboratory vortex experiments (Snow et al. 1980). We attribute this property to vase-like vortices (Fig. 1(b)), see also Snow (1987), and suggest that updraft motion predominates in tightly organized rope-like vortices.

In the forthcoming sections, we combine fluid dynamic, thermodynamic and numeric simulation approaches to discuss key properties of steady-state dust-devil vortices. The text addresses to a broad community of atmospheric and planetary scientists, including students participating in dust devil research. Simple analytical models of the steady-state structure of dust devil (Sect. 2), a thermophysical theory developed by Rennó et al. (Sect. 3), the structure of dust-devil-like vortices reproduced in LES (Sect. 4), and the impact of surface roughness on dust devil's corner flow structure (Sect. 5) are reviewed. Section 6 briefly discusses the feedbacks of lofted dust on dust-devil vortex flow, and Sect. 7 presents our conclusions and a list of open questions (challenges). Owing to the diversity of the review material, each section begins with a brief introduction.

2 Conceptual Fluid Dynamical Models of Dust-Devil Structure

2.1 A Gallery of Analytical Vortex Models

Among the aforementioned analytical models of vortices, the classical Rankine combined vortex model (Rankine 1901),

$$v(r) = \begin{cases} v_m r/r_m, & r \le r_m, \\ v_m r_m/r, & r > r_m, \end{cases} \tag{1}$$

remains one of the most popular because, perhaps first, it is the oldest (originating in the steam age) and secondly, it describes the broad features of observed vortices not significantly worse than more sophisticated theoretical models. A significant challenge in addressing the relative merits of different vortex models as representative of dust devils is that the stochastic component of flows in real dust devils and the challenges of acquiring measurements of rapidly-varying winds in field conditions mean that the noise in field data is comparable with the differences between different candidate vortex models. Many observations, including those in tornadoes (Wurman and Gill 2000) and dust devils (Bluestein et al. 2004), indicate that the tangential wind v grows linearly with radius r up to the radius r_m at which the wind is maximum v_m, in accord with (1). This can be explained by the vanishing of the small-scale turbulent viscosity in fluid's solid-body rotation. Rennó et al. (1998) state that "dust devils have tangential velocity profiles characteristic of a Rankine vortex". However, this statement of a Rankine vortex being a reasonable prima facie description of the velocity profile should not be taken as a statement that this model is the most accurate representation: this is especially true given that model fits are typically made to velocity profiles with the radius r_m as a free parameter, whereas ideally the vortex wall radius would be independently (e.g. optically) determined. More problematic is the fit of the tangential wind profile for radii greater than the radius of maximum wind. The classical Rankine vortex model assumes potential vortex flow at the vortex periphery, but field observations show examples where the tangential wind decays with radius much more slowly than it would follow from the hyperbolic law characteristic for the Rankine vortex. For example, Tratt et al. (2003) show a well-documented encounter with a devil in Arizona, and find that the tangential windspeed fall-off outside the wall is described as $r^{-1/2}$ rather than r^{-1}. A final deficiency of the Rankine vortex model compared with more sophisticated models is that it is a piecewise model, with an inelegant and unphysical step change in velocity gradient at the wall. Not only is this sharp change smoothed out by viscous effects in real vortices, but other analytic functions are more aesthetic and more convenient to implement, in part because (1) gives no information on the other than v velocity components, since they are ignored.

The Burgers–Rott (BR) vortex (Burgers 1948; Rott 1958; also see, e.g., Newton 2001),

$$\begin{gathered} u = -ar, \qquad v = \frac{\Gamma}{r}\left[1 - \exp\left(-\frac{ar^2}{2\nu_T}\right)\right], \qquad w = 2az, \\ \omega_z = \frac{1}{r}\left(\frac{\partial(rv)}{\partial r} - \frac{\partial u}{\partial \theta}\right) = \frac{2\Gamma}{a^2}\exp\left(-\frac{ar^2}{2\nu_T}\right), \end{gathered} \tag{2}$$

$(u, v, w) \equiv (u_r, u_\theta, u_z)$, with Γ and a as two (positive) constants of dimension L^2T^{-1} and T^{-1} (L is the unit of length and T of time), correspondingly, is often used to describe the (tangential) velocity field v in dust devils (e.g. Cantor et al. 2006). Polar cylindrical coordinates with r as the radius, θ as the azimuth, and z as the altitude are used to describe this model; the axis of symmetry of the vortex flow coincides with $r = 0$. This three-dimensional steady flow is an exact solution to the Navier–Stokes equations (NSE) in which diffusion of vorticity $\boldsymbol{\omega} = (\omega_r, \omega_\theta, \omega_z)$ counterbalances convection and stretching in the z direction. In the above formulas ν_T is the kinematic viscosity, which is molecular in the NSE setup, but taken as a small-scale turbulent eddy viscosity in the context of atmospheric vortices; to be consistent, velocity field should then be considered averaged in an appropriate sense, along with other kinematic parameters of the flow. Unlike the Rankine combined vortex, the BR vortex model tends to capture the mechanism of the vortex origination and maintenance. However, it describes the meridional circulation of the dust devil adequately only inside its core and a short distance beyond it. It has been argued in the literature (e.g. Fig. 8 of Kanak 2005) that the azimuthal velocity profile of BR model matches the high-resolution LES results for dust-devil vortices better than the Rankine combined model does; see more in Sect. 4.

The azimuthal component of the flow in the BR vortex corresponds to the Lamb–Oseen (LO) vortex

$$v = \frac{\Gamma}{r}\left[1 - \exp\left(-\frac{r^2}{4\nu_T t}\right)\right] \tag{3}$$

considered at fixed time t. A general Gaussian vortex model

$$v = \frac{v_m}{x}\left(\sigma^{-2} + 1\right)\left[1 - \exp\left(-\frac{x^2\sigma^2}{2}\right)\right] \tag{4}$$

can also be considered, where v_m is the maximum tangential wind velocity, $x \equiv r/r_m$, r_m is the radius of maximum wind, and $\sigma^2 = ar_m^2/\nu_T, r_m^2/2\nu_T t$, depending on whether BR (Eq. (2)) or LO (Eq. (3)) vortex is considered. In Eq. (4), it follows from the transcendental equation $1 + \sigma^2 = \exp(\sigma^2/2)$ that $\sigma \approx 1.5852$ (e.g. Mallen et al. 2005). In particular, for the BR solution, $r_m \approx 1.585\sqrt{\nu_T/a}$ and $v_m = (\sigma^{-2} + 1)^{-1}\Gamma/r_m \approx 0.715\Gamma/r_m$ (e.g. Kanak 2005).

An analog to the BR vortex with two cells has been found by Sullivan (1959). By "two cells" is meant that the flow does not simply spiral in toward the axis and out along it, as in the one-celled BR vortex, but has a region of reverse flow near the axis. Although many observed dust-devil vortices are two-celled, the Sullivan's vortex solution, due to its analytic complexity, is not widely used in practical applications. We also note here that Kuo's (1967) two-celled closed-form vortex solution has much formal similarity with the Sullivan's vortex.[6]

The radial pressure distribution for the vortex models (2)–(4), contains exponential integrals (e.g. Lewellen 1976), which requires the use of tables of integration. To avoid this difficulty, Vatistas et al. (1991) proposed an empirical formula for the tangential velocity

$$v(r) = \frac{\Gamma r/r_m^2}{[1 + (r/r_m)^{2n}]^{1/n}}, \quad n \geq 1 \tag{5}$$

[6]More detailed and comprehensive accounts of the classical analytical models can be found in reviews within the tornado literature (e.g. Lewellen 1976, 1993), but in this chapter we are focusing more on possible interests and demands of practitioners in dust devil research.

which fits well the observations and in certain particular cases provides a closed-form solution for the pressure (see below). In the limit of $n \to \infty$ Eq. (5) reduces to the Rankine combined vortex (1). The tangential velocity profile (5) is most close to the BR tangential velocity profile (2) for $n \approx 2$; see Fig. 1(a) in Vatistas et al. (1991).

For vortices (1) and (5), the latter for $n = 1, 2$, the pressure drop in the vortex center is twice the pressure drop at the radius of maximum azimuthal wind (e.g. Lorenz and Jackson 2015). More generally, the pressure deviation from the environmental value, as a function of radius r,

$$\Delta p(r) = \Delta P \left(1 + \frac{r^2}{r_m^2}\right)^{-1}, \tag{6}$$

$$\Delta p(r) = \Delta P \left(\frac{\pi}{2} - \arctan\left(\frac{r^2}{r_m^2}\right)\right), \tag{7}$$

is derived from Eq. (5) (for $n = 1, 2$, respectively) and the cyclostrophic balance equation; here, $\Delta P = -\rho \Gamma^2/2r_m^2$ and ρ is the air density.

The dust devil is embedded in a turbulent atmospheric environment. To adequately describe the effect of angular momentum radial mixing, partition of the turbulence into "relative vorticity and angular momentum diffusing elements" was proposed by Bretherton and Turner (1968); see also McEwan (1973). These authors considered the angular momentum balance equation

$$r^2 \frac{\partial v}{\partial t} = \frac{\partial}{\partial r}\left[\nu_T r^3 \frac{\partial}{\partial r}\left(\frac{v}{r}\right) + \chi_T r \frac{\partial}{\partial r}(vr)\right], \tag{8}$$

where $\nu_T > 0$ is the conventional turbulent eddy viscosity coefficient, whereas χ_T describes the angular momentum diffusion. Bretherton and Turner (1968) explain how the presence of preferred direction in the mixing process could induce an anisotropic mixing causing a mean radial flux of angular momentum, which homogenizes the fluid's angular momentum. These authors and McEwan (1973) considered $\chi_T > 0$; hereafter, an ad hoc generalization onto $\chi_T < 0$, i.e. the case of "negative" angular momentum diffusion (e.g. Lilly 1969), is made. To have a non-negatively defined kinetic energy dissipation rate, it should, however, be $\nu_T \geq \chi_T \geq -\nu_T$. In a steady case, the general solution to (8) is the circular motion described by the tangential velocity

$$v = ar^{s_1} + br^{s_2}, \tag{9}$$

where $s_1 = -1$ corresponds to a potential flow and $s_2 = (\nu_T - \chi_T)/(\nu_T + \chi_T)$ corresponds to a vorticity profile distinct from the solid body-like rotation; a and b are two integration constants determined by the boundary conditions. The vortex family under consideration is specified by the solution described by the second right-hand side term in (9) in the vortex core and a potential vortex described by the first right-hand side terms in (9) outside the core

$$v(r) = \begin{cases} v_m (r/r_m)^s, & r \leq r_m, \\ v_m (r_m/r), & r > r_m, \end{cases} \tag{10}$$

where $s \geq 1$. The radius of the vortex core is r_m; it is also the radius at which the maximum tangential velocity, v_m, occurs. The Rankine combined vortex (1) corresponds to $s = 1$ and is obtained for the turbulent vorticity mixing (diffusion) alone: $\chi_T = 0$. Barcilon (1967a), see also Fiedler and Rotunno (1986), considered an axial, steady, inviscid, cylindrical jet forced along the axis of a potential vortex flow in an unbounded fluid. Barcilon assumed that the vertical velocity in the jet is uniform across its cross-section and that at any such cross-section the pressure is a constant dictated by the value of the vortex pressure at the outer edge of the jet and at that same level. The Barcilon model corresponds to $s = \infty$ in (10). This "hollow-core" vortex solution (e.g. Snow and Pauley 1984), with irrotational (stagnant) core and all the vorticity being concentrated at the core edge, is obtained for $\chi_T = -\nu_T$. Dust devils (A) and (B) from Bluestein et al. (2004) shown in Figs. 2 and 3 are closer to a Rankine-like and a "hollow-core"-like vortex, correspondingly.

We have considered several analytical formulations for the dust-devil velocity structure ranging from the idealized Rankine combined vortex and the hollow-core vortex to more complex exact vortex solutions of NSE. Nevertheless, the reader may be left wondering "Which model is best?" The answer basically depends on the purpose for which the analytical model is intended to be used. For the purpose of working with experimental data the "middle-complexity" Vatistas et al. (1991) model (5) with $n = 1, 2$ can be recommended. Lorenz (2014) notes that the entire (inner and outer) velocity profile in observed dust devils is well-described by the Vatistas et al. (1991) model. The function (6) was extensively used by Ellehoj et al. (2010) to fit the observed pressure drop profiles for convective vortices and dust devils on Mars. An equally easily handled function (7) serves a good approximation to the pressure profile for the BR vortex (2), which is arguably the model with the strongest theoretical underpinning, since it corresponds to an exact solution of NSE. On the other hand, the Rankine combined vortex and the hollow-core vortex, as well as the broader vortex family (10), can be used in theoretical studies due to their simplicity and overall satisfactory fit to observations. The last remark concerns the irrotational character of peripheral flow (at $r \to \infty$) for all considered analytical vortices. It does not always correspond to observations which show examples of a slower than r^{-1} (usually as $r^{-1/2}$) fall-off of the tangential velocity, especially close to the ground (e.g. Balme and Greeley 2006),[7] but it secures the finiteness of the total pressure load—if one takes the analytic description of surface pressure drop and integrates to infinite radius—and of the power demands (integrating dissipation by surface drag) to sustain the vortex.

The above discussion mainly relates to analytical models of the swirl velocity in dust devil vortices. The question of the meridional circulation in dust devil vortices, including the axial velocity structure, is more complicated. Available exact analytical solutions (e.g. Sullivan 1959; Kuo 1967) are not easy to implement in practice, due to their complexity, and the relevant discussion should necessarily be focused more on results of numerical simulations (including LES) and laboratory experiments. Laboratory experiments with vortex chambers (Ward 1972; Church et al. 1979; etc.) and the supporting numerical simulations (Rotunno 1977, 1979; etc.) suggest that the single most important non-dimensional parameter that governs the tornado-like vortex structure/dynamics is the swirl ratio S; see also footnote 3 in Sect. 1 and Sects. 4, 5. It is the ratio of tangential velocity imposed at the fixed outer radius to the mean vertical velocity of the total flow-through imposed in the vortex chamber (Davies-Jones 1973, 1986). As the swirl ratio increases, simulated vortex structure changes (e.g. Davies-Jones 2015 and references therein). At low S there is a one-cell vortex with updraft flow in its center but when S exceeds a certain (for fixed geometrical parameters) threshold value, then according to the laboratory and numerical simulation results the vortex should possess a two-cell structure, with downdrafts in the vortex center (Fiedler and

[7] In his study of steady-state hurricanes, Riehl (1963) assumed conservation of potential vorticity in the vortex inflow region, i.e. vanishing of the azimuthal component of the frictional force, and on this basis concluded that $r\tau_{\theta z} =$ constant, where $\tau_{\theta z}$ is the shearing stress in the $\theta - z$ plane, taken on the ground surface. Application of the bulk formula then yields $v \propto r^{-1/2}$.

Rotunno 1986). The flow becomes highly asymmetric with formation of multiple vortices at large S. In general, this variety of vortex physical appearances (regimes) is consistent with the Doppler-radar observations of tornadoes and dust devils (e.g. Bluestein 2005). It is, however, difficult to relate the swirl ratio S defined for the vortex chambers and their numerical analogs to the swirl ratio in nature, which is most relevant to atmospheric vortices, in particular, tornadoes and dust devils (cf. Bluestein 2013). Lewellen et al. (2000) introduced the local corner flow swirl ratio S_c, the ratio of a characteristic swirl velocity to a characteristic flow-through velocity in the surface/corner/core portion of the flow in tornado-like vortices. The classic swirl ratio S tends to have smaller values than S_c for similar corner flow behavior, but there is no one-to-one correspondence between the two quantities in general. They measure different properties of the flow, and S_c is affected by other factors (for example, by diverse surface roughness; cf. Davies-Jones (2015)) in addition to the "outer" swirl ratio S (Lewellen et al. 2000).

So far, our discussion was focused on the velocity field. Illuminating is inspection of the vertical vorticity ω_z patterns in steady-state dust devil vortices; see also Figs. 2, 3 and the last formula of (2). Significant ω_z values are mainly explained by concentration of ambient vorticity, due to a converging airflow within the surface layer, and by tilting and/or stretching of vortex tubes in intense vertical motions within the vortex core. Sinclair (1973) calculated ω_z values of about 7–14 s^{-1} from his observations of a dust devil core. LES results by Kanak (2005) with 2-m horizontal grid size show that the order of magnitude of ω_z-extrema is about 1–3 s^{-1}, and peaks in ω_z are often associated with the time and location of a vertical vortex in the simulation. The simulated domain (global) extrema of ω_z are about an order of magnitude greater than observed background values of ω_z, but they are persistently smaller than calculated dust devil core values of ω_z based on observations.[8] Since the times of Sinclair's research of the 1960–1970s, it is suggested in the literature (e.g. Ryan and Carroll 1970; Hess and Spillane 1990; Kanak et al. 2000) that some top-down mechanism, related to convective cells developing in the CBL and transporting momentum downward, should be considered in addition to the regular bottom-up buoyancy forcing in dust devils to account for some of their observed features like, e.g., enhanced, as compared to model expectations, vorticity.

For dust devil vortices, which have cyclonic/anticyclonic sense of rotation in $\approx 50\ \%$ cases, respectively (e.g. Carroll and Ryan 1970), both positive and negative signs of ω_z are met with nearly equal probability. This contrasts to tornadoes which are cyclonic in the absolute majority of cases (e.g. Brooks 1951; Bluestein 1996). It reflects essentially different mechanisms of these two vortices formation. Concerning dust devils, we, first and mostly, refer the reader to the review article by Rafkin et al. (this issue). Second, we only mention that the origination of vorticity in dust devils can be related either to ground surface inhomogeneities, which are as a rule random and do not have a preferred orientation in space, or to vertices of Rayleigh–Bénard convective cells developing within the convective boundary layer (Kanak et al. 2000 and references therein), which generally have a similar (with respect to the initial vorticity sign) parity property.

2.2 Vortex Structure and Intensity

Despite their apparently different nature, size and intensity, dust devils have many constitutional commonalities with tornadoes. According to Davies-Jones (1986) (see also Lewellen

[8]These extrema are about an order of magnitude greater than those in Kanak et al. (2000) where 35-m horizontal grid spacing is used.

1976), there are several different flow regions in a tornado vortex. Two regions are immediately adjacent to the ground surface and affected by the latter. These are the surface boundary layer region (or the plate boundary layer in Barcilon 1967a) and the corner flow region. The main vortex flow region resides above these two regions. Its central part, the vortex core, is located just above the corner flow region; whereas the peripheral part is above the boundary layer region. The uppermost, or top, region of a tornado vortex joins the parent cloud.

Basically the same structure is characteristic for dust devil vortices (Cantor et al. 2006; see their Fig. 27 reproduced from Davies-Jones 1986). According to the thermal wind equation, in a warm-core dust devil vortex the core radius monotonically increases with height, which is accompanied by the vertical vorticity attenuation, and the vortex core region gradually transforms into the top, or exhaust, region (Cantor et al. 2006).

Hereafter, we concentrate mostly on the main vortex flow and exhaust regions in a dust-devil vortex and use inviscid solutions to one-component fluid dynamical Boussinesq equations with full neglect of Coriolis force. The effect of lofted (suspended) dust is considered in Sect. 6.

For a steady axisymmetric vertical vortex flow of a Boussinesq fluid, an integral formula

$$\frac{1}{2}w^2(0,z)\Big|_{z_1}^{z_2} - \int_0^\infty \frac{v^2}{r}dr\Big|_{z_1}^{z_2} = \int_{z_1}^{z_2} b(0,z)dz \tag{11}$$

follows from the momentum balance equations taken under the cyclostrophic approximation (e.g. Kurgansky 2015). Here z_1 and z_2 are two arbitrary altitudinal levels ($z_1 < z_2$) and it is assumed that the non-hydrostatic pressure vanishes at $r \to \infty$; $b(0,z)$ is the buoyancy at the axis of symmetry. Equation (11) remains valid when the turbulent diffusion of vertical momentum is accounted for, but the (radius-dependent) eddy diffusivity vanishes at the vortex axis; compare to the axial boundary layer in Barcilon (1967a). Otherwise, a correction term appears in (11), whose sign depends on whether a one-cell or, alternatively, two-cell vortex flow is considered (Lewellen 1976). In the following, the effect of viscous stresses along the axis $r = 0$ on the vortex net intensity is neglected (see also Davies-Jones 2015). If $z_2 >> z_1$ and the contribution of both v and w at the upper level is negligible compared with that coming from the lower level, it follows from (11) that

$$\frac{1}{2}w^2(0,z_1) + CAPE = \int_0^\infty \frac{v^2(r,z_1)}{r}dr, \qquad CAPE = \int_{z_1}^{z_2} b(0,z)dz. \tag{12}$$

Here, *CAPE* is the acronym for convective available potential energy. Davies-Jones (2015) gives a formula analogous to (12) for a compressible atmosphere, by approximating the potential temperature with its reference value in the radial pressure-gradient force; see also Lewellen (1976). For moist-convective tornadic vortices, the level z_2 coincides with the level of neutral buoyancy (LNB), and according to Davies-Jones (2015; p. 288): "A rotating updraft does not have to overshoot the LNB because the centrifugal force can accelerate the upper-level outflow (Lilly 1969)". In the dust devil framework, the level z_2 may either coincide with the main capping inversion atop a convective boundary layer (CBL), usually at a height of 3 to 4 km over deserts, or more likely, with one of possible multiple secondary inversions within the CBL bulk that limit the dust devil vortex height. While *CAPE* conventionally measures the energy available to lift air parcels, one might alternatively think of *CAPE* as being the energy capable of supporting a pressure deficit within the core of a dust devil. According to (12), the sum of the hydrostatic pressure deficit in a warm-core vortex, described by *CAPE*, and the pressure deficit due to the Bernoulli effect in a downward narrowing warm-core vortex equals the pressure deficit supported by the cyclostrophic balance in the radial direction.

Table 1 K-values for different vortex models considered in this review

	"Hollow-core" vortex	Rankine vortex	Gaussian vortex	Vatistas et al. vortex $n=1$	Vatistas et al. vortex $n=2$
K	1/2	1	$\approx 1.702^{*}$	2	$\pi/2 \approx 1.571$

*For the Gaussian vortex $K = (\ln 2/2)(\sigma^2+1)^2/\sigma^2$, $1+\sigma^2 = \exp(\sigma^2/2)$, $\sigma \approx 1.5852$

We further use (12) to estimate the peak azimuthal velocity at the top of the boundary layer taking $z_1 = \delta$ (e.g. Kurgansky 2005; Davies-Jones 2015); alternatively, $z_1 = \delta$ may correspond to the top of a corner flow region. We introduce short-hand notations $W \equiv w(0,\delta)$, $V(r) \equiv v(r,\delta)$ and write $\int_0^\infty (V^2/r)dr = KV_m^2$, where V_m is the maximum value of $V(r)$ and K is the functional of the chosen tangential velocity profile. To estimate $CAPE$ in (12), consider a vortex with constant buoyancy at the vortex center, $b(0,z) \equiv b(0,\delta) = b_0 = const$, in an altitudinal range $\delta \le z < z_2 = h < Z$. The vortex is embedded in a CBL of height Z. The mean potential temperature is height-constant in the main portion of the CBL; all superadiabatic temperature gradients are confined to a near-surface layer of thickness $\sim\delta$. This view of CBL corresponds to the so-called zero-order bulk model, see e.g. Fedorovich (1995). The vortex transports heat (without loss) from the surface, where heat is absorbed, toward the level $z = h$, where heat is rejected. By using the vortex Rossby number $\alpha = W/V_m$, one gets from (12)

$$V_m^2 = \frac{b_0(h-\delta)}{K - \alpha^2/2}, \tag{13}$$

where, e.g., $K = (s+1)/2s$ for the vortices (10). For other vortex models the K-value can also be computed; see Table 1. By K-metrics, too, the vortex in the last column of Table 1 is closest to the Gaussian vortex (cf. Vatistas et al. 1991).[9] In Eq. (13), $b_0 = g\Delta T/T_\infty$, where g is the acceleration due to gravity, ΔT the temperature increment in the vortex center and T_∞ the ambient air temperature; both ΔT and T_∞ correspond to the height δ.

Formula (13) is reminiscent of the thermodynamically derived formula in Rennó et al. (1998) (see also, Rennó 2008; Sect. 3, Eq. (19)) for the wind speed in the dust devil. More generally, the maximum wind speed, as shown in (13), may be referred to as the "thermodynamic speed limit" in intense atmospheric vortices, particularly in tornadoes (e.g. Fiedler and Rotunno 1986; Rotunno 2013; Davies-Jones 2015; also Sect. 3), if one disregards dependence of V_m on α. Given an observed $CAPE$-value in pre-tornado environmental atmospheric conditions, Eqs. (12) and (13) predict a substantially lower maximum velocity compared to Fujita scale (or, alternatively, enhanced Fujita scale, since 2007) rating. It is a well-known difficulty in tornado research (e.g. Fiedler and Rotunno 1986; Lewellen et al. 2000; Lewellen and Lewellen 2007). By matching the main vortex flow to a steady supercritical end-wall vortex which is adjacent to the Earth's surface, it is possible, in principle, to achieve that an end-wall vortex has a maximum azimuthal velocity approximately 1.7× the maximum azimuthal velocity in the subcritical vortex aloft (Fiedler and Rotunno 1986; see also Lewellen and Lewellen 2007). As explained in Zhao et al. (2004), see also Sect. 5, the lack, with rare exceptions, of a pronounced supercritical end-wall vortex is characteristic for dust

[9] Sinclair (1966, 1973) notes that the Rankine-combined model (1) is capable of explaining approximately 75 % of the total mean pressure drop from the environment to the dust devil center (see also Leverson et al. 1977). It follows from Table 1 that an application of the Gaussian and/or Vatistas et al. (especially for $n = 2$) vortex model gives better results, in this respect.

devils, and (13), as well as Eq. (19) of Sect. 3, provide a satisfactory estimate of their peak velocity. Equations (11)–(13) do not distinguish between the updraft and/or, alternatively, downdraft in the vortex center. They can be applied both to one-celled vortices with upflow along their axis and to two-celled vortices with downflow along their axis and upflow in a surrounding annular region.

For other parameters in (13) being the same, the "hollow-core" vortex ($K = 1/2$) attains a higher intensity than the Rankine vortex ($K = 1$). It might explain why the more intense and thick vortices need to be "hollow-core" ones (Snow et al. 1980; Snow and Pauley 1984). In the Gaussian vortex, all other parameters in (13) being the same, too, the maximum wind speed is less than in the Rankine vortex (in approximately 1.3 times for $\alpha = 0$; e.g., Cantor et al. 2006). By setting $b_0 = 9.81\ \mathrm{m\,s^{-2}} \times 6\ \mathrm{K}/305\ \mathrm{K} \approx 0.192\ \mathrm{m\,s^{-2}}$ (on Earth) and $\alpha = 0.25$ one gets for the Rankine vortex that $h - \delta \approx h \approx 505$ m, if $V_m = 10\ \mathrm{m\,s^{-1}}$ were substituted into (13). For the same vortex height, and other parameters being the same, too, the "hollow-core" vortex would have the peak rotational wind speed $V_m \approx 14.3\ \mathrm{m\,s^{-1}}$ and the Gaussian vortex $V_m \approx 7.6\ \mathrm{m\,s^{-1}}$, respectively (see Table 1). In the laboratory and in the field, secondary vortices form in the sheared upflow region of parent two-celled dust-devils, but the secondary vortices themselves appear to be one-celled (Snow and Pauley 1984). Evidence suggests (ibid) that secondary vortices may produce higher wind speeds than predicted based on an axisymmetric model of the parent dust devil vortex.

To infer the vertical structure of steady-state dust devils in the main vortex flow and exhaust regions, more assumptions are needed. Consider a special case when the relative distribution of velocity components is the same across the vortex at all altitudes (the similarity assumption). It is assumed hereafter that the tangential velocity v at each horizontal level is described by (10), where the radius of maximum winds $r_m = r_m(z)$ is a monotonic increasing function of z; the maximum tangential velocity equals $v_m(z) = \Gamma / r_m(z)$ and the specific angular momentum Γ is constant at the vortex core edge. The vertical velocity w is uniform inside the vortex core at each horizontal level and corresponds to an updraft flow; in the peripheral flow: $w = 0$. Because the dynamic stability caused by fluid rapid rotation inhibits mixing, the vertical volumetric flux $Q(z) = \pi r_m^2(z) w(z)$ is assumed height-independent. It follows from (12) ($z_1 = z, z_2 = h$) that

$$\frac{1}{2}\frac{Q^2}{\pi^2 r_m^4(z)} + b_0(h - z) = K\frac{\Gamma^2}{r_m^2(z)} \tag{14}$$

and for the less-than-unity vortex Rossby number α one has from (14) with good accuracy (which ever increases as $z \to h$)

$$r_m^2(z) \approx K\frac{\Gamma^2}{b_0(h - z)}. \tag{15}$$

For the considered idealized vortex, its core expands with height (see Fig. 5) concordant with many observed dust devils (e.g. Humphreys 1940) and due to Γ-constancy the vertical vorticity concentrates in the vicinity of the Earth's surface.

For a "hollow-core" vortex ($K = 1/2$) the solution (15) coincides with a simple exact solution for horizontal flow in Yih (2007), "with a stagnant funnel consisting of a lighter fluid and narrowing toward the ground, which exhibits the main features of real tornadoes" (ibid). Equation (15) is also applicable to a larger rotating thermal (or a "swirling rising plume", Zhao et al. (2004)), at the bottom of which the dust devil usually originates, regardless of

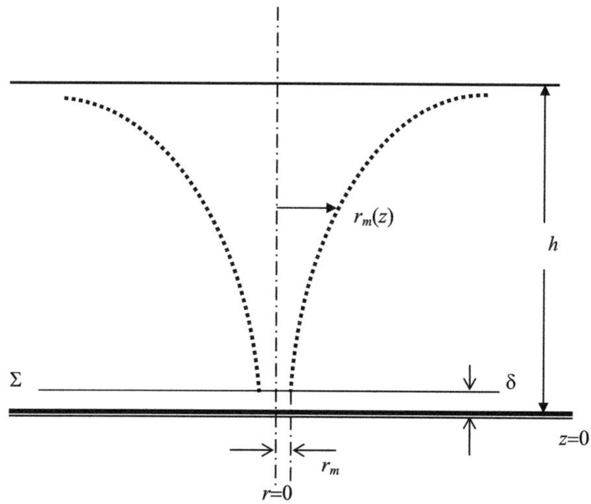

Fig. 5 Sketch of a dry convective Rankine-like vortex with the surface Σ which marks the top of the surface turbulent viscous boundary layer. Vortex cross-section in a vertical plane passing through $r = 0$ is depicted. *Dotted lines* represent the vortex core edge. Other notations are explained in the text. With regard to applications to atmospheric dust devils, the drawing is not in a proper scale, because in the real atmosphere $r_m/h \approx \delta/h \sim 10^{-2}$. Therefore, the *upper part* of the drawing, which is above Σ, must be stretched in vertical by a factor ~ 10 and the sketch actually corresponds to a nearly columnar dust devil vortex in, at least, the lower half of the convective boundary layer (adapted from Kurgansky 2005)

whether or not any dust devil vortex is actually embedded in the thermal.[10] In accordance with observations (Sinclair 1966, 1973; Hess and Spillane 1990) and simple theoretical arguments by Rennó et al. (1998, 2000) and Rennó and Bluestein (2001), Eq. (15) shows that in environments of high wind speed and large horizontal wind shears, with presumably large initial Γ-values, dust devil diameters are biased toward larger values. A detailed study of the statistical distribution of dust devils with respect to their size is the topic of the review by Lorenz and Jackson (2016, this issue).

In a one-cell vortex described by (10) with a uniform vertical flow within the vortex core and the vanishing vertical velocity beyond it (see above), the vertical flux of mechanical energy across the level $z_1 = \delta$,

$$J \equiv 2\pi\rho \int_0^{\infty} \left(\frac{V^2 + W^2}{2} + \varpi\right) W r\,dr = \frac{1}{2}\pi\rho r_m^2 W\left(W^2 - V_m^2\right), \tag{16}$$

is directed upward when $W > V_m$ and downward if $0 < W < V_m$ (Kurgansky 2013).[11] The last condition ensures positive values of V_m^2 in (13) for $K \geq 1/2$, and only this case is considered here. In Eq. (16), ϖ is the non-hydrostatic pressure divided by the air density ρ. The absolute value of flux J in (16) equals the kinetic energy production rate P in the main

[10] Regarding the possibility that the tallest dust-devil vortices form in the vertices of Rayleigh–Bénard convective cells within the convective boundary layer see, e.g., Willis and Deardorff (1979); Hess and Spillane (1990); Kanak et al. (2000).

[11] The first option is more relevant to mechanically driven vortices in vortex chambers (e.g. Ward 1972; Davies-Jones 1973, 1986; Rotunno 1984; and references therein). The second option refers more directly to the atmospheric dust devil.

vortex updraft flow due to positive correlation between the buoyancy and vertical velocity in the warm-core dust devil. For terrestrial dust devils, as a rough estimate, we accept $V_m = 10\ \mathrm{m\,s^{-1}}$, $r_m = 5$ m, $W = 0.25V_m$, $\rho = 1\ \mathrm{kg\,m^{-3}}$ in order to get $P = |J| \sim 9$ kW. For Martian dust devils, a similar, or even somewhat larger, estimate is obtained, because the smaller air density is overbalanced with the increased dust devil vortex size and intensity. Note that for a quasi-steady-state vortex the mechanical work produced must be less than that offered by the Carnot limit, given the heat flux in the dust devil and the temperature boundary conditions; see Sect. 3.

In this review we do not touch a difficult question of hydrodynamic stability of vortex solutions; concerning stability of the "hollow-core" vortex, see Rotunno (1978) and Drazin and Reid (2004). In one particular respect this question relates to the vortex breakdown phenomenon, somewhat analogous to the hydraulic jump. For the analytical vortex models (10) and under an assumption of uniform vertical flow within the vortex core, and of vanishing vertical velocity elsewhere, the condition for the vortex breakdown reads $\alpha^2 > 1/2$; this criterion was derived in Barcilon (1967a) and Fiedler and Rotunno (1986) (see also, Lewellen and Lewellen 2007) for the "hollow-core" vortex, and in Kurgansky (2013) for the Rankine vortex.

We also do not address the problem of detailed analytical treatment of the boundary layer beneath the dust devil's main vortex flow. Since pioneering efforts in this direction (e.g. Barcilon 1967a) progress has been relatively slow. For example, up to now the most comprehensive analytical study of the (laminar) boundary layer induced by a concentrated linear vortex filament with axis normal to a fixed plane (the so-called end-wall vortex) is due to Burggraf et al. (1971); see also Fiedler and Rotunno (1986) in the context of tornado research. An ever increasing emphasis is nowadays being placed on LES studies, and this also applies to the corner flow dynamics of dust devils, as discussed in Sects. 4 and 5 of this review.

Supplementing and elucidating fluid dynamical considerations of Sect. 2 are general thermodynamic arguments of Sect. 3.

3 A Thermophysical Scaling Theory for Dust Devils

3.1 Background

The thermodynamic speed limit, defined as the maximum flow speed V_m that would result from the conversion of convective available potential energy ($CAPE$) into kinetic energy (Bluestein et al. 1993; Fiedler 1994),

$$V_m \approx \sqrt{2 \times CAPE}, \tag{17}$$

has been studied intensively in the context of tornadoes. The thermodynamic pressure drop, defined as the difference between the static pressure in the vortex surrounding and that in its warm core (Fiedler 1994), has also been studied intensively in the context of tornadoes. Bluestein et al. (1993) show that tornadoes usually exceed the thermodynamic speed limit, while Fiedler (1994) shows that supercritical end-wall vortices and vortex breakdown can explain the fact that observed velocities exceed (17) (see also Sect. 2).

Rennó et al. (1998) and Rennó and Bluestein (2001) use a scaling theory based on the analytical Lagrangian theory developed for hurricanes by Emanuel (1986) to estimate the maximum pressure drop in small vortices such as dust devils and waterspouts. The maximum flow velocity is calculated assuming that the vortex is in dynamic (cyclostrophic)

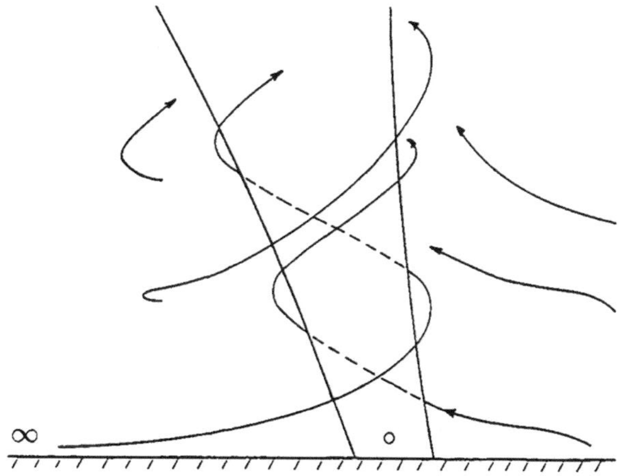

Fig. 6 Sketch of the circulation around a dust devil (Sinclair 1973). The circulation extends to the top of the convective layer (Rennó et al. 1998)

balance. In this approach, the pressure drop is the result of the work of expansion done on air parcels moving from the vortex surroundings towards a stagnation point at their center. Rennó and Ingersoll (1996) show that the work done around a convective circulation is to a first approximation equal to the total work done by buoyancy in the convective updrafts and downdrafts. Thus, the Lagrangian approach is equivalent to the classical approach used by Fiedler (1994) and Bluestein et al. (1993), but results in larger velocity limits mainly because the work done on the descending branch of the circulation (including downdrafts) is also accounted for. The Lagrangian approach developed by Rennó and Ingersoll (1996) and Rennó et al. (1998, 2000) is reviewed in this section.

3.2 Theory

The thermodynamic properties of convective circulations such as those associated with dust devils (see, e.g., Fig. 6) can be calculated using the scaling theories proposed by Rennó and Ingersoll (1996) and Rennó et al. (1998, 2000). According to these scaling theories, the bulk pressure drop from the ambient to the center of a convective vortex such as a dust devil is a measure of its intensity (see also Lorenz and Jackson 2015). These thermophysical scaling theories show that the pressure drop across a dust devil is

$$\Delta p \equiv (p_\infty - p_0) \approx \frac{\gamma \eta c_p p_\infty \Delta T}{R T_\infty}, \tag{18}$$

where p_0 is the surface pressure in the dust-devil center, p_∞ and T_∞ are the surface pressure and temperature at a large radial distance (∞) from the dust-devil center, γ is the fraction of the total dissipation of mechanical energy consumed by friction near the surface, η is the thermodynamic efficiency, c_p is the specific heat capacity of air at constant pressure, R is the gas constant of air, and ΔT is the effective temperature perturbation, defined below by (20).

An upper bound to the thermodynamic efficiency of a dust devil is the Carnot efficiency $\eta = (T_h - T_c)/T_h$, where $T_h \cong T_\infty$ and T_c are the entropy-weighted mean temperatures of the regions where heat is absorbed (the surface air) and rejected (the troposphere). For a dust devil $\eta \approx \Gamma_d Z/T_h$, where Γ_d is the dry adiabatic temperature lapse-rate and Z is the depth of the CBL (Souza et al. 2000).

It follows from (18) that the intensity of a dust devil depends on its height (via the thermodynamic efficiency) and the value of the temperature perturbation across it. The genesis and size of a dust devil, in turn, depends only on the establishment of a pressure drop and the magnitude of the vorticity (Rennó and Bluestein 2001). As air parcels move toward the center of an updraft in the presence of vorticity, they spin while attempting to conserve angular momentum. To a first approximation, the wind within a small convective vortex (with Rossby number $\gg 1$) is in cyclostrophic balance; that is, the pressure gradient force is balanced by the centrifugal force. Surface friction reduces the angular momentum of spinning air parcels moving toward the dust devil center and perturbs the balance between centrifugal and pressure gradient forces. The decrease in the centrifugal force makes the near-surface air to converge toward the dust devil center. A dust devil becomes visible when dust is entrained into the air rotating around it.

Assuming cyclostrophic balance, and using (18) and the ideal gas law, we find that the maximum steady-state tangential wind speed in a dust devil is

$$V_m \approx \sqrt{\gamma \eta c_p \Delta T}. \tag{19}$$

Equation (19) shows that the maximum wind speed in a dust devil depends on its thermodynamic efficiency and the near surface temperature increase across it. Indeed, it does not depend explicitly on the mechanisms responsible for the generation of vorticity. However, the dust devil size is proportional to the value of the background vorticity (Rennó and Bluestein 2001). The potential impact of boundary-layer effects on the maximum wind speed is included in the parameter γ. For larger convective vortices such as tropical cyclones and polar lows, planetary vorticity must be taken into account, but for dust devils it can be neglected (in fact, many early papers on dust-devil observations devote attention to the numbers of cyclonic versus anticyclonic vortices, which generally emerged to be roughly equal).

Rennó and Ingersoll (1996) show that the product of the buoyancy with the distance over which work is produced by rising air parcels is equal to the energy available for a unit mass air parcel to do work, $W = \eta c_p \Delta T$. They used this idea and the Newtonian cooling approximation, to show that the maximum temperature fluctuation associated with convective plumes over homogeneous surfaces is

$$\Delta T \approx \frac{\eta c_p F_{in}}{8 \varepsilon \sigma_R g Z T_c^2}, \tag{20}$$

where F_{in} is the surface heat flux, ε is the emissivity of the convective layer, σ_R is the Stefan–Boltzmann constant, and g is the gravity acceleration.

Souza et al. (2000) show that the intensity of convective circulations forced by surface inhomogeneities in sloping terrains depends on the near-surface nonadiabatic temperature difference across the ascending and descending branches of the circulation and the depth of the convective layer. In this case, ΔT is given by

$$\Delta T = \Delta T_{ac} + \Delta T_{ad} = \Delta T_{ac} - \Gamma_d \Delta z, \tag{21}$$

where the subscripts ac and ad stand for actual and adiabatic; $\Delta T_{ad} = -\Gamma_d \Delta z$ is the temperature drop following a dry adiabat, and Δz is the difference in elevation between the ascending and descending branches of the circulation. This result is consistent with observations that intense convective circulations such as dust devils and (non-rotating) convective plumes are more frequent in regions of large temperature gradients or sloping terrains.

Rennó and Ingersoll (1996) show that assuming that an ensemble of convective plumes and vortices is approximately in radiative-convective equilibrium, their updraft vertical velocity (e.g., the convective velocity) is

$$w \approx \sqrt{\frac{\eta c_p F_{in}}{8\mu\varepsilon\sigma_R T_c^3}}, \tag{22}$$

where $\mu \approx 10$–20 is a dimensionless coefficient of dissipation of mechanical energy.

Equations (18)–(22) provide only order-of-magnitude estimates similar to those that may be obtained from dimensional analysis. The goal of the thermodynamic theory discussed in this section is to provide insights into the physical process governing convective vortices, not to provide accurate predictions.

When the surface is composed of loose materials, dust particles might become airborne making dust devils visible and transporting dust from the surface all the way to the top of the convective layer. Dust is lifted from the surface by saltating dust particles (Bagnold 1941—see also the chapter on dust lifting by Neakrase et al., this issue). During saltation sand grains move in a skipping motion that propels dust particles a few microns in diameter into the air. It follows from Bagnold's findings that the minimum friction velocity u_* necessary to initiate saltation is

$$u_*^3 \approx A\frac{\rho_g g \nu}{\rho}, \tag{23}$$

where $A \approx 1.2 \times 10^2$ is a nondimensional parameter whose exact value generally depends on the angle of repose and the terrain slope, ρ_g is the sand-grain density, ν is the kinematic viscosity of the air, and ρ is the air density. Csanady (1967) derived a simple relationship between the friction and free-stream wind by assuming that the wind joins the frictional boundary layer through a logarithmic velocity profile. It follows from Csanady's study that the free-stream wind speed necessary to initiate saltation is

$$V_a \approx \frac{u_*}{C_D^{1/2}}, \tag{24}$$

where C_D is the surface's drag coefficient and u_* is given by (23). According to Bagnold (1941), C_D ranges from 5×10^{-4} for a smooth surface to 1.5×10^{-3} for a rough surface (see also, Rennó et al. 2004). Thus, dust is lifted more easily from rough than from smooth surfaces. This idea is consistent with the more precise experimental values obtained in wind tunnels (Greeley et al. 1980; Greeley and Iversen 1986).

A necessary condition for the formation of a visible dust devil is the availability of loose particles on the surface and the formation of a vortex strong enough to force saltation. The equations summarized above can be used to predict the possibility of saltation around a convective vortex. Saltation occurs whenever the tangential velocity predicted using (19) is larger than the free-stream wind speed necessary to initiate saltation where a vortex forms predicted using (23) and (24), that is where $V_m \geq V_a$. The collisions of salting particles with the surface and with dust particles cause the transfer of electric charges, while the relative motion of these particles in the convective updrafts causes charge separation. This leads to large electric fields in dust devils (e.g. Crozier 1964, 1970; see also the chapter in this issue by Harrison et al. 2016) that might play a significant role in dust lifting (Rennó et al. 2004; Wei and Gu 2015).

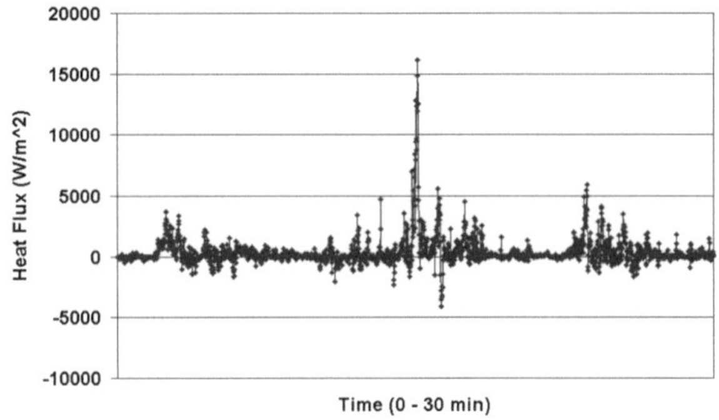

Fig. 7 Surface sensible heat flux for a period of 30 min that includes the passage of a large dust devil (at the center of the plot) (Rennó et al. 2004)

3.3 Measurements and Calculations

Rennó et al. (2004) report results of measurements in dust devils made near Eloy in Arizona, a region of intense boundary layer convection and dust devil activity. In particular, they used the eddy correlation technique to measure the sensible heat flux in convective plumes and dust devils.

The heat flux measured during the passage of a group of convective plumes containing a large dust devil is shown in Fig. 7 (Rennó et al. 2004). The dust devil (indicated by the peak heat flux at the center of the plot) had peak vertical velocity of $\sim 7\ \mathrm{m\,s^{-1}}$ (not shown), temperature perturbations of 3–5 K (not shown) and peak sensible heat flux of $\sim 15\ \mathrm{kW\,m^{-2}}$. Dust devils are extremely efficient in transporting heat (and dust) because the dynamic stability caused by rotation inhibits mixing.

Sinclair's (1973) measurements in dust devils indicate that in a typical day in Arizona's warm season $T_\infty \approx 320$ K and that in a typical large dust devil the pressure perturbation is $\Delta p \approx 3$ hPa, the temperature perturbation is $\Delta T \approx 5$ K, the near surface vertical velocity is $w \approx 10\ \mathrm{m\,s^{-1}}$ and that the tangential velocity is $V_m \approx 10$–$15\ \mathrm{m\,s^{-1}}$. Assuming that a typical value of the surface pressure near Eloy in Arizona where the measurements were done is $p_\infty \approx 925$ hPa, the depth of the convective layer is $Z \approx 3$–4 km, and taking $\gamma \approx 0.5$–1.0 (Rennó et al. 1998), it follows from (18), (19), and (22) that $\Delta p \approx 1.3$–2.7 hPa, $V_m \approx 10$–$15\ \mathrm{m\,s^{-1}}$ and $w \approx 8$–$16\ \mathrm{m\,s^{-1}}$. These values are similar to those measured by Sinclair.

The order of magnitude of the peak sensible heat flux in a dust devil can be estimated as

$$F = \rho c_p \Delta T w. \tag{25}$$

Since, for dry air $c_p = 1004\ \mathrm{J\,kg^{-1}\,K^{-1}}$ and $\rho = 1\ \mathrm{kg\,m^{-3}}$, it follows from (25) that $F \sim 25\ \mathrm{kW\,m^{-2}}$ if $w \sim 5\ \mathrm{m\,s^{-1}}$. This value is of the same order magnitude as the peak value shown in Fig. 7.

For a large dust devil shown in Fig. 7, the maximum kinetic energy production rate P equals the thermodynamic efficiency $\eta \cong gZ/c_pT_\infty \approx 0.09 - 0.12$ times the total sensible heat flux in the vortex and is on the order of 10^2 kW (for the observed vortex radius ~ 10 m); cf. P-estimate of Sect. 2.

4 Structure of Numerically Simulated Dust Devils

Dust devils on Earth and Mars have spatial scales on the order of 1 km or less (i.e., micro-scales); see Table 2 of Kanak (2006). Because of their micro-scales, the numerical simulation of dust devils requires high-resolution models. Since typical terrestrial dust devils have horizontal spatial scales on the order of 1 to 10 m (Kanak 2006), the grid spacing in numerical simulations of dust devils should be on the same order or finer. The LES approaches that have been used for studying dust devils are reviewed in detail by Spiga et al. (this issue). In this section, the structure of quasi-steady dust devils reproduced in LES is briefly described.

Before describing the structure of the simulated dust devils, the characteristics of their vortical structure reproduced in laboratory experiments are reviewed hereafter, by giving more details compared to what was already mentioned very briefly in Sect. 2. In general, dust devils and tornadoes have commonalities in the sense that they are both small-scale[12] vortical flows with concentrated, intense updrafts at their circulation centers. Depending on the intensity of the circulating motion, the form of the vortex changes, as described in Sect. 2. From a laboratory experiment targeted for tornadoes, Ward (1972) showed that the form of the simulated vortex changes from a one-celled, two-celled vortex to multiple vortices with the increase in swirl ratio (about swirl ratio see footnote 3 in Sect. 1 and also Sect. 2). In studying the dynamical aspects of tornadoes, it is convenient to focus on the central part of a vortex within a full system embodying the tornado and its parental convective storm. Using a Ward-type chamber, Church et al. (1979) demonstrated a sequence of transitions of the vortex structure from (i) a laminar one-celled vortex, (ii) a single vortex with breakdown separating upper levels and lower levels, (iii) a fully developed turbulent vortex with breakdown reaching the surface, to (iv) two intertwined helical vortices or (v) a circulating vortex with multiple vortices embedded. Monji (1985) summarized his experimental data obtained by a Ward-type chamber as a vortex regime diagram in the phase space spanned by the Reynolds number and the swirl ratio and indicated that the swirl ratio is more important than the Reynolds number in distinguishing one-celled, two-celled, and multiple vortices. Multiple vortices were also indicated in a larger-scale atmospheric vortex, such as an intense tropical cyclone (e.g. Mitsuta and Yoshizumi 1973), and it was suggested in Mitsuta et al. (1987) that the evolution of a vortex from one-celled, two-celled to multiple vortices is a common feature seen not only in tornadoes but also in dust devils, tropical cyclones, and other atmospheric vortices.

As far as dust-devil-like vortices are concerned, Mullen and Maxworthy (1977) investigated vortex formation and vortex-form transition in a Ward-type laboratory experiment with a heated lower plate and found that a vortex that undergoes transition from one-celled to two-celled structure terminates as a plume. A transition to multiple vortices was not observed by Mullen and Maxworthy. In other words, convective plumes forced by a heated surface become dominant motions that mask the development of multiple vortices. This demonstrates a major difference between dust devils and tornadoes, as far as it concerns the generation mechanism of updrafts: strong surface heating for the dust devil case versus middle-level intense diabatic heating in a parental thunderstorm for the tornado case. From the study of Mullen and Maxworthy (1977), it is suggested that one-celled and two-celled structures are the basic modes of dust devils. It is described hereafter how such structures can be observed in LES.

Examining the structure of dust devils in an LES requires not only a finer grid resolution but also a large computational domain sufficient to cover a CBL. Although an earlier LES study of Kanak et al. (2000) who used a grid spacing of 35 m was not able to sufficiently resolve the internal structure of dust devils, in later studies both the internal structure of dust devils and the spatial extent of the CBL have been resolved due to advances in computational capabilities.

[12] Where the radius is small compared with the atmospheric scale height, and in fact (for dust devils) with the CBL depth.

Toigo et al. (2003) conducted LES of Martian dust devils using a non-hydrostatic mesoscale model, i.e., the Mesoscale Model Version 5 (MM5) (Dudhia 1993), and successfully reproduced the detailed structure of vortices simulated in a 10-m resolution experiment. It was demonstrated that the tangential flow distribution agrees well with the Rankine model, cf. Eq. (1), the radius of maximum tangential wind (RMW) being about 200 m (at a height of 500 m above the ground). Interestingly, strong upward motion was found not at the circulation center but at around RMW. Although downward motion was not seen at the vortex center, the result seems to indicate an intensifying vortex that would evolve into a two-celled structure.

Kanak (2005) used an LES with the grid spacing of 2 m and the computational area of 740 m by 740 m, with extension in the vertical of 1200 m, to investigate the internal structure of terrestrial dust devils. Kanak compared the circulating flow fields of the simulated vortices with two types of theoretical vortex models: the Rankine combined and the Burgers–Rott vortex, see Sect. 2; Eqs. (1) and (2). It was found that the circulating flow of the simulated vortices in their early stages of development, which have no significant vertical motion associated with the vortices, agrees better with the Burgers–Rott vortex model than with the Rankine one (Kanak 2005; her Fig. 8). It was also indicated that the vortices reproduced in the coarser-resolution simulation of Kanak et al. (2000) agree better with the Rankine vortex (ibid; Fig. 7). Thus, it is suggested that the theoretical model to be considered as the best fit to the numerically simulated vortices depends on the horizontal resolution. It was further noted in Kanak (2005) that after evolving to a two-celled structure another analytical solution, like that of Sullivan (1959), is more appropriate.

Although the analyses of Kanak (2005) indicated that the Burgers–Rott vortex model is more appropriate for vortices in high-resolution simulations, the result of Kanak (2005) seems to demonstrate that the Rankine combined vortex model is a good approximation for the simulated vortices at least within the radius of maximum tangential wind (see Figs. 7 and 8 of Kanak 2005). This point is useful in examining the statistical characteristics of dust devils from a sufficient amount of samples. Ohno and Takemi (2010a) conducted an LES of the CBL in a 1 km by 1 km area with 3-m grid spacing to simulate the development and evolution of dust devils using the Weather Research and Forecasting (WRF) model (Skamarock et al. 2008) and investigated the common characteristics of a set of the simulated dust devils. Ohno and Takemi (2010a) used an analytical vortex model to identify a vortical structure within a turbulent, unsteady flow field by separating the vortex flow from the background flow field. They employed the Rankine combined vortex model for its simplicity and good performance to mimic the tangential wind distribution within RMW. In their analysis, the vortical flow was identified by fitting the circulating flow to the Rankine model with a least-square method after searching the local minimum of pressure and the local maximum of vertical vorticity. With this procedure, Ohno and Takemi were able to identify 225 vortices from the simulated outputs generated every 10 second during the time period of 30 minutes after the time when the CBL developed. By compiling a large number of the samples, they successfully demonstrated the statistical characteristics of the simulated dust devils. In this sense, theoretical vortex models are useful for automatically extracting dust-devil-like vortices from the simulated data, and in spite of its simplicity the Rankine combined vortex model performs well.

Ohno and Takemi (2010a) further examined the structure and evolution of intense dust devils. It was shown that the intense vortices are mostly strengthened by the merger of multiple vortices. One case examined in their analysis indicated that the downward motion at the center of the vortex developed after the vortex intensified through the merger of two vortices within the time period of 30 seconds (see their Fig. 3). At this intensified phase, the vortex became two-celled. This intensified vortex was maintained for another 300 seconds, through tilting horizontal vortices. At the bottom of this vortex the strong convergence of horizontal winds was seen, while just above this convergent layer the divergent regions were found. Such a convergence/divergence couplet seems to be a common feature for a well-developed vortex (Zhao et al. 2004). It was also noted that the vortex was elongated vertically up to the height of 300 m by strong updraft.

Ohno and Takemi (2010b) investigated the effects of background wind on the intensity and evolution of dust devils by performing LES in the model settings similar to those in Ohno and Takemi (2010a) but with a uniform background flow imposed and with a slightly coarser grid spacing of 10 m. With the use of the vortex identification procedure of Ohno and Takemi (2010a), the results of Ohno and Takemi (2010b) indicated that among the background winds examined in their LES experiments (0, 5, 10, and 15 $\mathrm{m\,s^{-1}}$) vertical vortices are generated predominantly in medium-wind conditions in terms of their number and intensity (i.e., the cases with the mean wind speed of 5 and 10 $\mathrm{m\,s^{-1}}$). As it follows from the results for the mean wind of 5 $\mathrm{m\,s^{-1}}$, the structure of a simulated dust devil is shown in Fig. 8. This dust devil moves rightward, see Fig. 8(c); asymmetric structure can be seen between the front and the rear side of the moving vortex, which is tilted in the vertical toward the direction of the movement. Upward motion around the vortex center is also asymmetric. Even with the asymmetric structure of the vortex, no upward motion is seen at the center of the vortex, while strong upward motion occurs surrounding the vortex core, which suggests a signature for the transition to a two-celled structure.

5 Impact of Varying Surface Roughness on Dust Devil Structure

Surface friction has profound and non-trivial effect on convective vortices. For example, tornado-like vortices over land are generally more severe over land than over the ocean, probably because of the effect of surface friction (see Dessens 1972 and references therein). Friction affects intensity by producing a central region of strong horizontal convergence and therefore rapidly rising air (ibid). Dessens (1972) and Wilkins et al. (1975) pioneered laboratory studies of the effects of surface friction on vortices. They found that friction reduces the maximum tangential velocity of the vortex and enlarges its diameter. This is in agreement with results of more recent laboratory experiments (e.g. Neakrase and Greeley 2010).[13] The non-monotonic dependence of the tangential wind profile on the surface roughness is less obvious. Low roughness increases the velocity maximum and makes the vortex more concentrated compared to zero-roughness experiments, whereas for medium and high roughness the aforementioned regular tendency dominates (ibid; their Fig. 5). Most importantly, surface roughness increases the difference between the pressure gradient force and the centrifugal force near the ground, thus enhancing air convergence in the surface boundary layer and, therefore, increasing the vertical velocity in the vortex core. Surface friction also makes the flow more turbulent, thus increasing exchange of momentum and, consequently,

[13] However, differently from Dessens (1972) and Wilkins et al. (1975), the vortex core diameter was found to decrease after introduction of surface roughness in the laboratory study of Diamond and Wilkins (1984). Also, PIV measurements by Zhang and Sarkar (2008) show that the magnitudes of maximum tangential velocity and vortex radius both reduce with increasing surface roughness. These and some similar LES results (e.g. Liu and Ishihara 2016, and references therein) show that there is not full consensus as to how the roughness affects the flow near the ground and the net effect may be sensitive to details of the laboratory and/or numerical setup.

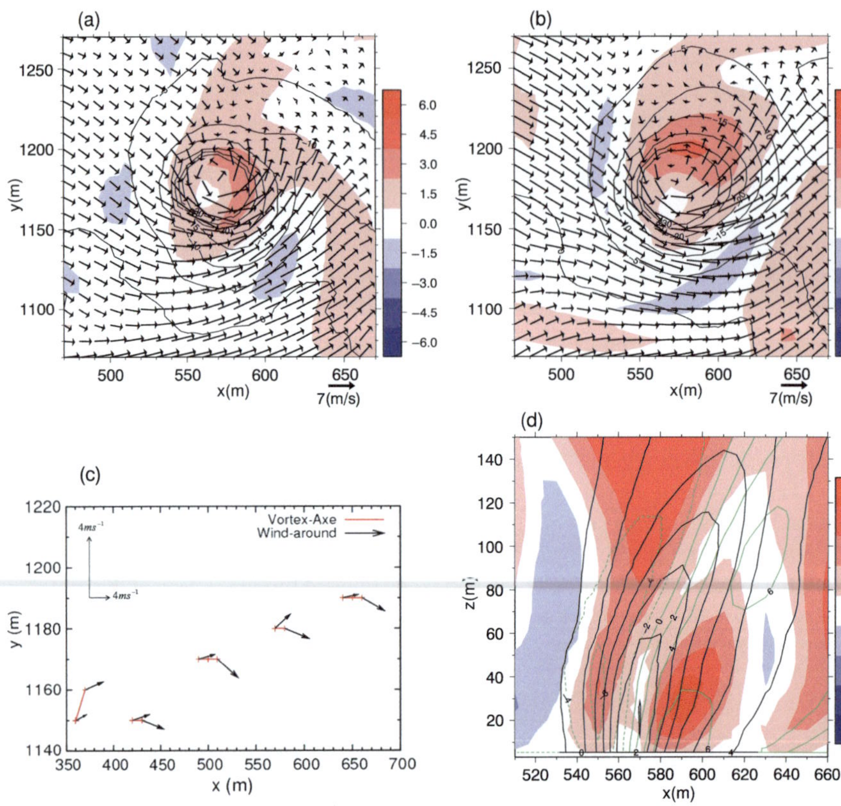

Fig. 8 The structure of a moving dust devil simulated with the background mean wind speed of 5 $m\,s^{-1}$. Horizontal cross sections of the vortex at the height of (**a**) 15 m and (**b**) 50 m above the surface at the time instant $t = 4800$ s. In (**a**) and (**b**), color shading indicates updraft (*red*) and downdraft (*blue*) speeds, *contour* denotes negative pressure perturbations at 5-hPa interval, and *vectors* show horizontal winds. (**c**) Tilted vertical structures of the vortex in the lowest 70 m, indicated by *red lines*, at the time instants $t = 4740$, 4760, 4780, 4800, and 4820 s (from *left* to *right*). A *longer* (*shorter*) *vector* for *each red line* denotes the horizontal wind at the height of 70 (5) m. (**d**) The vertical cross section of the vortex at $y = 1170$ m at $t = 4800$ s. The *color shading* and the *black contour* indicate the same as in (**a**) and (**b**). The *green contour* denotes the y-component of wind (*solid contour* means positive y-component values) (Ohno and Takemi 2010b)

decreasing the sharpness of velocity gradients (cf. Leslie 1977). Intuitively we expect dust devils to decay when they move from smoother to very rough surface. However, other processes must be taken into account, such as the increase in their updraft speed caused by the increase in the surface roughness. Interestingly, Kurgansky et al. (2011) witnessed the sudden intensification of a dust devil when it moved from a smooth sandy surface to a rougher surface (an area covered with dry shrubs). In this case, other processes such as the release of buoyant air trapped between the shrubs could also have caused the observed increase in the intensity of the dust devil.

Dust devils have a variety of corner flow profiles, such as inverse cone or column as shown in Fig. 9; compare Fig. 1 in Sect. 2. The dust devil shown in Fig. 9(a) occurs on a smooth ground surface, it looks like a column with no outward leaning updraft in the corner flow region. The dust devil shown in Fig. 9(b) occurs on a rough ground surface that is covered with bushes; it looks like an inverse cone with strong outward leaning updraft

Fig. 9 Different profiles of dust devils: (**a**) column-like dust devil; (**b**) inverse-cone-like dust devil (Zhao et al. 2004)

in the corner flow region. This suggests that surface roughness is an important factor that determines the profiles of dust devil corner flow.

In fact, the surface roughness is one factor influencing the shape of dust devils. There are others including particularly the vortex swirl ratio, which would affect their structure (see e.g. Lewellen et al. 2000 and references therein). The development stages of the structure of dust devil vortices are discussed in detail by Zhao et al. (2004). In summary (see also Sects. 2 and 4), as the swirl ratio S and hence the angular momentum input to the swirling flow are increased, the rotation core progressively evolves from a one- to a two-celled configuration. The vortex breakdown zone in dust devils (see also Sects. 2, 3 and 4) is, as a rule, positioned close to the ground surface, where the maximum of rotational velocity occurs (Zhao et al. 2004; their Fig. 8). This explains why Eq. (13) of Sect. 2 (derived for a subcritical main vortex flow) and Eq. (19) of Sect. 3 represent the maximum rotational wind speed in a dust devil.

It has been found experimentally that increasing surface roughness also causes transitions from a single vortex to multiple vortices to occur at higher swirl ratios (Leslie 1977), although it remains unclear to what extent the multiple vortex regime is applicable to dust devil vortices (see Sect. 4 and a reference to Mullen and Maxworthy (1977)). In the surface layer, with the increased surface roughness and, consequently, with the greater angular momentum loss, "a parcel of air must enter the convergence zone with a larger tangential component in order to possess the critical amount of swirl by the time it reaches the zone of maximum vorticity" (Leslie 1977). From a general perspective, now directly applicable to dust devils, it suggests that transition from one-celled to two-celled vortices occurs at a higher swirl ratio over the rougher surface.

Numerical simulations of dust devils on Earth (e.g. Smith and Leslie 1976; Kanak et al. 2000; Kanak 2005) and Mars (e.g. Rafkin et al. 2001; Toigo et al. 2003; Michaels and Rafkin 2004) revealed the similar evolutionary mechanism for dust devils on both planets. Since the simulation of dust devils on Earth can be more easily verified by field observations and in situ measurements, the simulation of terrestrial dust devils is more common in the literature. In their seminal study, Kanak et al. (2000) used LES for dust devil exploration. However, their early model is based on a CBL-scale simulation that uses 35-m horizontal grid spacing (see also Sects. 2 and 4). This model can only simulate vertical vortices of rather greater diameter than is typical for most observed dust devils.[14] In general, the CBL-scale simulation is carried out to investigate the genesis of dust devils, that is, to investigate how air flows give rise to and maintain the swirling, converging plume on the hundreds-of-meter scale, which embeds a further developing dust devil vortex. The dust-devil scale simulation is used to get the fine coherent structure of dust devil flow and the properties of velocity and pressure fields; the computational domain is usually confined to or is little larger than the size of a convection cell, setting the initial values of local vorticity via boundary conditions (Willis and Deardorff 1979; Moeng and Sullivan 1994; Zhao et al. 2004; Kanak 2005).

Leslie and Smith (1977) placed an emphasis on the choice of the outer radial boundary condition to simulate dust devils. The free-slip and no-slip lower boundary conditions were considered to investigate the dust-devil vortex strength. In fact, the free-slip boundary is not conceived as a physical possibility. However, from the viewpoint of numerical simulation, the free-slip boundary condition compensates the effect of extra ground torque to reduce the vortex strength. The relative simplicity of solutions obtained without the effects of the frictional lower boundary makes them a convenient starting point for analysis (Rotunno 2013). Zhao et al. (2004) compare the CBL-scale simulation for the convection cells with the dust devil-scale simulation for the airflow in dust devils. They find that given the appropriate initial conditions on the boundary of the computational domain, such as the heat flux and the initial temperature field, the convective cell can be initiated, which readily allows for the development of the airflow of dust devils. In Zhao et al.'s (2004) simulations, a three-dimensional, unsteady, high resolution (dust-devil scale) model utilizes Smagorinsky's sub-grid-scale (SGS) scheme and fine grids; the radial grid spacing smoothly stretches from 0.1 m at the center to 4 m at outer domain and the vertical grid spacing stretches from 0.1 m on the surface to about 10 m at the top.[15] In the simulations by Zhao et al. (2004), all initial and boundary conditions are kept unchanged, except the lower surface condition. The modeled dust devil over the free-slip surface and the no-slip surface is called hereafter for shortness FS and NS, respectively. FS is weak and laminar, whereas NS is violent and turbulent. The structure and intensity of the corner flow of FS and NS is also different. Rotunno (2013) shows that with free-slip lower boundary conditions, two-celled vortices occur in the range of swirl ratios, within which only one-celled vortices exist for no-slip lower boundary. The details of vortex shape difference between free-slip and no-slip lower boundary condition setup can be found in Rotunno (2013).

The tangential velocity on the lateral boundary characterizes the background vorticity in the dust devil-scale simulation (cf. Lewellen et al. 2000; Gu 2010). The reduction in the angular momentum values caused by friction at the surface (e.g., owing to grass, sand, bushes, etc.) results in the vertical distribution of the near-surface tangential velocity on the domain boundary (Leslie and Smith 1977; Vatistas et al. 1991). In Gu et al. (2008) the surface momentum impact height (SMIH) is defined as a parameter to account for the surface friction effect. Accordingly, the tangential velocity on the lateral boundary is a function of SMIH. The profiles of temperature, pressure and three components of the wind velocity (in polar cylindrical coordinates) through the base of a dust devil are obtained from Sinclair's (1973) measurements near Tucson, Arizona. In light of Sinclair's recorded profiles of tangential

[14] It has been observed that typical diameters and near-surface shapes of dust devils may depend strongly on the geographical position of an observational site (Balme and Greeley 2006).

[15] Lewellen et al. (1997) have performed the LES of turbulent transport in the tornado vortex for a particular set of physical boundary conditions that lead to several features resembling those ones found in radar observations by Wurman et al. (1996) and showed that for sufficiently fine grid resolution the simulation results are relatively independent of grid resolution and subgrid model modifications.

Table 2 Modeled dust devils and typical parameters (Gu et al. 2008)

Experiments	SMIH (m)	Ground temperature (K)	C_1, see Eq. (26) ($\mathrm{m\,s^{-1}}$)	Maximum tangential and updraft velocity, V_m and W, and pressure drop, Δp
A	4.0	343	2.5	$V_m = 12\ \mathrm{m\,s^{-1}}$, $W = 16\ \mathrm{m\,s^{-1}}$, $\Delta p = 2$ hPa
B	4.0	343	5.0	$V_m = 20\ \mathrm{m\,s^{-1}}$, $W = 22\ \mathrm{m\,s^{-1}}$, $\Delta p = 4.2$ hPa
C	4.0	323	2.5	$V_m = 10\ \mathrm{m\,s^{-1}}$, $W = 10\ \mathrm{m\,s^{-1}}$, $\Delta p = 1$ hPa
D	4.0	323	5.0	$V_m = 18\ \mathrm{m\,s^{-1}}$, $W = 17\ \mathrm{m\,s^{-1}}$, $\Delta p = 3.6$ hPa
E	2.0	343	2.5	$V_m = 12\ \mathrm{m\,s^{-1}}$, $W = 9.5\ \mathrm{m\,s^{-1}}$, $\Delta p = 3$ hPa
F	6.0	343	2.5	$V_m = 16\ \mathrm{m\,s^{-1}}$, $W = 38\ \mathrm{m\,s^{-1}}$, $\Delta p = 8.5$ hPa

velocity beyond the center of the dust devil (Sinclair 1966), a tangential velocity profile

$$v(z) = C_1\left(1 - e^{-z/C_2}\right) \tag{26}$$

was adopted for the swirling flow condition at the lateral boundary (Gu et al. 2008). In (26) C_1 is the initial tangential velocity, representative of the background vorticity, and C_2 reflects the vertical distribution of the tangential velocity, accounting for the influence of surface roughness on the near-ground flow. Distinct from the surface roughness height, C_2 is identified with SMIH. For the chosen incipient vortex, SMIH and surface temperature, as the parameters describing surface properties, dust-devil-scale LES can be carried out. Details of the LES model, boundary and initial conditions can be referred to Zhao et al. (2004) and Gu et al. (2008). In the computational domain of 200 m (radius) $\times 1200$ m (height), non-uniformly distributed grid points were used. The radial grid spacing stretched from 0.1 m at the center to 5 m on the lateral boundary, and the vertical grid spacing stretched from 0.1 m on the surface to around 15 m at the domain top. The time step is 0.02 s. By using different combinations of the incipient vortex, SMIH and surface temperature, a variety of flow patterns of modeled terrestrial dust devils in their (quasi-steady) matured stage have been reproduced in Gu et al. (2008) to unveil the different shapes and intensities of dust devils in nature. However, the different scales of coherent structures with differing spatial and temporal dimensions (Zurn-Birkhimer and Agee 2005), particularly the details of fine-flow field structure near ground surface, are not well identified for dust devils in Gu et al.'s study. The chosen values of the coefficients in (26), with C_2 representing SMIH are given in Table 2 (the second and fourth columns). The smaller SMIH means more flat and smooth surface, while the larger SMIH represents the uneven ground. Most values of V_m and Δp in the modeled dust devils are close to those in Rennó et al. (1998, 2004); see Gu et al. (2008).[16]

[16]The values of V_m, W and Δp in Table 2 approach the values observed by Sinclair (1973). Experiment F is an exception. Owing to high SMIH, the low (reduced) momentum layer near the ground might be unrealistically thick, which leads to simulation results far from Rennó et al. (2004) and the observations. However, the goal was to demonstrate the influence of SMIH.

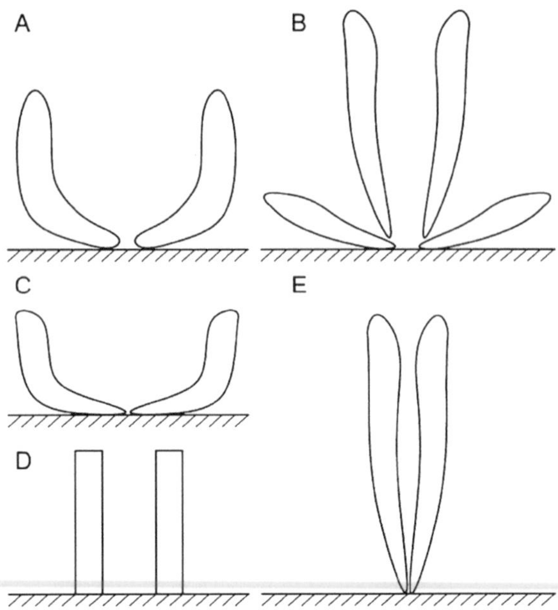

Fig. 10 Near-surface shapes of dust devils simplified by the periphery of the rotating velocity contour of their cores. (**a**) Upside-down cone-like dust devil. (**b**) Almost cylindrical vortex core surrounded by an upside down cone-like dust spray. (**c**) Upside-down cone-like dust devil with enlarged cone. (**d**) Regular cylinder-like dust devil. (**e**) Upside-down thin cone-like dust devil with very small cone angle (Gu et al. 2008)

From the LES results, it appears that the SMIH substantially affects the shape of terrestrial dust devils, changing it from a column on a very smooth surface (Fig. 10(d)), an upside-down cone-like dust devil near the ground (Fig. 10(a)), a bowl-like dust devil on the surface with small SMIH (Fig. 10(c)), to an upside-down thin cone-like dust devil with a very small cone angle on the surface with large SMIH (Fig. 10(e)), whilst the intensity of dust devils increases with a decrease in conic angle. Even for a smaller temperature difference between the local environment and the center of the dust devil, severe dust devils may be formed by strong incipient vortices. According to Gu et al. (2008), Figs. 10(a, b, c, e) illustrate schematically the near-surface shape of the modeled dust devils in Experiments A, D, E, F of Table 2, respectively.

6 Feedbacks of Lofted Dust on Dust Devils

It is important to describe the backward effect of lofted (suspended) dust on the steady-state structure of dust devils that we have so far neglected in this review. This possible influence can be roughly classified, in order of declining importance, as thermodynamic, dynamic (mechanic), and electric.

The thermodynamic effect is a potentially powerful positive feedback on dust devil intensity (Lorenz 2004; Fuerstenau 2006) resulting from direct solar heating of the whirling and rising dust column. It enhances buoyancy and serves as an analog to the latent heat release in ascending moist plumes. In effect, solar absorption by dust is communicated as sensible heat to the rising air in the column, injecting heat into the hot end of the heat engine, and (less efficiently) towards the cold end at higher altitudes. This effect is likely less important for terrestrial dust devils but may be crucial for Martian dust devils which occur in 100 times less dense Martian atmosphere, whereas the suspended dust load in the Martian dust devils may be comparable to that in their terrestrial counterparts (thus, Martian air is more easily heated). On Mars, dust particles over a wide range of sizes settle at the same rate as on Earth (Fuerstenau 2006), since "in the Stokes regime the only aspect of the atmosphere influencing particle drag is viscosity, a property that is independent of pressure and that varies in proportion to the square root of the gas temperature" (ibid). Fuerstenau (2006) estimates the direct solar heating rate for Martian dust devils up to $dT/dt \approx 0.05\ \mathrm{K\,s^{-1}}$. With the use of dust devil characteristic vertical velocity W and dust devil height h, the effective temperature rise along the vortex axis can be estimated as $\Delta T \sim (h/W)dT/dt$. Now, from the formulas of Sect. 2 an order-of-magnitude estimate follows for the maximum rotational velocity $V_m^2 \sim (\beta h^2/W)dT/dt$, $\beta = g/T_\infty$, with T_∞ as the constant reference temperature; compare Kurgansky (2006), where the proportionality factor in this formula, $2/\pi$, is given on the basis of a more detailed treatment of the problem. Upon substitution (for Martian conditions) $g = 3.7\ \mathrm{m\,s^{-2}}$, $T_\infty = 270$ K, $h = 5 \times 10^3$ m, $W = 0.25V_m$, one finds that the value of $V_m \sim 41\ \mathrm{m\,s^{-1}}$ can be explained by this thermodynamic effect. It may be noted that the effect could be quite powerful for devils late in the day, since with the sun low in the sky, a tall optically-thick dust devil will intercept sunlight over an area close to its diameter times its height, which can be a factor of 10–20 higher than its diameter squared (which a dust devil might intercept near noon with the sun overhead). The radiative effect of lofted dust on dust devils specifically has not been explored yet in LES, but related "dust rocket" plumes have been explored in connection with Mars (Spiga et al. 2013). For the temperature excess in the vortex center of $\Delta T_0 = 8$ K (near the ground), which roughly corresponds to Mars Pathfinder measurements (Schofield et al. 1997), the above-obtained value $V_m \sim 41\ \mathrm{m\,s^{-1}}$ can be deduced from (13) if $h \sim 15.3 \times 10^3$ m, which is an unrealistically large value for Martian conditions. In fact, for a considered hypothetical Martian dust devil the radiative effect of lofted dust mainly explains the dust devil high intensity.

In general, relative importance of (i) direct solar heating of dust aloft and (ii) surface heat fluxes can be quantified by a non-dimensional parameter $E = B/(\beta(q/c_p)h)$ (cf. Kurgansky 2006), where B is the buoyancy flux, q is the rate of (solar) heating per unit mass and c_p is the specific heat at constant pressure (i.e., $dT/dt = q/c_p$); two other parameters are explained above. When $E \gg 1$ then the heating q is much less important than the buoyancy flux B, as it occurs for convective vortices without dust. In contrast, in the limit case $E \ll 1$ of dust-laden vortices subject to direct sunlight, the heating q is dominant. For the above-taken example of a Martian dust devil $E \equiv \Delta T_0/\Delta T \approx 8\ \mathrm{K}/25\ \mathrm{K} \approx 0.32$, and it belongs to the latter limit case.

Regarding the mechanical effect of lofted dust we, first, mention the effect of dust centrifuging in the whirling column, second, possible surface drag reduction due to suspended dust in the boundary layer, and third, the frictional dissipation of "suspended" dust falling through the rising air in the column. The former effect is mentioned and discussed in some detail in Bluestein et al. (2004). The second effect is present in dust storms and manifests itself, inter alia, in reduction in the value of von Kármán constant in the Prandtl logarithmic law (e.g. Barenblatt and Golitsyn 1974; see citation in Barenblatt et al. 2005), i.e., at the same external conditions (the same friction velocity u_*) the flow moves faster in comparison with the flow of a pure air. However, upon our knowledge of literature, this effect still remains unexplored systematically in the context of the small-scale atmospheric dust devil. It will be noted that the effect of gravity on dust is to pull it downwards, and the upward drag by the rising air which allows dust to remain aloft for extended periods in a dust devil represents mechanical work done by the devil on that dust. The energy created due to that work is dissipated via viscosity into heat. While small overall, this energy loss term may need to be accounted for, especially on Mars where the terminal velocity of dust is somewhat higher than on Earth.

Lewellen et al. (2008) considered effects of finescale debris on near-surface tornado dynamics. They concluded that centrifuging, negative buoyancy, and angular momentum transport are important mechanisms through which debris can alter the tornado corner flow. These factors are especially essential when the debris mass loading D (in $\mathrm{kg\,kg^{-1}}$) approaches or even exceeds unity, as it can occur in tornadoes. Lewellen et al. (2008) identify three dimensionless parameters governing the debris dynamics in tornadoes. These are: (i) the corner flow swirl ratio S_c (the ratio of a swirl velocity to a flow-through velocity within the surface–corner–core flow (Lewellen et al. 2000; see also Sect. 2)), providing a measure of the "type" of tornado with respect to tornado's shape; (ii) a ratio of a characteristic tornado velocity scale to the terminal velocity of the debris in free fall, A_v, providing a measure of the ease of debris lofting; (iii) and a ratio of a characteristic centrifugal force in the core flow to gravity, A_a, providing a measure of the relative importance of debris centrifuging. The results by Lewellen et al. (2008) suggest that debris cloud mass, spatial dimensions, and the reduction in velocities due to debris loading within a strong, medium-swirl, quasi-steady tornado all rise monotonically with A_v. Concerning dust devils, we note that in the vortex core of terrestrial dust devils the dust mass loading D is usually on the order of $10^{-4}\ \mathrm{kg\,kg^{-1}}$. So, the corresponding negative buoyancy is by two orders of magnitude less than the positive thermal buoyancy acting there. For Martian dust devils, with comparable (to terrestrial ones) lofted dust content (e.g. Fuerstenau 2006), but the two orders of magnitude less dense Martian atmosphere, the lofted-dust-caused negative buoyancy may be on the order of thermal buoyancy but its effect could, in certain circumstances, be overbalanced by the additional positive buoyancy due to direct solar heating of a rising dust column (see above). A region where D-values can—due to the sand blasting—reach appreciable values is the corner flow where a devil's dust skirt is sometimes observed (see also Sect. 5 and an advanced LES study by Gu et al. (2006) of dust lifting within dust devils). We also mention an important difference between tornadoes and dust devils in the discussed context. Whereas, in tornadoes the threshold wind velocity for debris lifting is practically always exceeded, for dust devils (especially Martian ones) it is not necessarily the case, and alternative, more subtle mechanisms for dust lifting into the atmosphere should be accounted for; see more in the review by Klose et al. (2016, this issue) and also in a recent theoretical paper by Wang (2016).

We note the role of electrostatic forces on lofted dust. Electrodynamics of dust devils, including electric and magnetic ponderomotive forces, which act in them, is discussed in much detail in the review by Harrison et al. (2016, this issue). Here, we note that the by far greatest among these forces is the radially directed electric ponderomotive force in a negatively charged (due to the triboelectric effects, see, e.g., Farrell et al. (2004)) main vortex flow. We accept the electric charge vertical density $|q_-| \sim 2 \times 10^{-4}\ \mathrm{C\,m^{-1}}$ that can be inferred (see, Kurgansky et al. 2007) from the value of the macroscopic electric dipole moment, $\sim 1\ \mathrm{C\,m}$, of a dust devil (Farrell et al. 2003; see also Crozier 1964, 1970). Assuming that the electric charge is evenly distributed (well-mixed) across the horizontal cross-section of the vortex core, the repulsive electrostatic ponderomotive force (per unit volume) within the vortex core reads

$$F_E = \frac{|q_-|^2 r}{2\pi^2 \varepsilon_0 r_m^4(z)}, \tag{27}$$

where $\varepsilon_0 \cong 8.854 \times 10^{-12}\ \mathrm{F\,m^{-1}}$ is the electric permittivity of vacuum; $r_m(z)$ is the vortex core radius dependent on height (see Sect. 2). This force can be compared with the fluid dynamic centrifugal force (in notations of Sect. 2),

$$F_C = \frac{\rho \Gamma^2 r}{r_m^4(z)}, \tag{28}$$

for the Rankine combined vortex. By taking $\rho = 1\ \mathrm{kg\,m^{-3}}$, $\Gamma = 50\ \mathrm{m^2\,s^{-1}}$ in (27) and (28) one gets $F_E \approx 0.09 F_C$ for a typical terrestrial dust devil, i.e. the effect of electrostatic force is secondary, though not totally negligible as might seem. On Mars, the electric permittivity $\varepsilon \approx 1.6\varepsilon_0$ of the CO_2 atmosphere should be taken into account. The 100-times smaller value of Martian air density ρ is compensated by an order of magnitude greater Γ value; so, the ratio F_E/F_C remains small, as well. It cannot be excluded that together with the dust grain centrifuging the electrostatic repulsive force F_E favors dust concentration at the vortex core periphery and its lesser abundance in the vortex center, which is often so evident on dust devil images (e.g. Fig. 9(a)).

7 Conclusions and Open Questions (Challenges)

In this chapter we have focused on the review of the structure of dust devils in their mature stage when they can be treated as (quasi-)steady vortices. In Sect. 2, we provided an overview of existing analytic formulations for the tangential velocity field in steady axisymmetric vortex flows, ranging from the idealized Rankine model (which is conceptually useful, but not accurate) to more realistic models such as the Burgers–Rott or Vatistas model (Vatistas et al. 1991). In the current literature (e.g. Kanak 2005; Cantor et al. 2006) these analytic formulations have been shown to be in broad agreement with field measurements and the results of simulations with high-resolution numerical models, like those reported in Sects. 4 and 5. However, a definite set of tests of models with real-world dust devils is lacking so far, nor even exists a systematic study of relationship of dust devil morphology with environmental conditions (the discussion in Sect. 5 of corner flow structure and surface roughness, while not a rigorous test given only a couple of examples, is promising in this respect, however). We suggest that such datasets and analyses are needed in future.

An attempt has been made to estimate the power required to sustain a "typical" steady-state dust devil on Earth and Mars. Based both on the analytic model described in Sect. 2 and the "thermophysical" scaling theory summarized in Sect. 3 we estimate that the power required to sustain a typical dust devil is of the order of 10 kW or more.

An overview of dust-devil-like vortices structures reproduced in LES is provided in Sect. 4. The impact of varying surface roughness on dust devil's corner flow structure is discussed, in terms of LES findings, in Sect. 5.

The feedback of lofted (suspended) dust on the steady-state structure of dust devils is roughly classified in order of declining importance in Sect. 6, as thermodynamic, dynamic (mechanic), and electric. The thermodynamic effect is capable of augmenting the dust devil strength and seems most important in the gigantic Martian dust devils (visible, e.g., from orbit), which are intense and extend in the vertical direction onto distances comparable with the atmospheric scale height. In these gigantic dust devils, an additional, and potentially important, source of buoyancy is the direct absorption of solar radiation by the lofted dust. This effect is expected to be most pronounced in the late afternoon for large Sun zenith angles when the Sun rays are nearly perpendicular to the whirling dust column.

Three challenging tasks that theoretical meteorologists, fluid dynamicists and numerical modelers may face when studying dust devils are worth to be mentioned in the conclusion of this review.

First, contemporary theoretical models have difficulties reproducing the variety of morphological forms of observed dust devils. Indeed, this variety is even more striking for dust devils than for tornadoes, because tornadoes form under meteorological/physical conditions that never are identical, whereas, for example the two morphologically different dust devils illustrated in Fig. 4 occurred under virtually identical environmental conditions. Only small progress has been made so far in the understanding of dust devil morphologies and more attention should be devoted to this subject in the future.

In general, dust devil vortices must move in order to sustain. Otherwise, they would very soon exhaust the available supply of warm air in their neighborhood.[17] As a rule, dust devils travel with the mean wind (cf. Sinclair 1969, 1973; Balme et al. 2012), and this explains why light-moderate environmental winds favor the dust devil occurrence. The vortex movement explains the visible asymmetry between the forward (front) and rear (back) side of the vortex cone as illustrated on the left of Fig. 4 (cf. Bluestein et al. 2004, their Fig. 1). This asymmetry, in a "sloping dust devil", was described by Sinclair (1973, his Fig. 15), and is illustrated in Fig. 6. The asymmetric character of vortices being advected by horizontal wind is frequently observed in the field data (e.g. Fig. 11)—one sees that the negative "attack" slope in the pressure time series, i.e. in front of the dust devil, is shallower than the positive "decay" slope as the pressure recovers to the ambient value in the wake of the devil passage.

This peculiar behavior seems to occur frequently, and it can be speculated that the asymmetry may be stronger when the wind is stronger. Sinclair (1973) also noted this asymmetry, which is well captured by LES studies and is seen in more recent field observations by Lorenz (2012, see also Lorenz and Jackson 2015). Figure 8 demonstrates the structure of a rightward-moving dust devil simulated in ambient with the uniform background wind speed of 5 $\mathrm{m\,s^{-1}}$ by Ohno and Takemi (2010b). Asymmetry is evident, Fig. 8(d) being in accord with Fig. 11(a) showing that the surface pressure decline is shallower than its subsequent recovery. As explained above, this effect is attributed to the inclination of the vortex in the direction of its movement.

Maxworthy (1973) proposed that an essentially infinite source of vertical vorticity for dust devils is the tilting of horizontal vorticity in the atmospheric near-surface layer flow (which is typically a few m thick, i.e. an order of the absolute value of the Obukhov length scale (Hess and Spillane 1990; Kanak 1999; Kanak et al. 2000)) into the vertical direction. In order to acquire this boundary layer vorticity the (cyclonic) dust-devil vortex has to be curved (to resemble a fragment of the vortex ring) and travel to the right of the mean wind. Within this conceptual framework, the idea that such travelling vortices are axisymmetric needs to be abandoned and we need to realize that they are embedded in a convective boundary-layer wind field (see, Maxworthy 1973). As a whole, a satisfactory theoretical model for a moving dust devil-like vortex remains the second challenging task, emphasized in this review.

The third challenge relates to insufficient current understanding of the smaller vertical vorticity found in numerical and conceptual models compared to vorticity of observed dust devils (see also Sect. 2). It should also be noted in this connection that models/simulations of atmospheric CBL with imbedded dust devils usually produce many low-intense (in terms of vorticity) dust-devil-like vortices instead of one or two major intense vortices that would be expected for the same setting (the same domain size and forcing) in the atmospheric CBL. In addition to what has been said in Sect. 2 about the significance of "top-down" mechanism

[17] In this respect dust devils are more similar to hurricanes than tornadoes, the source of energy for the dust devil being the thermal disequilibrium between the surface and the overlying air, which is the same as that in hurricanes (Emanuel 1988; see also Kanak 1999). Emanuel (1988) suggests that dust devils may be maintained in a similar fashion as the hurricanes; that is, by a feedback process between the vortex circulation and surface heat fluxes. Sinclair (1969) came to a similar conclusion by suggesting that vortex translation permits the dust devil to continuously acquire new potential energy sources. Sinclair (1969) explains the observed decrease in dust devil activity directly following a maximum in occurrence, as due to depletion of the surface-adjacent superadiabatic layer, which must recover before dust devil frequency can again increase.

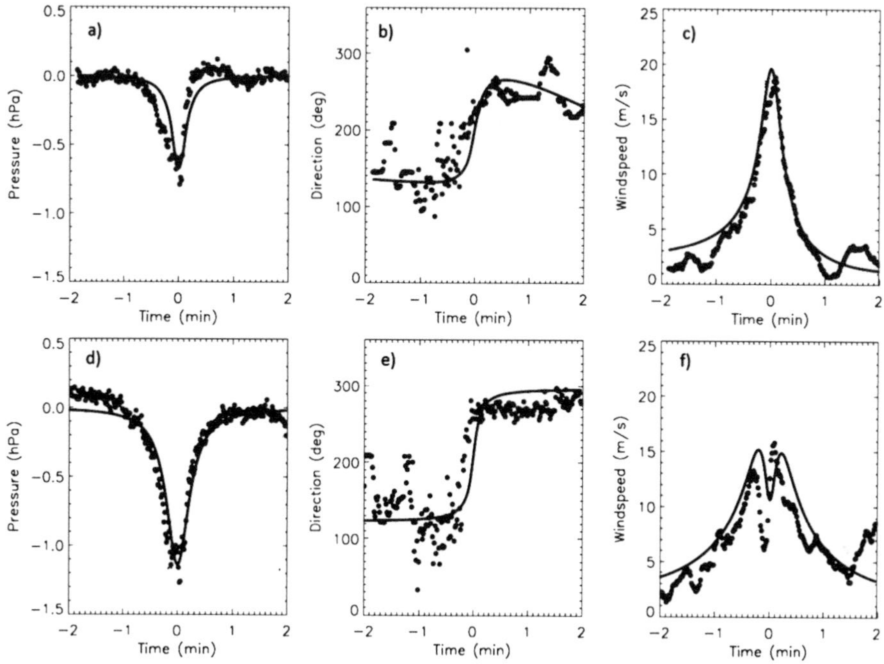

Fig. 11 Pressure (**a**), (**d**), wind direction (**b**), (**e**) and speed (**c**), (**f**) for two dust devil encounters in New Mexico during a month-long single-station survey in 2014. Data points are acquired at 1 second intervals, and the curves are a simultaneous fit to the three parameters by an analytic vortex model. Note the characteristic pressure dip, which in (**a**) clearly has a shallower onset slope than the recovery, probably related to the vortex advection. Note also in (**f**) the double-peaked wind speed profile, where winds are light in the central "eye" of the vortex (Lorenz 2016)

of momentum transport, an important role in resolving this paradox belongs to a clearer theoretical understanding (including through the use of a flow helicity concept (Moffatt 1969; Lilly 1986; Bluestein 1992, etc.)) and better reproduction in models/simulations of a "synergetic effect" of vortex merging (cf. Ohno and Takemi 2010a).

Acknowledgements We are thankful to two anonymous reviewers whose critical review and useful suggestions contributed tremendously to improvements in the content and in the style of this article.

Michael Kurgansky is grateful to Aldo Montecinos, Victor Villagran and Stephen M. Metzger with whom he carried out field studies of dust devils in Chile's Atacama Desert in 2008–2009, where the photos of Figs. 1 and 4 were taken. His work on this review was supported by the Presidium of the Russian Academy of Sciences, program no. 9 (Experimental and Theoretical Studies of Solar System Objects and Stellar Planetary Systems) and the Russian Foundation for Basic Research, projects 13-05-01025-a and 14-05-00847-a. Ralph Lorenz acknowledges the support of the NASA Mars Fundamental Research Program grant NNX12AI04G.

References

R.A. Bagnold, *The Physics of Blown Sand and Desert Dunes* (Methuen, New York, 1941)

M. Balme, R. Greeley, Rev. Geophys. **44**, RG3003 (2006)

M.R. Balme, A. Pathare, S.M. Metzger, M.C. Towner, S.R. Lewis, A. Spiga, L.K. Fenton, N.O. Renno, H.M. Elliott, F.A. Saca, T.I. Michaels, P. Russell, J. Verdasca, Icarus **221**, 632 (2012)

A. Barcilon, J. Fluid Mech. **27**, 155 (1967a)

A. Barcilon, J. Atmos. Sci. **24**, 453 (1967b)

A. Barcilon, P.G. Drazin, Geophys. Fluid Dyn. **4**, 147 (1972)

G.I. Barenblatt, G.S. Golitsyn, J. Atmos. Sci. **31**, 1917 (1974)

G.I. Barenblatt, A.J. Chorin, V.M. Prostokishin, Proc. Natl. Acad. Sci. USA **102**(32), 11148 (2005)

L.J. Battan, J. Meteorol. **15**, 235 (1958)

L. Bengtsson, J. Lighthill (eds.), Intense atmospheric vortices, in *Proceedings of the Joint Symposium*, IUTAM/IUGG Reading, UK, July 14–17, 1981 (Springer, Berlin, 1982)

H.B. Bluestein, *Synoptic-Dynamic Meteorology in Midlatitudes: Principles of Kinematics and Dynamics, vol. 1* (Oxford University Press, Oxford, 1992)

H.B. Bluestein, Tornadoes, in *Encyclopedia of Climate and Weather*, vol. 2, ed. by S.H. Schneider (Oxford University Press, Oxford, 1996), pp. 764–768

H.B. Bluestein, Dyn. Atmos. Ocean. **40**, 163 (2005)

H.B. Bluestein, *Severe Convective Storms and Tornadoes Observations and Dynamics* (Springer, Berlin, 2013)

H.B. Bluestein, J.G. Ladue, H. Stein, D. Speheger, W.F. Unruh, Mon. Weather Rev. **121**, 2200 (1993)

H.B. Bluestein, C.C. Weiss, A.L. Pazmany, Mon. Weather Rev. **132**, 209 (2004)

F.P. Bretherton, J.S. Turner, J. Fluid Mech. **32**, 449 (1968)

E.M. Brooks, Tornadoes and related phenomena, in *Compendium Meteorol.*, Boston, MA (1951), pp. 673–680

J.M. Burgers, A mathematical model illustrating the theory of turbulence, in *Advances in Applied Mechanics*, vol. 1 (Academic Press, New York, 1948), pp. 171–199

O.R. Burggraf, K. Stewartson, R.J. Belcher, Phys. Fluids **14**, 1821 (1971)

B.A. Cantor, K.M. Kanak, K.S. Edgett, J. Geophys. Res. **111**, E12002 (2006)

J.J. Carroll, J.A. Ryan, J. Geophys. Res. **75**, 5179 (1970)

C.R. Church, J.T. Snow, G.L. Baker, E.M. Agee, J. Atmos. Sci. **36**, 1755 (1979)

T. Cortese, S. Balachandar, Phys. Fluids A **5**, 3226 (1993)

W.D. Crozier, J. Geophys. Res. **69**, 5427 (1964)

W.D. Crozier, J. Geophys. Res. **75**, 4583 (1970)

G.T. Csanady, J. Atmos. Sci. **24**, 467 (1967)

R.P. Davies-Jones, J. Atmos. Sci. **30**, 1427 (1973)

R.P. Davies-Jones, Tornado dynamics, in *Thunderstorm Morphology and Dynamics*, ed. by E. Kessler (University of Oklahoma Press, Norman, 1986), pp. 197–388

R.P. Davies-Jones, Atmos. Res. **158–159**, 274 (2015)

J. Dessens, J. Appl. Meteorol. **11**, 72 (1972)

C.J. Diamond, E.M. Wilkins, J. Atmos. Sci. **41**, 2574 (1984)

P.G. Drazin, W.H. Reid, *Hydrodynamic Stability*, 2nd edn. (Cambridge University Press, Cambridge, 2004)

J. Dudhia, Mon. Weather Rev. **121**, 1493 (1993)

M.D. Ellehoj, H.P. Gunnlaugsson, P.A. Taylor, H. Kahanpää, K.M. Bean, B.A. Cantor, B.T. Gheynani, L. Drube, D. Fisher, A.-M. Harri, C. Holstein-Rathlou, M.T. Lemmon, M.B. Madsen, M.C. Malin, J. Polkko, P.H. Smith, L.K. Tamppari, W. Weng, J. Whiteway, J. Geophys. Res. **115**, E00E16 (2010)

K.A. Emanuel, J. Atmos. Sci. **43**, 585 (1986)

K.A. Emanuel, Am. Sci. **76**, 371 (1988)

W.M. Farrell, G.T. Delory, S.A. Cummer, J.R. Marshall, Geophys. Res. Lett. **30**(20), 2050 (2003)

W.M. Farrell, P.H. Smith, G.T. Delory, G.B. Hillard, J.R. Marshall, D. Catling, M. Hecht, D.M. Tratt, N. Rennó, M.D. Desch, S.A. Cummer, J.G. Houser, B. Johnson, J. Geophys. Res. **109**, E3 (2004)

E. Fedorovich, J. Appl. Meteorol. **34**, 1916 (1995)

L.K. Fenton, R. Lorenz, Icarus **260**, 246 (2015)

F. Ferri, P.H. Smith, M. Lemmon, N.O. Renno, J. Geophys. Res. **108**(E12), 5133 (2003)

B.H. Fiedler, Atmos.-Ocean **32**, 335 (1994)

B.H. Fiedler, R. Rotunno, J. Atmos. Sci. **43**, 2328 (1986)

S.D. Fuerstenau, Geophys. Res. Lett. **33**, L19S03 (2006)

B.T. Gheynani, P.A. Taylor, Bound.-Layer Meteorol. **137**, 223 (2010)

R. Greeley, J.D. Iversen, Aeolian processes and features at Amboy lava field, California, in *Physics of Desertification*, ed. by F. El-Baz, M.H.A. Hassan (Martinus Nijhoff Publishers, Dordrecht, 1986), pp. 290–317

R. Greeley, R. Leach, B. White, J. Iversen, J. Pollack, Geophys. Res. Lett. **7**, 121 (1980)

Z.L. Gu, *Wind-Blown Dust: Near Surface Gas-Solid Two-Phase Turbulent Flow* (Science Press, Beijing, 2010) (in Chinese)

Z. Gu, Y. Zhao, Y. Li, Y. Yu, X. Feng, J. Atmos. Sci. **63**, 2630 (2006)

Z.L. Gu, J. Qiu, Y.Z. Zhao, Y. Li, Adv. Atmos. Sci. **25**, 31 (2008)

L.N. Gutman, Bull. Acad. Sci. USSR Geophys. Ser. **1**, 87 (1957)

R.G. Harrison, E. Barth, F. Esposito, J. Merrison, F. Montmessin, K.L. Aplin, C. Borlina, J.J. Berthelier, G. Déprez, W.M. Farrell, I.M.P. Houghton, N.O. Rennó, K.A. Nicoll, S.N. Tripathi, M. Zimmerman, Space Sci. Rev. (2016, this issue). doi:10.1007/s11214-016-0241-8

G.D. Hess, K.T. Spillane, J. Appl. Meteorol. **29**, 498 (1990)
W.J. Humphreys, *Physics of the Air* (McGraw-Hill, New York, 1940)
J. Ito, H. Niino, SOLA **10**, 108 (2014)
J.C. Kaimal, J.A. Businger, J. Appl. Meteorol. **9**, 612 (1970)
K.M. Kanak, On the formation of vertical vortices in the atmosphere. Ph.D. Dissertation, University of Oklahoma (1999)
K.M. Kanak, Geophys. Res. Lett. **33**(1–4), L19S05 (2006)
K.M. Kanak, Q. J. R. Meteorol. Soc. **131**, 1271 (2005)
K.M. Kanak, D.K. Lilly, J.T. Snow, Q. J. R. Meteorol. Soc. **126**, 2789 (2000)
M. Klose, B.C. Jemmett-Smith, H. Kahanpää, M. Kahre, P. Knippertz, M.T. Lemmon, S.R. Lewis, R.D. Lorenz, L.D.V. Neakrase, C. Newman, M.R. Patel, D. Reiss, A. Spiga, P.L. Whelley, Space Sci. Rev. (2016, this issue). doi:10.1007/s11214-016-0261-4
H.L. Kuo, J. Atmos. Sci. **23**, 25 (1966)
H.L. Kuo, J. Atmos. Sci. **24**, 95 (1967)
M.V. Kurgansky, Dyn. Atmos. Ocean. **40**, 151 (2005)
M.V. Kurgansky, Geophys. Res. Lett. **33**, L19S06 (2006)
M.V. Kurgansky, Proc. IUTAM **7**, 193 (2013)
M.V. Kurgansky, Izv., Atmos. Ocean. Phys. **51**, 299 (2015)
M.V. Kurgansky, L. Baez, E.M. Ovalle, J. Geophys. Res. **112**, E11008 (2007)
M.V. Kurgansky, A. Montecinos, V. Villagran, S.M. Metzger, Bound.-Layer Meteorol. **138**, 285 (2011)
F.W. Leslie, J. Atmos. Sci. **34**, 1022 (1977)
L. Leslie, R. Smith, Q. J. R. Meteorol. Soc. **103**, 499 (1977)
V.H. Leverson, P.C. Sinclair, J.H. Golden, Mon. Weather Rev. **105**, 725 (1977)
W.C. Lewellen, Theoretical models of the tornado vortex, in *Proceedings of the Symposium on Tornadoes*, ed. by R.E. Peterson (Texas Tech. University, Lubbock, 1976), pp. 107–143
W.S. Lewellen, Tornado vortex theory, in *The Tornado; Its Structure, Dynamics, Prediction, and Hazards*, ed. by C. Church, D. Burgess, C. Doswell, R. Davies-Jones. Geophysical Monograph Series, vol. 79 (Am. Geophys. Union, Washington, 1993), pp. 19–39
D.C. Lewellen, W.S. Lewellen, J. Atmos. Sci. **64**, 2179 (2007)
W.S. Lewellen, D.C. Lewellen, R.I. Sykes, J. Atmos. Sci. **54**, 581 (1997)
D.C. Lewellen, W.S. Lewellen, J. Xia, J. Atmos. Sci. **57**, 527 (2000)
D.C. Lewellen, B. Gong, W.S. Lewellen, J. Atmos. Sci. **65**, 3247 (2008)
D.K. Lilly Tornado dynamics, NCAR Manuscript 69-117, 52 pp. (1969)
D.K. Lilly, J. Atmos. Sci. **42**, 126 (1986)
Z. Liu, T. Ishihara, J. Wind Eng. Ind. Aerodyn. **151**, 1 (2016)
R.R. Long, J. Meteorol. **15**, 108 (1958)
R.R. Long, J. Fluid Mech. **11**, 611 (1961)
R.D. Lorenz, J. Meteorol. **29**, 275 (2004)
R.D. Lorenz, Geosci. Instrum. Method. Data Syst. **1**, 209 (2012)
R.D. Lorenz, J. Atmos. Sci. **71**, 4461 (2014)
R.D. Lorenz, Icarus **271**, 326 (2016)
R.D. Lorenz, B.K. Jackson, Geo. Res. J. **5**, 1 (2015)
R.D. Lorenz, B.K. Jackson, Space Sci. Rev. (2016, this issue). doi:10.1007/s11214-016-0277-9
R.D. Lorenz, M.J. Myers, J. Meteorol. **30**, 178 (2005)
R.D. Lorenz, M.R. Balme, Z. Gu, H. Kahanpää, M. Klose, M.V. Kurgansky, M.R. Patel, D. Reiss, A.P. Rossi, A. Spiga, T. Takemi, W. Wei, Space Sci. Rev. (2016, this issue). doi:10.1007/s11214-016-0239-2
H.J. Lugt, *Vortex Flow in Nature and Technology* (Wiley-Interscience, New York, 1983)
K.J. Mallen, M.T. Montgomery, B. Wang, J. Atmos. Sci. **62**, 408 (2005)
T. Maxworthy, J. Atmos. Sci. **30**, 1717 (1973)
A.D. McEwan, Geophys. Fluid Dyn. **5**, 283 (1973)
S. Metzger, M. Kurgansky, A. Montecinos, V. Villagran, H. Verdejo, in *41st Lunar and Planetary Science Conference*. The Woodlands, Texas, USA, 2010. LPI Contribution, vol. 1533 (2010), p. 2564
T.I. Michaels, S.C.R. Rafkin, Q. J. R. Meteorol. Soc. **130**, 1251 (2004)
I. Michelson, I. On dust devils. II. Linearized theory of conical turbomachines. Ph.D. Thesis, California Institute of Technology, Pasadena (1951)
Y. Mitsuta, S. Yoshizumi, J. Meteorol. Soc. Jpn. **51**, 475 (1973)
Y. Mitsuta, N. Monji, H. Ishikawa, J. Geophys. Res. **92**(D12), 14827 (1987)
C.H. Moeng, P.P. Sullivan, J. Atmos. Sci. **51**, 999 (1994)
H.K. Moffatt, J. Fluid Mech. **35**(1), 117 (1969)
N. Monji, J. Meteorol. Soc. Jpn. **63**, 703 (1985)
J.B. Mullen, T. Maxworthy, Dyn. Atmos. Ocean. **1**, 181 (1977)

D.V. Neakrase, R. Greeley, J. Geophys. Res. **115**, E05003 (2010)
P.K. Newton, *The N-Vortex Problem. Analytical Techniques* (Springer, New York, 2001)
V.V. Nikulin, J. Appl. Mech. Tech. Phys. **21**(1), 62 (1980)
H. Ohno, T. Takemi, Atmos. Sci. Lett. **11**, 27 (2010a)
H. Ohno, T. Takemi, SOLA **6A**, 5 (2010b)
A.M.C. Oke, N.J. Tapper, D. Dunkerley, J. Arid Environ. **71**, 201 (2007)
O.G. Onishchenko, W. Horton, O.A. Pokhotelov, L. Stenflo, Phys. Scr. **89**, 075606 (2014)
S. Raasch, T. Franke, J. Geophys. Res. **116**, D16120 (2011)
S.C.R. Rafkin, R.M. Haberle, T.I. Michaels, Icarus **151**, 228 (2001)
W.J.M. Rankine, *A Manual of Applied Mechanics*, 16th edn. (Charles Griffin and Company Limited, London, 1901)
D. Reiss, M.I. Zimmerman, D.C. Lewellen, Earth Planet. Sci. Lett. **383**, 7 (2013)
N.O. Rennó, Tellus **60A**, 688 (2008)
N.O. Rennó, H.B. Bluestein, J. Atmos. Sci. **58**, 927 (2001)
N.O. Rennó, A.P. Ingersoll, J. Atmos. Sci. **53**, 572 (1996)
N.O. Rennó, M.L. Burkett, M.P. Larkin, J. Atmos. Sci. **55**, 3244 (1998)
N.O. Rennó, A.A. Nash, J. Lunine, J. Murphy, J. Geophys. Res. **105**, 1859 (2000)
N.O. Rennó, V.J. Abreu, J. Koch, P.H. Smith, O.K. Hartogenisis, H.A.R. De Bruin, D. Burose, G.T. Delory, W.M. Farrell, C.J. Watts, J. Garatuza, M. Parker, A. Carswell, J. Geophys. Res. **109**, E07001 (2004)
H. Riehl, J. Atmos. Sci. **20**, 276 (1963)
N. Rott, Z. Angew. Math. Phys. **9b**, 543 (1958)
R. Rotunno, J. Atmos. Sci. **34**, 1942 (1977)
R. Rotunno, J. Fluid Mech. **87**, 761 (1978)
R. Rotunno, J. Atmos. Sci. **36**, 140 (1979)
R. Rotunno, J. Atmos. Sci. **41**, 283 (1984)
R. Rotunno, Annu. Rev. Fluid Mech. **45**, 59 (2013)
J.A. Ryan, I.J. Carroll, J. Geophys. Res. **75**, 531 (1970)
J.T. Schofield, J.R. Barnes, D. Crisp, R.M. Haberle, S. Larsen, J.A. Magalhães, J.R. Murphy, A. Seiff, G. Wilson, Science **278**, 1752 (1997)
R.L. Schwiesow, J. Appl. Meteorol. **20**, 349 (1981)
A. Shapiro, Y. Kogan, Atmos. Res. **33**, 125 (1994)
P.C. Sinclair, A quantitative analysis of the dust devil. Ph.D. Dissertation, University of Arizona (1966)
P.C. Sinclair, J. Appl. Meteorol. **8**, 32 (1969)
P.C. Sinclair, J. Atmos. Sci. **30**, 1599 (1973)
P.C. Sinclair, in *Atmosphere-Surface Exchange of Particulate and Gaseous Pollutants*, ed. by R. Engelman, G. Sehmel (ERDA, Oak Ridge, 1976), pp. 497–527
W.C. Skamarock, J.B. Klemp, J. Dudhia, D.O. Gill, O.M. Barker, M.G. Duda, X.Y. Huang, W. Wang, J.G. Powers, A description of the Advanced Research WRF Version 3. NCAR Tech. Note 475 (2008), 113 pp.
R.K. Smith, L.M. Leslie, Q. J. R. Meteorol. Soc. **102**, 791 (1976)
J.T. Snow, Rev. Geophys. **25**, 371 (1987)
J.T. Snow, R.L. Pauley, J. Clim. Appl. Meteorol. **23**, 1465 (1984)
J.T. Snow, C.R. Church, B.J. Barnhart, J. Atmos. Sci. **37**, 1013 (1980)
E.P. Souza, N.O. Rennó, M.A.F. Silva Dias, J. Atmos. Sci. **57**, 2915 (2000)
A. Spiga, J. Faure, J.B. Madeleine, A. Määttänen, F. Forget, J. Geophys. Res., Planets **118**, 746 (2013)
R.D. Sullivan, J. Aerosp. Sci. **26**, 767 (1959)
A.D. Toigo, M.I. Richardson, S.P. Ewald, P.J. Gierasch, J. Geophys. Res. **108**, 5047 (2003)
D.M. Tratt, M.H. Hecht, D.C. Catling, E.C. Samulon, P.H. Smith, J. Geophys. Res. **108**(E11), 5116 (2003)
G.H. Vatistas, V. Kozel, W.C. Mih, Exp. Fluids **11**, 73 (1991)
Z.-T. Wang, Icarus **265**, 79 (2016)
N.B. Ward, J. Atmos. Sci. **29**, 1194 (1972)
W. Wei, Z. Gu, Phys. Rep. **600**, 1 (2015)
E.M. Wilkins, Y. Sasaki, H.L. Johnson, Mon. Weather Rev. **103**, 305 (1975)
G.E. Willis, J.W. Deardorff, J. Geophys. Res. **84**, 296 (1979)
J. Wurman, S. Gill, Mon. Weather Rev. **128**, 2135 (2000)
J. Wurman, J.M. Straka, E.N. Rasmussen, Science **272**, 1774 (1996)
C.-S. Yih, Phys. Fluids **19**, 076601 (2007)
W. Zhang, P.P. Sarkar, in *AAWE Workshop*, Vail, CA, USA (2008)
Y.Z. Zhao, Z.L. Gu, Y.Z. Yu, Y. Ge, Y. Li, X. Feng, Atmos.-Ocean **42**(1), 61 (2004)
S.M. Zurn-Birkhimer, E.M. Agee, J. Atmos. Sci. **62**, 2414 (2005)

Space Sci Rev (2016) 203:245–275
DOI 10.1007/s11214-016-0284-x

Large-Eddy Simulations of Dust Devils and Convective Vortices

Aymeric Spiga[1] · Erika Barth[2] · Zhaolin Gu[3] · Fabian Hoffmann[4] · Junshi Ito[5] · Bradley Jemmett-Smith[6] · Martina Klose[7,8] · Seiya Nishizawa[9] · Siegfried Raasch[4] · Scot Rafkin[2] · Tetsuya Takemi[10] · Daniel Tyler[11] · Wei Wei[12]

Received: 15 October 2015 / Accepted: 26 August 2016 / Published online: 27 September 2016

Abstract In this review, we address the use of numerical computations called Large-Eddy Simulations (LES) to study dust devils, and the more general class of atmospheric phenomena they belong to (convective vortices). We describe the main elements of the LES methodology. We review the properties, statistics, and variability of dust devils and convective vortices resolved by LES in both terrestrial and Martian environments. The current challenges faced by modelers using LES for dust devils are also discussed in detail.

Keywords Dust devils · Large-Eddy Simulations · Convective vortices · Convective boundary layer

✉ A. Spiga
aymeric.spiga@upmc.fr

1 Laboratoire de Météorologie Dynamique, UMR CNRS 8539, Institut Pierre-Simon Laplace, Sorbonne Universités, UPMC Univ Paris 06, Paris, France

2 SouthWest Research Institute, Boulder, CO, USA

3 Xi'an Jiaotong University, Xi'an, China

4 Institute of Meteorology and Climatology, Leibniz Universität Hannover, Hannover, Germany

5 Meteorological Research Institute, Ibaraki, Japan

6 Institute of Climate and Atmospheric Science, University of Leeds, Leeds, UK

7 Institute for Geophysics and Meteorology, University of Cologne, Cologne, Germany

8 Present address: USDA-ARS Jornada Experimental Range, Las Cruces, NM, USA

9 RIKEN Advanced Institute for Computational Science, Kobe, Japan

10 Disaster Prevention Research Institute, Kyoto University, Kyoto, Japan

11 Oregon State University, Corvallis, OR, USA

12 Wuhan University of Technology, Wuhan, China

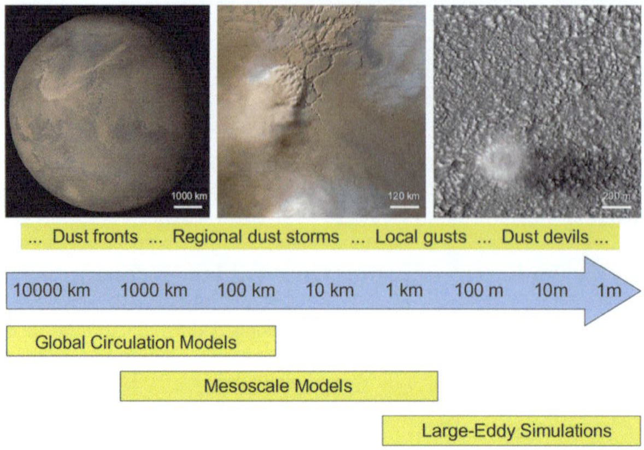

Fig. 1 *Typical dust lifting events (top images + yellow strip), relevant spatial scales (middle blue arrow) and meteorological models suitable for their analysis (bottom yellow strips). Left*: Dust front observed through Mars Orbiter Camera (MOC) imagery. *Middle*: Dust regional storm observed with MARs Color Imager (MARCI). *Right*: Dust devil observed with HIgh Resolution Imaging Science Experiment (HIRISE). Note that: 1. the lower bound of the GCM mesh spacing continues to decrease as computational capability and modeling techniques improve; 2. mesoscale modeling and (turbulence-resolving) Large-Eddy Simulations can be often carried out with the same non-hydrostatic dynamical core; 3. the horizontal scale is neither unique nor the best possible choice by which to classify dust events: obviously meteorological origin (fronts, deep convection, turbulence, ...) allows for a better understanding, although this criterion may be less straightforward for a broader audience to understand. Figure published in Spiga and Lewis (2010), used with permission

1 Introduction

In this chapter, we describe dust devils as events worthy of interest in atmospheric fluid dynamics. A dust devil is a particular case of a more general dynamical phenomenon, the convective vortex, occurring in the Planetary Boundary Layer (PBL), the part of the atmosphere directly influenced by the presence of the planetary surface. Convective vortices form during daytime hours when the warm surface causes strong convective instabilities to occur in the PBL, giving rise to the Convective Boundary Layer (CBL). A dust devil appears when one convective vortex has the ability (for a variety of reasons, one being the availability of dust particles at the surface) to lift and advect dust in its swirling circulation structure. A given area might be prone to the occurrence of convective vortices, but devoid of visible dust devils.

Herein we specifically describe the numerical modeling of dust devils by turbulence-resolving simulations called Large-Eddy Simulations (LES). In LES, the equations of motion for the atmospheric fluid are integrated on a gridded mesh with a typical spacing of tens of meters or less in order to resolve the larger turbulent eddies that are responsible for most of the energy transport within the CBL. Thus, LES are capable of resolving dust devils, which have diameters smaller than 1 km (see Murphy et al. 2016, Fenton et al. 2016, and Lorenz and Jackson 2016 in this issue), in contrast to simulations carried out by either Global Climate Models (GCM, in which mesh spacing is a couple hundreds of kilometers) or Regional Climate Models (RCM, in which mesh spacing is typically several kilometers) (Fig. 1). Furthermore, in LES, convective vortices possibly leading to dust devils are explicitly resolved. This is in contrast to other modeling work described in this special issue, where either dust transport by dust devils is predicted without resolving the individual structure (see Rafkin et al. 2016 in this issue), or pre-existing vortical structures are prescribed (see Kurgansky et al. 2016 in this issue). It shall be emphasized, however, that the smallest-diameter dust devils observed on both the Earth and Mars (a couple meters across) are still challenging to resolve with LES.

An important aspect to bear in mind in this chapter is that the occurrence of convective vortices is inseparable from the growth and mature stage of the CBL in daytime. Hence, reproducing convective vortices is only one of the overarching goals of LES which, in a broader sense, aim at exploring the whole range of atmospheric phenomena associated with the CBL growth: convective cells, plumes, vortices, and the resulting mixing of heat and momentum. In addition to CBL dynamics, LES offer interesting modeling perspectives for the stable and neutral PBL.

In this chapter, we review what LES brought to the study of dust devils and convective vortices. LES have been instrumental in broadening the knowledge of the properties, evolution, and statistics of dust devils. The LES modeling approach is highly complementary to measurements (see Murphy et al. 2016 and Fenton et al. 2016 in this issue) to study dust devils, since all properties of modeled vortices (wind speed, temperature, pressure, stability), as well as local environmental conditions, are accessible with LES.

We discuss LES performed in the context of both the Earth and Mars. Comparing the two environments gives insight into the mechanisms and characteristics of convective vortices in general, with Mars' strong near-surface gradients of temperature making the Red Planet an interesting extreme counterpart to the Earth for CBL studies.

2 LES Methodology

2.1 Dynamical Integrations

LES consist of four-dimensional (horizontal + vertical + temporal) numerical integrations of atmospheric fluid dynamics equations, usually the fully compressible, non-hydrostatic Navier–Stokes equations. While the vast majority of GCM simulations are performed under the approximation of hydrostatic balance between gravity and pressure forces in the vertical, such an approximation is not appropriate in LES because vertical wind accelerations in CBL turbulent structures might become comparable to the acceleration due to gravity. For hydrodynamical integrations, Martian LES are based on similar solvers as terrestrial LES: e.g., for their Martian studies, Rafkin et al. (2001) (see Sect. 4.1) used the RAMS terrestrial model (Pielke et al. 1992), and Spiga and Forget (2009) used the WRF model (Moeng et al. 2007; Skamarock and Klemp 2008), also used by Klose and Shao (2013) to simulate terrestrial CBL vortices (see Sect. 3 and Tables 1 and 2).

A remaining cornerstone for the proper simulation of dust devils, even in high-resolution LES, is the application of suitable numerical schemes. The numerical approximations used by the schemes cause characteristic numerical errors due to their specific construction (e. g., Durran 2010, Chap. 3). Numerical diffusion is usually caused by schemes of odd order of accuracy, whereas numerical dispersion is caused by schemes of even order. Numerical dispersion is triggered by strong gradients of the advected quantities, which typically exist in the center of dust devils. The resulting wiggles are cumbersome for the development and evolution of dust devils. By comparing a standard advection scheme to a scheme which is especially designed to avoid these errors, Weißmüller et al. (2016) showed that the average central pressure drop intensity increases by 8 % if the latter scheme is used.

Generally speaking, every LES model is capable of carrying out dynamical integrations at very high horizontal resolution. However, the increasing number of simulated grid points is accompanied by increasing demands on the model's high-performance computing capabilities, such as its efficient parallelization, which is necessary to distribute the simulation on up to thousands of computing nodes, and the optimized handling of data, simply because a higher number of grid points produces more data, which has to be processed and analyzed.

2.2 Domain Extent

LES integrations are carried out on a limited-area horizontal domain. LES are often qualified as idealized simulations, for this limited domain is associated with uniform surface conditions (e.g. topography, albedo, thermal inertia, solar forcing, roughness) and doubly periodic boundary conditions, effectively resulting in a simulation of an infinite flat plane. It is an open question whether this is a valid approach in the strictest sense to LES either for the Earth or for Mars: the approximation is probably satisfactory for a large enough flat area, but cannot properly address items such as the formation of dust devils near slightly larger-scale topographical or albedo contrasts (Renno et al. 2004). Furthermore, periodic lateral boundaries prevent LES from accounting for any gradients in regional or large-scale circulations.

The physical size of the LES domain is a crucial issue. For the mature stage of convection, the LES domain must be both deep enough and wide enough to contain at least several of the largest polygonal cells that develop in the CBL (an aspect ratio of three, width/depth, is minimally sufficient as is discussed in Mason 1989). Otherwise, a single (or, in the worst-case scenario, half a) polygonal cell might be aliased by the periodic boundary conditions, leading to unrealistic results. For Mars, following the deeper CBL mixing layer (up to 10 km deep vs. a few kilometers deep on the Earth, see Spiga 2011, and references therein), horizontal convective cells are larger in both dimensions by at least a factor of two, whereas the actual size of dust devils can be very similar to those on the Earth. This leads to a very large computational burden when performing LES on Mars, at least an order of magnitude larger cost when taking into consideration the lower timestep necessary to resolve the more vigorous vertical velocities encountered on Mars compared to the Earth. The state-of-the-art LES for the Martian CBL has not yet surpassed this obstacle, an aspect that should be kept in mind when further discussing convective vortices and dust devils in Martian LES.

This issue on the size of the LES domain, quite specific to the Martian environment, is also encountered in terrestrial LES when computationally demanding high-resolution LES are carried out. A proper representation of the fundamental boundary layer dynamics, i. e., the cellular pattern of convection, is as important as the highest possible resolution for the successful simulation of dust devils (e. g., Raasch and Franke 2011). Since the cells' vertices are one of the main dust devils' point of origin (see Sect. 3.2), dust-devil LES should not only feature a grid spacing in the order of meters for the successful simulation of the dust devil microstructure, but also a model domain adequate to well-represent larger-scale CBL dynamics—generally multiple kilometers or more in all directions, depending on the location and time-of-day.

2.3 Treatment of Sub-grid Scale Processes

Turbulence is the dominant mechanism of vertical atmospheric mixing. As mentioned in the introduction, the mesh spacing typically used for the LES horizontal grid is from a few tens of meters to a few meters, i.e. suitable to resolve the largest turbulent eddies responsible for about 80 %–90 % of the energy transport in the CBL (hence the name LES, cf. Lilly 1962; Sullivan and Patton 2011). Contrary to what is done in GCMs and RCMs, LES resolve the large turbulent eddies, i.e. convective plumes and vortices. Hence the mixing of heat and momentum they produce does not need to be parameterized. In other words, LES allow for a more fundamental study of the CBL. Still, LES lack the mixing produced by the unresolved small-scale eddies. This includes any turbulent phenomena at smaller scales than the adopted LES grid spacing, part of which can be the smallest convective vortices (and related dust devils). The treatment for this small-scale mixing in LES can be achieved by computing the eddy diffusivity (i.e., turbulent mixing coefficients) from either the local velocity and temperature gradients (Smagorinsky 1963), or subgrid Turbulent Kinetic Energy (TKE) prognosed from a high-order closure equation Deardorff (1972). The latter strategy is usually adopted in LES (see e.g. Moeng et al. 2007), while the former is more typical of GCM and RCM (though it is also used in LES, e.g. Kanak et al. 2000). The treatment of subgrid-scale mixing in Martian LES is similar to terrestrial LES, although the latest developments on TKE-based schemes for terrestrial LES Mirocha et al. (2010) are yet to be considered for use in Martian LES.

The Surface Layer (SL) is the part of the PBL immediately adjacent to the surface. The SL thickness is several meters to tens of meters on Earth (and on Mars, although this is yet to be confirmed by direct measurements Petrosyan et al. 2011). The whole SL thickness is thus usually enclosed in the 1–2 lowermost atmospheric levels in LES. Furthermore, turbulent motions in the SL develop at lower spatial and temporal scales than what is achievable by LES. Hence, SL processes are in need to be parameterized in LES, as is the case in GCMs and RCMs. This basically means developing a simple model for the surface-atmosphere transfer of heat, momentum, and tracers. For instance, the quantity related to heat transfer from the surface to the atmosphere via the SL is the sensible heat flux H_s. The simplest SL parameterization for H_s is a bulk parameterization in which H_s is simply proportional to the specific heat capacity of the surface, the atmospheric density, the wind stress exerted by the atmosphere on the surface, and the surface-atmosphere temperature contrast. Under simplifying assumptions, the last two quantities can be calculated respectively from the wind and temperature predicted at the lowermost atmospheric model layer. More accurate SL parameterizations account for the variation of the surface-atmosphere transfer coefficient (both for heat and momentum) with nighttime stable vs. daytime unstable conditions (e.g. Haberle et al. 1993).

2.4 Coupling Dynamical Integrations with Physical Processes

Within the CBL, the mixing of heat through turbulent motions (adiabatic heat transfer) typically dominates the atmospheric thermal structure. Diabatic processes (such as radiative transfer, molecular conduction with the surface, and volatile phase change) also modulate the static stability profile within and above the CBL, hence the intensity and vertical extent of convective motions. The phase change of volatiles is often insignificant for the arid areas/seasons that exhibit dust devil activity. Radiative forcing includes the absorption of incoming solar flux (by aerosols, i.e. dust particles in deserts) and the absorption/re-emission of infrared flux outgoing from the surface (by greenhouse gases and aerosols). Accounting for radiative forcing in LES may be separated into three categories, which illustrate major differences in the predominance of radiative processes between Mars and the Earth.

- *Category 1* If the radiative heating rates can be neglected, typically in hour-long terrestrial LES aimed at investigating a particular phenomenon, no radiative processes are included, and a fixed sensible heat flux value is imposed at the bottom of the atmosphere. For instance, Klose and Shao (2013) investigated CBL dynamical structures leading to particle lifting through 90-minute long LES runs using 200 × 200 horizontal grid points with

10 m mesh spacing. The combination of both idealized character and short duration for this LES setup reflects the limitations imposed by the timestep of 0.05 seconds; neither performing a longer run nor including radiative processes would have been crucial for the scientific objectives of Klose and Shao (2013) (other examples are e.g. Zhao et al. 2004; Kanak 2006).
- *Category 2* If radiative heating cannot be neglected, but radiative timescales are larger than typical convective timescales, precalculated (offline) time-varying radiative heating rates are employed. This setting is typically employed for daytime terrestrial LES—although Category 1 LES also yields correct results (to first order) in this configuration. Category 2 LES appears to yield acceptable results for Mars (Sorbjan 2007; Gheynani and Taylor 2011), although this remains to be tested more thoroughly.
- *Category 3* If radiative heating is significant, and radiative timescales are smaller than typical convective timescales, the dynamical integrations must be coupled with time-marching (online) radiative computations—usually associated with a ground model to account for subsurface heat transfers. This setting is typically employed for Martian LES because the thin CO_2 atmosphere loaded with dust on Mars leads to both fast radiative timescales and a predominance of the divergence of infrared radiative flux over sensible heat flux within near-surface atmospheric levels (Haberle et al. 1993; Sävijarvi 1999; Spiga et al. 2010). Category 3 LES also emerged recently for the Earth, to study areas particularly prone to the radiative influence of airborne dust, with a model coupling radiation and land-surface schemes (Shao et al. 2013).

Whether or not such modeling assumptions change the predictions for convective vortices in LES remains an open question, both for the Earth and for Mars.

Initializing LES is usually handled with a prescription of the same initial temperature and background wind profiles at every horizontal grid point. Temperature profiles are usually extracted from GCM or RCM simulations; on terrestrial LES, atmospheric soundings are also being used as initial conditions. Prescribing at all vertical levels the exact same temperature for the entire horizontal domain imparts a symmetry that does not exist in reality—small-scale, low-amplitude temperature fluctuations are ever-present in a real atmosphere. Random (noise) temperature perturbations of about 0.1 K are therefore added to the initial temperature field to break the horizontal symmetry of this initial field, and help trigger convective motions. Often LES are run without prescribing a background wind profile, in what is called “free convection” LES. Somewhat more realistic scenarios, where the horizontal wind is forced by the geostrophic wind derived from a mesoscale model, have been performed for Mars (e.g., Tyler et al. 2008). Such a strategy was primarily employed to account for the strong diurnal cycle of synoptic and regional winds on Mars, driven by thermal tides and slope flow. In terrestrial LES, experiments which include the Coriolis force to simulate the effect of background wind rotation were performed by Ito et al. (2011).

It is commonly assumed in the various published studies that the first half-hour/hour of a simulation should be considered as a spin-up from the initial state to the resolved LES state. Except for the finest resolutions, which clearly limit the total duration of the LES integrations, or LES driven by the same forcing throughout the simulation, most LES are usually started very early in the morning (local time 6 to 8 AM) to allow for this spin-up to occur before the CBL starts to develop and grow in earnest as the sun rises higher.

2.5 Dust Particles

Thus far, the vast majority of LES studies have been used to study convective vortices rather than true dust devils. In other words, the emphasis has been put on the dynamics and statistics of CBL vortices, with neither dust lifting nor dust transport processes included to form realistic dust devils. LES for Mars typically use a uniform background dust opacity; on the Earth, LES are often conducted without explicit dust treatment.

Progress has been made recently by including a representation for the full dust cycle (i.e. emission, transport, and deposition of dust particles) into terrestrial and Martian LES (Klose and Shao 2013). The first efforts towards this goal date back to the terrestrial LES study of Zhao et al. (2004), in which sand particles are injected into their dust-devil scale simulation to study dust transport and deposition, although no explicit formulation for dust entrainment is used. The first LES investigation to include a saltation bombardment scheme is the Martian study by Michaels (2006). The author uses the LES friction velocities to calculate bulk fluxes of dust and sand (partitioned according to a predefined particle-size distribution), that are then removed from the ground, transported within the modeled atmosphere, and ultimately fall back to the ground. The focus of Michaels (2006) was on the formation of dust devil tracks though—and not on the dust emission fluxes generated by dust devils. This was more the focus of Ito et al. (2010), who included dust lifting in their LES model by using the empirical formulation of Loosmore and Hunt (2000). According to this formulation, dust emission is proportional to friction velocity, u_*. However, Loosmore and Hunt (2000) have obtained their formulation for background dust lifting in the absence of saltation and without focus on convective turbulence. The fluxes obtained by Ito et al. (2010) in their simulation of the convective boundary layer are thus relatively low. To overcome this limitation, Klose and Shao (2013) included a physics-based dust emission scheme (Klose and Shao 2012) in their LES model and simulated dust emission due to convective turbulence (see Neakrase et al. 2016 in this issue). After further model developments and calibration against field measurements for dust emission, Klose et al. (2014) were the first to explicitly simulate size-resolved dust entrainment by dust devils.

This move towards more realistic prediction of the feedbacks between CBL dynamics and dust lifting—and thus the formation of spontaneous dust devils in LES—is still being more fully developed for both terrestrial and Martian LES. In the latter case, assessing the radiative effect of dust carried within a dust devil (as suggested by Fuerstenau 2006) would be an interesting open question to address through LES. This would relate to a new *Category 4* (see Sect. 2.4) where radiative transfer and dynamical transport are fully coupled.

3 LES for the Earth

3.1 First LES Studies for Convective Vortices

As pointed out by Kanak (2006), Mason (1989) may have been one of the first to present evidence of convective vortices in LES for the terrestrial CBL, but those were not the focus of his study, nor were they discussed. Experimental and computational efforts in the early 1990s allowed the gathering of robust evidence of spiraling updrafts in Rayleigh–Bénard convection (e.g. Cortese and Balachandar 1993), but the applicability of this body of work to CBL convective vortices and dust devils was only considered after the first LES studies on this topic (Fiedler and Kanak 2001).

It is generally accepted that the first work to report three-dimensional high-resolution LES capable of spontaneously generating convective vortices on scales similar to dust devils is Kanak et al. (2000), later complemented by Kanak (2005) (Fig. 2). The importance of convective tilting of horizontal vorticity in dust devil formation is demonstrated by simulations that show vortices forming within convergent branches of convective cells (and not

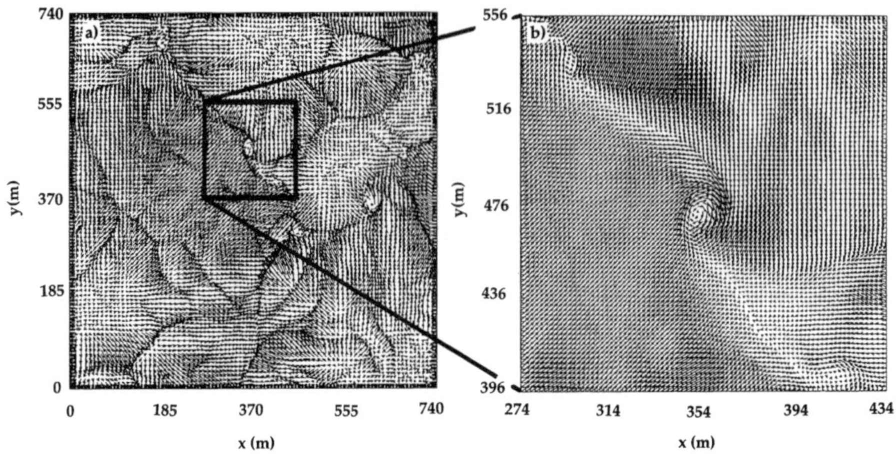

Fig. 2 *Convective vortices in terrestrial LES.* Horizontal cross-sections 2 m above the surface of velocity vectors from 2 m LES simulations after 1000 simulated seconds. *Left*: vectors at every fourth grid point over the total domain of 740 × 740 m^2. *Right*: expanded vortex of interest on 80 × 80 grid points, with a vector plotted at every grid point. Figure published in Kanak (2005), used with permission

Table 1 Properties of modeled convective vortices in published LES studies in terrestrial conditions. Maximum values are reported, and shall be compared only to the order of magnitude level, given the differences in grid settings and prescribed background winds (for the latter point, cf. text for a more extended discussion). dx is the horizontal mesh spacing employed in LES, $N_{x,y}$ is the number of grid points in the horizontal (x- and y-axis), Z is the altitude above the surface of the adopted model top, N_z is the number of grid points in the vertical (note that the vertical resolution changes with height, with higher resolution closer to the surface), $[w', v']$ are respectively the updraft velocity and tangential velocity within a simulated convective vortex, $[P', \zeta]$ are respectively the pressure drop and the magnitude of vertical vorticity associated with the vortex

Paper	Dyn. core	dx (m)	$N_{x,y}$	$\mathcal{Z}$ (km)	N_z	w' (m/s)	v' (m/s)	P' (Pa)	ζ (s^{-1})
Kanak et al. (2000)	Specific	35	86	2.1	48	0.15	5	18	0.1
Kanak (2005)	Specific	2	370	1.2	86	< 0.1	6	43	2.5
Ohno and Takemi (2010a)	WRF	3	333	1.5	185	1.5	–	64	2
Ohno and Takemi (2010b)	WRF	10	200	1.5	70	–	–	70	0.8
Ito et al. (2010)	Specific	50	90	3	60	4	6	40	0.3
Raasch and Franke (2011)	PALM	2	2000	1.7	449	2	4	72	6.8
Ito et al. (2013)	Specific	5	360	1.6	320	0.5	4.5	–	1.0

only at cell vertices). The convective vortices modeled in LES agree well with field measurements (see Murphy et al. 2016 in this issue), showing similar velocity structure and diurnal behavior as actual dust devils (Table 1). However, some differences from observations were noted in the vortex intensity in terms of the central pressure deficit, where the simulated vortices are weaker than the observations: Kanak (2005) found that the central pressure drop deviated from measured values in Sinclair (1973) by nearly a factor of 5. This difference may be due to the effects of the spatial resolutions used in the simulations. Nevertheless, even a horizontal mesh spacing of 2 m in Kanak (2005), a significant increase

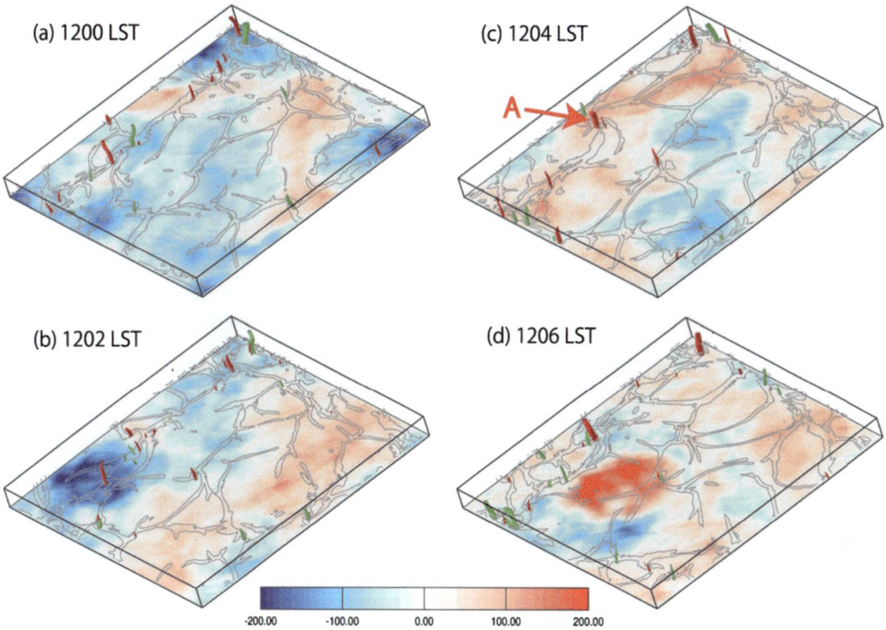

Fig. 3 *The origin of convective vortices resolved by terrestrial LES.* Time series of vertical vorticity iso-surfaces of 0.25 (*red*) and $-0.25\ \mathrm{s}^{-1}$ (*green*) from (**a**) 1200 to (**d**) 1206 local time, at 2-min intervals. The lower domain at heights < 100 m is displayed. Circulation, calculated only for downdraft regions within a horizontal circle of radius 300 m at each grid point at $z = 2.5$ m, is shown by *color shading* ($\mathrm{m}^2\,\mathrm{s}^{-1}$). Regions where vertical velocities are $> 0.1\ \mathrm{m\,s}^{-1}$ at the lowest model level are shown by *gray isolines*. The horizontal domain shown is 1.8 km across. Figure published in Ito et al. (2013), used with permission

from the resolution of 35 m in Kanak et al. (2000), did not improve the intensity of simulated dust devils because the diameters of the simulated vortices in Kanak (2005) were still larger than those observed (and thus insufficiently resolved). In spite of the deficiencies in LES pointed out in Kanak (2005), qualitative analyses on the dynamics and mechanisms of dust devils have been conducted in subsequent studies and have permitted the knowledge of those phenomena to progress.

3.2 Mechanisms for Vortex Formation in LES

Kanak (2006) emphasizes that LES demonstrate that no source of ambient angular momentum/wind shear is required to be imposed in order for vertical vortices to form: those occur even in LES devoid of background wind (cf. also Rafkin et al. 2016 in this issue). The mechanisms by which convective vortices and dust devils appear in LES are still being debated. Raasch and Franke (2011) studied the generation of dust devils by examining the vorticity budget, which has been directly derived from LES. They found that the vorticity of dust devils is mainly produced by twisting the vorticity generated by shear close to the surface from a horizontal to a vertical direction—a conclusion similar to the one reached by Kanak (2005) with coarser-resolution LES. In other words, those studies highlight the prominent role played by updrafts in CBL cells close to the surface. The LES study of Ito et al. (2013) indicated, however, that the origin of vorticity structures could also be traced back to downdrafts in CBL cells, and that the circulations can be originally generated by tilting of baroclinically generated horizontal vorticity within the CBL mixed layer, rather than

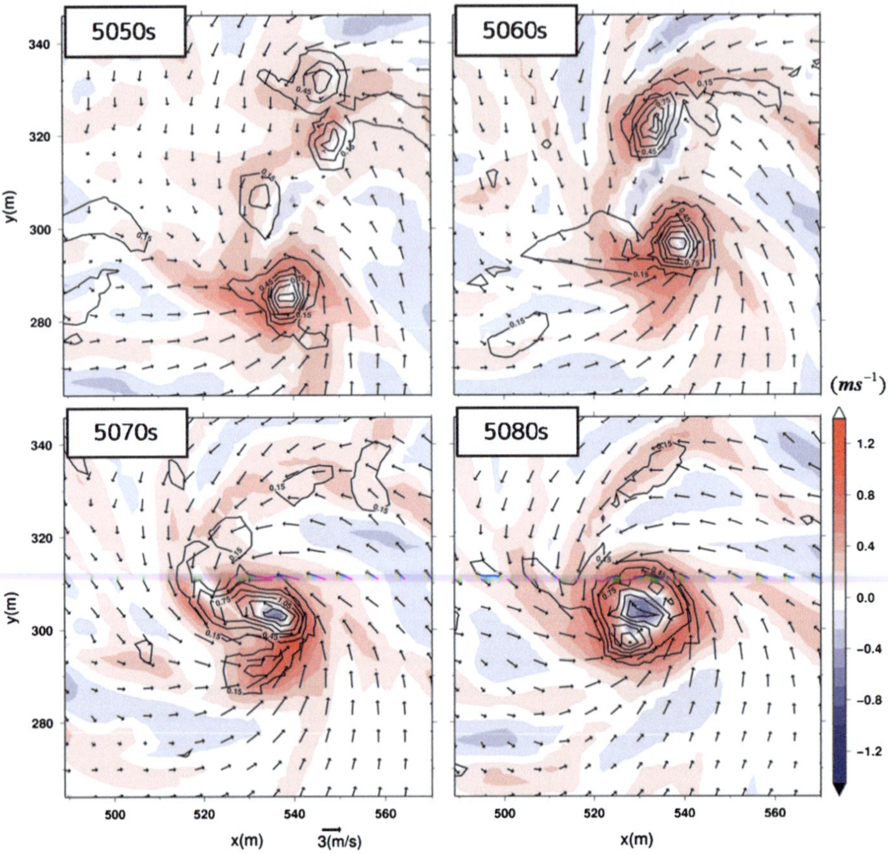

Fig. 4 *Merging of two convective vortices simulated by terrestrial LES.* Horizontal sections of wind vectors, vertical velocity (*shaded*) and vertical vorticity (*contoured* at 0.3 s^{-1} interval from 0.15 s^{-1}) at the 1.5 m height. Figure published in Ohno and Takemi (2010a), used with permission

closer to the surface (Fig. 3). This apparent discrepancy between the two studies (Raasch and Franke 2011; Ito et al. 2013) might simply reflect the variety of possible initial vorticity sources for the formation of dust devils in the CBL. At any event, Ohno and Takemi (2010a) demonstrated that, once formed, a vortex with a significant intensity will be further intensified through the merger with surrounding vortices (Fig. 4).

3.3 Influence of Environmental Conditions on Convective Vortices

How environmental conditions impact the properties of convective vortices is addressed in detail in Rafkin et al. (2016) in this issue. Here we review the main results obtained through LES on this topic. Takemi (2008) examined the influence of atmospheric stability on the development of daytime CBL and vertical dust transport by conducting high-resolution RCM simulations (with an horizontal resolution of 100 m, which makes it coarse-resolution LES) and found that an early morning stable stratification strongly regulates the temporal evolution and the vertical growth of convection and therefore the amount of dust emission and transport. Ohno and Takemi (2010b) investigated the effects of background wind speed on

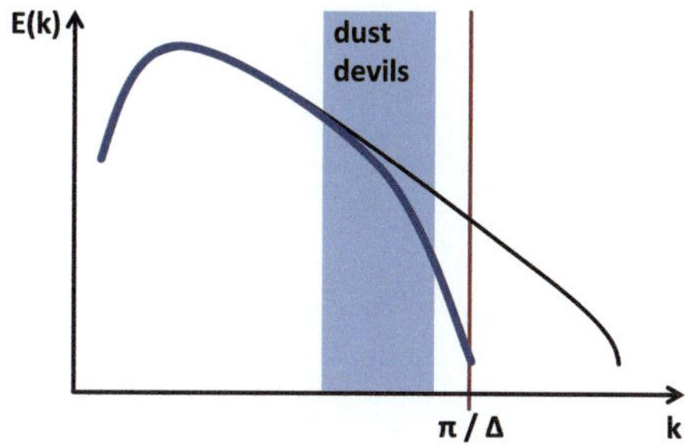

Fig. 5 *The required horizontal resolution for dust devils in LES.* Idealized spectrum of kinetic energy (E) as a function of wavenumber (k) as measured in the boundary layer (*black line*) or derived from LES (*blue line*). The *red line* denotes the wavenumber corresponding to the LES grid spacing Δ; the *shaded area* marks the typical size range of dust devils. Figure provided by co-authors Hoffmann and Raasch

the generation of vortices and found that vortices are generated most preferably in medium-wind conditions in terms of their number, intensity, and lifetime. Ohno and Takemi (2010b) also indicated that a stability condition controls the generation of dust devils, as was identified by Takemi (2008) and from field observations (see Murphy et al. 2016 in this issue).

Other LES modeling studies confirmed and complemented those findings. In the sensitivity experiment of Ito et al. (2010), the strong vorticity extrema corresponding to convective vortices—and thus possible dust devils—are more frequent when both the background wind is weaker (as is also found by Ohno and Takemi 2010b) and sensible heat flux is stronger. Those findings are consistent with observations (see Murphy et al. 2016 and Rafkin et al. 2016 in this issue). Ito et al. (2010) compared experiments with nearly calm environmental wind ($0.5\ \mathrm{m\,s^{-1}}$) and those with much stronger winds (5–$15\ \mathrm{m\,s^{-1}}$). Other studies (e.g., Raasch and Franke 2011) suggest that an intermediate, moderate background wind of a couple $\mathrm{m\,s^{-1}}$ is the most propitious condition for the formation of convective vortices.

3.4 Influence of Horizontal Resolution Adopted in LES

Although relatively coarse-resolution LES (e.g. 50 m horizontal resolution in Ito et al. 2010) are able to resolve the characteristic structure of dust devils at least qualitatively, there is still much room for improvements by performing high-resolution LES. When the mesh spacing is such that vortical structures are resolved by only a couple of grid points, the interpretation of LES results should consider the impact of numerical errors, which are known to affect, in particular, phenomena with a size close to the grid spacing (e. g., Sagaut 2006, Chap. 7). As is shown in Fig. 5, LES agree well with the observed spectrum of kinetic energy for the largest scales (small wavenumbers), but increasingly deviate from the observed spectrum as the wavenumber approaches the grid spacing—which is, incidentally, the spatial range where dust devils occur (a similar result is described in the broader study about CBL turbulence by Nishizawa et al. 2015). Hence the solution illustrated in Fig. 5 to further decrease the mesh spacing to move the poorly resolved scales to higher wavenumbers, away from the spatial scale of dust devils.

Refining the horizontal resolution in LES mostly impacts the size of simulated vortices (which become smaller), and the number of detected vortices (with more detections at higher resolutions; Raasch and Franke 2011). However, Ito et al. (2013) showed that the results of Ito et al. (2010) on the influence of background wind on dust devil formation, and the intensity of simulated dust devils, are basically left unchanged by refining the horizontal

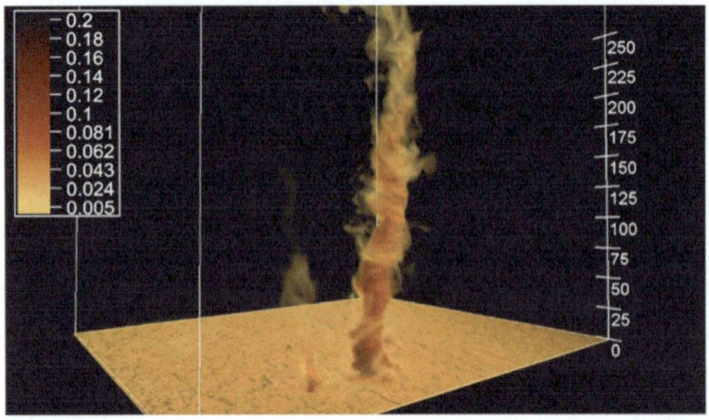

Fig. 6 *High-resolution terrestrial LES for dust devils.* Animation of dust devils in the convective boundary layer using a virtual dust source. LES of dust devils performed by Weißmüller et al. (2016) using the model of Raasch and Franke (2011). This figure is extracted from a movie which can be found by following the link http://dx.doi.org/10.5446/9352

resolution from 50 m to 5 m. Weißmüller et al. (2016), using the model of Raasch and Franke (2011), showed that LES with mesh spacing of 1 m (cf. Fig. 6) yielded dust devils with central pressure drops comparable to observed values. This result could not be obtained in the 2 m LES of Kanak (2005), and appears different from the results obtained by Ito et al. (2013). It is still unclear, nevertheless, if refining horizontal resolution alone is the only factor needed for obtaining realistic dust-devil-like vortices through LES. The simulations of Weißmüller et al. (2016) include a higher-order advection scheme and background winds, which might explain the differences with results reported in Kanak (2005). Furthermore, it still a matter of active research to determine the resolution at which the LES spectrum starts to significantly deviate from the theory, and what is the impact of the choice of the advection scheme on this resolution.

3.5 Considerations on Vortex Structure

The flow structure of well-developed dust devils was investigated in Kanak (2005) through theoretical models of vortices, i.e., the Rankine and the Burgers–Rott models which consist in linking tangential velocity in the vortex to the radial distance from the vortex center (see Kurgansky et al. 2016 in this issue for further details). It was found that the Burgers–Rott model matches the convective vortices arising spontaneously in LES better than the Rankine model, because the internal flow structure within the vortex core cannot be approximated as a simple solid body rotation within the core (which is assumed in the Rankine model). Nevertheless, the findings of Kanak (2005) seem to indicate that the Rankine vortex is a good approximation for the simulated vortices at least within the radius of maximum wind. Thus, Ohno and Takemi (2010a) employed the Rankine vortex model to identify the near-surface flow structure within the vortex core. Most of the radii of maximum wind are within 1–2 horizontal grid points (with a mesh spacing of 3 m), with major ones being greater than 3 grid points. These large vortices have more substantial pressure deficits and enhanced vorticities on the order of 1 s^{-1}. While the Burgers–Rott model appears more appropriate than the Rankine model to represent the flow structure within the vortex, Ohno and Takemi (2010a) shows that the latter is sufficient to conduct a statistical analysis from a large number of vortex samples.

The question of the applicability of the Rankine model was also addressed recently using the high-resolution LES by Weißmüller et al. (2016) (see Sect. 3.7). Figure 7 suggests that the tangential winds of vortices beyond the vortex wall are rarely as tangential, and the fall-off as steep or as uniform, as the Rankine model would suggest. Similar limitations of the Rankine model can also be diagnosed from in-situ observations (Lorenz 2014). Those are

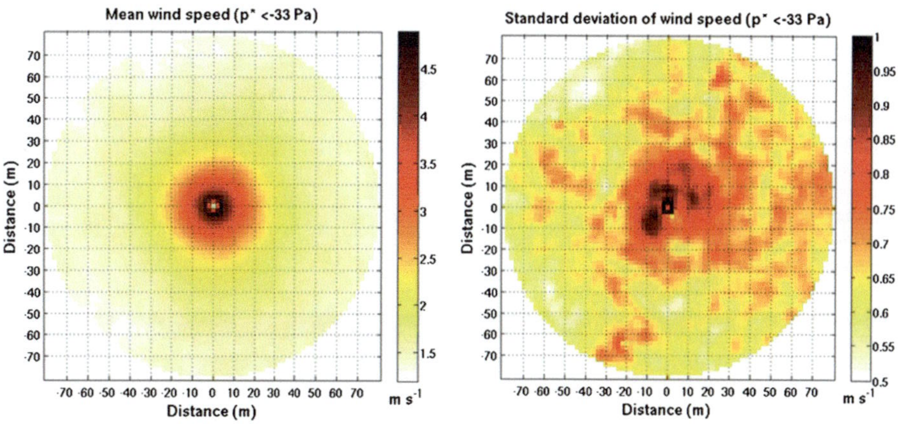

Fig. 7 *The structure of a convective vortex obtained with high-resolution terrestrial LES.* Composite of mean and standard deviation of 5 m wind speeds for 111 of the strongest vortices (central pressure < -33 hPa) using the high-resolution LES with no background wind in Weißmüller et al. (2016). Figure produced by co-author Jemmett-Smith (personal communication)

due to the interactions of the vortices with the convective ambient atmosphere, since convective vortices are mostly observed at the branches of the convective cell patterns (Kanak 2005; Gheynani and Taylor 2010), which generate local winds. Subsequently, only when vortices are mature (or on the rare occasions when they form in a locally weak wind region before moving away from those areas) are they likely to closely follow a Rankine model.

Interestingly, if the vortex intensity is increased, the vortex structure will change its form from a circular, single-celled flow to another regime (e.g. Rotunno 2013). In the LES of Ohno and Takemi (2010a) and Ito et al. (2010), it was demonstrated that an intense vortex has a double-celled structure which includes a downdraft at the vortex center—this type of double-celled structure was identified in observed dust devils (Bluestein et al. 2004, see also Murphy et al. 2016 and Fenton et al. 2016 in this issue). Thus, a transition from a single-celled to a double-celled structure is found for intense dust devils in both observations and LES—quite similar to what occurs in tornadoes, which are much stronger atmospheric vortices than dust devils (Rotunno 2013). This suggests a similarity between high-intensity dust devils and low-intensity tornadoes from a viewpoint of vortex properties. This analogy could help to explain the "spiral-arm" structure of some convective vortices resolved by LES, see e.g. Fig. 4, although this item is seldom discussed in the literature (Kanak 2005). However, the analogy does not extend to intense tornadoes which undergo an additional transition from double-celled to multiple vortices, which most likely does not occur in the weaker CBL vortices that become dust devils.

3.6 Dust Particles and Convective Vortices

Recently, the gap between simulating convective vortices versus predicting actual dust devils started to be overcome by LES coupled to dust emission schemes (Klose and Shao 2012, see Sect. 2.5). Using this new modeling perspective, Klose and Shao (2013) adopted a systematic approach with a set of 15 LES with various background wind (initialized with logarithmic wind profiles for $u_* = 0.15$, 0.3, and 0.5 $\mathrm{m\,s^{-1}}$) and atmospheric stratification settings (as specified by constant surface sensible heat fluxes, $H_s = -50$, 0, 200, 400, and 600 $\mathrm{W\,m^{-2}}$, with the highest value chosen to mimic strongly heated

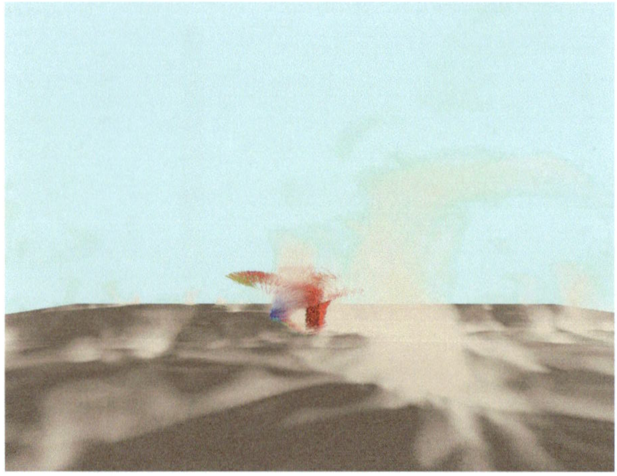

Fig. 8 *Turbulent dust emission simulated in a terrestrial LES, including the effect of a dust devil.* Modeled dust concentration is visualized with a volume rendering approach. A particle-tracing is applied to illustrate a dust devil occurring in the simulation. *Colors* indicate the direction of particle movement. Modeling results are described in Klose and Shao (2013). This figure is extracted from a movie which can be found by following the link http://dx.doi.org/10.5880/SFB806.5

desert surfaces). The model runs were not set up with a particular focus on dust devils, but with the broader goal to investigate convective turbulent dust emission, of which convective vortices are a component. Klose and Shao (2013) found that the largest CBL dust emissions occur in micro-convergence lines, micro-downdrafts, and in the area of convective vortices (Fig. 8). In a follow-up work, Klose and Shao (2016) repeated the model runs with a further developed and calibrated dust emission scheme (Klose et al. 2014), this time with a focus on dust entrainment by dust devils detected through a tracking algorithm based on a range of criteria for pressure drops ($\Delta p = 0.05$–0.25 hPa) and vorticity (0.1–1.0 s^{-1}). Statistical analysis of dust emission in dust devils shows that dust devils occurring in a similar setting (i.e. same model run) can exhibit a wide range of dust fluxes. The maximum values reached in dust devils are determined by atmospheric stability. Dust devils occur most frequently at highly negative Richardson numbers (e.g. unstable stratification and weak mean winds) and exhibit a relatively longer duration than under other conditions. Although of shorter duration and less frequent, dust devils at weakly negative Richardson numbers (i.e. unstable stratification and relatively strong mean winds) can have larger dust emissions, as the stronger mean wind shear contributes to the lifting (see Rafkin et al. 2016 in this issue).

The simulations show that turbulent dust emission in dust devils is typically of the order of magnitude $10^2\ \mu g\, m^{-2}\, s^{-1}$ (Fig. 9), but can reach the order of magnitude $10^3\ \mu g\, m^{-2}\, s^{-1}$, similar to that typically generated by saltation bombardment (Klose and Shao 2016) and at the lower edge of that measured in dust devils in the field (Metzger et al. 2011, and see Chapter Murphy et al. 2016 in this issue). The additional inclusion of intermittent saltation (e.g., Dupont et al. 2013) in the turbulent dust emission scheme would lead to even larger fluxes, closer to those measured in the field (see Klose et al. 2016 in this issue for further discussion on fluxes associated with dust devils). The recent study of Klose and Shao (2016) shows that modeling dust devils through LES is possible, provided an appropriate dust emission scheme is included.

To complement those LES studies of dust emission, steady-state vortex models can be employed to address the transport of dust particles within the convective vortex. For instance, Gu et al. (2006) studied the trajectories of sand particles (diameter 100 to 300 μm) injected within a mature-stage swirling flow bearing similarities with dust devils. The resulting spatial distributions for the different grain sizes are summarized in Fig. 10. The small particles move along with the main streamline of the updraft, while the large ones deflect outward from, and deposit beyond, the main core of the dust devil. In addition to the drag

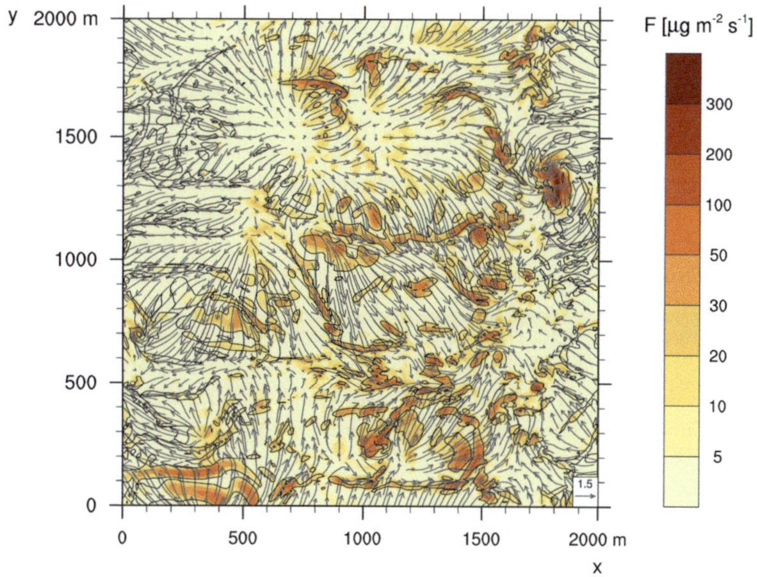

Fig. 9 *Flux of dust particles associated with turbulent structures resolved in terrestrial LES.* Turbulent wind speed (vectors, in $\mathrm{m\,s^{-1}}$) and instantaneous turbulent momentum flux (*black contour lines* at $1\ \mathrm{N\,m^{-2}}$) at 10 m height together with turbulent dust emission F (*shaded*, in $\mathrm{\mu g\,m^{-2}\,s^{-1}}$). The strongest turbulent dust emissions occur at micro-convergence lines (e.g. at coordinates $x = 850$ m, $y = 800$ m), micro-bursts (e.g. at coordinates $x = 800$ m, $y = 1700$ m), and vortices (e.g. at coordinates $x = 1800$ m, $y = 1400$ m). Figure published in Shao et al. (2015), used with permission

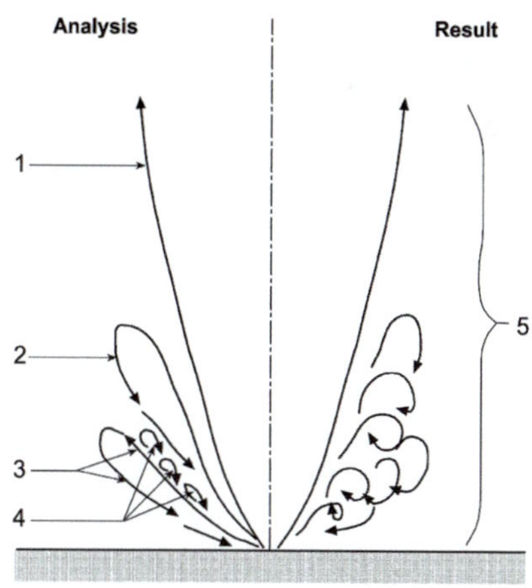

Fig. 10 *Dust lifting patterns in a dust devil. 1*-track of fine dust grains; *2*-track of medium grains; *3*-track of large grains; *4*-small vortices induced by the interaction of different sized grains; *5*-general pattern of interactions. Figure published in Gu et al. (2006), used with permission

in the updraft, the centrifugal force exerted on the solid particles in the swirling air flow of dust devils is another key force component on the horizontal plane, which might explain the deflection of larger particles. Some of the ejected larger particles may be re-entrained into the rapid near-surface inflow, and be returned to the vortex. In effect, such particles may move circularly on the flanks of the dust devil until the updraft decays and the drag is not enough to lift them. Large particles thus draw the "outer boundary" of the dust devil, while smaller-size particles spiral up within the dust devil core and draw an "inner boundary".

Medium-sized particles adopt an intermediate behavior. This implies that size discrimination is at play within dust devils (as was hypothesized by Ferrel in 1881, see Lorenz et al. 2016 in this issue), and that mostly smaller particles are being helically raised within (or even above) the CBL.

LES has not yet reached the level where high-resolution dynamical integration can be coupled to reliable dust emission scheme, in order to produce a very detailed simulation of the inner workings of dust devils. However, this goal is within reach in the near future given the recent body of work on this topic (e.g. Klose and Shao 2016). This will confirm the conclusions drawn from the aforementioned idealized vortex modeling with particle injection (Gu et al. 2006), which does not account for a complete distribution of particle sizes, and lacks smaller-scale turbulent components that can be resolved in high-resolution LES. Nevertheless, LES would still not be appropriate to resolve the individual dynamics of transported particles (and collisions between them), for which idealized modeling with a pre-existing vortex will still be of great value.

3.7 A Dust-Devil Parameterization Based on LES?

The preceding discussions make clear that LES studies for the terrestrial CBL recently progressed from reproducing spontaneous convective vortices towards obtaining predictions for real dust devils, i.e. convective vortices that have gained the capability to lift and transport dust particles (e.g. Klose and Shao 2013). Since obtaining in-situ field observations for a population of dust devils remains a difficult task (Lorenz 2014, and see Murphy et al. 2016 and Lorenz and Jackson 2016 in this issue), the relative success of LES in representing dust devil-like vortices in recent years highlights a unique point in dust devil research, where LES can be used as a tool to improve our understanding and representation of these phenomena.

It is now possible to envision the use of LES to design a physical model ("parameterization") to account for the effects of dust devils, notably on the dust cycle, in GCM and RCM which cannot resolve those phenomena—and are unlikely to do so within the foreseeable future. This approach is distinct from the earlier dust devil parameterizations based on thermodynamical arguments (Renno et al. 1998, see also Klose et al. 2016 in this issue). Using relatively large domains (on the order of 4 km in the horizontal) and high resolutions (down to 2 m in the horizontal), a population of dust devil-like vortices, with different sizes and strengths, can be investigated under different initial conditions (Weißmüller et al. 2016). Moreover, Klose and Shao (2016) showed that relating to the Richardson number the dust-devil emission and counts predicted through LES is a valuable means to estimate regional and global dust transport by dust devils. A fully mature dust devil parameterization based on LES has yet to emerge, however. One has to represent a full range of dust devil behaviour changes caused by different meteorological (i.e., ambient wind, temperature profile) and surface conditions (i.e, soil properties, terrain and roughness). This requires numerous, computationally expensive simulations to fully explore these effects. To further complicate matters, an interplay of mechanisms contribute to dust emission by dust devils: saltation bombardment, ΔP effect (see Neakrase et al. 2016 in this issue), direct emission by CBL gusts (see Sect. 3.6), and electrification of sand and dust particles (e.g., Huang et al. 2008, and see Harrison et al. 2016 in this issue for further details). Each of these phenomena warrants further investigation—some of them possibly with LES—before a dust devil parameterization could be used in

GCM and RCM, leading to improved estimates of their contribution to the global mineral dust budget, which is a key component of the Earth system (Knippertz and Stuut 2014).

4 LES for Mars

4.1 First LES Studies and Properties of Simulated Vortices

As is summarized in the introduction, studying convective vortices and dust devils on Mars through LES was part of a broader body of work aiming to understand CBL dynamics on Mars: daytime CBL growth, polygonal cells, thermal up/downdrafts and convective vortices (Rafkin et al. 2001; Toigo et al. 2003; Michaels and Rafkin 2004; Sorbjan 2007; Richardson et al. 2007; Tyler et al. 2008; Spiga and Forget 2009; Spiga et al. 2010; Gheynani and Taylor 2011). Here we only review LES studies where the properties of modeled convective vortices are discussed (Table 2). Those modeling studies of the CBL started when quantitative *in-situ* and orbital measurements complemented the early detection of dust devils structures in imagery (Thomas and Gierasch 1985; Schofield et al. 1997; Malin and Edgett 2001, see Murphy et al. 2016 and Fenton et al. 2016 in this issue for detailed discussions, and Figure 1 upper right for an example image of a dust devil).

The pioneering two-dimensional Martian LES by Odaka et al. (1998) provided insight into CBL growth and associated vertical velocities, but was not suitable to study the emergence of convective vortices (because, e.g., vortex stretching is not being permitted in two-dimensional dynamical integrations). The first successful attempt by Rafkin et al. (2001) to perform three-dimensional LES on Mars reports the emergence of convective vortices (Fig. 11) at the boundaries of open polygonal cells formed by narrow updrafts (and broad downdrafts in the middle of each cell, dictated by the conservation of mass). The behavior of convective vortices with time was discussed in more detail in Michaels and Rafkin (2004). Convective vortices are predicted from morning (local time 08:30) to late afternoon when the CBL collapses (local time 17:00). In the early afternoon, when the convection is at its most vigorous, LES show numerous convective vortices of diameter 100–1000 m in the first hundred meters above the surface. However, only a small fraction of these vortices resolved through LES evolved into structures in cyclostrophic equilibrium comparable to the observed dust devils, which have a vertical extent of 60 % of the CBL and exhibit no preferred direction of rotation.

Using the LES model of Michaels and Rafkin (2004), and accounting for dust lifting and aerosol transport (see Sect. 2.5), Michaels (2006) was able to show that convective vortices would indeed be conducive to the formation of dust devils, if dust material is available at the surface. This would indeed account for the observed tracks along the ground of darker material, where bright dust is removed by vortices (see Fig. 12, and Reiss et al. 2016 in this issue). Michaels (2006) relates the cycloidal path followed by the simulated dust devil in Fig. 12 to the dynamically-evolving convective motions being similar in magnitude to the weak background wind set in his simulation; LES with stronger background wind predict a more uniform rate of translational motion for the simulated convective vortices.

Toigo et al. (2003) dedicated their paper to an in-depth analysis of convective vortices in Martian LES (Fig. 13), by means of a model using a distinct dynamical core than Rafkin et al. (2001) and Michaels and Rafkin (2004). They noticed a great similarity with the equivalent terrestrial phenomena (Kanak et al. 2000; Kanak 2005), including the formation of vortices preferably at the intersection of the convective cells. A plausible scenario to account

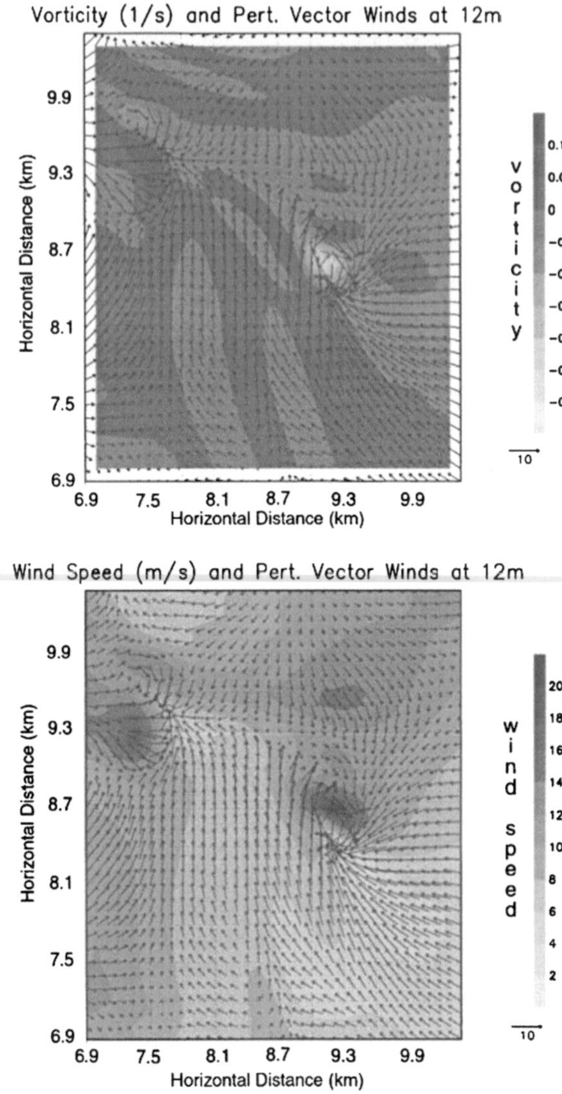

Fig. 11 *Convective vortices in Martian LES*. Dust-devil-like circulations captured during the Large-Eddy Simulation of Rafkin et al. (2001). Fields are shown on a portion of the horizontal domain at about 10 m above the surface: vertical vorticity and vector winds (*top*), wind speed and vector winds (*bottom*). The prescribed background wind velocity has been subtracted to obtain the displayed wind vectors. Figure published in Rafkin et al. (2001), used with permission

for this is the tilting into the vertical of the horizontal vorticity resulting from temperature contrasts in the lowermost levels of the CBL (the same diagnostic is proposed in Michaels and Rafkin 2004). The detailed LES analysis enabled Toigo et al. (2003) to distinguish the contribution of each term in the TKE equation. Convective vortices result from an equilibrium between production of TKE by buoyancy and removal of TKE via both advection and dissipation towards the smaller eddies. Toigo et al. (2003) found a high sensitivity of the activity of the dust devil-like vortices to the background wind. Once formed, a convective vortex in LES obeys a structure that can be described to first order by the thermodynamic theory of Renno et al. (1998) (see Rafkin et al. 2016 and Kurgansky et al. 2016 in this issue).

The properties of modeled convective vortices in the aforementioned LES studies are summarized in Table 2. Despite the range of simulated values obtained in the various stud-

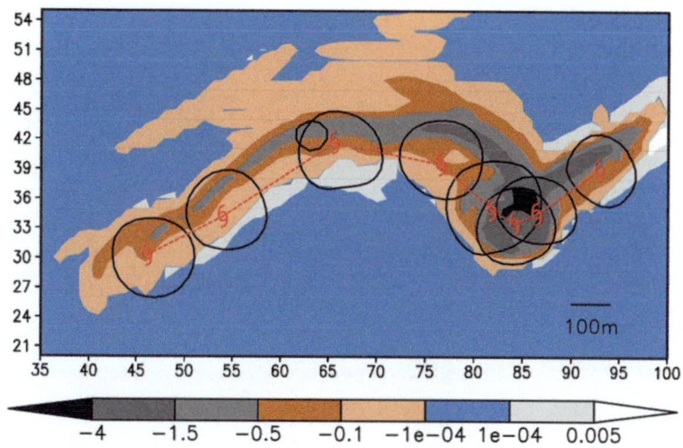

Fig. 12 *Dust devil track predicted by Martian LES.* Simulated track made by a clockwise rotating dust devil (movement is from *left* to *right*; approximately 1100 local time; the x-axis and y-axis represent the indices of horizontal grid points). Depth of net surface dust reservoir change is shaded (mm; negative values denote dust removal). *Red markers* indicate the approximate position of the surface center of the vortex (shown every 75 seconds), and *black contours* indicate the approximate surface radial extent of the vortex core walls (updrafts; the small, unassociated contour near the figure center is a separate, smaller vertical convective vortex). Figure published in Michaels (2006), used with permission

ies, they agree at the order of magnitude level: simulated convective vortices exhibit updraft velocities up to about 10 $m\,s^{-1}$, tangential velocities from 10 to 20 $m\,s^{-1}$ (compliant with the orbital measurements of Choi and Dundas 2011), and central pressure drops of about a couple Pascal (consistent with measurements on board the Martian landers, e.g. Schofield et al. 1997; Ellehoj et al. 2010, although the pressure drops > 5–10 Pa required to form dust devil tracks and to cause solar panel clearing events according to Lorenz and Reiss (2015) are not reproduced by existing LES). Comparing Table 2 with Table 1 appears to confirm that Martian dust devils are more intense than their terrestrial counterparts, when considering velocities and pressure drop scaled to the mean surface pressure on each planet. Those estimates have been bolstered by other LES studies performed with distinct models and in distinct locations (Toigo and Richardson 2003; Spiga and Forget 2009; Spiga and Lewis 2010; Gheynani and Taylor 2011). Not all the studies report the duration of convective vortices in LES, but the most distinctive and extreme predicted vortices are estimated to last about a few thousands of seconds (Toigo and Richardson 2003; Spiga and Forget 2009).

An interesting complement to this whole body of work based on Category 3 LES (see Sect. 2.4) is provided by the Category 1 LES for Mars reported by Kanak (2006) who investigated the minimum physical requirements for vortex formation in the Martian CBL. Despite the arguably very idealized setting compared to usual Mars LES (no radiative processes, simpler surface layer and sub-grid scale parameterizations, little to no background wind), Kanak (2006) showed that convective vortices still arise with similar properties as those predicted by more sophisticated LES—and that the mechanism by which they acquire rotation is, as discussed by others, deeply rooted in the CBL dynamics (i.e. dust devils do not develop independently from the CBL growth, but as a part of it). The fact that every Martian lander (see Murphy et al. 2016 in this issue) was able to detect pressure drops characteristic of convective vortices supports their ubiquitous presence in Mars LES. This confirms that studying convective vortices is a means to more generally study CBL processes, because those develop wherever the CBL grows deep enough.

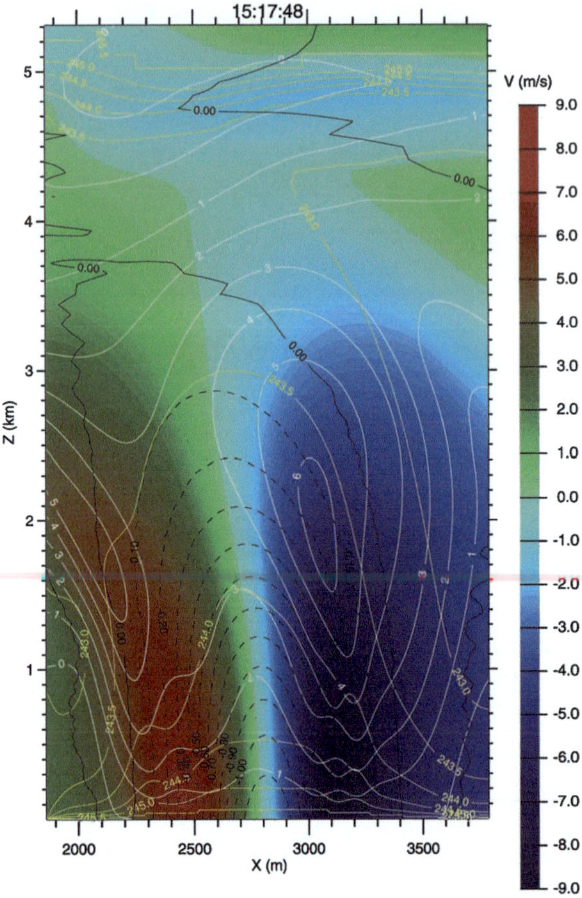

Fig. 13 *The vertical structure of one convective vortex obtained in Martian LES.* High-resolution Large-Eddy Simulation ($\Delta x = 10$ m) of the "no wind" simulation dust devil of Toigo et al. (2003). Here is plotted a vertical slice through the center of the dust devil. *Background color* shows the tangential wind speed. *Black contours* show the pressure perturbation in Pa, reaching a maximum difference near the surface of about 1 Pa less than the background. *Yellow contours* show potential temperature in K, and the warm core of the dust devil. *White contours* show upward wind velocity in $\mathrm{m\,s^{-1}}$. Upward wind velocity peaks at the walls of the dust devil, and the decrease in upward velocity can be seen in the center of the dust devil core. Figure published in Toigo et al. (2003), used with permission

4.2 Regional Variability

Determining tendencies for the regional variability of dust devils from the LES studies reported in Table 2 is difficult. First and foremost, as is indicated in Table 2, LES settings in various studies are too different (horizontal resolution, background wind). But even keeping this caveat in mind, the actual regional variability in dust devils—e.g. what makes Amazonis Planitia such a propitious place for dust devil occurrence (Fisher et al. 2005), or the fact that the Spirit rover witnessed far more dust devils than the Opportunity rover (Greeley et al. 2006)—remains an open question that LES has not addressed yet. A plausible explanation for this is the regional variability of dust availability, which is not accounted for in existing Martian LES. Another possibility is that the occurrence of dust devils, as well as their height and spacing Fenton and Lorenz (2015), somehow relates to the intensity of convective vortices—which makes it possible for existing LES to solve this scientific issue. Although this problem remains unsolved thus far, recent studies were able to draw a few preliminary conclusions or perspectives.

Radio-occultation measurements showed that the Martian CBL appeared to extend to higher altitudes over high plateaus than in lower altitude plains, despite similar surface temperatures in the considered regions (Hinson et al. 2008). Spiga et al. (2010) showed that these

Table 2 Same table as Table 1 for published LES studies in Martian conditions. The LES by Rafkin et al. (2001), Toigo et al. (2003), Michaels and Rafkin (2004), Michaels (2006) were run for environmental conditions typical of Meridiani Planum; Toigo and Richardson (2003) and Spiga and Forget (2009) of the plain within Gusev crater; Spiga and Lewis (2010) of Amazonis Planitia; Gheynani and Taylor (2011) of Vastitas Borealis; and LES by Nishizawa et al. (2016) were run with idealized Martian radiative forcing

Paper	Dyn. core	dx (m)	$N_{x,y}$	$\mathcal{Z}$ (km)	N_z	w' (m/s)	v' (m/s)	P' (Pa)	ζ (s^{-1})
Rafkin et al. (2001)	RAMS	100	180	9	36	10	10	2	0.3
Toigo et al. (2003)	MM5	10	200	6	140	7	9	1	0.06
Toigo and Richardson (2003)	MM5	100	100	7.5	57	7	5	–	–
Michaels and Rafkin (2004)	RAMS	30	200	2.3	61	5	10	2	0.2
Michaels (2006)	RAMS	25	102	8	120	12	19	6.5	–
Spiga and Forget (2009)	WRF	100	150	9.5	71	8	10	1.5	–
Spiga and Lewis (2010)	WRF	50	145	12	201	12	10	2	–
Gheynani and Taylor (2011)	NCAR	25	200	10	60	4	15	2.5	0.7
Nishizawa et al. (2016)	SCALE	5	4000	21	4000	10	20	4.2	2

dramatic regional variations of CBL depth are qualitatively and quantitatively predicted by LES by accounting for the regional differences in the surface properties and incoming sunlight. Those regional variations of CBL depth arise because the prominent radiative control of the Martian CBL (as opposed to sensible heat driving the terrestrial CBL) makes the intensity of convective motions correlated with the inverse surface pressure: the LES of Spiga et al. (2010) hence predict stronger near-surface horizontal winds and CBL vertical winds over high plateaus than over lower plains. This led to a plausible explanation as to how dust devils would be found at the Martian volcanoes' summit where atmospheric density would be too low to allow for dust to be lifted from the surface (Reiss et al. 2009).

Most Mars LES models are still rather idealized (see Sect. 2.2) and, without the forcing of the regional and large-scale circulation, should not be assumed too representative for any specific location on the planet—especially if the region is characterized by uneven topography or strong contrasts in surface thermophysical properties. For terrestrial LES, neglecting the larger-scale circulation is typically seen as a reasonable approach, although similar caveats apply if significant mesoscale circulations are present. For LES on Mars, although the CBL depth is primarily controlled by incoming sunlight and surface thermophysical properties, both the regional circulations (e.g. slope winds) and the strong atmospheric thermal tides can play very important roles in the diurnal growth of the CBL:

- Strong atmospheric thermal tides produce large diurnal and semidiurnal amplitudes in the surface pressure cycle. This leads to large and rapid variations in the atmospheric horizontal pressure gradient. When this rapidly varying forcing interacts with the growth of a deep Martian CBL, the result can vary greatly when compared to a case without such forcing. Using a prescribed time- and height-dependent geostrophic wind (a theoretical wind that would exist in an exact balance between the pressure gradient force and the Coriolis effect) computed from RCM for the Phoenix landing site, Tyler et al. (2008) showed that daytime convection was stronger than with a fixed uniform background wind, which leads to the formation of a deeper CBL. Regional circulation can produce sizeable differences in LES results, but the effect thereof on dust devil activity remains to be quantified.

- Whether convergence/divergence exists in the local circulation is another important aspect. In crater circulations, daytime convergence at the height of the surrounding terrain above the floor of the crater produces a temperature inversion due to the adiabatic warming of widespread subsidence (Tyler and Barnes 2013, 2015). This provides a plausible explanation for the small CBL depths assessed by imagery from the Curiosity rover operating within Gale Crater (Moores et al. 2015). Given the topography of Mars, and the strong slope flows that develop, this reality poses an additional problem to LES on Mars for specific locations. One investigative path forward may be to add a heating term in the LES that mimics the adiabatic warming of subsidence. As is the case for the preceding point, the impact of this on dust devil activity (and a possible explanation for the lack of observed dust devils by Curiosity, see Murphy et al. 2016 in this issue) is yet to be determined.

4.3 Sensitivity Exploration

Recent work on LES for Mars has explored the sensitivity of earlier results to various modeling parameters. This was discussed notably by Fenton and Michaels (2010), although the aim of their set of LES experiments at two distinct seasons, and using different conditions of background wind, was to determine the velocity distribution of near-surface CBL winds, rather than to focus on convective vortices alone. However, the tail of the wind distribution provides direct information on the strength of near-surface winds imposed by convective vortices, and possibly dust devils since maximum wind speeds in the LES of Fenton and Michaels (2010) do exceed the stress threshold to mobilize sand grains on Mars. Fenton and Michaels' LES show that maximum friction velocities increase with incoming sunlight and background wind, although simulations without background wind also produce a high speed wind tail that exceeds the local saltation stress threshold. Interestingly, the LES using strong uniform winds with superimposed wind shear yield a wind distribution which departs significantly from the Weibull distribution (unlike the LES with lesser background winds). Since convective vortices are associated with the tail of the wind distribution, those results could indicate, similarly to terrestrial studies (Sect. 3.3), that a significant rise in large-scale and regional winds could impact the occurrence of dust devils.

A notable sensitivity study on convective vortices predicted by Martian LES was carried out recently by Gheynani and Taylor (2011), who used a higher resolution of 25 m (see Table 2). Gheynani and Taylor (2011) performed LES with various background wind speeds $(0, 4, 8\ \mathrm{m\,s^{-1}})$ and two distinct assumptions on radiative forcing (comparing Category 1 LES vs. Category 2 LES, as defined in Sect. 2.4). Their study shows a decrease of vortex populations in windy conditions, finding that mean vortex tangential velocity increases accordingly with both pressure drop and vorticity. Gheynani and Taylor (2011) add to the current knowledge the result that smaller and more intense vortices form in LES with larger background wind. This indicates that background wind destroys weak vortices—or limits their formation, and only vortices with enough momentum will survive in windy conditions (see Rafkin et al. 2016 in this issue for detailed discussions on the relationship of background wind to dust devil size and number). The LES of Gheynani and Taylor (2011) with and without radiative forcing (Category 1 vs. Category 2 LES) indicate that vortices' vertical velocities slightly increase when radiative forcing is taken into account. The differences between both cases in vorticity, pressure drops, and tangential velocities is only about 1–5 % (Fig. 14). Thus, the effect of including radiative forcing in Gheynani and Taylor (2011) was relatively

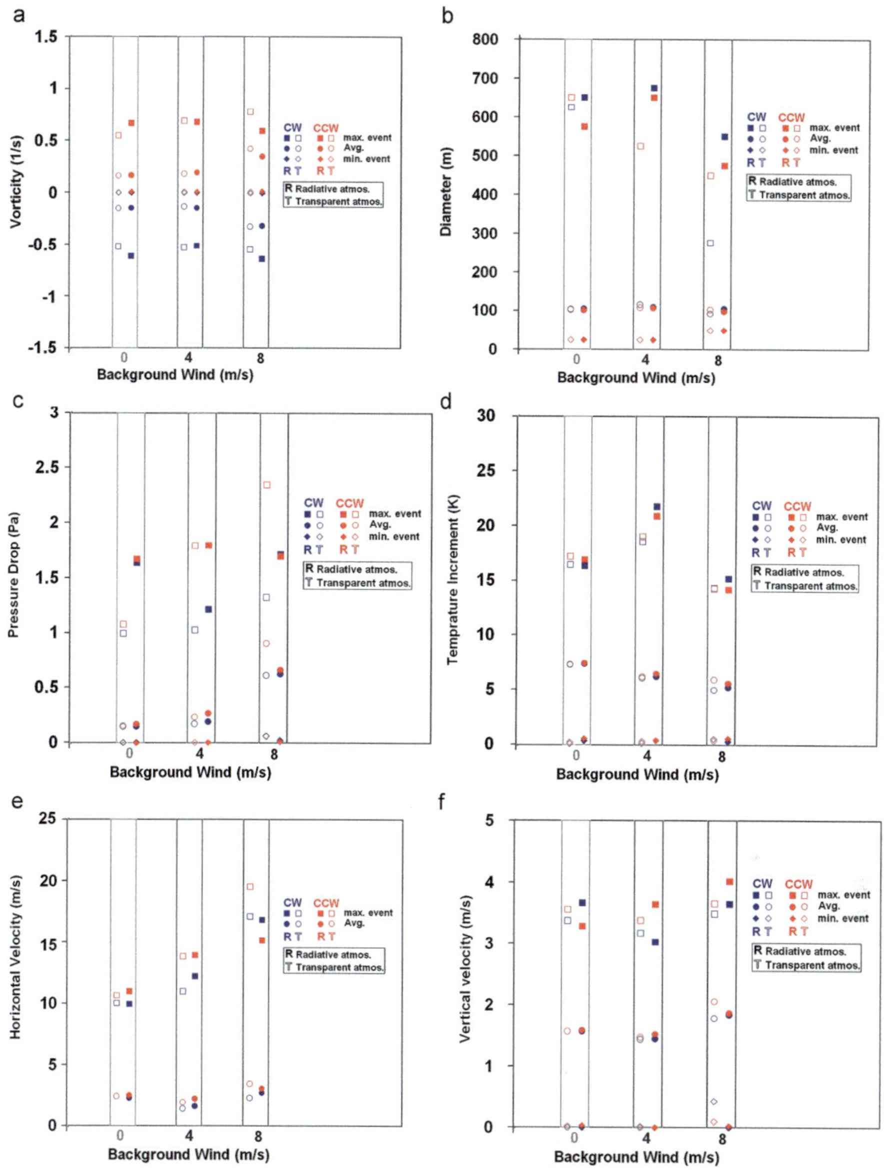

Fig. 14 *Properties of simulated convective vortices in Martian LES with distinct model assumptions.* Vortices' maximum (*square*) and minimum (*diamond*) property values, and the average (*circle*) of (**a**) vorticity (**b**) diameter (**c**) pressure drop (**d**) temperature increment (**e**) tangential velocity (**f**) vertical velocity, during the last 45 min (1100–3800 s) of simulation time at 20 m above the surface. *Blue stands* for clockwise vortices, *red stands* for counter-clockwise vortices. Figure published in Gheynani and Taylor (2011), used with permission

insignificant, if not negligible (confirming the results in Kanak 2006, cf. Sect. 4.1). This implies that, while radiative forcing is key on Mars to obtain appropriate CBL structure and dynamics (see e.g. Spiga et al. 2010), it is less of a requirement to obtain correct dust devil properties and statistics on Mars. This also suggests a closer similarity than expected with terrestrial LES (see also Gheynani and Taylor 2010). However, further investigation is warranted since the LES of Gheynani and Taylor (2011) were only hour-long runs which did not account for CBL growth during the day, and no radiatively active dust was being advected into the simulated vortices.

4.4 Perspectives and Challenges

The ability of Martian LES to reproduce convective vortices that bear resemblance to actual dust devils opens interesting perspectives. One of them is the possibility to simulate electrostatic disturbances and the electric field induced by dust devils through triboelectric charging of saltating and colliding sand and dust particles (e.g. Renno et al. 2003, see also Harrison et al. 2016 in this issue). Following Michaels (2006), Barth et al. (2014) coupled a detailed aerosol model (operating on discrete mass bins) to the LES model of Rafkin et al. (2001) run at an improved horizontal resolution of 33 m, and incorporated dynamic spatial redistribution of initially charged dust particles in order to evaluate the possibilities for electric fields generated within the dust devils in their LES. Using the Microscopic Triboelectric Simulation (MTS; see Harrison et al. 2016 in this issue for further description), they added tracers to represent the moments of a distribution of particle charges across each mass bin. As the dust was lifted and redistributed spatially by both convective vortices and gravitational settling, the charge density at each grid point would also evolve. They found consistency with observations of terrestrial dust devils in that the smaller, negatively charged particles tended to almost exclusively occupy the top portions of the vortex, while the larger, positively charged particles were more likely to fall to the surface. This charge separation allowed for the calculation of electric fields using Poisson's equation. In addition to electric fields generated by the dust devils, Barth et al. (2014) also noted that gust fronts along the boundaries of the convective cells could lift substantial amounts of dust and thus result in larger-scale electric fields. Given an initial grid box size of 100 m^2 in the horizontal, these gust front E-fields tended to overwhelm anything produced within the dust devil. The above mode of grain moving, mixing and collision could develop a small-size (lighter) grain domain aloft and a near-surface large-size (heavier) grain domain (i.e. grain mass stratification), and thus produce strong electric fields within dust devils. The dust cloud has a macroscopic electric dipole moment oriented in the nadir (downward) direction, which might play a significant role in dust sourcing.

In a more general scope, perspectives and challenges for Martian LES somewhat mirror those for terrestrial LES (Sect. 3.7), although observational limitations in the Martian case prevent dust emission schemes as sophisticated as the Earth's (Klose et al. 2014, see Sect. 3.6) to be calibrated. As is the case on the Earth, progress is being made that may soon lead to parameterizations in GCM and RCM for lifting by dust devils (and, more generally speaking, by CBL structures). These would be based on CBL statistics drawn from LES (Fenton and Michaels 2010; Mulholland et al. 2015) rather than thermodynamical arguments, as is the case in existing GCM parameterizations (Renno et al. 1998; Newman et al. 2002) (see Klose et al. 2016 in this issue for further discussions). Given that dust is a major driver of the Martian climatic system (e.g. Leovy 1982), improving dust devil parameterizations in GCM and RCM is an even more crucial goal for Mars than it is for the Earth.

Martian LES also share with terrestrial LES the quest for the combination of mesh resolutions and wide domains suitable to resolve a fair share of the size distribution of convective vortices and dust devils (cf. Sect. 3.4). This is an even more stringent issue in the

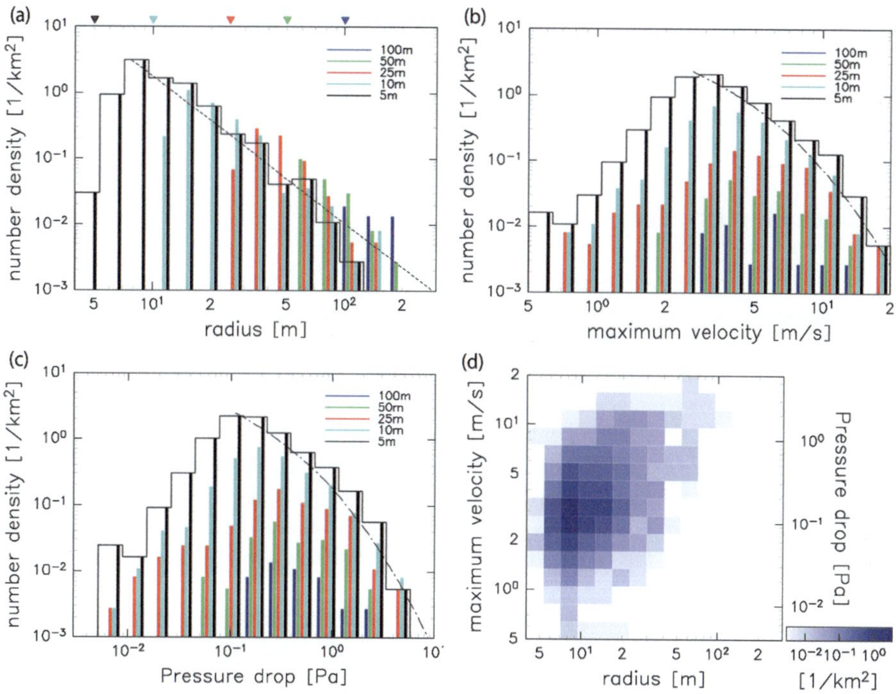

Fig. 15 *Statistics of convective vortices obtained in Martian LES while refining the horizontal resolution.* Histograms of (**a**) radius, (**b**) maximum velocity, (**c**) pressure drop, and (**d**) two-dimensional joint histogram of the radius and maximum velocity for convective vortices simulated at 14:30 LT around 62.5 m altitude in a 5 m resolution LES run. In Figures (**a–c**), the *thin lines* show the histogram of the 5 m resolution run, with the dependency of the count in each bin on the resolution shown by the bars. *Colors* represent the horizontal resolution: 100 m (*blue*), 50 m (*green*), 25 m (*red*), 10 m (*cyan*), 5 m (*black*). *Dotted lines* in Figure (**a**) represents the fitted power law distribution, and the *dash-dotted line* in Figures (**b**) and (**c**) represents the fitted Weibull distribution. *Triangles* in the *upper axis* of Figure (**a**) represent the grid spacing. Figure published in Nishizawa et al. (2016), used with permission

Martian case than in the terrestrial case (cf. Sect. 2.2). Recently, the potentialities of high-resolution LES performed on High-Performance Computing resources have been demonstrated by Nishizawa et al. (2016), who use a Martian version of Nishizawa et al. (2015)'s LES model with horizontal resolutions ranging from 100 m to an unprecedented 5 m, while keeping the domain size 20 km wide. The combination of high horizontal resolution with a wide domain allows Nishizawa et al. (2016) to gather a large sample of simulated vortices on which to compute statistics (Fig. 15). While a power law is the best fit distribution for the radii of simulated convective vortices (which is also suggested by observations, Lorenz 2011, although the sample size is too small to reach a robust conclusion), the maximum velocities and pressure drops follow a Weibull distribution (confirming the diagnostic of Fenton and Michaels 2010, when they performed LES with low background wind). The LES of Nishizawa et al. (2016) demonstrates that the transition to high-resolution LES not only affect the size of simulated vortices, but also the vertical vorticity which is nearly one order of magnitude larger than obtained in coarse-resolution LES (cf. Table 2), and the area fraction of vortices which is larger for high-resolution runs, despite a smaller number of simulated large vortices. Moreover, high-resolution LES suggest that an apparent linear relationship between the radius and velocity for large vortices seems to disappear for smaller vortices (Fig. 15d). Eventually, Nishizawa et al. (2016) note that the mode of the radii histogram still depends on horizontal resolution, even in the finest 5 m run. This suggests that further work is still needed with high-resolution Martian LES to predict the full population of convective vortices developing in the Martian CBL.

5 Concluding Remarks

The study of dust devils and convective vortices by Large-Eddy Simulations (LES) is at a relatively early stage of development since it started in the late 1990s/early 2000s. Nevertheless, the community has progressed significantly on this topic in the decade 2005–2015: the two short paragraphs mentioning LES for dust devils in the Balme and Greeley (2006) review paper has grown ten years later into a full chapter in this issue.

Using LES, modelers were not only able to simulate convective vortices with properties akin to the observed dust devils (without any *ad hoc* assumptions about those structures), but were also able to obtain crucial information on their statistics and variability. Those major LES achievements were obtained jointly for dust devils on the Earth and on Mars, with some studies addressing the two environments with similar modeling tools (Kanak 2006; Gheynani and Taylor 2010).

Numerous challenges still remain and are likely to arouse the interest of LES modelers in the next decades both for the terrestrial and Martian environments.

1. There is a strong need to run LES at finer horizontal mesh spacing than what has been done thus far—and solve the potential issues arising with this mesh refinement (including the need to use computational resources powerful enough to keep a horizontal domain size similar to LES with coarser mesh spacing). This is required to fully resolve the statistics of the (numerous) small vortices and to assess the fine scales of the vortex structure within dust devils.
2. LES resolve convective vortices that could give rise to dust devils, if dust particles were to be lifted and transported within the vortex. The latter element of dust devil formation has yet to be routinely included in Martian LES, and is still a nascent area of research in terrestrial LES. It would be exaggeration to state that LES have not reproduced any true dust devil yet, but work is still needed to obtain a complete numerical view of the formation of dust devils.
3. A follow-up challenge is to perform LES accounting for the radiative impact of the dust particles lifted and transported within the vortex. Could this radiative-dynamical coupling affect the characteristics and duration of dust devils?
4. LES for dust devils are still idealized numerical experiments: complex topography and albedo contrasts, as well as varying large-scale and regional meteorological conditions, have not yet been incorporated, although they must be included to make progress on the understanding of the variability of dust devil activity with environmental conditions, and on the comparative studies between LES and observations. Heterogeneities (heating, topography, roughness length) can already be studied within the existing LES modeling framework, although it has not been done for dust devil studies yet. As far as environmental meteorological conditions are concerned, coupling LES numerical integrations with those of a Global Climate Model (GCM) or a Regional Climate Model (RCM) is probably a very promising method to explore.
5. LES results remain to be employed to improve physical parameterizations of dust devils in GCM and RCM (where those phenomena are unresolved). Such an effort has been

recently started for the Earth, but is left to be initiated on Mars, where the question of whether or not dust devils play a major part in maintaining the permanent veil of dust particles in the Martian atmosphere is still open.

This list of challenges is only exhaustive to the extent of our current knowledge. Undoubtedly the list will grow longer as new observations, modeling tools, and computational resources become available. The former element is probably the most important: improving LES models is not only a matter of numerical techniques, but critically depends on better and more complete observations with which to validate the models.

Acknowledgements We acknowledge the logistic and financial help of the International Space Science Institute (ISSI, Bern, Switzerland) for the organization of a "dust devils" international workshop that led to the writing of this review chapter. We are indebted to two anonymous reviewers and associate editor Ralph Lorenz for constructive comments which helped to improve this review.

References

M. Balme, R. Greeley, Dust devils on earth and mars. Rev. Geophys. **44**(3) (2006)

E.L. Barth, W.M. Farrell, S.C.R. Rafkin, Electric fields in simulated martian dust devils, in *Mars Atmosphere: Modelling and Observation, 5th International Workshop*, ed. by F. Forget, M. Millour, 2014, p. 2204

H.B. Bluestein, C.C. Weiss, A.L. Pazmany, Doppler radar observations of dust devils in Texas. Mon. Weather Rev. **132**, 209 (2004). doi:10.1175/1520-0493(2004)132<0209:DROODD>2.0.CO;2

D.S. Choi, C.M. Dundas, Measurements of martian dust devil winds with HiRISE. Geophys. Res. Lett. **38**, 24206 (2011). doi:10.1029/2011GL049806

T. Cortese, S. Balachandar, Vortical nature of thermal plumes in turbulent convection. Phys. Fluids A, Fluid Dyn. (1989–1993) **5**(12), 3226–3232 (1993)

J.W. Deardorff, Numerical investigation of neutral and unstable planetary boundary layers. J. Atmos. Sci. **29**, 91–115 (1972). doi:10.1175/1520-0469(1972)029<0091:NIONAU>2.0.CO;2

S. Dupont, G. Bergametti, B. Marticorena, S. SimoëNs, Modeling saltation intermittency. J. Geophys. Res., Atmos. **118**, 7109–7128 (2013). doi:10.1002/jgrd.50528

D.R. Durran, *Numerical Methods for Fluid Dynamics: With Applications to Geophysics*. Texts in Applied Mathematics (Springer, Berlin, 2010)

M.D. Ellehoj, H.P. Gunnlaugsson, P.A. Taylor, H. Kahanpää, K.M. Bean, B.A. Cantor, B.T. Gheynani, L. Drube, D. Fisher, A.-M. Harri, C. Holstein-Rathlou, M.T. Lemmon, M.B. Madsen, M.C. Malin, J. Polkko, P.H. Smith, L.K. Tamppari, W. Weng, J. Whiteway, Convective vortices and dust devils at the Phoenix Mars mission landing site. J. Geophys. Res., Planets **115**(E14), 16 (2010). doi:10.1029/2009JE003413

L.K. Fenton, R. Lorenz, Dust devil height and spacing with relation to the martian planetary boundary layer thickness. Icarus **260**, 246–262 (2015). doi:10.1016/j.icarus.2015.07.028

L.K. Fenton, T.I. Michaels, Characterizing the sensitivity of daytime turbulent activity on Mars with the MRAMS LES: early results. Int. J. Mars Sci. Explor. **5**, 159–171 (2010). doi:10.1555/mars.2010.0007

L. Fenton, D. Reiss, M. Lemmon, B. Marticorena, S. Lewis, B. Cantor, Orbital observations of dust lofted by daytime convective turbulence. Space Sci. Rev. 1–54 (2016). doi:10.1007/s11214-016-0243-6

B.H. Fiedler, K.M. Kanak, Rayleigh–Bénard convection as a tool for studying dust devils. Atmos. Sci. Lett. **2**, 104–113 (2001). doi:10.1006/asle.2001.0043

J.A. Fisher, M.I. Richardson, C.E. Newman, M.A. Szwast, C. Graf, S. Basu, S.P. Ewald, A.D. Toigo, R.J. Wilson, A survey of martian dust devil activity using Mars global surveyor Mars orbiter camera images. J. Geophys. Res., Planets **110**(E9), 3004 (2005). doi:10.1029/2003JE002165

S.D. Fuerstenau, Solar heating of suspended particles and the dynamics of martian dust devils. Geophys. Res. Lett. **33**, 19 (2006). doi:10.1029/2006GL026798

B.T. Gheynani, P.A. Taylor, Large-eddy simulations of vertical vortex formation in the terrestrial and martian convective boundary layers. Bound.-Layer Meteorol. **137**, 223–235 (2010). doi:10.1007/s10546-010-9530-z

B.T. Gheynani, P.A. Taylor, Large eddy simulation of typical dust devil-like vortices in highly convective martian boundary layers at the Phoenix lander site. Planet. Space Sci. **59**, 43–50 (2011). doi:10.1016/j.pss.2010.10.011

R. Greeley, P.L. Whelley, R.E. Arvidson, N.A. Cabrol, D.J. Foley, B.J. Franklin, P.G. Geissler, M.P. Golombek, R.O. Kuzmin, G.A. Landis, M.T. Lemmon, L.D.V. Neakrase, S.W. Squyres, S.D. Thompson, Active dust devils in Gusev crater, Mars: observations from the Mars exploration rover spirit. J. Geophys. Res., Planets **111**(E10), 12 (2006). doi:10.1029/2006JE002743

Z. Gu, Y. Zhao, Y. Li, Y. Yu, X. Feng, Numerical simulation of dust lifting within dust devils simulation of an intense vortex. J. Atmos. Sci. **63**, 2630–2641 (2006). doi:10.1175/JAS3748.1

R.M. Haberle, H.C. Houben, R. Hertenstein, T. Herdtle, A boundary layer model for Mars: comparison with Viking lander and entry data. J. Atmos. Sci. **50**, 1544–1559 (1993)

R.G. Harrison, E. Barth, F. Esposito, J. Merrison, F. Montmessin, K.L. Aplin, C. Borlina, J.J. Berthelier, G. Déprez, W.M. Farrell, I.M.P. Houghton, N.O. Renno, K.A. Nicoll, S.N. Tripathi, M. Zimmerman, Applications of electrified dust and dust devil electrodynamics to martian atmospheric electricity. Space Sci. Rev. 1–47 (2016). doi:10.1007/s11214-016-0241-8

D.P. Hinson, M. Pätzold, S. Tellmann, B. Häusler, G.L. Tyler, The depth of the convective boundary layer on Mars. Icarus **198**, 57–66 (2008). doi:10.1016/j.icarus.2008.07.003

N. Huang, G. Yue, X. Zheng, Numerical simulations of a dust devil and the electric field in it. J. Geophys. Res., Atmos. **113**(D20) (2008)

J. Ito, H. Niino, M. Nakanishi, Large eddy simulation on dust suspension in a convective mixed layer. SOLA **6**, 133–136 (2010). doi:10.2151/sola.2010-034

J. Ito, R. Tanaka, H. Niino, M. Nakanishi, Large eddy simulation of dust devils in a diurnally-evolving convective mixed layer. J. Meteorol. Soc. Jpn. **88**, 64–77 (2010)

J. Ito, H. Niino, M. Nakanishi, Effects of ambient rotation on dust devils. SOLA **7**, 165–168 (2011). doi:10.2151/sola.2011-042

J. Ito, H. Niino, M. Nakanishi, Formation mechanism of dust devil-like vortices in idealized convective mixed layers. J. Atmos. Sci. **70**, 1173–1186 (2013). doi:10.1175/JAS-D-12-085.1

K.M. Kanak, Numerical simulation of dust devil-scale vortices. Q. J. R. Meteorol. Soc. **131**(607), 1271–1292 (2005)

K.M. Kanak, On the numerical simulation of dust devil-like vortices in terrestrial and martian convective boundary layers. Geophys. Res. Lett. **33**(19) (2006)

K.M. Kanak, D.K. Lilly, J.T. Snow, The formation of vertical vortices in the convective boundary layer. Q. J. R. Meteorol. Soc. **126**(569), 2789–2810 (2000)

M. Klose, Y. Shao, Stochastic parameterization of dust emission and application to convective atmospheric conditions. Atmos. Chem. Phys. **12**(12), 7309–7320 (2012). doi:10.5194/acp-12-7309-2012

M. Klose, Y. Shao, Large-eddy simulation of turbulent dust emission. Aeolian Res. **8**, 49–58 (2013). doi:10.1016/j.aeolia.2012.10.010

M. Klose, Y. Shao, A numerical study on dust devils with implications to global dust budget estimates. Aeolian Res. **22**, 47–58 (2016). doi:10.1016/j.aeolia.2016.05.003

M. Klose, Y. Shao, X.L. Li, H.S. Zhang, M. Ishizuka, M. Mikami, J.F. Leys, Further development of a parameterization for convective turbulent dust emission and evaluation based on field observations. J. Geophys. Res., Atmos. **119**, 10441–10457 (2014). doi:10.1002/2014JD021688

M. Klose, B.C. Jemmett-Smith, H. Kahanpää, M. Kahre, P. Knippertz, M.T. Lemmon, S.R. Lewis, R.D. Lorenz, L.D.V. Neakrase, C. Newman, M.R. Patel, D. Reiss, A. Spiga, P.L. Whelley, Dust devil sediment transport: from lab to field to global impact. Space Sci. Rev. 1–50 (2016). doi:10.1007/s11214-016-0261-4

P. Knippertz, J.-B.W. Stuut, *Mineral Dust: A Key Player in the Earth System* (Springer, Netherlands, 2014)

M.V. Kurgansky, R.D. Lorenz, N.O. Renno, T. Takemi, Z. Gu, W. Wei, Dust devil steady-state structure from a fluid dynamics perspective. Space Sci. Rev. ["Dust devils" special issue] (2016, submitted). doi:10.1007/s11214-016-0281-0

C.B. Leovy, Martian meteorological variability. Adv. Space Res. **2**, 19–44 (1982)

D.K. Lilly, On the numerical simulation of buoyant convection. Tellus **14**(2), 148–172 (1962)

G.A. Loosmore, J.R. Hunt, Dust resuspension without saltation. J. Geophys. Res. **105**(D16), 20663–20671 (2000). doi:10.1029/2000JD900271

R. Lorenz, On the statistical distribution of dust devil diameters. Icarus **215**, 381–390 (2011). doi:10.1016/j.icarus.2011.06.005

R.D. Lorenz, Vortex encounter rates with fixed barometer stations: comparison with visual dust devil counts and large-eddy simulations. J. Atmos. Sci. **71**(12), 4461–4472 (2014)

R.D. Lorenz, B.K. Jackson, Dust devil populations and statistics. Space Sci. Rev. 1–21 (2016). doi:10.1007/s11214-016-0277-9

R.D. Lorenz, D. Reiss, Solar panel clearing events, dust devil tracks, and in-situ vortex detections on Mars. Icarus **248**, 162–164 (2015). doi:10.1016/j.icarus.2014.10.034

R.D. Lorenz, M.R. Balme, Z. Gu, H. Kahanpää, M. Klose, M.V. Kurgansky, M.R. Patel, D. Reiss, A.P. Rossi, A. Spiga, T. Takemi, W. Wei, History and applications of dust devil studies. Space Sci. Rev. 1–33 (2016). doi:10.1007/s11214-016-0239-2

M.C. Malin, K.S. Edgett, Mars global surveyor Mars orbiter camera: interplanetary cruise through primary mission. J. Geophys. Res. **106**, 23429–23570 (2001). doi:10.1029/2000JE001455

P. Mason, Large-eddy simulation of the convective atmospheric boundary layer. J. Atmos. Sci. **46**(11), 1492–1516 (1989)

S.M. Metzger, M.R. Balme, M.C. Towner, B.J. Bos, T.J. Ringrose, M.R. Patel, In situ measurements of particle load and transport in dust devils. Icarus **214**(2), 766–772 (2011). doi:10.1016/j.icarus.2011.03.013

T.I. Michaels, Numerical modeling of Mars dust devils: albedo track generation. Geophys. Res. Lett. **33**(19) (2006). doi:10.1029/2006GL026268

T.I. Michaels, S.C.R. Rafkin, Large eddy simulation of atmospheric convection on Mars. Q. J. R. Meteorol. Soc. **130**, 1251–1274 (2004). doi:10.1256/qj.02.169

J.D. Mirocha, J.K. Lundquist, B. Kosovic, Implementation of a nonlinear subfilter turbulence stress model for large-eddy simulation in the advanced research WRF model. Mon. Weather Rev. **138**, 4212–4228 (2010)

C. Moeng, J. Dudhia, J. Klemp, P. Sullivan, Examining two-way grid nesting for large eddy simulation of the PBL using the WRF model. Mon. Weather Rev. **135**(6), 2295–2311 (2007)

J.E. Moores, M.T. Lemmon, H. Kahanpää, S.C.R. Rafkin, R. Francis, J. Pla-Garcia, K. Bean, R. Haberle, C. Newman, M. Mischna, A.R. Vasavada, M. de la Torre Juárez, N. Rennó, J. Bell, F. Calef, B. Cantor, T.H. Mcconnochie, A.-M. Harri, M. Genzer, M.H. Wong, M.D. Smith, F.J. Martín-Torres, M.-P. Zorzano, O. Kemppinen, E. McCullough, Observational evidence of a suppressed planetary boundary layer in northern Gale crater, Mars as seen by the Navcam instrument onboard the Mars Science Laboratory rover. Icarus **249**, 129–142 (2015). doi:10.1016/j.icarus.2014.09.020

D.P. Mulholland, A. Spiga, C. Listowski, P.L. Read, An assessment of the impact of local processes on dust lifting in martian climate models. Icarus **252**, 212–227 (2015). doi:10.1016/j.icarus.2015.01.017

J. Murphy, K. Steakley, M.B. Balme, G. Deprez, F. Esposito, H. Kahanpää, M. Lemmon, R.D. Lorenz, N. Murdoch, L.D.V. Neakrase, M. Patel, P. Whelley, Field measurements of terrestrial and martian dust devils. Space Sci. Rev. ["Dust devils" special issue] (2016, submitted). doi:10.1007/s11214-016-0283-y

L.D.V. Neakrase, M.B. Balme, F. Esposito, T. Kelling, M. Klose, J.F. Kok, B. Marticonera, J. Merrison, M.R. Patel, G. Wurm, Particle lifting processes in dust devils. Space Sci. Rev. ["Dust devils" special issue] (2016, submitted)

C.E. Newman, S.R. Lewis, P.L. Read, F. Forget, Modeling the martian dust cycle, 1. Representations of dust transport processes. J. Geophys. Res., Planets **107**, 5123 (2002). doi:10.1029/2002JE001910

S. Nishizawa, H. Yashiro, Y. Sato, Y. Miyamoto, H. Tomita, Influence of grid aspect ratio on planetary boundary layer turbulence in large-eddy simulations. Geosci. Model Dev. **8**(10), 3393–3419 (2015). doi:10.5194/gmd-8-3393-2015. http://www.geosci-model-dev.net/8/3393/2015/

S. Nishizawa, M. Odaka, Y.O. Takahashi, K. Sugiyama, K. Nakajima, M. Ishiwatari, S. Takehiro, H. Yashiro, Y. Sato, H. Tomita, Y.-Y. Hayashi, Martian dust devil statistics from high-resolution large-eddy simulations. Geophys. Res. Lett. **43**(9), 4180–4188 (2016). doi:10.1002/2016GL068896

M. Odaka, K. Nakajima, S. Takehiro, M. Ishiwatari, Y. Hayashi, A numerical study of the martian atmospheric convection with a two-dimensional anelastic model. Earth Planets Space **50**, 431–437 (1998)

H. Ohno, T. Takemi, Mechanisms for intensification and maintenance of numerically simulated dust devils. Atmos. Sci. Lett. **11**(1), 27–32 (2010a)

H. Ohno, T. Takemi, Numerical study for the effects of mean wind on the intensity and evolution of dust devils. SOLA **6**(1), 5–8 (2010b). doi:10.2151/sola.6A-002

A. Petrosyan, B. Galperin, S.E. Larsen, S.R. Lewis, A. Määttänen, P.L. Read, N. Renno, L.P.H.T. Rogberg, H. Savijärvi, T. Siili, A. Spiga, A. Toigo, L. Vázquez, The martian atmospheric boundary layer. Rev. Geophys. **49**, 3005 (2011). doi:10.1029/2010RG000351

R. Pielke, W. Cotton, R. Walko, C. Trembaek, W. Lyons, L. Grasso, M. Nieholls, M. Moran, D. Wesley, T. Lee, et al., A comprehensive meteorological modeling system-RAMS. Meteorol. Atmos. Phys. **49**, 69–91 (1992)

S. Raasch, T. Franke, Structure and formation of dust devil-like vortices in the atmospheric boundary layer: a high-resolution numerical study. J. Geophys. Res., Atmos. (1984–2012) **116**(D16) (2011)

S.C.R. Rafkin, R.M. Haberle, T.I. Michaels, The Mars regional atmospheric modeling system: model description and selected simulations. Icarus **151**, 228–256 (2001)

S. Rafkin, L. Fenton, R. Lorenz, B. Jemmett-Smith, N. Renno, T. Takemi, P. Knippertz, J. Ito, D. Tyler, Dust devil formation conditions and process. Space Sci. Rev. ["Dust devils" special issue] (2016, submitted)

D. Reiss, D. Lüsebrink, H. Hiesinger, T. Kelling, G. Wurm, J. Teiser, High altitude dust devils on Arsia Mons, Mars: testing the greenhouse and thermophoresis hypothesis of dust lifting, in *Lunar and Plane-*

tary Institute Science Conference Abstracts. Lunar and Planetary Inst. Technical Report, vol. 40, 2009, p. 1961

D. Reiss, P. Whelley, L.D.V. Neakrase, M. Zimmerman, L. Fenton, M. Balme, A.P. Rossi, T. Statella, Dust devil tracks and surface albedo changes. Space Sci. Rev. ["Dust devils" special issue] (2016, submitted)

N.O. Renno, M.L. Burkett, M.P. Larkin, A simple thermodynamical theory for dust devils. J. Atmos. Sci. **55**, 3244–3252 (1998)

N.O. Renno, A.-S. Wong, S.K. Atreya, I. de Pater, M. Roos-Serote, Electrical discharges and broadband radio emission by martian dust devils and dust storms. Geophys. Res. Lett. **30**(22), 220000-1 (2003)

N.O. Renno, V.J. Abreu, J. Koch, P.H. Smith, O.K. Hartogensis, H.A.R. De Bruin, D. Burose, G.T. Delory, W.M. Farrell, C.J. Watts, J. Garatuza, M. Parker, A. Carswell, MATADOR 2002: a pilot field experiment on convective plumes and dust devils. J. Geophys. Res. **109**(E18), 7001 (2004). doi:10.1029/2003JE002219

M.I. Richardson, A.D. Toigo, C.E. Newman, PlanetWRF: a general purpose, local to global numerical model for planetary atmospheric and climate dynamics. J. Geophys. Res. **112**(E09001) (2007). doi:10.1029/2005JE002636

R. Rotunno, The fluid dynamics of tornadoes. Annu. Rev. Fluid Mech. **45**, 59–84 (2013). doi:10.1146/annurev-fluid-011212-140639

P. Sagaut, *Large Eddy Simulation for Incompressible Flows: An Introduction* (Springer, Berlin, 2006)

H. Sävijarvi, A model study of the atmospheric boundary layer in the Mars Pathfinder lander conditions. Q. J. R. Meteorol. Soc. **125**(554), 483–493 (1999)

J.T. Schofield, D. Crisp, J.R. Barnes, R.M. Haberle, J.A. Magalhaães, J.R. Murphy, A. Seiff, S. Larsen, G. Wilson, The Mars Pathfinder atmospheric structure investigation/meteorology (ASI/MET) experiment. Science **278**, 1752–1757 (1997)

Y. Shao, S. Liu, J. Schween, S. Crewell, Large-eddy atmosphere—land-surface modelling over heterogeneous surfaces: model development and comparison with measurements. Bound.-Layer Meteorol. **148**(2), 333–356 (2013). doi:10.1007/s10546-013-9823-0

Y. Shao, W. Nickling, G. Bergametti, H. Butler, A. Chappell, P. Findlater, J. Gillies, M. Ishizuka, M. Klose, J. Kok, J. Leys, H. Lu, B. Marticorena, G. McTainsh, C. McKenna-Neuman, G. Okin, C. Strong, N. Webb, A tribute to M.R. Raupach for contributions to aeolian fluid dynamics. Aeolian Res. **19 Part A**, 37–54 (2015). doi:10.1016/j.aeolia.2015.09.004

P.C. Sinclair, The lower structure of dust devils. J. Atmos. Sci. **30**(8), 1599–1619 (1973)

W.C. Skamarock, J.B. Klemp, A time-split nonhydrostatic atmospheric model for weather research and forecasting applications. J. Comput. Phys. **227**, 3465–3485 (2008). doi:10.1016/j.jcp.2007.01.037

J. Smagorinsky, General circulation experiments with the primitive equations. I. The basic experiment. Mon. Weather Rev. **91**, 99–164 (1963)

Z. Sorbjan, Statistics of shallow convection on Mars based on large-eddy simulations. Part 1: shearless conditions. Bound.-Layer Meteorol. **123**, 121–142 (2007). doi:10.1007/s10546-006-9128-7

A. Spiga, Elements of comparison between martian and terrestrial mesoscale meteorological phenomena: katabatic winds and boundary layer convection. Planet. Space Sci. **59**, 915–922 (2011). doi:10.1016/j.pss.2010.04.025

A. Spiga, F. Forget, A new model to simulate the martian mesoscale and microscale atmospheric circulation: validation and first results. J. Geophys. Res., Planets **114**, 02009 (2009). doi:10.1029/2008JE003242

A. Spiga, S.R. Lewis, Martian mesoscale and microscale wind variability of relevance for dust lifting. Int. J. Mars Sci. Explor. **5**, 146–158 (2010). doi:10.1555/mars.2010.0006

A. Spiga, F. Forget, S.R. Lewis, D.P. Hinson, Structure and dynamics of the convective boundary layer on mars as inferred from large-eddy simulations and remote-sensing measurements. Q. J. R. Meteorol. Soc. **136**, 414–428 (2010). doi:10.1002/qj.563

P.P. Sullivan, E.G. Patton, The effect of mesh resolution on convective boundary layer statistics and structures generated by large-eddy simulation. J. Atmos. Sci. **68**, 2395–2415 (2011). doi:10.1175/JAS-D-10-05010.1

T. Takemi, An eddy-resolving simulation of the diurnal variation of fair-weather convection and tracer transport. Atmos. Res. **89**, 270–282 (2008). doi:10.1016/j.atmosres.2008.02.012

P.C. Thomas, P.J. Gierasch, Dust devils on Mars. Science **230**, 175–177 (1985)

A.D. Toigo, M.I. Richardson, Meteorology of proposed Mars exploration rover landing sites. J. Geophys. Res., Planets **108**(E12), 8092 (2003). doi:10.1029/2003JE002064

A.D. Toigo, M.I. Richardson, S.P. Ewald, P.J. Gierasch, Numerical simulation of martian dust devils. J. Geophys. Res., Planets **108**, 5047 (2003). doi:10.1029/2002JE002002

D. Tyler, J.R. Barnes, Mesoscale modeling of the circulation in the Gale crater region: an investigation into the complex forcing of convective boundary layer depths. Mars **8**, 58–77 (2013). doi:10.1555/mars.2013.0003

D. Tyler, J.R. Barnes, Convergent crater circulations on mars: influence on the surface pressure cycle and the depth of the convective boundary layer. Geophys. Res. Lett. **42** (2015). doi:10.1002/2015GL064957

D. Tyler, J.R. Barnes, E.D. Skyllingstad, Mesoscale and large-eddy simulation model studies of the martian atmosphere in support of Phoenix. J. Geophys. Res., Planets **113**(E12) (2008). doi:10.1029/2007JE003012

M. Weißmüller, F. Hoffmann, , S. Raasch, Towards large-eddy simulations of dust devils with observed intensity: effects of numerics and surface heterogeneities. J. Geophys. Res. (2016, submitted)

Y.Z. Zhao, Z.L. Gu, Y.Z. Yu, Y. Ge, Y. Li, X. Feng, Mechanism and large eddy simulation of dust devils. Atmos.-Ocean **42**(1), 61–84 (2004). doi:10.3137/ao.420105

Space Sci Rev (2016) 203:277–297
DOI 10.1007/s11214-016-0277-9

Dust Devil Populations and Statistics

Ralph D. Lorenz[1] · **Brian K. Jackson**[2]

Received: 15 February 2016 / Accepted: 22 July 2016 / Published online: 16 August 2016

Abstract The highly-skewed diameter and pressure drop distributions of dust devils on Earth and Mars are noted, and challenges of presenting and comparing different types of observations are discussed. The widely-held view that Martian dust devils are larger than Earth's is critically assessed: the question is confounded somewhat by different observation techniques, but some indication of a ∼3x larger population on Mars is determined. The largest and most intense (in a relative pressure sense) devils recorded are on Mars, although the largest reported number density is on Earth. The difficulties of concepts used in the literature of 'average' diameter, pressure cross section, and area fraction are noted in the context of estimating population-integral effects such as dust lifting.

Keywords Dust devils · Statistics · Power law

1 Introduction

Dust devils are a prominent feature of meteorology on Mars and in desert areas on the Earth. They are responsible for dust-lifting on both Earth and Mars, for surface albedo changes on Mars which may give rise to interannual climate variability, for the clearing of dust from the solar panels of landers and rovers, and for occasional death, injury and damage on Earth (e.g. Lorenz et al. 2016). The question of 'how many dust devils are there?' is therefore an important one but requires some rigor and subtlety in answering.

Clearly, the question of whether one location has more dust devil activity than another would be straightforward to answer if all dust devils were the same size and had the same lifetime. However, small dust devils are abundant and short-lived, while larger devils tend to last longer. A single photo from the field (or from a planetary lander) would be able to see devils that would be too small to observe from orbit, yet an orbital image covers a much

✉ R.D. Lorenz
ralph.lorenz@jhuapl.edu

1 Applied Physics Laboratory, Johns Hopkins University, Laurel, MD 20723, USA

2 Department of Physics, Boise State University, Boise, ID 83702, USA

wider area and is therefore able to count rare, large devils more effectively. Thus the data on which a comparison is drawn may sample quite different parts of the two populations, and so comparisons require some care. Furthermore, the enhanced longevity of large dust devils must be taken into account, in that it increases their probability of detection by snapshot observations (e.g. images once per day at a given region) or in-situ measurements at a fixed point, but does not affect continuous field observations.

It is commonly considered that dust devils are larger on Mars than on Earth – but how much of that perceived distinction is due to the very different instruments and techniques used to evaluate the populations? We address this question specifically in this paper.

Another motivation for proper statistical treatment is the nonlinear dependence of dust devil effects on their basic properties such as intensity (e.g. pressure drop ΔP at the core) or diameter. As a trivial example, consider dust lifting (a key effect of the dust devil population – see the chapter by Klose et al. 2016). Even if dust fluxes and longevity did not correlate with diameter (and there is evidence that they do), the dust raised by a single devil would be proportional to its area, or its diameter squared. In a skewed population, the mean value of (squared diameters) is not the same as the (mean diameter) squared (e.g. Lorenz 2011). Thus the population function needs to be taken into account when upscaling field observations to global effects.

In addition to discussing these questions, and the variability in dust devil occurrence, some rational approaches to the presentation of dust devil population data are suggested in this review to help manage the difficulties of evaluating and comparing highly-skewed populations.

2 Definition and Representation

Before discussing inferences about the population of dust devils in different environments, it is important first to establish a working definition of the objects under study (and later we will introduce quantitative criteria). We also lay some groundwork in this section on the styles of data presentation to be used later.

2.1 Dust Devil Definition

An Euler Diagram (Fig. 1) is helpful to describe various phenomena discussed in the context of dust devils. While there are many visible clouds of dust in desert areas, not all are dust devils. Similarly, there exist many dry columnar vortices in the planetary boundary layer, but not all carry dust. The two essential characters of dust devils, namely a rotation about a vertical axis and lofted dust, provide the definition. In this respect our definition is slightly wider than Oke et al. (2007) who suggest a specific criterion for circumferential windspeed, diameter, and duration: "*A vortex rotating in either direction faster than its motion downwind. It must have a coherent columnar or cone shape extending in a vertical direction above the surface. It must carry dust to a height of at least 1 m and last for at least 10 s. It is distinct from a dust blow in that it maintains a fast rotating coherent structure with vertical extension and distinct from a small dust whirl in its size and duration.*" It is useful to keep this alternative definition in mind when considering the rest of this paper in so far as it represents an example of a heuristic set of criteria for counting features observed in the field. We will discuss later other quantitative criteria in the context of automatic detection of features in meteorological time series, images, or simulated fields of meteorology data. We may note also in Fig. 1 that features of dust devil populations can be inferred from studies of dust devil tracks (discussed in a later section, and in the companion chapter by Reiss et al.), but such tracks are formed only by a subset of dust devils.

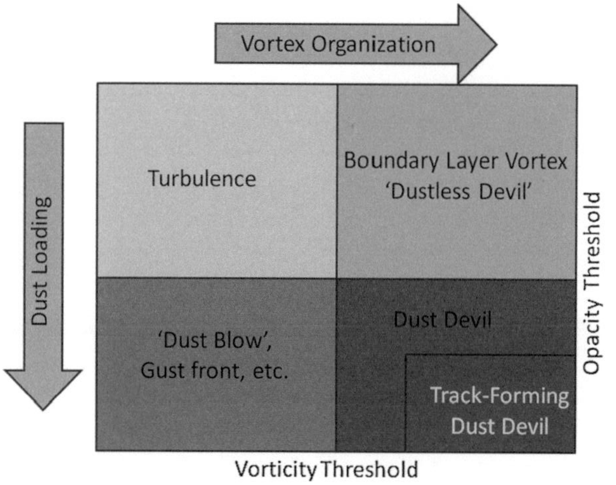

Fig. 1 Schematic categorization of events. 'Intensity' is in essence a combination of the organization and strength of a wind or pressure perturbation, and is a maximum in the *lower right part of the plot*. Small fluctuations always exist in the turbulent windfield, which can sometimes raise dust in a less organized way ('dust blows') in the *lower left*. Well-organized vortices often exist without lofted dust, where the dust-lifting threshold is high, for example, but can be detected with meteorological instrumentation, yielding events in the *upper right*. When dust is lifted by a vortex, we see the classic 'dust devil', a subset of which can form trails on the ground

2.2 Presentation of Population Data

Having established that some criterion is needed to 'count' entities such as dust devils, we now make some general remarks may be made on the presentation of dust devil counts, echoing discussion in Lorenz (2011, 2012).

Although historically visual counts in the field have of necessity been classified in three or four size bins, in general, binning is to be avoided ("Binning is sinning") since information is lost. If it must be done, then ideally an information-theoretic criterion might be used in bin-size selection (e.g. Jackson and Lorenz 2015), or at least binning be done in some manner well-suited to strongly skewed data e.g. logarithmically (bins of size ratio equal to the square root of two, as used often in the impact cratering community, is one approach). Cumulative plots have some virtues, not least in making noisy data look somewhat smoother – e.g. Fig. 2. Because cumulative plots can integrate out some details (since they have a wide dynamic range), a cumulative plot normalized to some function such as a power law can be useful (e.g. in the impact cratering literature, crater counts normalized to a -2 power law is referred to as an 'R-plot' (Melosh 1990) and indicates the relative area covered by craters of a given diameter).

The problem of how to estimate densities and bin data has a long history in the data science and machine-learning literature and comes down to striking the right balance between using few enough bins to smooth out statistical fluctuations without removing important trends in the data. Many solutions have been developed to address this problem, ranging from simple rules-of-thumb to more sophisticated procedures based on Bayesian analysis of the data. For instance, the recently developed Bayesian Blocks scheme (Scargle et al. 2013) aims for this balance by increasing the number of bins to minimize variations within a bin, while incorporating a penalty each time a new bin is added.

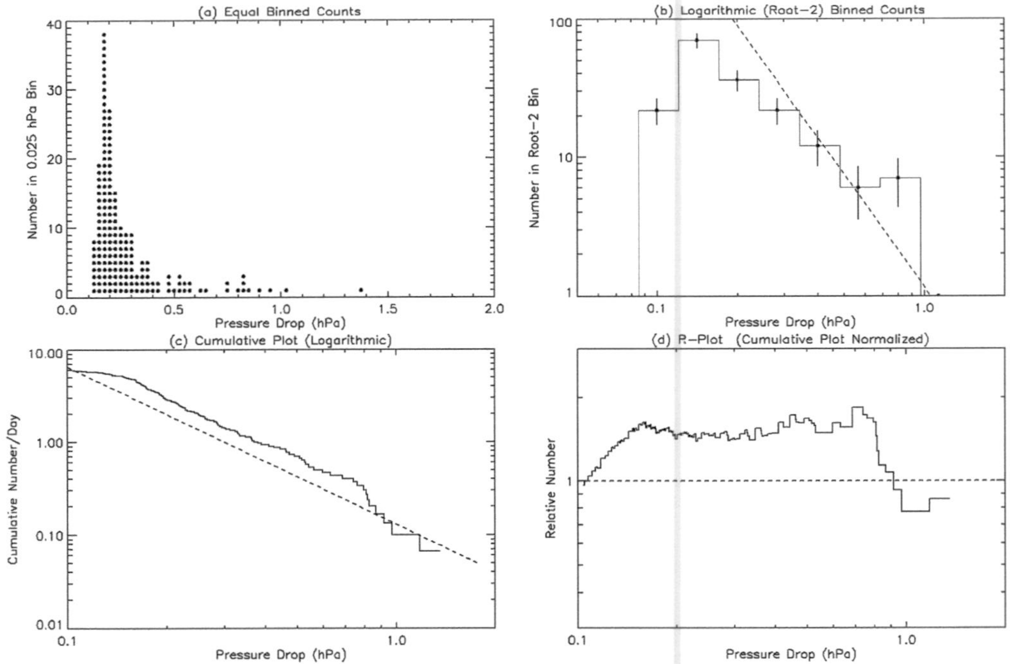

Fig. 2 Different ways of presenting the same data, the 'P28' dataset of dust devil pressure drops measured in a 30-day period at El Dorado by Lorenz and Lanagan (2014). Many visual surveys have presented data as simple linear equal-bin histograms or counts, similar to (**a**). The widely different number of counts at small and large sizes makes it difficult to assimilate the population function, and the statistics at the large end are poor. That issue can be addressed by using logarithmic binning, with bins having an equal ratio, rather than difference, in threshold. A $\sqrt{2}$ ratio is often a good compromise between retaining enough datapoints to define the function shape, while keeping enough counts in each bin to avoid large counting errors except in the last bin. However, by using a cumulative count, binning can be avoided altogether, as in (**c**). The staircase appearance is due to poor statistics at the large end, but is faithful to the data. For comparing different datasets, it may be useful to normalize the cumulative function by an analytic expression to obtain a relative rate, in this case an entirely empirical function $R(>\Delta P) = (\Delta P/0.3)^{-1.7}$, shown by the *dashed line* in (**c**) and (**d**) is used. This normalization (in the impact-cratering literature, referred to as an 'R-plot' where the normalization exponent is 2.0) reduces the dynamic range required in the y-axis, and therefore exposes differences more clearly

Many binning routines are readily available via open-source scientific computing modules such as scikit-learn (http://scikit-learn.org/stable/modules/density.html) and astroml (http://www.astroml.org/user_guide/density_estimation.html).

Whether binned or not, and shown in a differential or cumulative manner, a variety of analytic functions may be used to describe or parameterize the distribution. Kurgansky (2006) suggested an exponential model fitted the crudely-binned statistics of Sinclair's Tucson survey, whereas Lorenz (2009) showed that the Mars diameters measured by Greeley et al. (2006) on the Spirit rover were better-fit by a (truncated) power law. Further discussion by Pathare et al. (2010) advocated an exponential model for some terrestrial field data, but this was challenged by Lorenz (2011), who cautioned that the coarse binning and poor statistics challenged the significance of selection of one analytical model against another. The one exception to date appears to be the data from Ryan and Carroll (1970), where 5 size bins were populated. Those data, with bin thresholds of 1.5, 3, 6, 12 and 25 m, are 370, 165, 31, 17, 5 devils are quite accurately fit with a power law of −2.6 (i.e. a cumulative power law of −1.6). The question of best function for the Gusev data was revisited by Kurgansky (2012).

Pending the availability of better (unbinned) datasets for Earth, we suggest that the choice of diameter function is something of an aesthetic choice, with log-normal, Weibull, exponential or (truncated) power law functions all likely to give adequate fit to binned data. The power law enjoys some analytic convenience, however.

3 Population Assessment of Dust Devils

3.1 Optical Detections

The survey technique with the longest pedigree is naturally that of observing in the field. As noted by Balme and Greeley (2006), the observed formation rate of dust devils varies with the area surveyed – and the range of reported occurrence rates (from ~0.1 to some ~800 $km^{-2}\,day^{-1}$) defies straightforward interpretation. However, Lorenz (2009) showed that this variation is systematic, with detected rates N (devils $km^{-2}\,day^{-1}$) being inversely proportional to survey area A (km^2), with $N \sim 50/A$, and suggested that this could be explained by a power law diameter distribution and a detection threshold that varied with distance (and thus survey area – see Fig. 3). The detection efficiency of small and/or optically-thin devils is essential to take into account, for example in comparing populations observed at different sites, and to understand the shape of the actual dust devil population function (e.g. Lorenz 2011; Kurgansky 2012). It may be that for a given survey circumstance, e.g. camera location, there are temporal variations in detection efficiency owing to the different viewing/scattering geometry as a function of time of day, etc.

It is instructive to consider a recent observation in the context of this plot. A first detection of terrestrial dust devils in satellite imaging was described by Reiss et al. (2016) finding as many as 30–40 devils per 1000 km^2 in Landsat and ASTER data, corresponding to diameters of 20–155 m. Following the empirical longevity relationship of Lorenz (2013), dust devils in this size last 300–1200 s, or about one-sixtieth to one twentieth of a 'day' of dust devil activity (~6 hours). Thus the observed occurrence rate is 0.6–2.4 $km^{-2}\,day^{-1}$, corresponding to a survey area equal to the image size of ~1500 km^{-2}. It is seen (Fig. 3) that this datapoint plots just above a line of $N \sim 1000/A$ which defines roughly the upper values of vortex observations.

Figure 3 represents an update of the presentation first adopted in Lorenz (2009, 2013). In addition to the only terrestrial spaceborne imaging (Reiss 2016) described above, new

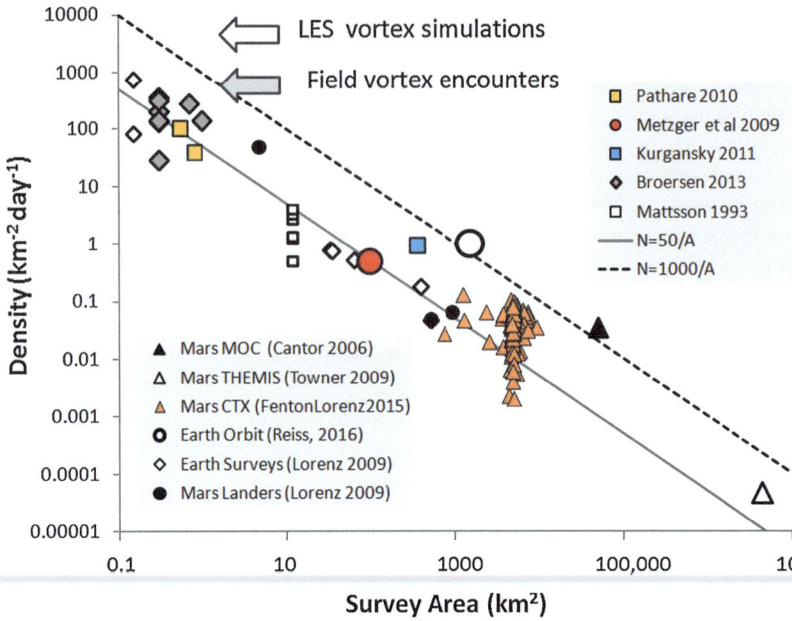

Fig. 3 A compilation of observed dust devil frequencies from visual surveys on Earth and Mars, extended from Lorenz (2009, 2013). There is a systematic variation of observed counts with area surveyed, presumably because large area surveys (especially those from orbit) only detect the larger, rarer dust devils. Also indicated with a *large grey arrow* is the vortex generation rate of 300 $km^{-2}\,day^{-1}$ required, with assumptions on longevity and advection speed, to reproduce in a Monte Carlo model (Lorenz 2014) the observed vortex pressure dips above 20 Pa seen in terrestrial field data. The vortex formation rate measured in two different terrestrial LES simulations, of 1500–4500 $km^{-2}\,day^{-1}$ of 10 and 4 Pa respectively are also shown with a *large arrow*. Most of these rather weak vortices will be dustless, however. The Mars and Earth rates appear, with survey area taken into account, to be broadly similar: most data fall on a line of $N = 50/A$, with the more active sites (and especially, more sensitive observations from orbit) falling on a line of $N = 1000/A$

datapoints shown here are a collection of individual Mars orbital image counts from Fenton and Lorenz (2015) described later in this paper, and field surveys in Jordan (Broersen 2013), Eldorado playa in Nevada (a ∼20 day, 300 km^2 survey by Metzger et al. 2009) and a report from Tunisia by Mattsson et al. (1993).

In this paradigm, generally the smallest areas will have the highest detection rates as 100 % of even the smallest devils will be detected and the inventory can be considered ‘complete’.[1] The highest observed rates reported in the field are those by Carroll and Ryan (1970) who encountered some 1151 devils in 10 days in a 0.15 km^2 area, or 767 devils $km^{-2}\,day^{-1}$. This count is especially high for a couple of reasons. First, the site was completely clear of vegetation and they ‘groomed’ the desert surface daily (breaking up the surface with a rake), to maximize the availability of dust. This likely reduced the threshold windspeed or pressure drop required for dust lifting (see later) and thereby increased the number of observed dev-

[1] ‘Complete’ in the sense of detecting all devils present in the survey area. Surveys in small areas may statistically encounter less than one devil for large sizes and so are not ‘complete’ in the sense of fully characterizing the population unless they are conducted for a long enough period – see Lorenz (2011). Note that a true power law has no limit, and so there are an infinite number of infinitely small dust devils, which is clearly not the case. In practice there is a lower limit, suggested to be the Obhukhov scale or ∼1 m for typical conditions on Earth.

ils. Second, as discussed in Lorenz (2013), counts of 'occurrence' in very small study areas are inflated over 'formation rates' because dust devils that form elsewhere but are advected into the site are counted – for the 0.15 km^2 area and the typical sizes (and thus longevities) and windspeeds encountered, the rate at which vortices form in the study area is probably lower by a factor of 2–3 (see Lorenz 2013, Fig. 6). Hence we consider their corrected result as a formation rate of ~300 devils $km^{-2}\,day^{-1}$, noting that the vortex intensity threshold for dust-lifting may have been especially low due to the grooming of the surface.

The highest count in more typical field conditions (indeed, at Eldorado playa in Nevada, where the pressure drop data in the next section was obtained) is that by Pathare et al. (2010), of 528 devils in 0.55 km^2 in 9 days. However, this period included some less favorable days, with daily counts varying between 18 and 83. Adopting the highest number yields $N = 151$ devils observed per km^2 per day, although again this should be reduced by a factor relating to the advection speed. Thus 100 devils $km^{-2}\,day^{-1}$ is probably about right for the formation rate, and certainly >75 devils $km^{-2}\,day^{-1}$. Other surveys (see Lorenz 2011, 2014) have lower counts. Snow and McClelland (1990) report mean counts of ~0.5 devils $km^{-2}\,day^{-1}$ at White Sands missile range, but close inspection of their Fig. 3 shows that these were very unevenly distributed in the survey area, and one 0.5×0.5 km survey plot saw 151 devils in 41 days, or 14 devils $km^{-2}\,day^{-1}$. This plot was some kilometers from the observer location, suggesting that this number is rather incomplete, and they note that 'it is likely that less than 10 % of whirlwinds in the small category (<3.3 m diameter) were counted, greater than 80 % of dust devils in the medium category (3–16 m) and all those in the large category … were counted by the observer'. This, incidentally, is the only completeness estimation known to the present authors for terrestrial visual surveys. Hence the overall dust devil rate was probably a factor of a few higher than the 14 devils $km^{-2}\,day^{-1}$ above, and seems consistent with the ~100 devils $km^{-2}\,day^{-1}$ indicated by Pathare et al. (2010).

A diameter criterion is somewhat straightforward to implement with appropriate preparation – visual data gives the angular width of a target, thus its range must be known by stereo imaging (Balme et al. 2012), by geometry (e.g. exploiting a raised imaging platform to derive range from elevation e.g. Greeley et al. 2006) or by comparison with known landmarks (e.g. Sinclair 1969).

The radiometric criterion of detecting dust loading is rarely defined, and is surely variable depending on illumination conditions and visibility (e.g. the background atmospheric dust loading). It also depends on observing wavelength or whether human observers use polarizing sunglasses, for example. It follows that, for a given dust loading of particles/m^3 (assumed uniform within a dust devil; note that the extinction or area/volume depends also on the particle size), the opacity, or column optical depth, will depend on the size of the dust devil. Thus for a given optical contrast detection criterion, larger devils are easier to detect.

It would be desirable in stating observed dust devil counts to document an absolute contrast threshold used in the detection, that can at least crudely be related to dust loading. In practice, a relative contrast threshold may be used, which may depend on the image noise properties and/or the variation in the sky or ground background. In this connection it may be noted that dust devils are often – perhaps counterintuitively – easier to detect against a terrain background than against the sky.

It may be noted in passing that dust devil shadows are instrumental in Mars orbital studies of dust devils. Not only can the height of dust devils be measured this way, and in some cases the opacity can be measured via the shadow contrast, but many surveys rely on the shadows to detect dust devils as a pair of features (a bright circle, with a quasi-linear dark shadow). In other words, the presence of a shadow adjacent to a dust devil feature enhances the probability of detecting the feature (or equivalently lowers the size or opacity threshold for detection). This effect, plus the increased line-of-sight contrast for dust devil columns seen from above (Lorenz 2014) may explain in part the slightly higher-than-trend counts of Mars dust devils seen from orbit (Fig. 3).

3.2 Vortex Detections in Meteorological Time Series Data

Dust devil passage manifests itself in meteorological data by several clear signatures (e.g. Sinclair 1973; Ryan and Lucich 1983; Ringrose et al. 2003, 2007; Tratt et al. 2003) – most notably a pressure drop ΔP lasting a few seconds to hundreds of seconds and a rotation of the wind direction. Wyett (1954) noted a dust devil passage in a recording barograph at an airport meteorology station, and Lambeth (1966) documented about 20 pressure drops with an array of six stations in a 4-month period at White Sands, New Mexico. The low detection efficiency is likely due to a high threshold for triggering the chart recording (i.e. while the effort was devoted to acquiring in-situ data, the detection was optical).

For comparison with Mars in-situ detections, where meteorological data were recorded quasi-continuously and examined post-hoc for vortex signatures (Ryan and Lucich 1983; Murphy and Nelli 2002; Ellehoj et al. 2010), a comparable measurement approach is needed for Earth, with high-cadence meteorological data acquired at a fixed station for weeks or months.

After some initial experiments (yielding a 0.6 mb devil in a 3-day survey reported in Lorenz 2010) new measurements of this type have been made, enabled by the availability of small pressure loggers using flash memory to store the ~Gbits of data required for multi-week multi-Hz sampling (Lorenz 2012, 2013). A preliminary inspection of data from a trial field campaign at El Dorado Playa in Nevada (Lorenz 2013) suggested detectable encounter rates (>0.3 mbar) on average of 1.3–2.6 per day. More systematic evaluations recording more than a hundred candidate events (Lorenz and Lanagan 2014) and indeed some thousands of events (Lorenz and Jackson 2015) have been conducted.

There has been some debate regarding the optimal way in which to parameterize or even how display the distributions of dust devil pressure parameters, with different groups advocating power-law or exponential fits, cumulative or differential histograms (Pathare et al. 2010; Lorenz 2011; Kurgansky 2012). Since different processes may give rise to different populations, answering these questions is not merely an academic subject.

With 252 million pressure readings, Jackson and Lorenz (2015) recovered more than 1000 putative dust devil pressure drops during a two-year terrestrial field campaign and were able to compare different parametric fits to the histogram of pressure drops. Figure 5 shows data from that study which illustrates the pressure drop histogram, with bin sizes determined using Knuth's Rule, a binning scheme that uses a Bayesian framework to determining bin size. Neglecting small drops (with $\Delta P < 0.2$ hPa), that study found that a power-law fit to the histogram (with an index of -2.59 ± 0.11) gave a reduced chi-squared nearly half that for an exponential fit (with an e-folding pressure of 0.15 ± 0.01 hPa), which argues that the power-law is statistically a better fit.

Jackson and Lorenz (2015) also leveraged their large number of recovered pressure drops to investigate how the number of detections influenced the resulting fits and determine how robustly a given power-law index could be recovered if fewer detections were made. Randomly selecting sub-samples of the full, recovered population and fitting a power-law to the each resulting histogram, the study found that smaller populations of drops could give wildly different power-law fits and, to retrieve the final best-fit to the full sample, at least 75 % of the full sample would have had to be recovered. This result shows that small number statistics could easily explain why analyses of smaller samples give different functional fits to the dust devil distributions.

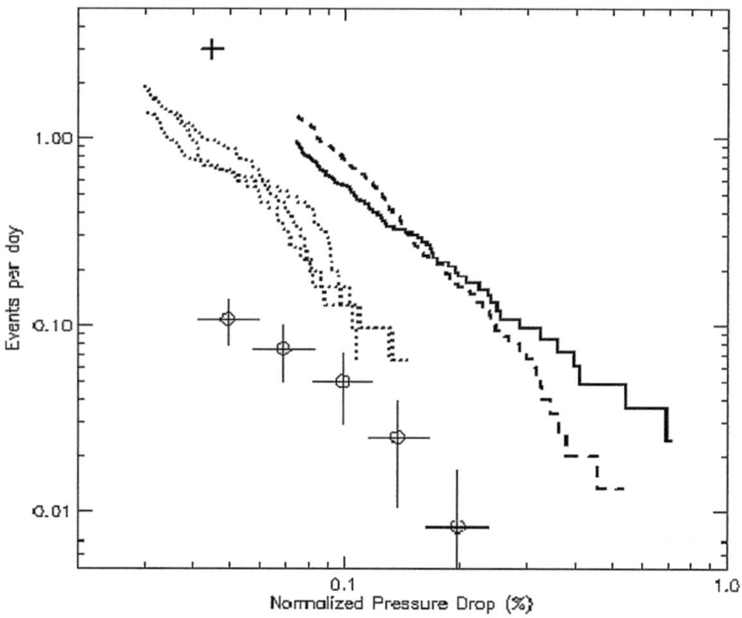

Fig. 4 The intensity of vortex encounters at fixed stations on Mars (Pathfinder – *solid line*; Phoenix – *dashed line* and *cross*) and on Earth (El Dorado playa – Lorenz and Lanagan 2014 – *dotted line*). The double-logarithmic axes show power-law populations as *straight lines*. The likely incomplete survey by Lambeth (1966) at White Sands is shown by the *circles* with 1–σ error bars. Populations are expressed as cumulative numbers of events per 100 days with amplitude greater than the abscissa. Since Mars ambient pressures are ~100x smaller than Earth's, the population is shown as a normalized pressure. This metric is roughly proportional to vortex speed

Fig. 5 The aggregated counts of Jackson and Lorenz (2015) shown as a differential population function with bins sized to maximize information content. A power law fit is shown with the *solid line*, and an exponential with a *dashed line*. Both candidate functions deviate from observations below 0.2 mb, either due to an intrinsic minimum intensity of the population, or low detection efficiency for events near the noise level

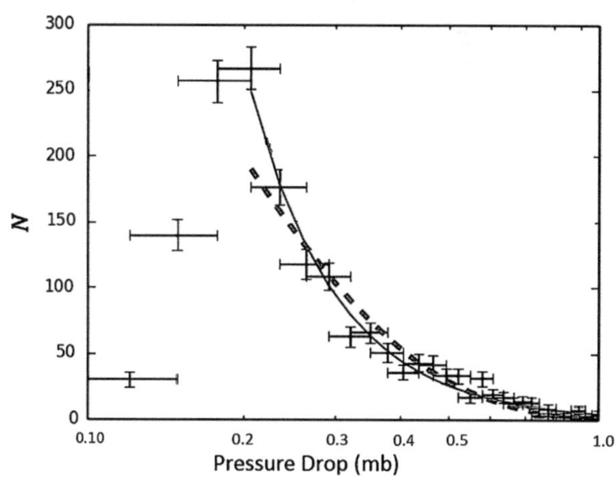

Finally on this topic, we may remark that the cumulative plot has the virtue that even very meagre datasets can be meaningfully portrayed and compared in this manner (e.g. Lorenz and Radebaugh 2016) report data from a pocket pressure logger at a high-elevation site (~3600 m) in the Argentine Puna. Even though only three events were detected, the cumulative counts suggested a higher level of activity, with 95 % Poisson error bars barely overlapping with the benchmark Eldorado counts of Lorenz and Lanagan (2014).

3.3 Vortex Detection in Simulations

Large Eddy Simulations (LES – see chapter on LES modeling by Spiga) can reproduce vortex structures under conditions similar to those in which they are seen in nature. Simulations expose the full fields of (computed) pressure and velocity, so while vortices can be detected by inspection by a human operator, for efficient gathering of statistics, an automatic vortex detection criterion must be defined. Practically, as for field data, a pressure drop makes a useful first-order criterion, but in addition a vorticity criterion is employed. Vorticity is defined as the curl of the flow velocity vector u, i.e. $\bar{\omega} = \nabla \times \bar{u}$. The vorticity in a surface element with area δA is related to the flow's circulation Γ (the line integral of the velocity around that element) simply as $\omega \delta A = \Gamma$. For the 'wall' of a dust devil, with diameter D and circumferential windspeed u, the circulation is $\Gamma = \pi u D$, and the vorticity $\omega = 4u/D$. Note that the vorticity defined at the wall this way not only has the same dimensions as the spin rate, but a value that differs only by a factor (the spin rate R is simply $u/\pi D$, thus $R \sim \omega/12$).

Ohno and Takemi (2010) report vortex formation and evolution in an LES with a 1×1 km domain and a horizontal resolution of 3 m, with a CBL height of 800 m and a surface heat forcing of 0.24 $\mathrm{K\,m\,s^{-1}}$ (i.e. ~290 $\mathrm{W\,m^{-2}}$, typical of desert conditions on Earth). They find after model convergence that 225 vortices were detected (with a 10 Pa pressure drop threshold, and a vorticity threshold ζ of 0.15 $\mathrm{s^{-1}}$) in a 2000 s period. If we extrapolate to a typical sunny day (i.e. ~4 hrs or 15,000 s of such strong heating), then this corresponds to 1600 vortices per $\mathrm{km^2}$ per day. Raasch and Franke (2011) show results for a 4×4 km domain and 2m resolution (as well as a smaller run at 1 m resolution) with a CBL of 700 m and 0.2 $\mathrm{K\,m\,s^{-1}}$ surface heating. They used a vortex detection threshold of $\zeta > 1\ \mathrm{s^{-1}}$ and $\Delta P >$ 4 Pa and found 25,000 vortex tracks in a 5400 s run. This corresponds, given their 16 $\mathrm{km^2}$ model domain, to ~1500 tracks/$\mathrm{km^2}$ in the run, or ~4500 vortices per $\mathrm{km^2}$ per day. These LES vortex results are plotted in Fig. 2 – because the core pressure drops may correspond to intensities below typical field dust-lifting thresholds, many of these vortices would not be visible in nature, and so the frequencies are somewhat higher than field measurements of dust devil occurrence. The simulations appear broadly consistent with the highest counts of visible dust devils (i.e. those showing the smallest dust devils).

Some of the caveats associated with field observations may be repeated here in the context of deriving vortex statistics from LES studies. Typically the domains have periodic boundary conditions, so can observe the full life cycle of dust devil vortices. However, subtleties may emerge associated with tracking vortices that leave (and re-enter) the domain, analogous to the formation rate vs occurrence rate problem of small-area field observations. Many interpretations (e.g. the generalization of ΔP statistics throughout the domain over a short period, to ΔP measured at a point measured over a much longer period, typical of field measurements) rely on the assumption of ergodicity, which may or may not hold.

3.4 Reconciling Vortex Detections and Observed Dust Devils

Lorenz (2014) showed with a Monte Carlo model that the occurrence frequency of in-situ vortex detections (via pressure drop) measured on El Dorado playa (Lorenz and Lanagan 2014; Jackson and Lorenz 2015) could be related to the observed number of dust devils seen in visual surveys (e.g. Pathare et al. 2010). The model introduced random vortices into an ‘arena’ containing a virtual pressure sensor. The vortices were drawn from a specified (truncated power law) diameter and core pressure drop distribution, with an assumed velocity distribution and a longevity related to diameter (see later), and the detection process (pressure drop threshold exceedance) was simulated to recover the observed (‘sample’) population as a function of the underlying complete population. Unsurprisingly, the visual detections had to be less abundant than the full vortex population, indicating that only a fraction of vortices form visible dust devils. In the case simulated, the vortex population was considered with core pressure drops as small as 0.1 hPa, but only those of 0.3 hPa or higher were assumed to be dusty. (The simulation assumed that all vortices with intensity above this threshold would have been detected optically – this makes some assumptions about the acuity and efficiency of the observing process.) The dust devil generation rate consistent with the observed detections ($\sim 3/\text{day}$ at 0.2 hPa) is shown in Fig. 2, and also appears consistent with an ‘asymptotic’ underlying vortex population of a few hundred per km^2 per day.

Some insight into the relative numbers of dusty and dustless devils is afforded by field measurements at El Dorado. Lorenz and Jackson (2015) and Lorenz et al. (2016) report using pressure loggers with solar cells to detect dust devils – the vortex signature is indicated as a pressure drop, which is accompanied in about 40 % of cases by a drop in measured sunlight due to shadowing by lofted dust. As shown in Lorenz and Jackson (2015)’s field data, and modeled by Mason et al. (2013), sometimes a double-shadow is seen in near-diametric encounters with devils with well-defined cylindrical ‘walls’ of dust. Lorenz and Jackson (2015) also note an occasional increase in solar flux, due to dark sky in the anti-sun direction being occupied by a scattering dust plume. Lorenz and Jackson (2015) found that about 40 % of detected vortices (using a 10-s average pressure drop of 0.2 hPa as a detection threshold) had no measurable (~ 0.2 %) accompanying drop in sunlight, and 60 % had a drop less than 1 % (which approximately corresponds to a human visual detection threshold). This does not mean, however, that 40 % of vortices have no dust. For example, no shadow was measured for near-misses with the devil to the antisunward direction of the sensor.

We will for the moment introduce the concept of a pressure-detection cross section (see later). Vortex models (see chapter by Kurgansky et al.) indicate pressure falls off from the core as $\Delta P(x)/\Delta P_{\circ} \sim 1/([2x/D]^2+1)$, where $\Delta P_{\circ}$ is the drop at the core ($x=0$). If the pressure drop detection threshold is ~ 0.2 hPa, and the core pressure drop is ~ 0.7–2.7 hPa, then the dust devil can be detected in the pressure time series at a miss distance of $x < 0.7D$ to $1.7D$ in any direction. In contrast, while in typical conditions a shadow might be detected at a sunward distance of as far as $\sim 6D$ (e.g. for a height/diameter of ~ 5 and a sun elevation a little less than 45°). Thus all sunward pressure-detected vortices would have detectable shadows, but not all shadow-detected devils would have a measureable pressure signature. On the other hand, the shadow is detected only for miss distances less than $0.5D$ in the antisunward direction. Thus for uniformly-distributed miss distances, the entire sunward half of the pressure-dip events are measured to have shadows if dust is present, whereas only a fraction ($0.5/0.7$ to $0.5/1.7 = 30$–70 %) of the antisunward half do, or 65–85 % of the total. Thus for the observed 60 % of shadows when vortices were detected, this geometric correction suggests that in fact 70–95 % of detected vortices should have had observable amounts of dust.

3.5 Dust Devil Trails as a Population Metric

Although dust devil tracks or trails have been observed on Earth (see review chapter by Reiss et al. 2016), they are comparatively rare, and so we consider here only trails at Mars where their quantitative treatment may be useful.

It is commonly observed on Mars that dust falls out of the atmosphere onto surfaces, altering their color or brightness with a typical characteristic timescale of the order of 100 sols (although there are of course regional, seasonal and interannual variations). Arvidson et al. (1983) noted that dark material exhumed by the Viking lander sampling arm brightened on this timescale. Kinch et al. (2007) suggested a monolayer of dust accumulated on rover camera calibration targets in about 100 sols. The Pathfinder rover solar panels lost capacity at a rate of about 0.3 % per sol (Landis and Jenkins 2000) or an e-folding timescale of about 200 sols, but considering that the dust is partly reflective in the solar array bandpass, this timescale is probably larger than that to accumulate unity optical depth.

Thus to a first order, the observation of a dust devil track on the surface of Mars indicates a vortex passage intense enough to remove dust in the last 100 sols. Dust devil tracks can in fact be recognized (after the fact, Lorenz and Zimbelman 2014) in the first Mariner 9 images from 1972 that show sand dunes on Mars. However, their detailed quantification awaited higher-resolution images that became available after 1998.

In Hellas, the number of observed dust devil tracks in a survey by Balme et al. (2003) increased from ~0.6 to 3.5 km^{-2} over about 30 sols (between the intervals of $L_s = 285$–300 and $L_s = 300$–315). This corresponds to a track generation rate of about 0.1 km^{-2} per day. More typically the generation rate was about a fifth of this rate. Verba et al. (2012) find similarly at Gusev in 2006–2008 a peak formation rate of 0.103 tracks km^{-2}/sol, although as low as 0.0011. At Russell crater, track formation was rather more frequent, 0.04–0.95 tracks km^{-2}/sol.

Verba et al. (2012) report dimensions of dust devil tracks in Gusev crater, finding average widths of 40 m and lengths of 2.5 km. The area swept by an 'average' track is therefore 0.1 km^{-2}. The formation of a 0.1 km^{-2} trail at a rate of 0.1 per km^2 per day implies a typical interval between cleaning events seen by a given point of 100 days during the most intense part of dust devil season, and 500 days more typically.

This 100–500 day interval is reassuringly close to that of dust clearing events indicated at Spirit. This agreement between track formation rate, dust clearing, and observed vortex distributions again supports the notion that dust devils are responsible for clearing events.

While the notion of 'average' for a highly skewed population requires considerable care, it is instructive to compare this average trail area of 0.1 km^2 with the area swept by a typical dust devil. Lorenz (2013) determines a longevity of a dust devil diameter d (in m) as $\sim 40d^{0.66}$ s, or for a 40 m dust devil, about 8 min. With a 5–15 m s^{-1} ground speed typical of dust devils (e.g. Reiss et al. 2014), a devil would migrate 2–6 km, spanning the average length noted by Verba et al. (2012). This suggests some overall consistency, and circumstantially implies that dust devils may be able to form trails for most of their observed lifetime. In this connection, Raasch and Franke (2011, their Fig. 6) plot the time histories of two vortices in a Large Eddy Simulation (LES) model of terrestrial dust devils: the two vortices were tracked over 500 and 600 s respectively, and had core pressure drops exceeding 20 Pa (a likely dust-lifting threshold) for over 80 % of their lifetime.

Verba et al. (2012) note that the population of observed tracks are wider than observed devils. As Fig. 6 shows, HiRISE observations of track widths peak at around 40–60 m across, with a mean of 56 m ($N = 640$), whereas the dust devils observed by the Spirit rover ($N = 480$) are generally between 10 and 20 m in diameter (see e.g. Greeley et al.

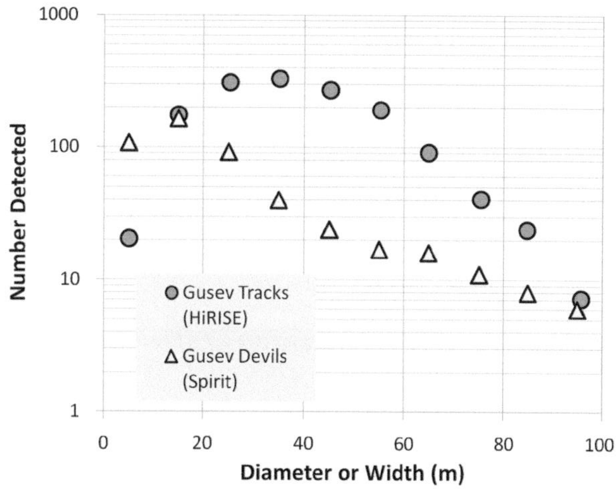

Fig. 6 Gusev dust devil diameters compared with Gusev dust devil track widths. Since the HiRISE imaging resolution is of the order of 1 m or better, image resolution should not be the main reason for the different distributions: it seems larger dust devils may be more effective at forming trails at this site, or that the path over which dust is removed is three times larger than the visible dust column

2006). While there may be some systematic differences in detection efficiencies, this difference is suggestive that track formation is favored for large devils. While it is possible that larger devils are more intense (i.e. ΔP and tangential windspeed correlate with diameter, such that larger vortices are better able to lift dust), modeling of terrestrial dust devils finds no evidence of correlation (Lorenz 2014).

4 Statistics on Earth and Mars

Here we address the question of whether Martian dust devils are larger than those on Earth. As noted earlier, the skewed size distribution makes this a subtle question. A formal assessment can in principle be made non-parametrically (i.e. without fitting a model function), by the application (for example) of a Kolmogorov-Smirnov test, which measures the probability that two different populations are drawn from distinct underlying distributions. However, the combination of coarse binning, and the lack of truly equivalent datasets (i.e. observations with the same survey areas, contrast criteria etc.) means only a somewhat informal answer (yes) can be given in practice.

The reason for caution will be apparent in Fig. 7, which shows convincingly that the Ryan and Carroll (1970) dust devil population indicates much smaller dust devils than the Gusev survey by Greeley et al. (2006) – the peak in the function (the modal diameter) is about 20 times larger in the Gusev survey. But the terrestrial measurements were made over a short period in a very small survey area, so preferentially detect small devils. This survey area/size/number dependence is the foundation for the variations in detected numbers shown in Fig. 3, and is evident when an orbital survey at Mars is considered, also in Fig. 9. It might be concluded that dust devils seen from orbit at Mars are much larger than those seen from the Spirit rover – and while that is true, it is merely a result of the poor efficiency of detection of small devils from long ranges. In other words, Mars dust devils typically look larger, because we typically are looking from further away.

This distance bias could be addressed with a model of the observing process to correct for detection thresholds, but to a first order can be examined by comparing (Fig. 9) two datasets with somewhat similar conditions, namely the Tucson survey by Sinclair (1969) – over a ~500 km^2 area, and that at Gusev, over a similar area. Unfortunately the coarse binning

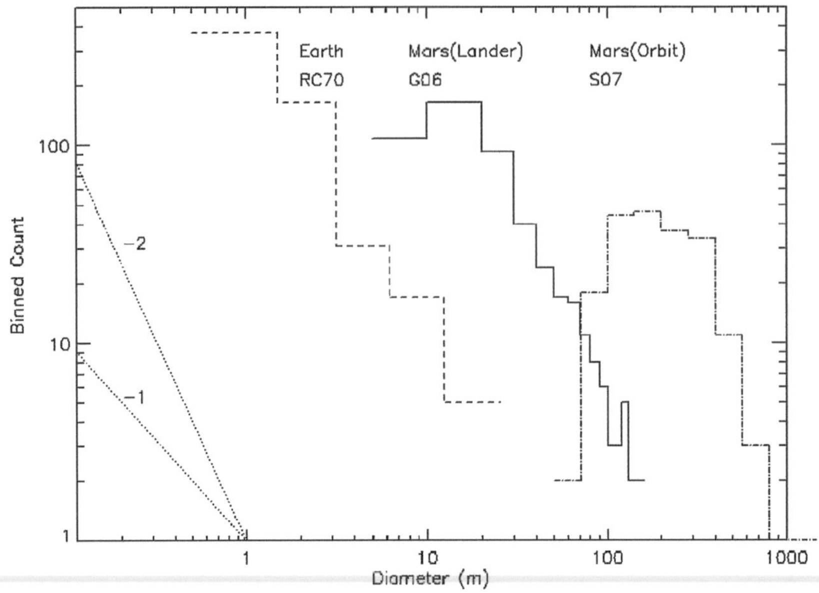

Fig. 7 Plotted are the binned counts of dust devils on Earth (Ryan and Carroll 1970), and on Mars from the Spirit rover (Greeley et al. 2006) and from Mars Express (Stanzel et al. 2008). Differential power law slopes of −1 and −2 are shown at right. From this plot one might conclude that Martian dust devils are larger (the modal diameter on Mars is ∼15 m, while more like ∼1 m on Earth), although it is also clear that one might also conclude that dust devils seen from orbit are much larger than those seen from the ground, by a similar factor. Thus it should similarly be noted that the Ryan and Carroll study was over a very small (<1 km^2) area, and so observed many small dust devils

of the terrestrial data makes the comparison challenging (the curves could be moved up or down if the observing period were different), but the modal diameter appears to be ∼3 times larger on Mars. An unbinned (or at least finely-binned) terrestrial dataset may permit a more confident answer of this question.

While the diameter distributions shown in Fig. 8 are merely suggestive of a larger modal diameter, it is certainly the case that the largest devil diameters reported are larger on Mars – see Table 1. Those summary statistics also suggest that Martian dust devils have a higher intensity (in the sense of relative pressure drop, expressed as a percentage of ambient) – again, the highest values recorded by surveys are at Mars, although possibly some targeted encounters on Earth that penetrate the very core of the devil may be higher yet. Note that the normalized pressure drop corresponds to a vortex wind speed – the atmospheric density contribution to dynamic pressure is proportional to ambient pressure. Encounters with a given threshold of normalized pressure drop (0.1 %, 0.5 %) appear to occur more frequently on Mars than at typical terrestrial locations. The maximum number of dust devils per km^2 per day on Mars, from the Spirit rover during its first season (Greeley et al. 2006) appears lower than the Ryan and Carroll (1970) value, although this may be an observational limit since the area surveyed was rather larger, exposing fewer small devils, than that terrestrial survey (see Sect. 3.1).

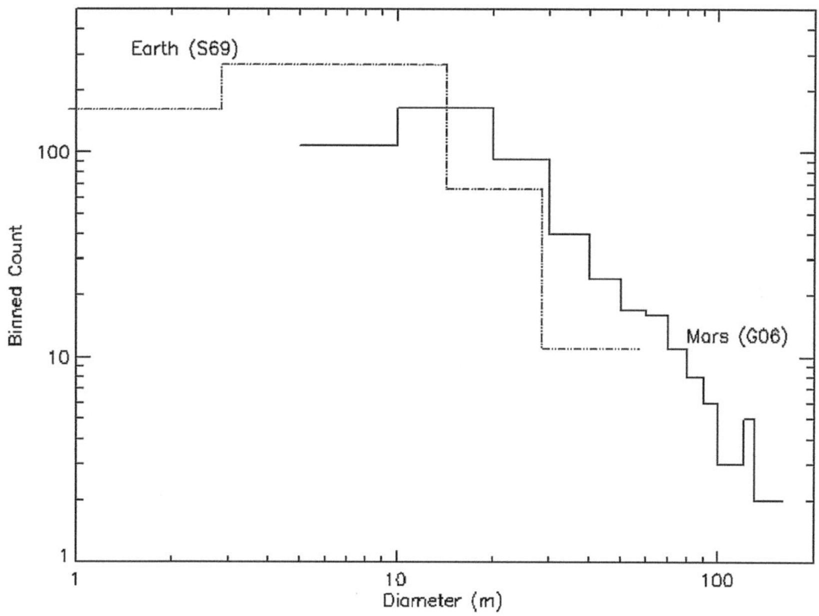

Fig. 8 A plot to examine the possibility that Mars and Earth devils are actually the same size. These data, from Sinclair's (1969) study at Tucson, and the Spirit observations of Greeley et al. (2006) involve roughly comparable survey areas, and so the bias in the terrestrial survey to smaller devils is not as severe as in the Ryan and Carroll example: in this instance the modal diameter on Earth is more like 5 m, only a factor 3 smaller than on Mars. It is notable that the largest devil seen in the Mars set is larger than that at Earth, but perhaps a longer survey on Earth (pushing the curve upwards) would allow the expected number of detections of larger devils on Earth to rise above 1. It is evident that the terrestrial data in particular suffers loss of information by binning which frustrates a more robust comparison

Table 1 Properties of dust devil populations

Characteristic	Unit	Earth	Mars	Notes
Largest Diameter Observed	(m)	>100	1650	1,2
Tallest Height Observed	(m)	~2500	20,000	1,3
Maximum Field Frequency	(km^{-2} day^{-1})	300	51	4,5
Largest Pressure Drop Observed	(Pa)	450*	4.8	6,7
(Normalized to Background)	(%)	0.5	0.7	6,7
Interval for 0.5 % Encounters	(days)	>300	20–100	8,9
Interval for 0.1 % Encounters	(days)	~10	1	8,9

1. Snow and McClelland (1990); 2. Stanzel et al. (2008); 3. Fenton and Lorenz (2015); 4. Carroll and Ryan (1970); 5. Greeley et al. (2006); 6. Sinclair (1973); 7. Murphy and Nelli (2002); 8. Lorenz and Lanagan (2014); 9. Lorenz and Reiss (2015). * this is the largest reliable value in a refereed journal. See Balme and Greeley (2006) who note theses by Ringrose and by Metzger which report 1000 and 1500 Pa respectively for Earth

5 Heuristic Metrics

Although our philosophy is that wherever possible, the effects of a dust devil population or the magnitude of a meteorological perturbation should be evaluated explicitly, it is worth discussing a couple of heuristic quantities used in the literature as approximations.

5.1 Pressure Cross Section

Some attempts to compare visual dust devil counts with in-situ meteorological detections (e.g. Moores et al. 2015) have invoked a 'pressure radius' distinct from an 'optical radius'. While in principle one can assess detection cross-sections in this way, we discourage this approach as it is only meaningful when integrated across a whole population. The problem is that the detection distance of a vortex (the 'pressure radius') depends not only on the actual vortex radius (i.e. that of the 'wall', the optical radius, where the windspeed is a maximum and the pressure drop is half that at the center) but also on the pressure detection threshold, and on the core pressure drop. The core pressure drop and the diameter are not necessarily correlated, and thus the average over the full population of these properties is difficult to estimate accurately.

As discussed in Sect. 3.4, for terrestrial surveys a pressure radius of ~1.5–3.5 optical radii is a reasonable range which might be considered for the most crude calculations. However, the correlation of diameter with longevity needs to be considered in any attempt at an accurate reconciliation of visual and in-situ records (e.g. Lorenz 2014).

5.2 Area Fraction of Dust Devils

A common metric in the literature is the 'area fraction' – the instantaneous or averaged area covered by dust devils (presumably defined by their optical diameter). This is a more useful metric in some senses than a simple count of dust devils, in that larger dust devils are given larger 'weighting'. Thus a total dust flux derived from multiplying an 'average' flux per unit area in dust devils and the area fraction should be more representative than some 'average' dust flux per devil multiplied by the number of devils. In an ideal world, of course, these averages would be accurately computed and the results would be equivalent, but this is rarely the case and indeed it is all but impossible to know how well such quantities are evaluated.

If we adopt a value of 100 km^{-2} day^{-1} with a minimum diameter of 1 m, and assume that a dust devil has a longevity $t = 40d^{0.66}$ s (per Lorenz 2013), then the fractional area covered by dust devils can be calculated by integrating the time-area product $t\pi d^2/4$ over the population. If all dust devils were only 1m in diameter (i.e. a power law with an exponent of 0), then there are 100 dust devils per day per km^2 (10^6 m^2), each lasting 40 seconds, and each with an area of 0.75 m^2. Then the fractional area covered is $100 \cdot 40 \cdot 0.75/(86{,}400 \cdot 10^6) = 3 \times 10^{-8}$.

If, instead, devil diameters follow a power law with exponents -1, -1.6 and -2, the resultant area fractions are 3×10^{-4}, 2.3×10^{-5} and 4×10^{-6}. It can be seen that these values are much higher than if all dust devils were small. As noted by Lorenz (2011), population-integral quantities such as total dust lifting or area fraction are dominated by the largest dust devils. Thus the notion of an 'average' dust devil requires exceptional caution, and perhaps should be avoided altogether, since such an average cannot be accurately determined from a small sample of the population.

For the -1.6 cumulative power law (favored by the Mojave data), the 2.3×10^{-5} fraction is reduced to 1.1×10^{-5} if the power law is truncated at 100 m diameter (i.e. if one assumes there are no dust devils larger than this). Evidently, the fraction also scales directly with the density of devils (100 km^{-2} day^{-1}).

It is striking that the area fraction computed by integrating the observed parameters (-1.6 cumulative power law, 100 km^{-2} day^{-1}) of 1–2 $\times$ 10^{-5} is very similar to the area fraction computed by Koch and Renno (2005). It follows, perhaps, that the -1.6 power law, truncated at $\sim$1 m at the small end, and some diameter of the order of 100–300 m at the high end, is a reasonably good representation of the terrestrial population.

Greeley et al. (2006) report dust devils observed on Mars by the Spirit rover at Gusev crater over years: the densities observed were 51, 11 and 20 per km^2/sol with median diameters of 19, 24 and 39 m respectively, average lifetimes of 129–170 s and average speeds of 1.5–2.5 $m\,s^{-1}$. The Gusev data for 2006 was fit with a -2 differential power law (-1 cumulative), although the following two years appear to have steeper fall-offs.

The smallest size bin was 0–10 m. If we adopt 5m as the minimum size, and a -1.6 cumulative power law as for Earth, but truncated at 500 m, then we find for $N(>5\text{ m}) = 20\text{ dd}\,km^{-2}/\text{sol}$ that the area fraction occupied by dust devils is some 1.6×10^{-4}, essentially an order of magnitude higher than for Earth. The distinction appears largely due to the larger size of dust devils on Mars. This in turn is likely related to the thicker planetary boundary layer.

6 Statistical Properties of Dust Devil Counts

In so far as dust devils are relatively rare events, it is tempting to consider (and may be convenient to model) their occurrence as a Poisson process. For example, the P28 Eldorado dataset shown in Fig. 2, also presented in Fig. 4) previously suggests a 1 mbar dust devil encounter may occur once every ten days or so, and thus has an occurrence frequency of ~ 0.1/day. This is of course not the same as a rate of 0.04 /hr, since dust devils typically do not occur during most of a 24 day period. Thus we implicitly consider a periodic or conditional factor in dust devil frequency – most usually by considering an 'active' period of only 6–8 hours per day. Thus during the active part of the day, 10–11 am local solar time to 5–7 pm or so, the recurrence rate is about 0.015/hr. This is quite a small value, and thus one would not expect two devils to occur in the same hour if they were truly independent events. Similarly, one would expect images of large areas showing multiple dust devils to be progressively less and less common as the number of dust devils increases.

This, however, is not what is observed. A survey of Mars Reconnaissance Orbiter Context Imager (CTX) images of Amazonis Planitia on Mars (Fenton and Lorenz 2015) shows not the monotonically-declining frequency of higher-populated images, but rather a bimodal distribution, with either a handful of dust devils, or about 30 (see Fig. reffig9). The images cover areas of 4000–8000 km^2 at a resolution of 6m/pixel, with dust devils being recognized for minimum diameters of 30–60 m.

Since dust devil vortices occur in a set of favorable meteorological conditions (convective heating, with moderate wind shear), simultaneous dust devils are in fact quite common. Furthermore, occurrence of dust devils (or dustless vortices) at regular intervals – typically 5–20 minutes – associated with the advection of a quasi-regular cellular pattern of boundary layer convection over a meteorological station have been noted (e.g. Lorenz and Christie 2015). Trains of regularly-spaced dust devils are sometimes reported (see Fenton and Lorenz 2015 and references therein).

The correlated occurrence of dust devils makes population measurements challenging, in that a short observing interval may encounter a 'lucky' day with many events, or a quiet

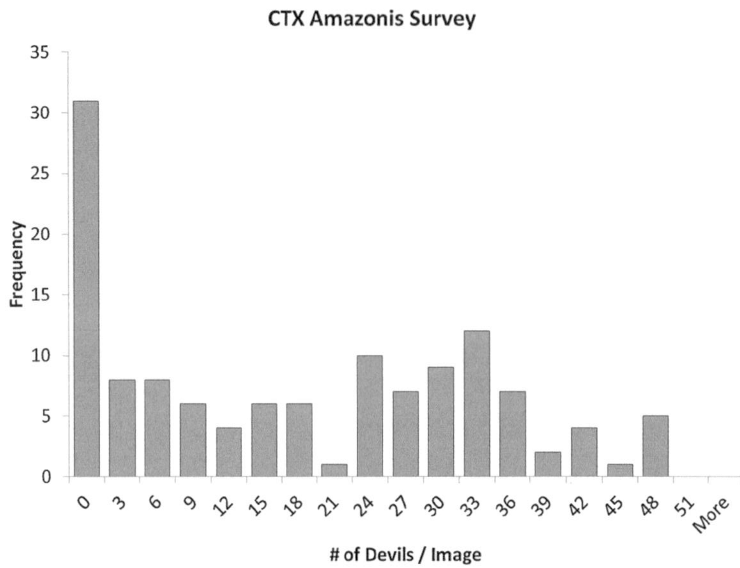

Fig. 9 The number of dust devils seen in the very active Amazonis Planitia area of Mars with the CTX camera (Fenton and Lorenz 2015). CTX images span an area of 4000–8000 km^2 and the corresponding devil diameters are some tens of meters. A purely Poisson process would have a systematic decline in frequency towards the right. Here the clear bulge in the histogram for 24–36 devils/image is striking, and shows that dust devil appearances over $\sim$100 km ranges are correlated, in that favorable conditions must simultaneously apply over such a large region

day when conditions are for some reason unfavorable. Put another way, dust devils are a nonstationary process in a statistical sense, in that mean values (e.g. of formation rate) are not a constant.

7 Summary and Conclusions

From this review of population studies, we can conclude the following:

When evaluating the broad influence of dust devils on the martian or terrestrial atmosphere, population-weighted statistics should be considered to the extent possible. The apparently skewed nature of the dust devil population and the steep dependence of their dust-lifting capacity on their physical properties mean that 'typical' or 'average' values may be grossly inadequate – the mean diameter dust devil does not loft a mean amount of dust. For instance, lab experiments reported in Neakrase et al. (2006) suggested that the dust flux from a vortex scales exponentially with a vortex's pressure drop, meaning the deepest (largest) but rarest devils dominate the martian dust flux, and the flux estimated by considering the "average" devil could be incorrect by orders of magnitude.

The *observed* populations of Martian dust devils indicate a larger diameter than for Earth. This is reflected not only in the largest diameters (and heights) reported in various datasets, but also in sample statistics such as the median. It is not, however, firmly established whether either Mars or Earth samples are 'complete' in the sense of efficiently detecting the smallest devils. Thus these statistics should be used with some caution, although examination of somewhat-comparable surveys suggest Martian dust devils may have a modal diameter a factor of $\sim$3 larger than Earth's.

If dust devil height and diameter are correlated (as they appear to be), and if the ultimate limit on dust devil height is the thickness of the planetary boundary layer, then it follows from measurements of the latter (usually several km, in some cases more than 10 km, compared with a maximum of 2–4 km on Earth) that the upper limit on dust devil diameter will be larger on Mars than Earth by a factor of $\sim$4. Systematic observations that 'completely' survey the largest devils are not presently available for Earth, although it is perhaps reasonable to assume that anecdotal reports (see Lorenz et al. 2016) may shed light on the question.

Available data do not yet indicate fundamental differences between Earth and Mars on the correlation between diameter and longevity. This correlation is important in that it influences the extent to which dust-lifting may be dominated by the largest, rarest dust devils.

Some recommendations in future analysis or re-analysis of observational datasets (field or remote sensing) are as follows:

1. The 'stopping rule' in surveys should be reported. As an example, did field observations extend for a predetermined period (e.g. 'stop at 6 pm, no matter what') or did they terminate when there appeared to be nothing happening (e.g. an hour since the last dust devil was seen)? Similarly the criteria for selecting remote images for study should be given (and the number of images examined in which dust devils were not detected should be reported). Detection criteria should be quantified as well as possible (e.g. contrast threshold) and methodology in visual surveys reported (e.g. counts made in a 1-minute horizon sweep at 15 minute intervals, or continuous surveillance, etc.).
2. Survey completeness should also be assessed to the extent possible. For surveys using barometric time series, injection and recovery of synthetic dust devil signals can reveal the efficiency of a chosen detection scheme and its dependence on the signal's properties. Similar injection efforts may be possible for surveys involving orbital or in-situ visual data. For both types of surveys, images of identified dust devils could be mixed together with control images without identified devils and contrast ratios adjusted. The workers surveying the visual data for devils could then attempt to sift the real devils from the control images, allowing the detection efficiency to be evaluated.
3. Binning into size categories may be the only practical approach in some cases, but in general binning is to be avoided as information is lost. Tabulation of raw times, longevities, diameters, speeds etc. for individual devils is most useful. In particular, data should be archived and made available to other workers, a requirement in other fields.[2] In graphical presentation, a cumulative plot on logarithmic axes can be useful.

References

R.E. Arvidson, E.A. Guinness, H.J. Moore, J. Tillman, S.D. Wall, Three Mars years: Viking lander 1 imaging observations. Science **222**, 463–468 (1983)

M. Balme, R. Greeley, Dust devils on Earth and Mars. Rev. Geophys. **44**(3) (2006). doi:10.1029/2005RG000188

M.R. Balme, P.L. Whelley, R. Greeley, Mars: Dust devil track survey in Argyre Planitia and Hellas Basin. J. Geophys. Res. **108**, 5086 (2003). doi:10.1029/2003JE002096

M.R. Balme, A. Pathare, S.M. Metzger, M.C. Towner, S.R. Lewis, A. Spiga, L.K. Fenton, N.O. Renno, H.M. Elliott, F.A. Saca, T.I. Michaels, Field measurements of horizontal forward motion velocities of terrestrial dust devils: towards a proxy for ambient winds on Mars and Earth. Icarus **221**(2), 632–645 (2012)

[2]The Planetary Atmospheres Node of NASA's Planetary Data System provides support for researchers who wish to archive their dust devil data – http://atmos.nmsu.edu/atmos-home.html.

T. Broersen, Quantification of soil erosion by dust devil in the Jordan Badia. MSc thesis, University of Utrecht, The Netherlands (2013)

J.J. Carroll, J.A. Ryan, Atmospheric vorticity and dust devil rotation. J. Geophys. Res. **75**(27), 5179–5184 (1970)

M.D. Ellehoj, H.P. Gunnlaugsson, P.A. Taylor, H. Kahanpa, K.M. Bean, B.A. Cantor, B.T. Gheynani, L. Drube, D. Fisher, A.-M. Harri, C. Holstein-Rathlou, M.T. Lemmon, M.B. Madsen, M.C. Malin, J. Polkko, P.H. Smith, L.K. Tamppari, W. Weng, J. Whiteway, Convective vortices and dust devils at the Phoenix Mars mission landing site. J. Geophys. Res. **115**, E00E16 (2010)

L. Fenton, R. Lorenz, Dust devil height and spacing with relation to the Martian planetary boundary layer thickness. Icarus **260**, 246–262 (2015)

R. Greeley, P.L. Whelley, R.E. Arvidson, N.A. Cabrol, D.J. Foley, B.J. Franklin, P.G. Geissler, M.P. Golombek, R.O. Kuzmin, G.A. Landis, M.T. Lemmon, Active dust devils in Gusev crater, Mars: observations from the Mars Exploration Rover Spirit. J. Geophys. Res., Planets **111**(E12) (2006). doi:10.1029/2006JE002743

B.K. Jackson, R.D. Lorenz, A multi-year dust devil vortex survey using an automated search of pressure time-series. J. Geophys. Res., Planets **120** (2015). doi:10.1002/2014JE004712

K.M. Kinch, J. Sohl-Dickstein, J.F. Bell III., J.R. Johnson, W. Goetz, G.A. Landis, Dust deposition on the Mars Exploration Rover Panoramic Camera (Pancam) calibration targets. J. Geophys. Res. **112**, E06S03 (2007). doi:10.1029/2006JE002807

M. Klose, B.C. Jemmett-Smith, H. Kahanpaa, M. Kahre, P. Knippertz, M.T. Lemmon, S.R. Lewis, R.D. Lorenz, L.D.V. Neakrase, C. Newman, M.R. Patel, D. Reiss, A. Spiga, P.L. Whelley, Dust devil sediment transport: from lab to field to global impact. Space Sci. Rev. (2016). doi:10.1007/s11214-016-0261-4

J. Koch, N.O. Renno, The role of convective plumes and vortices on the global aerosol budget. Geophys. Res. Lett. **32**(18), L18806 (2005)

M.V. Kurgansky, Steady-state properties and statistical distribution of atmospheric dust devils. Geophys. Res. Lett. **33**, L19S06 (2006)

M.V. Kurgansky, Statistical distribution of atmospheric dust devils. Icarus **219**, 556–560 (2012)

R.L. Lambeth, On the measurement of dust devil parameters. Bull. Am. Meteorol. Soc. **47**, 522–526 (1966)

G.A. Landis, P.J. Jenkins, Measurement of the settling rate of atmospheric dust on Mars by the MAE instrument on Mars Pathfinder. J. Geophys. Res. **105**, 1855–1857 (2000)

R.D. Lorenz, Power law of dust devils on Earth and Mars. Icarus **203**, 683–684 (2009)

R.D. Lorenz, Studies of desert dust devils with a sensor network, timelapse cameras and thermal imaging, in *Aerospace Conference*, 6 March 2010 (IEEE Press, New York, 2010), pp. 1–7

R.D. Lorenz, On the statistical distribution of dust devil diameters. Icarus **215**, 381–390 (2011)

R.D. Lorenz, Pressure drops in dust devils: Earth and Mars. Planet. Space Sci. **60**, 370–375 (2012)

R.D. Lorenz, The longevity and aspect ratio of dust devils: effects on detection efficiencies and comparison of landed and orbital imaging at Mars. Icarus **226**, 964–970 (2013)

R.D. Lorenz, Vortex encounter rates with fixed barometer stations: comparison with visual dust devil counts and large eddy simulations. J. Atmos. Sci. **71**, 4461–4472 (2014)

R.D. Lorenz, D. Christie, Dust devil signatures in infrasound records of the international monitoring system. Geophys. Res. Lett. **42**, 2009–2014 (2015). doi:10.1002/2015GL063237

R.D. Lorenz, B.K. Jackson, Dust devils and dustless vortices on a desert playa observed with surface pressure and solar flux logging. J. Geophys. Res. **5**, 1 (2015)

R.D. Lorenz, P.D. Lanagan, A barometric survey of dust devil vortices on a Desert Playa. Bound.-Layer Meteorol. **53**, 555–568 (2014). doi:10.1007/s10546-014-9954-y

R. Lorenz, J. Radebaugh, Dust devils in thin air: vortex observations at a high elevation Mars analog site in the Argentinian Puna. Geophys. Res. Lett. (2016). doi:10.1002/2015GL067412

R.D. Lorenz, D. Reiss, Solar panel clearing events, dust devil tracks, and in-situ vortex detections on Mars. Icarus **248**, 162–164 (2015)

R.D. Lorenz, J. Zimbelman, *Dune Worlds: How Wind-Blown Sand Shapes Planetary Landscapes* (Springer, Berlin, 2014)

R.D. Lorenz, M.R. Balme, Z. Gu, H. Kahanpää, M. Klose, M. Kurgansky, M.R. Patel, D. Reiss, A.P. Rossi, A. Spiga, T. Takemi, W. Wei, History and applications of dust devil studies. Space Sci. Rev. (2016). doi:10.1007/s11214-016-0239-2

J.P. Mason, M.R. Patel, S.R. Lewis, Radiative transfer modelling of dust devils. Icarus **223**(1), 1 (2013)

J.O. Mattsson, T. Nihlén, W. Yue, Observations of dust devils in a semi-arid district of southern Tunisia. Weather **48**(11), 359–363 (1993)

H.J. Melosh, *Impact Cratering: A Geologic Process* (Oxford Univ. Press, London, 1990)

S. Metzger, M. Balme, A. Pathare, Meteorologic conditions and the formation of terrestrial dust devils, in *40th Lunar and Planetary Science Conference*, Houston, TX (2009)

J.E. Moores, M.T. Lemmon, H. Kahanpää, S.C. Rafkin, R. Francis, J. Pla-Garcia, K. Bean, R. Haberle, C. Newman, M. Mischna, A.R. Vasavada, Observational evidence of a suppressed planetary boundary layer in northern Gale Crater, Mars as seen by the Navcam instrument onboard the Mars Science Laboratory rover. Icarus **249**, 129–142 (2015)
J.R. Murphy, S. Nelli, Mars pathfinder convective vortices: frequency of occurrence. Geophys. Res. Lett. **29**(23) (2002). doi:10.1029/2002GL015214
L.D.V. Neakrase, R. Greeley, J.D. Iversen, M.R. Balme, E.E. Eddlemon, Dust flux within dust devils: preliminary laboratory simulations. Geophys. Res. Lett. **33**, L19S09 (2006). doi:10.1029/2006GL026810
H. Ohno, T. Takemi, Mechanisms for intensification and maintenance of numerically simulated dust devils. Atmos. Sci. Lett. **11**, 27–32 (2010). doi:10.1002/asl.249
A.M.C. Oke, N.J. Tapper, D. Dunkerley, Willy-willies in the Australian landscape: the role of key meteorological variables and surface conditions in defining frequency and spatial characteristics. J. Arid Environ. **71**, 201–215 (2007)
A.V. Pathare, M.R. Balme, S.M. Metzger, A. Spiga, M.C. Towner, N.O. Renno, F. Saca, Assessing the power law hypothesis for the size-frequency distribution of terrestrial and martian dust devils. Icarus **209**, 851–852 (2010)
S. Raasch, T. Franke, Structure and formation of dust devil-like vortices in the atmospheric boundary layer: a high-resolution numerical study. J. Geophys. Res., Atmos. **116**(D16) (2011). doi:10.1029/2011JD016010
D. Reiss, First observations of terrestrial dust devils in orbital image data: comparison with dust devils in Amazonis Planitia, Mars, in *Lunar and Planetary Science Conference*, Houston, TX, 21–25 March (2016)
D. Reiss, A. Spiga, G. Erkeling, The horizontal motion of dust devils on Mars derived from CRISM and CTX/HiRISE observations. Icarus **227**, 8–20 (2014)
D. Reiss, L. Fenton, L. Neakrase, M. Zimmerman, T. Statella, P. Whelley, A.P. Rossi, M. Balme, Dust devil tracks. Space Sci. Rev. (2016 this issue)
T.J. Ringrose, M.C. Towner, J.C. Zarnecki, Convective vortices on Mars: a reanalysis of Viking Lander 2 meteorological data, sols 1–60. Icarus **163**, 78–87 (2003)
T.J. Ringrose, M.R. Patel, M.C. Towner, M. Balme, S.M. Metzger, J.C. Zarnecki, The meteorological signatures of dust devils on Mars. Planet. Space Sci. **55** (14), 2151–2163 (2007)
J.A. Ryan, J.J. Carroll, Dust devil wind velocities: mature state. J. Geophys. Res. **75**, 531–541 (1970)
J.A. Ryan, R.D. Lucich, Possible dust devils, vortices on Mars. J. Geophys. Res. **88**, 11005–11011 (1983)
J. Scargle, J. Norris, B. Jackson, J. Chiang, Studies in astronomical time series analysis, VI: Bayesian block representations. Astrophys. J. **764**(2), 167 (2013). doi:10.1088/0004-637X/764/2/167
P.C. Sinclair, General characteristics of dust devils. J. Appl. Meteorol. **8**(1), 32–45 (1969)
P.C. Sinclair, The lower structure of dust devils. J. Atmos. Sci. **30**(8), 1599–1619 (1973)
J.T. Snow, T. McClelland, Dust devils at White Sands Missile Range, New Mexico, 1: temporal and spatial distributions. J. Geophys. Res. **95**, 13,707–13,721 (1990)
C. Stanzel, M. Pätzold, D.A. Williams, P.L. Whelley, R. Greeley, G. Neukum (HRSC Co-Investigator Team), Dust devil speeds, directions of motion and general characteristics observed by the Mars Express High Resolution Stereo Camera. Icarus **197**(1), 39–51 (2008)
D.M. Tratt, M.H. Hecht, D.C. Catling, E.C. Samulon, P.H. Smith, In situ measurements of dust devil dynamics: toward a strategy for Mars. J. Geophys. Res. **108**(E11), 5116 (2003). doi:10.1029/2003JE002161
C.A. Verba, P. Geissler, T. Titus, D. Waller, Observations from the High Resolution Imaging Science Experiment (HiRISE): martian dust devils in Gusev and Russell craters. J. Geophys. Res. **115**, E09002 (2012). doi:10.1029/2009JE003498
R.E. Wyett, Pressure drop in a dust devil. Mon. Weather Rev. **82**, 7 (1954)

Space Sci Rev (2016) 203:299–345
DOI 10.1007/s11214-016-0241-8

Applications of Electrified Dust and Dust Devil Electrodynamics to Martian Atmospheric Electricity

R.G. Harrison[1] · E. Barth[2] · F. Esposito[3] · J. Merrison[4] · F. Montmessin[5] · K.L. Aplin[6] · C. Borlina[7] · J.J. Berthelier[5] · G. Déprez[5] · W.M. Farrell[8] · I.M.P. Houghton[6] · N.O. Renno[7] · K.A. Nicoll[1] · S.N. Tripathi[9] · M. Zimmerman[10]

Received: 28 August 2015 / Accepted: 15 February 2016 / Published online: 12 April 2016

Abstract Atmospheric transport and suspension of dust frequently brings electrification, which may be substantial. Electric fields of 10 kV m^{-1} to 100 kV m^{-1} have been observed at the surface beneath suspended dust in the terrestrial atmosphere, and some electrification has been observed to persist in dust at levels to 5 km, as well as in volcanic plumes. The interaction between individual particles which causes the electrification is incompletely understood, and multiple processes are thought to be acting. A variation in particle charge with particle size, and the effect of gravitational separation explains to, some extent, the charge structures observed in terrestrial dust storms. More extensive flow-based modelling demonstrates that bulk electric fields in excess of 10 kV m^{-1} can be obtained rapidly (in less than 10 s) from rotating dust systems (dust devils) and that terrestrial breakdown fields can be obtained. Modelled profiles of electrical conductivity in the Martian atmosphere suggest the possibility of dust electrification, and dust devils have been suggested as a mechanism of charge separation able to maintain current flow between one region of the atmosphere and another, through a global circuit. Fundamental new understanding of Martian atmospheric electricity will result from the ExoMars mission, which carries the DREAMS (Dust characterization, *Risk Assessment, and Environment Analyser on the Martian Surface*)—

✉ R.G. Harrison
r.g.harrison@reading.ac.uk

1 Department of Meteorology, University of Reading, Reading, UK

2 Southwest Research Institute, Boulder, CO, USA

3 INAF – Osservatorio Astronomico di Capodimonte, Naples, Italy

4 University of Aarhus, Aarhus, Denmark

5 Laboratoire Atmosphères, Milieux, Observations Spatiales (LATMOS), Guyancourt, France

6 Department of Physics, University of Oxford, OX1 3RH, Oxford, UK

7 Atmospheric, Oceanic and Space Sciences, University of Michigan, Ann Arbor, USA

8 NASA Goddard, Greenbelt, USA

9 Centre for Environmental Science and Engineering, Indian Institute of Technology, Kanpur, India

10 Johns Hopkins University, Baltimore, USA

MicroARES (*Atmospheric Radiation and Electricity Sensor*) instrumentation to Mars in 2016 for the first in situ electrical measurements.

Keywords Planetary electrostatics · Lightning discharge · Particle electrification · Global circuit

1 Charge Separation in Dust

It has long been known that dust lofted or transported can become highly electrified. Strong effects were reported on a gold leaf electrometer connected to a roof-level electrode which was exposed to a dust storm in Lahore in 1847, together with the generation of sparks (Baddeley 1860). An accompanying characteristic of the Dust Bowl disaster in the US Great Plains during the 1930s was severe static electricity, audible on domestic medium wave radios and able to cause electric shocks, with earthing chains necessary on automobiles. Under fair weather conditions, the magnitude of the vertical terrestrial atmospheric electric field is about 100 V m^{-1}, sustained by distant thunderstorm and disturbed weather electrification. Dust storms are considerably electrified in comparison. Quantitatively, electric fields exceeding 100 kV m^{-1} have been measured during blowing sand, dust storms and dust devils (Rudge 1913; Demon et al. 1953; Freier 1960; Crozier 1964; Harris 1967; Stow 1969; Kamra 1972; Schmidt et al. 1998; Renno et al. 2004; Jackson and Farrell 2006; Kok and Renno 2006, 2008a, 2008b; Williams et al. 2009). As an example, Fig. 1 shows the variation in electric field during the passage of a dust devil in the Sahara desert, opportunistically obtained during measurements undertaken for another purpose (Freier 1960).

As discussed later in this paper (Sect. 3), it is generally accepted that dust electrification is due to contact and triboelectric charging between blowing particles during these phenomena. The details of the charge transfer mechanism remain incomplete, but several experiments and observations (Freier 1960; Inculet et al. 2006; Duff and Lacks 2008) suggest that on average, during collisions, the smallest grains acquire negative charge with respect to larger particles. In general, the smallest particles are brought into suspension and transported aloft into the atmosphere by local turbulence, whereas the larger particles stay close to the surface. This gravitational charge separation is consistent with increase of the atmospheric electric field observed during dusty phenomena.

Laboratory and theoretical work (Kok and Renno 2006, 2008a, 2008b) suggest that intense electric fields can influence charged particle dynamics through changing their trajectories and reducing the threshold friction speed necessary for initiating their motion. This facilitates the lifting of particles from the surface, so increasing the concentration of saltating grains at a given wind speed. Electric forces may therefore play a further role in the evolution of dust events, including dust devils.

Observations of dust electrification are sparse in literature, both because this field of study is relatively young and because of the difficulty in sampling stochastic dust activity. There is renewed interest in dust electrification due to its consequences for human activities such as the breakdown of power transmission lines, electrical alignment of atmospheric particles (Ulanowski et al. 2007), volcanic plume electrification (Mather and Harrison 2006) and planetary exploration (Helling et al. 2016). In this latter case, the presence of electric discharges or electromagnetic noise can potentially affect communications or communications equipment, but substantial electric fields have also been argued to affect atmospheric chemistry and planet habitability (Atreya et al. 2006; Delory et al. 2006) and the possible development of life (Miller 1953), in particular for Mars. New measurements from

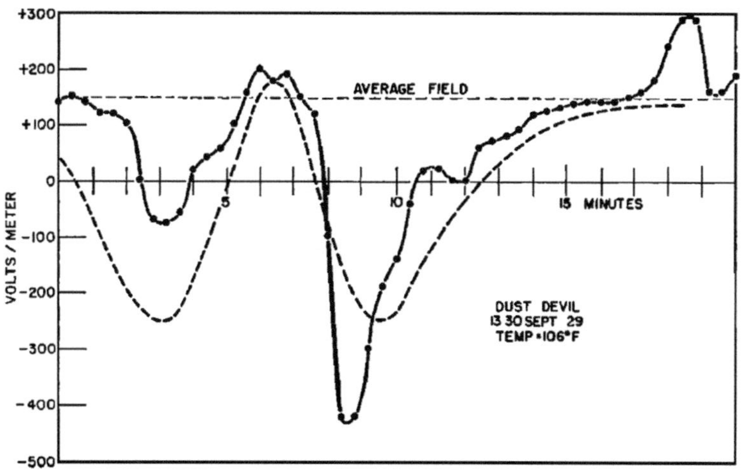

Fig. 1 Variation of electric field with time beneath the passage of a dust devil in the Sahara desert. (Reproduced from Freier 1960 with permission of Wiley)

the ExoMars 2016 space mission (Bettanini et al. 2014) using the DREAMS-MicroARES (acronym formed from Dust characterization, *Risk Assessment, and Environment Analyser on the Martian Surface*, and *Atmospheric Radiation and Electricity Sensor*) instrument will provide the first direct measurements of the electric properties of Martian atmosphere. This is likely to mark a hugely important step change in the knowledge of Martian atmospheric electricity.

This paper therefore presents a pre-MicroARES view of the electrical properties of dusts and dust devils, with particular reference to Mars. To do this, Sect. 2 first briefly summarises knowledge on Martian atmospheric electricity. Section 3 then provides an overview of particle electrification mechanisms concerning dust. Section 4 reports the results of some of the most recent observations of dust electrification during dust storms/devils in terrestrial deserts both close to the surface and aloft. Section 5 provides an overview of modelling of dust devils, and Sect. 6 describes the DREAMS-MicroARES sensor to be used on the forthcoming ExoMars mission.

2 The Martian Electrical Environment

2.1 Atmospheric Electricity on Mars

Atmospheres electrify to varying extents, depending on the charged particles they contain and whether winds, convection or other meteorological processes can actively separate local charge (Harrison et al. 2008; Aplin 2006). The consequences of electrification include the motion of charged particles and ions under the action of electric fields, and electrical discharges where substantial accumulations of charge occur. In the Martian atmosphere, cosmic rays and ultra violet radiation generate molecular cluster ions which, together with free electrons, make the atmosphere electrically conductive. With active charge separation therefore, current flow can potentially occur, and a planetary electrical system analogous to the terrestrial global atmospheric electric circuit has accordingly been considered (Fillingim 1998;

Farrell and Desch 2001; Aplin 2006). The abundance of dust in Mars' atmosphere (and remote sensing of its dust devils), combined with knowledge that terrestrial dust electrification is common has led to the expectation that Martian dust devils can become electrified.

The basis for electrical activity on Mars has been assumed to be similar to electrical activity in Earth's atmosphere. This has been considered through the global circuit concept introduced for the Earth in the early 20th century by Wilson (1921), Harrison (2011). A global circuit relies on the existence of several key characteristics: a global generator (lightning, discharges, etc.), conductive surfaces (the ionosphere and the rocky surface) overall forming a spherical capacitor system (Rycroft et al. 2008, 2012). Lightning is a key indicator of planetary atmospheric electricity which can be remotely sensed. For example, at Earth, VLF signals propagating in the surface-ionosphere waveguide allow location of lightning events, and high frequency "sferic" (>3 MHz) emissions can be detected by Earth-orbiting spacecraft (Herman et al. 1973). Global circuits may prevail on a number of solar system bodies including Mars (Aplin 2006; Aplin et al. 2008).

The case for Martian atmospheric electricity has taken a unique path motivated by a strong astrobiological context. After the initial tribo-electricity experiments of Eden and Vonnegut (1973), Mills (1977) mixed grains in a low pressure CO_2 gas and created impressive electrical activity (glow and spark discharges) in order to gain insights into the environment that may have created ambiguities in the Viking lander biological experiments. It was concluded that this active dust-created electrical environment could make Martian dust storms an effective 'scavenger' of organic material. To date though, there have been no direct measurements of the atmospheric electrical environment on Mars. Indirect evidence of electrification however exists from the apparent adhesion of dust to the wheels of the Mars Pathfinder and Sojourner rovers, suggested to be electrostatic in origin (Farrell et al. 1999; Ferguson et al. 1999) and from laboratory measurements using Martian analogue materials (e.g. Krauss et al. 2003; Aplin et al. 2012).

Given the lack of in situ measurements, there are three strong but circumstantial arguments that dust storms in the low pressure CO_2 atmosphere of Mars may be electrical in nature:

(1) Laboratory experiments (Farrell et al. 2015) of the breakdown from mixing sand grains all display measurable electrical effects. These not only include the early glow-creating laboratory experiments by Eden and Vonnegut (1973) and Mills (1977), but also a later set of dust mixing studies in a low pressure CO_2 gas performed by Krauss et al. (2003, 2006) and pre-glow and spark discharge studies by Farrell et al. (2015).
(2) Modelling studies (see Sect. 5) suggest that an electron avalanche and collisional plasma can be created when dust storm E-fields stress the low pressure CO_2 gas.
(3) Measurements in terrestrial desert environments reveal that dust devils and dust features can generate large electric fields (Freier 1960; Crozier 1964; Farrell et al. 2004; Renno et al. 2004; Delory et al. 2006; Jackson and Farrell 2006; Seran et al. 2013; Esposito et al. 2015, 2016). If this same electrical generator process occurred in a lower pressure atmosphere (like that at Mars), there would be the initiation of enhanced electron impact ionization and atmospheric breakdown.

In terms of arguments against Martian atmospheric electricity, Ruf et al. (2009) reported the possible remote-sensed detection of lightning emission at 8 GHz from Mars using a terrestrial-based radio telescope. Given the detection, the electric dipole moment to account for the discharge was found to be relatively large. However, subsequent work (Anderson et al. 2012) could not confirm the initial ~8 GHz observation. Also, a sensitive 4–5 MHz radio system aboard Mars Express carried out an unsuccessful search for lightning RF discharges, in close proximity to the planet (Gurnett et al. 2010).

These remote sensing observations do not confirm impulsive lightning events from Mars. However, instead of intense impulsive cloud-to-ground discharges known at Earth, laboratory experiments suggest the effects may be more subtle, with mixing of particles creating dark Townsend (gas ionization) discharges, low current glows, and local weaker but numerous spark discharges. In other words, the Martian system may be continually discharging (or 'leaking') at low current levels into the low-pressure atmosphere, thereby avoiding the excessively large accumulation of charge which leads to impulsive lightning events. In situ measurement systems offer the possibility to resolve this apparent fundamental difference in our understanding of the terrestrial and Martian systems.

2.2 Ion Balance in Atmospheres

On Mars, the main permanent ionization source is the bombardment by cosmic rays and the cascade of ions and electrons generated when they collide with neutral atoms and molecules. The maximum ion production rate by cosmic rays is at the surface because the atmospheric thickness of Mars is not sufficient to allow cosmic rays to deposit all their energy in the atmosphere, hence most of the energy reaches the surface. During day-time photo-ionization by solar EUV and soft X-rays can also contribute with an efficiency increasing with altitude while, at night, energetic electron precipitation from the magnetic tail of the planet may also provide a supplementary source of ionization.

Since the initial work of Whitten et al. (1971), progresses in laboratory measurements of ion-neutral reactions as well as radio-electric observations of the electron density profiles (Patzold et al. 2005) have allowed several authors to build more detailed models of ionized species in the Martian lower atmosphere. According to Molina-Cuberos et al. (2002) the most abundant ions below ~50 km are hydronium ions $H_3O^+(H_2O)_n$ while O_2^+ dominates above ~70 km. The main negative ion species are water clusters of CO_3^-, NO_2^-, NO_3^- reaching a density of about 4500 ions cm^{-3} at the surface. At 35 km the electron density reaches a peak of $\sim 10^2\ e\,cm^{-3}$ and after a decrease with altitude, a secondary peak of $\sim 70\ e\,cm^{-3}$ appears at ~5 km. This secondary peak may significantly impact the negative conductivity at ground level. Although not indicated in the paper, this secondary peak close to the surface might be due to photo-electron production from UV impacts on the surface (Grard 1995).

An atmosphere's electrical conductivity depends on the local concentrations of ions and electrons and ion species it contains. The total conductivity σ is given by

$$\sigma = e(\mu_+ n_+ + \mu_- n_- + \mu_e n_e) \tag{1}$$

where n_+, n_-, and n_e are the number concentrations per unit volume of positive ions, negative ions and free electrons respectively, and μ_+, μ_-, and μ_e are their associated electrical mobility (the drift speed per unit electric field). The concentrations of ions and electrons are determined by the production rates by ionization and photoelectron emission, and their losses by recombination and attachment to other particles. In the presence of a dust-laden atmosphere, the concentration of ions and electrons (and therefore the electrical conductivity) will be reduced. The ratio of ions to electrons depends on the concentration of atmospheric electrophilic species; on Mars both negative CO_2^- ions and electrons are thought to be present (e.g. Aplin 2006). The ion and electron concentrations are governed by a set of rate equations for ion production and loss, written separately for positive ions, negative ions and electrons by Tripathi et al. (2008) as

$$\frac{dn_+}{dt} = q - \alpha n_+ n_- - \alpha_e n_+ n_e - n_+ \sum_j \beta_j^+ N_j, \tag{2}$$

$$\frac{dn_-}{dt} = q - \alpha n_+ n_- - n_- \sum_j \beta_j^- N_j - F n_-, \tag{3}$$

$$\frac{dn_e}{dt} = q_e + F n_- - \alpha_e n_+ n_e - n_e \sum_j \beta_j^e N_j, \tag{4}$$

assuming no transport of ionization. In Eqs. (2) to (4), the volumetric ion and electron production rates are q and q_e respectively, the ion-ion and ion-electron recombination coefficients are α and α_e, and the negative ion detachment rate to yield electrons is F. The positive ion-particle attachment coefficient to particles of concentration N carrying j elementary charges is β_j^+, that for negative ion attachment to particles β_j^-, and that for electron attachment to particles β_j^e. Theoretical formulations for recombination and attachment coefficients exist (e.g. Harrison and Tammet 2008; Tripathi et al. 2008), and the electron detachment rate can be found from thermodynamic considerations.

2.3 Martian Conductivity Profiles and Relaxation Timescales

Solution of the ion balance equations with time gives the variations in ion and electron concentrations, which, when the production and loss rates are equal, become steady-state concentrations. If the ion production rates are known or can be calculated as a function of height (for example from information about the cosmic ray energies and the atmospheric density profile), and assumptions are made about the vertical profile of particles, the steady state ion and electron concentrations can be found. In turn, this allows calculation of the electrical conductivity, σ. An important parameter which can be determined from σ is the charge relaxation timescale, τ, given by

$$\tau = \frac{\varepsilon_0}{\sigma} \tag{5}$$

where ε_0 is the permittivity of free space. This defines the timescale for the decay of charge on a particle in a medium with conductivity σ, as if the particle were considered to be of a finite capacitance and connected to a distant fixed potential through the medium. It therefore provides a characteristic timescale for charge to be sustained on a particle. In the Earth's lower atmosphere, timescales of 10 s to 100 s are typical, which allows electrification to develop with moderately active electrical processes such as those in terrestrial dust storms, thunderstorms or volcanic plumes.

Figure 2(a) shows calculations from Tripathi et al. (2008) for the vertical profile of electrical conductivity in the Martian atmosphere, under different assumptions of aerosol loading. It is immediately apparent that the electrical conductivity is usually two orders of magnitude greater than that for Earth. Only when the Martian atmosphere is assumed to contain a dust storm does its lower atmosphere have an electrical conductivity comparable to that of Earth. (This has practical consequences for testing Martian atmospheric electrical instrumentation on Earth, see Sect. 6). The associated relaxation times are shown

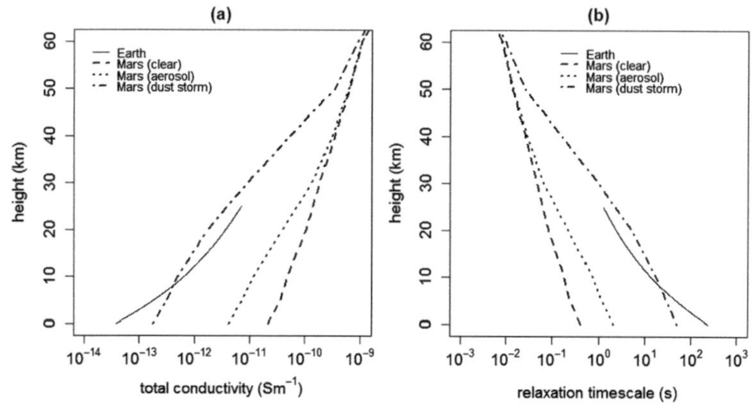

Fig. 2 Profiles of (**a**) atmospheric electrical conductivity σ and (**b**) the associated charge relaxation timescale ε_0/σ, for Mars (from Tripathi et al. 2008) and Earth (from Nicoll 2012). In the Martian case, three scenarios of conductivity profile are considered, for a clear atmosphere, an aerosol-laden atmosphere, and a dust storm of opacity 5. In the terrestrial case, the total conductivity is considered to be twice the positive conductivity, under the quiet solar conditions observed by Gringel (1978)

in Fig. 2(b). This indicates that, for charge to be sustained on particles in the Martian atmosphere for timescales appreciably greater than one second, a dust storm conductivity environment would be required. The combination presented by a dust devil generating particle electrification in a low conductivity environment may therefore offer unique circumstances on Mars for active charge generation.

The effect of including the aerosols in the electrical model of the lower Martian atmosphere very substantially reduces the free electron concentration at low altitude. It decreases by more than two orders of magnitude compared to the simple model of ion-neutral chemistry of Molina-Cuberos et al. (2002) to be less than $\sim 0.5\ \mathrm{e\,cm^{-3}}$ at ground with the $\sim 100\ \mathrm{e\,cm^{-3}}$ peak at 35 km lifted to ~45 km, and also reduced by two orders of magnitude at $\sim 1\ \mathrm{e\,cm^{-3}}$.

Michael et al. (2007, 2008) comprehensively modelled the role of aerosols on the charged particle distribution and conductivity in the lower atmosphere using aerosol properties (Chassefière et al. 1995). These studies were performed for both night-time and daytime conditions when dust particles are photo-ionized by low energy (~6 eV) solar UV that reach the Martian surface and produce a photo-electron population whose properties were taken from the work of Grard (1995). A role for dust particles, in particular close to the surface, is clearly emphasized by this study.

At night aerosols become charged by attachment of positive or negative ions. More aerosols become charged close to the surface than at higher altitudes, and the dust content from Phobos 2 spacecraft data indicates an average decrease of conductivity by a factor of two over the whole range of altitudes (Chassefière et al. 1995).

2.3.1 Tribo-Electric Charging Processes

The build-up of intense electric fields in the Martian atmosphere is expected to result from charge exchange between particles colliding in a turbulent environment (see Fig. 3).

In order to understand the differences between Earth and Mars charging processes, the combined system must be considered, where the E-field is generated by the competing tribo-

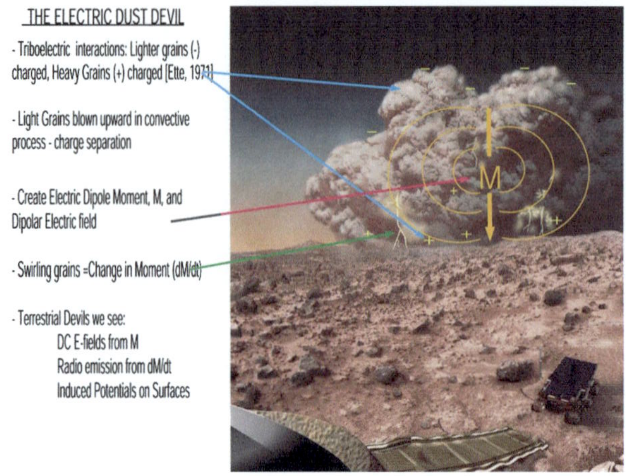

Fig. 3 An artist's impression of an electrified dust devil at Mars (from Farrell et al. 2004)

electric charging currents and atmospheric dissipation currents. This can be expressed as

$$\varepsilon_0 \frac{dE}{dt} = -J_{\text{tribo}} + J_{\text{diss}} = -n_{\text{dust}} q_{\text{dust}} v_{\text{dust}} + \left[\sigma E + n_0 \exp(\alpha_T d) e \mu_e E\right] \tag{6}$$

where n_0 is the nominal near-surface electron content, α_T is Townsend's first coefficient representing the number of electron impact ionizations per unit distance, d is the extent of the driving E-field in the dust devil, and μ_e is the electron mobility. The last two terms in square brackets represent the atmospheric dissipation currents, the first of which is the nominal dissipation current and the second an enhanced current associated with the electron avalanche process for a gas under electrical stress. In the following subsections, these different terms are compared for both Earth and Mars.

2.3.2 *Tribo-Electric Current Generation,* J_{tribo}

The mixing of dust grains creates electricity via tribo-electric processes with the grain polarity being a function of the particle mass and composition. Forward et al. (2009a, 2009b) found, for particles of identical composition, that smaller grains tend to charge negatively and larger grains tend to charge positively. As discussed more extensively in Sect. 3, they argued that the mass-size dependency involves the collisional transfer of electrons trapped in defect-created meta-states upon contact or rubbing, with the net exchange of electrons statistically favouring the smaller grains due to their increase likelihood of defect region 'rubbing' with larger grains (see Fig. 3 in Forward et al. 2009a, 2009b). In the centre of a convective feature, the grains become stratified in the upward flow of warm air. Small negative grains tend to lift upward and large positive grains remain lower in such features, giving rise to the large-scale dipole moment and E-field like that measured within terrestrial dust devils (Freier 1960; Crozier 1964; Farrell et al. 2004; Renno et al. 2004; Delory et al. 2006; Jackson and Farrell 2006).

The dust load for a nominal terrestrial dust devil has been estimated to be up to $\sim 10^{-3}$ kg m^{-3} (Metzger et al. 2011). Assuming a dust devil primarily contains particles of ~ 5 μm diameter carrying -5 fC per particle with an upward flux of $\sim 2 \times 10^9$ m^2 s^{-1} (Farrell et al. 2004; 2006b), an estimate of the associated upward tribo-charging current source would be $\sim 10^{-5}$ A m^{-2}. Similar mass-dependent tribo-charging and mass stratification process are anticipated for Mars. However, given that dust devils on Mars can be larger, larger tribo-charging current sources can be anticipated.

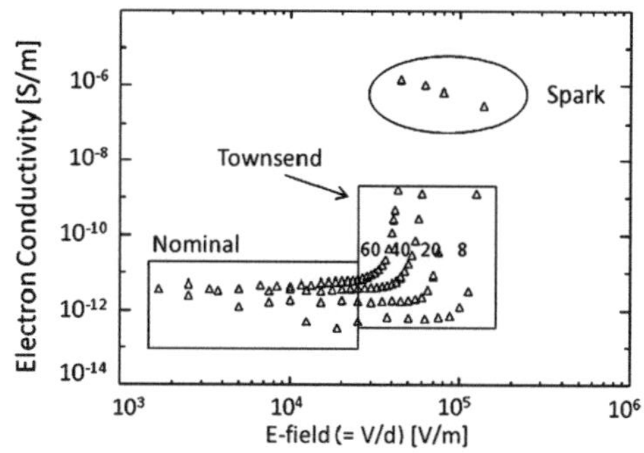

Fig. 4 Effective electron conductivity measured in a representative Martian environment, as a function of an electric field maintained between two plates with a varying separation (from left to right in the figure) of 60, 40, 20, and 8 mm. The different current regimes (nominal, Townsend, and spark) are delineated. (Reprinted from Farrell et al. 2015, with permission of Elsevier)

2.3.3 *Atmospheric Dissipation Currents, J_{diss}*

The large contrast between terrestrial and Martian dust devil electrification lies in the nature of the response by the atmosphere. At Earth, for typical disturbed electric field values, the atmosphere does not initiate the electron avalanche process, and the dissipation currents are defined primarily by conduction (i.e. proportional to σE). For the terrestrial atmosphere with conductivity ~50 $\mathrm{fS\,m^{-1}}$ only a modest removal of current will occur in a dust devil having $E \sim 100\ \mathrm{kV\,m^{-1}}$, with the competing dissipation current at $J_{diss} \sim 10^{-8}\ \mathrm{A\,m^{-2}}$. This current is far less than the driving tribo-electric current, and is not the process limiting the development of dust devil electric field. It is likely that it is the efficiency of the dust lifting process itself which limits the electric fields generated.

In contrast, in a 6 hPa CO_2 atmosphere like that at Mars, the nominal value of the atmospheric conductivity is predicted to be about ~100 times greater than that at Earth (Farrell et al. 2015), with $\sigma \sim 10^{-12}\ \mathrm{S\,m^{-1}}$ (see Fig. 2). However, even at modest E-field values of 10 $\mathrm{kV\,m^{-1}}$, electron avalanche processes will develop, with the exponential increase of electron currents directed along the driving E-field (Delory et al. 2006; Jackson et al. 2010). In the low-pressure Martian atmosphere, the mean free path for electrons under the driving E-field is larger, thereby creating more energetic electron-CO_2 impacts compared to the case of the denser Earth-like pressure. This greater number of electron impact ionizations will lead to the exponential increase in electron content.

In essence, the stress of even a modest E-field at Mars initiates the electron avalanche process, creating a substantial dissipation current that increases itself exponentially with increasing driving E-field. As found in recent laboratory work (Farrell et al. 2015; see Fig. 4), the dissipation current can become comparable to the driving J_{tribo}, creating a situation where the E-field becomes limited by the atmospheric dissipation (i.e. $dE/dt \sim 0$ in Eq. (6)). These dissipation currents can take the form of the Townsend dark discharge or a glow or even a spark if breakdown is reached. This visual effect was reported by Eden and Vonnegut (1973), Mills (1977) and Krauss et al. (2003). The atmosphere at Mars therefore creates a substantial and significant competing dissipation current that act to deplete the development of large separated charge centres.

2.3.4 *Lightning*

Given the possibility of strong dissipation currents that limit the growth of storm charge centres, it seems that large impulsive 'dissipation' currents via lightning would be unlikely. However, Jackson et al. (2008) and Kok and Renno (2009) found that the same dust grains responsible for the charging current, J_{tribo}, can also absorb electrons from the dissipation currents. This electron absorption would then reduce dissipation currents and thus could lead to anomalously large charge centres in the convective feature, thus requiring impulsive breakdown as a means to remediate excessive charge. Clearly, resolving which of these competing electrical processes dominates requires in situ measurements.

2.3.5 *Implications for Dust and Climate*

The existence of a planetary electric field with very intense local enhancements and possibly significant breakdown currents may have a significant influence on the physics and chemistry of the surface material. A detailed description of the expected impacts of electrification on processes relevant to the climate of Mars is presented in Kok and Renno (2008b).

Dust is the most important source of local heating in the Martian atmosphere and is known to control in large part the thermal structure of the troposphere. Many studies in the past have focused on identifying and further characterizing the processes at work in levitating dust from the surface of Mars. Well-established work (Bagnold 1962) indicated that direct wind lifting on Mars typically occurs at wind speeds exceeding 25 ms^{-1} near the surface (Greeley et al. 1992), a condition that is difficult to reach, according to current general circulation models. Instead, several other mechanisms have been proposed and explored in laboratories that provide the necessary lifting strength without requiring winds as strong as the standard theory predicts. The proposed mechanisms comprise sand blasting process and levitation induced by electrostatic forces. Considering the potential electrical force exerted on micron size dust particles by electric fields predicted inside dust devils or dust storms, electrification may potentially contribute to the global transport of dust at a level comparable as the drag force due to the wind (Berthelier et al. 2000). Studies have explored the more specific impacts of electrification on the lifting processes of dust from the surface of Mars, unveiling the unexpected role of electrification in the trajectories of saltating particles, an effect first suggested by Schmidt et al. (1998) and Zheng et al. (2003). A major effect concerns the limitation imposed by electric forces on the height of the saltation layer. Contrary to the standard theory, the observed height does not increase with increasing wind speed, as it simultaneously leads to development of a downward-pointing electric force exerted over the saltating particles (Kok and Renno 2008a, 2008b). A related effect is the emergence of a positive lifting feedback loop as a charge gradient exists between the positively charged particles lying at the surface and the negatively charged saltating particles. This phenomenon leads to a significant reduction of the shear velocity and thus aids hydrodynamic forces in lifting particles from the surface (Kok and Renno 2008a, 2008b). Direct lifting by the electric force may also occur in parallel, and thus establishes a combination of effects that can contribute significantly to dust lifting and potentially ultimately atmospheric dust loading on Mars. This remains something which global modelling of the Martian atmosphere has yet to consider quantitatively.

3 Contact and Tribo-Electrification of Environmental Dust and Sand

The importance of particle–particle interactions in the electrification of dust clouds and structures such as dust devils highlights the understanding needed in contact and tribo-electrification. This section now reviews understanding of the physical process(es) of contact and tribo-electrification. The discussion is based principally on laboratory studies, but also presents contemporary models and theories considering mineral dust and sand, because of their relevance to dust devils.

3.1 Definitions, History and Background

It is worth discussing the terminology regarding particulate electrification, since even this can be a source of confusion. *Contact electrification* was originally used as a broad term for electrostatic charge transfer resulting from contact, including contact modes such as detachment, sliding, rolling, impact, etc. *Tribo-electrification* was later used to describe this phenomenon resulting from 'rubbing'. Currently both contact- and tribo-electrification are widely and interchangeably used in the literature. Conventionally (though not formally) contact electrification has since become associated with 'contact and macroscopic separation leading to charge transfer' (e.g. McCarty and Whiteside 2008), whereas tribo-electrification is typically used where emphasis is placed upon (empirical) dependence on composition (i.e. the triboelectric series) and impact velocity, and it is also interchangeably used with 'frictional electrification'.

Despite the importance of contact electrification and the amount of research in this field, there is little agreement upon the core mechanism of charge transfer. Generally it is now thought that there are several competing process occurring, rather than one single mechanism being responsible, and that this is dependent upon the nature of the interacting surfaces and especially the surface chemistry. On the most fundamental level one may consider contact electrification to occur via charge exchange with the transfer of electrons and/or ions. There are several models involving each of these processes which have been successfully applied in respective experimental cases. A complete picture for all materials in all cases still eludes researchers and has been a source of active debate for decades (McCarty and Whiteside 2008; Harper 1998; Lowell and Rose-Innes 1980).

3.2 Experimental Studies of the Phenomenon

Numerous contact or tribo-electrification studies have used macroscopic contacting surfaces, often involving an insulator and a metal electrode (McCarty and Whiteside 2008). Here, however, focus will be placed on granular materials (sand/dust). Typical methods for quantifying electrification of coarse granular material (sand) involve removal and direct measurement of the electric charge, often using a combination of a Faraday cup and electrometer. The specifics of transport and collection vary, including; cascades, fountains, fluidized beds, blow-away experiments, aerosolizers or single particle impact studies (Matsusaka et al. 2010; Sickafoose et al. 2001; Poppe et al. 2000; Kok and Lacks 2009; Merrison et al. 2012).

3.2.1 Surface Charge Density

Techniques involving direct electric charge extraction are ineffective when dealing with dust sized (micrometer) particulates since they are typically well suspended and difficult to extract (e.g. into a Faraday cup). Suspended particles may also only be slightly charged, and, in contact with a surface, they may not be electrically conductive enough to allow rapid charge extraction. For fine suspended dust other techniques have typically been used to quantify particulate electrification. They rely upon the application of an electric field and the drift of electrified dust grains. The field induced drift velocity can be determined using optical or laser systems to study particle trajectories (Kunkel 1950; Merrison et al. 2012; Mazumder et al. 1991) or by extracting dust onto a surface (Merrison et al. 2012). The field-induced drift velocity will be proportional to the electrical force and therefore the electric charge and polarity of the dust grains.

In determining order of magnitude limits for expected values of contact electrification there has been considerable success using the ratio of charge to surface area. In a variety of experimental techniques and over a broad range of grain sizes (μm–mm) the electrification has been seen to be of order 0.1 $\mathrm{mC\,m^{-2}}$ (Poppe et al. 2000; Lowell 1986; Merrison et al. 2004; Nieh and Nguyen 1988). In many granular electrification studies the electrified surface area is taken as the (total) surface area of the grain; since the contact area is not known, this possibly explains the lower level of electrification seen in such cases e.g. $\sim 10^{-3}\ \mathrm{mC\,m^{-2}}$ (Sickafoose et al. 2001; Gross et al. 2001).

3.2.2 *Specific Charge*

Many workers, however, still express electrification in terms of the specific charge (i.e. the charge per unit mass, Q/m), due often to experimental convenience. There is less success in establishing expected/limiting values. In aeolian sand transport studies values have been quoted of order 60 $\mathrm{\mu C\,kg^{-1}}$, (Schmidt et al. 1998; Zheng et al. 2003, 2004; Qu et al. 2004; Bo et al. 2014). In gas-solid pipe flows continuous sand transport is performed while measuring the current to metal pipe, here values of order 3 $\mathrm{mC\,kg^{-1}}$ have been measured (Matsusaka et al. 2010). In metal oxides charges after blow-off exceeding $\pm 200\ \mathrm{mC\,kg^{-1}}$ have been seen (Oguchi and Tamatani 1993) (see the section on wind tunnel experiments).

3.2.3 *Practical Experimental Limitations*

Another common cause for observing reduced electrification in charge collection systems is charge leakage via surface water, which is related to ambient humidity (Nieh and Nguyen 1988). In dust collection experiments electrostatic aggregation can also de-electrify suspended dust, especially when in high concentrations and after long suspension times (Merrison et al. 2012). In relatively recent studies using Atomic Force Microscopy (AFM) techniques and well prepared surfaces, significantly higher values were reported, although in these no account was made for lateral spreading of charge (Horn et al. 1993).

In many cases electrical breakdown can limit the charge on grains during separation, e.g. dielectric breakdown in terrestrial surface air is at an electric field of around $3 \times 10^5\ \mathrm{V\,m^{-1}}$. (The lower surface pressure on Mars yields a smaller breakdown voltage there (Laughton and Warne 2004)). This could provide a quantitative explanation for the experimentally observed upper surface charge concentration Q and has been supported experimentally in detailed investigation using polymer micro-spheres (50–500 μm radius r) where Q/r^2 is seen to be constant, of order 0.1 $\mathrm{mC\,m^{-2}}$. Electrical breakdown has also been directly observed in some experiments (McCarty et al. 2007; Harper 1998; Matsuyama and Yamamoto 1997; Horn et al. 1993; Matsusaka et al. 2010). The observed charging will therefore in many cases be limited (by dielectric breakdown) to values less than 0.1 $\mathrm{mC\,m^{-2}}$.

Even lacking a detailed physical understanding of the electrification process, there is therefore general agreement on the order of magnitude of the electrification and this is sufficient in most cases to quantify the effect of electrification on, for example, entrainment and transport of sand and dust.

3.3 Particulate Size Dependence

It is widely accepted that particulates of the same composition and of differing size will show a tendency for larger particulates to electrify positively and smaller ones to electrify negatively, upon contact and separation. In experimental studies the electric fields generated by terrestrial dust devils support the idea that suspended dust becomes electrified negatively with respect to the sand/sand-bed (Schmidt et al. 1998; Zheng et al. 2003; Qu et al. 2004). This size dependence has been best demonstrated in laboratory experiments involving sand cascading (Lacks et al. 2008; Kok and Lacks 2009; Lacks and Levandovsky 2007, Forward et al. 2009a, 2009b; Bilici et al. 2014). However, not all experimental studies reproduce this behaviour (Trigwell et al. 2003; Sowinski et al. 2010; Kunkel 1950). It should also be noted that in many of these electrification studies multiple particle interactions are involved, including particle-wall interactions (Aplin et al. 2012), and material purity (surface composition) is not well controlled. A more complex charge exchange behaviour can therefore often not be ruled out.

In laboratory experiments of sand transport in the absence of dust, it has been seen that sand may electrify either positively or negatively depending on the size distribution within the sand bed (Bo et al. 2014; Zheng et al. 2003, Kok and Renno 2008a, 2008b). Laboratory experiments have also shown that dust re-suspension in the absence of sand shows little net (size dependent) dust electrification (Merrison et al. 2012), however the role of dust aggregates (acting like large sand-sized particles) could possibly complicate this behaviour. This complexity makes describing/predicting contact electrification for example within a Martian dust devil problematic. Modelling has been unsuccessful in satisfactorily explaining this size dependence in contact electrification, although several promising models are being pursued. For example, one model is based upon electron transfer through so-called high energy electron surface states (Lacks and Sankaran 2011; Apodaca et al. 2010; Bo et al. 2014). It would be extremely useful here to experimentally identify the precise charge/polarity dependence of contact electrification with grain size: this has yet to be done.

3.4 Material and Humidity Dependence

Attempts to apply a single model (i.e. involving either electron or ion transfer) for all cases of contact electrification have failed. Although the use of various so called tribo-electric series have been empirically useful (and widely published) there are many cases where electrification does not follow such tribo-series, and examples exist of circular tribo-series (Harper 1998) wherein a repeating sequence of materials paradoxically generates increasing values on the tribo-series. Also the tribo-series cannot explain electrification of like materials with different sized particles (Lacks and Sankaran 2011). This suggests that there is not one single electrification process occurring (McCarty and Whiteside 2008).

It is however apparent that material properties are crucial in applying models of contact electrification. For contacting metals, electron transfer models have been extremely successful in reproducing observed electrification based upon differences in electron work function (e.g. Matsusaka et al. 2010). Similarly for insulators containing mobile ions, modelling involving ion exchange has provided a consistent picture and even allowed control in contact electrification (Diaz and Felix-Navarro 2004; Law et al. 1995; Mizes et al. 1990). Contact electrification in (non-ionic) insulating materials though, remains problematic. Currently most researchers assume that contacting (non-ionic) insulators involve electron transfer (Lowell and Rose-Innes 1980; Grzybowski et al. 2005).

Experimentally, a typical problem is purity, especially the presence of surface contamination. Since the number concentration of charges at the surface involved in electrification is low compared to the concentration of molecules/atoms, even low concentrations of impurities (of order ppm) can in principle dominate the electrification process. This makes studies of contact electrification technically challenging.

Under ambient terrestrial conditions layers of water molecules are invariably present at surfaces and can play role in electrification by enabling ion or electron transport. Since humidity and surface ions such as H^+ and OH^-, are ubiquitous, models have been proposed to describe electrification of insulating materials (not containing ions) in which for example OH^- accumulate on surfaces with water layers. In experiments using polymers and bound/unbound ions (Diaz and Felix-Navarro 2004) contact electrification was seen to be eliminated at 0 % RH (relative humidity) and rise to a maximum at 30–40 % RH, supporting this ion exchange model. Again, however, there are experimental cases in which contact electrification has been observed despite the absence of surface water (low humidity/vacuum) (Harper 1998; Lowell and Rose-Innes 1980) and also with well-prepared dry surfaces (e.g. Horn et al. 1993; Gady et al. 1998). Hence, despite advances made (Diaz and Felix-Navarro 2004; Law et al. 1995; Mizes et al. 1990) involving ion transfer (and the use of charge control agents), this model still cannot be applied in all cases of contact electrification.

In the case of real planetary silicate minerals, ions will likely be present at the surfaces (e.g. OH^-, alkali metals, halide ions, etc.), and, terrestrially, water vapour (and therefore H^+, OH^-) will also be present and allow ion-transfer models to be applied. Generally, however, most current modelling of aeolian (contact/tribo-) electrification involves electron transfer (only) (Lacks and Sankaran 2011).

3.5 Models and Theories for Dust and Sand Electrification

For mineral dust and sand it is likely that both electron and/or ion transfer mechanisms may be relevant. Here material transfer is considered to be a form of ion transfer (Tanoue et al. 1999). For non-ionic insulators (e.g. polymers and possibly including mineral sand or dust) most researchers attempt to employ electron exchange models (Lowell and Rose-Innes 1980). However contact electrification does not correlate with surface or bulk electron properties such as dielectric constant, atomic properties, ionization energy, electron affinity or electro-negativity (Wiles et al. 2003). Despite this, some success has been achieved with electron transfer models involving electron donor/acceptors or so called high energy electron states in insulator-insulator contact, (also known as a molecular ion state model) (Lowell and Rose-Innes 1980; Bailey 2001; Duke and Fabish 1978). Similar electron transfer models involving so-called high energy surface electron states (likely due to impurities/contamination) are used in describing size dependent electrification (Lacks and Sankaran 2011; Bo et al. 2014).

Specific electron transfer models have been developed for electrification in metal-insulator contact, for example; involving an effective potential difference (Davies 1969) or performing quantum chemical (electron state) calculations (Shirakawa et al. 2008). Other models have also studied multiple surface impacts; one involved a capacitance charging model including charge relaxation (electrical discharge) (Matsusyama and Yamamoto 2006).

Despite these advances in electron transfer models, similar success and progress has also been achieved using ion exchange models specifically involving (proton exchange) which not only works well for ionic or ion doped materials, but also where water layers may be involved relating to chemical properties such as pH and/or zeta potential (Diaz and Felix-Navarro 2004; Law et al. 1995; Mizes et al. 1990). It has even been argued that there is 'never electron transfer with insulators' (Harper 1998), however this has not been demonstrated experimentally and is not widely accepted.

Based on an ion transfer (entropy driven) model, an order of magnitude calculation can be performed assuming a concentration of mobile ion groups (e.g. H^+, OH^-, metal ions, halide ions, etc.) of, for example, around 0.1 nm^2 generating around 3000 $e\,\mu m^{-2}$, this is in reasonable agreement with the expected upper limit (Diaz and Felix-Navarro 2004). This still corresponds to a surface concentration of <1 %. Experimental tests have yet to be performed of the temperature dependence in electrification which could verify the entropy driven model.

3.6 Wind Tunnel Experiments

On the scale of laboratory wind tunnels it has proved difficult to generate and therefore measure electric fields produced by sand/dust transport. It has been more practical to measure and determine the electrification of single grains (and distributions) and then apply modelling to predict electric field generation.

The electrification of saltating sand grains has been studied using wind tunnels in an extensive series of experiments, which also combined modelling with theory, involving determining the effects on grain trajectories of applied electric fields. In these experiments sand grain electrification was measured in a conventional manner using an Faraday cup type electrode to collect the charge of impacting grains and an electrometer to quantify the discharging current (Bo et al. 2014).

Grain electrification distribution was typically seen to be bipolar and broadly in the range from -300 to $+600$ $\mu C\,kg^{-1}$. In the earliest wind tunnel studies the average electrification was seen to be negative (Zheng et al. 2003; Qu et al. 2004; Zhang et al. 2004), however in a more recent study the net charge is seen to be positive and also a higher degree of electrification (agreeing with field experiments of Schmidt et al. 1998), Bo et al. (2014). This discrepancy was interpreted as due to interaction (charge exchange) with finer grains (non-saltating) within the wind tunnel sand bed. Notably a (broad) grain size distribution was used, which was not representative of those seen in the field. From this work it seems that there is a dependence of the measured electrification on the grain size of both the saltating and non-saltating grains, as well as on wind speed and height (increasing strongly with height).

Based on this sand electrification work electric fields have been predicted to be generally less than 50 $kV\,m^{-1}$ and typically a few $kV\,m^{-1}$ (Bo et al. 2013a, 2013b), this is below values at which significant effects would be expected on saltation and below (upper limit for) value measured in the field (Farrell et al. 2004; Schmidt et al. 1998; Renno et al. 2004). It should be noted however, that the models applied did not include dust entrainment, i.e. fine, suspended fraction which may have enhanced the electric field.

For fine (suspended) dust grains the techniques based on measuring their electrical discharge current to an electrode is difficult and ineffective due to: the small charge degree, the particles not being ballistic (having low impact velocity) and not liberating electrical charge rapidly. An alternative technique has been applied in a series of wind tunnel studies of dust electrification. Here an electric field is applied in order to drift dust grains out of suspension and collect them on electrodes. The quantity of dust collected can then be quantified either optically or by mass. By determining the amount of collected dust as a function of applied electric field, the average electrification per grain can be established (Merrison et al. 2012; Merrison et al. 2004).

It was found that the degree of electrification of re-suspended dust is typically of the order of 10^3 to 10^4 e/grain for μm scale particles (or around 0.1 $mC\,m^{-2}$). In the absence of sand the net electrification of the dust was seen to be close to zero (therefore not expected to generate an electric field). However, dust aggregates (agglomerates) are also seen to become electrified and may show the electrostatic behaviour of real dust clouds, even in the absence of larger sand particles. Specifically in wind tunnel studies, it was found that electrification is affected by competing processes of aggregation (electrostatic self-assembly), causing reduced charge and aggregate dispersion causing electrification.

Xie and Han (2012) performed experiments in a wind tunnel to investigate the effect of relative humidity on aeolian E-field. The working section of the chamber was 20 m × 1.3 m × 1.3 m. A sand bed of 8 m × 12 cm was deposited in the chamber. Sand particles were collected in the Badain Jaran desert in China. The electric field was measured with a KDY-IV field mill. At a given wind speed, they measured the aeolian electric fields at different air relative humidity (RH) and sand relative moisture conditions. They observed the electric field linearly increased with increasing relative humidity up to a critical value and then exponentially decrease. This critical RH value was observed to increase with the wind speed. For a wind speed of $14\ \mathrm{m\,s^{-1}}$ the critical RH was 32.7 %.

Wind tunnels have been more extensively used to investigate the influence of applied electric fields upon sand transport (saltation). Several groups have reported reduction in the threshold for saltation and enhanced transport rates of sand due to applied electric fields (Kok and Renno 2006, Rasmussen et al. 2009), typically at electric fields above around $100\ \mathrm{kV\,m^{-1}}$. This work has mainly been based upon the assumption of a conductive sand bed (surface). However, recently it has been experimentally demonstrated that, for the case in which the sand bed is insulating, the application of an electric field in fact significantly increases the threshold shear stress for saltation (Holstein-Rathlou 2012). This is due to dielectric attraction, entirely analogous in its operation to diamagnetic chain formation in the presence of an applied magnetic field.

3.7 Charging and Radioactivity in Volcanic Plumes

Theoretical studies of triboelectric charging of single particle systems demonstrate that the particle size distribution determines the magnitude and size of charging (Lacks and Levandovsky 2007). Laboratory studies of electrical charging of volcanic ash systems confirm that the particle size distribution plays an important role in the electrical charging of terrestrial volcanic plumes (Houghton et al. 2013; Cimarelli et al. 2014) and indicate that the composition of the particulate matter making up the plume also affects electrical charging. The factors that affect triboelectric charging in terrestrial volcanic plumes, namely the particle size distribution, particle composition and relative humidity, are expected to have similar effects in the aeolian Martian environment. While recent work has contributed significantly to the understanding of triboelectric charging in terrestrial volcanic plumes, further investigation is required to fully understand this process and the effects of the particle size distribution, material composition and ambient conditions on charging.

Alpha and beta particles are emitted following the radioactive decay of the uranium, thorium and potassium radioisotopes, leaving a residual charge on the particle from which they have been emitted (Clement and Harrison 1992). Studies of ash collected from the 2010 Eyjafjallajökull and 2011 Grimsvötn eruptions demonstrate that radioactive decay is only seen in larger particles (Aplin et al. 2014). This size dependence means that for terrestrial volcanic plumes, charging associated with radioactive decay is only likely to occur near the vent itself, as the larger particles containing radioisotopes are rapidly removed by gravitational settling. The composition and size of particles in Martian dust storms are expected to determine whether or not radioactive decay contributes to charging in Martian dust storms; larger particles are expected to be lost first from the dust cloud.

During volcanic eruptions, magma fragmentation can contribute to plume charging near the vent (Gilbert et al. 1991; James et al. 2008). Charged species are ejected from cracked surfaces following material fracture, and the loss of these charged species results in the formation of charged fragments. However, particle fragmentation is not expected to occur in Martian dust storms.

Clearly there is much more research needed in this area before we reach an understanding of these disparate electrification processes. For wind tunnel investigations it would be informative to perform electrification studies which involve both sand and dust grains (i.e. using an extremely broad size distribution from 1 μm–1 mm) in order to gain an understanding of the interplay of electrification and field generation processes.

4 Atmospheric Observations of Dust Electrification

Atmospheric electrical parameters can be sensed in a variety of ways. The most commonly measured quantity is the vertical electric field,[1] but the air conductivity, charge density and vertical conduction current have also been routinely observed at some measurement sites. Many of these techniques have developed rapidly as electronic systems have improved, as exceptionally good performance—very often ultra-low leakage current—is required from the signal processing electronics across a wide range of conditions. A widely used fast-response instrument for electric field measurement is the field mill, which operates by the alternate exposure and screening of a sensing electrode using a rotating shutter followed by phase-sensitive detection. A field mill can be orientated with its sensing surface upwards or downwards for the vertical component of the field, although spherical geometries have also been used in dust devil studies, to obtain the horizontal field component (Ravichandran and Kamra 1999). Vertical electric fields can also be measured using passive (long horizontal wire) antennas or other electrodes which come into electrical equilibrium with the potential of the surrounding air. These can have a slow response because of the low conductivity of atmospheric air: a radioactive coating can increase the air conductivity and the associated time response. The position of a sensing instrument mounted above the surface acts to distort the local electric field, and calibration to the electric field over a flat undisturbed surface is needed. The negligible distortion of a long wire antenna provides one possible reference technique to calibrate atmospheric electric field measurements, which has been used for over a century (Harrison 1997, 2013; Bennett and Harrison 2006).

Another approach used in high field conditions is to measure the so-called point discharge current, which is the current flowing into a vertically facing sharpened vertical point under the influence of the atmospheric electric field (Marlton et al. 2013). Techniques for measurement of air conductivity and conduction current density are not discussed further here, but summaries of the associated methods required are given in Harrison (2004), Aplin and Harrison (2001) and Bennett and Harrison (2008).

4.1 Surface Measurements

As introduced in Sect. 1, Freier (1960) was one of the first researchers to measure electric fields in dust devils. It was found that the electric field was upward pointing in dust devils. Freier's findings were later confirmed by Crozier (1964, 1970). These pioneering measurements of dust devil electric fields are consistent with results of more recent measurements

[1]Note than a convention in fair weather atmospheric electricity is to report the Potential Gradient (PG) rather than the vertical electric field E_z. In fair weather, the potential increases positively with height, and the PG is typically $+120\ \mathrm{V\,m^{-1}}$ near to the surface. (E_z in these circumstances would be $-120\ \mathrm{V\,m^{-1}}$, downward pointing).

Fig. 5 ACI Project site at the Owens Lake playa with electric field sensors mounted at 0.5, 1.5, 2, and 3 m above the surface. The results reported here are from measurements of the electric field at 3 m above the ground. The PM2.5 aerosol concentration is measured at approximately 3.5 m above the ground, and saltation is measured at 7 cm above the ground

also indicating upward pointing electric fields in dust devils (e.g., Farrell et al. 2004; Renno et al. 2004).

In dust storms the near surface electric field is stronger and in the opposite direction to the downward pointing fair weather electric field (e.g., Latham 1964). This suggests that dust particles become negatively charged after colliding with the larger sand particles and the surface (e.g., Schonland 1953; Latham 1964; Kok and Renno 2008a, 2008b; Renno et al. 2003). Schmidt et al. (1998) reported that the electric field is also upward pointing in saltation alone (that is, without dust lifting). Indeed, Schmidt et al. (1998) reported upward pointing electric fields of 160 $\mathrm{kV\,m^{-1}}$ at a few cm above the surface during saltation in sand dunes. Electric fields of this magnitude could have significant effects on saltation and therefore dust lifting.

4.1.1 Recent Measurements of Electric Fields in Dust-Devils

Except for the measurements of Schmidt et al. (1998), all previous measurements summarized above were made using grounded instruments. Since grounded instruments typically measure the difference in potential between the ground and their sensing elements, rather than the electric field at the sensor height, measurements with these grounded instruments must be interpreted carefully. The presence of grounded instruments distorts the electric fields being measured and the instrument configuration must be calibrated on site (e.g. Sullivan 2013). In addition, the impact of charged dust or sand particles introduces noise on grounded instruments by the effect of electric currents between the instruments and the ground (Sullivan 2013).

Schmidt et al. (1998) used a cylindrical field mill isolated from the ground to measure the electric fields in the saltation layer directly (Johnston et al. 1986). Since charge transfer by the collision of charged particles with isolated field mills vanishes as soon as the field mills reach the same potential as the charged particles, the errors produced by these collisions also vanish. Renno et al. (2008) developed an isolated cylindrical field mill for making measurements of the electric fields in saltation, dust storms and dust devils. This new instrument is capable of measuring the electric fields accurately even when the instrument is subject to the impact of charged particles (Renno and Rogacki 2013).

Here we report results of measurements with this instrument in the Aerosols-Climate Interaction (ACI) project site at the Owens Lake salty playa, in California (Halleaux and Renno 2014). Figure 5 shows four electric field sensors installed at the ACI project site at the Owens Lake playa. Figure 6 indicates that, in mid-March, a saltation event triggered an

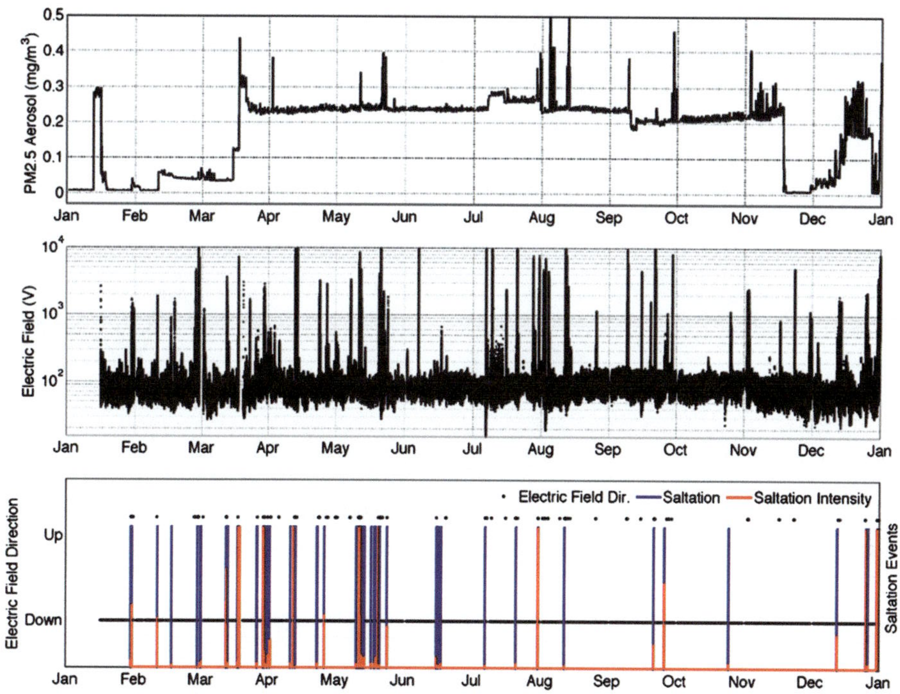

Fig. 6 PM2.5 aerosol concentration (*upper panel*), near surface electric field strength (*middle panel*), as well as electric field direction and indication of saltation events in blue (*lower panel*) at ACI project site in the Owens Lake in 2014. The measurements of the electric field strength and direction presented in Fig. 6 were made at 10 Hz. (Upward and downward pointing electric fields are defined as electric fields within approximately 30° of the vertical. Saltation is measured with a Sensit sensor placed at about 7 cm above the surface e.g. Halleaux and Renno (2014). Saltation events are defined as those time periods in when the impact rate at the sensor is greater than 10^{-4} particles/second. Saltation intensity refers to the normalized number of impacts on the Sensit sensor per unit time.)

order-of-magnitude increase in PM2.5 aerosol concentration. The hourly mean value of the electric field exceeded 10 kV m^{-1} during this event. This value is two orders of magnitude larger than the fair weather electric field.

Figure 6(c) indicates that, in general, the electric field is downward pointing, but that when saltation and therefore dust lifting occurs the electric field points upward. The few exceptions occur when saltation is weak and during stormy weather when charged clouds are also present.

Figure 7 shows the results of measurements of PM2.5 aerosol concentration, electric field and saltation during a day of intense saltation and dust lifting. The measurements indicate that, in general, the electric field changes its polarity to upwards when saltation occurs. The exceptions to this occurred during periods when saltation was weak or when charged clouds were also present above the site.

4.1.2 Measurements of Electric Fields in Dust Storms

Measurements of electric fields during dust storms are sparse in the literature but show both upward (Rudge 1913; Harris 1967; Stow 1969; Zhang et al. 2004) and downward (Demon et al. 1953) electric field directions.

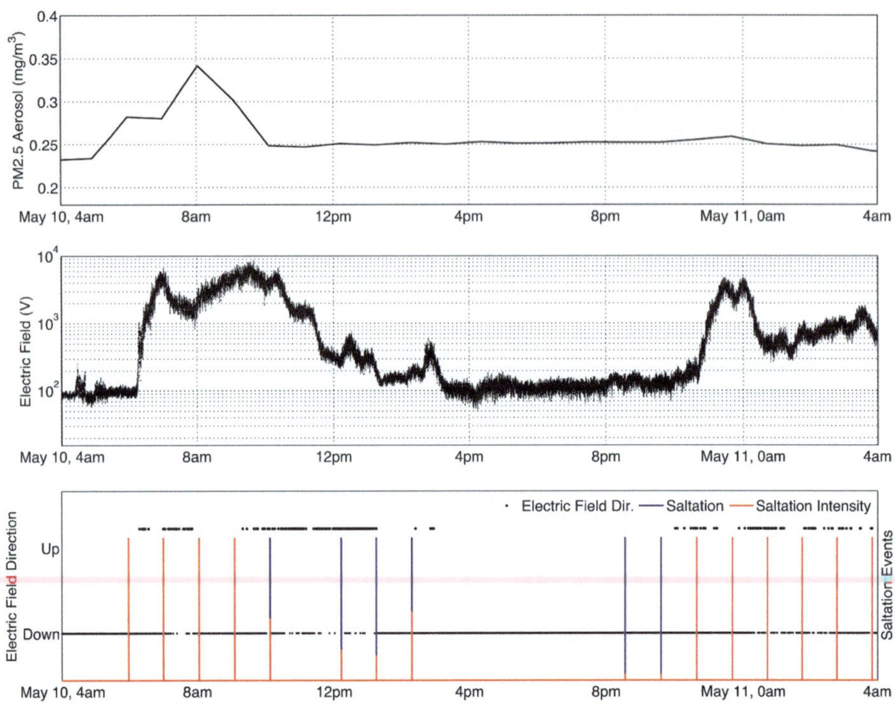

Fig. 7 PM2.5 aerosol concentration (*upper panel*), near surface electric field strength (*middle panel*), as well as electric field direction and saltation events (*lower panel*) at the ACI project site in the Owens Lake on a day of active saltation and dust lifting events (May 10, 2014)

Kamra (1972) performed measurements of the potential gradient between the surface and at 1 m above, in several US locations using a 500 μCi Polonium probe as a potential sensing probe. (The use of radioactivity enhances the time response of the sensor, without adding the mechanical complexity of a field mill.) The potential of the radioactive probe was measured by a Keithley model 600B electrometer, a laboratory grade device. During dust storms, this instrument recorded both upward and downward E-fields depending on the nature of soil in the measurement site. In particular, Kamra observed that dust storms dominated by clay minerals produced only negative potential gradients, while, when silica dominated in the dust storm, both polarities of potential gradients were produced.

During measurements in the Sahel with a Mission Instruments field mill, Williams et al. (2009) observed E-fields directed both downward and upward. Very recently, Esposito et al. (2016) carried out several measurements in the West Sahara (Merzouga region, Morocco) during the dust storm season in 2013 and 2014. They performed simultaneous measurements of atmospheric electric field, using a Campbell CS110 field mill, mounted at 2 m and directed toward the surface, atmospheric dust abundance and size distribution (in the range 0.265–34 μm), sand saltation rate (with the Sensit sensors), wind speed and direction at six different heights, air and soil temperature, pressure, relative humidity, soil moisture content and solar irradiance. They observed predominantly negative electric fields, i.e. with the same sign of fair weather E-field, with some sporadic and short inversions in polarity (Fig. 8). This was observed both during dust storms and dust devils (Fig. 9).

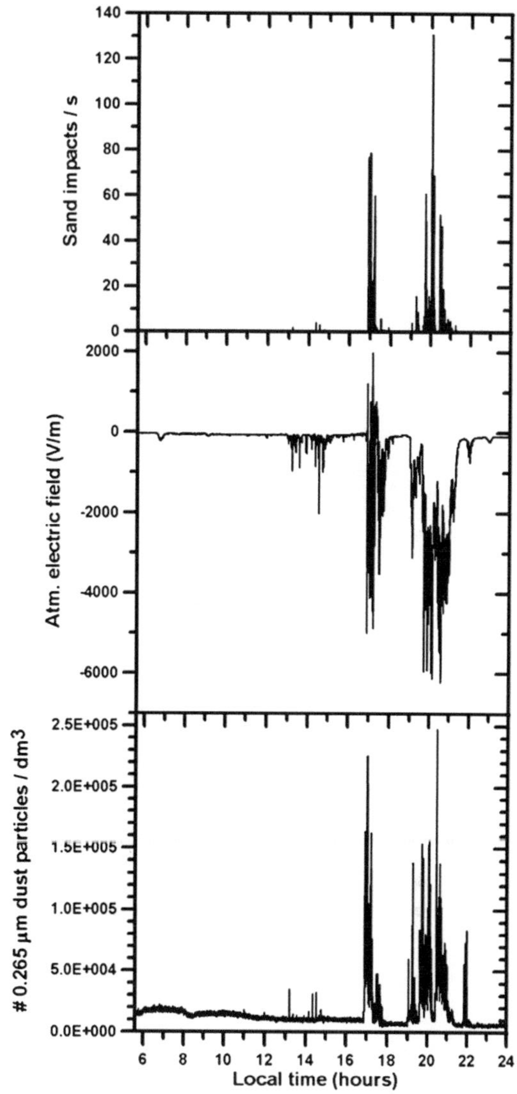

Fig. 8 Typical dust storm measured in the Sahara desert in July 2013 (Esposito et al. 2016)

The relation between humidity and particle charging mentioned in Sect. 3.4 was observed by Esposito et al. (2016) during dust storms monitored in the Sahara desert. By a statistical analysis of the data, they found a linear relationship between the concentration of dust entered in suspension and atmospheric electric field. The slope of this linear trend was observed to increase when relative humidity reached a critical value. This value seemed to depend also on the content of soil moisture and was around 20 % in the 2013 campaign and 30 % in 2014 where the soil was drier (Fig. 10).

The increase in the slope indicates that, for each value of dust concentration a lower E-field is observed for relative humidity larger than the critical value. Taken together, all these observations suggest there are complex mechanisms underlying natural dust electrification, and that different effects should be expected in different regions on Earth and Mars.

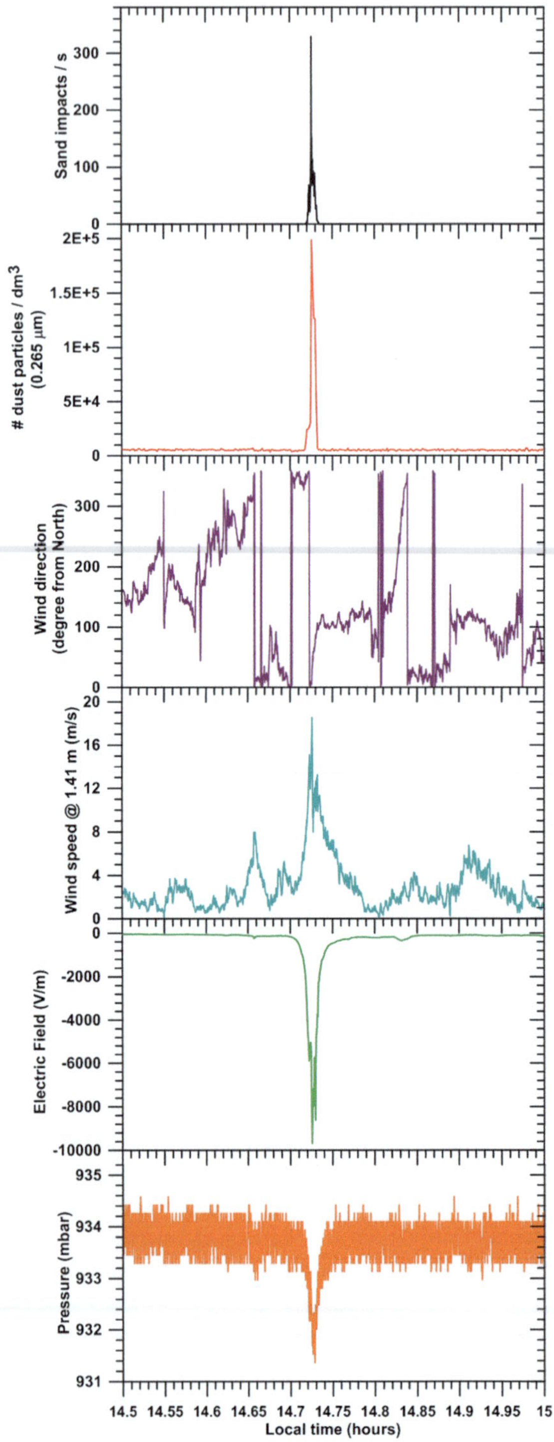

Fig. 9 Properties of a dust devil observed in the Sahara desert in July 2014. *Upper panel*: rate of impact of particles. *Second panel*: dust particle concentration. *Third panel*: wind direction. *Fourth panel*: wind speed. *Fifth panel*: electric field. *Bottom panel*: atmospheric pressure

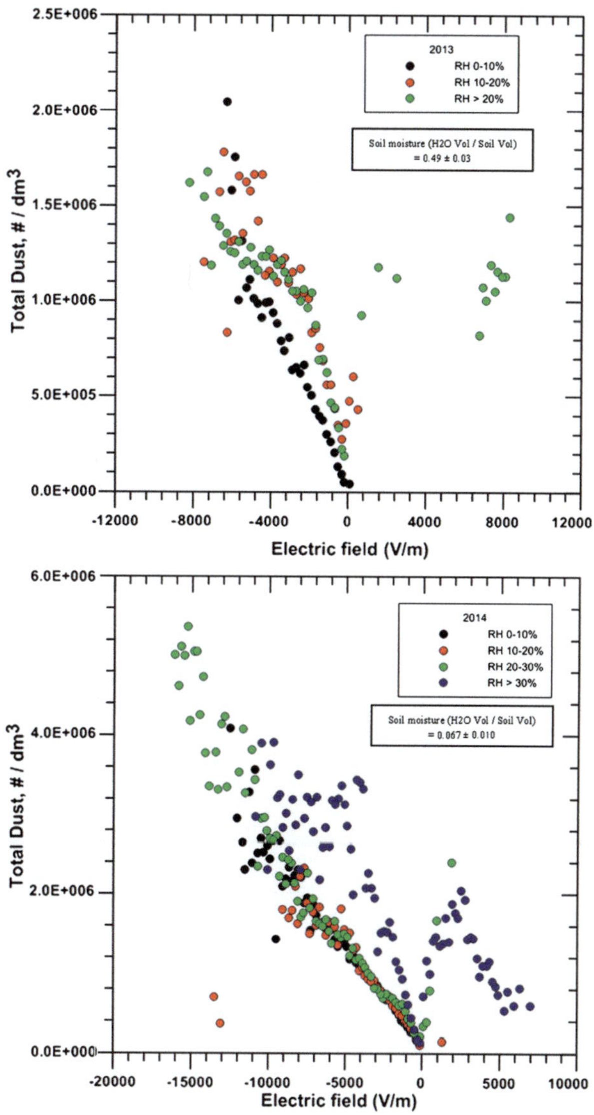

Fig. 10 Relationship between atmospheric electric field and concentration of dust emitted during several dust storms observed in the West Sahara in 2013 (*upper*) and 2014 (*lower*) field campaigns by Esposito et al. (2016). The slope of the linear trend increases when relative humidity exceeds a critical value. This value depends also on the moisture content of the soil emitting dust grains

4.2 Measurements of Particle Charging Aloft

Although measurements of dust electrification in different circumstances have been made at and near Earth's surface, measurements made aloft are rare in comparison, due to the difficulties associated with measurement platforms and lack of suitable instrumentation. Of the handful of airborne studies which do exist, all have focused on passive dust layers, which are typically isolated from the surface, travelling many hundreds of km from their source regions. As yet, no attempt has been made to measure electrification in dust storms or dust devils above the surface, which is certainly something that should be attempted in future to establish the in situ charge structure. Of the measurements that do exist, Gringel and Mühleisen (1978) were amongst the first to publish measurements of electrical conductivity, σ,

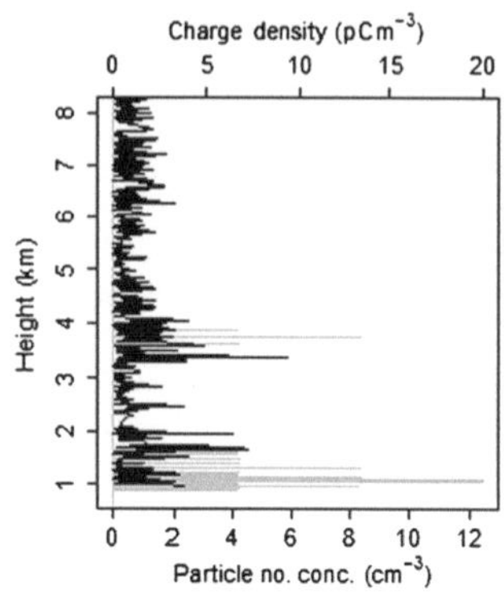

Fig. 11 Vertical profile of total aerosol particle concentration (*grey*) and magnitude of space charge density (*black*) measured on a specially instrumented radiosonde flight through Saharan dust layers from Sal, Cape Verde. Adapted from Nicoll et al. (2011) "© IOP Publishing Ltd. CC BY-NC-SA"

measured by a Gerdien condenser, made from a free balloon platform through a 2 km thick layer of Saharan dust off the west coast of Africa. Their data show a clear decrease in the conductivity (caused by the attachment of ions to the dust particles) within the layer, dropping by a factor of two compared to its clean air value at the same altitude. These findings are to be expected from a passive dust layer where no active charge generation or charge separation occurs. This is in accordance with Ohm's law, which governs charge conservation in the fair weather atmosphere, and states that the vertical conduction current, J_c, is equal to the product of the electrical conductivity and the local atmospheric electric field E_z, ($J_c = \sigma E_z$). In the case of no active charge generation, J_c through the dust layer should be continuous, therefore any decrease in conductivity (caused by ion-aerosol attachment) will lead to an increase in the local electric field within the dust layer. This hypothesis is supported by measurements through aerosol layers (e.g. Markson 2007), which are similar to dust layers in terms of the presence of large particles, and show clear increases in E_z inside the layer, as a result of the conductivity decrease.

Following on from the work of Gringel and Mühleisen (1978), Nicoll et al. (2011) made a series of instrumented balloon flights from the Cape Verde Islands, near the west coast of Africa, which frequently experiences elevated layers of dust transported from the Sahara in the so-called "Saharan Air Layer", which can reach altitudes of 5–7 km. Simultaneous measurements of dust particle concentration and space charge, ρ, were obtained by two small sensors attached to standard balloon-carried meteorological radiosondes. The space charge sensor measures the net charge (i.e. the difference between positive and negative charge) carried by the dust particles and comprises a small spherical electrode connected to a sensitive electrometer circuit, measuring the voltage on the electrode.

Figure 11 shows measurements from one of the instrumented balloon flights from Sal, Cape Verde, which shows two distinct layers of dust particles at 2 km and 4 km (with particle diameters up to 2.6 μm). Measurements from the charge sensor show weak charge on the dust particles in both layers (up to 10 pC m^{-3}). Particle–particle charging (i.e. triboelectric charging) of dust grains is likely to have originated near the surface during the initial lofting of the dust, but this charge will decay quickly (within several minutes) due to recombination with ions. This therefore suggests the presence of a continual charging (albeit weak) mechanism which is active in elevated dust layers, likely to be related to tribo-electrification of the dust particles (Houghton et al. 2013). Such an effect has also been observed in elevated layers of volcanic ash from the eruption of Eyjafjallajökull in Iceland in 2010 (Harrison et al. 2010). The effect of such charge on long range transport of dust and ash is as yet unknown, but may have implications for modification of deposition speed as well as particle–particle agglomeration rates and particle wet removal by droplets.

4.3 Particle Electrification Observed in Volcanic Plumes

A special case of particle electrification aloft is that associated with charge separation in volcanic plumes. Observations have shown that charging of terrestrial volcanic plumes can occur up to hundreds of kilometres from the eruption site (Harrison et al. 2010; Hatakeyama 1949). The presence of charge in the Earth's electrically conductive atmosphere indicates that self-charging occurs in the plume and is attributed to triboelectric charging. Observations of volcanic lightning during the Eyjafjallajökull eruption in 2010 demonstrate that ambient atmospheric conditions also contribute to electrical charging of terrestrial volcanic plumes (Bennett et al. 2010), known as the ash-rich ice electrification system (ARIES) mechanism (Aplin et al. 2016). Convective processes, as in the ARIES mechanism, are expected to contribute to charging of dust storms on Mars, but without the enhancement associated with the presence of ice.

5 Models of charge exchange between particles

Simulations of the particle charging mechanism derive from laboratory measurements and have been developed in order to understand a number of different phenomena. The basic studies are described below and applications to dust devils follow.

Melnik and Parrot (1998) were interested in the conditions required for electric discharge in dust disturbances on Mars. Their charging mechanism assumed that the average charge on a particle was proportional to the particle size. They adopted the relation that each collision resulted in charge exchange between particles of $\Delta q = 1$ fC/μm, with the larger particle losing electrons to the smaller particle.

Desch and Cuzzi (2000) proposed an alternate mechanism of charging where both particle size and composition influenced the amount of charge exchange between the colliding particles. While their simulations were conducted in order to understand the mechanisms for lightning generation in the solar nebula, the concepts are transferrable to particle collisions within dust storms or dust devils. The important parameter in their study is the contact potential difference between the two colliding particles. The contact potential of a material is the energy needed to remove an electron from that material. When two particles of differing material collide, the electrons will migrate to the material with the higher contact potential. For metals, this quantity is directly related to the work function. For insulators, the contact potential energy is more difficult to determine from intrinsic properties of the material, but potential differences between two materials can be found from the triboelectric series (Faraday 1855). A formula was derived to calculate the charge exchange, which is a function of potential difference and the mutual capacitance of the two materials. See Sect. 3.2 of Desch and Cuzzi (2000) for the full derivation, resulting in

$$\Delta q = f_1 \Delta \Phi - (1 - f_2) q_{\mathrm{tot}} \tag{7}$$

where $f_1 = (c_{12}c_{21} - c_{11}c_{22})/(c_{11} + c_{12} + c_{21} + c_{22})$ and $f_2 = (c_{11} + c_{12})/(c_{11} + c_{12} + c_{21} + c_{22})$ are functions of the mutual capacitances, c_{ij}. (Φ is the potential.) When mixing large insulator particles with small metallic particles, f_2 approaches unity and the charge exchange can be rewritten as a function of potential difference and particle size as

$$\Delta q \sim 2668(\Delta \Phi / 2\ \mathrm{V})(r_f / 0.5\ \mu\mathrm{m})e, \tag{8}$$

where $r_f = (r_L^{-1} + r_S^{-1})^{-1} \sim r_S$ is the reduced radius, and e is the elementary charge. (The approximate potential difference between a metal and an insulator is 2 V.)

Kok and Renno (2008a, 2008b) modelled charge exchange between sand particles and the Earth's surface employing the Desch and Cuzzi charging mechanism, with an additional term for collisions between particle pairs of similar composition,

$$\Delta\Phi_{\text{eff}} = (r_L - r_S)/(r_L + r_S), \tag{9}$$

where S is a scaling factor determined by calibrating the model with field measurements ($S = 6 \pm 4$ V). Their study was applied to reconcile discrepancies between classical saltation theory and terrestrial observations of the phenomenon. By including the electric force in the equation of motion for particles in saltation, they found that the wind stress needed to lift particles is reduced and the particle trajectories remain closer to the surface with a reduced horizontal speed.

Kok and Renno (2009) applied their terrestrial saltation model to explore saltation with charged particles on Mars. They concluded that the breakdown E-field in the Martian saltation layer is ~43 kV m^{-1}, but their modelling could only achieve fields of ~15–20 kV m^{-1}. Therefore, electric discharges in the saltation layer seem unlikely, but this does not prohibit discharges from occurring in larger scale phenomena such as dust storms and dust devils.

5.1 Applications of the Charge Exchange Mechanism to Dust Devils

The Melnik-Parrot and Desch-Cuzzi grain charging processes described above were applied to a comparative analytical study of dust devils on Earth and Mars in the work of Farrell et al. (2003). They created an electrodynamic model of a dust devil by applying the methodology used for modelling terrestrial thunderstorms, e.g. see Mathpal et al. (1980) to the induction electrification process between graupel and water/ice in terrestrial thunderstorms. The time derivative of the electric field, E, is related to rate of charging on the larger grains, Q'_L, by

$$E'' + \sigma E'/\varepsilon_0 = -n_L \Delta v Q'_L/\varepsilon_0 \tag{10}$$

where σ is the local atmospheric conductivity, n_L is the particle number density, Δv is the relative velocity between the large/small grains ($\Delta v = v_L - v_S < 0$), and ε_0 is the permittivity of free space. Here Q'_L is based on triboelectric processes (for dust devils) rather than induction processes (for thunderstorms) and is proportional to the charge exchange per collision, Δq, as $Q'_L = \upsilon \Delta q$, where υ is the grain collision frequency. The model is applied to both Earth and Mars by varying the atmospheric conductivity appropriately. Results from their simulations are shown in Fig. 12.

A number of aspects become evident from Fig. 12. First, exponentially growing E-fields develop in both the terrestrial and Martian dust devils due to the tribocharging and subsequent mass (and charge) separation within the convective cloud. The electric fields in the Martian case are always lower than in the terrestrial case due to greater dissipation leakage current which removes charge from the stratified grains. Additionally, the Desch and Cuzzi charging results in electric fields consistently lower than those produced using Melnik and Parrot. The two cases are approximately equal for the Desch and Cuzzi $\Delta\Phi = 2$ V which represents the compositional difference of a metal and an insulator (i.e., simulating the collision between a metallic grain and a basaltic grain).

A follow-up study by Farrell et al. (2006a) added the acceleration term to Eq. (10), such that the right hand size becomes: $-n_L(Q'_L \Delta v + Q_L \Delta v')/\varepsilon_0$. The $Q'_L \Delta v$ acceleration term depends on the lifting process associated with the fluid, and the addition of this term links

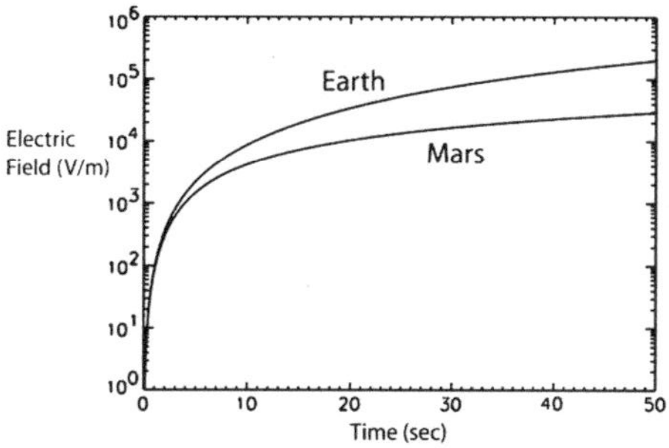

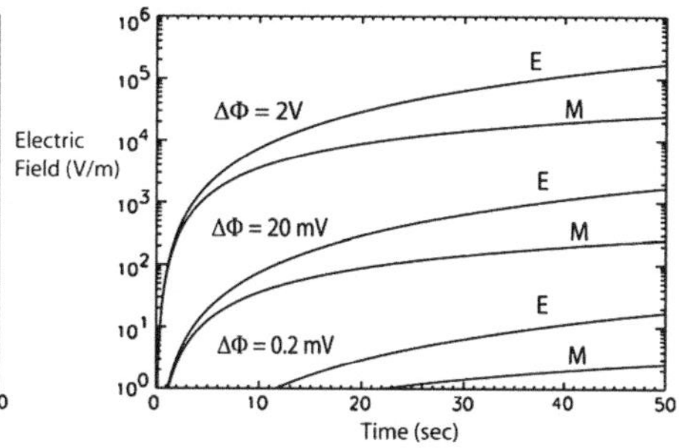

Fig. 12 Comparison of the size-only grain charging mechanism from (**a**) Melnik and Parrot with the (**b**) Desch and Cuzzi grain charging mechanism which also includes the effects of grain composition (from Figs. 2 and 3 in Farrell et al. 2003)

the electrostatic effects to the fluid properties of the medium, particularly the wind speed. The effects are seen only during the early dust devil formation period, but are evident as a rapid rise in E-field value by as much as a factor of 10 over that shown in Fig. 12 in the first ten seconds. As time progresses, the $Q'_L \Delta v$ term primarily drives the system and the E-field asymptotes to approach results similar to the cases shown in Fig. 12.

5.2 Simulation Codes Integrating Particle Charging with Vortex Dynamics

Melnik & Parrot (1998) conducted a simulation where a number of particles of like composition but differing sizes were placed in a box and subjected to wind movement. These particles were then placed in a larger simulation domain (50 m × 100 m) where the wind vortex was halted and further particle movement was controlled by the forces of gravity, air drag, and the electrostatic field. The redistribution of charge was tracked over time until an electric field built up to 20 kV m^{-1}, the breakdown field for the Martian atmosphere. The maximum electric field seen in the simulation reached ~10 kV m^{-1} in the first 2 s; the breakdown field was reached in 6.6 s. A second set of simulations tracking the movement of the electric field as particles roll up a hill (under the influence of a wind parallel to the surface) were also conducted. The electric field at the top of the hill reached the Martian breakdown value in about 13.5 s; growth of the electric field was exponential and much more regular than in the wind vortex case. In both wind environments, the amount of time to reach the breakdown field could be considerably lengthened (by greater than a factor of 10) by decreasing the number of particles involved, the amount of charge exchange upon collision, or the wind velocity. Over these longer timescales other processes (e.g. recombination of charges through subsequent collisions) are likely to occur, which could prevent the electric field increasing to breakdown strength.

Huang et al. (2008) numerically simulated a dust devil in a cylindrical domain of radius 100 m and height 200 m, in a terrestrial environment. They modelled the entire process iteratively beginning with development of the convective vortex from local surface heating, lifting of sand grains and their subsequent movement due to gravity and vortex winds, and applying a specified charge-to-mass ratio (Huang and Zheng 2001) on each grain in order to calculate the resulting electric field from Coulomb's law. The closest fit to the observations of Farrell et al. (2004) employed charge-to-mass ratios of −120 μC kg^{-1}, 60 μC kg^{-1}, and 57 μC kg^{-1} for sand grains of diameter 0.15 mm, 0.2 mm, and 0.25 mm, respectively. The maximum E-field within the dust devil was reached in about 80 seconds of simulation time.

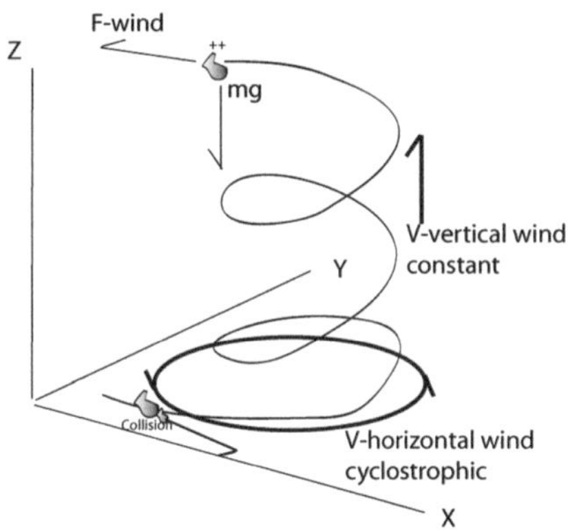

Fig. 13 Illustration of the forces on the grains in the Macroscopic Triboelectric Simulation model. Along the z-direction, grains are subject to an imposed constant upward wind and the downward gravitational force. In the horizontal plane, the prevailing winds are defined by a pattern of cyclostrophic balance between the outward centripetal force and inward pressure gradient force. (The z and xy motions are not coupled.) (From Farrell et al. 2006b)

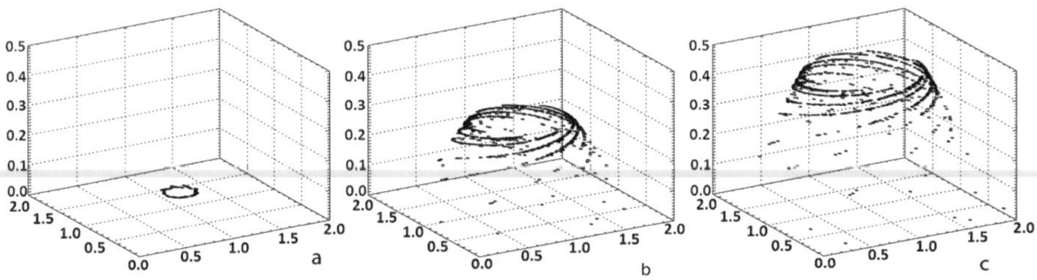

Fig. 14 3-D perspective of the grain motion in an MTS simulation. The plots correspond to (**a**) 10 time steps, (**b**) 300 time steps, and (**c**) 499 time steps, where each timestep is approximately 1 ms (adapted from Fig. 4 in Farrell et al. 2006b)

Motivated by the detection of ULF magnetic radiation emitted in terrestrial dust devils (Houser et al. 2003; Farrell et al. 2004), Farrell et al. (2006b) developed the Macroscopic Triboelectric Simulation (MTS). MTS is a 3-D particle code which quantifies charging associated with swirling, mixing dust grains. Grains of pre-defined sizes and compositions are placed in simulation box (2 m × 2 m × 1 m) and allowed to move under the influence of winds and gravity. The model tracks the movement of individual grains in the prevailing winds, and charge exchange upon grain-grain collision. An illustration of the model and a 3-D perspective of the grains during a simulation are shown in Fig. 13 and Fig. 14.

The grains are initially placed in distinct clumps in close proximity to one another surrounding the box centre. Grain sizes are randomly generated but constrained such that the radius ranges 0.05–10 μm with a median value near 0.1 μm. To explore the Desch and Cuzzi compositional charging effects, the smaller grains ($r < 2$ μm) are given a metallic composition (conductors) and the less abundant larger grains are modelled as silicates (insulators), with the potential difference between the two populations of $\Delta\Phi = 1.6$ V. The motion of the smaller grains traces the wind pattern, while gravitational effects on the larger grains force them into a less circular wind pattern and they move towards the walls of the box as they fall. The distribution of charge on each of the grains at the point in the simulation before the grains reach the walls of the box ($t = 0.5$ s) is shown in Fig. 15(a).

The spatial distribution of the charged grains establishes a distinct current system involving the relative displacement of negative and positive charge in the dust devil; the transverse part of this current is responsible for a time-variable magnetic field, and the vertical part is responsible for a large-scale vertical electrostatic E-field. The assumption of solenoidal

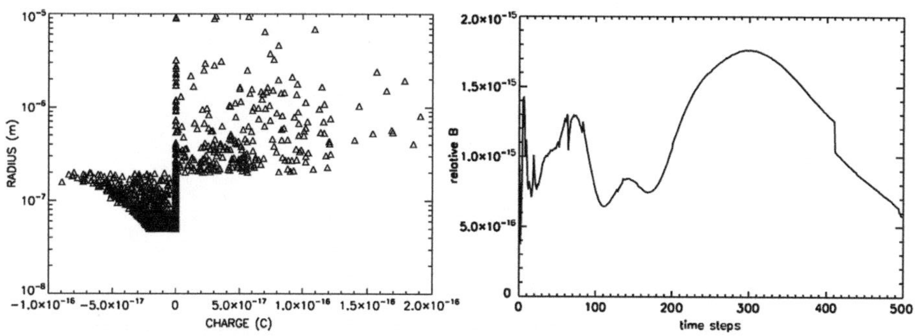

Fig. 15 (**a**) Distribution of grain charge as a function of grain radius from an MTS simulation. The smaller metallic grains charge negatively and the large silicate grains charge positively as prescribed by the D&C charging mechanism. Additionally, and particularly evident with the metallic grains is the increase in grain capacity to hold charge with increasing particle radius. (**b**) The magnetic field generated as a function of time over the course of the same simulation (from Farrell et al. 2006b)

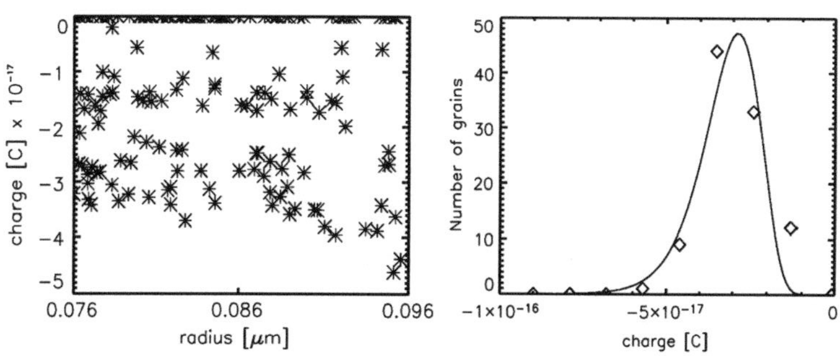

Fig. 16 The MTS model returns charge values on individual dust grains. For each size-bin in MRAMS (Mars Regional Atmospheric Modeling System) the charge values were binned and fit to a log-normal function (from Barth et al. 2015, with permission of Elsevier)

current flow has allowed the magnetic effects of dust devils to be modelled (Kurgansky et al. 2007), which can explain some aspects of the ULF magnetic observations.

The MTS work described above used prescribed winds and the results were scaled up to the size of a real (terrestrial size) dust devil. Barth et al. (2015) applied the MTS simulations to Martian dust devils by incorporating the MTS-charged dust grains into their Mars Regional Atmospheric Modeling System (MRAMS; see Chap. 7 for additional description). The charged grains shown in Fig. 15 were partitioned into the MRAMS radius bins and a log-normal function was fitted to the charge values in each radius bin as shown in Fig. 16. Then the first and second moments of the charge distribution were added as tracers so that the charge on the dust particle tracers could be reconstructed at any MRAMS grid point.

As dust devils developed in the MRAMS Large Eddy Simulations (LES), the temporal/spatial change in charge distribution within the simulated Martian dust devil could be tracked and resulting E-fields were calculated. They found that the magnitude of the resulting E-field had strong dependence on particle size, amount of charge, and amount of dust lifted; the range of E-field values ($\mathrm{mV\,m^{-1}}$ to $\mathrm{kV\,m^{-1}}$) was in general, more modest than in previous studies (an example is shown in Fig. 17). As this study was a first look at the charge environment, a number of processes were not yet included. However, with this code one can examine the dust devil E-field in three dimensions, make estimates of the overall

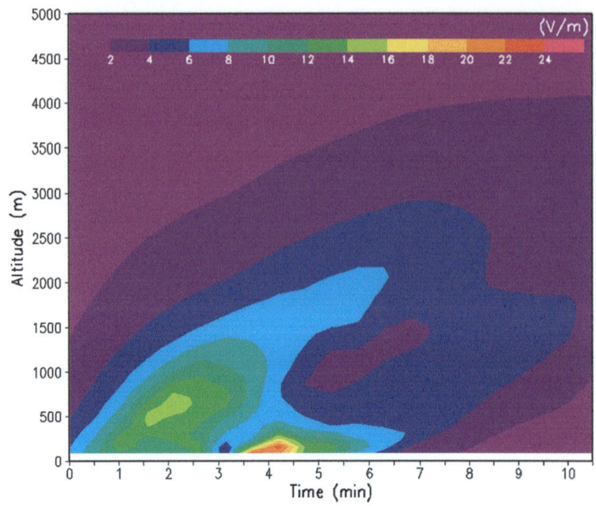

Fig. 17 Time and vertical variation of the electric field (V/m) an observer would see if traveling within the dust devil at the minimum pressure point, as generated by the MRAMS-MTS model (from Barth et al. 2015, with permission of Elsevier)

dipole moment, and get new insights on the development of higher order moments that form in regions of inhomogeneous fluid flows.

5.3 Breakdown Effects

Large E-fields like those measured in situ in terrestrial dust devils have been found which exceed 50 kV m^{-1} (Farrell et al. 2004; Renno et al. 2004; Jackson and Farrell 2006). These same electrical convective features placed in the low pressure Martian atmosphere are expected to initiate atmospheric breakdown. Specifically, as the E-fields grow in the Martian dust devil, the atmospheric currents transition from nominal flows described by $J = \sigma E$ to the creation of a current-enhancing electron avalanche from Eq. (6) as

$$J_a = n_0 \exp\big(\alpha_{\mathrm{T}}(E)d\big) e v_e E \tag{11}$$

where n_0 is the initial electron density, α_{T} is Townsend's first coefficient defining the number of ionizing events per unit length, and v_e is the electron drift velocity in the gas. In the electron avalanche, those electrons with energy greater than 14.5 eV can ionize a CO_2 molecule via electron impact ionization to thus create extra electrons that get accelerated by the driving E-field. The population of electrons above 14.5 eV and their overall drift speed increases with electric field (Nighan 1970).

Farrell et al. (2015) noted that the electron avalanche represents a substantial modification of the atmospheric conductivity. At low E-fields, the atmospheric conductivity is represented as a bulk isotropic quantity, σ, where electron and ion mobility are assumed to be nearly equal. In contrast, within the electron avalanche itself, the electron conductivity is exponentially greater than that from the positive ion component, and is highly directional. As such, electrical dissipation of any charged object has to be examined based on the flow of fast beam-like electrons vs slow ions reaching the surface. Negatively charged dust grains immersed within the electron avalanche will not quickly dissipate their surface charge but instead reach a negative charge equilibrium state consistent with current balance to the grain surfaces. The interruption of dissipation of the negative dust grains is likely to act to reinforce the overall dust charge build-up in the convective feature.

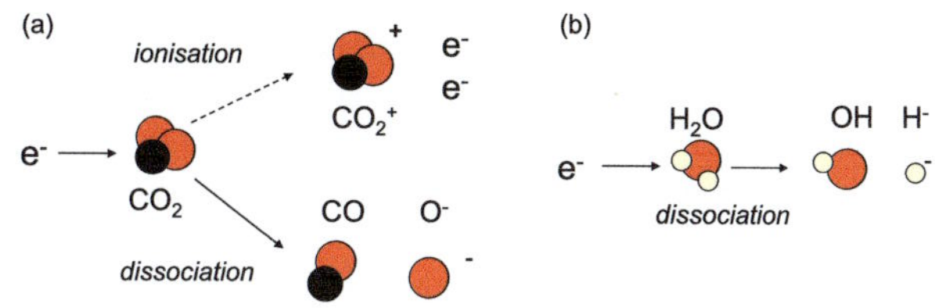

Fig. 18 Electron chemistry associated with (**a**) CO_2 and (**b**) H_2O, in the vicinity of an electrified dust devil or storm (after Delory et al. 2006)

Delory et al. (2006) modelled the electron avalanche process and growth in electron density in a dust feature and found that in dipole electric fields greater than 15 kV m^{-1} the mean free path for electron impact ionization of less than a metre leads to substantial ionization. These energized electrons can then interact with CO_2 and H_2O, via dissociative attachment creating CO, O^-, OH, and H^- (see Fig. 18). The dust devil-created CO and OH interactions are then suspected to generate hydrogen peroxide at concentrations well above those predicted via photochemistry (Atreya et al. 2006). These excesses of hydrogen peroxide become adsorbed onto aerosols and ultimately end up in the regolith, giving the H_2O_2 a longer lifetime than the chemically formed H_2O_2, which is photochemically dissociated during the day with a mean lifetime of 6 hours. Implications of this longer lifetime are that H_2O_2 could be an effective oxidiser (Encrenaz et al. 2012). The energetic electrons in the avalanche also can possibly create methane loss via dissociative attachment (Farrell et al. 2006c).

5.3.1 *Atmospheric Chemistry Effects and Breakdown*

Delory et al. (2006) were the first to suggest that catalytic production of hydrogen peroxide can be triggered by electrostatic fields generated in the Martian dust devils and dust storms as well as near the surface within the saltation layer. The origin of this hydrogen peroxide enhancement lies in the possibility for electrons, after acceleration by large electric fields, to ionize CO_2 and H_2O molecules, dissociating them and thereby releasing hydroxyl radicals as oxidizers of the Martian atmosphere. Atreya et al. (2006) predict that the HO_x enhancement factor can be up to 100 in electric fields reaching the breakdown limit ($\sim$20 kV m^{-1}). However, no dedicated laboratory measurement to date has been performed in support of these assertions and the chemical implications for such catalytic production of oxidizer remain difficult to reconcile with our present understanding of Martian chemistry (Lefèvre and Forget 2009). The electron avalanche process including dust/electron absorption losses and the active conversion to O^- via electron dissociative attachment have been the focus of several later modelling studies (Jackson et al. 2008, 2010; Kok and Renno 2009). All of these models indicate that electron avalanche processes could develop, leading to electron impact excitation and new chemical products. However, each varied in the intensity of the electron impact ionization and chemical production. Delory et al. (2006) and Atreya et al. (2006) predicted an electron avalanche and plasma content having values below 1 part in 10^{14}. Kok and Renno (2009) suggested that the activity could be of lower intensity when including grain charge dissipation effects. Jackson et al. (2010) found that the final equilibrium values for the plasma (electrons, O^-, CO_2^+) depend upon the assumed saturation process, with the most optimistic case to be 1 part in 10^4, and in the least ideal case to be 1 part in 10^{13}. These models all uniformly suggest the electron avalanche and ionization process is very mild, producing ions at concentrations $\ll$1 % of atmospheric gas densities. While electron dissociative processes are considered a loss process for methane (Farrell et al. 2006c), laboratory work has suggested that electrical activity associated with CO_2 and water could stimulate methane formation (Robledo-Martinez et al. 2012). Specifically, the electrons dissociate CO_2 and H_2O into CO/O^- and OH/H^-, with the CO and H_2 then create methane and water. Farrell et al. (2015) concluded that the added energetic electrons in the atmosphere associated with relatively modest dust-generated E-fields act to create new chemistry, and in the case of the methane, can act to both destroy and create this critical biomarker. An obvious remaining question regarding methane is which process dominates at any given time, and how do these competing processes act to affect the overall global methane stability. To date, models have described the development of a bulk E-field, and its effect on storm-size chemistry. However, one can imagine the E-fields near individual dust particles to be amplified via the converging field near a local point source. A 5 micron particle charged to ~ -2.5 fC will have a local surface potential of -5 V (Farrell et al. 2015). However, in its immediate vicinity the local E-field values can become very large. For example, within 30 μm the E-fields exceed about 27 kV m^{-1}, and within 10 μm in excess of 250 kV m^{-1}. Thus very active local chemistry in the vicinity of these particles could be expected. Specifically, the electrons will initially collect or be removed on an individual small lofted particle depending upon the particle charge state and current balance to the particle in the electron avalanche flow. However, as these particles reach their negative equilibrium charge state, the electrons in the flow will be locally accelerated (i.e., redirected) about the particle. This effect will stimulate local chemistry in the near-grain environment. In equilibrium, the negative charged particles will also have a cloud of positive CO_2^+ ions about it, but will tend to reject the O^- ions. The O^- ions should then congregate near the positively charged surface and in the vicinity of the larger positively charged grains near the bottom of the dust devil. The O^- ion is highly reactive and should modify near-surface chemistry. These avenues have yet to be fully examined.

6 Measurements of Martian Atmospheric Electricity

To provide a thorough picture of the atmospheric electricity on Mars, three principal parameters need to be measured in situ. These are the electrical conductivity, the DC atmospheric electric field and the AC electric fields associated with electromagnetic wave emissions. The atmospheric electric fields are directly related to atmospheric charging and discharging processes, and are characterized by large amplitudes, extending from DC to frequencies representative of the dynamics of charging and discharging processes.

The electrical conductivity of an atmosphere is an essential parameter of its global electric circuit. It also provides information on the ionization processes that control the electrical state of the atmosphere and on the major charged particles species, free electrons, positive and negative ions that are present in the medium. The conductivities associated with either positively or negatively charged species have to be measured separately to gain the relevant information. From model calculations (Sect. 2.3), the anticipated conductivities range from 10^{-12} S m^{-1} to 10^{-10} S m^{-1}.

The quasi-DC electric fields provide direct evidence of the charging mechanisms and of their temporal variations. Based on the properties of the Martian atmosphere, measurements of the quasi DC electric fields should ideally cover the wide range from a few mV m^{-1} to $\sim$10 kV m^{-1} in order to encompass all situations of interest, from the faint electrification due to photo-electrons charging to potential electrical breakdown. Impact charging processes are directly associated with winds and thus their temporal variations at a fixed location should arise from e.g. the motion of dust devils and/or turbulence in the wind flow with a frequency of the order of 10 Hz. On the other hand, following observations in thunderstorms, discharge processes and the resulting redistribution of charges give rise to large amplitude variations as rapid as 0.5 to 1 ms. Ideally, therefore, in situ measurements should cover a frequency range up to ~2 kHz.

Measuring the small amplitude AC electric fields will provide information on the electromagnetic wave emissions and the impact of charged dust grains on the lander. Electromagnetic waves can originate from atmospheric processes, such as the EM waves emitted by filamentary discharge currents in case of atmospheric breakdown or the Schumann resonances that result from the trapping of these waves in the surface-ionosphere resonator. They may also arise from totally different phenomena in the distant ionized environment of Mars through interactions with the solar wind. To achieve these observations AC electric fields have to be measured in the frequency range from a few Hz to more than 3 kHz, with a sensitivity of better than $\mu V\,m^{-1}\,Hz^{-1/2}$. This upper limit of 3 kHz is estimated from the greatest frequency signals expected, those of transverse resonances at ~1500 Hz (Simões et al. 2009), and consideration of sampling requirements.

Several proposals have been made for an atmospheric electricity package on the surface of Mars (Berthelier et al. 2000; Farrell et al. 2000, 2004; Renno et al. 2004). The upcoming ExoMars Entry Descent Module (EDM) will carry the DREAMS package to the surface, which includes the MicroARES sensor that will provide an opportunity for the first-ever in situ measurements of Martian atmospheric electricity. These observations are likely to provide great insights into the Mars atmospheric electricity-chemistry-astrobiology interconnection. MicroARES is motivated by two fundamental science questions which are well addressed by in situ measurements: (1) is there an electric field on Mars?, and (2) are Martian dust devils electrified?

Ideally, an array of field mills could be deployed to answer these questions, operating indefinitely to sample a wide range of conditions. However, the very tight constraints imposed on the dimensions, mass and power of the sensors on the EDM surface payload restrict the electric field measurements to a short time window of only a few days, during which there is some expectation of dust storms. As on Earth, the planetary large scale electric field of atmospheric origin is essentially vertical and this component is of primary interest. MicroARES is based on a simplified version of the standard double probe technique used to measure electric field in space. Rather than deploying a boom with two sensors, MicroARES was designed to measure the potential with respect to the lander ground, using a single spherical electrode positioned 30 cm above the lander's upper surface. Previous balloon flights of double-probe instruments together with numerical models have shown that the intended MicroARES configuration provides reliable measurements of the local atmospheric potential at the position of the electrode and higher frequency variations of electrostatic or electromagnetic origin.

6.1 MicroARES Electric Sensor

The MicroARES electric field sensor integrated in the ExoMars/DREAMS experiment (Esposito et al. 2014, 2015; Bettanini et al. 2014) will measure electrical properties of the Martian atmosphere. A modified version of this sensor has also been developed and tested in desert conditions, demonstrating its capability to measure electric fields found in terrestrial conditions, and under a simulated Martian conductivity environment. Because of the fundamental scientific results in planetary atmospheric electricity anticipated from the ExoMars/MicroARES measurements, the sensor to be deployed is described in some detail here.

In its final configuration, the sensor consists of a 27 cm mast supporting a 3 cm spherical electrode (see Fig. 19). The electrode is free to float electrically to a local potential, from

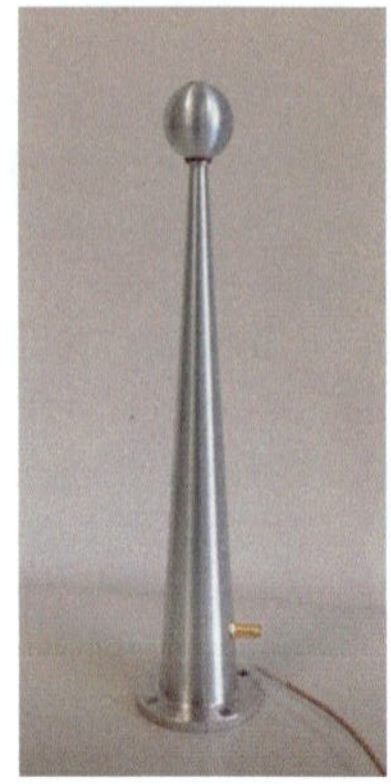

Fig. 19 Photograph of electrode

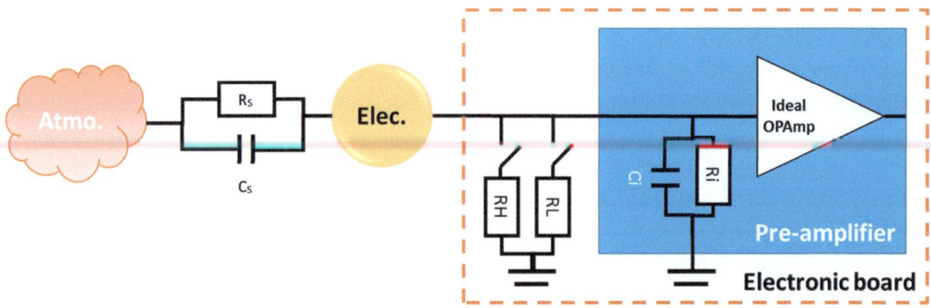

Fig. 20 Concept of the electrode—atmosphere interaction used in the MicroARES measurement system. The sheath around the electrode is represented by the Resistance Capacitance combination of R_S and C_S, and the input parameters of the electronic pre-amplifier by R_i and C_i. RL (Relay Low range: 25 GΩ) and RH (Relay High range: 10 GΩ) are range setting resistances, which can be switched in successively (RL first, then RH) as necessary as the electric field increases

which the local electric field can be derived using known electrical characteristics of the input stage. For weight and power consumption reasons, the reference electrode is the lander chassis. The spherical sensing electrode system has been designed upon the presumption that the Martian atmospheric conductivity lies between 10^{-10} and 10^{-13} $\mathrm{S\,m^{-1}}$. Measurements of the conductivity itself can also be made, from determining the charging time to restore the equilibrium potential of the electrode after bringing it to a fixed potential (e.g. Aplin 2005; Bennett and Harrison 2006).

The configuration used for the electrode-atmosphere interaction and associated electronic system is summarised in Figure 20. The interaction of the sensing electrode with the atmosphere can be modelled by a very simple RC circuit characterized by the sheath impedance (R_s and C_s). Accurate modelling of the sheath around the electrode is therefore important in order to derive the conductivity from the measurement of R_s alone and vice-versa. Two methods can be used to achieve this, firstly a mathematical model approach (Molina-Cuberos et al. 2010, §2.2; Berthelier et al. 2000, §3.2), where the spherical electrode is considered immersed in a collisional and weakly ionized medium, and secondly an empirical approach of modelling of the equipotential lines around the electrode in order to retrieve the coefficients k_e and k_ϕ which link the sheath resistance to the atmospheric conductivity (Seran et al. 2013).

For the MicroARES sensor, the first approach is used, using

$$R_s \cdot C_s = \frac{\varepsilon_0}{\sigma} \quad \text{with}$$
$$C_s = 4\pi r_e \varepsilon_0 \quad \text{and} \quad R_s = \frac{1}{4\pi\sigma r_e} \tag{12}$$

where r_e is the electrode radius, ε_0 is the permittivity of free space and σ is the local atmospheric conductivity. Further simulations of the electrode coupling with its environment will enable a comparison of the two approaches. With the Martian atmospheric conductivity assumptions, R_s will be between $10^{10}\ \Omega$ and $10^{13}\ \Omega$.

The input of the preamplifier is also modelled using a RC circuit. The R_i value is around $10^{14}\ \Omega$ (based on the input impedance of the LMC6041 operational amplifier used combined with some leakage across the board), and the circuit has been optimised to maximize this parameter since R_s and R_i form a potential divider at the input. To allow measurement of the DC potential of the electrode regardless of the atmospheric conductivity, R_s has to be negligible compared to R_i. For large electric fields, dividers formed by the R_L and R_H grounding resistors can be activated using controllable high impedance relays, similar to the method used for terrestrial measurements by Harrison and Aplin (2000), using the same model of operational amplifier. The R_L and R_H values of 25 GΩ and 10 GΩ respectively have been chosen to suit the Martian atmospheric conductivity assumption and the corresponding sheath resistance R_s.

The range of measurable frequencies extends from DC to 3.2 kHz for sensor potential values ranging across ±100 V, corresponding to vertical electric fields of $\pm 300\ \mathrm{V\,m^{-1}}$. At low frequencies, the electrode is coupled to the undisturbed plasma through the resistance of its surrounding sheath, to measure the DC electric fields. Above a cut-off frequency, which depends on the sheath resistance and electrode capacitance, the preamplifier is capacitance coupled to the ambient atmosphere (see Fig. 20). The high sheath resistance makes it necessary to use high input impedance ($>10^{14}\ \Omega$), low leakage current ($<10^{-14}$ A) electronic preamplifiers.

To cope with possible large electric fields exceeding the normal range of voltages applied to the electrode, the high impedance relay can switch in a parallel input resistance to the preamplifier (Fig. 20). This resistance is significantly smaller than the anticipated value of the electrode sheath resistance, which will therefore produce a potential divider, lowering the potential to lie within the input voltage range of the preamplifier. In practice there are two different shunt resistances that can be switched independently, 10 GΩ and 25 GΩ. The measurements performed with these two shunt resistances allows the sheath resistance to be determined, which in disturbed conditions, may differ from that obtained in quiet conditions.

A further measurement possible is that of the atmospheric conductivity through the relaxation technique: the sensor potential is displaced from the local floating potential and then allowed to recover its equilibrium potential. The time constant for this charging (or discharging) process $\tau = \varepsilon_0/\sigma$ readily provides the local electric conductivity σ, provided that the conductive area of the lander is much larger than that of the sensor. This condition ensures that the return current collected by the lander to counterbalance the current collected by the polarized sensor does not entail any significant variation of the lander potential with respect to the atmosphere, and ensures that the lander ground stays constant with respect to the ambient medium.

Fig. 21 Terrestrial adaptation of MicroARES antenna as used in the Sahara desert

6.2 MicroARES Tests in the Terrestrial Atmosphere

MicroARES can also be used for the measurement of the electric properties of Earth's atmosphere, and this provides a test opportunity. A terrestrial version of MicroARES (Fig. 21) was developed and operated in the Sahara desert (at Merzouga, Morocco) for a week in July 2014 during the local dust storm season. It operated during the field campaign described in Sect. 4.1 and fully reported in Esposito et al. (2016). As Fig. 2 shows, however, the atmospheric conductivity profiles are very different between Earth and Mars, with the surface atmospheric conductivity at Mars comparable with that of the terrestrial stratosphere. Although space instrumentation can be tested using stratospheric balloons (Fulchignoni et al. 2004), use at this altitude would not be practical, not least as there would be no prospect of obtaining a ground connection. Since the Earth's atmospheric conductivity[2] at sea-level is about 100 times smaller than that near the surface on Mars, the antenna-atmosphere coupling will generate a greater sheath resistance of up to $10^{15}\ \Omega$, which will cause the signal to be almost fully attenuated and the electrode will not be able to maintain itself at the floating local potential because of the small bias current flow to the input amplifier.

Since the standard MicroARES configuration will not properly measure the Earth's DC electric field, possible solutions for a test system are to either change the relay resistances and the pre-amplifier input resistance or to change the antenna. Given the fact that the input resistance is an intrinsic property of the design, increasing it is clearly not possible. Therefore, the antenna needs to be adapted to the Earth atmospheric properties instead.

For atmospheric conductivity roughly 100 times smaller than that on Mars, an electrode roughly 100 times larger is needed in order to keep R_s at the same order of magnitude. For the tests undertaken in the Sahara, for which there were numerous practical issues, a 60 cm side copper cube constructed from 0.8 mm thickness sheets was used. Its 2.16 m^2 surface area was equivalent to a 41.6 cm radius sphere, around 30 times larger than the original antenna. Thus, by calculation, the associated C_s becomes around 46 pF, and R_s (with assumed atmospheric conductivity of 50 $fS\ m^{-1}$) will be around 3.5 $T\ \Omega$ instead of 100 $T\ \Omega$. With this modified sheath resistance, the input relays are expected to provide switching of the input stage as originally planned, and the pre-amplifier input resistance should not

[2]The conductivity considered here as the mean conductivity is the mobility weighted average of the bipolar conductivities.

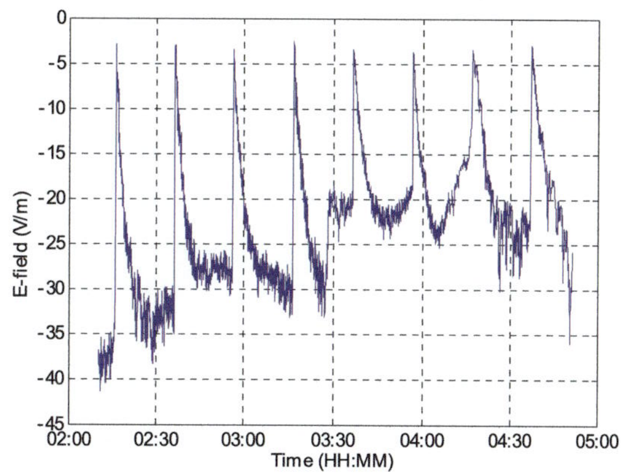

Fig. 22 Time series of the MicroARES unprocessed signal. Measurements were acquired in the Sahara desert, Morocco in July 2014

significantly affect the measurements. The mast height was also increased to 50 cm, which lowers the parasitic capacity to a few pF. A further consequence is that the centre of the cube is now located at 80 cm from the ground, which will lead to a greater measured potential (Fig. 21).

As the cube antenna and Earth atmosphere have different electrical properties from the anticipated Martian input configuration, some different post-processing is required. The response time of the instrument input is inversely proportional to the conductivity ($\tau = \varepsilon_0/\sigma = 1/R_p C_p$), hence the expected response time on Earth with this cube antenna will be ~200 s. The two main consequences of this are that the signal is highly deformed, which can be corrected numerically, and that, in the conductivity measurement mode, a switching sequence as rapid as 4 s is no longer possible. Since the post-processing correction depends on σ, its determination is performed with the signal correction. The instrument can only record 20 min worth of data, so it is restarted, with the signal correction applied every 20 min.

To reconstruct the input response, the instrument input stage parameters are needed accurately. The model employed for the electronics is that presented in

Figure 20, with $R_i = 10\ \mathrm{T\Omega}$, $C_i = 2$ pF (preamplifier) + 6 pF (antenna), $R_L = 25\ \mathrm{G\Omega}$ and $R_H = 10\ \mathrm{G\Omega}$. The antenna-atmosphere coupling depends on two parameters, C_s and R_s, but the measurements only provide one parameter, the atmospheric conductivity σ. Since the equations describing the instrument input depend on these variables separately, an equivalent radius to a spherical electrode (41.6 cm) is assumed to derive C_s and R_s from the conductivity.

The signal distortion at each zeroing of the instrument is clearly visible on Fig. 22. Note that at every restart (switching on and off of the signal processing board), the input of the preamplifier is connected to ground and then consequently discharged, so the slow response time of the atmospheric coupling is exaggerated.

In order to correct the signal the response is modelled with the assumptions made previously. The data reconstruction depends on the determination of σ (for the values of R_s and C_p) and the initial potential condition i.e. $U_e(0)$. On Mars, σ would have been provided by the conductivity measurement mode, during calm weather, and the relay operation during stormy weather (the quick response time allows computation of the ratio before and after relay operations to easily determine R_s and σ), but for Earth measurements it is not available directly, without the post processing. The data reconstruction also relies on the assumption that, during the 6 s reset time, the electric field is steady, thus allowing a good approximation of $U_e(0)$ during each 20 min of measurement.

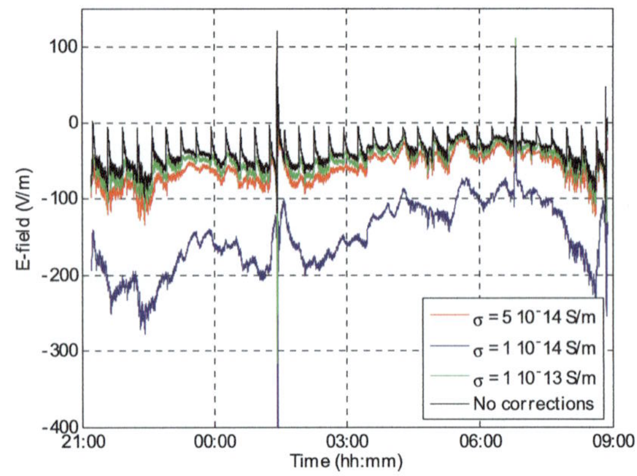

Fig. 23 Discontinuity smoothing with simultaneous determination of conductivity. The time axis is local time

Determination of σ has been achieved by iterating its value in the input inversion equations, at every switching cycle of the instrument visible on Fig. 22, so that the derivative and potential changes before and after the switch are minimized (see Fig. 23). Comparison of the reconstructed electric field with that obtained from an adjacent commercial field-mill (Fig. 24) demonstrates the effectiveness of this technique, with a correlation of 0.9 during calm weather and 0.75 during stormy weather.

As explained, the MicroARES measurement sequence and post-processing also allow the retrieval of conductivity, which is generally not possible with commercial field-mills alone running in a standard mode. Figure 25 shows the retrieved conductivity for 9–13th July 2014; the scale chosen emphasises the variability. The increase of conductivity during dust events implies that the dust is contributing to the conductivity, as discussed for dense aerosol concentrations by Dhanorkar and Kamra (1997).

7 Conclusions

Dusty environments on Earth can, as shown here, readily become electrified, and, by analogy and theoretical consideration, the Martian atmosphere also seems likely to contain electrified dust. Central to developing an improved understanding of this dust electrification is the understanding of charge exchange between the substances concerned, and possible breakdown effects. Ultimately, bulk measurements made in natural dust storms on Earth and Mars need to be reconciled with laboratory measurements of electrification at a particle level and the modelled electric field structures they present under the appropriate electrodynamic conditions.

Measurements of terrestrial dust electrification at the surface and aloft indicate that particle charging in dust storms is a common, and, perhaps, even a universal phenomenon to a greater or lesser extent. The use of field mill measurements in dust devils shows that electric fields of order 10 kV m^{-1} are readily observed, which are associated with the pressure minimum and wind speed maximum (e.g. Fig. 8). Larger values to $\sim$100 kV m^{-1} have also been reported. The magnitude of the field appears to be reduced with increasing humidity (Fig. 10), which is likely to be related to the properties of the contact electrification process when water layers at the molecular level are present (Sect. 3.4).

Several modern experimental techniques are proving to be extremely informative with regard to contact electrification processes, notably, as mentioned, atomic force microscopy

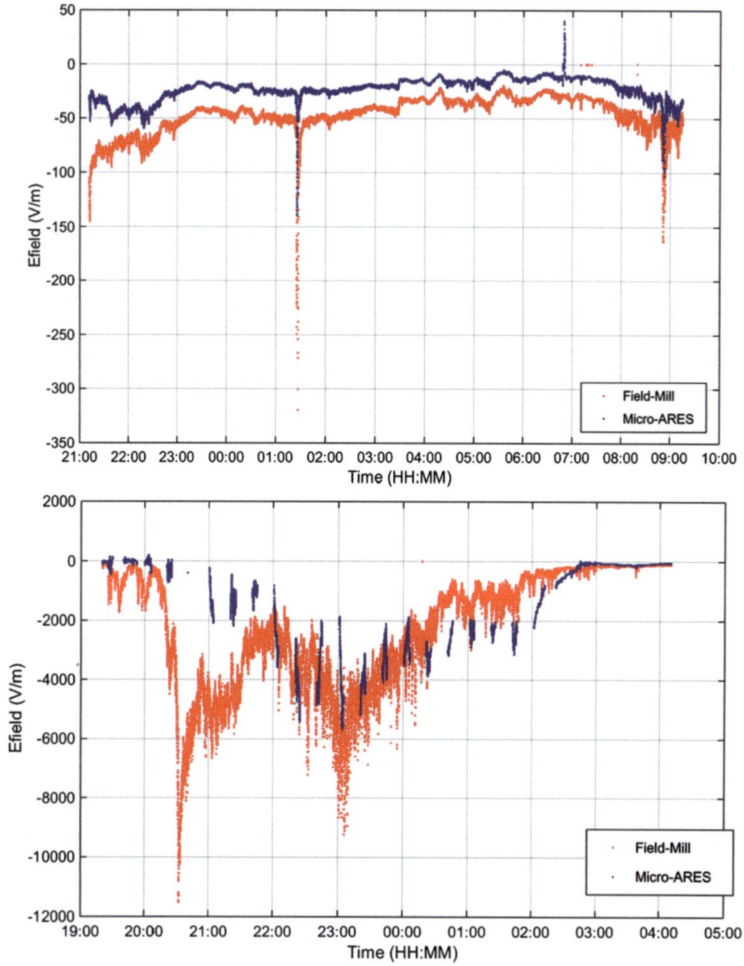

Fig. 24 Electric field samples made at 1 Hz: comparison during calm (*upper panel*: 10th July 2014) and disturbed (*lower panel*: 12th July 2014) weather. Note that the electric field intensity measured during fair (calm) weather by MicroARES at 80 cm from the ground is smaller than the value measured by the reference field mill that was placed at 2 m from the ground. This is in agreement with the expected atmospheric electric field vertical profile

(AFM) (Matsusaka et al. 2010; Horn et al. 1993, Gady et al. 1998). This approach promises to provide a more detailed (atomic scale) understanding of electrification and has already led to the discovery that net surface electrification in fact is made up of a 'mosaic' of highly electrified regions of positive and negative patches (Baytekin et al. 2011). Both processes of electron and ion transfer can be involved in contact electrification, depending on the surface conditions or composition. In the terrestrial environment, surface water layers will typically be expected on particles facilitation ion exchange irrespective of surface composition.

For the electrostatics in dust devils, specific studies aimed at understanding and quantifying the dust or sand electrification interactions are of direct practical use. Several laboratory (and wind tunnel) based studies are currently being performed (Bo et al. 2014; Lacks and Sankaran 2011, Merrison et al. 2012). To allow modelling of electrification

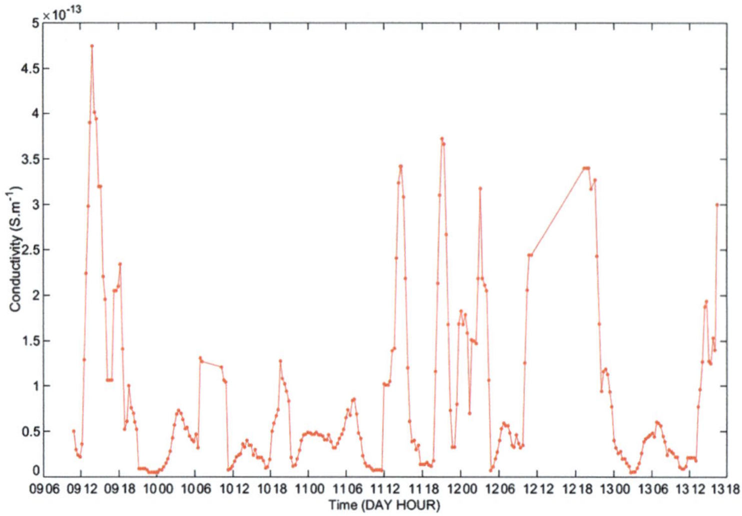

Fig. 25 Conductivity inferred with MicroARES instrument in terrestrial desert conditions from 9–13th July 2014

within a dust devil, a deeper understanding is needed in the size dependence of contact electrification. This is probably only likely to be achieved by pursuing more advanced laboratory experiments, applying detailed physical models (possibly involving both electron and ion transfer processes). Detailed and novel field studies could also contribute here.

Modelling of dust devils has shown that, by linking the flow dynamics produced by the balance between centripetal and pressure gradient forces (i.e. cyclostrophic balance) with theoretically derived inter-particle charge exchange processes, leads to electric fields comparable with those observed (Fig. 12). This includes a rapid initial rise in the electric field. At extreme electric fields, the dynamics of the charged particles can be influenced by the field (Kok and Renno 2008a, 2008b), providing a feedback on the electrification or a means by which the effectiveness of dust uplift is enhanced. Ultimately the growth of the field is limited by breakdown processes, which are dependent on the chemical composition of the atmosphere concerned.

Within the Martian atmosphere there are good arguments to expect electrification of dust devils and dust storms, in the absence of any direct electrical measurements to date. The DREAMS-MicroARES instrument due to land on Mars in 2016 will provide these first direct measurements, and with them, an important and substantial change in the knowledge of its planetary electrical environment.

Acknowledgements KAN acknowledges the support of the UK's Natural Environment Research Council through an Independent Research Fellowship (NE/L011514/1). This work was facilitated by a workshop on dust devils hosted by the International Space Science Institute in Bern.

References

M.M. Anderson et al., The Allen Telescope Array search for electrostatic discharges on Mars. Astrophys. J. **744**, 15 (2012). doi:10.1088/0004-637X/744/1/15

K.L. Aplin, Aspirated capacitor measurements of air conductivity and ion mobility spectra. Rev. Sci. Instrum. **76**, 104501 (2005). doi:10.1063/1.2069744

K.L. Aplin, Atmospheric electrification in the solar system. Surv. Geophys. **27**, 63–108 (2006).

K.L. Aplin, R.G. Harrison, A self-calibrating programable mobility spectrometer for atmospheric ion measurements. Rev. Sci. Instrum. **72**(8), 3467–3469 (2001)

K.L. Aplin, R.G. Harrison, M.J. Rycroft, Investigating Earth's atmospheric electricity: a role model for planetary studies. Space Sci. Rev. **137**, 11–27 (2008). doi:10.1007/s11214-008-9372-x

K.L. Aplin, T. Goodman, K.L. Herpoldt, C.J. Davis, Laboratory analogues of Martian electrostatic discharges. Planet. Space Sci. **69**, 100–104 (2012). doi:10.1016/j.pss.2012.04.002

K.L. Aplin, I.M.P. Houghton, K.A. Nicoll, M. Humphries, A. Tong, Electrical charging of volcanic ash, in *Proc. ESA Annual Meeting on Electrostatics* (2014)

K.L. Aplin, A.J. Bennett, R.G. Harrison, I.M.P. Houghton, Electrostatics and in situ sampling of volcanic plumes, in *Volcanic Ash: Methods of Observation and Monitoring*, ed. by S. Mackie (Elsevier, Amsterdam, 2016)

M.M. Apodaca, P.J. Wesson, K.J.M. Bishop, M.A. Ratner, B.A. Grzybowski, Contact electrification between identical materials. Angew. Chem., Int. Ed. Engl. **49**, 946–949 (2010)

S.K. Atreya, A.-S. Wong, N.O. Renno, W.M. Farrell, G.T. Delory, D.D. Sentman, S.A. Cummer, J.R. Marshall, S.C.R. Rafkin, D.C. Catling, Oxidant enhancement in Martian dust devils and storms: implications for life and habitability. Astrobiology **6**(3), 439–450 (2006) doi:10.1089/ast.2006.6.439

P.F.H. Baddeley, *Whirlwinds and dust-storms of India* (Bell and Daldy, London, 1860)

R.A. Bagnold, Auto-suspension of transported sediment; turbidity currents. Proc. R. Soc. Lond. Ser. A **265**(1322), 315–319 (1962)

A.G. Bailey, The charging of insulator surfaces. J. Electrost. **51/52**, 82–90 (2001)

E.L. Barth, W.M. Farrell, S.C.R. Rafkin, Electric field generation in Martian dust devils. Icarus **268**, 253–265 (2015). doi:10.1016/j.icarus.2015.12.020

H.T. Baytekin, A.Z. Patashinski, M. Branicki, B. Baytekin, S. Soh, B.A. Grzybowski, The mosaic of surface charge in contact electrification. Sci. Express (2011). doi:10.1126/science.1201512

A.J. Bennett, R.G. Harrison, In situ calibration of atmospheric air conductivity measurements. Rev. Sci. Instrum. **77**, 016103 (2006)

A.J. Bennett, R.G. Harrison, Surface measurement system for the atmospheric electrical vertical conduction current density, with displacement current correction. J. Atmos. Sol.-Terr. Phys. **70**, 1373–1381 (2008). doi:10.1016/j.jastp.2008.04.014

A.J. Bennett, P. Odams, D. Edwards, TH. Arason, Monitoring of lightning from the April–May 2010 Eyjafjallajökull volcanic eruption using a very low frequency lightning location network. Environ. Res. Lett. **5**(4), 044013 (2010)

J.-J. Berthelier, R. Grard, H. Laakso, M. Parrot, ARES, Atmospheric relaxation and electric field sensor, the electric field experiment on NETLANDER. Planet. Space Sci. **48**, 1193–1200 (2000). doi:10.1016/S0032-0633(00)00103-3

C. Bettanini, F. Esposito, S. Debei, C. Molfese, I. Arruego Rodríguez, G. Colombatti, A.-M. Harri, F. Montmessin, C. Wilson, A. Aboudan, S. Abbaki, V. Apestigue, G. Bellucci, J.-J. Berthelier, J.R. Brucato, S.B. Calcutt, F. Cortecchia, G. Di Achille, F. Ferri, F. Forget, G.P. Guizzo, E. Friso, M. Genzer, P. Gilbert, H. Haukka, J.J. Jiménez, S. Jiménez, J.-L. Josset, O. Karatekin, G. Landis, R. Lorenz, J. Martinez, L. Marty, V. Mennella, D. Möhlmann, D. Moirin, R. Molinaro, E. Palomba, M. Patel, J.-P. Pommereau, C.I. Popa, S. Rafkin, P. Rannou, N.O. Renno, P. Schipani, W. Schmidt, S. Silvestro, F. Simoes, A. Spiga, F. Valero, L. Vázquez, F. Vivat, O. Witasse, R. Mugnuolo, S. Pirrotta E:Marchetti, The DREAMS experiment on the ExoMars 2016 mission for the study of Martian environment during the dust storm season, in *Proceedings of IEEE Metrology for Aerospace (MetroAeroSpace)* (2014), pp. 167–173. doi:10.1109/MetroAeroSpace.2014.6865914

M.A. Bilici, J.R. Toth II., R.H. Sankaran, D. Lacks, Particle size effects in particle-particle triboelectric charging studied with an integrated fluidized bed and electrostatic separator. Rev. Sci. Instrum. **85**, 103903 (2014)

T. Bo, H. Zhang, W. Hu, X. Zheng, The analysis of electrification in windblown sand. Aeolian Res. **11**, 15–21 (2013a)

T.L. Bo, H. Zhang, W. Zhu, X.J. Zheng, Theoretical prediction of electric fields in wind-blown sand. J. Geophys. Res. **118**, 4494–4502 (2013b)

T. Bo, H. Zhang, X. Zheng, Charge-to-mass ratio of saltating particles in wind-blown sand. Sci. Rep. **4**, 5590 (2014)

E. Chassefière, P. Drossart, O. Korablev, Post-Phobos model for the altitude and size distribution of dust in the low Martian atmosphere. J. Geophys. Res. **100**(E3), 5525–5539 (1995)

C. Cimarelli, M.A. Alatorre-Ibargüengoitia, U. Kueppers, B. Scheu, D.B. Dingwell, Experimental generation of volcanic lightning. Geology **42**(1), 79–82 (2014)

C.F. Clement, R.G. Harrison, The charging of radioactive aerosols. J. Aerosol Sci. **23**(5), 481–504 (1992)

W.D. Crozier, The electric field of a New Mexico dust devil. J. Geophys. Res. **69**, 24 (1964). doi:10.1029/JZ069i024p05427

W.D. Crozier, Dust devil properties. J. Geophys. Res. **75**, 4583–4585 (1970). doi:10.1029/JC075i024p04583

D.K. Davies, Charge generation on dielectric surfaces. Br. J. Appl. Phys. **2**, 1533–1537 (1969)

G.T. Delory, W.M. Farrell, S.K. Atreya, N.O. Renno, A.-S. Wong, S.A. Cummer, D.D. Sentman, J.R. Marshall, S.C.R. Rafkin, D.C. Catling, Oxidant enhancement in Martian dust devils and storms: storm electric fields and electron dissociative attachment. Astrobiology **6**(3), 451–462 (2006). doi:10.1089/ast.2006.6.451

L. Demon, P. DeFelice, H. Gondet, Y. Kast, L. Pontier, Premiers résultats obtenus au cours du printemps J. Rech. Cent. Natl Rech. Sci. **24**, 126–137 (1953)

S.J. Desch, J.N. Cuzzi, The generation of lightning in the solar nebula. Icarus **143**, 87–105 (2000)

S. Dhanorkar, A.K. Kamra, Calculation of electrical conductivity from ion aerosol balance equations. J. Geophys. Res. **102**(D25), 30147–30159 (1997)

A.F. Diaz, R.M. Felix-Navarro, A semi-quantitative tribo-electric series for polymeric materials: the influence of chemical structure and properties. J. Electrost. **62**, 277 (2004)

N. Duff, D.J. Lacks, Particle dynamics simulations of triboelectric charging in granular insulator systems. J. Electrost. **66**, 51 (2008). doi:10.1016/j.elstat.2007.08.005

C.B. Duke, T.J. Fabish, Contact electrification of polymers: a quantitative model. J. Appl. Phys. **49**, 315–321 (1978)

H.F. Eden, B. Vonnegut, Electrical breakdown caused by dust motion in low pressure atmospheres: considerations for Mars. Nature **280**, 962–963 (1973)

T. Encrenaz, T.K. Greathouse, F. Lefèvre, S.K. Atreya, Hydrogen peroxide on Mars: observations, interpretation and future plans. Planet. Space Sci. **68**(1), 3–17 (2012)

F. Esposito, S. Debei, C. Bettanini, C. Molfese, I. Arruego Rodriguez, G. Colombatti, A.M. Harri, F. Montmessin, C. Wilson, A. Aboudan, S. Abbaki, V. Apestigue, G. Bellucci, J.J. Berthelier, J.R. Brucato, S.B. Calcutt, F. Cortecchia, F. Cucciarrè, G. Di Achille, F. Ferri, F. Forget, E. Friso, M. Genzer, H. Haukka, J.J. Jimènez, S. Jimènez, J.L. Josset, O. Karatekin, G. Landis, R. Lorenz, E. Marchetti, J. Martinez, L. Marty, V. Mennella, D. Möhlmann, D. Moirin, R. Molinaro, E. Palomba, M. Patel, J.P. Pommereau, C.I. Popa, S. Rafkin, P. Rannau, N.O. Renno, P. Schipani, W. Schmidt, E. Segato, S. Silvestro, F. Simoes, A. Spiga, F. Valero, L. Vázquez, F. Vivat, O. Witasse, R. Mugnuolo, S. Pirrotta, in *The DREAMS Experiment of the ExoMars 2016 Mission for the Study of Martian Environment During the Dust Storm Season*, Eighth International Conference on Mars, LPI Contribution No. 1791, July 14–18, 2014, Pasadena, California (2014), p. 1246

F. Esposito, S. Debei, C. Bettanini, C. Molfese, I. Arruego Rodríguez, G. Colombatti, A.-M. Harri, F. Montmessin, C. Wilson, A. Aboudan, S. Abbaki, V. Apestigue, G. Bellucci, J.-J. Berthelier, J.R. Brucato, S.B. Calcutt, F. Cortecchia, F. Cucciarrè, G. Di Achille, F. Ferri, F. Forget, E. Friso, M. Genzer, P. Gilbert, H. Haukka, J.J. Jiménez, S. Jiménez, J.-L. Josset, O. Karatekin, G. Landis, R. Lorenz, M. Mannetta, J. Martinez, L. Marty, V. Mennella, D. Möhlmann, D. Moirin, R. Molinaro, E. Palomba, M. Patel, J.-P. Pommereau, C.I. Popa, S. Rafkin, P. Rannou, N.O. Renno, P. Schipani, W. Schmidt, E. Segato, S. Silvestro, F. Simoes, E. Simoncini, A. Spiga, F. Valero, L. Vázquez, F. Vivat, O. Witasse, R. Mugnuolo, S. Pirrotta, E. Marchetti, in *The DREAMS Experiment On-Board the Schiaparelli Lander of ExoMars Mission*. EPSC Abstracts, vol. 10 (2015), EPSC2015-364

F. Esposito, R. Molinaro, C.I. Popa, C. Molfese, F. Cozzolino, L. Marty, K. Taj-Eddine, G. Di Achille, S. Silvestro, G.G. Ori, The role of atmospheric electric field in the dust lifting process. Geophys. Res. Lett. (2016, in revision)

M. Faraday, *Experimental Researches in Electricity* (Dover, New York, 1855)

W.M. Farrell, M.D. Desch, Is there a Martian atmospheric electric circuit. J. Geophys. Res. **106**(E4), 7591–7595 (2001)

W.M. Farrell, M.L. Kaiser, M.D. Desch, J.D. Houser, S.A. Cummer, D.M. Wilt, G.A. Landis, Detecting electrical activity from Martian dust storms. J. Geophys. Res. **104**(2), 3795–3801 (1999)

W.M. Farrell et al., Radio and optical detection of Martian dust storm discharges. Acta Astronaut. **46**, 25–36 (2000)

W.M. Farrell, G.T. Delory, S.A. Cummer, J.R. Marshall, A simple electrodynamic model of a dust devil. Geophys. Res. Lett. **30**, 2050 (2003)

W.M. Farrell et al., Electric and magnetic signatures of dust devils from the 2000–2001 MATADOR desert tests. J. Geophys. Res. **109**, E03004 (2004). doi:10.1029/2003JE002088

W.M. Farrell, N. Renno, G.T. Delory, S.A. Cummer, J.R. Marshall, Integration of electrostatic and fluid dynamics within a dust devil. J. Geophys. Res. **111**, E01006 (2006a)
W.M. Farrell, J.R. Marshall, S.A. Cummer, G.T. Delory, M.D. Desch, A model of the ULF magnetic and electric field generated from a dust devil. J. Geophys. Res. **111**, E11004 (2006b)
W.M. Farrell et al., Martian dust storms as a possible sink to atmospheric methane. Geophys. Res. Lett. **33**, L21203 (2006c)
W.M. Farrell, J.L. McLain, M.R. Collier, J.W. Keller, T.J. Jackson, G.T. Delory, Is the electron avalanche process in a Martian dust devil self-quenching? Icarus **254**, 333–337 (2015)
D.C. Ferguson, J.C. Kolecki, M.W. Siebert, D.M. Wilt, J.R. Matijevic, Evidence for Martian electrostatic charging and abrasive wheel wear from the Wheel abrasion experiment on the Pathfinder Sojourner rover. J. Geophys. Res. **104**(E4), 8747–8759 (1999)
M. Fillingim, Global electric circuit of Mars. Unpublished report, Florida Institute of Technology (1998)
K.M. Forward, D.J. Lacks, R.M. Sankaran, Particle-size dependent bipolar charging of Martian regolith simulant. Geophys. Res. Lett. **36**, L13201 (2009a). doi:10.1029/2009GL038589
K.M. Forward, D.J. Lacks, R.M. Sankaran, Charge segregation depends on particle size in triboelectrically charged granular materials. Phys. Rev. Lett. **102**, 028001 (2009b)
G.D. Freier, The electric field of a large dust devil. J. Geophys. Res. **65**(10), 3504 (1960). doi:10.1029/JZ065i010p03504
M. Fulchignoni, A. Aboudan, F. Angrilli, M. Antonello, S. Bastianello, C. Bettanini, G. Bianchini, G. Colombatti, F. Ferri, E. Flamini, V. Gaborita, N. Ghafoor, B. Hathi, A.-M. Harri, A. Lehto, P.F. Lion Stoppato, M.R. Patel, J.C. Zarnecki, A stratospheric balloon experiment to test the Huygens atmospheric structure instrument (HASI). Planet. Space Sci. **52**, 867–880 (2004)
B. Gady, R. Reifenberger, D.S. Rimai, Contact electrification studies using atomic force microscope techniques. J. Appl. Phys. **84**(1), 219–322 (1998)
R. Greeley, N. Lancaster, S. Lee, P. Thomas, Martian aeolian processes, sediments, and features. Mars **1**, 730–766 (1992)
J.S. Gilbert, S.J. Lane, R.S.J. Sparks, T. Koyaguchi, Charge measurements on particle fallout from a volcanic plume. Nature **349**(6310), 598–600 (1991)
R. Grard, Solar photon interaction with the Martian surface and related electrical and chemical phenomena. Icarus **114**, 130–138 (1995)
W. Gringel, Untersuchungen zur elektrischen Luftleitfahigkeit unter Berucksichtihung der Sonnenaktivitat und der Aerosolteilchenkonzentration bis 35 km Höehe. PhD thesis, Universität Tübingen (1978)
W. Gringel, R. Mühleisen, Sahara dust concentration in the troposphere over the North Atlantic derived from measurements of air conductivity. Beitr. Phys. Atmos. **51**(2), 121–128 (1978)
F.B. Gross, S.B. Grek, I.C. Calle, R.U. Lee, JSC Mars-1 Martian regolith simulant particle charging experiments in low pressure environment. J. Electrost. **53**, 257–266 (2001)
B.A. Grzybowski, M. Fialkowski, J.A. Wiles, Kinetics of contact electrification between metals and polymers. J. Phys. Chem. B **109**, 20511–20515 (2005)
D.A. Gurnett, D.D. Morgan, L.J. Granroth, B.A. Cantor, W.M. Farrell, J.R. Espley, Non-detection of impulsive radio signals from lightning in Martian dust storms using the radar receiver on the mars express spacecraft. Geophys. Res. Lett. **37**, L17802 (2010). doi:10.1029/2010GL044368
D. Halleaux, N.O. Renno, Aerosols–climate interactions at the Owens "Dry" Lake, California. Aeolian Res. **15**, 91–100 (2014). doi:10.1016/j.aeolia.2014.08.004
W.R. Harper, *Contact and Frictional Electrification* (Laplacian Press, Morgan Hill, 1998)
D.J. Harris, Electrical effects of the Harmattan dust storms. Nature **214**, 585 (1967)
R.G. Harrison, An antenna electrometer system for atmospheric electrical measurements. Rev. Sci. Instrum. **68**(3), 1599–1603 (1997)
R.G. Harrison, The global atmospheric electrical circuit and climate. Surv. Geophys. **25**(5–6), 441–484 (2004). doi:10.1007/s10712-004-5439-8
R.G. Harrison, The cloud chamber and CTR Wilson's legacy to atmospheric science. Weather **66**(10), 276–279 (2011). doi:10.1002/wea.830
R.G. Harrison, The Carnegie curve Surv. Geophys. **34**(2), 209–232 (2013). doi:10.1007/s10712-012-9210-2
R.G. Harrison, K.L. Aplin, A multimode electrometer for atmospheric ion measurements. Rev. Sci. Instrum. **71**(12), 4683–4685 (2000)
R.G. Harrison, H. Tammet, Ions in the terrestrial atmosphere and other solar system atmospheres. Space Sci. Rev. **137**, 107–118 (2008). doi:10.1007/s11214-008-9356-x
R.G. Harrison, K.L. Aplin, F. Leblanc, Y. Yair, Planetary atmospheric electricity. Space Sci. Rev. **137**, 5–10 (2008). doi:10.1007/s11214-008-9419-z
R.G. Harrison, K.A. Nicoll, Z. Ulanowski, T.A. Mather, Self-charging of the Eyjafjallajökull volcanic ash plume. Environ. Res. Lett. **5**(2), 024004 (2010). doi:10.1088/1748-9326/5/2/024004

H. Hatakeyama, On the disturbance of the atmospheric potential gradient caused by the smoke-cloud of the Volcano Yake-Yama. J. Geomagn. Geoelectr. **1**(2), 48–51 (1949)
C. Helling, R.G. Harrison, F. Honary, D.A. Diver, K.L. Aplin, I. Dobbs-Dixon, U. Ebert, S. Inutsuka, F.J. Gordillo-Vazquez, S. Littlefair, Atmospheric electrification in dusty, reactive gases in the solar system and beyond. Surv. Geophys. (2016, in press). doi:10.1007/s10712-016-9361-7
J.R. Herman, J.A. Caruso, R.G. Stone, Radio astronomy explorer (RAE)—1. Observations of terrestrial radio noise. Planet. Space Sci. **21**, 443–461 (1973)
R.G. Horn, D.T. Smith, A. Grabbe, Contact electrification induced by monolayer modification of a surface and relation to acid-base interactions. Nature **366**, 442–443 (1993)
I.M.P. Houghton, K.L. Aplin, K.A. Nicoll, Triboelectric charging of volcanic ash from the 2011 Grímsvötn eruption. Phys. Rev. Lett. **111**, 118501 (2013). doi:10.1103/PhysRevLett.111.118501
J.G. Houser et al., ULF and ELF magnetic activity from a terrestrial dust devil. Geophys. Res. Lett. **30**, 1027 (2003)
N. Huang, X. Zheng, A laboratory test of the electrification phenomenon in wind-blown sand flux. Chin. Sci. Bull. **46**, 417–420 (2001)
N. Huang, G. Yue, X. Zheng, Numerical simulations of a dust devil and the electric field in it. J. Geophys. Res. **113**, D20203 (2008)
I.I. Inculet, G.S.P. Castle, G. Aartsen, Generation of bipolar electric fields during industrial handling of powders. Chem. Eng. Sci. **61**, 2249–2253 (2006)
T.L. Jackson, W.M. Farrell, Electrostatic fields in dust devils: an analog to Mars. IEEE Trans. Geosci. Remote Sens. **44**(10), 2942–2949 (2006). doi:10.1109/TGRS.2006.875785
T.L. Jackson, W.M. Farrell, G.T. Delory, J. Nithianandam, Effect of dust absorption on the electron avalanche process occurring within Martian dust storms. Geophys. Res. Lett. **35**, L16201 (2008). doi:10.1029/2008GL034523
T.L. Jackson et al., Martian dust devil electron avalanche process and associated electrochemistry. J. Geophys. Res. **115**, E05006 (2010)
M.R. James, L. Wilson, S.J. Lane, J.S. Gilbert, T.A. Mather, R.G. Harrison, R.S. Martin, Electrical charging of volcanic plumes, in *Planetary Atmospheric Electricity* (Springer, New York, 2008), pp. 399–418
A.R. Johnston, H. Kirkham, B. Eng, DC electric field meter with fiber optic readout. Rev. Sci. Instrum. **57**, 2746–2753 (1986)
A.K. Kamra, Measurements of the electrical properties of dust storms. J. Geophys. Res. **77**(30), 5856 (1972)
J.F. Kok, D.J. Lacks, Electrification of granular systems of identical insulators. Phys. Rev. E **79**, 051304 (2009)
J.F. Kok, N.O. Renno, Enhancement of the emission of mineral dust aerosols by electric forces. Geophys. Res. Lett. **33**, L19S10 (2006)
J.F. Kok, N.O. Renno, Electrostatics in wind-blown sand. Phys. Rev. Lett. **100**, 014501 (2008a)
J.F. Kok, N.O. Renno, Electrical activity and dust lifting on Earth, Mars, and beyond. Space Sci. Rev. **137**, 419–434 (2008b)
J.F. Kok, N.O. Renno, Electrification of wind-blown sand on Mars and its implications for atmospheric chemistry. Geophys. Res. Lett. **36**, L05202 (2009). doi:10.1029/2008GL036691
C.E. Krauss, M. Horanyi, S. Robertson, Experimental evidence for electrostatic discharging of dust near the surface of Mars. New J. Phys. **5**, 70.1–70.9 (2003)
C.E. Krauss, M. Horanyi, S. Robertson, Modeling the formation of electrostatic discharges on Mars. J. Geophys. Res. **111**, E02001 (2006). doi:10.1029/2004JE002313
W.B. Kunkel, The static electrification of dust particles on dispersion into a cloud. J. Appl. Phys. **21**, 820–832 (1950)
M.V. Kurgansky, L. Baez, E.M. Ovalle, A simple model of the magnetic emission from a dust devil. J. Geophys. Res. **112**, E11008 (2007). doi:10.1029/2007JE002952
D.J. Lacks, A. Levandovsky, Effect of particle size distribution on the polarity of triboelectric charging in granular insulator systems. J. Electrost. **65**, 107–112 (2007)
D.J. Lacks, R.M. Sankaran, Contact electrification of insulating materials. J. Phys. D. **44**, 45 (2011)
J.D. Lacks, N. Duff, S.K. Kumar, Nonequilibrium accumulation of surface species and triboelectric charging in single component particulate systems. Phys. Rev. Lett. **100**, 188305 (2008)
J. Latham, The electrification of snowstorms and sandstorms. Q. J. R. Meteorol. Soc. **90**(383), 91–95 (1964)
M.A. Laughton, D.F. Warne (eds.), *Electrical Engineer's Reference Book*, 16th edn. (2004), Newnes
K.Y. Law, I.W. Tarnawskyj, D. Salamida, T. Debies, Investigation of the contact charging mechanism between an organic salt doped polymer surface and polymer-coated metal beads. Chem. Mater. **7**(11), 2090–2095 (1995)
F. Lefèvre, F. Forget, Observed variations of methane on Mars unexplained by known atmospheric chemistry and physics. Nature **460**(7256), 720–723 (2009). doi:10.1038/nature08228

J. Lowell, Constraints on contact charging of insulators: I. Spatial localisation of insulator states. J. Phys. D, Appl. Phys. **19**, 95–104 (1986)

J. Lowell, A.C. Rose-Innes, Contact electrification. Adv. Phys. **29**(6), 947–1023 (1980)

R. Markson, The global circuit intensity: its measurement and variation over the last 50 years. Bull. Am. Meteorol. Soc. **88**(2), 223–241 (2007)

G.J. Marlton, R.G. Harrison, K.A. Nicoll, Atmospheric point discharge current measurements using a temperature-compensated logarithmic current amplifier. Rev. Sci. Instrum. **84**, 066103 (2013). doi:10.1063/1.4810849

T.A. Mather, R.G. Harrison, Electrification of volcanic plumes. Surv. Geophys. **27**(4), 387–432 (2006)

K.C. Mathpal, N.C. Varshneya, N. Dass, Precipitation-powered mechanisms of cloud electrification. Rev. Geophys. **18**, 361 (1980)

S. Matsusaka, H. Maruyama, T. Matsuyama, A. Ghadiri, Chem. Eng. Sci. **65**, 5781 (2010)

T. Matsusyama, H. Yamamoto, Impact charging of particulate materials. Chem. Eng. Sci. **61**, 2230–2238 (2006)

T. Matsuyama, H. Yamamoto, Charge relaxation process dominates contact charging of a particle in atmospheric conditions II. General model. J. Phys. D, Appl. Phys. **30**, 2170–2175 (1997)

M.K. Mazumder, R.E. Ware, T. Yokoyama, B.J. Rubin, D. Kamp, Measurement of particle size and electrostatic charge distributions on toners using E-SPART analyzer. IEEE Trans. Ind. Appl. **27**, 611–619 (1991)

L.S. McCarty, G.M. Whiteside, Electrostatic charging due to separation of ions at interfaces: contact electrification of ionic electrets. Angew. Chem., Int. Ed. Engl. **47**(12), 2188–2207 (2008)

L.S. McCarty, A. Winkleman, G.M. Whiteside, Ionic electrets: electrostatic charging of surfaces by transferring mobile ions upon contact. J. Am. Chem. Soc. **129**, 4075–4087 (2007)

O. Melnik, M. Parrot, Electrostatic discharge in Martian dust storms. J. Geophys. Res. **103**, 29107–29118 (1998)

J. Merrison, J. Jensen, K. Kinch, R. Mugford, P. Nørnberg, The electrical properties of Mars analogue dust. Planet. Space Sci. **52**, 279–290 (2004)

J.P. Merrison, H.P. Gunnlaugsson, M.R. Hogg, M. Jensen, J.M. Lykke, M. Bo Madsen, M.B. Nielsen, P. Nørnberg, T.A. Ottosen, R.T. Pedersen, S. Pederse, A.V. Sørensen, Factors affecting the electrification of wind-driven dust studied with laboratory simulations. Planet. Space Sci. **60**, 328–335 (2012)

S.M. Metzger, M.R. Balme, M.C. Towner, B.J. Bos, T.J. Ringrose, M.R. Patel, In situ measurements of particle load and transport in dust devils. Icarus (2011). doi:10.1016/j.icarus.2011.03.013

M. Michael, M. Barani, S.N. Tripathi, Numerical predictions of aerosol charging and electrical conductivity of the lower atmosphere of Mars. Geophys. Res. Lett. **34**, L044201 (2007). doi:10.1029/2006GL028434

M. Michael, S.N. Tripathi, S.K. Mishra, Dust charging and electrical conductivity in the day and nighttime atmosphere of Mars. J. Geophys. Res. **113**, E07010 (2008). doi:10.1029/2007JE003047

S.L. Miller, A production of amino acids under possible primitive Earth conditions. Science **117**(3046), 528–529 (1953). doi:10.1126/science.117.3046.528

A.A. Mills, Dust clouds and frictional generation of glow discharges. Nature **268**, 614 (1977). doi:10.1038/268614a0

H.A. Mizes, E.M. Conwell, D.P. Salamida, Direct observation of ion transfer in contact charging between a metal and a polymer. Appl. Phys. Lett. **56**, 1597–1599 (1990)

G.J. Molina-Cuberos, H. Lichtenegger, K. Schwingenschuh, J.J. Lopez-Moreno, R. Rodrigo, Ion-neutral chemistry model of the lower ionosphere of Mars. J. Geophys. Res. **107**(E5), 5027 (2002). doi:10.1029/2000JE001447

G.J. Molina-Cuberos, R. Godard, J.J. López-Moreno, M. Hamelin, R. Grard, F. Simões, K. Schwingenschuh, V.J.G. Brown, P. Falkner, F. Ferri, I. Jernej, J.M. Jerónimo, R. Rodrigo, R. Trautner, M.J. Núñez, N. Ibrahim, C. Groth, M. Fulchignoni, A new approach for estimating Titan's electron conductivity based on data from relaxation probe sensors on the Huygens experiment. Planet. Space Sci. **58**(14–15), 1945–1952 (2010)

K.A. Nicoll, Measurements of atmospheric electricity aloft. Surv. Geophys. (2012). doi:10.1007/s10712-012-9188

K.A. Nicoll, R.G. Harrison, Z. Ulanowski, Observations of Saharan dust electrification. Environ. Res. Lett. **6**, 014501 (2011)

S. Nieh, T. Nguyen, Effects of humidity, conveying velocity, and particle size on electrostatic charges of glass beads in a gaseous suspension flow. J. Electrost. **21**, 99–114 (1988)

W.L. Nighan, Electron energy distribution and collision rates in electrically excited N_2, CO, and CO_2. Phys. Rev. A **2**, 1989–2000 (1970)

T. Oguchi, M. Tamatani, Contact electrification phenomena and powder surface treatments. Wear **i68**, 91–98 (1993)

M. Patzold, S. Tellmann, B. Hausler, D. Hinson, R. Schaa, G.I. Tyler, A sporadic third layer in the ionosphere of Mars. Science **310**(5749), 837-9 (2005)

T. Poppe, J. Blum, T. Henning, Experiments on collisional grain charging of micron-sized preplanetary dust. Astrophys. J. **533**, 472–480 (2000)

J.J. Qu, M.H. Yan, G.R. Dong, H.F. Zhang, R.P. Zu, W.Q. Tuo, A.G. Zhao, Z.H. Xiao, F. Li, B. Yang, Wind tunnel simulation experiment and investigation on the electrification of sandstorms. Sci. China, Ser. D **47**, 529–539 (2004)

K.R. Rasmussen, J.F. Kok, J.P. Merrison, Enhancement in wind-driven sand transport by electric fields. Planet. Space Sci. **57**(7), 804–808 (2009)

M. Ravichandran, A.K. Kamra, Spherical field meter to measure the electric field vector—measurements in fair weather and inside a dust devil. Rev. Sci. Instrum. **70**, 2140–2149 (1999). doi:10.1063/1.1149727

N.O. Renno, S.A. Rogacki, US Patent No. 8,536,879. Washington, DC: US Patent and Trademark Office

N.O. Renno, A.S. Wong, S.K. Atreya, I. de Pater, M. Roos-Serote, Electrical Discharges and broadband radio emission by Martian dust devils and dust storms. Geophys. Res. Lett. **30**, 2140 (2003). doi:10.1029/2003GL017879

N.O. Renno, V.J. Abreu, J. Koch, P.H. Smith, o.K. Hartogensis, H.A.R. De Bruin, D. Burose, G.T. Delory, W.M. Farrell, C.J. Watts, J. Garatuza, M. Parker, A. Carswell, MATADOR 2002: a pilot field experiment on convective plumes and dust devils. J. Geophys. Res. **109**, E07001 (2004). doi:10.1029/2003JE002219

N.O. Renno, J.F. Kok, H. Kirkham, S. Rogacki, A miniature sensor for electrical field measurements in dusty planetary atmospheres. J. Phys. Conf. Ser. **142**, 012075 (2008)

A. Robledo-Martinez, H. Sobral, A. Ruiz-Meza, Electrical discharges as a possible source of methane on Mars: lab simulation. Geophys. Res. Lett. **39**, L17202 (2012)

W.A.D. Rudge, Atmospheric electrification during South African dust storms. Nature **91**, 31–32 (1913). doi:10.1038/091031a0

C. Ruf, N.O. Renno, J.F. Kok, E. Bandelier, M.J. Sander, S. Gross, L. Skjerve, B. Cantor, Emission of non-thermal microwave radiation by a Martian dust storm. Geophys. Res. Lett. **36**, L13202 (2009). doi:10.1029/2009GL038715

M.J. Rycroft, R.G. Harrison, K.A. Nicoll, E.A. Mareev, An overview of Earth's global electric circuit and atmospheric conductivity. Space Sci. Rev. **137**, 83–105 (2008). doi:10.1007/s11214-008-9368-6

M.J. Rycroft, K.A. Nicoll, K.L. Aplin, R.G. Harrison, Recent advances in global electric circuit coupling between the space environment and the troposphere. J. Atmos. Sol.-Terr. Phys. **90–91**, 198–211 (2012). doi:10.1016/j.jastp.2012.03.015

D.S. Schmidt, R.A. Schmidt, J.D. Dent, Electrostatic force on saltating sand. J. Geophys. Res. **103**(D8), 8997–9001 (1998)

B.F.J. Schonland, *Atmospheric Electricity*, (Methuen, New York, 1953)

E. Seran, M. Godefroy, N. Renno, H. Elliott, Variations of electric field and electric resistivity of air caused by dust motion. J. Geophys. Res. Space Phys. **118**, 5358–5368 (2013). doi:10.1002/jgra.50478

Y. Shirakawa, N. Ii, M. Yoshida, R. Takashima, A. Shimosaka, J. Hidaka, Quantum chemical calculation of electron transfer at metal/polymer interfaces. Adv. Powder Technol. **45**, 366–372 (2008)

A.A. Sickafoose, J.E. Colwell, M. Horanyi, S. Robertson, Experimental investigations on photoelectric and triboelectric charging of dust. J. Geophys. Res. **106**, 8343–8356 (2001)

F. Simões, J.J. Berthelier, M. Godefroy, S. Yahi, Observation and modeling of the Earth-ionosphere cavity electromagnetic transverse resonance and variation of the D-region electron density near sunset. Geophys. Res. Lett. **36**(14), L14816 (2009)

A. Sowinski, L. Miller, P. Mehrani, Investigation of electrostatic charge distribution in gas-solid fluidized beds. Chem. Eng. Sci. **65**, 2771–2781 (2010)

C.D. Stow, Dust and sand storm electrification. Weather **24**, 134–140 (1969)

V. Sullivan, Variations of the vertical electric field and wind speed on days with airborne dust in Lubbock, Texas, Master thesis, Texas Tech University, Lubbock, Texas (2013)

K. Tanoue, A. Ema, H. Masuda, Effect of material transfer and work hardening of metal surface on the current generated by impact of particles. J. Chem. Eng. Jpn. **32**, 544–548 (1999)

S. Trigwell, N. Grable, C.U. Yurteri, R. Sharma, M.K. Mazumder, Effects of surface properties on the tribocharging characteristics of polymer powder as applied to industrial processes. IEEE Trans. Ind. Appl. **39**, 79 (2003)

S.N. Tripathi, M. Michael, R.G. Harrison, Profiles of ion and aerosol interactions in planetary atmospheres. Space Sci. Rev. **137**, 193–211 (2008). doi:10.1007/s11214-008-9367-7

Z. Ulanowski, J. Bailey, P.W. Lucas, J.H. Hough, E. Hirst, Alignment of atmospheric mineral dust due to electric field. Atmos. Chem. Phys. **7**, 6161–6173 (2007)

C. von Holstein-Rathlou, J.P. Merrison, C.F. Brædstrup, P. Nørnberg, The effects of electric fields on wind driven particulate detachment. Icarus **220**(1), 1–5 (2012)

R.C. Whitten, I.G. Poppoff, J.S. Sims, The ionosphere of Mars below 80 km altitude. Planet. Space Sci. **17**, 243–250 (1971)
J.A. Wiles, B.A. Grzybowski, A. Winkleman, G.M. Whitesides, A tool for studying contact electrification in systems comprising metals and insulating polymers. Anal. Chem. **75**, 4859–4867 (2003)
E. Williams, N. Nathou, E. Hicks, C. Pontikis, B. Russel, M. Miller, M.J. Bartholomew, The electrification of dust-lofting gust fronts ('haboobs') the Sahel. Atmos. Res. **91**, 292–298 (2009)
C.T.R. Wilson, Investigations on lightning discharges and on the electric field of thunderstorms. Philos. Trans. R. Soc. Lond. A **221**, 73–115 (1921)
L. Xie, K. Han, Influence of relative humidity on the aeolian electric field. Aeolian Res. **7**, 45–50 (2012)
H.F. Zhang, T. Wang, J.J. Qu, M.H. Yan, An experimental and observational study on the electric effect of sandstorms. Chin. J. Geophys. **47**, 53–60 (2004)
X.J. Zheng, N. Huang, Y.H. Zhou, Laboratory measurement of electrification of wind-blown sands and simulation of its effect on sand saltation movement. J. Geophys. Res. **108**, 4322 (2003)

Space Sci Rev (2016) 203:347–376
DOI 10.1007/s11214-016-0296-6

Particle Lifting Processes in Dust Devils

L.D.V. Neakrase[1] · M.R. Balme[2] · F. Esposito[3] · T. Kelling[4] · M. Klose[5,6] · J.F. Kok[7] · B. Marticorena[8] · J. Merrison[9] · M. Patel[2,10] · G. Wurm[4]

Received: 18 November 2015 / Accepted: 1 October 2016 / Published online: 17 October 2016

Abstract Particle lifting in dust devils on both Earth and Mars has been studied from many different perspectives, including how dust devils could influence the dust cycles of both planets. Here we review our current understanding of particle entrainment by dust devils by examining results from field observations on Earth and Mars, laboratory experiments (at terrestrial ambient and Mars-analog conditions), and analytical modeling. By combining insights obtained from these three methodologies, we provide a detailed overview on interactions between particle lifting processes due to mechanical, thermal, electrodynamical and pressure effects, and how these processes apply to dust devils on Earth and Mars. Experiments and observations have shown dust devils to be effective lifters of dust given the proper conditions on Earth and Mars. However, dust devil studies have yet to determine the individual roles of each of the component processes acting at any given time in dust devils.

Keywords Dust devils · Mars · Earth · Particle lifting · Aeolian processes

✉ L.D.V. Neakrase
lneakras@nmsu.edu

1 Department of Astronomy, New Mexico State University, Las Cruces, NM, USA

2 Open University, Walton Hall, Milton Keynes, MK7 6AA, UK

3 INAF—Osservatorio Astronomico di Capodimonte, Napoli, Italy

4 Faculty of Physics, University of Duisburg-Essen, Duisburg, Germany

5 Institute for Geophysics and Meteorology, University of Cologne, Cologne, Germany

6 Present address: USDA-ARS Jornada Experimental Range, Las Cruces, NM, USA

7 Department of Atmospheric and Oceanic Sciences, UCLA, Los Angeles, CA, USA

8 Laboratoire Interuniversitaire des Systèmes Atmosphériques, Université Paris, Créteil, France

9 Institute of Physics and Astronomy, Aarhus University, Aarhus, Denmark

10 Space Physics, Rutherford Appleton Laboratory, Oxfordshire, UK

1 Introduction

Dust devils are vertical convective vortices that are intense enough to erode surficial particulate materials and perhaps even weakly-indurated sediments. They are of scientific interest because of their potential to lift considerable amounts of surface material into the atmosphere. In contrast to regular boundary layer winds, dust particles ($D_p <\sim 63$ μm) are more easily removed from the surface by dust devil action than by boundary-layer wind shear alone (Greeley et al. 2003). This has been proposed to be due to a combination of influences dubbed the "ΔP-*effect*" by Greeley et al. (2003), in reference to the idea that the low–pressure core of a dust devil might "suck up" material.

Understanding this effect is important for Mars because of the many dust devils seen there (e.g., Metzger et al. 1999; Cantor et al. 2006; Greeley et al. 2006, 2010), and the likelihood that the low atmospheric density makes the lifting of small particles by boundary layer winds alone difficult (Greeley and Iversen 1985). The dust cycle on Mars (Kahn et al. 1992; Zurek and Martin 1993; McKim 1996; Newman et al. 2002) dominates much of the weather on Mars, and has many sources and sinks responsible for regulating the global movement of fine-grained material. The martian dust cycle includes interactions between multiple drivers of dust entrainment into the atmosphere (such as dust storms, dust devils, polar outflow, etc.) and dust removal processes (primarily, dust settling out from the atmosphere over time). Dust devils and dust storms are the major contributing sources for atmospheric dust on Mars (e.g., Newman et al. 2002; Kahre et al. 2006). Beyond understanding the nature of martian weather, the dust cycle plays an integral role in the continued exploration of Mars: dust constantly settles out of the atmosphere onto spacecraft sent to Mars to explore the surface, affecting surface mobility and solar-powered power generation (Greeley et al. 2006, 2010). It is possible that the electrostatic properties of fine grains can wreak havoc with moving parts (e.g., Wagner 2004). However, the enhanced ability of dust devils to lift fine grains can also be beneficial by sometimes removing dust from spacecraft solar panels as the dust devils pass over them (Greeley et al. 2006, 2010; Lorenz and Reiss 2015).

Dust devils, although localized, small-scale meteorological phenomena, seem to be more significant sources of atmospheric aerosols on Mars than on Earth (e.g., Metzger et al. 1999; Balme and Greeley 2006; Greeley et al. 2010; Allen et al. 2015; Jemmett-Smith et al. 2015, see also Klose et al., *this issue*). The presence of dust devils on Mars has led to many dust devil laboratory and field studies aimed at investigating their ability to lift fine particles, which explore the idea that dust devils might be the cause of pervasive background dust concentration required to maintain the ongoing dustiness of the martian atmosphere (Montabone et al. 2015). Exactly how dust devils lift fine-grained sand and dust is not well understood, being a complex set of interactions within a transient, micrometeorological process. However, significant work has been performed to investigate the conditions responsible for the enhanced lifting capabilities of dust devils compared to simple boundary layer flow.

On Earth, dust devils share many attributes of their martian counterparts, but are smaller in size, perhaps due to Earth's shallower planetary boundary layer (Fenton and Lorenz 2015). Dust devils occur frequently in arid and semi-arid regions of the Earth, where sensible heat fluxes are large, and loose sediment is available (e.g. Sinclair 1969; Snow and McClelland 1990; Hess and Spillane 1990; Oke et al. 2007; Kurgansky et al. 2011; Lorenz et al. 2015). They rarely form in other regions, where aeolian sediment transport is severely hampered by the presence of soil moisture and vegetation, which prevents the lifting of soil particles and reduces sensible heat fluxes (and thus reduces the likelihood of dust devil formation). Dust devils tend to be most active during the middle of the day starting in the mid-morning and tapering off in the late afternoon, and have a seasonal dependence ('dust devil season') with minimum activity in winter and maximum activity in summer. Due to their localized extent, less attention has been paid to the impact of dust devils in dust modeling than to the impact of regional or continental-scale dust storms (Klose et al., *this issue*). In addition, turbulence in the atmospheric boundary layer (ABL) is highly stochastic and, although bulk properties of the ABL are relatively well understood (e.g., Kaimal and Finnigan 1994; Smits et al. 2011), microscale meteorological phenomena such as dust devils remain difficult to represent in atmospheric circulation models.

In this paper we review our current understanding of the processes involved in dust devil sediment lifting, including a discussion of possible contributing factors to the "ΔP-effect". After reviewing the general physics of aeolian particle lifting (Sect. 2), we discuss insights into particle lifting specifically by dust devils with respect to: (i) results from Mars and Earth field observations (Sect. 3.1), (ii) laboratory experiments (Sect. 3.2), and (iii) analytical modeling (Sect. 3.3). Examining all three methodologies together provides a more complete picture of the underlying physics of dust lifting within dust devils, including exploring why not all vertical thermal convective vortices are necessarily capable of lifting dust, and what the dust lifting potential on Earth and Mars implies for the global and regional dust budgets of both planets.

2 Physics of Particle Lifting

There is a long history of the study of aeolian particle movement dating back to the early 20th century, including wind tunnel experiments and observations pertaining to the movement of sand in dunes and ripples (Bagnold 1941; Chepil 1945). Observations showed that aeolian sediment transport is initiated if the wind stress τ at the surface exceeds a threshold value τ_t. Since $\tau = \rho u_*^2$, where u_* is the friction velocity, this threshold stress is usually expressed in terms of the threshold friction velocity, u_{*t}. For a given pressure, u_{*t} is a property of the surface, describing its resistance to wind erosion (Shao and Lu 2000). Theoretical considerations on the balance of forces acting on a particle at rest lead to estimates on u_{*t}. By accounting for aerodynamic drag and gravity forces, Bagnold (1941) obtained

$$u_{*t} = A_B(Re_{*t})(\sigma_p g d)^{1/2} \tag{1}$$

where σ_p is the ratio of particle to air density, g is the gravitational acceleration, d is the particle diameter, and A_B is a coefficient that depends on the particle friction Reynolds number, Re_{*t}. According to (1), u_{*t} increases monotonically with particle diameter. However, observations have shown that there exists a minimum of u_{*t} for particles with diameters of $\sim$100 μm, with u_{*t} increasing with decreasing particle size below that value (e.g. Iversen et al. 1976) (Fig. 1a). By further including aerodynamic lift and interparticle cohesion in addition to drag and gravity, u_{*t} including this minimum can be predicted (Iversen and White 1982; Greeley and Iversen 1985) using

$$u_{*t} = A_1 F(Re_{*t}) G(d) \tag{2}$$

where A_1 is an empirical constant and F and G are relatively complex empirical functions. Shao and Lu (2000) found that these functions can be replaced by a simple theory-based expression if interparticle cohesion is treated explicitly, and if the Reynolds number dependence of aerodynamic forces is neglected. They obtained:

$$u_{*t} = A_N(\sigma_d g d + \gamma/\rho d)^{1/2} \tag{3}$$

where A_N is a dimensionless coefficient, and γ [N/m] scales the interparticle force.

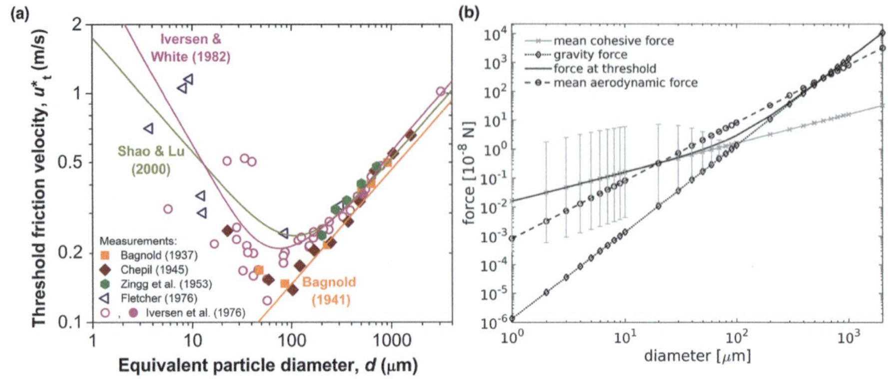

Fig. 1 (**a**) Theoretical relations for the threshold friction velocity from Bagnold (1941), Greeley and Iversen (1985) and Shao and Lu (2000) (*colored lines*), with experimental measurements using sand and dust (*filled symbols*) and other materials (*open symbols*). After Kok et al. (2012). (**b**) Relative importance of forces relevant for dust emission (from Klose 2014, modified from Shao 2008). The aerodynamic force is shown for $u_* = 0.4\ \mathrm{m\,s^{-1}}$. Error bars indicate the standard deviation of the cohesive force as given in Klose et al. (2014, Eq. (8)). The stochastic nature of the cohesive force may lead to particle uplift at a lower threshold

If roughness elements on the soil surface are present or if the soil is moist, the shear stress required to mobilize particles is larger than predicted based on u_{*t} for idealized (smooth and dry) surfaces. Drag partition is used to describe the influence of roughness elements on the surface momentum flux. In drag partition, the total drag is decomposed into a pressure-drag on the roughness elements, a skin-drag on the roughness element surface, and a skin-drag on the ground surface (e.g. Schlichting 1936; Arya 1975; Raupach 1992; Raupach et al. 1993; Marticorena and Bergametti 1995; Shao and Yang 2008). The latter is what drives dust emission. The ratio of surface to total drag is used to correct either u_{*t} (Marticorena and Bergametti 1995; Marticorena et al. 1997; Shao 2008), or u_* (Kok et al. 2014b), depending on the theoretical approach used for parameterization of the effect in models. For Earth, the moisture cohesion must be additionally accounted for in parameterizations of u_{*t}. Such parameterizations have been developed by McKenna Neuman and Nickling (1989), Fécan et al. (1999), McKenna Neuman (2003), and Cornelis et al. (2004a, 2004b).

Due to the minimum in u_{*t}, sand particles or particle aggregates with diameters $\sim$100 μm are most readily entrained (Fig. 1a). Corresponding to their large sizes, such grains are often too heavy to remain suspended. Rather, they hop along the surface horizontally in a process called saltation. During saltation, dust can be emitted either by the impacts of saltating particles overcoming the energetic interparticle bonds between soil particles, which is called saltation bombardment or sandblasting (Gillette 1974; Shao et al. 1993), or by the physical break-up of saltating dust aggregates (Shao 2001; Kok et al. 2014b) (Sect. 2.3).

Dust particles generally need stronger aerodynamic forces for direct aerodynamic entrainment than sand-sized particles, due to the larger entrainment thresholds at smaller diameters (Fig. 1). Since particle weight scales with d^3, whereas most cohesion forces are thought to scale with d (Hamaker 1937; Johnson et al. 1971), the interparticle cohesion force f_i is the dominant retarding force acting on small particles. Interparticle cohesive forces include Van der Waals forces, electrodynamic forces, and chemical forces (e.g., Castellanos 2005). These are governed in turn by a variety of factors such as particle shape, particle surface roughness, and chemical composition. Although theory suggests that f_i is, on average, proportional to d (Hamaker 1937; Johnson et al. 1971), the influencing factors suggest a stochastic behavior as has been observed by Zimon (1982). This stochastic behavior due to the various controls on f_i is more relevant for dust-sized particles than for sand-sized particles, because gravity dominates over interparticle cohesion for larger particles (Fig. 1b). This means that in the dust-size range, cohesion can be well below average and the lifting of dust-sized particles is thus also possible through direct aerodynamic entrainment and without saltation as an intermediate process (e.g. Loosmore and Hunt 2000; Roney and White 2004; Macpherson et al. 2008; Chkhetiani et al. 2012; Klose et al. 2014). The assumption of a stochastic behavior of interparticle cohesion suggests that the entrainment threshold friction velocity for the fraction of dust particles with smaller-than-average cohesion may be as small as that for particles in the saltation size range. The entrainment of this fraction is most likely strongly supply limited (Shao and Klose 2016).

Although the flux of directly entrained dust is typically smaller (for most natural surface types) than the dust amount lifted by the much more efficient saltation-bombardment mechanism, it can become a dominant mechanism in the absence of saltation. For example, while mean wind shear stresses may be below entrainment threshold during a calm summer day, organized atmospheric turbulence can produce shear stresses strong enough to lift sand-sized particles, leading to intermittent saltation and associated dust emission (Stout and Zobeck 1997; Dupont et al. 2013) and to aerodynamically entrain dust particles directly (Klose and Shao 2013). This is especially relevant for dust devils due to intense shear stresses associated with their rotation (e.g. Balme et al. 2003).

3 Dust Devil Sediment Entrainment

Dust devils contain complicated wind fields that include significant rotational components, near-surface inflows, and vertical updrafts and downdrafts within the core. In addition, many dust devils vary rapidly in intensity, and even when the dust devil as a whole is in a steady state, the wind field can change significantly at the small scale as the vortex traverses the surface. Nevertheless, by using a combination of approaches, there has been significant progress in understanding how, and how much, material is lifted by dust devils.

Dust devils may contain a variety of structural and morphological features including all or some of the following depending on the ambient conditions, as well as on the internal pressure structure. Visualization of each of these conditions is dependent on the availability of sediment to trace the dust devil's helical flow. The most dramatic feature of a fully developed dust devil is the dust column (Fig. 2, number 1), which comprises dust lofted from the surface wrapped up around a central thermal updraft. Individual dust grains and/or aggregates of dust are lifted into the helical flow by a combination of the shear stress from air inflow at the base of the dust devil and other, poorly understood mechanisms, such as low-pressure core effects (Sinclair 1969; Balme and Greeley 2006; Balme and Hagermann 2006). The column may extend beyond the top of the visible dust devil, which is highly dependent on the ambient conditions and the strength of the vortex itself. In many weak cases, dust devils may only produce a partially visible column due to inadequate flow velocities, small core pressure gradients, or sediment availability.

Another common feature in many dust devils is a saltation skirt. Saltation skirts (Fig. 2, number 2) can be seen around the base of a dust devil, demarking sediment load that was initially lofted by the dust devil but was comprised of particles too large or heavy to be suspended in the column. The presence of this saltation skirt is highly dependent on the particle size distribution at the surface and the strength of each individual dust devil (Balme and Greeley 2006). In some cases the larger grains can be thrown out of the skirt producing a cycloidal deposition track. This is often seen on Mars but is less common on Earth

Fig. 2 Schematic diagram showing various parts of a dust devil. (*Left*) Unprocessed, unannotated image of the dust devil to clearly see structure (*Center*) Annotated, 1 indicates the main column of the dust devil, 2 the "saltation skirt" where larger and heavier particles fall out of the column, and 3 the dust-free core downdraft that sometimes occurs. (*Right*) The individual velocity components of dust devil vortical flow where u_θ, u_z, and u_r, are respectively the tangential, vertical and radial velocity components. U_∞ represents the background boundary layer wind field that increases with height, z, causing dust devils to tilt by a proportional angle, α. (Photo credit: S.P. Idso, Arizona, c. 1975)

(Greeley et al. 2004; Neakrase 2009; Reiss et al. 2010; Neakrase et al. 2012; Reiss et al., *this issue*).

In many of the strongest dust devils, a clear central core may be observed (Fig. 2, number 3). This feature requires a strong pressure gradient within the dust devil allowing a central, dust-free, downdraft to form (Sinclair 1969; Balme and Greeley 2006). Cooler air in this downdraft is brought into the structure from outside the dust devil at the top: warm, dust-laden air is rising and rotating upwards around the downdraft.

Once a stable dust devil has formed, it can behave as a single structure. The column can be pushed by the ambient winds, with the ambient wind speed controlling the lateral motion of the dust devil (Balme et al. 2012). Because the ambient boundary layer winds increase with height above the surface, often dust devils will tilt in the direction of motion (Sinclair 1969; Kaimal and Businger 1970; Balme and Greeley 2006). The flow within a dust devil, resulting in a variety of dust devil morphologies, is controlled by complex interactions between all these processes. Figure 3 shows an idealized two-dimensional view of the dust devil flow structure. Stagnation points where flow reverses within the dust devil can occur near the surface, where the strength of the updraft can allow a region of no flow near the center. If the dust devil is strong enough to have a clear core, another stagnation point can form at the interface of where the downdraft meets the rising, warm, dust-laden air.

3.1 Field Observations

Earth Various field investigations of dust devils have been made in arid environments on Earth, but only a few have attempted to measure the flux of dust emitted from the surface beneath a dust devil. Renno et al. (2004) used LIDAR to remotely estimate a terrestrial dust devil column peak loading of 100 $\mathrm{mg\,m^{-3}}$ in a dust devil with a peak vertical velocity of 10 $\mathrm{m\,s^{-1}}$ to estimate a dust flux of 1000 $\mathrm{mg\,m^{-2}\,s^{-1}}$. The largest data set, though, comes

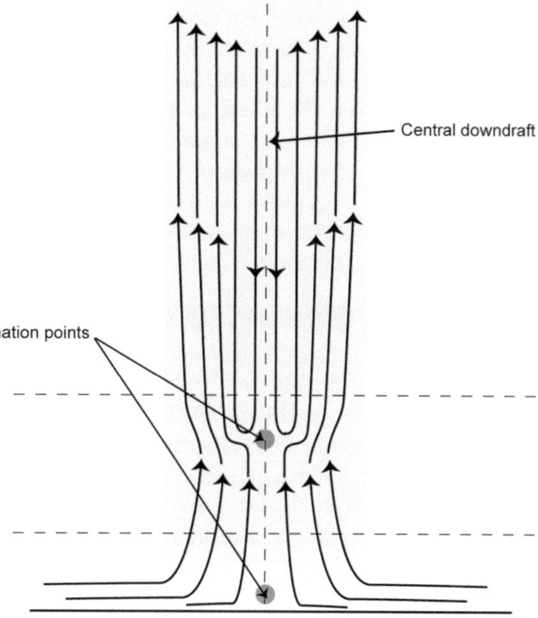

Fig. 3 Sketch showing vertical flow within an idealized dust devil. Stagnation points indicate where possible flow reversal occurs in some dust devils. The upper stagnation point can be due to the presence of a core downdraft, as observed in some dust devils, and could progress all the way to the surface within the core depending on the strength of the vortex. When an elevated stagnation point is not present, flow within the dust devil is dominated by only vertical flow. (After Balme and Greeley 2006, and Neakrase 2009)

from several field campaigns made between 1996 and 2005 at the El Dorado Playa, Nevada, and near Eloy, Arizona, both USA (Metzger et al. 2011). Metzger et al. (2011) made measurements using two types of commercial sensors, PM10 (particulate matter with diameter <10 μm) dust sensors, and TSP (Total Suspended Particle Load) aerosol monitors, as well as piezoelectric saltation-impact detectors being tested for the Beagle2 Mars mission (Towner et al. 2004). In many cases, these dust concentration data were acquired at high sample rates allowing particle-load profiles through the dust devil to be made. By combining these data with simultaneously acquired vertical wind speed profile data, they estimated the integrated upwards flux of dust or particulates. Metzger et al. (2011) found that the TSP concentrations in the dust devils they sampled ranged from 6 to 875 $\mathrm{mg\,m^{-3}}$, whereas the PM10 concentration varied from 1.3 to 162 $\mathrm{mg\,m^{-3}}$. From this they extrapolated TSP fluxes of between about 600 to 4375 $\mathrm{mg\,m^{-2}\,s^{-1}}$ and PM10 fluxes of between 1.3 and 162 $\mathrm{mg\,m^{-2}\,s^{-1}}$. The Metzger et al. (2011) flux estimates are about an order of magnitude lower than those obtained by Renno et al. (2004) using LIDAR remote sensing. This probably reflects the fact that the Metzger et al. (2011) data are integrated over profiles of dust load and wind speed, whereas the Renno et al. (2004) flux is calculated from peak vertical wind speed multiplied by peak dust concentration. Metzger et al. (2011) also found that the PM10 concentration was around 10 % of the TSP value thus about 90 % of sediments entrained by dust devils in their field areas were larger than 10 μm and unlikely to be carried into suspension but instead form the 'sand-skirt' seen in many dust devils. The presence of a sand-skirt (saltation-skirt) indicates that wind speeds have exceeded threshold conditions to entrain larger particles in the dust devil and is dependent on the sediment distribution for each field area. Skirtless dust devils could indicate either dust devil conditions allow only dust lifting and the largest grains are not sufficiently lofted to form the skirt, or the sediment fraction contains fewer coarse grains available for lofting.

While these data can inform about the amount of dust contained in a dust devil, and how rapidly it is lifted, they do not inform about the processes by which the dust is entrained.

While the ΔP effect could play a role, Balme et al. (2003) showed that the intense winds contained in dust devils generate shear stresses sufficient to lift most natural materials on their own, so it seems likely that, on Earth, shear stress dominates any other effect within dust devils. However, as outlined by Balme and Hagermann (2006), the suction effect created by the low-pressure core of a dust devil passing over a particulate surface is likely to be more effective when particle size is small, and when the dust devil is itself small and traveling quickly across the surface (i.e., when the pressure change occurs suddenly). Hence there might be occasions, even on Earth, when dust devils can suck material up rather than or in addition to blowing it into motion.

Even on Earth, quantitatively measuring dust vertical lifting is a difficult task and dust emission flux data sets are not numerous. A large part of this difficulty comes from the fact that it is difficult to properly sample the large size-range of mineral dust, especially with a high temporal frequency. Dust emission fluxes are usually estimated using the so-called "gradient method" (e.g., Gillette 1977; Nickling 1983; Nickling and Gillies 1993, Nickling et al. 1999; Gomes et al. 2003; Rajot et al. 2003). This method is based on the assumption that particles with aerodynamic diameters smaller than 20 μm are light enough to follow air movements perfectly and that the dust flux is constant with height (Gillette et al. 1972). In such conditions the causal relationship of air momentum and particle mass exchanges allows the upward vertical flux of particles to be expressed as a function of the wind friction velocity (u_*) and of the difference of dust concentration measured at two vertical levels (Gillette 1977). As noted by Shao et al. (2011), however, even in emission conditions the diffusive flux is not constant with height due to gravitational settling and a correction may be applied to the computed flux as a function of particle size. For particles in the range of 2–3.5 μm, this correction is estimated as being negligible (2 %), but it is 15 % for particle sizes from 6 to 8 μm, and should be higher for larger particle sizes (Shao et al. 2011).

From field measurements, the computation of u_* is usually performed over durations of at least 15 min, in order to integrate over the major time scales of turbulence occurring in the atmospheric boundary layer (Wieringa 1993). In addition, if dust concentration is measured with filter-based instruments a minimum integration time is required to get measurable dust loads. For measurements within dust devils, exposure times may be too short to obtain meaningful results depending on the method of instrument deployment (stationary versus mobile platforms). Dust sampling also involves the use of inlets with a specified size range. The most commonly used is the PM_{10} inlet that allows an efficient collection of particles up to 10 μm. More recently, the gradient method was applied to size-resolved dust measurements performed in Australia (Shao et al. 2011) during the Japanese Australian Dust Experiment (JADE, Ishizuka et al. 2008) and in Niger (Sow et al. 2009), as part of the AMMA (African Monsoon Multidisciplinary Analysis) international project. In both cases, significant, large, vertical dust fluxes were measured only during periods where saltation occurred, confirming the efficiency of the sand-blasting process to release fine dust particles.

Mars Both in situ meteorological measurements (landers/rovers) and remote sensing estimations (landers/rovers/orbiters) have provided a consistent picture of dust devils for different regions on Mars. Of particular interest to examine the processes of particle entrainment in dust devils are exact measurements and estimations of the surface conditions. However, observations of the surface conditions and their variation throughout the year have been limited by instrument availability on landers/rovers/orbiters. The most active dust devil period, or "dust devil season", on Mars is roughly defined as local spring and summer—i.e., for solar longitudes, L_s, of 0 to 180° in the northern hemisphere and 180 to 360° in the southern hemisphere. It varies for individual locations predominantly due to surface geometry with respect to solar illumination. Therefore, the dust devil season is typically defined as the time of the year where dust devil activity is at its maximum for a given area, which usually coincides with maximum insolation of the surface. Dust devil activity ceases almost completely in autumn, partly due to seasonal variability in regional to global dust storm activity reducing solar irradiation through increased atmospheric dust load, and partly due to the change in solar radiance associated with the transition from summer to winter (Greeley et al. 2006, 2010).

Dust devils were never imaged by the *Viking* landers in the late 1970s (see also Murphy et al., *this issue*). However, the detection of short isolated pressure drops in the meteorological data (Ryan and Lucich 1983; Ringrose et al. 2003) suggest that convective vortices did pass over or close to the lander, though it is not possible to tell whether they were dust loaded or not. Further information from these early surface vortex detections was limited due to the low resolution and sampling rate, and the sheltered position of the pressure sensors on the *Viking* landers. Diurnal and seasonal relationships can be seen as similar to terrestrial cases with peak activity occurring near midday and more detections occurring during the summertime (Ringrose et al. 2003).

Visual surface observations of dust devils began with *Mars Pathfinder* (MPF), in 1997, with the first images of dust devils from a lander in Ares Vallis. Metzger et al. (1999) discovered dust devils around the lander after employing a filter subtraction method to make the dust devils visible in the enhanced images. They reported the first estimates of dust devil properties on Mars, including sizes, speeds, and dust concentrations made with assumptions based on terrestrial field studies. Assuming vertical velocities of $\sim 7\ \mathrm{m\,s^{-1}}$ and measured dust concentrations of $\sim 70\ \mathrm{mg\,m^{-3}}$, from MPF dust opacity studies, Metzger et al. (1999) estimated the average dust devil dust flux to be $\sim 500\ \mathrm{mg\,m^{-2}\,s^{-1}}$. They also estimated a range in dust devil diameter of 14–79 m and estimated heights of 46–350 m. Dust devil traverse speeds were estimated at ~ 0.5–$4.6\ \mathrm{m\,s^{-1}}$. Ferri et al. (2003) performed a reanalysis of the MPF dust devil data substituting vertical velocities of up to $20\ \mathrm{m\,s^{-1}}$ (based on martian thermal convection studies), and making more detailed estimates of the distances to the dust devils and hence the dust devil diameters. The largest dust devils were determined to be around 100–200 m in diameter with a dust loading of ~ 700 times the background dust concentration. Average dust devil dust fluxes were determined to be $\sim 70\ \mathrm{mg\,m^{-2}\,s^{-1}}$ for a 200 m diameter dust devil. Local activity at MPF was observed at $L_s =\sim 143$–188, toward the end of northern summer, suggesting that *Pathfinder* potentially observed only the end of the dust devil season in Ares Vallis.

Further visual observations during the course of the *Mars Exploration Rover, Spirit* campaign in Gusev Crater (2004–2010) suggested that, as is the case for Earth, dust devil activity and dust lifting capability is highly variable from region to region and from year to year (Greeley et al. 2006, 2010). Three dust devil seasons were visually observed from *Spirit* in Gusev Crater, where over 700 individual dust devils were cataloged corresponding to local southern hemisphere spring to autumn (for Gusev $L_s =\sim 173$–340). Due to a favorable viewing geometry, namely the location of the rover on the Columbia Hills for each of the dust devil seasons, a more accurate estimate of the distances from the rover to the dust devils was possible. Gusev Crater contained relatively few large dust devils—their sizes ranged from a few meters up to ~ 276 m—but over the three seasons about six dust devils out of 761 were calculated as being larger than 160 m in diameter. Dust fluxes of between 0.004 and $460\ \mathrm{mg\,m^{-2}\,s^{-1}}$ were estimated, suggesting lower values than the estimates from Ares Vallis, but the frequency of dust devils was higher during the three dust devil seasons observed in Gusev Crater than in Ares Vallis (Greeley et al. 2006, 2010). The use of higher resolution cameras in conjunction with the better viewing angles is likely the reason so many more smaller dust devils were observed (as many as 51 dust devils per sol (martian day)) at Gusev compared to Ares Vallis. The other improvements used in this mission included "movie" sequences taken by *Spirit*, in which rapidly acquired sequences of images were used to capture dust devil movement. The same dust devil could be captured in many frames in a time sequence. The benefits of these sequences included being able to monitor how dust devils changed with time as they traversed the floor of Gusev Crater and the ability to estimate their translation speeds and hence, indirectly, the instantaneous near-surface wind speeds in the study area. Estimations of vertical velocities were also made within some of the dust devils. Traverse speeds were consistent with 1–3 $m\,s^{-1}$ ambient winds, and where clumps of sediment could be identified in subsequent frames, vertical velocities were estimated as being between 1–2 $m\,s^{-1}$ (Greeley et al. 2006, 2010). Other observations from the Mars Exploration Rovers suggest that dust devil induced removal of dust from solar panels occurred, and that this coincided with the peak diurnal dust devil activity when boundary layer winds alone probably would not have been as effective an agent (Greeley et al. 2006, 2010; Lorenz and Reiss 2015).

During the course of the 2008 Mars Phoenix Lander (PHX) mission several dust devils were detected at high northern latitudes. The meteorological package recorded $\sim$502 pressure drops >0.3 Pa and lasting at least 20 s, presumably associated with the nearby passage of dust devil-like vortices (Ellehoj et al. 2010). Using a method similar to that used by Murphy and Nelli (2002) to estimate vortex populations at MPF from the meteorological instruments, Ellehoj et al. (2010) examined the frequency of convective vortices at the PHX landing site (68.2°N, 234.3°E). Their results showed that convective vortices at high latitudes on Mars might have complicated initiation mechanisms, including sources of vorticity such as local weather systems and topographic controls. For example, a few vortex signatures were detected at night, when there is no solar input. These nighttime vortices are hypothesized to be due to vortex shedding from surrounding topographic features. Vortex detections at the PHX site peaked with the typical midday activity for dust devil populations, albeit having a slightly earlier peak than previously measured on Mars, but the seasonal activity was different due to the high latitude and interactions with weak frontal systems and topographic controls (Ellehoj et al. 2010). Measured ambient winds during the PHX mission were on average 4–5 $m\,s^{-1}$ in the NE–SE direction with estimated vortex wind speeds of 5–10 $m\,s^{-1}$ (Ellehoj et al. 2010; Holstein-Rathlou et al. 2010), well below the boundary layer wind tunnel estimates of $\sim$30 $m\,s^{-1}$ necessary for dust entrainment (e.g., Greeley and Iversen 1985) but similar to values estimated from other landing sites.

Surface observations of martian dust devils have all suffered from the same issues that hinder good correlative studies: firstly, visual detections of dust devils (e.g., Greeley et al. 2010) are generally not available with correlated meteorological information (i.e., pressure, temperature, wind speed, etc.). Secondly, meteorological detections of convective vortices, as in the *Viking*, *Pathfinder*, and *Phoenix* cases, generally lack simultaneous images. This dichotomy illustrates the difficulties in documenting such a transient natural phenomenon, compounded by remote sensing limitations of exploring a planet other than Earth.

3.2 Laboratory Investigations of Particle Lifting

3.2.1 Wind Stress Entrainment

Numerous terrestrial boundary layer wind tunnel studies of sediment lifting have been performed under varying planetary conditions, such as using different ambient temperatures and pressures, and particle sizes and densities to mimic the effects of lower gravity (e.g., Bagnold 1941; reviewed in Greeley and Iversen 1985, etc.) These experiments all suggest that threshold wind speed curves for wind stress lifting are likely to be similar but shifted when extrapolated to other planetary bodies; specifically, that a minimum in the u_{*t} curve should occur at a particular particle diameter (see Sect. 2), with that diameter and minimum u_{*t} varying from planet to planet.

One possible explanation for easier entrainment of dust-sized particles presented and discussed in the 1970s (e.g., reviewed in Greeley and Iversen 1985), was that aggregated dust could be removed as 'low density sand particles'. Such aggregates would have the lower cohesion forces of large particles yet be lighter than solid particles of the same diameter. In support of this claim dust aggregates are seen to be ubiquitous on Mars (e.g., Sullivan et al. 2008). Experiments in wind tunnels have also recently demonstrated that the threshold wind speed falls significantly for increased deposited dust depth (i.e., greater than 1 mono layer), due to the formation of low-density dust particle aggregates (Rondeau et al. 2015; Merrison et al. 2007; Merrison 2012). The detailed micro-scale dynamics of aggregate removal/breakup has yet to be studied in detail; however, the experimental techniques necessary for such studies (e.g., the use of high-speed cameras or laser-based velocimetry) are now readily available.

It is widely accepted that the simple boundary layer model typically used in sand/dust entrainment is incomplete and that in most cases turbulence plays a significant role in aeolian mobilization (Ibrahim et al. 2008; Reeks and Hall 2001; Ziskind et al. 1995; Dupont et al. 2013; Klose et al. 2014; Carneiro et al. 2015). The role of turbulence has been widely discussed in experiments and modeling of dust (and sand) entrainment, though there has been limited success in implementing a quantitative model to reproduce observed dust re-suspension (e.g., Rondeau et al. 2015).

It is typical in wind tunnel studies of sand transport that mobilization (detachment) occurs at significantly lower wind speeds (shear stress) than entrainment (lift). This type of transport, thought to result from rolling detachment, is a pre-cursor to entrainment. It has been shown that rolling detachment can have a qualitatively different behavior than the conventional aerodynamic model (e.g., De Vet et al. 2014; Reeks and Hall 2001), and could lead to particles on Mars being transported by winds at speeds below the usually considered aerodynamic threshold.

3.2.2 *Thermo-luminescent Lifting*

Atmospheric pressure has a significant influence on the particle entrainment threshold. Small particles such as fine sand and dust are more difficult to lift by aerodynamic forces because of stronger interparticle forces acting on small particles. At lower atmospheric pressures, changes in the flow regime near the surface can have significant influence on particle entrainment. Cohesion and electrical effects are generally described as retarding forces. In typical wind tunnel experiments, particle entrainment is explained through fluid (wind) shear stress providing a lift component to the particles that will induce motion through rolling, saltation, secondary impacting, and eventually some amount of suspension. Vortex studies (see Sect. 3.2.4) demonstrated that dust devil passage overcomes retarding forces more efficiently than typical boundary layer winds. However, wind shear stress may only be one piece of the total effect. Another potential lifting effect involves exchange of gas flows between sub-soil overpressures and the near-surface boundary layer flow resulting from temperature gradients within the soil.

Gas exchange in this context is a function of the ratio between the mean free path of the gas molecules (λ) and the dust or pore diameter (d) within the soil which defines the

Knudsen number ($Kn = \lambda/d$). If Kn increases or approaches 1, the gas can no longer be considered an ideal gas and the relationship between pressure, temperature, and gas density can no longer be applied. If the connection between the soil and the atmosphere is smaller than the mean free path ($Kn \gg 1$), the pressure is not equal across the interface. In this case, gas—somewhat counterintuitively—flows from the cold to warm side until the flow balances if $n_1 v_1 = n_2 v_2$, where n is the particle number density and v the thermal velocity. This corresponds to $P_1/P_2 = (T_2/T_1)^{1/2}$. This type of gas flow along a surface (here the walls of the connecting capillary) is known as thermal creep.

At martian conditions the mean free path is on the order of 10 μm, within the range where low pressure physics ($Kn \geq 1$) becomes important for dust and sand-sized structures. Therefore, if temperature gradients are present within the martian soil, thermal creep gas flow, as above, dictates flow from cold to warm. Under these conditions, the martian soil could act as a natural pump. Drop tower experiments used by De Beule et al. (2015) demonstrated gas flow within soils. Their experiments were not buoyancy driven as thermal convection does not exist in microgravity. Using tracer particles for visualization, the gas flow was found to be into cool areas and out of warm areas.

For comparison, the mean free path for air on Earth at 1 bar is only 67 nm. (Because typical pore-diameters for sand and soil are much larger than this, thermal creep is not important in natural terrestrial settings.)

The question is if and how this process can lead to dust or sand lifting. Thermal creep gas flow does not directly provide lift within the pumping soil because it requires a temperature gradient to exist. The top layer of the soil does not show a strong temperature variation because of insolation, which can cool by thermal radiation. The thickness of this insolated layer depends on the duration of illumination. De Beule et al. (2015) found thicknesses of 100 to 200 μm in laboratory settings for seconds of observation. More extended illumination times over a day on the surface of Mars, might lead to mm-thick layers of constant temperature. Gas flow in the layers below this constant temperature cap would be subject to thermal creep. Continuity requires that the gas pumped from below passes through the top layer. In the absence of thermal creep, gas flow can only be produced by a pressure-driven gas flow. Therefore, a sub-surface overpressure is established, with a mass flow rate described by Muntz et al. (2002). Two flow types contribute to mass flow: thermal creep flow characterized by the non-dimensional gas flow, and the pressure difference driven flow. The pressure difference (ΔP) driven flow provides a lift for particles on an illuminated surface on Mars, which is set by the temperature difference along capillaries or pores, ΔT.

This thermal-creep lifting force can be strong, but as outlined above is generally limited to low, millibar-pressure for soils. Wurm and Krauss (2006) found that particle lifting can easily overcome gravity and interparticle forces if essentially any light-absorbing dust sample was illuminated with a strong laser beam. Wurm et al. (2008) studied the gravity dependence and applied the results to Mars. An example of the gravity dependence is show in Fig. 4 taken from De Beule et al. (2013). The light flux used in the experiments ($12\ \mathrm{kW/m^2}$) is an order of magnitude larger than natural insolation at the martian surface.

On Earth typically at least a few $\mathrm{kW\,m^{-2}}$ are needed to lift dust in a millibar-pressure environment. In this case no other lifting force is needed, but if the light flux is lower, thermal creep and the pressure gradient are less intense though still present (Fig. 5). At lower light fluxes, below $1\ \mathrm{kW\,m^{-2}}$, the flow is not strong enough to lift particles against gravity (terrestrial or martian) or interparticle forces. De Beule et al. (2015) conducted experiments where a short vibration removed cohesive forces between the particles. As a result, the topmost layers were lifted by the pressure gradient created by thermal pumping. De Beule et al. (2015) called this an insolation-activated layer. This shows that the pressure support is still present at low light flux.

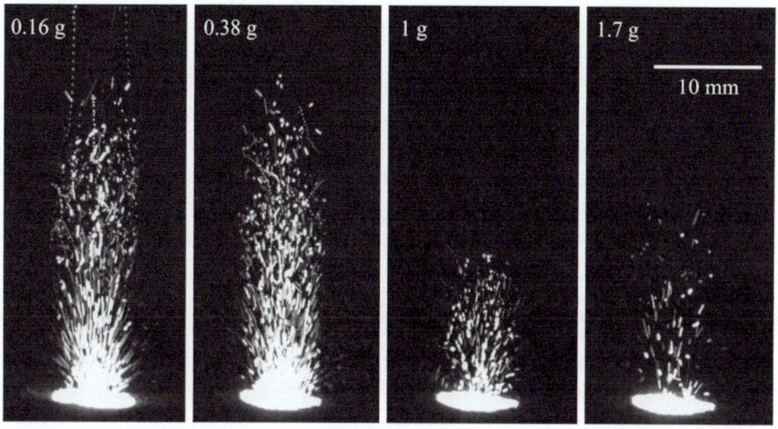

Fig. 4 Examples of particle ejections due to illumination at different g-levels (De Beule et al. 2013)

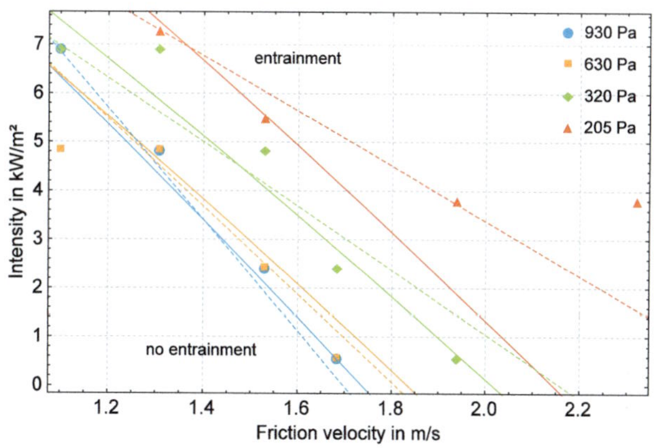

Fig. 5 Threshold friction velocities at given light flux and for different pressures (Küpper and Wurm 2015)

Estimates for martian conditions imply that insolation activation could reduce the threshold velocity for gas drag to lift dust (De Beule et al. 2015). This is consistent with current wind tunnel experiments by Küpper and Wurm (2015) where the threshold friction velocity was dependent on the insolation for Mars-analog pressures. Figure 6 shows the results of their laboratory measurements of threshold friction velocities depending on the insolating flux.

Scaled to Mars, Küpper and Wurm (2015) found that illumination can reduce the threshold velocity by between 5–18 %. Applied to dust devils this means that threshold velocities for translational or rotational motion of the vortex could also be smaller, making them more capable of lifting dust. Kelling et al. (2011) and Kocifaj et al. (2010) also demonstrated that up to 100 times more particles could be released counterintuitively after the light source was shut off. The implication is that as a dust devil traverses and blocks the sunlight from the surface, the shadowed dust could be more easily lifted as a result of similarly forcing a change in the soil temperature gradient producing outgassing.

3.2.3 *The Role of Electrodynamics in Natural Particle Lifting*

Electrification of particles is another candidate for enhanced lifting in martian dust devils. Although not specific to dust devils, grains moving via saltation favor particle electrification. In ballistic trajectories, particles bounce along the surface colliding with other grains and exchanging electric charge (Harper 1967; Renno et al. 2004; Kok and Renno 2008). The exact mechanism governing this charge exchange is not yet completely understood, but some theoretical models and some experiments suggest this is a size dependent process (Freier 1960; Schmidt et al. 1998; Inculet et al. 2006; Duff and Lacks 2008). During dust events, as local turbulence transports the finest particles higher into the atmosphere leaving the more massive grains closer to the surface, the atmosphere experiences a charge separation that produces an enhancement of its electric field. Moreover, some theoretical models and laboratory experiments suggest that the electric forces produced this way can be of the same order as gravitational forces. Thus, electric forces could be able to influence the trajectories of charged particles (Schmidt et al. 1998; Zheng et al. 2003) and also to reduce the threshold friction velocity u_{*t} necessary to initiate saltation and dust lifting. Electrical forces could thus play a major role in enhancing the particle lifting process especially in phenomena such as dust devils that are efficient at segregating particle sizes (Kok and Renno 2006).

Recent field experiments (Esposito et al. 2016) monitored the atmospheric electric field during dust events, including saltation, dust storms and dust devils. The experiments were conducted during an intense field campaign in the West Sahara. Esposito et al. performed simultaneous measurements of atmospheric parameters (pressure, wind, relative humidity, temperature, solar irradiance, electric field), soil parameters (temperature, moisture), and sand and dust parameters (dust size distribution and abundance, sand saltation rate and flux) during the dust storm seasons in 2013 and 2014 in the desert region of Merzouga in Morocco. They observed a very strong correlation between saltation rate and dust emission, and between these two parameters and the intensity of the atmospheric electric field. This demonstrated both the role of saltation in the dust emission and the particle electrification processes acting during these phenomena (Fig. 6).

Esposito et al. (2016) observed E-fields up to 15 kV/m during the strongest dust events as measured by a field mill mounted two meters above the ground, similar to what has been reported by other observers (Rudge 1913; Demon et al. 1953; Stow 1969; Kamra 1972; Williams et al. 2009). E-fields were generally directed in the downward direction, the same as the fair-weather field. A linear trend was observed between dust abundance and the E-field shown in Fig. 7.

The observed trend was almost the same in both the 2013 and 2014 campaigns even if the two measurement sites were different in terms of soil moisture content and composition. This demonstrates the effect of the electrification of fresh lifted dust on the electric properties of the atmosphere. A similar trend was observed in dust devils (see Murphy et al., *this issue*) suggesting that the electrification process is similar and pertinent to particle entrainment in dust devils. Although it is thought that electrification of dust is typical during mobilization and that it can be instrumental in the formation of dust aggregates (see Harrison et al., *this issue*) it has yet to be implemented into a predictive model.

3.2.4 *The "ΔP-Effect"*

Laboratory studies of the mechanical properties of dust devils were conducted via a series of experiments using the Arizona State University Vortex Generator (ASUVG) (Greeley et al. 2003; Greeley et al. 2004; Neakrase et al. 2006, Neakrase and Greeley 2010a; 2010b). These

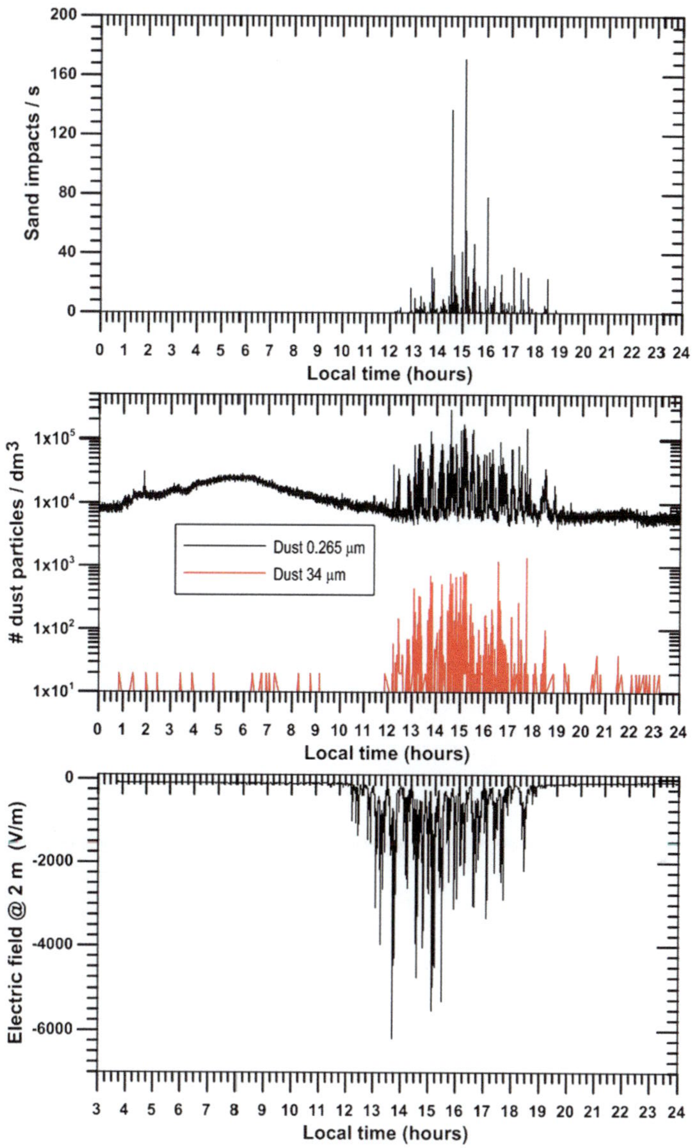

Fig. 6 Typical dust storm observed in the Moroccan desert (Esposito et al. 2016): sand saltation rate, atmospheric dust concentration and electric field at 2 m from the ground versus time

experiments explored dust devil particle entrainment threshold, sediment flux and basic interactions with surface roughness. Initial threshold studies (Greeley et al. 2003) demonstrated that vortices were very efficient at lifting small particles (Fig. 8). This suggested that a rotationally induced pressure drop in the vortex core provided an additional lift component that was substantial enough to overcome the interparticle forces acting on the small grains allowing them to be more easily entrained in the vortex flow (Fig. 9).

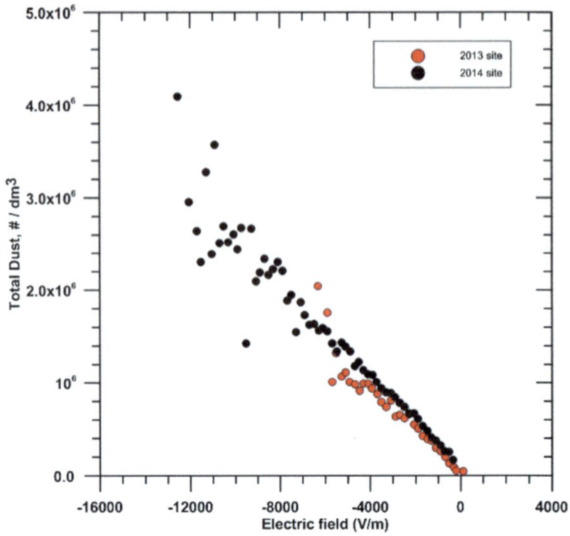

Fig. 7 Atmospheric dust concentration as a function of the electric field. Data acquired in two different sites in the desert region of Merzouga (Morocco) respectively in 2013 and 2014 (Esposito et al. 2016) in dry conditions (relative humidity below 10 %)

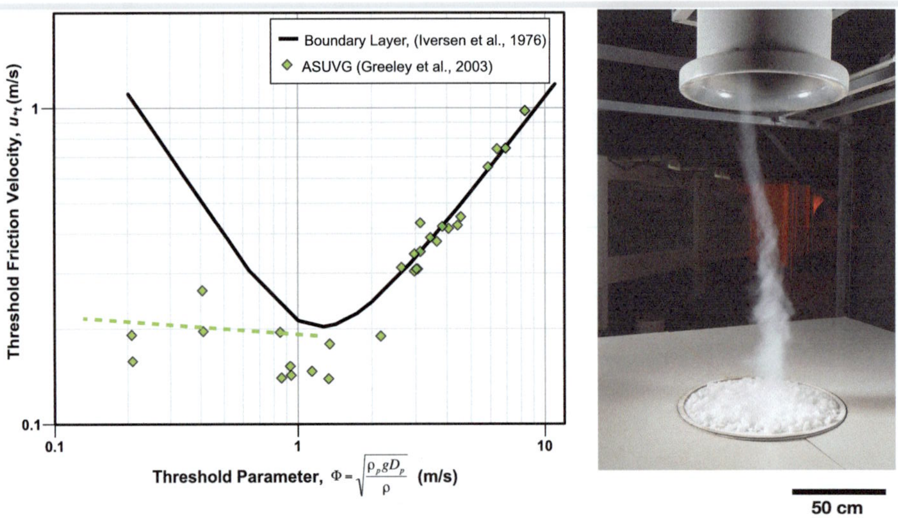

Fig. 8 Plot of the threshold parameter (*left*), Φ, versus threshold friction speed, u_{*t}, showing differences in laboratory data comparing boundary layer versus vortex induced particle thresholds, where ρ_p is particle density, g is gravity, D_p is particle diameter, ρ is atmospheric density. Image of the ASUVG with CO_2 sublimate for flow visualization (*right*). (After Greeley et al. 2003, and Neakrase and Greeley 2010a)

Further laboratory characterization of this "ΔP-*effect*" examined sediment flux for vortex flow on both aerodynamically smooth and rough surfaces (Neakrase et al. 2006; Neakrase and Greeley 2010a; 2010b). Expanding on the threshold experiments of Greeley et al. (2003), sediment flux was examined for a variety of materials at a range of tangential velocities (1–45 $\mathrm{m\,s^{-1}}$), which were above the measured threshold for both terrestrial ambient and martian analog conditions. The parameterized sediment flux was demonstrated to be a function of the strength of the vortex, described by the "lifting parameter", $\Delta P/u_\theta$. The use of this parameter allowed comparison of sediment lifting in Earth

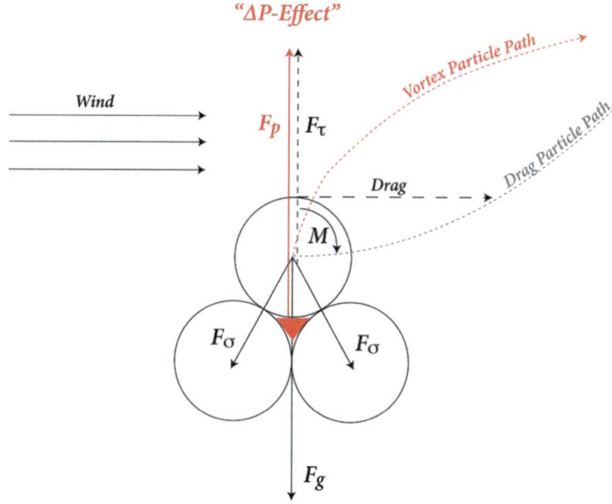

Fig. 9 Simplified force diagram showing idealized forces acting on a particle as it is lifted in boundary layer flow (*gray*) and vortex flow with ΔP-effect (*red*). F_g represents gravity, F_σ, interparticle forces, M, the Magnus rotational force, F_τ, the lift component from boundary layer shear stress, and F_p, lift component from the change in pressure possibly resulting from pore space pressure equalization creating a pressure differential across the particle. (After Greeley et al. (2003), Neakrase and Greeley (2010a), and Balme and Hagermann (2006))

and Mars environments, and accommodated the non-coupled nature of the core pressure differential and the tangential velocity. In other words, the tangential velocity is not determined solely by the size of the vortex, meaning that for the same sized vortex, there could exist multiple configurations of ΔP and tangential velocity (Neakrase et al. 2006; Neakrase and Greeley 2010a). The empirical results of their work showed that dust devil sediment flux is proportional to the lifting parameter to the fourth power.

Balme and Hagermann (2006) sought to examine the nature of the "ΔP-*effect*" and explain the wind-stress-entrainment laboratory results by using two different simple numerical models to explore how vertical pressure gradients (caused by, for example, the passage of the low-pressure core of a dust devil over a surface) might lift loose particles. The two potential mechanisms described by Balme and Hagermann (2006) included (1) an impermeable bed of particles with lifting by vertical pressure gradients only, and (2) a bed in which lifting is controlled by the drag forces on particles as gas is sucked out from the interparticle pore spaces. Their conclusion was that the vertical pressure-gradient process alone was the more efficient and effective process and that equilibration of pore space pressures with the transient passage of the localized low-pressure system led to a decrease in the lifting potential of the vortex. Hence, for a permeable bed, the lifting potential would be largest when the pressure gradient is applied suddenly, and when the particle sizes of the sediment are small, such that 'degassing' from the interparticle pore-spaces happens slowly compared to the application of the pressure gradient. As a consequence, the ΔP effect would be most effective in small, fast-moving (across the surface) dust devils with intense low-pressure cores. This might explain why the ΔP effect was seen so clearly in the small, laboratory-scale vortices, but is yet to be seen in the natural dust devil size regime. It is also possible that this effect is negligible in real dust devils (Shao and Klose 2016) or currently too difficult to assess in the natural case.

As has been shown through the laboratory studies throughout this section, this added lift component from vortex action, the "ΔP-*effect*", is most likely a complex interaction of several factors resultant from the passage of the dust devil over the surface. As such, it is perhaps better to designate this combined lifting effect the "enhanced vortex entrainment effect" (EVE-effect). Any or all of the contributing factors could be active in any given dust devil, contributing to rates of change for each process. Pressure-excursion effects such as those proposed by Greeley et al. (2003) could be explained in part by Balme and

Hagermann (2006) as mechanical pressure equalization, where the 'trapped' pockets of gas beneath a low pressure zone within a dust devil lifts particles into the flow. Furthermore, vortex-induced particle lifting could occur by some combination of those mechanical processes with varying contributions from thermo-luminescent (e.g., Küpper and Wurm 2015) or electrodynamical processes (e.g., Esposito et al. 2016). The EVE-effect is thus not only a function of the vortex properties, but also in part, the surface materials and the ambient atmospheric conditions. Availability of sediment, pore space within the sediment, surface roughness, and sediment intrinsic properties (e.g., particle size and density distributions, moisture content, electrical properties, etc.) all influence how the surface responds to the passage of a convective vortex and whether or not that vortex becomes a dust devil by entraining sand and dust. Background ambient atmospheric conditions add to this effect by externally controlling the formation of the convective vortex. All of these factors are important to consider for designing new multi-tiered, co-registered observations in the field and laboratory to separate individual factors better for consideration within analytical models.

In summary, although wind tunnel and vortex generator studies demonstrated that small vortices are particularly efficient at lifting particles of all sizes under both terrestrial and martian conditions, they have not been able to explain the process by which this occurs specifically. As wind tunnel studies progress, there has been new recognition of the effects of saltation triggering, rolling, and turbulence on the lifting of dust and sand. All of these mechanisms are probably important for dust devils on Mars, but, as shown above, other, perhaps less intuitive, effects might also play a role.

3.3 Analytical Parameterizations of Dust Emission

As outlined in Sect. 2, aerodynamic entrainment, saltation bombardment, and aggregate disintegration processes can contribute to dust emission in dust devils. On Earth, saltation bombardment and aggregate disintegration are the most effective dust emission mechanisms (Gillette et al. 1974; Shao et al. 1993) and have thus been the main focus of research. Both require saltation as an intermediate process. Parameterizations of saltation commonly used for dust emission modeling (e.g. Owen 1964; Kawamura 1964; White 1979) assume that the downward flux of horizontal fluid momentum, which sustains saltation against dissipation due to friction with the bed, remains approximately constant over a time period of a few to a few tens of minutes, and that saltation is in equilibrium with the atmospheric forcing (Barchyn et al. 2014). This assumption may be especially inappropriate in the case of dust devils due to their short duration and small spatial extent. However, although the indiscriminate application of existing parameterizations of saltation bombardment and aggregate disintegration to dust entrainment by dust devils is problematic, both processes likely contribute to dust emission in dust devils. Their parameterizations are thus useful starting points for understanding dust entrainment in dust devils, and are discussed in the following.

On Mars, the mechanics of dust lifting are quite different from those on Earth. In particular, because of Mars' low atmospheric density, formidable wind speeds of the order of $u_{*t} \approx 2\ \text{m/s}$ are needed to initiate saltation (Iversen and White 1982). However, theory and numerical modeling indicate that, once initiated, saltation (and thus dust emission) can be sustained by wind speeds that are an order of magnitude less intense (Claudin and Andreotti 2006; Almeida et al. 2008; Kok 2010a, 2010b; Pähtz et al. 2012). This occurs because the lower gravity and vertical drag on Mars makes saltating particles travel higher and longer trajectories, causing them to be accelerated by wind for a much longer duration during a single hop than on Earth. Furthermore, since wind speed increases away from the surface, the higher particle trajectories increase the acceleration on the particles (Kok 2010a). These two factors compensate for the lower atmospheric density on Mars, causing Martian saltators to relatively easily pick up large speeds. Upon impact, these particles then eject ('splash') surface particles into the air stream, thereby sustaining saltation at relatively low wind speeds. Consequently, the 'impact threshold' shear velocity down to which saltation can be sustained by splashing is predicted to be well below the 'fluid threshold' required to initiate saltation. The theoretical and numerical prediction that the Martian impact threshold is well below the Martian fluid threshold is supported by the Earth-like size of ripples encountered by Mars rovers (e.g., Sullivan et al. 2005), which are consistent with a low Martian impact threshold (Yizhaq et al. 2014). Nonetheless, the extent of the role that saltation plays in dust lifting in martian dust devils and dust storms remains unclear (Kok 2012). Indeed, observations by Sullivan et al. (2008) indicate that dust emission regularly occurs without saltation, namely by the aerodynamic entrainment of low-density sand-sized aggregates of dust that has settled from the sky (also see Merrison et al. 2007).

Below, we discuss parameterizations of dust emission through either saltation bombardment and aggregate fragmentation (Sect. 3.3.1) or direct aerodynamic entrainment (Sect. 3.3.2). These parameterizations have been developed for Earth, where observations are available to test them. Some of these process-based parameterizations were applied to estimate the dust emission by a "typical" dust devil (Wang 2016) or by a dust devil population (Klose and Shao 2016). Because of the dearth of experimental constraints, dust devil parameterizations for Mars are generally less based on the microphysics of dust emission, and instead aim to quantify dust devil emission from thermodynamics-based estimates of dust devil activity (Newman et al. 2002; Basu et al. 2004; Kahre et al. 2006; Newman and Richardson 2015). The parameterization of dust lifting by dust devils in martian atmospheric circulation models is thus generally more simplistic, in part because of the lack of observations, and in part because of the aforementioned remaining fundamental questions regarding the mechanics of martian dust lifting (Sullivan et al. 2008; Kok 2010b; Mulholland et al. 2015).

3.3.1 Parameterizations of Dust Emission Due to Saltation Bombardment and Aggregate Disintegration

Due to their hopping motion, the effective movement of saltating grains is in the horizontal wind direction and is expressed as a horizontal saltation flux, Q. For smaller dust devils, saltating particles might be expelled from the bottom of the vortex quite rapidly, but in larger ones, wind speeds are intense enough to make sand grains follow broad curving paths within the dust devil. In parameterization schemes, saltation is typically described as a uniform equilibrium process based on the momentum balance in the saltation layer, a layer close to the surface covering the height of most saltation trajectories (Owen 1964), though the accuracy of this simplifying assumption has been questioned by recent studies (e.g., Barchyn et al. 2014). Most studies, starting with Bagnold (1941), have suggested that Q is proportional to the cube of the friction velocity (u_*) (Kawamura 1951; Owen 1964; Lettau and Lettau 1978; Raupach and Lu 2004). However, more recent theoretical, numerical, and experimental studies have suggested that the linear scaling of particle speed with wind speed assumed in these models is incorrect (Ungar and Haff 1987; Rasmussen and Sørensen 2008; Creyssels et al. 2009; Kok 2010a, 2010b; Ho et al. 2011; Martin and Kok 2016). Models that do not assume that particle speed scales with wind speed instead conclude that sand flux scales with the square of u_* (Duran et al. 2011; Kok et al. 2012; Pähtz et al. 2012; Rasmussen et al. 2015). A recent comprehensive set

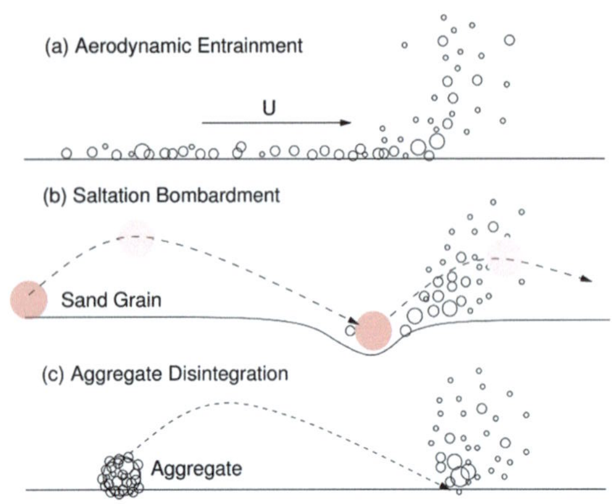

Fig. 10 Illustration of (**a**) dust emission by aerodynamic lift, (**b**) by saltation bombardment, and (**c**) through aggregate disintegration. Modified from Shao (2008)

of field campaigns directly supports the scaling of sand flux with the square, not the cube, of u_* (Martin and Kok 2016).

The vertical dust emission flux, F, generated by saltation bombardment (also known as abrasion) or aggregate disintegration (auto-abrasion) is dependent on the horizontal saltation flux, Q, as the energy transferred to the surface by saltating grains at impact determines the energy available for dust lifting (Fig. 10) (e.g., Shao et al. 1993; Marticorena and Bergametti 1995). Thus, F is often expressed as

$$F(d_i, d_s) = \alpha(d_i, d_s)Q(d_s) \tag{4}$$

where d_i denotes the diameter of dust particles, and d_s that of saltating grains (Shao 2004). The determination of α is the major challenge addressed in saltation-based dust emission parameterizations.

Marticorena and Bergametti (1995) proposed a semi-empirical dust emission parameterization. The authors suggest that the magnitude of F depends on dust particle abundance at the surface and relate α to soil clay content. Other studies propose physically-explicit parameterizations. For example, Shao et al. (1993, 1996) and Alfaro and Gomes (2001) relate α to the binding energy, ψ, of the particles at the surface, $\alpha \propto \beta/\psi$, where β is a parameter.

Shao (2001, 2004) explicitly consider both saltation bombardment and aggregate disintegration. The bombardment process is described in terms of the volume removal at the soil surface through impacting grains based on the work of Lu and Shao (1999) (Fig. 11a). Whether dust is emitted through abrasion or aggregate disintegration is determined by the mass fraction of free/aggregated dust available. The size of the volume that can be removed depends on the resistance the saltating grain experiences at impact. As a consequence, α is based on soil texture and soil plastic pressure in the scheme of Shao (2001, 2004).

More recently, Kok (2011b) proposed that most dust aerosol emission results from the fragmentation of aggregates of dust particles, which can occur either from impacts of saltators onto these aggregates, or from the saltation of the aggregates themselves (Fig. 11b). This analogy with the well-studied phenomenon of brittle material fragmentation (Astrom 2006) allowed the derivation of a simple analytical expression for the size distribution of emitted dust aerosols that eliminates some of the complexities of previous schemes, while being in good agreement with available measurements (also see Mahowald et al. 2014). Kok et al. (2014b) then built on this hypothesis that most dust emission results from aggregate fragmentation to propose a new theoretical model for how the dust flux emitted by an eroding soil depends on wind and soil properties. In this model, the efficiency with which the

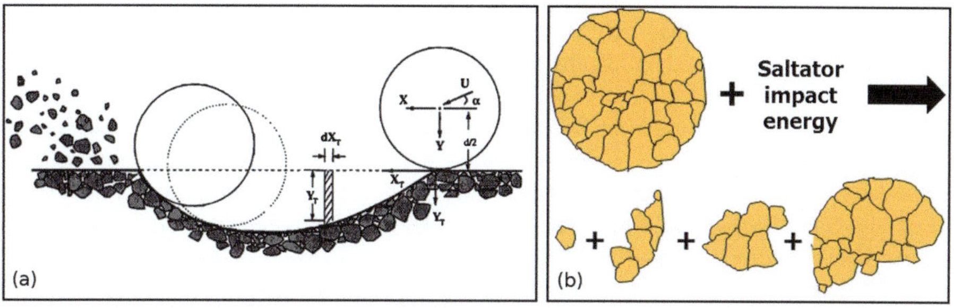

Fig. 11 (**a**) Illustration of dust emission via soil volume removal by impacting saltating grains during saltation bombardment as described by Shao (2001, 2004) (from Shao 2001); (**b**) Illustration of dust emission through soil or saltation aggregate disintegration in analogy with brittle fragmentation as described by Kok (2011b) and Kok et al. (2014b)

horizontal saltation flux produces a vertical dust flux depends primarily on soil clay content, which determines the dust mass available for disaggregation, and on the soil binding energy, which determines the resistance of the soil to fragmentation. This new dust emission model improves the representation of the global dust cycle in the Community Earth System Model (Kok et al. 2014a).

All schemes have been proven valuable for the modeling of regional and/or global-scale dust patterns, such as dust storms (e.g. Cavazos et al. 2009; Shao et al. 2010; Kok et al. 2014b), although differences between model predictions and observations of the dust cycle remain substantial (Huneeus et al. 2011; Evan et al. 2014). These differences can be due to a number of problems, including sub-gridscale variability and errors in wind speeds (Ridley et al. 2013), uncertainties in soil properties and moisture, sub-gridscale variability in the threshold wind speed due to spatial heterogeneity in soil properties and moisture, and deficiencies in parameterizations of the effects of differences in soil properties on the dust flux (Kok et al. 2014b).

Some of these problems are mitigated when applying the parameterizations on local scales for which soil properties can be defined with greater accuracy. However, all these parameterizations are formulated in terms of the friction velocity, which is by its nature a time-averaged property of the turbulent wind. Applying these parameterizations on smaller time scales, such as necessary for dust devils, can thus produce errors (Martin et al. 2013; Barchyn et al. 2014). Nonetheless, dust emission parameterizations such as the above are still a viable option for describing saltation-induced dust fluxes in dust devils. This can for instance be done by simulating (using high-resolution modeling frameworks such as large-eddy simulation; see Spiga et al., *this issue*) or measuring the spatial distribution of the instantaneous wind stress on the soil surface, and using the above parameterizations to estimate the dust emission flux. This ignores the fundamental problem in applying friction velocity-based parameterizations on short time scales (Barchyn et al. 2014), but the sensitivity study of Namikas et al. (2003) suggests that the effect of this discrepancy on predicted fluxes might be limited. Further work is needed to better resolve this issue.

3.3.2 *Parameterizations of Dust Emission Due to Aerodynamic Entrainment*

When the averaged wind stress is below the threshold stress, i.e. in the situation of $u_* < u_{*t}$, the above-mentioned dust emission schemes do not predict any dust emission. However, wind gusts can cause the instantaneous wind stress on the surface to exceed the threshold required to initiate saltation, causing dust emission to occur intermittently even though

u_* is on average smaller than u_{*t} (Stout and Zobeck 1997; Cameiro et al. 2015). Furthermore, dust can also be aerodynamically entrained when $u_* < u_{*t}$ (Loosmore and Hunt 2000; Macpherson et al. 2008).

Intermittently large momentum fluxes that can cause aerodynamic dust entrainment are due to atmospheric turbulence, which is most coherent in the case of unstable atmospheric stratification or in the presence of roughness elements (Stull 1988; Raupach et al. 1996). Klose and Shao (2012) and Klose et al. (2014) developed a parameterization for the direct aerodynamic entrainment of dust with a focus on convective turbulent conditions. Dust emission is described as a stochastic process by representing both interparticle cohesive forces and aerodynamic lifting forces, i.e. momentum fluxes due to atmospheric turbulence, as probability distributions. Dust emission fluxes produced by the scheme are mostly two to three orders of magnitude smaller than fluxes typically obtained for saltation-generated dust emission. However, under favorable conditions such as in dust devils, fluxes can be of similar magnitude and have been found to be close to those observed in the field (Klose and Shao 2016). Turbulent gusts also entrain sand-sized particles. This intermittent saltation is not yet included in the scheme and would further increase modeled turbulent dust emission. Klose and Shao (2016) applied the scheme in the framework of large-eddy simulation (LES) to investigate the dust entrainment associated with dust devils. They found that fluxes due to aerodynamic entrainment can be as large as $\sim 1000\ \mu g\,m^{-2}\,s^{-1}$ in dust devils and thus can explain large parts of dust fluxes measured in dust devils. Based on LES experiments for a wide range of atmospheric conditions, Klose and Shao (2016) related dust devil dust emission to Richardson number and suggested a simple parameterization for use in regional and global models. See Klose et al., *this issue* for more detail.

Wang (2016) compared the friction and horizontal pressure forces acting in a dust devil and found that friction forces are strong enough to lift dust particles while pressure forces are not. Wang concluded that dust entrainment in dust devils is likely caused by friction rather than pressure forces. The author derived a theoretical expression for the direct aerodynamic entrainment of dust in dust devils based on the Navier–Stokes equations for a rotating flow and obtained larger emissions for direct aerodynamic entrainment than for dust emission generated by saltation bombardment. However, interparticle cohesion has been neglected in the approach of Wang (2016), so his estimate of aerodynamic dust emission is likely an overestimation.

4 Conclusions and Future Work

Particle lifting and entrainment within dust devils are complicated processes that result from the internal balancing of multiple subsidiary processes, all of which may or may not be active in any given dust devil. The complex nature of these interactions means they are difficult to separate and study in the context of dust devils on Earth and Mars. Nevertheless, interactions between the various dust devil communities has led to advances that incorporate field observations on both planets, laboratory experiments (at terrestrial ambient and Mars analog conditions), and analytical modeling, clarifying many of the component processes for sediment transport within dust devils. The so-called ΔP-effect, describing the 'suction' of a low-pressure core within a dust devil, is perhaps better described as the EVE-effect (enhanced vortex entrainment), as it is actually a combination of many processes resulting from the conditions associated with the passage of a dust devil over the surface. Initial laboratory work characterized the wind shear component of the lift, without consideration of other processes tied to the material properties of the soil and how airflow in and around the grains affects their entrainment. Dust devil entrainment is intimately tied to the rate of change in each of the processes interacting with the surface, which leads to varying amounts of sediment lifting depending on the strength and translation speed of the dust devil. For example, electrification processes could affect the amount of dust lifted, but is dependent on the strength and speed of the dust devil to allow enough interaction time to be effective. So, dust lifting within dust devils is resultant from both the combined effects of the ambient conditions necessary for producing the vorticity and pressure structure within the dust devil. Furthermore, the instantaneous interactions between the lifting potential of the flow and the heterogeneous characteristics of sediment on the surface also strongly control the amount of dust seen lofted from the surface. The dust load within the dust devil, and even the orientation of the dust devil with respect to the sun, might have an additional effect, as the thermo-luminescent effect might be enhanced by sudden shadowing of the surface as the dust devil passes. The dominant form of lifting within a single dust devil could therefore be dominated by pore-space pressure equalization or thermo-luminescent enhanced lifting or mechanical wind shear, but it is likely the feedback of each of these within the dust devil which helps fuel many of the other processes ongoing within the vortex, and ultimately leading to dust lifting and the production of the 'classical' dust devil column.

Dust devils have been demonstrated to be intrinsically different lifting phenomena (Lorenz et al., *this issue*) compared to boundary layer winds. Through experiments and observations dust devils have been identified as effective lifters of dust given the proper conditions on Earth and Mars, and have been shown to be capable of removing dust on spacecraft solar panels. One of the more difficult remaining questions for dust devil studies involves discerning the individual roles of each of the component processes acting at any given time in dust devils. The difficulty arises in the ability to simultaneously collect bulk data on both the internal properties via in situ sensor suites (e.g., how pressure, wind velocities and particle counts are changing with time) and external measurements of how the dust composition, size distributions, and flux are changing over the lifetime of a dust devil. Future field observations will need to make better use of combined networks of sensors and camera systems to better characterize dust devil populations by attempting to get both types of data simultaneously. Laboratory studies can aid in development of instrumentation for these types of networks. Laboratory studies can also continue to refine our knowledge about the individual component processes by further isolation of specific variables including mechanical versus thermal versus electrodynamical dust lifting mechanisms. Analytical parameterizations and numerical dust lifting simulations that predict dust emission fluxes can continue to evolve with better computational ability and improved inputs from both observations and laboratory experiments. Such theoretical and modeling work serves the important role of connecting what we know about the specific processes to the larger regional and global systems and helps to define how dust devils result from and are tied to the bigger picture.

Acknowledgement This work was supported by grant National Science Foundation grant AGS-1358621 to J. K.

References

S.C. Alfaro, L. Gomes, Modeling mineral aerosol production by wind erosion: emission intensities and aerosol size distributions in source areas. J. Geophys. Res. **106**(D16), 18,075–18,084 (2001)

C.J.T. Allen, R. Washington, A. Saci, Dust detection from ground-based observations in the summer global dust maximum: results from Fennec 2011 and 2012 and implications for modeling and field observations. J. Geophys. Res. **120**, 897–916 (2015). doi:10.1002/2014JD022655

M.P. Almeida, E.J.R. Parteli, J.S. Andrade, H.J. Herrmann, Giant saltation on Mars. Proc. Natl. Acad. Sci. USA **105**, 6222–6226 (2008)

S.P.S. Arya, A drag partition theory for determining the large-scale roughness parameter and wind stress on the Arctic pack ice. J. Geophys. Res. **80**(24), 3447–3454 (1975). doi:10.1029/JC080i024p03447

J.A. Astrom, Statistical models of brittle fragmentation. Adv. Phys. **55**, 247–278 (2006)

R.A. Bagnold, *The Physics of Blown Sand and Desert Dunes* (Methuen, London, 1941). 265 p.

M.R. Balme, R. Greeley, Dust devils on Earth and Mars. Rev. Geophys. **44**, RG3003 (2006). doi:10.1029/2005RG000188

M. Balme, A. Hagermann, Particle lifting at the soil-air interface by atmospheric pressure excursions in dust devils. Geophys. Res. Lett. **33**, L19S01 (2006). doi:10.1029/2006GL026819

M. Balme, S. Metzger, M. Towner, T. Ringrose, R. Greeley, J. Iversen, Friction wind speeds in dust devils: a field study. Geophys. Res. Lett. **30**(16), 1830 (2003). doi:10.1029/2003GL017493

M.R. Balme, A. Pathare, S.M. Metzger, M.C. Towner, S.R. Lewis, A. Spiga, L.K. Fenton, N.O. Renno, H.M. Elliott, F.A. Saca, T.I. Michaels, P. Russell, J. Verdasca, Field measurements of horizontal forward motion velocities of terrestrial dust devils: towards a proxy for ambient winds on Mars and Earth. Icarus **221**(2), 632–645 (2012). doi:10.1016/j.icarus.2012.08.021

T.E. Barchyn, R.L. Martin, J.F. Kok, C.H. Hugenholtz, Fundamental mismatches between measurements and models in aeolian sediment transport prediction: the role of small-scale variability. Aeolian Res. **15**, 245–251 (2014)

S. Basu, M.I. Richardson, R.J. Wilson, Simulation of the Martian dust cycle with the GFDL Mars GCM. J. Geophys. Res. **109**, E11006 (2004)

M.V. Cameiro, K.R. Rasmussen, H.J. Herrmann, Bursts in discontinuous Aeolian saltation. Sci. Rep. **5**, 11109 (2015). doi:10.1038/srep11109

B.A. Cantor, K.M. Kanak, K.S. Edgett, Mars Orbiter Camera observations of Martian dust devils and their tracks (September 1997 to January 2006) and evaluation of theoretical vortex models. J. Geophys. Res. **111**, E12002 (2006). doi:10.1029/2006JE002700

M.V. Carneiro, K.R. Rasmussen, H.J. Herrmann, Bursts in discontinuous Aeolian saltation. Sci. Rep. **5**, 11109 (2015)

A. Castellanos, The relationship between attractive interparticle forces and bulk behaviour in dry and uncharged fine powders. Adv. Phys. **54**, 4 (2005). doi:10.1080/17461390500402657

C. Cavazos, M.C. Todd, K. Schepanski, Numerical model simulation of the Saharan dust event of 6–11 March 2006 using the Regional Climate Model version 3 (RegCM3). J. Geophys. Res. **114**, D12109 (2009). doi:10.1029/2008JD011078

W.S. Chepil, Dynamics of wind erosion. 2. Initiation of soil movement. Soil Sci. **60**, 397–411 (1945)

O.G. Chkhetiani, E.B. Gledzer, M.S. Artamonova, M.A. Iordanskii, Dust resuspension under weak wind conditions: direct observations and model. Atmos. Chem. Phys. **12**, 5147–5162 (2012). doi:10.5194/acp-12-5147-2012

P. Claudin, B. Andreotti, A scaling law for aeolian dunes on Mars, Venus, Earth, and for subaqueous ripples. Earth Planet. Sci. Lett. **252**, 30–44 (2006)

W.M. Cornelis, D. Gabriels, R. Hartmann, A conceptual model to predict the deflation threshold shear velocity as a affected by near-surface soil water: I. Theory. Soil Sci. Soc. Am. J. **68**, 1154–1161 (2004a)

W.M. Cornelis, D. Gabriels, R. Hartmann, A parameterisation for the threshold shear velocity to initiate deflation of dry and wet sediment. Geomorphology **59**, 43–51 (2004b). doi:10.1016/j.geomorph.2003.09.004

M. Creyssels, P. Dupont, A. Ould el Moctar, A. Valance, I. Cantat, J.T. Jenkins, J.M. Pasini, K.R. Rasumssen, Saltating particles in a turbulent boundary layer: experiment and theory. J. Fluid Mech. **625**, 47–74 (2009). doi:10.1017/S0022112008005491

C. De Beule, T. Kelling, G. Wurm, J. Teiser, T. Jankowski, From planetesimals to dust: low gravity experiments on recycling solids at the inner edge of protoplanetary disks. Astrophys. J. **763**(11), 1–8 (2013)

C. De Beule, G. Wurm, T. Kelling, M. Köster, M. Kocifaj, An insolation activated dust layer on Mars. Icarus **260**, 23–28 (2015)

S.J. De Vet, J.P. Merrison, M.C. Mittelmeijer-Hazeleger, E.E. van Loon, L.H. Cammeraat, Effects of rolling on wind-induced detachment thresholds of volcanic glass on Mars. Planet. Space Sci. **103**, 205–218 (2014)

L. Demon, P. Defelice, H. Gondet, Y. Kast, L. Pontier, J. Rech. C.N.R.S. **24**, 126 (1953)

N. Duff, D.J. Lacks, Particle dynamics simulations of triboelectric charging in granular insulator systems. J. Electrost. **66**, 51 (2008). doi:10.1016/j.elstat.2007.08.005

S. Dupont, G. Bergametti, B. Marticorena, S. Simoëns, Modeling saltation intermittency. J. Geophys. Res., Atmos. **118**, 7109{7128 (2013). doi:10.1002/jgrd.50528

O. Duran, P. Claudin, B. Andreotti, On aeolian transport: grain-scale interactions, dynamical mechanisms and scaling laws. Aeolian Res. **3**, 243–270 (2011)

M.D. Ellehoj, H.P. Gunnlaugsson, P.A. Taylor, H. Kahanpää, K.M. Bean, B.A. Cantor, B.T. Gheynani, L. Drube, D. Fisher, A.-M. Harri, C. Holstein-Rathlou, M.T. Lemmon, M.B. Madsen, M.C. Malin, J. Polkko, P.H. Smith, L.K. Tamppari, W. Weng, J. Whiteway, Convective vortices and dust devils at the Phoenix Mars mission landing site. J. Geophys. Res. **115**, E00E16 (2010). doi:10.1029/2009JE003413

F. Esposito, R. Molinaro, C.I. Popa, C. Molfese, F. Cozzolino, L. Marty, K. Taj-Eddine, G. Di Achille, G. Franzese, S. Silvestro, G.G. Ori, The role of the atmospheric electric field in the dust-lifting process. Geophys. Res. Lett. **43**(10), 5501–5508 (2016). doi:10.1002/2016GL068463

A.T. Evan, C. Flamant, S. Fiedler, O. Doherty, An analysis of Aeolian dust in climate models. Geophys. Res. Lett. **41**, 5996–6001 (2014)

F. Fécan, B. Marticorena, G. Bergametti, Parameterization of the increase of the aeolian erosion threshold wind friction velocity due to soil moisture for arid and semi-arid areas. Ann. Geophys. **17**, 149–157 (1999)

L.K. Fenton, R. Lorenz, Dust devil height and spacing with relation to the martian planetary boundary layer thickness. Icarus **260**, 246–262 (2015). doi:10.1016/j.icarus.2015.07.028

F. Ferri, P.H. Smith, M. Lemmon, N.O. Rennó, Dust devils as observed by Mars Pathfinder. J. Geophys. Res. **108**(E12), 5133 (2003). doi:10.1029/2000JE001421

G.D. Freier, The electric field of a large dust devil. J. Geophys. Res. **65**, 3504 (1960). doi:10.1029/JZ065i010p03504

D.A. Gillete, Fine particulate emissions due to wind erosion. Trans. Amer. Soc. Agric. Eng. **20**(5), 0890 (1977). doi:10.13031/2013.35670

D.A. Gillette, On the production of soil wind erosion aerosols having the potential for long range transport. J. Rech. Atmos. **8**, 735–744 (1974)

D. Gillette, I.H. Blifford, C.R. Fenster, Measurements of aerosol size distributions and fluxes of aerosols on land subject to wind erosion. J. Appl. Meteorol. **11**, 977–987 (1972)

D.A. Gillette, I.H. Blifford, D.W. Fryrear, Influence of wind velocity on size distributions of aerosols generated by wind erosion of soils. J. Geophys. Res. **79**, 4068–4075 (1974)

L. Gomes, J.L. Rajot, S.C. Alfaro, A. Gaudichet, Validation of a dust production model from measurements performed in semi-arid agricultural areas of Spain and Niger. Catena **52**, 257–271 (2003). doi:10.1016/S0341-8162(03)00017-1. Wind Erosion in Europe

R. Greeley, M.R. Balme, J.D. Iversen, S. Metzger, R. Mickelson, J. Phoreman, B. White, Martian dust devils: laboratory simulations of particle threshold. J. Geophys. Res. **108**(E5), 5041 (2003). doi:10.1029/2002JE001987

R. Greeley, J.D. Iversen, *Wind as a Geologic Process on Earth, Mars, Venus, and Titan* (Cambridge Univ. Press, New York, 1985)

R. Greeley, D.A. Waller, N.A. Cabrol, G.A. Landis, M.T. Lemmon, L.D.V. Neakrase, M. Pendleton Hoffer, S.D. Thompson, P.L. Whelley, G. Crater, Mars: observations of three dust devil seasons. J. Geophys. Res. **115**, E00F02 (2010). doi:10.1029/2010JE003608

R. Greeley, P.L. Whelley, R.E. Arvidson, N.A. Cabrol, D.J. Foley, B.J. Franklin, P.G. Geisler, M.P. Golombek, R.O. Kuzmin, G.A. Landis, M.T. Lemmon, L.D.V. Neakrase, S.W. Squyres, S.D. Thompson, Active dust devils in Gusev crater, Mars: observations from the Mars exploration rover spirit. J. Geophys. Res. **111**, E12S09 (2006). doi:10.1029/2006JE002743

R. Greeley, P.L. Whelley, L.D.V. Neakrase, Martian dust devils: directions of movement inferred from their tracks. Geophys. Res. Lett. **31**, L24702 (2004). doi:10.1029/2004GL021599

H.C. Hamaker, The London—Van Der Waals attraction between spherical particles. Physica **4**, 1058–1072 (1937)

W.R. Harper, *Contact and Frictional Dissipation* (Clarendon Press, Oxford, 1967)

G.D. Hess, K.T. Spillane, Characteristics of dust devils in Australia. J. Appl. Meteorol. **29**, 498–507 (1990). doi:10.1175/1520-0450(1990)029<0498:CODDIA>2.0.CO;2

T.D. Ho, A. Valance, P. Dupont, A.O. El Moctar, Scaling laws in aeolian sand transport. Phys. Rev. Lett. **106**, 094501 (2011)

C. Holstein-Rathlou, H.P. Gunnlaugsson, J.P. Merrison, K.M. Bean, B.A. Cantor, J.A. Davis, R. Davy, N.B. Drake, M.D. Ellehoj, W. Goetz, S.F. Hviid, C.F. Lange, S.E. Larsen, M.T. Lemmon, M.B. Madsen, M. Malin, J.E. Moores, P. Nørnberg, P. Smith, L.K. Tamppari, P.A. Taylor, Winds at the Phoenix landing site. J. Geophys. Res. **115**, E00E18 (2010)

N. Huneeus, M. Schulz, Y. Balkanski, J. Griesfeller, J. Prospero, S. Kinne, S. Bauer, O. Boucher, M. Chin, F. Dentener, T. Diehl, R. Easter, D. Fillmore, S. Ghan, P. Ginoux, A. Grini, L. Horowitz, D. Koch, M.C. Krol, W. Landing, X. Liu, N. Mahowald, R. Miller, J.J. Morcrette, G. Myhre, J. Penner, J. Perlwitz, P. Stier, T. Takemura, C.S. Zender, Global dust model intercomparison in AeroCom phase I. Atmos. Chem. Phys. **11**, 7781–7816 (2011)

A.H. Ibrahim, P.F. Dunn, M.F. Qazi, Experiments and validation of a model for microparticle detachment from a surface by turbulent air flow. J. Aerosol Sci. **39**, 645–656 (2008)

I.I. Inculet, G.S. Peter Castle, G. Aartsen, Generation of bipolar electric fields during industrial handling of powders. Chem. Eng. Sci. **61**, 2249–2253 (2006). doi:10.1016/j.ces.2005.05.005

M. Ishizuka, M. Mikami, J. Leys, Y. Yamada, S. Heidenreich, Y. Shao, G.H. McTainsh, Effects of soil moisture and dried raindroplet crust on saltation and dust emission. J. Geophys. Res. **113**, D24212 (2008). doi:10.1029/2008JD009955

J. Iversen, J. Pollack, R. Greeley, B. White, Saltation threshold on Mars: the effect of interparticle force, surface roughness, and low atmospheric density. Icarus **29**(3), 381–393 (1976). doi:10.1016/0019-1035(76)90140-8

J.D. Iversen, B.R. White, Saltation threshold on Earth, Mars and Venus. Sedimentology **29**, 111–119 (1982)

B.C. Jemmett-Smith, J.H. Marsham, P. Knippertz, C.A. Gilkeson, Quantifying global dust devil occurrence from meteorological analyses. Geophys. Res. Lett. **42**, 1275–1282 (2015). doi:10.1002/2015GL063078

K.L. Johnson, K. Kendall, A.D. Roberts, Surface energy and contact of elastic solids. Proc. R. Soc. Lond. Ser. A, Math. Phys. Sci. **324**, 301–313 (1971)

R.A. Kahn, T.Z. Martin, R.W. Zurek, S.W. Lee, The Martian dust cycle, in *Mars*, ed. by H. Kieffer et al.(Univ. of Ariz. Press, Tucson, 1992), pp. 1017–1053. Ch. 29

M.A. Kahre, J.R. Murphy, R.M. Haberle, Modeling the Martian dust cycle and surface dust reservoirs with the NASA Ames general circulation model. J. Geophys. Res. **111**, E06008 (2006). doi:10.1029/2005JE002588

J.C. Kaimal, J.A. Businger, Case studies of a convective plume and a dust devil. J. Appl. Meteorol. **9**, 612–620 (1970)

J.C. Kaimal, J.J. Finnigan, *Atmospheric Boundary Layer Flows: Their Structure and Measurement* (Oxford Univ. Press, New York, 1994)

A.K. Kamra, Measurements of the electrical properties of dust storms. J. Geophys. Res. **77**(30), 5856 (1972)

R. Kawamura, Study on sand movement by wind (relationship between sand flow and wind friction, and vertical density distribution of sand). Tokyo Daigaku Rikogaku Kenkyusho Hokoku, Tokyo **5**(3), 95–112 (1951)

R. Kawamura, Study of sand movement by wind, Hydraulic Eng. Lab. Tech. Rep., University of California, Berkeley, CA, HEL-2-8, pp. 99–108 (1964)

T. Kelling, G. Wurm, M. Kocifaj, J. Klačka, D. Reiss, Dust ejection from planetary bodies by temperature gradients: laboratory experiments. Icarus **212**, 935–940 (2011)

M.R. Klose, Convective Turbulent Dust Emission: Process, parameterization, and relevance in the Earth system, Dissertation, University of Cologne (2014). urn:nbn:de:hbz:38-58264

M. Klose, Y. Shao, Stochastic parameterization of dust emission and application to convective atmospheric conditions. Atmos. Chem. Phys. **12**(12), 7309–7320 (2012). doi:10.5194/acp-12-7309-2012

M. Klose, Y. Shao, Large-eddy simulation of turbulent dust emission. Aeolian Res. **8**, 49–58 (2013). doi:10.1016/j.aeolia.2012.10.010

M. Klose, Y. Shao, A numerical study on dust devils with implications to global dust budget estimates. Aeolian Res. **22**, 47–58 (2016)

M. Klose, Y. Shao, X.L. Li, H.S. Zhang, M. Ishizuka, M. Mikami, J.F. Leys, Further development of a parameterization for convective turbulent dust emission and evaluation based on field observations. J. Geophys. Res., Atmos. **119**, 10,441–10,457 (2014). doi:10.1002/2014JD021688

M. Kocifaj, J. Klačka, G. Wurm, T. Kelling, I. Kohút, Dust ejection from (pre-)planetary bodies by temperature gradients: radiative and heat transfer. Mon. Not. R. Astron. Soc. **404**, 1512–1518 (2010)

J.F. Kok, An improved parameterization of wind-blown sand flux on Mars that includes the effect of hysteresis. Geophys. Res. Lett. **37**, L12202 (2010a)

J.F. Kok, Difference in the wind speeds required for initiation versus continuation of sand transport on Mars: implications for dunes and dust storms. Phys. Rev. Lett. **104**, 074502 (2010b)

J.F. Kok, A scaling theory for the size distribution of emitted dust aerosols suggests climate models underestimate the size of the global dust cycle. Proc. Natl. Acad. Sci. USA **108**(3), 1016–1021 (2011b)

J. Kok, Planetary science: Martian sand blowing in the wind. Nature **485**, 312–313 (2012). doi:10.1038/nature11193

J.F. Kok, S. Albani, N.M. Mahowald, D.S. Ward, An improved dust emission model—part 2: evaluation in the community Earth system model, with implications for the use of dust source functions. Atmos. Chem. Phys. **14**, 13043–13061 (2014a)

J.F. Kok, N.M. Mahowald, G. Fratini, J.A. Gillies, M. Ishizuka, J.F. Leys, M. Mikami, M.S. Park, S.U. Park, R.S. Van Pelt, T.M. Zobeck, An improved dust emission model—part 1: model description and comparison against measurements. Atmos. Chem. Phys. **14**, 13023–13041 (2014b)

J.F. Kok, E.J.R. Parteli, T.I. Michaels, D. Bou Karam, The physics of wind-blown sand and dust. Rep. Prog. Phys. **75**, 106901 (2012)

J.F. Kok, N.O. Renno, Enhancement of the emission of mineral dust aerosols by electric forces. Geophys. Res. Lett. **33**, L19S10 (2006)

J.F. Kok, N.O. Renno, Electrostatics in wind-blown sand. Phys. Rev. Lett. **100**, 014501 (2008)

M. Küpper, G. Wurm, Thermal creep assisted dust lifting on Mars: wind tunnel experiments for the entrainment threshold velocity. J. Geophys. Res. **120**, 1346–1356 (2015)

M.V. Kurgansky, A. Montecinos, V. Villagran, S.M. Metzger, Micrometeorological conditions for dust-devil occurrence in the Atacama Desert. Bound.-Layer Meteorol. **138**(2), 285–298 (2011). doi:10.1007/s10546-010-9549-1

K. Lettau, H. Lettau, Experimental and micrometeorological field studies of dune migration, in *Exploring the World's Driest Climate*, ed. by H.H. Lettau, K. Lettau (Center for Climatic Research, Univ. Wisconsin, Madison, 1978)

G.A. Loosmore, J.R. Hunt, Dust resuspension without saltation. J. Geophys. Res. **105**(D16), 20,663–20,671 (2000). doi:10.1029/2000JD900271

R.D. Lorenz, L.D. Neakrase, J.D. Anderson, In-situ measurement of dust devil activity at La Jornada experimental range, New Mexico, USA. Aeolian Res. (2015). doi:10.1016/j.aeolia.2015.01.012

R.D. Lorenz, D. Reiss, Solar panel clearing events, dust devil tracks and in-situ vortex detections on Mars. Icarus **248**, 162–164 (2015). doi:10.1016/j.icarus.2014.10.034

H. Lu, Y. Shao, A new model for dust emission by saltation bombardment. J. Geophys. Res. **104**, 16,827–16,842 (1999)

T. Macpherson, W.G. Nickling, J.A. Gillies, V. Etyemezian, Dust emissions from undisturbed and disturbed supply-limited desert surfaces. J. Geophys. Res., Earth Surf. **113**, F02S04 (2008). doi:10.1029/2007JF000800

N. Mahowald, S. Albani, J.F. Kok, S. Engelstaeder, R. Scanza, D.S. Ward, M.G. Flanner, The size distribution of desert dust aerosols and its impact on the Earth system. Aeolian Res. **15**, 53–71 (2014)

B. Marticorena, G. Bergametti, Modeling the atmospheric dust cycle: 1. Design of a soil-derived dust emission scheme. J. Geophys. Res. **100**(D8), 16,415–16,430 (1995)

B. Marticorena, G. Bergametti, D.A. Gillette, J. Belnap, Factors controlling threshold friction velocity in semi-arid and arid areas of the United States. J. Geophys. Res. **102**, 23277–23287 (1997)

R.L. Martin, T.E. Barchyn, C.H. Hugenholtz, D.J. Jerolmack, Timescale dependence of aeolian sand flux observations under atmospheric turbulence. J. Geophys. Res., Atmos. **118**, 9078–9092 (2013). doi:10.1002/jgrd.50687

R.L. Martin, J.F. Kok, Linear scaling of wind-driven sand flux with shear stress. (2016), in review

C. McKenna Neuman, Effects of temperature and humidity upon the entrainment of sedimentary particles by wind. Bound.-Layer Meteorol. **108**, 61–89 (2003)

C. McKenna Neuman, W.G. Nickling, A theoretical and wind tunnel investigation of the effect of capillary water on the entrainment of sediment by wind. Can. J. Soil Sci. **69**, 79–96 (1989)

R.J. McKim, The dust storms of Mars. J. Br. Astron. Assoc. **106**, 185–200 (1996)

J.P. Merrison, Sand transport, erosion and granular electrification. Aeolian Res. **4**, 1–16 (2012)

J.P. Merrison, H.P. Gunnlaugsson, P. Nørnberg, A.E. Jensen, K.R. Rasmussen, Determination of the wind induced detachment threshold for granular material on Mars using wind tunnel simulations. Icarus **191**, 568–580 (2007)

S.M. Metzger, M.R. Balme, M.C. Towner, B.J. Bos, T.J. Ringrose, M.R. Patel, In situ measurements of particle load and transport in dust devils. Icarus **214**, 766–772 (2011). doi:10.1016/j.icarus.2011.03.013

S.M. Metzger, J.R. Carr, J.R. Johnson, T.J. Parker, M.T. Lemmon, Dust devil vortices seen by the Mars Pathfinder camera. Geophys. Res. Lett. **26**(18), 2781–2784 (1999). doi:10.1029/1999GL008341

L. Montabone, F. Forget, E. Millour, R.J. Wilson, S.R. Lewis, B. Cantor, D. Kass, A. Kleinböhl, M.T. Lemmon, M.D. Smith, M.J. Wolff, Eight-year climatology of dust optical depth on Mars. Icarus **251**, 65–95 (2015). doi:10.1016/j.icarus.2014.12.034

D.P. Mulholland, A. Spiga, C. Listowski, P.L. Read, An assessment of the impact of local processes on dust lifting in martian climate models. Icarus **252**, 212–227 (2015). doi:10.1016/j.icarus.2015.01.017

E.P. Muntz, Y. Sone, K. Aoki, S. Vargo, M. Young, J. Vac. Sci. Technol. A **20**, 214 (2002)

J.R. Murphy, S. Nelli, Mars Pathfinder convective vortices: frequency of occurrence. Geophys. Res. Lett. **29**(23), 2103 (2002). doi:10.1029/2002GL015214

S.L. Namikas, B.O. Bauer, D.J. Sherman, Influence of averaging interval on shear velocity estimates for aeolian transport modeling. Geomorphology **53**, 235–246 (2003). doi:10.1016/S0169-555X(02)00314-8

L.D.V. Neakrase, *A Laboratory Study of Sediment Flux Within Dust Devils on Earth and Mars (Doctoral Dissertation)* (Arizona State University, Tempe, Arizona, USA, 2009)

L.D.V. Neakrase, R. Greeley, Dust devil sediment flux on Earth and Mars: laboratory simulations. Icarus **206**, 306–318 (2010a). doi:10.1016/j.icarus.2009.08.028

L.D.V. Neakrase, R. Greeley, Dust devils in the laboratory: effects of surface roughness on vortex dynamics. J. Geophys. Res. **115**, E05003 (2010b). doi:10.1029/2009JE003465

L.D.V. Neakrase, R. Greeley, J.D. Iversen, M.R. Balme, E.E. Eddlemon, Dust flux within dust devils: preliminary laboratory simulations. Geophys. Res. Lett. **33**, L19S09 (2006). doi:10.1029/2006GL026810

L.D.V. Neakrase, J. McHone, P.L. Whelley, R. Greeley, Terrestrial analogs to Mars: East-central Saharan dust devil tracks (Abstract), in *Proceedings of the 43rd Lunar and Planetary Science Conference*, The Woodlands, Texas (2012)

C.E. Newman, S.R. Lewis, P.L. Read, F. Forget, Modeling the Martian dust cycle 1. Representations of dust transport processes. J. Geophys. Res. **107**(E12), 5123 (2002). doi:10.1029/2002JE001910

C.E. Newman, M.I. Richardson, The impact of surface dust source exhaustion on the martian dust cycle, dust storms and interannual variability, as simulated by the MarsWRF General Circulation Model. Icarus **257**, 47–87 (2015). doi:10.1016/j.icarus.2015.03.030

W.G. Nickling, Grain-size characteristics of sediment transported during dust storms. J. Sediment. Petrol. **53**(3), 1011–1024 (1983)

W.G. Nickling, J.A. Gillies, Dust emission and transport in Mali, West Africa. Sedimentology **40**, 859–868 (1993)

W.G. Nickling, G.H. McTainsh, J.F. Leys, Dust emissions from the Channel Country of western Queensland, Australia. Z. Geomorphol., Suppl.bd **116**, 1–17 (1999)

A.M.C. Oke, N.J. Tapper, D. Dunkerley, Willy-willies in the Australian landscape: the role of key meteorological variables and surface conditions in defining frequency and spatial characteristics. J. Arid Environ. **71**, 201–215 (2007). doi:10.1016/j.jaridenv.2007.03.008

R.P. Owen, Saltation of uniform grains in air. J. Fluid Mech. **20**, 225–242 (1964)

T. Pähtz, J.F. Kok, H.J. Herrmann, The apparent roughness of a sand surface blown by wind from an analytical model of saltation. New J. Phys. **14**, 043035 (2012)

J.L. Rajot, S.C. Alfaro, L. Gomes, A. Gaudichet, Soil crusting on sandy soils and its influence on wind erosion. Catena **53**, 1–16 (2003). doi:10.1016/S0341-8162(02)00201-1

K.R. Rasmussen, M. Sørensen, Vertical variation of particle speed and flux density in aeolian saltation: measurement and modeling. J. Geophys. Res. **113**(F2), 2156–2202 (2008). doi:10.1029/2007JF000774

K.R. Rasmussen, A. Valance, J. Merrison, Laboratory studies of aeolian sediment transport processes on planetary surfaces. Geomorphology **244**, 74–94 (2015). doi:10.1016/j.geomorph.2015.03.041

M. Raupach, Drag and drag partition on rough surfaces. Bound.-Layer Meteorol. **60**(4), 375–395 (1992). doi:10.1007/BF00155203

M. Raupach, J. Finnigan, Y. Brunei, Coherent eddies and turbulence in vegetation canopies: the mixing-layer analogy. Bound.-Layer Meteorol. **78**(3–4), 351–382 (1996). doi:10.1007/BF00120941

M.R. Raupach, D.A. Gilette, J.F. Leys, The effect of roughness elements on wind erosion threshold. J. Geophys. Res. **98**(D2), 3023–3029 (1993). doi:10.1029/92JD01922

M.R. Raupach, H. Lu, Representation of land-surface processes in aeolian transport models. Environ. Model. Softw. **19**(2), 93–112 (2004). Modelling of Wind Erosion and Aeolian Processes. doi:10.1016/S1364-8152(03)00113-0

M.W. Reeks, D. Hall, Kinetic models for particle resuspension in turbulent flows: theory and measurement. J. Aerosol Sci. **32**, 1–31 (2001)

D. Reiss, J. Raack, A.P. Rossi, G. Di Achille, H. Hiesinger, First in-situ analysis of dust devil tracks on Earth and their comparison with tracks on Mars. Geophys. Res. Lett. **37**, L14203 (2010). doi:10.1029/2010GL044016

N.O. Renno, V.J. Abreu, J. Koch, P.H. Smith, o.K. Hartogensis, H.A.R. De Bruin, D. Burose, G.T. Delory, W.M. Farrell, C.J. Watts, J. Garatuza, M. Parker, A. Carswell, MATADOR 2002: a pilot field experiment on convective plumes and dust devils. J. Geophys. Res. **109**, E07001 (2004). doi:10.1029/2003JE002219

D.A. Ridley, C.L. Heald, J.R. Pierce, M.J. Evans, Toward resolution-independent dust emissions in global models: impacts on the seasonal and spatial distribution of dust. Geophys. Res. Lett. **40**, 2873–2877 (2013). doi:10.1002/grl.50409

T.J. Ringrose, M.C. Towner, J.C. Zarnecki, Vortices on Mars: a reanalysis of Viking Lander 2 meterological data, sols 1–60. Icarus **163**, 78–87 (2003). doi:10.1016/S0019-1035(03)00073-3

A. Rondeau, J. Merrison, J.J. Iversen, S. Peillona, J.-C. Sabroux, P. Lemaitre, F. Gensdarmes, E. Chassefièrec, First experimental results of particle re-suspension in a low pressurewind tunnel applied to the issue of dust in fusion reactors. Fusion Eng. Des. (2015). doi:10.1016/j.fusengdes.2014.12.038

J.A. Roney, B.R. White, Definition and measurement of dust aeolian thresholds. J. Geophys. Res., Earth Surf. **109**, F01013 (2004). doi:10.1029/2003JF000061

W.A.D. Rudge, Atmospheric electrification during South African dust storms. Nature **91**, 31 (1913)

J.A. Ryan, R.D. Lucich, Possible dust devils, vortices on Mars. J. Geophys. Res. **88**, 11,005–11,011 (1983)

H. Schlichting, Dusty ice clouds over Alaska. Ing.-Arch. **7**, 1–34 (1936). English translation, NACA Technical Memorandum, No. 823-1936

D.S. Schmidt, R.A. Schmidt, J.D. Dent, Electrostatic force on saltating sand. J. Geophys. Res. **103**(D8), 8997–9001 (1998)

Y. Shao, A model for mineral dust emission. J. Geophys. Res. **106**(D17), 20,239–20,254 (2001)

Y. Shao, Simplification of a dust emission scheme and comparison with data. J. Geophys. Res. **109**, D10202 (2004). doi:10.1029/2003JD004372

Y. Shao, *Physics and Modelling of Wind Erosion*, 2nd edn. (Springer, Berlin, 2008). 452 p.

Y. Shao, A.H. Fink, M. Klose, Numerical simulation of a continental-scale Saharan dust event. J. Geophys. Res. **115**, D13205 (2010). doi:10.1029/2009JD012678

Y. Shao, M. Ishizuka, M. Mikami, J.F. Leys, Parameterization of size-resolved dust emission and validation with measurements. J. Geophys. Res. **116**, D08203 (2011). doi:10.1029/2010JD014527

Y. Shao, M. Klose, A note on the stochastic nature of particle cohesive force and implications to threshold friction velocity for aerodynamic dust entrainment. Aeolian Res. **22**, 123–125 (2016). doi:10.1016/j.aeolia.2016.08.004

Y. Shao, H. Lu, A simple expression for wind erosion threshold friction velocity. J. Geophys. Res. **105**, 22,437–22,443 (2000)

Y. Shao, M.R. Raupach, P.A. Findlater, The effect of saltation bombardment on the entrainment of dust by wind. J. Geophys. Res. **98**, 12,719–12,726 (1993)

Y. Shao, M.R. Raupach, J.F. Leys, A model for predicting aeolian sand drift and dust entrainment on scales from paddock to region. Aust. J. Soil Res. **34**, 309–342 (1996)

Y. Shao, Y. Yang, A theory for drag partition over rough surfaces. J. Geophys. Res. **113**, F02S05 (2008). doi:10.1029/2007JF000791

P.C. Sinclair, General characteristics of dust devils. J. Appl. Meteorol. **8**, 32–45 (1969). doi:10.1175/1520-0450(1969)008<0032:GCODD>2.0.CO;2

A.J. Smits, B.J. McKeon, I. Marusic, High-Reynolds number wall turbulence. Annu. Rev. Fluid Mech. **43**, 353–375 (2011)

J.T. Snow, T.M. McClelland, Dust devils at white sands missile range, New Mexico: 1. Temporal and spatial distributions. J. Geophys. Res., Atmos. **95**(D9), 13,707–13,721 (1990). doi:10.1029/JD095iD09p13707

M. Sow, S.C. Alfaro, J.L. Rajot, B. Marticorena, Size resolved dust emission fluxes measured in Niger during 3 dust storms of the AMMA experiment. Atmos. Chem. Phys. **9**(12), 3881–3891 (2009)

J.E. Stout, T.M. Zobeck, Intermittent saltation. Sedimentology **44**, 959–970 (1997)

C.D. Stow, Dust and sand storm electrification. Weather **24**, 134–137 (1969)

R.B. Stull, *An Introduction to Boundary Layer Meteorology* (Kluwer Academic Publishers, Norwell, 1988). 666 p.

R. Sullivan, R. Arvidson, J.F. Bell III., R. Gellert, M. Golombek, R. Greeley, K. Herkenhoff, J. Johnson, S. Thompson, P. Whelley, J. Wray, Wind-driven particle mobility on Mars: insights from Mars exploration rover observations at 'El Dorado' and surroundings at Gusev Crater. J. Geophys. Res. **113**, E06S07 (2008). doi:10.1029/2008JE003101

R. Sullivan, D. Banfield, J.F. Bell III., W. Calvin, D. Fike, M. Golombek, R. Greeley, J. Grotzinger, K. Herkenhoff, D. Jerolmack, M. Malin, D. Ming, L.A. Soderblom, S.W. Squyres, S. Thompson, W.A. Watters, C.M. Weitz, A. Yen, Aeolian processes at the Mars exploration rover Meridiani Planum landing site. Nature **436**, 58–61 (2005)

M.C. Towner, M.R. Patel, T.J. Ringrose, J.C. Zarnecki, D. Pullan, M.R. Sims, S. Haapanala, A.-M. Harri, J. Polkko, C.F. Wilson, R.C. Quinn, F.J. Grunthaner, M.H. Hecht, J.R.C. Garry, The Beagle 2 environmental sensors: science goals and instrument description. Planet. Space Sci. **52**(13), 1141–1156 (2004). doi:10.1016/j.pss.2004.07.015

J.E. Ungar, P.K. Haff, Steady-state saltation in air. Sedimentology **34**, 289–299 (1987)

S. Wagner, An Assessment of Dust Effects on Planetary Surface Systems to Support Exploration Requirements, *NASA Technical Report*, CTSD-AIM-0029, JSC-62198, 20080047665, p. 23 (2004)

Z.-T. Wang, A theoretical note on aerodynamic lifting in dust devils. Icarus **265**, 79–83 (2016). doi:10.1016/j.icarus.2015.10.016

B.R. White, Soil transport by winds on Mars. J. Geophys. Res. **84**(B9), 4643–4651 (1979). doi:10.1029/JB084iB09p04643

J. Wieringa, Representative roughness parameters for homogeneous terrains. Bound.-Layer Meteorol. **63**, 323–363 (1993)

E. Williams, N. Nathou, E. Hicks, C. Pontikis, B. Russel, M. Miller, M.J. Bartholomew, The electrification of dust-lofting gust fronts ('haboobs') in the Sahel. Atmos. Res. **91**, 292–298 (2009)

G. Wurm, O. Krauss, Dust eruptions by photophoresis and solid state greenhouse effects. Phys. Rev. Lett. **96**, 134301 (2006)

G. Wurm, J. Teiser, D. Reiss, Greenhouse and thermophoretic effects in dust layers: the missing link for lifting dust on Mars. Geophys. Res. Lett. **35**, L10201 (2008)

H. Yizhaq, J.F. Kok, I. Katra, Basaltic sand ripples at Eagle crater as indirect evidence for the hysteresis effect in martian saltation. Icarus **230**, 143–150 (2014)

X.J. Zheng, N. Huang, Y.H. Zhou, Laboratory measurement of electrification of wind-blown sands and simulation of its effect on sand saltation movement. J. Geophys. Res. **108**, 4322 (2003)

A.D. Zimon, *Adhesion of Dust and Powder* (Consultants Bureau, New York, 1982). 438 p.

G. Ziskind, M. Fichman, C. Gutfinger, Resuspension of particulates from surfaces to turbulent flows—review and analysis. J. Aerosol Sci. **26**, 613–644 (1995)

R.W. Zurek, L.J. Martin, Interannual variability of planet-encircling dust activity on Mars. J. Geophys. Res. **98**, 3247–3259 (1993)

Space Sci Rev (2016) 203:377–426
DOI 10.1007/s11214-016-0261-4

Dust Devil Sediment Transport: From Lab to Field to Global Impact

Martina Klose[1,2] · Bradley C. Jemmett-Smith[3] · Henrik Kahanpää[4] · Melinda Kahre[5] · Peter Knippertz[6] · Mark T. Lemmon[7] · Stephen R. Lewis[8] · Ralph D. Lorenz[9] · Lynn D.V. Neakrase[10] · Claire Newman[11] · Manish R. Patel[8] · Dennis Reiss[12] · Aymeric Spiga[13] · Patrick L. Whelley[14]

Received: 26 October 2015 / Accepted: 23 May 2016 / Published online: 4 July 2016

Abstract The impact of dust aerosols on the climate and environment of Earth and Mars is complex and forms a major area of research. A difficulty arises in estimating the contribution of small-scale dust devils to the total dust aerosol. This difficulty is due to uncertainties in the amount of dust lifted by individual dust devils, the frequency of dust devil occurrence, and the lack of statistical generality of individual experiments and observations. In this paper, we review results of observational, laboratory, and modeling studies and provide an overview of dust devil dust transport on various spatio-temporal scales as obtained

✉ M. Klose
mklose@nmsu.edu

B.C. Jemmett-Smith
b.jemmett-smith@leeds.ac.uk

H. Kahanpää
henrik.kahanpaa@fmi.fi

M. Kahre
melinda.a.kahre@nasa.gov

P. Knippertz
peter.knippertz@kit.edu

M.T. Lemmon
lemmon@tamu.edu

S.R. Lewis
stephen.lewis@open.ac.uk

R.D. Lorenz
Ralph.Lorenz@jhuapl.edu

L.D.V. Neakrase
lneakras@nmsu.edu

C. Newman
claire@aeolisresearch.com

M.R. Patel
manish.patel@open.ac.uk

D. Reiss
dennis.reiss@uni-muenster.de

with the different research approaches. Methods used for the investigation of dust devils on Earth and Mars vary. For example, while the use of imagery for the investigation of dust devil occurrence frequency is common practice for Mars, this is less so the case for Earth. Modeling approaches for Earth and Mars are similar in that they are based on the same underlying theory, but they are applied in different ways. Insights into the benefits and limitations of each approach suggest potential future research focuses, which can further reduce the uncertainty associated with dust devil dust entrainment. The potential impacts of dust devils on the climates of Earth and Mars are discussed on the basis of the presented research results.

Keywords Dust devils · Dust emission · Lab experiments · Field measurements · Modeling · Dust environmental impact · Sediment transport · Earth · Mars · Planetary atmospheres

1 Introduction

Dust devils are a common occurrence on Mars and in semi-arid and arid regions on Earth when the surface is heated by insolation and convective turbulence is sufficiently developed (Chap. 5 [Rafkin et al. 2016]). The appearance of dust devils can be spectacular, but they are typically of small spatial extent (diameters of the order of magnitude $\sim 10^0 - 10^1$ m on Earth, $\sim 10^1 - 10^2$ m on Mars) and short duration ($\sim 10^0 - 10^1$ min), although exceptional larger and longer-lived cases have been reported (Balme and Greeley 2006; Greeley et al. 2010). The exact occurrence time and location of dust devils is hard to predict and makes

A. Spiga
spiga@lmd.jussieu.fr

P.L. Whelley
patrick.l.whelley@nasa.gov

[1] Institute for Geophysics and Meteorology, University of Cologne, Cologne, Germany

[2] Present address: USDA-ARS Jornada Experimental Range, Las Cruces, NM, USA

[3] Institute for Climate and Atmospheric Science, School of Earth and Environment, University of Leeds, Leeds, West Yorkshire, UK

[4] Finnish Meteorological Institute, Helsinki, Finland

[5] NASA Ames Research Center, Moffett Field, CA, USA

[6] Institute of Meteorology and Climate Research, Karlsruhe Institute of Technology, Karlsruhe, Germany

[7] Department of Atmospheric Sciences, Texas A&M University, College Station, TX, USA

[8] Department of Physical Sciences, The Open University, Milton Keynes, UK

[9] Johns Hopkins University Applied Physics Lab, Laurel, MD, USA

[10] Department of Astronomy, New Mexico State University, Las Cruces, NM, USA

[11] Aeolis Research, Suite 205, 600 North Rosemead Boulevard, Pasadena, CA 91107, USA

[12] Institut für Planetologie, Westfälische Wilhelms-Universität, Münster, Germany

[13] Laboratoire de Météorologie Dynamique, Université Pierre et Marie Curie, Paris, France

[14] NASA Goddard Space Flight Center, Greenbelt, MD, USA

in situ observations of dust devils a challenge. However, long-term monitoring by stationary instruments, as well as portable instruments that are transported into active dust devils, have provided valuable insights into dust devil characteristics on Earth (e.g. Sinclair 1969; Metzger et al. 2011; Lorenz et al. 2015). On Mars, cameras and meteorological in situ/remote sensing instrumentation on landers/orbiters have substantially advanced knowledge about Martian dust devils, thereby inspiring further research (e.g. Ryan and Lucich 1983; Thomas and Gierasch 1985; Schofield et al. 1997; Metzger et al. 1999; Cantor et al. 2006; Greeley et al. 2010; Ellehoj et al. 2010; Reiss et al. 2011; Moores et al. 2015). In addition, laboratory experiments and numerical modeling are powerful techniques to investigate dust devils (e.g. Neakrase et al. 2006; Kanak 2005; Gu et al. 2008; Ohno and Takemi 2010; Raasch and Franke 2011; Klose and Shao 2016). In all approaches, i.e. field, laboratory, and modeling, the estimate of total dust transport by dust devils and consequently the assessment of their environmental impact remain difficult due to the variety of dust devil sizes and intensities. Additionally, the dust devil dust load is ultimately dependent on the conditions of the land-surface in its path, which determines the abundance of soil dust particles available for entrainment.

The major impact of dust aerosols on the terrestrial and martian climates is through aerosol-radiation and aerosol-cloud interactions (e.g. Michelangeli et al. 1993; Rosenfeld et al. 2001; Bangert et al. 2012). Airborne dust particles scatter and absorb shortwave radiation and absorb and re-emit longwave radiation, leading to thermodynamic responses in the atmosphere (e.g. Sokolik and Toon 1996; Miller et al. 2014). Dust particles can also act as efficient ice nuclei both on Earth and Mars, thereby impacting cloud formation and cloud-climate feedback (e.g. Isono 1955; DeMott et al. 2003; Boucher et al. 2013). In the terrestrial environment, dust transport has further effects on ecosystems due to the transport of nutrients, minerals, and carbon, potentially acting as fertilizers (e.g. Bristow et al. 2010) or through deposition of pathogens or chemical contaminants (Shinn et al. 2000; Garrison et al. 2003). Human and animal health is influenced through inhalation of dust particles penetrating into the lung or carrying bacteria, fungi, pathogens, or allergens (Kellogg and Griffin 2006; Derbyshire 2007). Not least, reduced visibilities during dust events can cause severe traffic accidents.

Although the impact of dust aerosols on the global climate system has become a research focus in recent decades, the particular role and significance of dust devils compared to other meteorological dust injection processes is not clear. In this paper, we review estimates of dust transport in dust devils from in situ, remote sensing, laboratory, and modeling studies (Sect. 2). We summarize the occurrence frequency of dust devils on different spatial and temporal scales (Sect. 3) to provide a comprehensive picture of dust transported by dust devils on Earth and Mars (Sect. 4). The results are used to assess the impact of dust devils on climate and environment on local, regional, and global scales (Sect. 5). Findings for Earth and Mars are compared, indicating potential future directions for dust devil studies (Sect. 6).

2 Dust Transport in Dust Devils on Earth and Mars

The mechanisms by which dust is entrained into vortices to form dust devils are still not fully understood. In general, three dust emission mechanisms are commonly referred to in aeolian research: saltation bombardment, aggregate disintegration, and aerodynamic entrainment (Shao 2008). Saltation bombardment and aggregate disintegration are considered to be the most efficient dust emission mechanisms. Dust emission by both mechanisms is generated by saltation, the hopping motion of sand-sized particles or particle aggregates,

Table 1 Dust transport in individual dust devils

Dust flux [$kg\,m^{-2}\,s^{-1}$]			Method	Reference
Earth				
2.2×10^{-5}	–	3×10^{-3}	in situ	Gillette and Sinclair (1990)
1.4×10^{-6}	–	1.6×10^{-4}	in situ (PM25)	Gillette and Sinclair (1990)
0.6×10^{-3}	–	4.4×10^{-3}	in situ	Metzger (1999)
1×10^{-4}	–	1×10^{-3}	in situ	Renno et al. (2004)
4.6×10^{-6}	–	100	laboratory	Neakrase and Greeley (2010b)
4×10^{-7}	–	1.1×10^{-4}	in situ (PM10)	Metzger et al. (2011)
1×10^{-9}	–	4×10^{-7}	large-eddy simulation	Klose and Shao (2016)
Mars				
5×10^{-4}			lander images	Metzger et al. (1999)
4×10^{-9}	–	1.6×10^{-4}	lander images	Greeley et al. (2010)
2×10^{-5}	–	0.5	laboratory	Neakrase and Greeley (2010b)
3.8×10^{-7}	–	1.2×10^{-3}	orbital images	Reiss et al. (2014)

causing localized impacts strong enough to inject dust particles into the atmosphere. Saltation is initiated as soon as the mean surface drag exceeds a threshold which depends on the surface conditions (Bagnold 1941; Greeley and Iversen 1985; Shao and Lu 2000). While this threshold is easily exceeded during dust storms, it remains controversial whether or not the drag in dust devils is sufficiently strong to initiate saltation. The aerodynamic entrainment of dust particles, i.e. dust uplift due directly to strong aerodynamic drag, has been neglected for a long time due to the on-average stronger inter-particle cohesive forces acting on dust particles compared to those acting on sand-sized particles. However, measurements show that inter-particle cohesion can vary over orders of magnitude even for particles of the same size, thereby allowing for stronger aerodynamic dust emission than previously thought (Zimon 1982; Klose et al. 2014; Shao and Klose 2016).

In addition to these three dust emission mechanisms, further mechanisms apply in the special case of dust devils. The small-scale vortices exhibit a substantial pressure drop in their center. The suction effect associated with this pressure drop may facilitate dust entrainment (Balme and Hagermann 2006). In a low-pressure environment such as on Mars, dust particles can also experience thermal lifting (Wurm et al. 2008; de Beule et al. 2014; Küpper and Wurm 2015). In the presence of a temperature gradient, thermal lifting is generated by thermal creep, a gas flow in the pores of a dust bed arising at pressures of few millibars. During an ongoing emission event, particle collisions can result in the development of an electrostatic field, which can reduce the entrainment threshold and enhance further emission (Kok and Renno 2006). A more detailed discussion of dust lifting mechanisms in dust devils is given in Chap. 10 (Neakrase et al. 2016).

This section describes different approaches (laboratory based, observational, and theoretical) to estimate the dust amount entrained by individual dust devils. Table 1 summarizes the values obtained by a range of such methods in several different studies for Earth and Mars.

2.1 Laboratory Experiments on Dust Transport in Dust Devils

A series of laboratory experiments were conducted between 2000 and 2009 at Arizona State University to try to understand fundamental controls on dust devil sediment lifting poten-

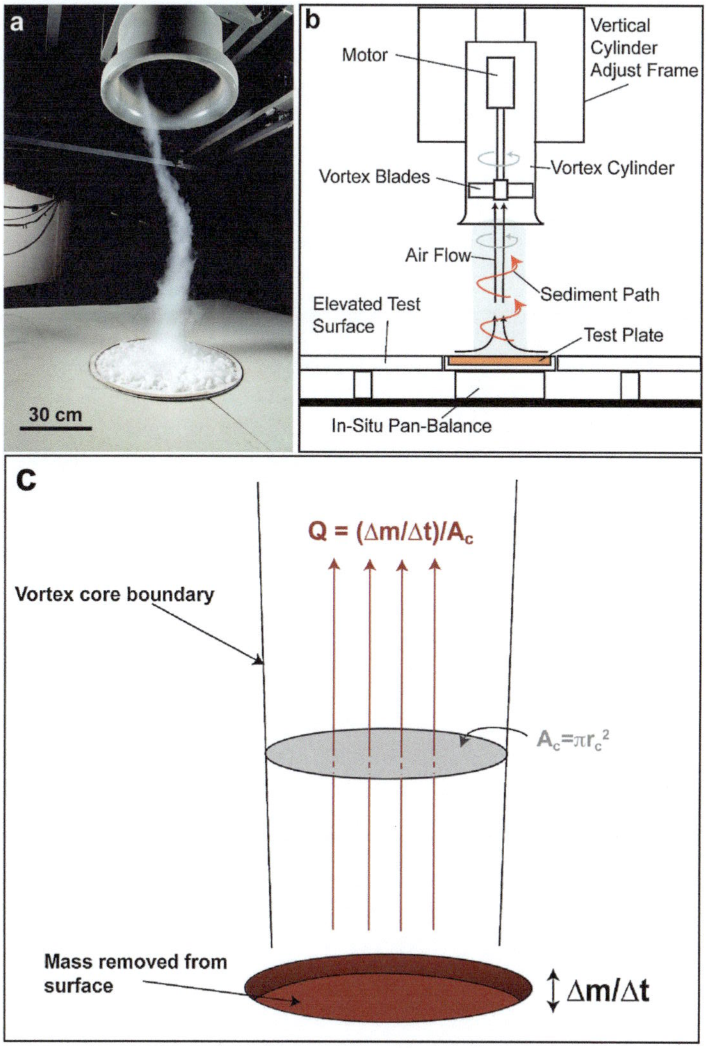

Fig. 1 Summary of Arizona State University Vortex Generator (ASUVG) sediment flux results. (a) Photo of the ASUVG; (b) Schematic of the ASUVG sediment flux experiments; (c) Cartoon of the mass calculation for vortex sediment flux (Reprinted from Neakrase and Greeley 2010b, with permission from Elsevier)

tial. The laboratory setup utilized the Arizona State University Vortex Generator (ASUVG), which consists of a $2.4 \times 2.4\ m^2$ translatable table (both in the vertical and horizontal directions), below a cylinder assembly that houses the motorized fan that can be moved to different heights above the test surface controlling the diameter of the vortex on the test surface (Fig. 1a) (Greeley et al. 2003; Neakrase et al. 2006; Neakrase and Greeley 2010a,b). The earliest experiments examined vortex threshold velocities for a range of particle sizes and densities both at terrestrial ambient and Mars-analog pressures (Greeley et al. 2003). Mars-analog atmospheric pressures ($\sim$10 hPa, though with terrestrial ambient air composition) were attained using the Planetary Aeolian Laboratory at NASA Ames Research Center.

The initial ASUVG experiments showed that for dust-sized particles (diameters <63 μm) the vortex threshold occurred at a lower value than the entrainment threshold in comparable boundary layer wind tunnel studies. This deviation from the standard boundary layer threshold suggests that sediment mobility under vortex conditions is subject to an additional lift component, which was dubbed the 'Δp-effect' by Greeley et al. (2003). This result implied that dust devils could be particularly efficient at lifting smaller particles (see also Chap. 10 [Neakrase et al. 2010]).

Building on the results of Greeley et al. (2003), sediment flux experiments were designed to investigate how dust devils lift sand and dust at terrestrial ambient and Mars-analog conditions using the ASUVG (Neakrase et al. 2006; Neakrase and Greeley 2010b). Using the results of the threshold experiments, the sediment flux study aimed to investigate the mass loss over time for conditions above threshold (Neakrase et al. 2006; Neakrase and Greeley 2010b). The sediment flux experiments consisted of an in situ pan balance beneath a 5 mm-deep plate flush with the elevated test bed (Fig. 1b). The plate was filled with different sediments and the ASUVG was set for different sized vortices (vortex diameter) and tangential velocities above the determined threshold. As the ASUVG was run for a set amount of time, mass measurements were made before and after each run to determine the change in mass per unit time ($\Delta m/\Delta t$). Knowing the vortex parameters for each setup as determined by the height above the test surface and speed of the fan-blade assembly, the effective cross-sectional area of the vortex could be determined. Sediment flux was then determined to be

$$Q = \frac{\Delta m}{\Delta t} \frac{1}{A_c} \tag{1}$$

where Q is the sediment flux through the area, A_c, of the vortex core (Fig. 1c). Results from these experiments yielded empirical relationships for dust devil sediment flux as a function of a parameterized lift ratio ($\Delta p/u_\theta$), represented by the average magnitude of the core pressure drop, Δp, and the maximum tangential velocity, u_θ, as determined by curve-fitting the initial pressure data as a function of radius (Neakrase and Greeley 2010b). The sediment flux relationship as provided by Neakrase and Greeley (2010b) was

$$Q \propto k \left(\frac{\Delta p}{u_\theta} \right)^4 \tag{2}$$

with k being a parameter and was shown to be independent of ambient atmospheric pressure under this parameterization. This result averaged all sediment used in their experiments, which included both sand (particle diameter $D_p >$ 63 μm) and dust ($D_p \leq 63$ μm). Average laboratory flux values were between 4.0×10^{-6} and 100 $\mathrm{kg\,m^{-2}\,s^{-1}}$ for terrestrial ambient conditions and 2.0×10^{-5} to 0.5 $\mathrm{kg\,m^{-2}\,s^{-1}}$ for martian analog conditions (Fig. 2). Neakrase and Greeley (2010b) also investigated whether the individual curves for dust flux were lower than for sand flux. The results confirmed the initial threshold results by Greeley et al. (2003) in that dust particles are more effectively lifted in the presence of small Δp than without, while more intensive Δp have a stronger effect on sand-sized particles. This can likely be attributed to the larger tangential velocities in the latter case.

Further experiments in sediment flux with the ASUVG investigated how vortex flow is affected by surface roughness elements. Neakrase and Greeley (2010a) used large roughness elements (i.e. sizes and spacing on the order of the size of the vortex core). Roughness is determined by the size and spacing of the nonerodible roughness elements and can be characterized by the roughness density, λ, also known as frontal-area index, defined by Lettau

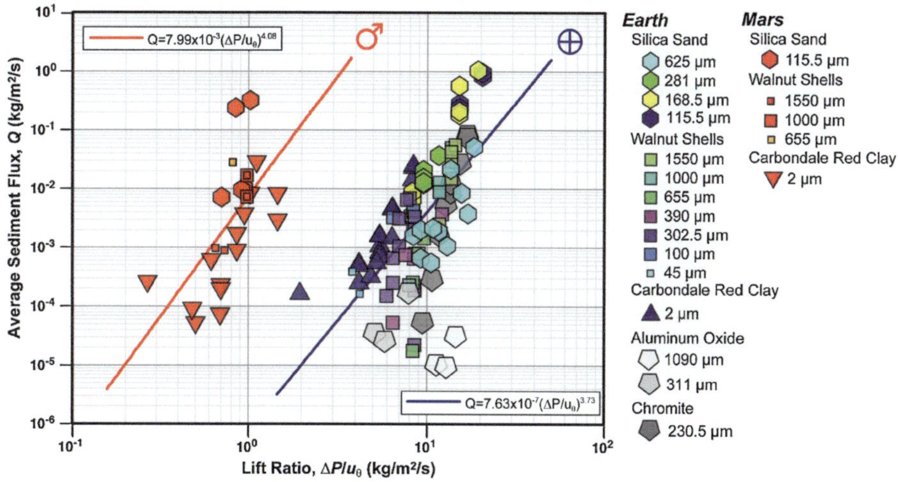

Fig. 2 Average sediment flux vs. lift ratio obtained for terrestrial ambient and Mars analog conditions for the entire tested range of sediments (Reprinted from Neakrase and Greeley 2010b, with permission from Elsevier)

(1969) as

$$\lambda = \frac{nbh}{S} \tag{3}$$

where n is the number of nonerodible roughness elements, b is the element width, h is the height of the element and S is the total surface area over which the elements exist. Neakrase and Greeley (2010a) showed that for small roughness at this scale ($\lambda \approx 0.01$) there seemed to be an optimal roughness, capable of enhancing vortex flow and, as a result, also the sediment flux. Neakrase and Greeley (2010a) suggested that this was a capability of the vortex of a given size and speed to adjust to the surface by conservation of angular momentum to fit between the roughness elements (enhancing flow) or expanding to incorporate the roughness elements (reducing flow). Both cases may be pertinent to the natural world where dust devils encounter large boulders or vegetation. Depending on the size and spacing of the roughness elements, dust devil sediment flux could be increased or decreased as a response to the vortex gaining or losing vorticity.

2.2 Field Observations of Dust Devil Dust Transport—In Situ Measurements and Remote Sensing

Direct measurements of sediment fluxes within dust devils are difficult to obtain, hence most fluxes are calculated using measurements of dust concentrations and assumed or simultaneously obtained measurements of the vertical velocity within the dust devil. Sediment or dust fluxes are then calculated as vertical wind speed multiplied by particle concentration. It must be noted, however, that the so obtained fluxes are likely different to surface dust emission fluxes (compare Sect. 4.1). Gillette and Sinclair (1990) estimated particle fluxes based on in situ aircraft measurements of particle concentration and vertical wind speed in the updrafts of several dust devils at altitudes of 142 and 330 m. Although the corresponding method paper was never published, Gillette and Sinclair (1990) summarized mean dust flux estimates for different dust devil sizes ranging from (a) 2×10^{-5} kg m^{-2} s^{-1} for dust devils <3 m in diameter to (b) 3×10^{-3} kg m^{-2} s^{-1} for dust devils >30 m in diameter.

Table 2 Calculated dust devil parameters including dust opacity from Mason et al. (2014)

Encounter	r [m]	z_{dd} [m]	τ_a at 450 nm	Δp [hPa]
E1	11	>130	0.8	0.30
E2	4	>100	0.6	0.42
E3	5	>18	0.2	0.29
E4	7	>200	2.3–2.7	–
E5	8	>200	1.6–2.0	–

However, the contribution of finer particles (<25 μm diameter) to these total dust flux estimates ranged only from 1.4×10^{-6} to 1.6×10^{-4} kg m^{-2} s^{-1} for (a) and (b), respectively. For dust devils of all sizes, Gillette and Sinclair (1990) obtain a total dust flux (<25 μm diameter) of 2.8×10^{-4} kg m^{-2} s^{-1} for their test area in the southwestern US. Renno et al. (2004) estimated a particle flux of 1×10^{-3} kg m^{-2} s^{-1} based on LIDAR measurements of particle concentration in one dust devil at 100 m height and using the peak vertical wind speed of 10 m s^{-1} measured in the dust devil using a sonic anemometer. The most reliable terrestrial in situ measurements were made by Metzger (1999) and Metzger et al. (2011). Metzger (1999) measured sediment loads of total suspended particles (TSP) in dust devils and calculated sediment fluxes in the range of 0.6×10^{-3}–4.4×10^{-3} kg m^{-2} s^{-1} for five dust devils near the surface using simultaneously obtained measurements of vertical speeds within the dust devils. Near-surface measurements of twelve dust devils by Metzger et al. (2011) gave suspended particle loads of particulate matter <10 μm in diameter (PM10). The authors calculated maximum dust fluxes ranging from 4×10^{-7}–1.1×10^{-4} kg m^{-2} s^{-1} using again simultaneously obtained measurements of vertical speeds within the dust devils. The authors found that the mean flux in dust devils is about one third of the measured maximum flux.

Mason et al. (2014) performed a field study on dust devils in 2009 in the Southern Nevada Desert (Eldorado Valley, USA) in a closed playa basin, using a chase vehicle-encounter approach in which instruments were driven into existing dust devils. Results were presented for 5 well-defined dust devil encounters, E1–E5. It should be noted that the chase-encounter approach is a targeted approach to observation, and thus introduces a selection effect into the results, biased towards larger, more defined dust devils that are more easily tracked. By using optical instrumentation fixed to the top of the vehicle, and ensuring that the dust devil trajectory passes over the instrumentation, the dust loading of the dust devils was recorded. For the encounters E1–E5, the radius of the dust devil (r), dust devil height (z_{dd}), apparent dust optical depth (τ_a) and pressure drop (Δp) were defined, shown in Table 2. The physical dimensions of the dust devils were calculated from images containing known reference lengths. The high dust loading and size of encounters E4 and E5 led to an undefined vortex core, and therefore, a range of τ_a is specified.

Table 2 shows five encounters that were studied in detail for dust optical property retrieval, however a larger data set of encounters was recorded for investigating the relation between Δp and dust loading. Greeley et al. (2003) and Neakrase et al. (2006) suggested that vortices are more efficient at dust lifting than non-rotating boundary layer flows due to the pressure deficit in the vortex centers. This would indicate that encounters with more intense pressure drops may have higher dust loadings. The extinction coefficient k_{ext} (attenuation per unit length) was calculated for 23 dust devils using τ_a and r and plotted against Δp, shown in Fig. 3. There is no unique relationship between dust loading and Δp as the dust load depends not only on entrainment threshold, but also on the dust devil dimensions

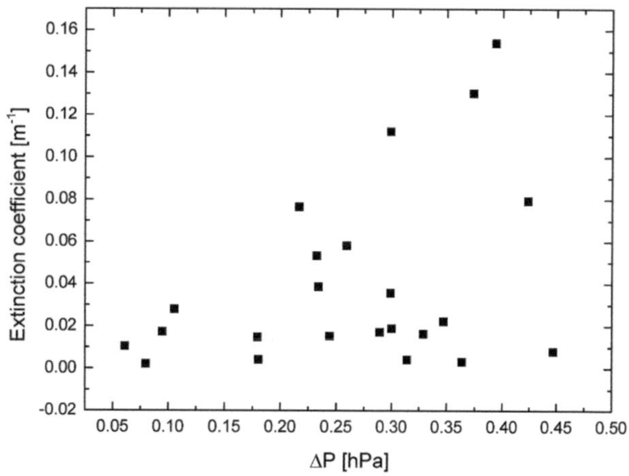

Fig. 3 Extinction coefficient [m^{-1}] (representing dust loading) as a function of Δp [hPa] for 23 dust devil encounters (modified from Mason et al. 2014)

and shape, tilt, and in a natural environment on dust supply. It is clear, however, that as Δp increases so does the maximum dust loading (and thus the range of dust loadings) that can be produced.

Lorenz and Jackson (2015) observed dust devil activity in summer 2013 at four spots in the same Eldorado field site as studied by Mason et al. (2014). Lorenz and Jackson (2015) used data loggers that recorded pressure and incident sunlight at a high cadence for over a month. Direct encounters and near-misses of boundary layer vortices are indicated by a sharp drop in pressure: when these vortices are dust-laden, there may also be a drop in incident sunlight. Not all dust devil encounters lead to such a drop – if the dust devil passes in the anti-sun direction (e.g. north or east during the afternoon), the dust column may not block the line-of-sight to the sun. Nonetheless, this measurement approach gave a census of dust associated with dust devil activity. Of 50–80 pressure dips (>0.3 hPa) detected per 100 station-days at each of four locations, about 40% had no detectable solar attenuation. Some fraction of these were simply anti-sun misses, but some are doubtless close encounters with devils with undetectably small amounts of dust. About 40% of the total had obscuration of 1% or more (indicating a line-of-sight opacity of ~ 0.01) and about 10% had 5% obscuration or more. While large pressure drops (indicating close distances and/or intense vortices) could have large or small obscuration, it was noticed that large obscuration was only found with the more intense pressure drops: the scatter of attenuation A [%] against Δp [hPa] points appeared to have an upper limit of $A \approx 50\Delta p$. This exact relationship may be site-dependent, however; indeed, differences in the populations between the four stations, separated by only $\sim$1 km, were noted, making evident the significance of small-scale variations that are not practically captured in numerical models except in a statistical manner.

Metzger et al. (1999) measured dust column opacities for a dust devil observed by the lander camera at the Mars Pathfinder landing site and used a terrestrial vertical velocity estimate of 7 $\mathrm{m\,s^{-1}}$, calculating a vertical flux of $5 \times 10^{-4}\ \mathrm{kg\,m^{-2}\,s^{-1}}$. Reiss et al. (2014) measured dust column opacities for three dust devils using orbiter images and calculated their dust loads. Minimum and maximum calculated dust fluxes were in the range between 3.8×10^{-7} and $1.2 \times 10^{-3}\ \mathrm{kg\,m^{-2}\,s^{-1}}$ assuming vertical velocity ranges between 0.1 and 10 $\mathrm{m\,s^{-1}}$.

From the surface of Mars, statistically significant datasets of dust devil observations have been limited to imaging campaigns from the Mars Exploration Rover *Spirit* (Greeley et al. 2006, 2010). Over three martian years the cameras onboard *Spirit* routinely imaged dust

Fig. 4 Dust devil in Gusev: A navigation camera image of a dust devil in Gusev crater. Taken on Mars Exploration Rover *Spirit* on 22 August 2004 mission day (sol) 581 (Product ID: 2N 177950967 RAD AD ND P0645 L0 C1)

devils on the floor of Gusev crater (Fig. 4). Greeley et al. (2010) determined the density of dust within vortices detected by *Spirit* by deriving atmospheric opacities from the images and comparing values obtained within the dust devils against background values. The obtained dust concentrations varied from 2.1×10^{-9}–$2.5 \times 10^{-4}\ \mathrm{kg\,m^{-3}}$. Further, vertical wind speeds inside vortex cores were determined by identifying features such as apparent "clots" of dust within the dust devils and tracking them in sequential images. The result was that the distribution of vertical speeds ranged from 0.04 to $17.0\ \mathrm{m\,s^{-1}}$. From these observations Greeley et al. (2010) calculated that the vertical dust flux for individual vortices ranged from 4.0×10^{-9}–$1.6 \times 10^{-4}\ \mathrm{kg\,m^{-2}\,s^{-1}}$.

Unfortunately there was no meteorological instrumentation onboard the *Spirit* rover. However, three other Mars landers have observed dust devils by both imaging and meteorological measurements. These landers are Mars Pathfinder (Murphy and Nelli 2002; Ferri et al. 2003), Phoenix (Ellehoj et al. 2010) and MSL Curiosity (Moores et al. 2015; Kahanpää et al. 2016; Steakley and Murphy 2016). Pathfinder and Phoenix operated for only a few months and took a small number of images suitable for a dust devil survey. Thus only 14 and 37 dust devils were identified from the images taken by these landers, respectively (Ferri et al. 2003; Ellehoj et al. 2010). Curiosity has operated to date for more than one martian year, but landed on an area where the dust devil activity is low and was thus able to detect only one plausible dust devil during its first Earth year on Mars (first 360 sols), despite an intensive imaging campaign of "dust devil search movies" (Moores et al. 2015). No more dust devils have been detected between sols 361 and 681 (J. Moores, 2016, personal communication). All of these three landers were equipped with high-resolution pressure sensors and detected numerous transient pressure drops apparently caused by convective vortices. A distinctive feature of these pressure drops is that they are small: less than 4.8 Pa, 3.6 Pa, and 3.0 Pa for Pathfinder, Phoenix, and Curiosity (first 681 sols), respectively (Murphy and Nelli 2002; Ellehoj et al. 2010; Kahanpää et al. 2016; Steakley and Murphy 2016).

Moores et al. (2015) compared the statistics of dust devil detections in the image data and pressure data of Pathfinder. A prediction of the number of dust devils that should be visible in the images was calculated based on the assumption that all identified pressure drops with magnitude greater than a preset threshold, Δp_t, were caused by dust-lifting vortices. The result was that with $\Delta p_t = 2.0$ Pa the number of visually detected dust devils matched the prediction. This is only a rough estimate given that it is based on a statistically small number of both pressure drops and dust devils detected by Pathfinder. Also, the detected pressure drops are lower than the central pressure drops of the vortices as they do not pass right over the sensor, so this result should be taken as a lower limit of the central pressure drops occurring in Martian dust devils. However, this calculation indicates that it is improbable that all dust devils seen by Pathfinder were caused by vortices much stronger than what was detected in the pressure data.

Assuming that a dust devil is in cyclostrophic balance, the tangential wind speed, u_θ, is proportional to the square root of the central pressure drop, Δp (Sinclair 1973; Renno et al. 2000). By making use of the ideal gas law, u_θ can be expressed as

$$u_\theta = \sqrt{RT\frac{\Delta p}{p_s}} \tag{4}$$

where R is the specific gas constant, T is air temperature, and p_s is surface pressure. The magnitudes of the largest pressure drops detected by Pathfinder, Phoenix and Curiosity can be transferred into tangential wind speeds using this relationship. The results are 17, 14 and 13 $\mathrm{m\,s^{-1}}$ for Pathfinder, Phoenix and Curiosity (first 681 sols), respectively. These values can be compared to threshold wind speeds for dust lifting by vortices at martian pressure, determined in laboratory experiments (Sect. 2.1). Greeley et al. (2003) concluded that the threshold for fine dust (2 μm) was ~20–30 $\mathrm{m\,s^{-1}}$, and Neakrase and Greeley (2010b) found that some fine dust was lifted already at 18 $\mathrm{m\,s^{-1}}$. Taking into account (a) that the central pressure drops of the vortices will generally be deeper than the detected pressure drops, unless the vortices pass directly over the sensor, (b) that the translational motion of the vortices also affects their ability to lift dust, and (c) the lower gravity of Mars, the tangential wind speeds calculated for the strongest vortices detected by Pathfinder agree with the minimum requirements for dust lifting as determined in the laboratory. The non-detection of practically any dust devils during Curiosity's first 681 sols, despite plenty of dust to lift, suggests that 3.0 Pa or 13 $\mathrm{m\,s^{-1}}$ is a lower limit for the dust lifting threshold on Mars. This limit is roughly in agreement with the threshold of $\Delta p_t = 2.0$ Pa obtained by Moores et al. (2015), which was based on comparing the statistics of vortex detections in the image and pressure data measured by Pathfinder.

Choi and Dundas (2011) used images taken from orbit to study the wind fields inside martian dust devils. They reported wind velocities within four dust devils imaged by the High Resolution Imaging Science Experiment (HiRISE) onboard Mars Reconnaissance Orbiter (MRO). HiRISE images the surface through three color filters with ~0.1 s intervals. Wind velocities were derived by automated tracking of dust devil cloud contrast features between the frames imaged through the different color filters. The determined composite wind velocities, i.e. the sum of translational and tangential wind, had typically magnitudes between 20 and 30 $\mathrm{m\,s^{-1}}$. When assuming cyclostrophic balance, the detected tangential wind speed profiles corresponded to 0.25–1% pressure decreases in the dust devil cores relative to ambient pressure. This agrees with the strongest pressure drops detected by Pathfinder and Phoenix.

2.3 Numerical Modeling of Dust Devils and Associated Dust Transport—Large-Eddy Simulation

Large-eddy simulation (LES) is a term used to describe the numerical simulation of atmospheric flow at horizontal grid spacings of $\sim 10^0$–10^2 m (Deardorff 1970). At such high spatial and associated high temporal resolution, most turbulence structures are resolved, making LES a powerful tool to study turbulent phenomena (see Chap. 7 [Spiga et al. 2016] for more details on LES). To date, most LES studies related to terrestrial dust devils have focused on vortex formation conditions and/or their physical characteristics (Kanak et al. 2000; Kanak 2005; Zhao et al. 2004; Gu et al. 2008; Ito et al. 2010b; Ohno and Takemi 2010; Raasch and Franke 2011; Klose and Shao 2013), with only a few such studies examining the fluxes of sand or dust.

Zhao et al. (2004) injected sand particles into their terrestrial LES to study particle trajectories in a dust devil. The authors found that particles with diameters up to 160 μm could be transported in their simulated dust devils, with the largest particles being carried in the periphery of the vortices. Ito et al. (2010a) included the empirical formulation for dust lifting of Loosmore and Hunt (2000) into their LES and studied dust suspension in a convective planetary boundary layer (PBL) on Earth. Loosmore and Hunt (2000) had measured dust flux in a wind tunnel for wind speeds below the threshold for saltation, but without particular focus on convective turbulence. They related dust emission flux to friction velocity only and therefore the dust fluxes simulated by Ito et al. (2010a) for a convective boundary layer at low mean wind speed (and thus on average small friction velocity) are relatively low (10^{-8} kg m^{-2} s^{-1}). Klose and Shao (2013) coupled their terrestrial LES with a size-resolved dust emission scheme representing the aerodynamic entrainment of dust particles by atmospheric turbulence (Klose and Shao 2012) and also employed parameterizations for dust transport and deposition. The dust transport by dust devils was thus explicitly simulated. Although turbulent dust emission is generally weak and of the orders of magnitude 10^{-9}–10^{-7} kg m^{-2} s^{-1}, Klose and Shao (2016) showed in a follow-up work, using an upgraded version of their scheme which was calibrated against field measurements of convective dust emission (Klose et al. 2014), that turbulent dust emissions can reach an order of magnitude 10^{-6} kg m^{-2} s^{-1} under favorable conditions such as in dust devils or dust plumes. Figure 5 shows a linear cross-section of dust emission through an individual dust devil detected in an LES run together with the corresponding vertical cross-section of PM20 dust concentration, perturbation pressure, and turbulent wind vectors. Dust emission reached values of up to 8×10^{-7} kg m^{-2} s^{-1} and the maximum dust concentration in this dust devil was $\sim 5 \times 10^{-7}$ kg m^{-3}. The dust emission in dust devils varied strongly from case to case but the maximum emissions varied only to an upper-limit envelope, which was related to atmospheric stability by Klose and Shao (2016). The finding is consistent with observations of solar attenuation in dust devils, which indicated that dust devil intensity does not uniquely determine the dust load in individual dust devils, but that the maximum possible dust load increases with dust devil intensity (Mason et al. 2014, Lorenz and Jackson 2015, see Sect. 2.2).

Assessing dust devil particle fluxes using LES for Mars, both for individual dust devils and for an area prone to dust devil occurrence, has proven difficult. The main difficulty is that the current martian literature focuses more on convective vortices than on actual dust devils—the subset of convective vortices capable of lifting dust (see Chap. 7 [Spiga et al. 2016] and references therein). However, vortices strong enough to potentially lift dust and become dust devils (and LES tends to show this could be as low as a few percent of the actual vortices arising from PBL convection) show characteristics similar to those observed

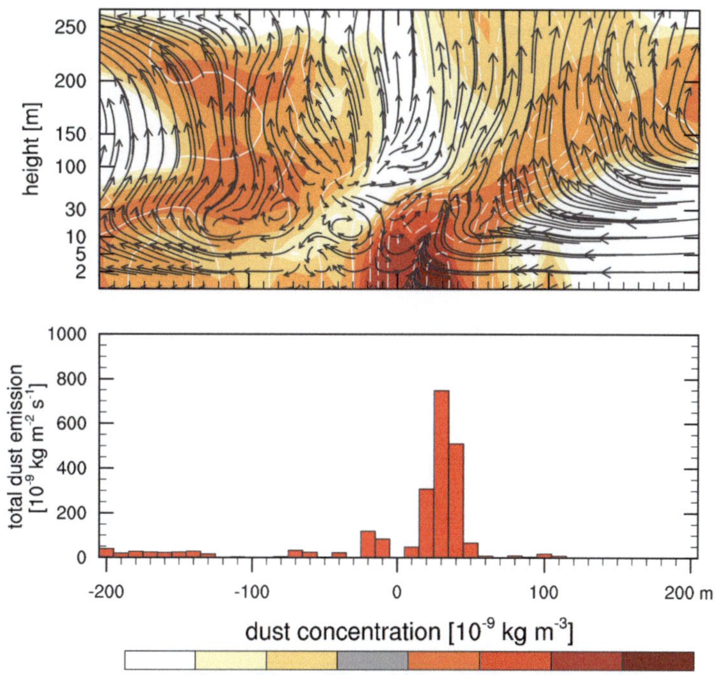

Fig. 5 (*Top*) Vertical cross section of dust concentration (*shaded*), perturbation pressure (*contour lines*), and turbulent wind vectors through a dust devil identified in LES; (*Bottom*) linear cross section of dust emission corresponding to the top plot (Reprinted from Klose and Shao 2016, with permission from Elsevier)

for dust devils by landers and orbiters. These characteristics include: pressure drops of a few Pascals, sizes from a few tens to hundreds of meters, vertical extension of about a kilometer, and durations from a few tens to at most a few thousands of seconds (Rafkin et al. 2001; Toigo and Richardson 2003; Michaels and Rafkin 2004; Spiga and Forget 2009; Gheynani and Taylor 2011). Those parameters are generally all larger for martian dust devils than for their terrestrial counterparts, which is consistent with the generally larger dust flux supposedly originating from Mars's dust devils compared to Earth's.

Michaels (2006) was the first to include a saltation bombardment scheme in a Mars LES, calculating bulk dust fluxes partitioned according to a predefined particle size distribution. While Michaels focused on the formation of dust devil tracks and did not aim to calculate dust fluxes, his simulation showed that the modeled dust devil track reached a maximum depth of about 8 μm only, with the majority of the track being less than 1.5 μm deep. This contrast results from "pulses" in vortex intensity, likely caused by the fluctuations of convective cells within which the vortex is embedded. Michaels (2006) noted that particle redeposition is prominent for the largest particles (about 100 μm radius) but negligible for particles smaller than 10 μm. Overall, the net redeposition of dust was very small anywhere in the model domain while the dust devil was active, and was slightly larger after its cessation. The LES results demonstrate that a dust devil should be able to transport dust particles rapidly from the surface to a height of several kilometers. Along the majority of its track, the modeled vortex exhibited a central pressure drop of ∼4.5 Pa, which compares well with the strongest vortices observed by Mars Pathfinder (Murphy and Nelli 2002). However, dust lifting in the vortex was probably even stronger than could be expected for such a pressure drop for two reasons: (1) the pulses in vortex intensity led to a pressure drop which could temporarily rise up to 6.5 Pa; and (2) the vortex was not symmetrical and not completely in cyclostrophic balance so that the vortex wind speed sometimes exceeded $19\ \mathrm{m\,s^{-1}}$ at certain locations.

Despite the success in reproducing dust-devil like vortices, LES models still need further improvement before they can be used to assess dust fluxes generated by dust devils. As a result, authors were cautious to present dust flux estimates in the existing martian LES literature. Several difficulties remain, such as LES models being highly idealized and lacking the influence of e.g. rugosity caused by small-scale topography or small-scale albedo contrasts. The parameterization of dust lifting by resolved turbulent winds on Mars is also prone to many shortcomings—including the major paucity in measurements on Mars to validate models derived from terrestrial measurements. Another major complication is a lack of knowledge of the availability of dust in a given area, which might explain the pronounced contrasts of observed dust devil activity between various regions on Mars, while LES modeling, along with in-situ pressure measurements, indicates that the formation of convective vortices is widespread.

3 Dust Devil Occurrence Frequency on Earth and Mars

To assess the relevance of dust devils, it is necessary to estimate their occurrence frequency. Dust devil occurrence varies considerably on different spatio-temporal scales and in different regions on Earth and Mars. Three approaches are examined in this section. Dust devil occurrence frequency is evaluated based on: (i) their formation conditions—i.e., the local meteorology and surface characteristics (Sect. 3.1; see also Chap. 5 [Rafkin et al. 2016]); (ii) statistical analysis of observations such as dust devil tracks (Sect. 3.2, see also Chap. 4 [Reiss et al. 2016]); and (iii) regional and global model results (Sect. 3.3).

A summary of dust devil occurrence densities as identified from visual surveys on Earth and Mars, augmented by results from modeling studies, is shown in Fig. 6. The largest densities of vortices detected in LES and inferred from field pressure measurements correspond to a couple of thousand vortices per $\mathrm{km^2}$ per day. Most of these are too weak to lift dust, and the largest densities of observed dust devils (from the surveys observing the smallest area and thus efficiently detecting the most abundant small devils) are of the order of $100\ \mathrm{km^{-2}\,day^{-1}}$. Other visual surveys on both Earth and Mars yield densities that fall off with survey area A roughly as $1/A$, since only scarce large devils are seen in wide area surveys. For more details on dust devil populations from the perspectives of modeling, remote sensing, and in situ observations, see Chap. 8 (Lorenz and Jackson 2016).

3.1 Dust Devil Occurrence Frequency Based on Meteorological and Land-Surface Conditions

The controls on the dust devil frequency of occurrence on Earth and Mars can broadly be determined by the availability of sand, dust or debris, the complexity of the terrain, and the meteorological conditions that prevail. The general characteristics for dust devil formation on Earth are summarized by the existence of the following conditions: (1) intense surface heating through insolation and a strong superadiabatic atmospheric temperature lapse rate, which implies a low soil moisture and a very high Bowen ratio (ratio of sensible to latent heat flux); (2) relatively weak ambient winds, i.e. generally less than $10\ \mathrm{m\,s^{-1}}$; (3) smooth arid terrain with some rock cover but few trees, buildings, or grassy areas; (4) relatively level

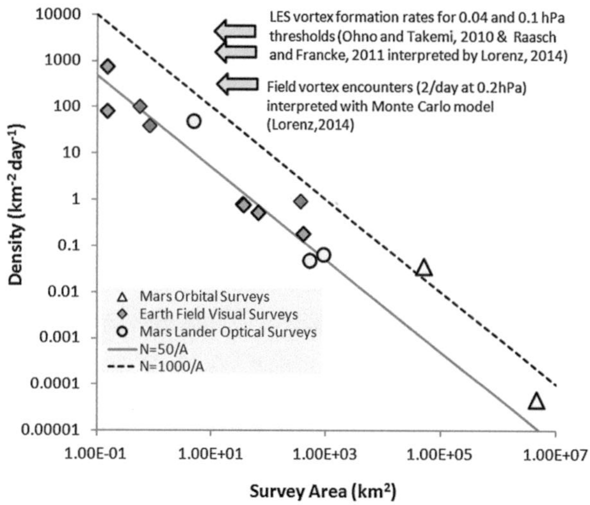

Fig. 6 A compilation of observed dust devil frequencies from visual surveys on Earth and Mars, derived from Lorenz (2013). There is a systematic variation of observed counts with area surveyed, presumably because large area surveys (especially those from orbit—*triangles*) only detect the larger, rarer dust devils. *Lines* show linear relationships between dust devil density and survey area, A, as a reference. Also indicated is the vortex generation rate of 300 $km^{-2}\,day^{-1}$ required, with assumptions on longevity and advection speed, to reproduce in a Monte Carlo model (Lorenz 2014) the observed pressure dips above 20 Pa seen in terrestrial field data to indicate vortices (not all of which may be dust-laden). The vortex formation rate measured in two different terrestrial LES simulations, of 4500 $km^{-2}\,day^{-1}$ and 1500 $km^{-2}\,day^{-1}$ for vortices with core pressure drops of more than 4 Pa and 10 Pa and vorticities of more than 1 s^{-1} and 0.15 s^{-1}, respectively, are also shown. Most of these rather weak vortices will be dustless, however. The Mars and Earth rates appear, with survey area taken into account, to differ only by a factor of a few at most

to gently sloping topography (Balme and Greeley 2006; Oke et al. 2007b; Kurgansky et al. 2011). See Chap. 5 (Rafkin et al. 2016) for a detailed discussion of dust devil formation conditions.

Dust devil activity is known to vary within seasonal, daily, and episodic cycles and across terrain types (Sinclair 1969; Oke et al. 2007b). On Earth, dust devils have a diurnal cycle that closely resembles the most active period of the convective boundary layer cycle, with activity generally observed between 10:00–17:30 local time (Balme and Greeley 2006). Until recently no global quantitative estimate of dust devil occurrence frequency had been conducted. The study by Jemmett-Smith et al. (2015) was the first to quantify regional and global potential of dust devil occurrence frequency on Earth (where the actual occurrence depends on the availability of dust), with results shown to be broadly consistent with observations (see Balme and Greeley 2006). This was achieved by using meteorological constraints to identify potential dust devil hours from global model outputs (operational analyses from the European Centre for Medium-Range Weather Forecasts, ECMWF; analyses data are model data with assimilation of observational data to provide the most realistic representation of a given time), with ~20 km horizontal resolution and hourly temporal resolution. The relatively high spatio-temporal resolution made it possible to sufficiently resolve the diurnal cycle as well as the conditions required for dust devils to exist, which was not possible before. Grid-box potential dust devil active hours were defined as hours when thresholds of w_*/u_* (ratio of convective velocity scale, w_*, and friction velocity, u_*; or equivalently convective boundary layer height over Obukhov length, $-h/L$, as used by

Deardorff (1978), Hess and Spillane (1990), Kurgansky et al. (2011)) and near-surface temperature lapse rate (Ryan 1972; Oke et al. 2007b; Ansmann et al. 2009) were exceeded; in this case the 8.5 $\mathrm{K\,m^{-1}}$ lapse rate criterion from Ansmann et al. (2009) and the $w_*/u_* > 5$ criterion from Lyons et al. (2008) were used. The combination of these two criteria provided a framework that incorporates all known measures of local meteorology that determine when and where dust devils are likely to occur, providing a dust source is available. Measures of atmospheric stability are also used for dust devil dust lifting parameterizations in martian global models (see Sects. 3.3.2 and 4.2.2).

Results show a diurnal cycle as well as geographical, seasonal and annual variations in dust devil frequency of occurrence, with a clear indication that arid areas, in both hemispheres, have the highest potential dust devil activity (Fig. 7). As expected, minima occur in winter and maxima in summer, showing a clear latitudinal dependence (related to solar insolation) with low latitude regions (generally between 10° and 30° North and South) experiencing higher total values and smaller seasonal variations. The total area of potential dust devil activity within a year is $\sim 3.7 \times 10^7\ \mathrm{km^2}$, with a mean active period of $\sim 205\ \mathrm{h\,yr^{-1}}$ (Fig. 7a). Potential occurrence tends to be restricted to "hot spot" regions, with dust devil active hours on the order of 2500 $\mathrm{h\,yr^{-1}}$ (equivalent to ~ 7 h of activity on average per day); an indication that these regions are active year round (Jemmett-Smith et al. 2015). Notable hot spots include parts of the Atacama and Sechura deserts of South America, which are known dust devil breeding grounds (see Kurgansky et al. 2011; Hesse 2012; Reiss et al. 2013). Many dust devil hot spots, including those mentioned, are located at or near coastal areas. When relatively cool water bodies exist near hot land-surface areas, onshore land-sea breezes can lead to extremely unstable atmospheric stratifications over land, thereby providing one of the key dust devil formation conditions.

Dust devils require not only vortices strong enough to raise dust but also the presence of a dust source at the surface. A known dust devil active area can therefore serve as a proxy for dust sources. As well as with local variations, dust sources can change with the seasons (Ginoux et al. 2012). Decadal changes in dust sources can be attributed to changes in climate or by anthropogenic means, such as land-use change (Ginoux et al. 2012). In that case, aridity can be affected over vast regions when sensitive to vegetation changes, such as in the Sahel (Cowie et al. 2013). Reduced vegetation can also decrease surface roughness, which in turn can lead to an increase in dust devil frequency of occurrence under certain conditions (Lyons et al. 2008). The uncertainty of dust devil occurrence on Earth caused by uncertainty of dust sources is discussed in Sect. 4.1.

3.2 Dust Devil Occurrence Frequency Based on Observations

Lorenz (2014) showed with a Monte Carlo model that field meteorological observations of vortices on a terrestrial desert playa (about 300 $\mathrm{km^{-2}\,day^{-1}}$, for a vortex detection threshold of 20 Pa) are rather consistent with the number of vortices detected in LES simulations by Raasch and Franke (2011) and Ohno and Takemi (2010) when a -1 cumulative power-law (-2 differential) in core pressure drop was assumed. Specifically, those simulations indicated about 4,500 and 1,600 vortices per $\mathrm{km^2}$ per day, but with core pressure drops of more than 4 and 10 Pa and maximum vorticities of more than 1 $\mathrm{s^{-1}}$ and 0.15 $\mathrm{s^{-1}}$, respectively. The Monte Carlo results indicate that vortices are more abundant than visually-observed dust devils, which at that site (Eldorado Playa) occurred with a frequency of about 100 $\mathrm{km^{-2}\,day^{-1}}$. This is readily understood as a dust devil intensity threshold for dust lifting to occur, expressed in pressure terms as about 0.3–0.8 hPa (regardless of the actual dust-lifting mechanism and its dependence on shear stress, pressure drop or other factors,

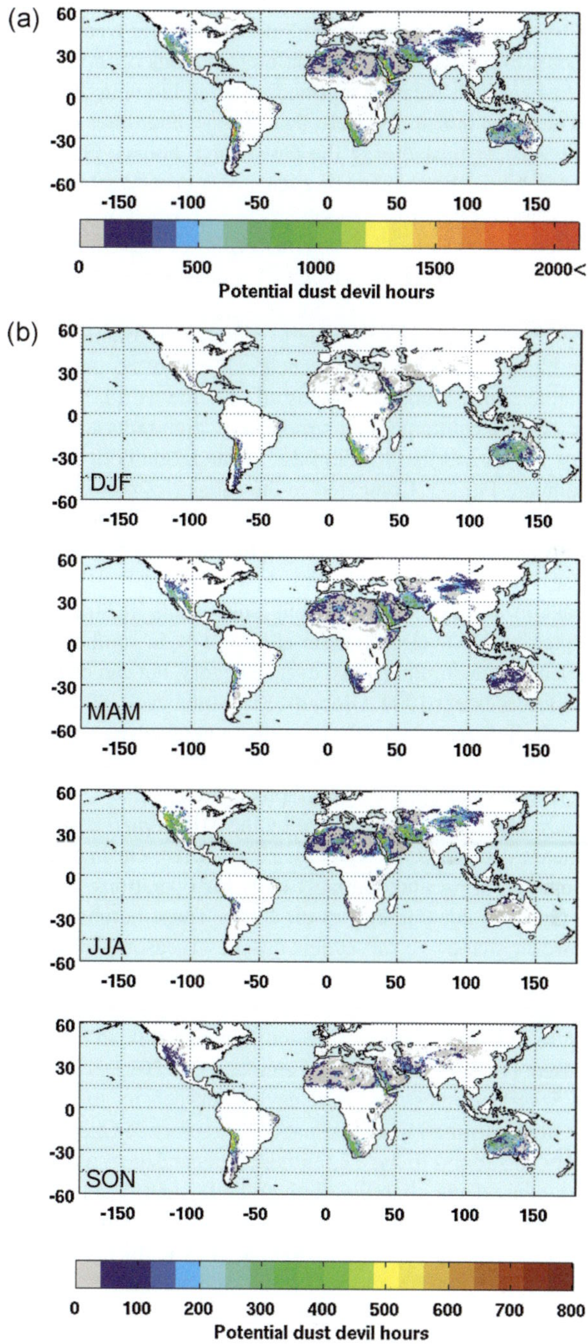

Fig. 7 Climatology of potential dust devil hours ($PDDP_{hours}$; mean of 2012 and 2013) using criteria of $w_*/u_* > 5$ and near-surface lapse rate $>8.5\ K\,m^{-1}$ from respectively Lyons et al. (2008) and Ansmann et al. (2009) for (a) annual total, and (b) seasonal totals (modified from Jemmett-Smith et al. 2015)

it is convenient to assume these factors are proportional to pressure drop), not too different from results of laboratory measurements.

Shao et al. (2013) analyzed global synoptic weather reports and calculated time series of dust event occurrence for different regions of the world for the categories of weak dust events (present weather reported as being blowing dust or dust in suspension), strong dust events (dust storm or severe dust storm), and other dust events (dust devils, thunderstorm with dust or dust during the past 3 or 6 hours). Occurrence [%] is defined as N_w/N_{obs}, where N_w is the number of events of a dust weather category (e.g. dust storm) and N_{obs} is the total number of synoptic records. Only present and past weather records of manned stations are considered (for more details see Shao et al. 2013). Klose (2014) augmented this analysis by treating the category for dust devils separately. Figure 8 shows the global occurrence of dust devils for the time period of 1984–2014. The highest frequencies of dust devil occurrence are restricted to remarkably few regions. Australia stands out with by far the highest occurrence frequencies. It cannot be guaranteed that observers report present weather consistently at stations all over the world, however, so the results should be interpreted with caution. Other regions with high reported dust devil occurrence frequencies are East Africa and the Middle East, Iceland, Southern America, and southern North America (compare Fig. 7). Compared to the categories of weak, strong, and other dust events, only Australia and southern North America show significant dust devil frequencies (Fig. 9). For Australia, the dust devil frequency mostly exceeds that of strong and other dust events. Only dust in suspension and blowing dust (weak dust events) are reported more often. Particular peaks in the frequency of strong dust events illustrate continent-wide dust storms, e.g. in September 2009 (De Deckker et al. 2014). In southern North America, dust devils are not generally the dominant dust event category, but can be in particular years, such as in 1999 or to a lesser extent 1994. Globally, a seasonal cycle of dust devil frequency can be recognized, which shows regular peaks in southern hemispheric spring and summer and is thus probably strongly related to the Australian records. See also Chap. 1 (Lorenz et al. 2016) for reports on dust devil activity by geographical area.

On Mars, the images taken by the MER *Spirit* rover are again the only statistically significant data set (Greeley et al. 2006, 2010). During the first full "dust devil season" (i.e., southern spring and summer) of *Spirit* operations, there were 51 dust devils $km^{-2} sol^{-1}$ in Gusev crater (Greeley et al. 2006). These numbers dropped to respectively 11 and 20, however, in the two subsequent dust devil seasons. Inter-annual differences in dust devil frequency are attributed to variations in atmospheric opacity due to regional dust storms that change the local thermodynamic conditions, with increased atmospheric dust content tending to reduce e.g. the surface sensible heat flux and thus reduce the number of dust devils present (Greeley et al. 2010; Lemmon et al. 2015).

Several studies used dust devil track density to infer global dust devil frequency (for detailed information about dust devil tracks, see Chap. 4 [Reiss et al. 2016]). Dust devil activity was found to have a latitudinal dependence, with the southern hemisphere producing more dust devils than the north (Whelley and Greeley 2006). Dust devil track (ddt) densities are 9×10^{-5} and 9×10^{-4} ddt $km^{-2} sol^{-1}$ for the northern and southern hemisphere, respectively. Furthermore, 55% of activity in the northern hemisphere is limited to between 45 and 75°N during northern spring and summer; while 65% of the dust devil activity during southern spring and summer is limited to between 45 and 75°S (Whelley and Greeley 2008).

Regional studies in Argyre and Hellas Planitiae by Balme et al. (2003) show dust devil track densities of 1.2×10^{-3} and 7×10^{-4} ddt $km^{-2} sol^{-1}$, respectively. These regional dust devil track densities are in good agreement with densities of 9×10^{-4} ddt $km^{-2} sol^{-1}$ for the southern hemisphere measured by Whelley and Greeley (2006). Local studies using multi-temporal high resolution remote sensing data reveal higher dust devil track densities. Verba

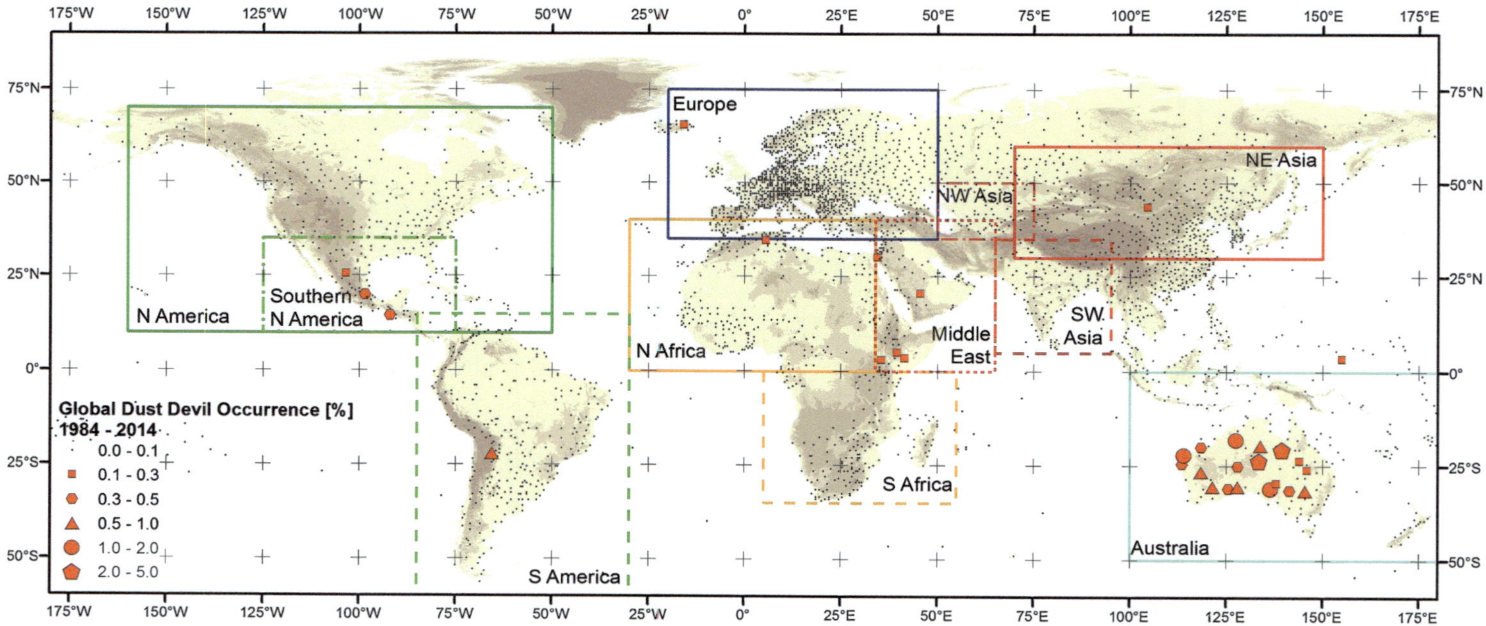

Fig. 8 Global dust devil occurrence for 1984–2014. Occurrence [%] is defined as number of events of a dust weather category (e.g. dust devils) divided by the total number of synoptic records (for more details see Shao et al. 2013). *Rectangles* mark areas of time series analyses (compare Fig. 9)

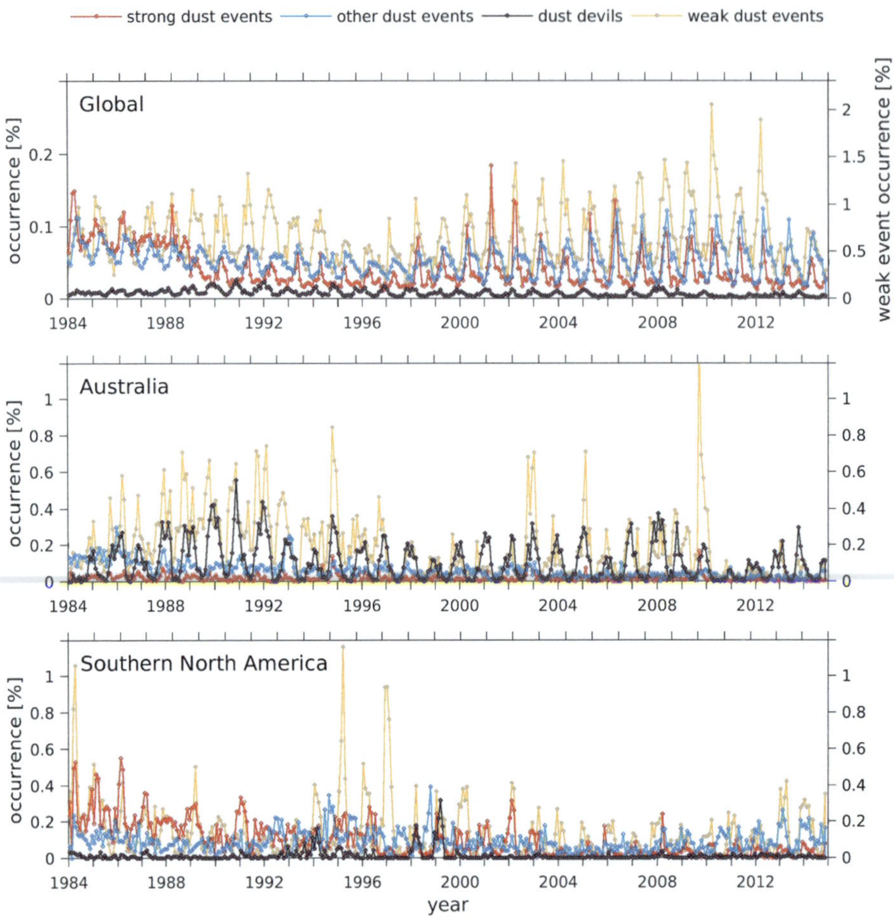

Fig. 9 Time series of dust event occurrence globally, for Australia, and southern North America (analysis domains are shown in Fig. 8). Note the different axes scales for the global time series (after Shao et al. 2013)

et al. (2010) analyzed dust devil tracks in Gusev and Russell crater and found averaged dust devil track formation rates of 3.4×10^{-2} and 2.5×10^{-1} ddt km^{-2} sol^{-1}. Formation rates at the future InSight landing site region in Elysium Planitia are with 4.6×10^{-2} ddt km^{-2} sol^{-1} similar to those at Gusev crater (Reiss and Lorenz 2015). Figure 10 shows ephemeral dust devil tracks imaged by HiRISE in Terra Sirenum in early southern summer and gives an impression on how numerous dust devils occur on Mars. The study locations given in this paragraph are shown in Fig. 11.

The dust devil frequencies inferred from dust devil tracks (Balme et al. 2003; Whelley and Greeley 2006, 2008; Verba et al. 2010; Reiss and Lorenz 2015) are much lower in comparison with the statistically significant data obtained by the MER *Spirit* rover in Gusev crater (Greeley et al. 2006, 2010). In general, inferring dust devil frequency from dust devil tracks is problematic. Global remote sensing observations of active dust devils leaving tracks (Cantor et al. 2006) as well as the comparison of dust devil track with active dust devil frequencies (Verba et al. 2010; Greeley et al. 2010) in Gusev crater showed that only a fraction of dust devils leave surface tracks. The formation of dust devil tracks depends on

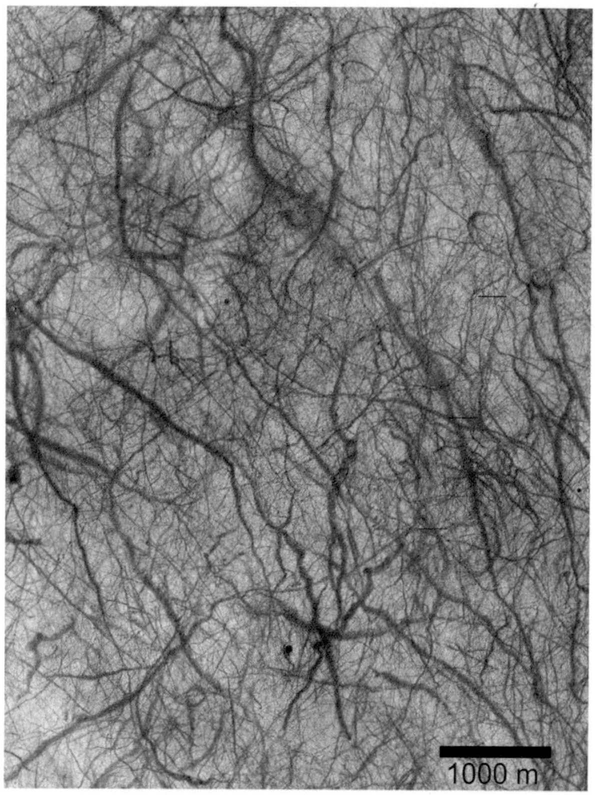

Fig. 10 Hundreds of ephemeral dust devil tracks (widths up to 100 m) left by passages of dust devils, giving an impression on how numerous dust devils occur on Mars. HiRISE image ESP_013538_1230 at 56.8°S and 198.1°E

several parameters such as dust availability, dust cover thickness and substrate properties (Balme et al. 2003; Fisher et al. 2005; Whelley and Greeley 2006, 2008). More detailed discussion is provided in Chap. 4 (Reiss et al. 2016).

The occurrence frequency of martian dust devils has also been studied by identifying dust devils in images taken from orbit (Cantor et al. 2006; Reiss et al. 2014). In contrast to the surveys based on dust devil tracks, these surveys found the dust devil activity to be stronger in the northern hemisphere. For example, 88.5% of the dust devils identified in the comprehensive survey by Cantor et al. (2006), reporting 11,456 dust devil observations, occurred in the northern hemisphere. The difference is probably explained by detection bias. Only the largest dust devils can be directly seen from orbit, but tracks left by all sizes are visible. Some areas in the northern hemisphere, especially Amazonis Planitia, are known as breeding grounds of dust devils with monumental dimensions (Fenton and Lorenz 2015). These areas probably distort the statistic of dust devils directly observed from orbit. Based on the surveys of Fisher et al. (2005) and Cantor et al. (2006) in Amazonis Planitia, dust devil frequencies are around 6.3×10^{-4} and 5.2×10^{-4} $\text{dd}\,\text{km}^{-2}\,\text{sol}^{-1}$, respectively. However, these very low dust devil frequencies do not represent real dust devil frequencies and are orders of magnitude underestimations, because satellite imagery only provides snapshots of the daily activity and image resolutions do not resolve smaller dust devils, which occur much more frequently.

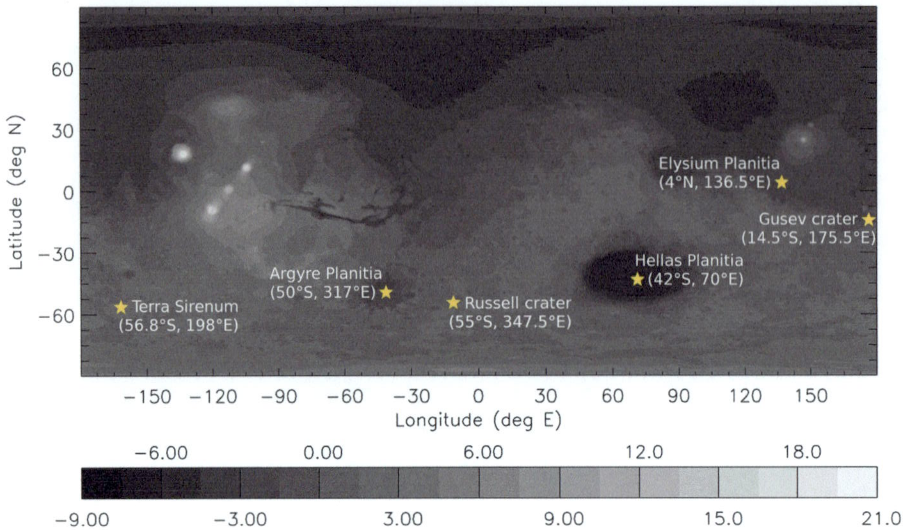

Fig. 11 Map of Mars showing topographic height above the geoid [km] as measured using the Mars Orbiter Laser Altimeter on board the Mars Global Surveyor (MGS) spacecraft. Locations of dust devil track studies are marked

3.3 Dust Devil Occurrence Frequency Based on Numerical Models

3.3.1 Large-Eddy Simulation

In LES, dust devils can be identified from their characteristic properties such as pressure drop, Δp, and vorticity, ζ, and are detected using threshold criteria for the latter quantities. Ohno and Takemi (2010) (OT10) and Raasch and Franke (2011) (RF11) conducted LES simulations for terrestrial conditions with a surface sensible heat flux of $H = 290\ \mathrm{W\,m^{-2}}$ and zero mean wind using 3 m and 2 m horizontal resolution, respectively. Klose and Shao (2016) (KS16) conducted a set of terrestrial simulations with varying surface heat fluxes and wind speeds at 10 m horizontal resolution. During 2000 s of simulation time in a 1 km^2 domain, OT10 identified 225 dust devils with pressure drops exceeding a threshold value of 10 Pa and maximum vorticities of more than 0.15 s^{-1} at 10 m height, corresponding to a dust devil number of about 400 km^{-2} h^{-1}. RF11 found 25,000 dust devils during 5400 s in a 16.8 km^2 domain, or about 1000 km^{-2} h^{-1}, by applying a perturbation pressure threshold of 4 Pa and a vorticity threshold of 1 s^{-1} at about 1 m height. Varying threshold criteria have been tested by KS16. With thresholds close to those applied by OT10 (10 Pa and 0.1 s^{-1} at 10 m), KS16 found 25 dust devils during 1 hour in a 4 km^2 domain for a simulation with $H = 200\ \mathrm{W\,m^{-2}}$ and weak mean wind ($u_* = 0.15\ \mathrm{m\,s^{-1}}$), and 100 for $H = 400\ \mathrm{W\,m^{-2}}$, yielding 6 to 25 km^{-2} h^{-1}. KS16 found no dust devils when using criteria of 5 Pa and 1 s^{-1} at 2 m height, close to those used by RF11. It is likely that the lower number of detected dust devils by KS16 is due to their coarser model resolution. Vorticities as large as 1 s^{-1} are thus barely reached. For example, RF11 identified twice as many dust devils in a simulation with 1 m horizontal resolution rather than 2 m. Additionally, KS16 have applied a post processing on the identified dust devil tracks. In this post processing, short (<30 s) tracks were deleted and tracks belonging to the same vortex were connected by removing gaps arising from values of Δp and ζ that intermittently did not satisfy the threshold criteria.

This further reduced the number of dust devil counts in the study of KS16. Despite being lower than those found by OT10 and RF11, the change in the number of detections as the threshold criteria are varied is consistent in the study of KS16 with the power-law relationship between dust devil frequency and pressure drop suggested by Lorenz (2014) (compare Sect. 3.2).

Rafkin et al. (2001) presented the first LES results for Mars using their Mars Regional Atmospheric Modeling System (MRAMS). The model was set up with a 100 m horizontal resolution for a $18 \times 18\ \text{km}^2$ domain. Rafkin et al. (2001) initialized the LES with a pre-sunrise thermodynamic sounding obtained from a 2D model version and vertically increasing background wind profile. By detecting local vorticity maxima and pressure minima, Rafkin et al. (2001) found a density of dust-devil like vortices of 0.03–0.1 km^{-2} at any given time during the simulation time of one half sol. Gheynani and Taylor (2011) conducted two-hour simulations for the Phoenix lander site using the NCAR LES model (Sullivan et al. 1994) with a horizontal resolution of 25 m, a horizontal domain size of $5 \times 5\ \text{km}^2$, and a vertically stretched grid. Vortices were identified in their simulations by searching for corner velocity vectors that follow a circular motion. Using a surface sensible heat flux of 24 $\text{W}\,\text{m}^{-2}$, Gheynani and Taylor (2011) found that dust devil density decreased with increasing background wind speed and obtained densities of 2.32, 0.88, and 0.08 km^{-2} for geostrophic winds of 0, 4, and 8 $\text{m}\,\text{s}^{-1}$ without and 1.96, 0.92, and 0.08 km^{-2} with radiative forcing included in their simulations.

3.3.2 Regional and Global Models

Due to the coarser spatial resolution of 10^0–10^2 km, dust devils are not resolved in regional and global atmospheric models of Earth and Mars. As the focus of terrestrial dust modeling has so far almost exclusively been on regional- and large-scale events such as dust storms, no studies exist that provide an explicit estimate of dust devil frequency using meso- and large scale numerical models. However, it is possible to estimate dust devil frequency from regional and global model data based on LES results. Klose and Shao (2016) conducted experiments for 15 different atmospheric stability and background wind conditions and obtained a relationship between the number density of dust devils, $n = N/(AT)$, where N is the number of dust devils detected in an area A during a time period T, and Richardson number, Ri (calculated for 10 m height):

$$n = \beta \cdot Ri^2 \quad (\text{with } Ri < 0) \tag{5}$$

with $\beta \approx 5.8\ \text{km}^{-2}\,\text{h}^{-1}$. The coefficient β was found to decrease rapidly for the weakest dust devil identification criteria (i.e. small Δp_t and ζ_t), but to be approximately constant for a range of threshold criteria with $\Delta p_t > 20$ Pa or equivalently about $\zeta_t > 0.5\ \text{s}^{-1}$. Equation (5) is a robust average of the relationships obtained based on the various identification threshold and is thus more representative of the larger dust devils. With Equation (5), it is possible to estimate the number of dust devils occurring in a given area, e.g. a grid box of a regional or global model, only from modeled Ri.

Applying this approach to results of regional or global simulations yields maps of dust devil occurrence numbers. Klose (2014) conducted simulations for Australia for July 2007–June 2008, a time period with high dust devil and low dust storm occurrence (see Fig. 9b). Equation (5) was obtained from LES results using an idealized setup, and does not account for how the relationship might vary with changes in vegetation cover. Further LES simulations would be required to investigate this relationship. The number of dust devils is in general negatively correlated with fractional vegetation cover, σ, although small

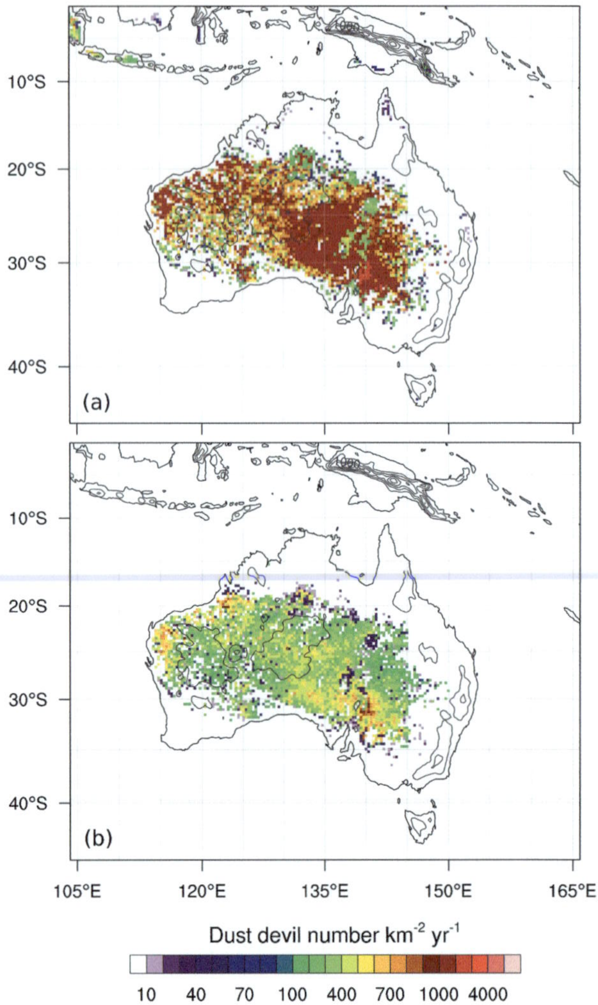

Fig. 12 (a) Estimated total number of dust devils occurring in Australia from July 2007 to July 2008 based on Equation (5) with correction for vegetation cover: $N_\sigma = N(1-\sigma)$; (b) as (a), but using N_L, i.e. numbers are only computed at times when the lapse rate between the surface and 2 m height exceeds $8.5\ \mathrm{K\,m^{-1}}$

cover fractions do not preclude and might even enhance dust devil development (Balme and Greeley 2006; Oke et al. 2007b; Neakrase and Greeley 2010a). To account for vegetation cover, a preliminary approximation of $N_\sigma = N(1-\sigma)$ was made by Klose and Shao (2016). In addition, N_σ was only computed for areas with less than 50% vegetation cover as it is assumed that areas with larger cover fractions are no dust sources. A clear seasonal dependence of the number of dust devils was found, with the largest numbers occurring in southern hemispheric spring and summer. Without the vegetation correction, dust devil numbers of about 100–200 $\mathrm{km^{-2}\,mon^{-1}}$ were predicted over wide regions of central Australia from October to January, with maxima of up to 500 $\mathrm{km^{-2}\,mon^{-1}}$ in the western part of Western Australia and in a region stretching from eastern South Australia (SA) to western New South Wales (NSW). After accounting for vegetation cover, the predicted dust devil numbers are generally smaller than 100 $\mathrm{km^{-2}\,mon^{-1}}$ in central Australia, with maxima of 300 $\mathrm{km^{-2}\,mon^{-1}}$ at few locations in the aforementioned areas. N_σ for the whole year of simulation is shown in Fig. 12a. Dust devil numbers in the areas of high activity integrate

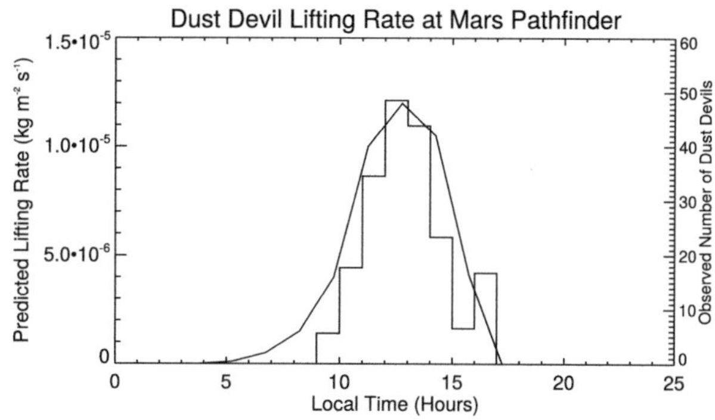

Fig. 13 Modeled diurnal dust devil lifting rate shown for the model grid point nearest the Mars Pathfinder landing site (*solid curve*) and histogram of observed diurnal occurrence of dust devils during the Pathfinder mission (Murphy and Nelli 2002). (Reprinted from Kahre et al. 2006, with permission from John Wiley and Sons)

to about 900–1000 $km^{-2}\,yr^{-1}$. Very few grid points in eastern SA and western WA show numbers of up to 4000 $km^{-2}\,yr^{-1}$.

Oke et al. (2007b) conducted a dust devil census at Fowlers Gap research station in western NSW, close to the dust devil hot spot in eastern SA identified here. During their 20 day observation period in January 2001, Oke et al. (2007b) counted 557 dust devils in a 35 km^2 area, translating to about 25 $km^{-2}\,mon^{-1}$. For January 2008, the approach described above yielded $N_\sigma \approx 32$ $km^{-2}\,mon^{-1}$ in the corresponding model grid cell, slightly larger than the number observed by Oke et al. (2007b). More dust devils have been reported in Australia in 2008, the time of simulation shown here, compared to 2001, when the census was conducted by Oke et al. (2007b) (Fig. 9), so the results are plausible. However, N_σ is larger in the surrounding grid cells. Oke et al. (2007b) found that no dust devils occurred for near surface temperature lapse rates lower than $\sim$1 $K\,m^{-1}$. If we apply a minimum lapse rate of 1 $K\,m^{-1}$ as an additional criterion to our estimates, we find negligible difference in our results, thus showing the applicability of Ri as sole indicator for dust devil occurrence. However, if a value of 8.5 $K\,m^{-1}$ is applied as suggested by Ansmann et al. (2009), then the dust devil number with lapse rate criterion, N_L, is smaller with 5 $km^{-2}\,mon^{-1}$ in the model grid cell corresponding to Fowlers Gap, again with larger values in the surrounding. A case-based comparison would be necessary to further investigate the criteria and validate the approach. On an annual basis, N_L yields values of mostly below 400 $km^{-2}\,yr^{-1}$ and maxima of about 2000 $km^{-2}\,yr^{-1}$ in the westernmost part of WA and in the area surrounding Fowlers Gap (Fig. 12b).

Global models for Mars have a much stronger focus on dust devils than is the case for terrestrial models. Martian global climate models (GCMs) contain separate dust devil parameterizations (Newman et al. 2002a). In these parameterizations, a positive surface heat flux and a nonzero PBL thickness is required for dust devils to occur (see Sect. 4.2.2). Models therefore predict that dust devils occur only during the day, with a peak activity during the early afternoon (Newman et al. 2002a; Basu et al. 2004; Kahre et al. 2006). In particular, Kahre et al. (2006) demonstrated that the predicted diurnal peak in dust devil emission rates at the Mars Pathfinder site during the season of the mission should have occurred between 12:00 and 13:00 LTST (Fig. 13), which is consistent with the observed peak in pressure dips from the pressure sensor on Mars Pathfinder (Murphy and Nelli 2002). Greeley et al. (2010) analyzed dust devils over three consecutive years as observed by the *Spirit* lander at Gusev Crater. They found that dust devils peaked over the entire period 12:00–15:00 LTST, in the second year featuring a later peak from 14:00–15:00 LTST, in the presence of higher dust loading, and in the third year there were significant numbers

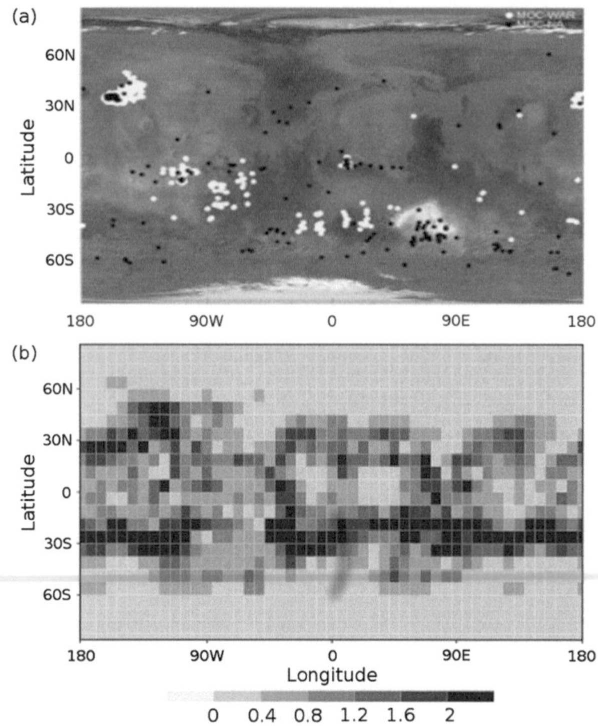

Fig. 14 (a) Afternoon spatial distribution of dust devils as observed by Mars Orbiter Camera plotted on a cylindrical map (Reprinted from Cantor et al. 2006, with permission from John Wiley and Sons). *White* and *black dots* indicate dust devils detected with two different cameras. (b) Annual dust devil dust lifting (arbitrary units) obtained from a Mars global model with only parameterized dust devil lifting (set proportional to the dust devil activity defined in Eq. (6); see Sect. 4.2.2 and radiatively active dust (Reprinted from Newman et al. 2002a, with permission from John Wiley and Sons)

as late as 16:00–17:00 LTST. These were consistent with the diurnal surface heating cycle, although selection issues should also be noted in sampling.

A seasonal trend in dust devil activity for Mars is also predicted in global models because both the surface-to-atmosphere temperature difference (and thus the heat flux) and the depth of the PBL are maximized near the sub-solar point. Predicted peak dust emission rates occur during local spring and summer, which is consistent with surveys of observed dust devils and dust devil tracks (Newman et al. 2002a; Basu et al. 2004; Fisher et al. 2005; Kahre et al. 2006; Cantor et al. 2006; Greeley et al. 2010). Both lander-based (Greeley et al. 2010) and orbital (e.g. Cantor et al. 2006) surveys have shown that the dust devil activity ceases almost completely during local fall and winter. By contrast, most dust devil parameterizations used in Mars climate models predict some lifting whenever thermal conditions are favorable for dust devils to occur, hence lifting occurs year-round. This suggests such parameterizations do not include all the physics required to predict dust devil occurrence and/or their ability to lift dust, such as the correct formulation of a threshold intensity that must be achieved before lifting can occur. However, the total dust lifting produced by such parameterizations, which is typically tuned to match the observed background dust loading, may be accounting for other small scale lifting phenomena not represented in the model, as discussed further in Sects. 4.2.2 and 5.3.2. The spatial distribution of dust devil activity predicted by the Mars climate models over a full Mars year is rather similar to orbital observations (Fig. 14), particularly in terms of the overall hemispheric asymmetry (more in the southern hemisphere, which has the more intense summer) and the basic latitudinal distribution (though the peak activity is observed at slightly higher latitudes than predicted). The model also captures some of the regional trends, e.g. the greater dust devil intensity in Amazonis Planitia, although the agreement is not as good in some

Table 3 Total dust transport by dust devils

Dust transport [$kg\,yr^{-1}$][a]	Method	Note	Reference
Earth			
566×10^9	field/theory	global	Koch and Renno (2005)
59×10^9	field/theory	global	Jemmett-Smith et al. (2015)
7×10^9	field	USA	Gillette and Sinclair (1990)
$0.2–1.1 \times 10^8$	modeling	AUS	Klose and Shao (2016)
Mars			
5.8×10^{11}	orbiter	global	Cantor et al. (2006)
2.3×10^{11}	orbiter/lander	global	Whelley and Greeley (2008)

[a]Note that a martian year is approximately twice as long as a terrestrial year

regions, for example the model does not capture the peak between Solis and the Tharsis Montes.

4 Estimates of Total Dust Transport by Dust Devils on Earth and Mars

The previous sections provided insights into the dust transport associated with individual dust devils and the statistics of dust devil occurrence. In this section, these insights are combined to obtain estimates of the contributions of dust devils to regional and global dust budgets, as listed in Table 3. These estimates are either based purely on observations of dust fluxes (Sect. 4.1) or are based on (to a varying extent) some consideration of the physical mechanisms behind dust devil occurrence and consequently their dependence on the spatio-temporally varying state of the atmosphere in which they form (Sect. 4.2).

4.1 Total Dust Transport Estimated from in Situ and Remote Sensing Observations

No global estimate of dust devil sediment transport on Earth is available that is based on observations only. Although regional dust devil censuses have been conducted, e.g. in Australia by Hess and Spillane (1990) and Oke et al. (2007b), in the USA by Sinclair (1969), Carroll and Ryan (1970), and Snow and McClelland (1990), or in South America by Kurgansky et al. (2011), no information on dust flux in the counted dust devils was available and thus no area estimate on transport could be obtained. The only estimate on regional dust devil transport is that of Gillette and Sinclair (1990). The authors conducted aircraft measurements in a test region near Tucson, Arizona, USA. Using these fluxes as a reference, Gillette and Sinclair (1990) assigned annual dust flux values to areas based on their climatic and vegetative conditions (Fig. 15). To account for non-erodible elements on the soil surface, they further assumed a reduction of the dust flux by 50%. On this basis, Gillette and Sinclair (1990) estimated that dust devils contribute $\sim 7 \times 10^9$ kg to the total mass of mineral dust aerosol in the contiguous USA.

On Mars, the most reliable estimates of the total amount of dust lifted on a regional scale are those based on the observations of the MER *Spirit* rover in Gusev grater (Sects. 2.2 and 3.2). Greeley et al. (2006, 2010) combined the values of dust fluxes (Sect. 2.2), diameters and lifetimes of the detected dust devils to calculate total amounts of dust injected by single

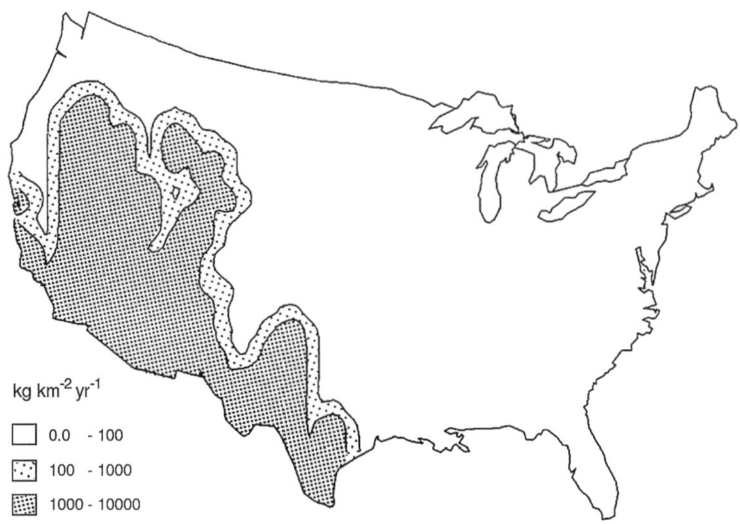

Fig. 15 Annual dust devil dust fluxes ($D_p < 25$ μm) estimated by Gillette and Sinclair (1990) for the USA. (Reprinted from Gillette and Sinclair (1990), with permission from Elsevier)

dust devils into the atmosphere. The results varied from 1.9×10^{-4} to 29 kg per dust devil. Amounts of dust lifted per unit area were then estimated by using the average of the masses lifted by individual dust devils and the number of dust devil observations extrapolated both temporally and spatially. The value for *Spirit*'s first dust devil season, given in Greeley et al. (2006), was $\sim$19 kg km^{-2} sol^{-1}. Instead of exact values, wide error ranges for the amount of dust lifted per unit area during the whole dust devil season were given in Greeley et al. (2010): 2.6 to 3.0×10^5 kg km^{-2} for season one, 4.4 to 5.3×10^3 kg km^{-2} for season two, and 1.5×10^2 to 1.6×10^5 kg km^{-2} for season three.

Whelley and Greeley (2008) estimated the total amount of dust lifted by dust devils on Mars using the value given by Greeley et al. (2006) for Gusev crater as ground truth and extrapolating it globally using dust devil track density observations (Sect. 3.2). The result was $2.3 \pm 1.0 \times 10^{11}$ kg per martian year, approximately half as much as local and regional dust storms.

Another estimate of the annual dust devil flux was calculated by Cantor et al. (2006), who estimated the occurrence frequency of dust devils using orbital observations of active vortices (Sect. 3.2). The dust lifting rate was calculated by determining dust devil optical depths from orbital images and assuming a 2 m s^{-1} vertical wind speed, an early result of Greeley et al. (2006). The estimated global mean dust flux was 4×10^3 kg km^{-2} per martian year. Multiplying this with the surface area of Mars yields 5.8×10^{11} kg per martian year. This is surprisingly close to the above estimate by Whelley and Greeley (2008), considering the difference in methodology, but this may of course be coincidental.

4.2 Total Dust Transport Estimated Based on the Atmospheric State

The studies presented in this section vary greatly, but all include at least some dependence on the atmospheric (and surface) state – and its control on dust devil occurrence and dust flux— in making their estimates of regional or global dust transport by dust devils. Conceptually, they all follow the same formula: *total dust devil lifting* = *a scaling factor* × *a function of the atmospheric (and surface) state*. In some studies, the scaling factor is based on upscaling field measurements of individual dust devil fluxes (e.g. Koch and Renno 2005; Jemmett-Smith et al. 2015), while in others it is based on requiring that the total global dust loading due to dust devils matches the observed background dust load (e.g. for Mars Basu et al. 2004; Kahre et al. 2005), or based on explicit modeling of dust lifting by individual vortices (Klose and Shao 2016). In some studies, the function is continuous, implying a dust devil strength or intensity (e.g. Koch and Renno 2005; Newman et al. 2002a), while in others it is discrete, predicting only that dust devils would or would not occur (e.g. Jemmett-Smith et al. 2015). Finally, in some studies, the atmospheric state is obtained from observations (e.g. Koch and Renno 2005), while in others it is taken from a model (e.g. Jemmett-Smith et al. 2015; Klose and Shao 2016). While the studies overlap considerably in terms of their various inherent assumptions and simplifications, this section is divided up according to the scaling factor used.

4.2.1 *Scaling Dust Fluxes Based on Field Measurements of Individual Vortices (Earth)*

The first global estimate for dust devil dust transport on Earth was achieved by Koch and Renno (2005). They assumed that individual dust devils lift dust at a rate of $0.7 \times 10^{-3}\ \mathrm{kg\,m^{-2}\,s^{-1}}$, based on field measurements of dust concentration and vertical wind speed in dust devils (Kaimal and Businger 1970; Renno et al. 2004). They then used the thermodynamic theory for dust devils of Renno and Ingersoll (1996) and Renno et al. (1998) to estimate the fractional area over which dust devils should be active. The theory developed by Renno et al. (1998) describes a convective vortex (dust devil) as a heat engine that performs mechanical work against frictional dissipation. Driven by this convective heat engine, the dust devil activity, F_{av}, is approximated by

$$F_{av} \approx \eta H \tag{6}$$

where η is the thermodynamic efficiency of the heat engine and H is the surface sensible heat flux. H is approximately proportional to the temperature difference between the surface and the near-surface air. The thermodynamic efficiency, η, is approximately $\eta = 1 - b$, with

$$b = \frac{p_s^{\chi+1} - p_{\mathrm{top}}^{\chi+1}}{(p_s - p_{\mathrm{top}})(\chi+1)p_s^{\chi}} \tag{7}$$

where p_{top} is the pressure at the top of the PBL, p_s is the surface pressure, and χ is the specific gas constant divided by the specific heat capacity at constant pressure. Based on these expressions, the dust devil activity (and thus the amount of dust transported by dust devils) increases with both increasing boundary layer thickness and increasing sensible heat flux.

Koch and Renno (2005) used atmospheric sounding data from 9 locations in Algeria, China, Iran, Saudi Arabia, and the USA to calculate the fractional area, σ, covered by dust devils (Renno and Ingersoll 1996):

$$\sigma \approx \left(\frac{\mu}{\eta}\right)^{1/2} \left(\frac{\Delta p_a}{\rho g \tau_R}\right)^{3/2} \left(\frac{F_{\mathrm{in}}}{\rho}\right)^{-1/2} \tag{8}$$

where μ is a dimensionless coefficient, Δp_a is the pressure decrease from the surface to the top of the convective PBL, ρ is air density, g is gravitational acceleration, τ_R is the radiative time scale, and F_{in} is the heat input into the vortex. τ_R was estimated using a theoretical approximation and F_{in} was obtained from measurements. Koch and Renno (2005) found that σ varied only slightly for the locations and time periods considered ($\sigma \approx 3 \times 10^{-5}$). Based on observations, Koch and Renno (2005) further assumed that dust devils likely occur 8 h per day at 72 days per year (80% of a three-month period) and that 40% of global arid and semi-arid areas are dust sources, leading to a global dust source area of $(1.3 \pm 0.2) \times 10^7$ km^2. Using these area and time fractions of dust devil occurrence, Koch and Renno (2005) estimated that dust devils annually transport $\sim 566 \times 10^9$ kg of dust with an uncertainty of 18%.

Jemmett-Smith et al. (2015) used the same individual dust devil fluxes (0.7 g m^{-2} s^{-1}) derived from field measurements as Koch and Renno (2005). Rather than using a fixed annual occurrence time, they instead determined when and where dust devils should (or should not) occur based on the following constraints: locations with high convective buoyancy and low frictional dissipation, using a criterion suggested by Lyons et al. (2008) and a high lapse rate criterion as suggested by Ryan (1972), Oke et al. (2007b), and Ansmann et al. (2009) (see Sect. 3.1 for more details). Jemmett-Smith et al. (2015) applied their criteria truly globally using high-resolution ECMWF operational analyses (see Sect. 3.1). They also used a global dust source region mask as opposed to a global dust source fraction as used by Koch and Renno (2005). Jemmett-Smith et al. (2015) assumed the same constant fractional area to be covered by dust devils within each active region (3×10^{-5}) as calculated by Koch and Renno (2005). Overall, Jemmett-Smith et al. (2015) proposed that only 59×10^9 kg of dust is transported by dust devils globally per year, with an uncertainty range for their estimates of ~ 2 to 66×10^9 kg, depending on which dust source mask was applied.

In addition to their differences in approach, the studies described above are both highly sensitive to uncertainties in the estimated individual dust devil dust fluxes, which are key to upscaling their results to produce regional and global estimates. Typical dust devil flux values given by Metzger et al. (2011) are about 2 orders of magnitude less than the values proposed by Renno et al. (2004), used by Koch and Renno (2005) and Jemmett-Smith et al. (2015) to gain their global estimates. Additionally, the fluxes calculated from measured dust concentrations and vertical wind speeds at a given height (as made in field observations of dust devils, see Sect. 2.2) are likely different to surface dust emission fluxes. Convergence in the dust devil vortex may lead to an overestimation of the dust flux at small heights ($\sim$2 m), whereas turbulent diffusion may lead to an underestimation if measurements are taken at larger heights ($\sim$100 m). If sediment coarser than dust is considered, then gravitational settling may lead to an underestimation of the coarse size-fraction even at small heights. The dust fluxes at altitude may thus not be directly transferable to surface sediment loss at a given time and location.

The explicit assumption that the characteristics of individual ('typical' or 'average') dust devils, such as the dust flux, may be 'scaled up' to provide regional and global estimates is problematic. For example, taking one field measurement of a 'typical' diameter and multiplying by some observed number of dust devils ignores the dependence of a number of parameters that scale with diameter. As noted by Lorenz (2011), for a fixed dust-lifting rate per unit area, the area occupied by a single devil varies as the square of diameter, or if advected at a constant speed, the area swept will vary with diameter raised to the nth power (with n between 1.5 and 1.75). This exponent arises because empirically dust devil longevity varies with diameter. For example, Lorenz (2014) suggested that duration varies as roughly $40d^{0.66}$, where d is the diameter in meters and duration is measured in seconds, yielding an exponent of $n \approx 1.66$ for a fixed wind speed. Additional scaling factors may further increase this exponent, however; for example, if larger dust devils are also more intense, then their dust-lifting rate is also expected to be larger as laboratory experiments have shown (Sect. 2.1). Thus despite the strongly skewed size distribution of dust devils (lots of small ones, few large ones) (Kurgansky 2006; Lorenz 2011), the dust lifting may in fact be dominated by the rarest, largest dust devils. Hence any approach that relies on multiplying up the dust lifting by a single 'typical' (i.e. rather smaller) dust devil is unlikely to accurately represent the overall lifting by the full population (see Chap. 8 [Lorenz and Jackson 2016] for a more detailed discussion).

Using different dust source maps, Jemmett-Smith et al. (2015) also showed that uncertainties in global dust sources affect global estimates by 1 order of magnitude. This combines with the roughly 2 orders of magnitude maximum uncertainty in the dust flux, giving up to 3 orders of magnitude uncertainty in the total estimate of global dust emission. Continued advances in satellite observations will reduce uncertainty in dust sources (Ginoux et al. 2012; Schepanski et al. 2012), which will subsequently reduce uncertainty in dust devil contributions. However, the inclusion of improved estimates of dust devil dust fluxes will lead to more substantial advances in total transport estimates given the larger uncertainties involved.

The large difference in the global estimates obtained by Koch and Renno (2005) and Jemmett-Smith et al. (2015) ($\sim 600 \times 10^9$ versus 60×10^9 kg yr^{-1}) demonstrate the effect of the variability in surface and atmospheric conditions on dust devils and dust lifting. The major reason for the difference is likely the better spatial and/or temporal representation of both dust source areas and meteorological conditions by Jemmett-Smith et al. (2015) compared to Koch and Renno (2005). While the latter used a specified proportion of the global area to scale dust devil lifting (1.3×10^7 km^2), Jemmett-Smith et al. (2015) used a spatially resolved dust source description. Both, Koch and Renno (2005) and Jemmett-Smith et al. (2015) used a fixed thermodynamic efficiency, obtained by Koch and Renno (2005) using atmospheric soundings for convectively active days from key dust regions, to deduce a (constant) fractional dust devil updraft area. However, while Koch and Renno (2005) used a fixed temporal fraction to correct their result for days during the active dust season which were without dust devils, Jemmett-Smith et al. (2015) used spatially and temporally varying criteria to constrain their annual estimate. Consequently, the dust devil dust fluxes resulting from both studies differ by one order of magnitude although both employ the same dust flux value for individual dust devils.

4.2.2 *Scaling Dust Fluxes Based on Global Observational Constraints on Dust Loading (Mars)*

Due to the greater radiative impact of atmospheric dust in the thin martian atmosphere and the lack of oceans or large amounts of water to collect dust rapidly once lifted, dust raised into the atmosphere on Mars has a far more direct and larger scale impact on the circulation than on Earth. This likely explains why studies of dust lifting on Mars have tended to focus on representing large-scale seasonal and spatial variations in dust load and their impact on the circulation, rather than on individual dust devils or small dust lifting events. For this reason, all current estimates of transport due to dust devils on Mars—other than the purely observational estimates given in Sect. 4.1—are derived from global atmospheric models, in which dust lifting parameterizations are 'tuned' such that the spatial and seasonal variation of dust loading (and/or the resultant mid-level temperatures) in the model matches that observed.

Dust devils are a larger component of the dust cycle on Mars than on Earth, because almost the entire martian planet is desert-like and typically has large near-surface lapse rates and PBL depths. As a result, lifting by dust devils has long been treated separately to lifting via near-surface wind stress in atmospheric models (e.g. Newman et al. 2002a,b). This is in contrast to the longstanding situation for Earth, as discussed in Sect. 4.2.3. The net result is that all estimates for dust devil transport on Mars given below have a scaling factor derived from global observational constraints on the total dust load (or its climatological effect), and depend on some function of the atmospheric and surface state as simulated by a Mars global circulation model.

To date, dust devil lifting in Mars climate models has been parameterized using functions ranging from a simple dependence on the near-surface lapse rate (e.g. Basu et al. 2004), to the 'dust devil activity' metric defined by Renno et al. (1998) or a combination of the Renno et al. (1998) thermodynamic theory with measurements of dust lifting by a laboratory-produced vortex (e.g. Newman et al. 2002a; Kahre et al. 2006). This first type of parameterization schemes assumes that the dust emission flux is proportional to respectively the ground-to-air temperature difference or the dust devil activity (given in Sect. 4.2.1). An implicit 'threshold' is applied, in that dust devil lifting shuts off when the surface is cooler than the atmosphere above it (switching off the sensible heat flux driving convection). Figure 16 shows an example of model predicted surface sensible heat flux, thermodynamic efficiency, and dust devil activity, which is proportional to dust lifting in the parameterization scheme.

The second parameterization scheme involves a threshold based on a semi-empirical formula for the tangential wind speed around the convective core that is needed for dust lifting to occur, derived from laboratory measurements of dust raising by a vortex (Greeley and Iversen 1985). The tangential wind speed around the vortex core is calculated from the model's atmospheric state via the convective heat engine model, in which the pressure drop to the vortex core, Δp, is given by

$$\Delta p = p_s \left\{ 1 - \exp\left[\left(\frac{\gamma \eta}{\gamma \eta - 1} \right) \left(\frac{\eta_H}{\chi} \right) \right] \right\} \tag{9}$$

where γ is the fraction of the total dissipation of mechanical energy that is consumed by friction at the surface (a free parameter often set to 0.5), and η_H is the horizontal thermodynamic efficiency of the dust devil, given by $\eta_H = (T_0 - \overline{T}_s)/\overline{T}_s$. In this expression, T_0 is the temperature at the center of the vortex at the surface (typically set to the ambient surface temperature) and $\overline{T}_s$ the temperature of the near-surface air outside the vortex.

By substituting Eq. (9) into Eq. (4), an expression for the tangential wind speed around a vortex is obtained:

$$u_\theta = \sqrt{RT \left\{ 1 - \exp\left[\left(\frac{\gamma \eta}{\gamma \eta - 1} \right) \left(\frac{\eta_H}{\chi} \right) \right] \right\}}. \tag{10}$$

Newman et al. (2002b) chose a threshold based on a semi-empirical expression for the tangential wind speed required to lift a single layer of dust in laboratory experiments by Gree-

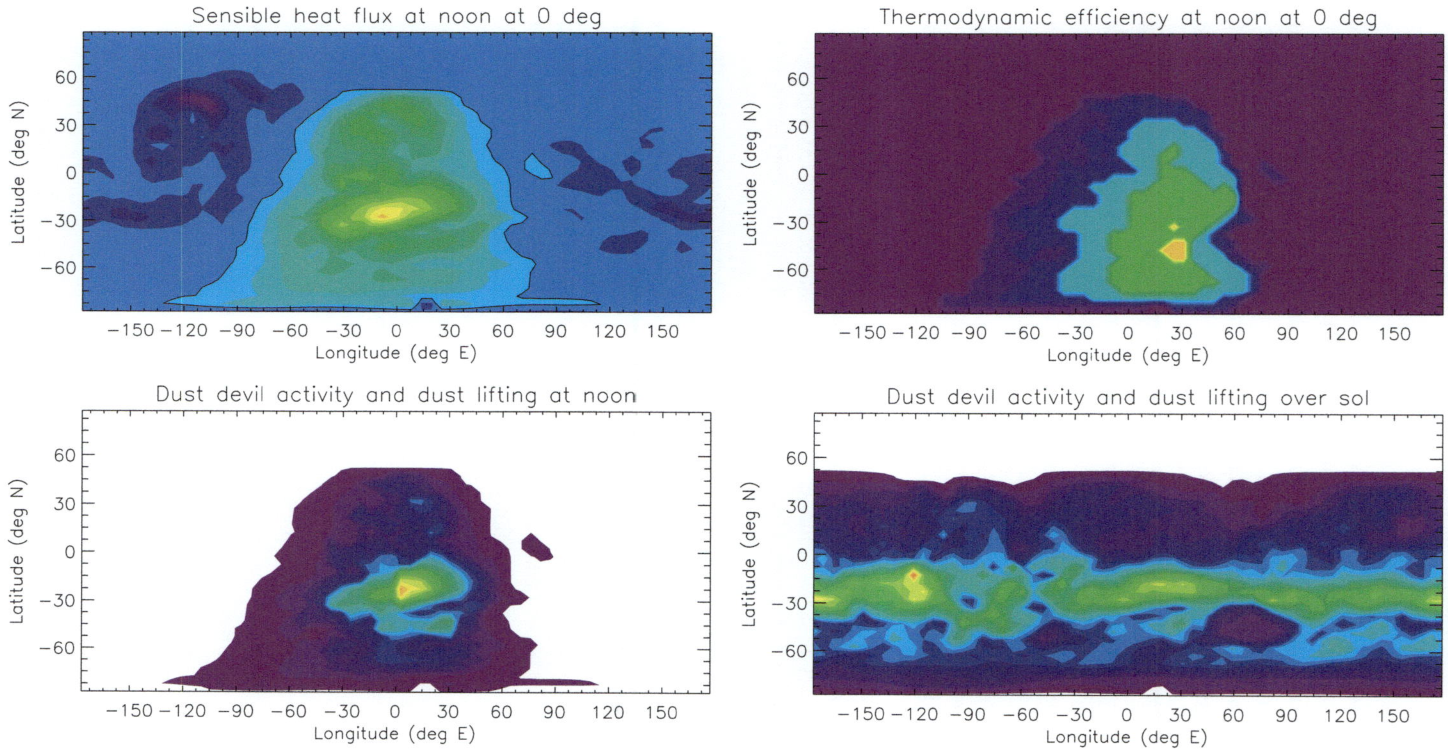

Fig. 16 Predicted surface sensible heat flux, thermodynamic efficiency, and dust devil activity (proportional to dust devil lifting in the parameterization scheme) for noon at 0° longitude in southern hemisphere summer on Mars (Ls ~ 270°). Also shown is dust devil activity averaged over 8 sols

ley and Iversen (1985). This threshold could be chosen using another method, based on the same principles as those used for general saltation processes. However, although it is expressed as a threshold wind speed, this laboratory-based threshold also accounts for other lifting effects associated with dust devils, such as the Δp-effect and influence of electric fields (Sect. 2), in addition to the saltation threshold for particle movement (i.e. it is lower than the value of the saltation threshold alone). Balme and Hagermann (2006) investigated the relative importance of the Δp-effect, e.g. lifting from the reduced pressure at the core of dust devils, compared to the winds. They found that the Δp-effect was most significant when the pressure change occurred rapidly, i.e. for quickly-moving dust devils and for the most intense vortices. In general, lifting based solely on wind shear might therefore be an under-prediction, although the significance of the Δp-effect has not been quantified yet. The relatively higher importance of additional lifting mechanisms specific to dust devils, such as the Δp-effect, is another motivation for parameterizing dust devils separately from other sub-grid or grid-scale winds for Mars (see Chap. 10 [Neakrase et al., 2016] for more details).

Figure 17 shows the depth of dust removed by dust devils according to the first ("dust devil activity"-based) parameterization and its seasonal evolution for regions included in the dust devil survey by Fisher et al. (2005). While both of the dust devil parameterizations described above have been implemented into and used in global dust cycle studies for Mars, the first scheme has been used more consistently because it is simpler and it provides a smoothly varying background dust loading during northern hemisphere spring and summer (Newman et al. 2002b; Basu et al. 2004; Kahre et al. 2006; Mulholland et al. 2013). Additionally, it has been shown to reproduce the observed seasonal behavior of dust devil activity in both the northern and southern hemispheres (Fisher et al. 2005; Kahre et al. 2006). However, we note that both parameterizations are simplifications of the actual situation. The dust devil activity, F_{av}, is a measure of the energy available for dust devils, but in itself it tells nothing about the intensity distribution of the vortex population. As weak vortices lift no dust and the amount of dust that can be lifted depends strongly on the central pressure drop (Sect. 2), two dust devil populations with the same F_{av}, but different pressure drop distributions, would lift different amounts of dust. The second method, involving a semi-empirical formula derived from lab vortex experiments, attempts to take into account the intensity of the vortices, but it assumes that all vortices within a region have equal central pressure drops, which is unrealistic.

On Mars, dust devils are believed to be critical for maintaining the background dust loading (and thus the correct atmospheric temperatures) throughout the year, but particularly during northern hemisphere spring and summer, when very few regional and no global dust storms occur. Most models that employ both a dust devil parameterization and a saltation bombardment scheme for larger-scale dust events first tune the dust devil scheme to roughly match the northern hemisphere summer global dust loading, before tuning the saltation scheme to achieve reasonable dust storm behavior in the northern hemisphere winter (e.g. Basu et al. 2004; Kahre et al. 2006; Newman and Richardson 2015). While saltation typically contributes slightly to the background dust loading too in these models (requiring further tuning of the dust devil contribution), the grid spacings used in most current global Mars models mean that many small-scale winds are not resolved. Thus it should be noted that the 'dust devil' parameterization may be used to account for dust in the global budget that is really lifted by other sub-grid scale circulation features (e.g. small-scale slope or ice cap edge winds, or convective gusts) which are also not represented explicitly, and hence the dust lifted by such a scheme is an over-estimate of the true dust devil lifting if the total budget is correct. It is likely, however, that as Mars models increase in resolution the dust

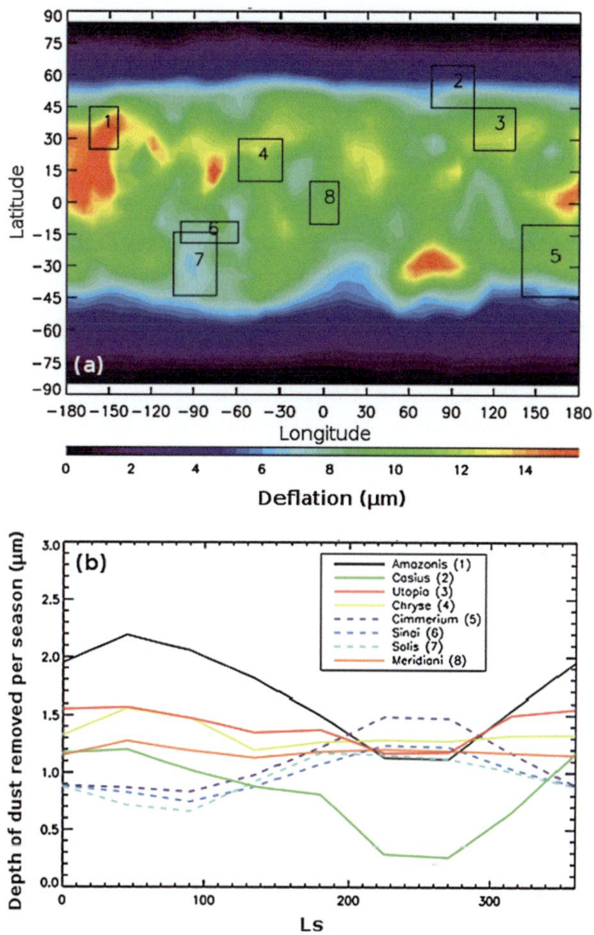

Fig. 17 (a) Annual surface dust removal by dust devils on Mars as obtained by Kahre et al. (2006) using a threshold-independent dust devil parameterization. *Boxes* indicate areas surveyed by Fisher et al. (2005). (b) Seasonal dust removal by dust devils for the areas labeled in (a). (Reprinted from Kahre et al. (2006), with permission from John Wiley and Sons)

lifting generated by the saltation bombardment scheme will become increasingly realistic, thus leading to a more accurate estimate of dust devil lifting.

Kahre et al. (2006) estimated that dust devils contribute 50% of the annual atmospheric dust loading during a year without a global-scale dust storm. This is broadly consistent with figures published in Newman et al. (2005) for the present orbital epoch, who show the total amount of dust lifted from the surface by each mechanism. Also, this fits with the observational estimates on the annual dust devil flux mentioned in Sect. 4.1 (Whelley and Greeley 2008; Cantor et al. 2006). The dust devil lifting has peak dust depletion values which are roughly a factor of two less than the peaks of saltation lifting from explicit wind stress, but the dust devil lifting occurs over a wider area of the planet and is broadly comparable in total effect.

4.2.3 *Scaling Dust Fluxes Based on Large Eddy Simulations of a Dust Devil Population (Earth)*

On Earth, particular attention has been given to the development of schemes that represent dust emission during large-scale dust events such as dust storms. Dust emission in regional and global models has typically been incorporated as a single dust emission parameterization, irrespective of the meteorological processes leading to dust emission. Dust emission schemes are usually based on the process of saltation bombardment (e.g. Shao et al. 1993; Marticorena and Bergametti 1995; Shao 2004; Kok et al. 2014). Dust emission estimates in most models for Earth are thus based on friction velocity, u_*, obtained from a surface layer parameterization, and on a threshold friction velocity, u_{*t}, for saltation (see Chap. 10 [Neakrase et al. 2016]). The dust emission flux is thus only dependent on bulk properties of the atmosphere and of the surface. Progress has been achieved by accounting for subgrid-scale winds in regional and global simulations induced by, for example, dry and/or moist convection (Lunt and Valdes 2002; Cakmur et al. 2004; Takemi et al. 2006; Pantillon et al. 2015). These studies assess the change in modeled dust emissions due to small- and meso-scale winds, but it is not possible to separately assess sediment transport from dust devils alone from such results. However, although convective vortices are small-scale events, they can be frequent and widespread and thus may lift sediment amounts that accumulate to a significant portion of the dust budget, especially in areas where dust devil occurrence frequency is higher than on global average.

This motivated Klose and Shao (2012) and Klose et al. (2014) to develop a parameterization scheme that would represent the direct aerodynamic entrainment of dust by atmospheric turbulence in the absence of strong mean winds leading to systematic saltation. The scheme was shown to be able to simulate the dust lifting in dust devils in the framework of LES (Klose and Shao 2013, 2016). Klose and Shao (2016) analyzed the dust emission generated by dust devils occurring in their LES and proposed a method to estimate dust devil dust flux per unit area and unit time, $\tilde{F}$, as

$$\tilde{F} = n\langle M_{DD}\rangle \tag{11}$$

where $\langle M_{DD}\rangle$ [kg] is the ensemble average of the dust mass transported by individual dust devils. By using $\langle M_{DD}\rangle$, the use of 'typical' values for dust devil size, duration, and intensity can be avoided (see Sect. 4.2.1). From their simulations, Klose and Shao (2016) obtain

$$\langle M_{DD}\rangle = \begin{cases} 0.17 \times \exp(8.15Ri) + 0.015 & \text{for } Ri < 0 \\ 0 & \text{otherwise.} \end{cases} \tag{12}$$

Together with Equation (5), Equations (11) and (12) allow to readily estimate dust devil dust transport in global models based on Richardson number. This enables the separate estimation of dust devil dust transport, for the first time using surface dust emission fluxes.

It must be noted that Equation (11) was obtained based on idealized simulations, i.e. using a homogeneous land-surface, and effects of vegetation cover, soil moisture, or soil type are not yet included. To account for changes in vegetation cover, a preliminary correction for N was applied (Sect. 3.3.2). The estimated dust fluxes do not account for variations in soil type at the moment, but are based on calculations for a loam soil. Further LES runs would be required to quantify the effects of changes in vegetation cover and soil type on $\langle M_{DD}\rangle$ and incorporate them into Eq. (12). This is feasible as the LES model and the dust emission scheme used by Klose and Shao (2016) are capable of accounting for flow changes due to vegetation or other roughness elements (Shao et al. 2013) and changes in dust emissions due to different soil types.

Applying Equations (5), (11), and (12) to the model results of Klose (2014) for Australia (see Sect. 3.3.2) yielded an estimate of the contribution of dust devils to the Australian dust budget. In dependence on the number of dust devils, the amount of dust transported by dust devils varies with season. Based on N_σ as the number of dust devils, the largest

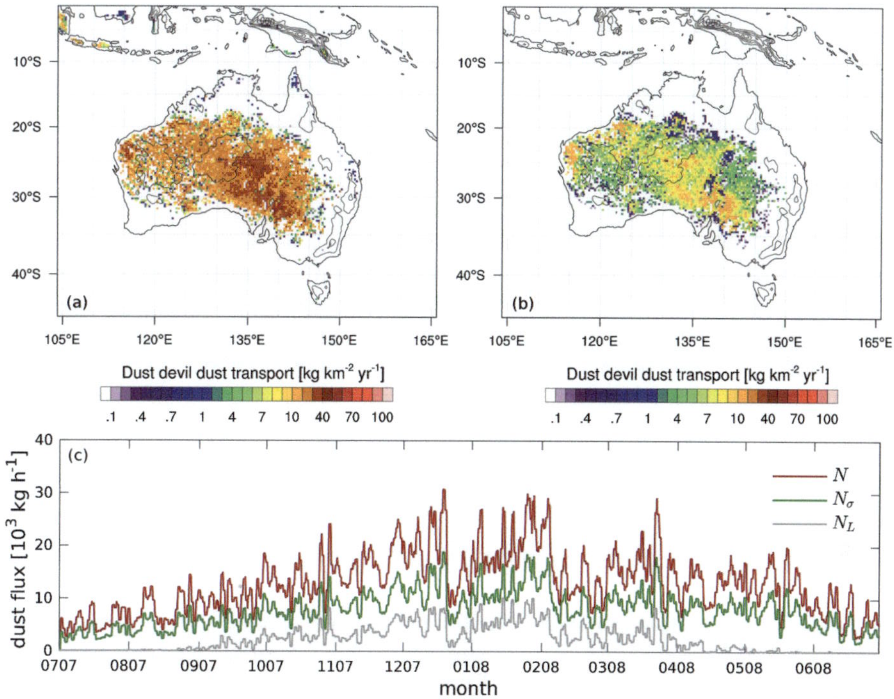

Fig. 18 (a) Estimate of the total dust transport by dust devils in Australia from July 2007 to July 2008 based on Eqs. (11), (12), and N_σ (Eq. (5) with correction for vegetation cover); (b) as (a), but using N_L (Eq. (5) with vegetation correction and lapse rate criterion). (c) Time series of total dust devil dust transport in the simulation domain. *Lines* show 24-h running means of hourly dust transport estimated based on N (Eq. (5), *red*), N_σ (*green*), and N_L (*gray*)

fluxes occurred in southern hemispheric summer with 1–3 $\text{kg}\,\text{km}^{-2}\,\text{mon}^{-1}$ in wide areas of central Australia, and up to about 10 $\text{kg}\,\text{km}^{-2}\,\text{mon}^{-1}$ at particular locations. During the year of simulation, totally between 10 and 50 $\text{kg}\,\text{km}^{-2}\,\text{yr}^{-1}$ of dust were transported by dust devils in central Australia based on this approach. Larger values only occurred at particular grid points. Almost identical results were obtained if a minimum lapse rate of 1 $\text{K}\,\text{m}^{-1}$ was used as additional criterion for the number of dust devils, but smaller fluxes of predominantly 5–30 $\text{kg}\,\text{km}^{-2}\,\text{yr}^{-1}$ were obtained during the simulation year using a minimum lapse rate of 8.5 $\text{K}\,\text{m}^{-1}$ (Fig. 18b). This estimate is smaller than that obtained by Jemmett-Smith et al. (2015), which shows values for dust devil dust transport in the category of 0–2×10^3 $\text{kg}\,\text{km}^{-2}\,\text{yr}^{-1}$ for most regions of Australia, but up to $\sim 20 \times 10^3$ $\text{kg}\,\text{km}^{-2}\,\text{yr}^{-1}$ at particular locations.

Figure 18c shows time series of domain integrated dust transport by dust devils during the investigation period. While dust fluxes of up to about 5×10^3 $\text{kg}\,\text{h}^{-1}$ were determined for the winter months June and July when using N_σ, no significant dust devil dust transport is estimated based on N_L (minimum lapse rate of 8.5 $\text{K}\,\text{m}^{-1}$ required). Dust fluxes remain below 20 and 10×10^3 $\text{kg}\,\text{h}^{-1}$ for N_σ and N_L, respectively, during the whole year. Without correction for vegetation cover, i.e. using N as in Eq. (5), dust fluxes can exceed 10×10^3 $\text{kg}\,\text{h}^{-1}$ in winter and reach maxima of about 27×10^3 $\text{kg}\,\text{h}^{-1}$ in summer.

In total, a dust mass of about 0.11, 0.07, or 0.02×10^9 $\text{kg}\,\text{yr}^{-1}$ was lifted by dust devils when using N, N_σ, or N_L as the number of dust devils. Estimates of total annual dust emissions for Australia vary largely (Huneeus et al. 2011; Shao et al. 2011). For particles with diameters of up to 20 μm, estimates range from 14.9 to $106 \times 10^9\ \mathrm{kg\,yr^{-1}}$ with an average of $59 \times 10^9\ \mathrm{kg\,yr^{-1}}$ (Tanaka and Chiba 2006; Huneeus et al. 2011). Compared to this, the dust devil contribution to the Australian dust budget would be very small (<1%). It must be noted, however, that the dust emission fluxes obtained by Klose and Shao (2016) are smaller or on the lower edge of those obtained in laboratory or field (Neakrase and Greeley 2010b; Metzger et al. 2011) as only aerodynamic entrainment is accounted for in the model and intermittent saltation is not yet included. Additionally, large parts of Australia have soil types with a larger fraction of dust-size particles than loam, e.g. sandy loam or clay, thus leading to larger dust emission fluxes. An increase in dust flux by 1–2 orders of magnitude to be closer to the values estimated by Renno et al. (1998) and Metzger et al. (2011), would yield a total contribution of only between 0.3–19%. On the contrary, surface crusting especially in clay-rich areas might again reduce dust emission and is not yet accounted for in any of the estimates given in this paper. Not least, the choice of the reference value (here $59 \times 10^9\ \mathrm{kg\,yr^{-1}}$) is naturally pivotal to the resulting dust devil contribution.

5 Impact on Local, Regional and Global Scales

The frequency of dust devils varies strongly from region to region and from season to season both on Earth and Mars. Dust devils are most significant on local and regional scales, less so on global scales, as has been shown in the previous sections. Nevertheless, the impact of dust devils on all spatial scales in the Earth system and in the martian environment can be manifold. The effects extend from property/instrument damage on local scales, to air quality on regional scales, to climate feedbacks on global scales. Despite their relatively small extent and short duration, their frequency of occurrence in particular areas make them a fundamental mechanisms for dust uplift with important impacts on daily life on Earth as well as on the climate and environment of Earth and Mars.

5.1 Local Scale: Incidents Caused by Dust Devils

At a local level, the winds associated with dust devils can be disruptive to picnics, sports etc., although occasionally can cause damage to light structures and even occasional deaths. These effects are due to the high winds within dust devils and are summarized briefly in Chap. 1 (Lorenz et al. 2016). The dust lifted by dust devils can present health and more general air quality issues, as is discussed at the regional level in Sect. 5.2. Local dust-lifting can also have a local impact on visibility: this may have been a factor in at least one vehicular fatality attributed to a dust devil (a South African astronomer—see Lorenz 2013) and the loss of pilot situational awareness in a hovering helicopter due to the dust in a dust devil was a stated factor in the crash of that aircraft (Lorenz and Myers 2005). On Mars, *Spirit* rover measurements of background dust opacity show 4–6 times larger opacities during the dust devil season compared to other seasons at Gusev crater on Mars (Greeley et al. 2010). The large number of dust devils on Mars during southern hemispheric spring and summer compared to other seasons suggests a strong contribution of dust devils to the generation of local dust haze at that time of year. However, a larger dust load decreases dust devil activity due to its negative feedback on surface sensible heat flux. In particular, regional-scale dust storms can suddenly increase dust opacity and thus shut down dust devil activity (Greeley et al. 2010).

The local impact of dust transport by dust devils is not always negative, however. Dust devils appear to be responsible for the 'dust-clearing events' noted on the Mars Exploration Rover *Spirit* (Lorenz and Reiss 2015) which restored the electrical output of its solar panels. Steady accumulation of dust on the panels leads to a progressive drop in power, which would eventually end the mission, but at the start of dust devil season, dust was suddenly removed and power levels improved, allowing the mission to operate much longer than originally planned.

5.2 Regional Scale: Dust Devil Feedback to the Surface and the Atmosphere

5.2.1 Earth

As shown in this paper, the dust devil contribution to the continental or global dust budgets on Earth seems to be small. However, dust devils can be a major dust event type on regional scales depending on the season, such as in the Eldorado (Nevada) or Avra Valleys (Arizona), USA (Sinclair 1969; Metzger et al. 2011), in parts of the Atacama and Sechura deserts in South America (Kurgansky et al. 2011; Hesse 2012; Reiss et al. 2013; Jemmett-Smith et al. 2015), or at locations in southeastern Australia (Oke et al. 2007b; Klose and Shao 2016). As a large part of the dust lifting in dust devils occurs through aerodynamic entrainment, dust devils can still cause emissions from soil surfaces which contain little or no particles in the saltation size range. This may for example be the case on playas, where the surface is crusted but may be covered by a thin dust layer, or on loosely packed silt- and/or clay-rich soils, such as on some cultivated agricultural fields. On such surfaces, dust devils may become a significant emission process.

In regions that favor dust devil development, the frequent occurrence of dust devils leads to a persistent removal of the top-soil layer. Although dust devils do not travel long distances, the vertical mixing in the convective boundary layer is intense. As a consequence, suspended particles are likely transported farther than the location of dust devil cessation might suggest, thus potentially leading to a net soil transport away from the typical formation regions of dust devils. As discussed by Oke et al. (2007a), this might lead to local topographic changes between non- or weakly vegetated areas where dust devils occur frequently, and strongly vegetated areas, thus potentially affecting the hydrological cycle in the area. Not least, the loss of top-soil particles is associated with a loss of soil nutrients, minerals, and carbon, thus impacting on soil productivity (e.g. Sterk et al. 1996; Webb et al. 2013). The degree to which these effects occur in a particular area and the feedback they have on dust devil formation, i.e. smaller availability of loose soil particles or changes in water flow, would need to be investigated on a case by case basis.

The strong vertical mixing of particles entrained by dust devils also has an effect on air quality. Due to the small terminal velocities of particles in the lower dust-size range, e.g. below 10 μm in diameter, these particles, once entrained, have long atmospheric residence times (Shao 2008). The frequent occurrence of dust devils in a particular area might thus lead to a significant increase of suspended particulate matter (PM) in the atmosphere. Effects are reduced visibilities (not for very low PM levels), changes in atmospheric thermodynamic properties through radiative interactions, increased availability of cloud condensation and ice nuclei, and damage to human and animal health through inhalation of pathogens or chemical contaminants transported with dust particles (Rosenfeld et al. 2001; Kellogg and Griffin 2006; Boucher et al. 2013; Miller et al. 2014).

5.2.2 *Mars*

When dust devils inject dust to great heights, through the deep planetary boundary layer (Petrosyan et al. 2011) and into the free atmosphere, it can be advected to much greater heights by the global circulation and reach 80 km altitude or more. This dust is rapidly mixed and thus is indistinguishable from that injected by any other source. Dust absorption is a major source of internal heating in the martian atmosphere and a primary component of the radiative-dynamic feedback and hence interannual variability (e.g. Read and Lewis 2004).

On a smaller scale, the direct radiative effects of dust associated with dust devil circulations have yet to be studied in great detail. Fuerstenau (2006) suggested that dust devils might be larger on Mars than on Earth as a result of solar heating via dust absorption. Internally-heated parcels of air would become more positively buoyant as a result of the warming and so ascend to greater heights, enhancing the dust devil. A similar radiative feedback that could lead to convective instability within larger, but still local dust storms has been demonstrated in a mesoscale model (Spiga et al. 2013). These calculations were, however, still for circulations on much larger scales than a typical individual dust devil that would have a lower dust load. Whether or not an individual dust devil would be prone to radiative feedback is yet to be investigated. The coupled modeling needed to establish this has not been performed for Mars, although there has been some initial progress with Monte Carlo simulations of radiative transfer within idealized dust devil-like cylindrical dust distributions (Mason et al. 2013).

It is clear that, although the amount of dust lifted by dust devils on Mars is difficult to quantify, the dust that is lifted can have a profound impact on atmospheric circulations through radiative feedback. It is less clear that dust devil lifting can have a profound effect upon the surface of the planet in turn. The most obvious change seen is the presence of 'graffiti-like' dust devil tracks that can be seen from orbit (see Fig. 10). The tracks appear dark (sometimes bright) compared to their surroundings (see Chap. 4 [Reiss et al. 2016]). This is because the bedrock itself is darker than the fine layer of dust which is deposited over it, and which tends to settle slowly from the atmosphere so forming a smooth surface with slightly higher albedo. Where the dust devils pass they lift some of the fine material and also act to redistribute grains of various sizes, changing the surface roughness and leaving a darker track. It may not be the case that all the fine dust is removed, leaving clean rock, but rather that the distribution of granular particle sizes is mixed and altered (for the different formation processes, see Chap. 4 [Reiss et al. 2016] and Reiss et al. 2010, 2011, 2013). Verba et al. (2010) show some more examples of dust devil tracks on Mars. Michaels (2006) provided an example simulation of a dust devil generating a track (see Chap. 7 [Spiga et al. 2016]).

Although dust devils clearly change the albedo of the martian surface on small scales (the tracks are of the order of magnitude 10 m wide and may be coherent over distances of kilometers or spiral and fade rapidly) there is presently little evidence that they change the landscape in a consistent way on longer timescales, for example by preferentially stripping dust from certain regions of the planet on an annual-mean basis. There are very many dust devils that occur over large regions of the surface. Dust is lifted from a variety of sites in an apparently stochastic pattern and eventually settles back down from the atmosphere. If dust devils were equally common everywhere, then such a process would produce no long term trend in dust distribution. Dust devils do, however, seem to be less common in certain localities, such as Gale Crater (Moores et al. 2015). If dust devils are a major component of the equatorial dust lifting and are not counter-balanced by enhancements in other small-scale effects, such as slope winds, then such regions should preferentially acquire fine dust over long timescales. Although relatively small masses of surface material are moved by each dust devil, any consistent bias in lifting might make significant changes to the size-distribution of dust in the upper layers of the regolith. Any consistent change to surface albedo over larger areas affects the radiative budget at the surface and so impacts on the martian circulation and climate in turn.

5.3 Global Scale: Interactions with Climate and Environment

5.3.1 Earth

For Earth, only two studies (Koch and Renno 2005; Jemmett-Smith et al. 2015) have attempted to upscale dust devil dust emissions to global scales to get an estimate of their contribution to the total emissions. Best estimates given by these studies vary by one order of magnitude, illustrating the substantial multiplicative uncertainties related particularly to dust flux and source area. Koch and Renno (2005) estimated a contribution of dust devils to the global dust budget of $26\% \pm 18\%$ based on a global annual dust emission of 2150×10^9 kg (IPCC 2001). Recent results by Jemmett-Smith et al. (2015) estimated that dust devils contribute $\sim 3\%$ (uncertainty 1 to 25%) to the total mineral dust budget on Earth, when assuming an annual global dust emission of 2000×10^9 kg (Griggs and Noguer 2002; Shao et al. 2011; Huneeus et al. 2011). In addition, if the dust devil flux values given by Metzger et al. (2011) were applied to the results by Koch and Renno (2005) and Jemmett-Smith et al. (2015), both would give dust devil contributions of less than 1% to the total mineral dust budget on Earth.

The continental-scale estimate for Australia shown in Sect. 4.2.3 seems to support that dust devils are of little importance from the perspective of global-scale dust emissions. With an estimated contribution of dust devils of 0.03–0.19% to the Australian dust budget, the dust transported by dust devils is smaller than one would think given the high frequency of dust devil reports on the continent (Fig. 9). For the USA, Gillette and Sinclair (1990) found that dust devils accounted for about 66% of the total lifted dust mass on the continent based on aircraft measurements conducted near Tucson, Arizona. The estimate obtained by Gillette and Sinclair (1990) is grounded on general assumptions on the relative magnitude of the dust flux in areas other than the test area that are based on climate and vegetation. The estimate does thus likely contain large uncertainty. In addition, no seasonal variation has been considered, so the percentage resulting from a dust devil census conducted during the dust devil season is probably an overestimation. However, the study by Gillette and Sinclair (1990) was the first that aimed to upscale dust devil measurements to a continental scale. Overall, the later results suggest that dust devils are unlikely to be a big player in the global mineral dust cycle on Earth.

5.3.2 Mars

As has been noted throughout this paper, observations of martian dust devils are incomplete, with significant detections at only a handful of lander sites for limited periods (see Sect. 3.2) and through orbital imaging of the largest dust devils and their surface tracks. Any quantitative estimates of global impacts are therefore reliant to some extent on extrapolation and modeling, with considerable uncertainties remaining in both.

The impact of dust devils on the global atmospheric dust loading is still highly uncertain. As noted in Sect. 4.2.2, many Mars models employ specific parameterizations for dust devils (e.g. Newman et al. 2002a,b; Basu et al. 2004; Kahre et al. 2006; Mulholland et al. 2013;

Newman and Richardson 2015). In these models the dust devils parameterization contributes up to one half of the total annual dust budget (Newman et al. 2005; Kahre et al. 2006). Such lifting tends to peak around both solstices in the summer hemisphere, though with greater lifting in the southern hemisphere during summer. However, the spatial and temporal variation in dust devil lifting is relatively smooth, and there is at least weak dust devil lifting predicted over most of the planet at most times of year. This is in direct contrast to lifting from explicit near-surface wind stress schemes, which tend to peak in restricted locations and times (particularly during the dust storm season, solar longitude $L_s \sim 180°–360°$), although they may lift several times more dust at these peak locations (e.g. Cantor et al. 2001; Cantor 2007; Wang and Richardson 2015).

Although dust may remain suspended for many tens of days in the martian atmosphere, even global dust events decay more rapidly than the seasonal timescale. The result is that the dust devil parameterization is required to support the background loading of dust in the model, as observed throughout the year, particularly in the northern hemisphere summer ($L_s = 90°–180°$), when dust loadings are lower, but still significant and repeatable (Smith 2004; Montabone et al. 2015). Section 4.2.2 has already questioned whether the total dust lifting that is accounted for by the dust devil parameterization in a model is truly the result of lifting by dust devils alone or may be used to account for other small scale lifting, from mesoscale and microscale winds, that is not represented in the model. The parameterization is likely tuned to account for all missing processes. On the other hand, such schemes have been shown to reproduce broadly the observed seasonal behavior of dust devil activity in both the northern and southern hemispheres (Fisher et al. 2005; Kahre et al. 2006) and so are not totally misleading.

Newman et al. (2002b) show that a dust devil parameterization, in contrast to the near-surface wind stress parameterization, typically has a negative feedback: a clearer atmosphere results in a greater atmosphere-surface thermal contrast and so increases dust devil activity and vice versa. Thus the dust devil lifting scheme will tend to prevent the model atmosphere from reaching unrealistically low dust loadings, even in northern hemisphere summer when winds are relatively weak on average. Even at these lower levels (typically visible total opacities of 0.1–0.2), atmospheric dust still plays a major role in the martian radiation budget.

6 Conclusions

This paper provides a review of dust devil dust transport on Earth and Mars based on state-of-the-art methods in laboratory-based and field studies, remote sensing, and modeling. Technological advances have led to significant advances during the recent years, providing means for further investigations of dust devils.

Laboratory studies on dust devils (Neakrase et al. 2006; Neakrase and Greeley 2010a,b) allowed for the detailed investigation of individual vortices and thus provided insights into dust devil dynamics. Further developed instruments, increasing data storage capacity, and higher-resolved images led to more comprehensive (in situ and remote) field observations of dust devils (e.g. Metzger et al. 2011; Mason et al. 2014; Whelley and Greeley 2008; Lorenz and Jackson 2015; Reiss and Lorenz 2015). High-performance computing environments allow for the use of high-resolution large-eddy simulations—with and without coupled dust emission schemes—to study the characteristics of individual dust devils. New methods to estimate dust devil occurrence frequency and dust devil dust transport (Jemmett-Smith et al. 2015; Klose and Shao 2016) yield the opportunity to study the effect of dust devils on a significantly higher spatio-temporal resolution, thus hopefully leading to more accurate results. In global models, the use of dust devil parameterizations enabled the reproduction of the martian background annual dust cycle (e.g. Newman et al. 2002a; Kahre et al. 2006).

Naturally, the overlap in methods used for Earth and Mars is small in the field of direct observations. On Earth, in situ observations are more easy, than on Mars, as the areas are more accessible. Owing to the stochastic nature of dust devils, their in situ measurement remains challenging, however. On Mars, the observations have a stronger focus on dust devil detections from imagery, which may bias them more toward larger events. However, while the use of imagery data for dust devil analyses is common practice for Mars, it is not as well established for Earth. This leads to more comprehensive observational datasets on the spatial distribution of dust devils for Mars, while datasets for Earth focus more on the properties of individual dust devils, which vary significantly. Consequently, it is important to ensure that general statements regarding dust devil properties are based on a sufficiently large dust devil population (see also Chap. 8 [Lorenz and Jackson 2016]). If such care is not taken, then conclusions—on topics such as the contribution of dust devils to the total dust budget—may not be representative.

Large-eddy simulation models can be applied similarly effective for Earth as for Mars, although domain extents need to be larger on Mars due to a deeper planetary boundary layer. Hence, more computing power and memory would be required if simulations were to be conducted at a similar spatial resolution. Comparison of the model results to observations is again more difficult for Mars due to limited data availability. This complicates the application of dust emission schemes developed and calibrated for Earth to the simulation of martian dust devils. For both, Earth and Mars, large-eddy simulation is usually applied in an idealized setup. The investigation of the effects of non-ideal conditions on dust devil properties, such as surface heterogeneity, vegetation cover, etc., requires additional modeling efforts in the future. Some of the models available to date already provide the necessary framework to conduct such modeling studies.

Different methods are used in terrestrial and global Mars models. While the focus of dust modeling on Earth has until recently almost exclusively been on large-scale dust events such as dust storms, separate parameterization schemes for dust devils have existed for a while for use in martian global models. This originates from the important role dust devils play in the martian climate, whereas they likely have smaller effects on the terrestrial climate. The dust devil parameterizations in models account for roughly half the total martian dust lifting, although it is arguable whether all this lifting is truly a result of dust devils or other small lifting events not captured in the models. These parameterizations are constrained only to reproduce the global atmospheric dust budget over annual timescales, acting in combination with saltation lifting by explicitly resolved winds. Although there is some broad agreement in terms of the time of day and year for peak lifting between model lifting schemes and observations of dust devil occurrence, the details of peak location differ and the actual lifted dust flux can only be roughly estimated.

The results of this paper suggest that dust devils on Earth are more likely to have local and regional effects on air quality and potentially on geomorphology, than global effects on climate. These potential environmental impacts of dust devils have not yet been quantified, however. On Mars, the radiative feedback of dust haze generated by dust devils may be significant on regional and possibly also on global scales, although uncertainties exist concerning the accuracy of estimated dust devil contributions to global dust aerosol. Due to the wide distribution of regions prone to dust devil occurrence, a strong effect on martian geomorphology appears unlikely. However, areas that exhibit persistent large numbers of dust devils may experience a significant sorting of top-soil particles both on Earth and Mars.

The methods and tools summarized in this paper provide the necessary means to quantify dust devil sediment transport on different spatial scales and for different locations, thus enabling the investigation of the effect of dust devils on local climate and geomorphology. It is now possible to conduct high-resolution studies of dust devils for focus areas with regional models, where in-situ field observations and remote sensing data can be compared to model outputs to gain knowledge of the meteorological conditions controlling their occurrence, and to further improve their representation in model parameterizations. At the moment, the discrepancies between individual research outcomes are large. Understanding the benefits and limitations of the different approaches is essential to interpret the results. A closer linkage between research disciplines, approaches, and study areas would help to relate individual research outcomes and be highly beneficial for the study of dust devils in the future, as would more interaction between those working on Earth and Mars studies, especially in remote sensing an modeling.

Acknowledgements We wish to thank Luca Montabone, one anonymous reviewer, and two editors for their careful review and valuable comments, and Bruce Cantor for his permission to reuse Fig. 4 of Cantor et al. (2006) in this paper. Bradley Jemmett-Smith and Peter Knippertz would like to acknowledge funding from the European Research Council Grant 257543 "Desert Storms". Ralph Lorenz acknowledges the support of NASA Mars Fundamental Research Program grant NNX12AI04G. Not least, we are grateful to the International Space Science Institute (ISSI), Bern, Switzerland, and to the conveners for organizing the workshop "Dust Devils on Mars and Earth" (www.issibern.ch/workshops/dustdevils/).

References

A. Ansmann, M. Tesche, P. Knippertz, E. Bierwirth, D. Althausen, D. Müller, O. Schulz, Vertical profiling of convective dust plumes in southern Morocco during SAMUM. Tellus B **61**(1), 340–353 (2009)

R.A. Bagnold, *The Physics of Blown Sand and Desert Dunes* (Methuen, London, 1941), p. 265

M. Balme, R. Greeley, Dust devils on Earth and Mars. Rev. Geophys. **44** (2006). doi:10.1029/2005RG000188

M. Balme, A. Hagermann, Particle lifting at the soil-air interface by atmospheric pressure excursions in dust devils. Geophys. Res. Lett. **33**, L19S01 (2006). doi:10.1029/2006GL026819

M.R. Balme, P.L. Whelley, R. Greeley, Mars: dust devil track survey in Argyre Planitia and Hellas Basin. J. Geophys. Res. **108**(E8), 5086 (2003). doi:10.1029/2003JE002096

M. Bangert, A. Nenes, B. Vogel, D. Barahona, V.A. Karydis, P. Kumar, C. Kottmeier, U. Blahak, Saharan dust event impacts on cloud formation and radiation over Western Europe. Atmos. Chem. Phys. **12**, 4045–4063 (2012). doi:10.5194/acp-12-4045-2012

S. Basu, M.I. Richardson, R.J. Wilson, Simulation of the Martian dust cycle with the GFDL Mars GCM. J. Geophys. Res., Planets **109**(E11), E11006 (2004). doi:10.1029/2004JE002243

O. Boucher, D. Randall, P. Artaxo, C. Bretherton, G. Feingold, P. Forster, V.-M. Kerminen, Y. Kondo, H. Liao, U. Lohmann, P. Rasch, S.K. Satheesh, S. Sherwood, B. Stevens, X.Y. Zhang, Clouds and Aerosols, in *Climate Change 2013: the Physical Science Basis. Contribution of Working Group I to the Fifth Assessment Report of the Intergovernmental Panel on Climate Change*, ed. by T.F. Stocker, D. Quin, G.-K. Plattner, M. Tignor, S.K. Allen, J. Boschung, A. Nauels, Y. Xia, V. Bex, P.M. Midgley (Cambridge University Press, Cambridge/New York, 2013)

C.S. Bristow, K.A. Hudson-Edwards, A. Chappell, Fertilizing the Amazon and equatorial Atlantic with West African dust. Geophys. Res. Lett. **37** (2010). doi:10.1029/2010GL043486

R.V. Cakmur, R.L. Miller, O. Torres, Incorporating the effect of small-scale circulations upon dust emission in an atmospheric general circulation model. J. Geophys. Res., Atmos. **109**(D7), D07201 (2004). doi:10.1029/2003JD004067

B.A. Cantor, K.M. Kanak, K.S. Edgett, Mars Orbiter Camera observations of Martian dust devils and their tracks (September 1997 to January 2006) and evaluation of theoretical vortex models. J. Geophys. Res. **111**, E12002 (2006). doi:10.1029/2006JE002700

B.A. Cantor, MOC observations of the 2001 Mars planet-encircling dust storm. Icarus **186**(1), 60–96 (2007). doi:10.1016/j.icarus.2006.08.019

B.A. Cantor, P.B. James, M. Caplinger, M.J. Wolff, Martian dust storms: 1999 Mars Orbiter Camera observations. J. Geophys. Res. **106**(E10), 23653–23687 (2001). doi:10.1029/2000JE001310

J.J. Carroll, J.A. Ryan, Atmospheric vorticity and dust devil rotation. J. Geophys. Res. **75**(27), 5179–5184 (1970). doi:10.1029/JC075i027p05179

D.S. Choi, C.M. Dundas, Measurements of Martian dust devil winds with HiRISE. Geophys. Res. Lett. **38**(24), L24206 (2011). doi:10.1029/2011GL049806

S.M. Cowie, P. Knippertz, J.H. Marsham, Are vegetation-related roughness changes the cause of the recent decrease in dust emission from the Sahel? Geophys. Res. Lett. **40**(9), 1868–1872 (2013)

C. de Beule, G. Wurm, T. Kelling, M. Küpper, T. Jankowski, J. Teiser, The Martian soil as a planetary gas pump. Nat. Phys. **10**, 17–20 (2014). doi:10.1038/nphys2821

P. De Deckker, C.I. Munday, J. Brocks, T. O'Loingsigh, G.E. Allison, J. Hope, M. Norman, J.-B.W. Stuut, N.J. Tapper, S. van der Kaars, Characterisation of the major dust storm that traversed over eastern Australia in September 2009; a multidisciplinary approach. Aeolian Res. **15**, 133–149 (2014). doi:10.1016/j.aeolia.2014.07.003

J.W. Deardorff, A numerical study of three-dimensional turbulent channel flow at large Reynolds numbers. J. Fluid Mech. **41**(2), 453–480 (1970). doi:10.1017/S0022112070000691

J. Deardorff, Observed characteristics of the outer layer. Short course on the planetary boundary layer (1978)

P.J. DeMott, K. Sassen, M.R. Poellot, D. Baumgardner, D.C. Rogers, S.D. Brooks, A.J. Prenni, S.M. Kreidenweis, African dust aerosols as atmospheric ice nuclei. Geophys. Res. Lett. **30**(14) (2003)

E. Derbyshire, Natural minerogenic dust and human health. Ambio **36**(1), 73–77 (2007). doi:10.1579/0044-7447(2007)36[73:NMDAHH]2.0.CO;2

M.D. Ellehoj, H.P. Gunnlaugsson, P.A. Taylor, H. Kahanpää, K.M. Bean, B.A. Cantor, B.T. Gheynani, L. Drube, D. Fisher, A.-M. Harri, C. Holstein-Rathlou, M.T. Lemmon, M.B. Madsen, M.C. Malin, J. Polkko, P.H. Smith, L.K. Tamppari, W. Weng, J. Whiteway, Convective vortices and dust devils at the Phoenix Mars mission landing site. J. Geophys. Res. **115**(E4) (2010). doi:10.1029/2009JE003413

L.K. Fenton, R. Lorenz, Dust devil height and spacing with relation to the martian planetary boundary layer thickness. Icarus **260**, 246–262 (2015). doi:10.1016/j.icarus.2015.07.028

F. Ferri, P.H. Smith, M. Lemmon, N.O. Rennó, Dust devils as observed by Mars Pathfinder. J. Geophys. Res., Planets **108**(E12) (2003). doi:10.1029/2000JE001421

J.A. Fisher, M.I. Richardson, C.E. Newman, M.A. Szwast, C. Graf, S. Basu, S.P. Ewald, A.D. Toigo, R.J. Wilson, A survey of Martian dust devil activity using Mars Global Surveyor Mars Orbiter Camera images. J. Geophys. Res. **110**(E3), E03004 (2005). doi:10.1029/2003JE002165

S.D. Fuerstenau, Solar heating of suspended particles and the dynamics of Martian dust devils. Geophys. Res. Lett. **33**(19) (2006). doi:10.1029/2006GL026798

V.H. Garrison, E.A. Shinn, W.T. Foreman, D.W. Griffin, C.W. Holmes, C.A. Kellogg, M.S. Majewski, L.L. Richardson, K.B. Ritchie, G.W. Smith, African and Asian dust: from desert soils to coral reefs. Bioscience **53**(5), 469–480 (2003). doi:10.1641/0006-3568(2003)053[0469:AAADFD]2.0.CO;2

B.T. Gheynani, P.A. Taylor, Large Eddy Simulation of typical dust devil-like vortices in highly convective Martian boundary layers at the Phoenix lander site. Planet. Space Sci. **59**, 43–50 (2011). doi:10.1016/j.pss.2010.10.011

D.A. Gillette, P.C. Sinclair, Estimation of suspension of alkaline material by dust devils in the United States. Atmos. Environ. **24**(5), 1135–1142 (1990)

P. Ginoux, J.M. Prospero, T.E. Gill, N.C. Hsu, M. Zhao, Global-scale attribution of anthropogenic and natural dust sources and their emission rates based on MODIS Deep Blue aerosol products. Rev. Geophys. **50**(3) (2012)

R. Greeley, J.D. Iversen, *Wind as a Geological Process on Earth, Mars, Venus and Titan* (Cambridge University Press, New York, 1985), p. 333

R. Greeley, M.R. Balme, J.D. Iversen, S. Metzger, R. Mickelson, J. Phoreman, B. White, Martian dust devils: laboratory simulations of particle threshold. J. Geophys. Res. **108**(E5), 5041 (2003)

R. Greeley, P.L. Whelley, R.E. Arvidson, N.A. Cabrol, D.J. Foley, B.J. Franklin, P.G. Geissler, M.P. Golombek, R.O. Kuzmin, G.A. Landis, M.T. Lemmon, L.D.V. Neakrase, S.W. Squyres, S.D. Thompson, Active dust devils in Gusev crater, Mars: observations from the Mars Exploration Rover Spirit. J. Geophys. Res. **111**(E12), E12S09 (2006). doi:10.1029/2006JE002743

R. Greeley, D.A. Waller, N.A. Cabrol, G.A. Landis, M.T. Lemmon, L.D.V. Neakrase, M. Pendleton Hoffer, S.D. Thompson, P.L. Whelley, Gusev Crater, Mars: observations of three dust devil seasons. J. Geophys. Res., Planets **115**(9), 1–18 (2010). doi:10.1029/2010JE003608

D.J. Griggs, M. Noguer, Climate change 2001: the scientific basis. Contribution of working group I to the third assessment report of the intergovernmental panel on climate change. Weather **57**(8), 267–269 (2002)

Z. Gu, J. Qiu, Y. Zhao, Y. Li, Simulation of terrestrial dust devil patterns. Adv. Atmos. Sci. **25**(1), 31–42 (2008). doi:10.1007/s00376-008-0031-7

G. Hess, K. Spillane, Characteristics of dust devils in Australia. J. Appl. Meteorol. **29**(6), 498–507 (1990)

R. Hesse, Short-lived and long-lived dust devil tracks in the coastal desert of southern Peru. Aeolian Res. **5**, 101–106 (2012)

N. Huneeus, M. Schulz, Y. Balkanski, J. Griesfeller, J. Prospero, S. Kinne, S. Bauer, O. Boucher, M. Chin, F. Dentener, T. Diehl, R. Easter, D. Fillmore, S. Ghan, P. Ginoux, A. Grini, L. Horowitz, D. Koch, M.C. Krol, W. Landing, X. Liu, N. Mahowald, R. Miller, J.-J. Morcrette, G. Myhre, J. Penner, J. Perlwitz, P. Stier, T. Takemura, C.S. Zender, Global dust model intercomparison in AeroCom phase I. Atmos. Chem. Phys. **11**(15), 7781–7816 (2011). doi:10.5194/acp-11-7781-2011. http://www.atmos-chem-phys.net/11/7781/2011/

IPCC, Clouds and aerosols, in *Climate Change 2001: The Scientific Basis. Contribution of Working Group I to the Third Assessment Report of the Intergovernmental Panel on Climate Change*, ed. by J.T. Houghton, Y. Ding, D.J. Griggs, M. Noguer, P.J. van der Linden, X. Dai, K. Maskell, C.A. Johnson (Cambridge University Press, Cambridge/New York, 2001), p. 881

K. Isono, On ice-crystal nuclei and other substances found in snow crystals. J. Meteorol. **12**, 456–462 (1955). doi:10.1016/j.aeolia.2013.11.002

J. Ito, H. Niino, M. Nakanishi, Large eddy simulation on dust suspension in a convective mixed layer. SOLA **6**, 133–136 (2010a). doi:10.2151/sola.2010-034

J. Ito, R. Tanaka, H. Niino, M. Nakanishi, Large eddy simulation on dust devils in a diurnally-evolving convective mixed layer. J. Meteorol. Soc. Jpn. **88**(1), 63–77 (2010b). doi:10.2151/jmsj.2010-105

B.C. Jemmett-Smith, J.H. Marsham, P. Knippertz, C.A. Gilkeson, Quantifying global dust devil occurrence from meteorological analyses. Geophys. Res. Lett. **42**(4), 1275–1282 (2015)

H. Kahanpää, C. Newman, J. Moores, M.-P. Zorzano, J. Martín-Torres, S. Navarro, A. Lepinette, M.T. Lemmon, B. Cantor, P. Valentín-Serrano, A. Ullán, W. Schmidt, Convective vortices and dust devils at the MSL landing site: annual variability. J. Geophys. Res. (2016, submitted)

M.A. Kahre, J.R. Murphy, R.M. Haberle, F. Montmessin, J. Schaeffer, Simulating the Martian dust cycle with a finite surface dust reservoir. Geophys. Res. Lett. **32**(20), L20204 (2005). doi:10.1029/2005GL023495

M.A. Kahre, J.R. Murphy, R.M. Haberle, Modeling the Martian dust cycle and surface dust reservoirs with the NASA Ames general circulation model. J. Geophys. Res., Planets **111**(E6), E06008 (2006). doi:10.1029/2005JE002588

J.C. Kaimal, J.A. Businger, Case studies of a convective plume and a dust devil. J. Appl. Meteorol. **9**, 612–620 (1970). doi:10.1175/1520-0450(1970)009<0612:CSOACP>2.0.CO;2

K.M. Kanak, Numerical simulation of dust devil-scale vortices. Q. J. R. Meteorol. Soc. **131**(607), 1271–1292 (2005). doi:10.1256/qj.03.172

K.M. Kanak, D.K. Lilly, J.T. Snow, The formation of vertical vortices in the convective boundary layer. Q. J. R. Meteorol. Soc. **126**(569), 2789–2810 (2000). doi:10.1002/qj.49712656910

C.A. Kellogg, D.W. Griffin, Aerobiology and the global transport of desert dust. Trends Ecol. Evol. **21**(11), 638–644 (2006). doi:10.1016/j.tree.2006.07.004

M. Klose, Y. Shao, Stochastic parameterization of dust emission and application to convective atmospheric conditions. Atmos. Chem. Phys. **12**(12), 7309–7320 (2012). doi:10.5194/acp-12-7309-2012

M. Klose, Y. Shao, Large-eddy simulation of turbulent dust emission. Aeolian Res. **8**, 49–58 (2013). doi:10.1016/j.aeolia.2012.10.010

M. Klose, Y. Shao, A numerical study on dust devils with implications to global dust budget estimates. Aeolian Res. **22**, 47–58 (2016). doi:10.1016/j.aeolia.2016.05.003

M. Klose, Y. Shao, X.L. Li, H.S. Zhang, M. Ishizuka, M. Mikami, J.F. Leys, Further development of a parameterization for convective turbulent dust emission and evaluation based on field observations. J. Geophys. Res., Atmos. **119**, 10441–10457 (2014). doi:10.1002/2014JD021688

M.R. Klose, Convective Turbulent Dust Emission: Process, parameterization, and relevance in the Earth system, Dissertation, Universität zu Köln, 2014. http://kups.ub.uni-koeln.de/id/eprint/5826

J. Koch, N.O. Renno, The role of convective plumes and vortices on the global aerosol budget. Geophys. Res. Lett. **32** (2005). doi:10.1029/2005GL023420

J.F. Kok, N.M. Mahowald, G. Fratini, J.A. Gillies, M. Ishizuka, J.F. Leys, M. Mikami, M.-S. Park, S.-U. Park, R.S. Van Pelt, T.M. Zobeck, An improved dust emission model—Part 1: model description and comparison against measurements. Atmos. Chem. Phys. **14**(23), 13023–13041 (2014). doi:10.5194/acp-14-13023-2014

J.F. Kok, N.O. Renno, Enhancement of the emission of mineral dust aerosols by electric forces. Geophys. Res. Lett. **33**(19), 2–6 (2006). doi:10.1029/2006GL026284

M. Küpper, G. Wurm, Thermal creep-assisted dust lifting on Mars: wind tunnel experiments for the entrainment threshold velocity. J. Geophys. Res. **120**(7), 1346–1356 (2015). doi:10.1002/2015JE004848

M.V. Kurgansky, Steady-state properties and statistical distribution of atmospheric dust devils. Geophys. Res. Lett. **33**(19) (2006). doi:10.1029/2006GL026142

M.V. Kurgansky, A. Montecinos, V. Villagran, S.M. Metzger, Micrometeorological conditions for dust-devil occurrence in the Atacama Desert. Bound.-Layer Meteorol. **138**(2), 285–298 (2011)

M.T. Lemmon, M.J. Wolff, J.F. Bell III, M.D. Smith, B.A. Cantor, P.H. Smith, Dust aerosol, clouds, and the atmospheric optical depth record over 5 Mars years of the Mars Exploration Rover mission. Icarus **251**, 96–111 (2015). doi:10.1016/j.icarus.2014.03.029. Dynamic Mars

H. Lettau, Note on aerodynamic roughness-parameter estimation on the basis of roughness element description. J. Appl. Meteorol. **8**, 828–832 (1969). doi:10.1175/1520-0450(1969)008<0828:NOARPE>2.0.CO;2

G.A. Loosmore, J.R. Hunt, Dust resuspension without saltation. J. Geophys. Res. **105**(D16), 20663–20671 (2000). doi:10.1029/2000JD900271

R. Lorenz, On the statistical distribution of dust devil diameters. Icarus **215**(1), 381–390 (2011). doi:10.1016/j.icarus.2011.06.005

R. Lorenz, The longevity and aspect ratio of dust devils: effects on detection efficiencies and comparison of landed and orbital imaging at Mars. Icarus **226**(1), 964–970 (2013). doi:10.1016/j.icarus.2013.06.031

R.D. Lorenz, Vortex encounter rates with fixed barometer stations: comparison with visual dust devil counts and large-eddy simulations. J. Atmos. Sci. **71**, 4461–4472 (2014). doi:10.1175/JAS-D-14-0138.1

R.D. Lorenz, B.K. Jackson, Dust devils and dustless vortices on a desert playa observed with surface pressure and solar flux logging. GeoResJ **5**, 1–11 (2015). doi:10.1016/j.grj.2014.11.002

R.D. Lorenz, M.J. Myers, Dust devil hazard to aviation: a review of United States air accident reports. J. Meteorol. **30**(298), 178–184 (2005)

R.D. Lorenz, D. Reiss, Solar panel clearing events, dust devil tracks, and in-situ vortex detections on Mars. Icarus **248**, 162–164 (2015). doi:10.1016/j.icarus.2014.10.034

R.D. Lorenz, L.D. Neakrase, J.D. Anderson, In-situ measurement of dust devil activity at La Jornada Experimental Range, New Mexico, USA. Aeolian Res., 1–12 (2015). doi:10.1016/j.aeolia.2015.01.012

D.J. Lunt, P.J. Valdes, The modern dust cycle: comparison of model results with observations and study of sensitivities. J. Geophys. Res., Atmos. **107**(D23), 4669 (2002). doi:10.1029/2002JD002316

T. Lyons, U. Nair, I. Foster, Clearing enhances dust devil formation. J. Arid Environ. **72**(10), 1918–1928 (2008)

B. Marticorena, G. Bergametti, Modeling the atmospheric dust cycle: 1. Design of a soil-derived dust emission scheme. J. Geophys. Res. **100**(D8), 16415–16430 (1995)

J.P. Mason, M.R. Patel, S.R. Lewis, Radiative transfer modelling of dust devils. Icarus **223**(1), 1–10 (2013). doi:10.1016/j.icarus.2012.11.018

J.P. Mason, M.R. Patel, S.R. Lewis, The retrieval of optical properties from terrestrial dust devil vortices. Icarus **231**(0), 385–393 (2014). doi:10.1016/j.icarus.2013.12.013

S.M. Metzger, M.R. Balme, M.C. Towner, B.J. Bos, T.J. Ringrose, M.R. Patel, In situ measurements of particle load and transport in dust devils. Icarus **214**(2), 766–772 (2011). doi:10.1016/j.icarus.2011.03.013

S.M. Metzger, Dust devils as aeolian transport mechanisms in the southern Nevada and in the Mars Pathfinder landing site, PhD thesis, University of Nevada, 1999

S.M. Metzger, J.R. Carr, J.R. Johnson, T.J. Parker, M.T. Lemmon, Dust devil vortices seen by the Mars Pathfinder Camera. Geophys. Res. Lett. **26**(18), 2781–2784 (1999). doi:10.1029/1999GL008341

T.I. Michaels, Numerical modeling of Mars dust devils: Albedo track generation. Geophys. Res. Lett. **33**(19) (2006). doi:10.1029/2006GL026268

T.I. Michaels, S.C.R. Rafkin, Large eddy simulation of atmospheric convection on Mars. Q. J. R. Meteorol. Soc. **130**, 1251–1274 (2004). doi:10.1256/qj.02.169

D.V. Michelangeli, O.B. Toon, R.M. Haberle, J.B. Pollack, Numerical simulations of the formation and evolution of water ice clouds in the martian atmosphere. Icarus **102**(2), 261–285 (1993). doi:10.1006/icar.1993.1048

R.L. Miller, P. Knippertz, C. Pérez García-Pando, J.P. Perlwitz, I. Tegen, Impact of dust radiative forcing upon climate, in *Mineral Dust*, ed. by P. Knippertz, J.-B.W. Stuut (Springer, Netherlands, 2014), pp. 327–357. ISBN 978-94-017-8977-6. doi:10.1007/978-94-017-8978-3_13

L. Montabone, F. Forget, E. Millour, R.J. Wilson, S.R. Lewis, B. Cantor, D. Kass, A. Kleinböhl, M.T. Lemmon, M.D. Smith, M.J. Wolff, Eight-year climatology of dust optical depth on Mars. Icarus **251**, 65–95 (2015). doi:10.1016/j.icarus.2014.12.034. Dynamic Mars

J.E. Moores, M.T. Lemmon, H. Kahanpää, S.C.R. Rafkin, R. Francis, J. Pla-Garcia, K. Bean, R. Haberle, C. Newman, M. Mischna, A.R. Vasavada, M. de la Torre Juárez, N. Renno, J. Bell, F. Calef, B. Cantor, T.H. Mcconnochie, A.-M. Harri, M. Genzer, M.H. Wong, M.D. Smith, F.J. Martín-Torres, M.-P. Zorzano, O. Kemppinen, E. McCullough, Observational evidence of a suppressed planetary boundary layer in northern Gale Crater, Mars as seen by the Navcam instrument onboard the Mars Science Laboratory rover. Icarus **249**(15), 129–142 (2015). doi:10.1016/j.icarus.2014.09.020

D.P. Mulholland, P.L. Read, S.R. Lewis, Simulating the interannual variability of major dust storms on Mars using variable lifting thresholds. Icarus **223**(1), 344–358 (2013). doi:10.1016/j.icarus.2012.12.003

J.R. Murphy, S. Nelli, Mars Pathfinder convective vortices: frequency of occurrence. Geophys. Res. Lett. **29**(23), 18–1184 (2002). doi:10.1029/2002GL015214

L.D.V. Neakrase, R. Greeley, Dust devils in the laboratory: effects of surface roughness on vortex dynamics. J. Geophys. Res. **115**, E05003 (2010a). doi:10.1029/2009JE003465

L.D.V. Neakrase, R. Greeley, Dust devil sediment flux on Earth and Mars: laboratory simulations. Icarus **206**(1), 306–318 (2010b)

L.D.V. Neakrase, R. Greeley, J.D. Iversen, M.R. Balme, E.E. Eddlemon, Dust flux within dust devils: preliminary laboratory simulations Geophys. Res. Lett., **33**, L19S09, (2006). doi:10.1029/2006GL026810

C.E. Newman, S.R. Lewis, P.L. Read, The atmospheric circulation and dust activity in different orbital epochs on Mars. Icarus **174**(1), 135–160 (2005). doi:10.1016/j.icarus.2004.10.023

C.E. Newman, M.I. Richardson, The impact of surface dust source exhaustion on the martian dust cycle, dust storms and interannual variability, as simulated by the MarsWRF General Circulation Model. Icarus **257**, 47–87 (2015). doi:10.1016/j.icarus.2015.03.030

C.E. Newman, S.R. Lewis, P.L. Read, F. Forget, Modeling the Martian dust cycle, 1. Representations of dust transport processes. J. Geophys. Res., Planets **107**(E12), 6-1–6-18 (2002a). doi:10.1029/2002JE001910. 5123

C.E. Newman, S.R. Lewis, P.L. Read, F. Forget, Modeling the Martian dust cycle 2. Multiannual radiatively active dust transport simulations. J. Geophys. Res., Planets **107**(E12), 7-1–7-15 (2002b). doi:10.1029/2002JE001920. 5124

H. Ohno, T. Takemi, Mechanisms for intensification and maintenance of numerically simulated dust devils. Atmos. Sci. Lett. (2010). doi:10.1002/asl.249

A.M.C. Oke, D. Dunkerley, N.J. Tapper, Willy-willies in the Australian landscape: sediment transport characteristics. J. Arid Environ. **71**(2), 216–228 (2007a). doi:10.1016/j.jaridenv.2007.03.014

A. Oke, N. Tapper, D. Dunkerley, Willy-willies in the Australian landscape: the role of key meteorological variables and surface conditions in defining frequency and spatial characteristics. J. Arid Environ. **71**(2), 201–215 (2007b)

F. Pantillon, P. Knippertz, J.H. Marsham, C.E. Birch, A parameterization of convective dust storms for models with mass-flux convection schemes. J. Atmos. Sci. **72**, 2545–2561 (2015). doi:10.1175/JAS-D-14-0341.1

A. Petrosyan, B. Galperin, S.E. Larsen, S.R. Lewis, A. Määttänen, P.L. Read, N. Renno, L.P.H.T. Rogberg, H. Savijärvi, T. Siili, A. Spiga, A. Toigo, L. Vázquez, The martian atmospheric boundary layer. Rev. Geophys. **49**(3) (2011). doi:10.1029/2010RG000351

S. Raasch, T. Franke, Structure and formation of dust devil-like vortices in the atmospheric boundary layer: a high-resolution numerical study. J. Geophys. Res. **116**, D16120, (2011). doi:10.1029/2011JD016010

S.C.R. Rafkin, R.M. Haberle, T.I. Michaels, The Mars regional atmospheric modeling system: model description and selected simulations. Icarus **151**, 228–256 (2001)

P.L. Read, S.R. Lewis, *The Martian Climate Revisited: Atmosphere and Environment of a Desert Planet* (Springer, Berlin/New York, 2004)

D. Reiss, N.M. Hoekzema, O.J. Stenzel, Dust deflation by dust devils on Mars derived from optical depth measurements using the shadow method in HiRISE images. Planet. Space Sci. **93–94**, 54–64 (2014). doi:10.1016/j.pss.2014.01.016

D. Reiss, J. Raack, H. Hiesinger, Bright dust devil tracks on Earth: implications for their formation on Mars. Icarus **211**(1), 917–920 (2011). doi:10.1016/j.icarus.2010.09.009

D. Reiss, A. Spiga, G. Erkerling, The horizontal motion of dust devils on Mars derived from CRISM and CTX/HiRISE observations. Icarus **227**, 8–20 (2014). doi:10.1016/j.icarus.2013.08.028

D. Reiss, M. Zanetti, G. Neukum, Multitemporal observations of identical active dust devils on Mars with the High Resolution Stereo Camera (HRSC) and Mars Orbiter Camera (MOC). Icarus **215** (2011). doi:10.1016/j.icarus.2011.06.011

D. Reiss, J. Raack, A.P. Rossi, G. Di Achille, H. Hiesinger, First in-situ analysis of dust devil tracks on Earth and their comparison with tracks on Mars. Geophys. Res. Lett. **37**(14), L14203 (2010). doi:10.1029/2010GL044016

D. Reiss, R.D. Lorenz, Dust devil track survey at Elysium Planitia, Mars: implications for the InSight landing sites. Icarus **266**, 315–330 (2015). doi:10.1016/j.icarus.2015.11.012

D. Reiss, M.I. Zimmerman, D.C. Lewellen, Formation of cycloidal dust devil tracks by redeposition of coarse sands in southern Peru: implications for Mars. Earth Planet. Sci. Lett. **383**, 7–15 (2013)

N.O. Renno, A.P. Ingersoll, Natural convection as a heat engine: a theory for CAPE. J. Atmos. Sci. **53**, 572–585 (1996). doi:10.1175/1520-0469(1996)053<0572:NCAAHE>2.0.CO;2

N.O. Renno, M.L. Burkett, M.P. Larkin, A simple thermodynamical theory for dust devils. J. Atmos. Sci. **55**, 3244–3252 (1998). doi:10.1175/1520-0469(1998)055<3244:ASTTFD>2.0.CO;2

N.O. Renno, V.J. Abreu, J. Koch, P.H. Smith, O.K. Hartogensis, H.A.R.D. Bruin, D. Burose, G.T. Delory, W.M. Farrell, C.J. Watts, J. Garatuza, M. Parker, A. Carswell, MATADOR 2002: a pilot field experiment on convective plumes and dust devils. J. Geophys. Res. **109** (2004). doi:10.1029/2003JE002219

N.O. Renno, A.A. Nush, J. Luninne, J. Murphy, Martian and terrestrial dust devils: test of a scaling theory using Pathfinder data. J. Geophys. Res. **105**, 1859–1866 (2000). doi:10.1029/19999JE001037

D. Rosenfeld, Y. Rudich, R. Lahav, Desert dust suppressing precipitation: a possible desertification feedback loop. Proc. Natl. Acad. Sci. USA **98**(11), 5975–5980 (2001). doi:10.1073/pnas.101122798

J.A. Ryan, R.D. Lucich, Possible dust devils, vortices on mars. J. Geophys. Res. **88**(C15), 11005–11011 (1983). doi:10.1029/JC088iC15p11005

J. Ryan, Relation of dust devil frequency and diameter to atmospheric temperature. J. Geophys. Res. **77**(36), 7133–7137 (1972)

K. Schepanski, I. Tegen, M. Todd, B. Heinold, G. Bönisch, Meteorological processes forcing Saharan dust emission inferred from MSG-SEVIRI observations of subdaily dust source activation and numerical models. J. Geophys. Res. **114**, 10201 (2012)

J.T. Schofield, J.R. Barnes, D. Crisp, R.M. Haberle, S. Larsen, J.A. Magalhaes, J.R. Murphy, A. Seiff, G. Wilson, The Mars Pathfinder atmospheric structure investigation meteorology (ASI/MET) experiment. Science **278**(5344), 1752–1758 (1997). doi:10.1126/science.278.5344.1752

Y. Shao, Simplification of a dust emission scheme and comparison with data. J. Geophys. Res. **109** (2004). doi:10.1029/2003JD004372

Y. Shao, *Physics and Modelling of Wind Erosion*, 2nd edn. (Springer, Berlin, 2008), p. 452

Y. Shao, M. Klose, A note on the stochastic nature of particle cohesive force and implications to threshold friction velocity for aerodynamic dust entrainment. Aeolian Res. (2016, in revision)

Y. Shao, H. Lu, A simple expression for wind erosion threshold friction velocity. J. Geophys. Res. **105**, 22437–22443 (2000)

Y. Shao, M. Klose, K.-H. Wyrwoll, Recent global dust trend and connections to climate forcing. J. Geophys. Res., Atmos. **118**, 1–12 (2013). doi:10.1002/jgrd.50836

Y. Shao, M.R. Raupach, P.A. Findlater, The effect of saltation bombardment on the entrainment of dust by wind. J. Geophys. Res. **98**, 12719–12726 (1993)

Y. Shao, K.-H. Wyrwoll, A. Chappell, J. Huang, Z. Lin, G.H. McTainsh, M. Mikami, T.Y. Tanaka, X. Wang, S. Yoon, Dust cycle: an emerging core theme in Earth system science. Aeolian Res. **2**, 181–204 (2011). doi:10.1016/j.aeolia.2011.02.001

Y. Shao, S. Liu, J. Schween, S. Crewell, Large-eddy atmosphere-land-surface modelling over heterogeneous surfaces: model development and comparison with measurements. Bound.-Layer Meteorol. **148**(2), 333–356 (2013). doi:10.1007/s10546-013-9823-0

E.A. Shinn, G.W. Smith, J.M. Prospero, P. Betzer, M.L. Hayes, V. Garrison, R.T. Barber, African dust and the demise of Carribbean coral reefs. Geophys. Res. Lett. **27**(19), 3029–3032 (2000). doi:10.1029/2000GL011599

P.C. Sinclair, General characteristics of dust devils. J. Appl. Meteorol. **8**, 32–45 (1969). doi:10.1175/1520-0450(1969)008<0032:GCODD>2.0.CO;2

P.C. Sinclair, The lower structure of dust devils. J. Atmos. Sci. **30**, 1599–1619 (1973). doi:10.1175/1520-0469(1973)0302.0.CO;2

M.D. Smith, Interannual variability in TES atmospheric observations of Mars during 1999–2003. Icarus **167**(1), 148–165 (2004). doi:10.1016/j.icarus.2003.09.010. Special Issue on DS1/Comet Borrelly

J.T. Snow, T.M. McClelland, Dust devils at White Sands Missile Range, New Mexico: 1. Temporal and spatial distributions. J. Geophys. Res., Atmos. **95**(D9), 13707–13721 (1990). doi:10.1029/JD095iD09p13707

I.N. Sokolik, O.B. Toon, Direct radiative forcing by anthropogenic airborne mineral aerosols. Nature **381**, 681–683 (1996)

A. Spiga, F. Forget, A new model to simulate the Martian mesoscale and microscale atmospheric circulation: validation and first results. J. Geophys. Res. **114**, 02009 (2009). doi:10.1029/2008JE003242

A. Spiga, J. Faure, J.-B. Madeleine, A. Määttänen, F. Forget, Rocket dust storms and detached dust layers in the Martian atmosphere. J. Geophys. Res. **118**(4), 746–767 (2013). doi:10.1002/jgre.20046

K. Steakley, J. Murphy, A year of convective vortex activity at Gale crater. Icarus (2016, accepted). doi:10.1016/j.icarus.2016.06.010

G. Sterk, L. Herrmann, A. Bationo, Wind-blown nutrient transport and soil productivity changes in southwest Niger. Land Degrad. Dev. **7**(4), 325–335 (1996). doi:10.1002/(SICI)1099-145X(199612)7:4<325::AID-LDR237>3.0.CO;2-Q

P.P. Sullivan, J.C. McWilliams, C.-H. Moeng, A subgrid-scale model for large-eddy simulation of planetary boundary-layer flows. Bound.-Layer Meteorol. **71**(3), 247–276 (1994). doi:10.1007/BF00713741

T. Takemi, M. Yasui, J. Zhou, L. Liu, Role of boundary layer and cumulus convection on dust emission and transport over a midlatitude desert area. J. Geophys. Res., Atmos. **111**(D11), D11203 (2006). doi:10.1029/2005JD006666

T.Y. Tanaka, M. Chiba, A numerical study of the contributions of dust source regions to the global dust budget. Glob. Planet. Change **52**, 88–104 (2006)

P. Thomas, P.J. Gierasch, Dust devils on Mars. Science **230**(4722), 175–177 (1985). doi:10.1126/science.230.4722.175

A.D. Toigo, M.I. Richardson, Meteorology of proposed Mars Exploration Rover landing sites. J. Geophys. Res. **108**(E12), 8092 (2003). doi:10.1029/2003JE002064

C.A. Verba, P.E. Geissler, T.N. Titus, D. Waller, Observations from the High Resolution Imaging Science Experiment (HiRISE): Martian dust devils in Gusev and Russell craters. J. Geophys. Res. **115**(E9), E09002 (2010). doi:10.1029/2009JE003498

H. Wang, M.I. Richardson, The origin, evolution, and trajectory of large dust storms on Mars during Mars years 24–30 (1999–2011). Icarus **251**, 112–127 (2015). doi:10.1016/j.icarus.2013.10.033. Dynamic Mars

N.P. Webb, C.L. Strong, A. Chappell, S.K. Marx, G.H. McTainsh, Soil organic carbon enrichment of dust emissions: magnitude, mechanisms and its implications for the carbon cycle. Earth Surf. Process. Landf. **38**(14), 1662–1671 (2013). doi:10.1002/esp.3404

P.L. Whelley, R. Greeley, Latitudinal dependency in dust devil activity on Mars. J. Geophys. Res., Planets **111**(E10), E10003 (2006). doi:10.1029/2006JE002677

P.L. Whelley, R. Greeley, The distribution of dust devil activity on Mars. J. Geophys. Res. **113**(E7) (2008). doi:10.1029/2007JE002966

G. Wurm, J. Teiser, D. Reiss, Greenhouse and thermophoretic effects in dust layers: the missing link for lifting of dust on Mars. Geophys. Res. Lett. **35**, L10201 (2008). doi:10.1029/2008GL033799

Y.Z. Zhao, Z.L. Gu, Y.Z. Yu, Y. Ge, Y. Li, X. Feng, Mechanism and large eddy simulation of dust devils. Atmos.-Ocean **42**(1), 61–84 (2004). doi:10.3137/ao.420105

A.D. Zimon, *Adhesion of Dust and Powder* (Consultants Bureau, New York, 1982), p. 438

MIX
Papier aus verantwortungsvollen Quellen
Paper from responsible sources
FSC® C105338

FSC
www.fsc.org

If you have any concerns about our products,
you can contact us on
ProductSafety@springernature.com

In case Publisher is established outside the EU,
the EU authorized representative is:
Springer Nature Customer Service Center GmbH
Europaplatz 3, 69115 Heidelberg, Germany

Printed by Libri Plureos GmbH
in Hamburg, Germany